SYMBOLS

n	An integer (1, 2, 3, . . .)		w	Energy density
n	Turns per unit length		w	Weight
N	Number of turns		W	Work
N	Normal (perpendicular) force		X_C	Capacitive reactance
p	Object distance		X_L	Inductive reactance
p	Pressure		Y	Young's Modulus
P	Power		Z	Atomic number
PE	Potential Energy		Z	Impedance
q	Image distance		α (alpha)	Angular acceleration
Q	Electric charge		α (alpha)	Temperature coefficient of resistivity
Q	Heat		β (beta)	Sound intensity level
r	Angle of reflection, refraction		γ (gamma)	Electromagnetic photon
r	Internal resistance		γ (gamma)	Surface tension
r, R	Radius		Δ (delta)	"Change in"
R	Range		η (eta)	Viscosity
R	Rate of flow		θ (theta)	An angle
R	Electrical resistance		λ (lambda)	Wavelength
s	Displacement from equilibrium position		μ (mu)	Coefficient of friction
s	Distance; displacement		μ (mu)	Magnetic permeability
S	Shear modulus		ρ (rho)	Density
t	Time		ρ (rho)	Resistivity
T	Period		σ (sigma)	Stefan-Boltzmann constant
T	Temperature		Σ (sigma)	"Sum of"
T	Tension		τ (tau)	Torque
v	Speed		φ (phi)	Phase angle
v	Velocity		Φ (phi)	Magnetic flux
V	Potential difference		ω (omega)	Angular speed
V	Volume			

POWERS OF TEN

10^{-10}	= 0.000,000,000,1	
10^{-9}	= 0.000,000,001	
10^{-8}	= 0.000,000,01	
10^{-7}	= 0.000,000,1	
10^{-6}	= 0.000,001	
10^{-5}	= 0.000,01	
10^{-4}	= 0.000,1	
10^{-3}	= 0.001	
10^{-2}	= 0.01	
10^{-1}	= 0.1	
10^{0}	= 1	

10^{0}	= 1
10^{1}	= 10
10^{2}	= 100
10^{3}	= 1000
10^{4}	= 10,000
10^{5}	= 100,000
10^{6}	= 1,000,000
10^{7}	= 10,000,000
10^{8}	= 100,000,000
10^{9}	= 1,000,000,000
10^{10}	= 10,000,000,000

MULTIPLIERS FOR SI UNITS

a	atto-	10^{-18}	da	deka-	10^{1}
f	femto-	10^{-15}	h	hecto-	10^{2}
p	pico-	10^{-12}	k	kilo-	10^{3}
n	nano-	10^{-9}	M	mega-	10^{6}
μ	micro-	10^{-6}	G	giga-	10^{9}
m	milli-	10^{-3}	T	tera-	10^{12}
c	centi-	10^{-2}	P	peta-	10^{15}
d	deci-	10^{-1}	E	exa-	10^{18}

SIXTH EDITION

MODERN TECHNICAL PHYSICS

SIXTH EDITION

MODERN TECHNICAL PHYSICS

ARTHUR BEISER

ADDISON-WESLEY PUBLISHING COMPANY

Reading, Massachusetts • Menlo Park, California • New York
Don Mills, Ontario • Wokingham, England • Amsterdam • Bonn
Sydney • Singapore • Tokyo • Madrid • San Juan • Milan • Paris

Beiser, Arthur.
    Modern technical physics / Arthur Beiser.—6th ed.
        p. cm.
    Includes index.
    ISBN 0-201-57899-9
    1. Physics.   I. Title
QC23.R36   1991
530—dc20

90–32028
CIP

11 12 13 14 15 16 DOC 10 09 08 07 06 05 04

PREFACE

Physics is concerned with the fundamental structure, properties, and behavior of matter. All technology, from the design of the simplest building to that of the most advanced computer, rests on physical principles. Students who plan careers in technology, whatever the branch, must have some competence in physics, and the purpose of this book is to help provide it.

Modern Technical Physics is a straightforward presentation of the laws of motion, the properties of matter in bulk, vibrations and waves, heat and thermodynamics, electromagnetism, optics, and atomic and nuclear physics. The mathematical level has been kept as low as possible.

Student Aids

A number of aids are provided to help the student master the text:

EXAMPLES   Nearly three hundred worked examples show how physical principles are put into practice.

IMPORTANT TERMS AND FORMULAS   The meanings of important terms are given at the end of each chapter, together with the formulas needed to solve problems based on the chapter material. These lists act as chapter summaries.

MATH APPENDIXES   Only basic algebra and simple trigonometry are used in the book, and these are reviewed in two appendixes to the extent required. Powers-of-ten notation for small and large numbers is carefully explained, and the use of electronic calculators for finding powers and roots is described. The appendixes are self-contained and include exercises to sharpen rusty skills (or to practice new ones).

EXERCISES   Each chapter ends with exercises of four kinds, for a total of over 2,500.

Multiple Choice.   These exercises, an average of thirty per chapter, serve as a quick, painless check on the student's grasp of the chapter contents. They serve as a warm-up for the regular exercises and have proved to be an effective learning tool. Correct answers provide reinforcement and encouragement; incorrect ones identify areas of weakness. The answers are given at the end of the exercise section.

Questions.   A typical chapter has over a dozen questions that test the student's general understanding of the ideas in the chapter. Answers to the odd-numbered questions are given at the back of the book.

Exercises.   An average of forty exercises per chapter (some have over sixty) are arranged in groups that correspond to sections in the text. They range from quite easy to moderately challenging. The simpler exercises need only one or two steps to solve and are meant to provide practice in manipulating symbols and numbers, to build confidence, and to promote an intuitive feeling for the magnitudes of the quantities involved. Other exercises are more complex, but most students should be able to handle them.

Problems.   The problems, an average of ten per chapter, are more difficult and more time-consuming than the exercises. Some require real thought and are therefore more fun for the abler student.

OUTLINE SOLUTIONS   All the odd-numbered exercises and problems have outline solutions at the back of the book. Each outline solution shows the method of attack and gives the answer, but some of the intermediate steps are omitted. In this way a puzzled student can find out how to proceed and still have something left to do on his or her own. The outline solutions supplement the worked examples in the text to furnish nearly a thousand models to follow in dealing with a wide variety of problems.

Changes to this Edition

In revising *Modern Technical Physics* for its sixth edition, one of the chief aims was to make the book more accessible to its users. Much of the text was rewritten in a less formal style for easier reading; few paragraphs escaped at least some change. Many arguments were recast for greater clarity and interest. A new design was adopted.

ORGANIZATION   The organization of *Modern Technical Physics* was slightly altered at the request of users of the previous edition. Vectors are now introduced before the laws of motion, and equilibrium has been shifted to follow rotational motion in a chapter that also contains a briefer treatment of mechanical advantage. The contents of the former chapter on heat transfer have been incorporated into two earlier chapters. Capacitance and inductance are now included respectively in the chapters on electric energy and on electromagnetic induction. Throughout the book, although some discussions were expanded and a few new topics were added, enough material that experience showed to be of marginal interest was shortened or deleted to give a net reduction in the length of the text.

ILLUSTRATIONS   The illustrations are full partners to the text, and much attention has been given to them in preparing this edition. There are about a hundred more figures than in the previous edition, and many of the old figures have been revised or entirely redrawn for greater effectiveness. Sequences of figures accompany complicated derivations, a figure for each step. In this way students unaccustomed to abstract arguments have a visual pathway as well as a formal one to appreciate how an important formula is obtained. In addition, figures are used liberally with key concepts and equations to show, as well as tell, the point being made. A selection of photographs is included as part of the effort to bring alive the ideas of physics.

Acknowledgments

In revising *Modern Technical Physics* I have had the benefit of constructive criticism from the following reviewers of both the previous edition and the manuscript of this one. Their generous help was of great value.

Duane Carmony, Purdue University
John Danna, Erie Community College
Hayes Everett, Gulf Coast Community College
John Mycroft, New England Institute of Technology
James Nolta, Lawrence Technical Institute

Finally, I want to thank my friends at Addison-Wesley, especially Stuart Johnson, Peggy J. Flanagan, and Laurie Sciuto, who did so splendid a job of turning the manuscript into a book.

ARTHUR BEISER

CONTENTS

1 Describing Motion   1

1.1 Units   2
1.2 Speed   4
1.3 Velocity   9
1.4 Acceleration   10
1.5 Velocity and Acceleration   12
1.6 Displacement, Time, and Acceleration 14
1.7 Displacement, Velocity, and Acceleration 16
1.8 How to Solve Problems   18
1.9 Acceleration of Gravity   21
1.10 Free Fall   24

2 Vectors   35

2.1 Vector Quantities   36
2.2 Vector Addition   37
2.3 Vector Subtraction   39
2.4 Resolving a Vector   40
2.5 Vector Addition by Components   43
2.6 Motion in a Vertical Plane   45
2.7 Projectile Flight   48

3 Force and Motion   59

3.1 First Law of Motion   60
3.2 Mass   60
3.3 Force   62
3.4 Second Law of Motion   64
3.5 Weight   69
3.6 Customary Units of Mass and Force 73
3.7 Third Law of Motion   75
3.8 Friction   78
3.9 Coefficient of Friction   80

4 Energy   93

4.1 Work   94
4.2 Work Done Against Gravity   97
4.3 Power   99
4.4 Energy   101
4.5 Kinetic Energy   102
4.6 Potential Energy   105

4.7 Conservation of Energy   109
4.8 Conservative and Nonconservative Forces 112

5 Momentum   121

5.1 Linear Momentum   122
5.2 Impulse   123
5.3 Conservation of Momentum   125
5.4 Rocket Propulsion   128
5.5 Inelastic Collisions   131
5.6 Elastic Collisions   133
5.7 Energy Transfer   136
5.8 Coefficient of Restitution   138

6 Circular Motion and Gravitation 147

6.1 Centripetal Force   148
6.2 Centripetal Acceleration   149
6.3 Magnitude of Centripetal Force   152
6.4 Motion in a Vertical Circle   156
6.5 Gravitation   161
6.6 Earth Satellites   164
6.7 Weightlessness   167

7 Rotational Motion   173

7.1 Angular Measure   174
7.2 Angular Speed   176
7.3 Rotational Kinetic Energy   178
7.4 Moment of Inertia   179
7.5 Combined Translation and Rotation 181
7.6 Angular Acceleration   183
7.7 Torque   186
7.8 Work and Power   189
7.9 Angular Momentum   191

8 Equilibrium and Mechanical Advantages 201

8.1 Translational Equilibrium   202
8.2 Rotational Equilibrium   208
8.3 Center of Gravity   211
8.4 Finding a Center of Gravity   216

8.5   Mechanical Advantage   220
8.6   Some Simple Machines   224

9   Mechanical Properties of Matter 242

9.1   Density   243
9.2   Elasticity   246
9.3   Structure of Solids   250
9.4   Young's Modulus   251
9.5   Shear   256
9.6   Bulk Modulus   258
9.7   Building over Space   260

10   Fluids   267

10.1   Pressure   268
10.2   Hydraulic Press   272
10.3   Pressure and Depth   276
10.4   Buoyancy   278
10.5   Fluid Flow   281
10.6   Bernoulli's Equation   285
10.7   Applications of Bernoulli's Equation   288
10.8   Viscosity   292
10.9   Surface Tension   295

11   Harmonic Motion   304

11.1   Elastic Potential Energy   305
11.2   Simple Harmonic Motion   309
11.3   A Model of Simple Harmonic Motion   313
11.4   Position, Speed, and Acceleration   316
11.5   Trigonometric Notation   320
11.6   The Simple Pendulum   321
11.7   The Physical Pendulum   322
11.8   Damped Harmonic Oscillator   325

12   Waves   331

12.1   Pulses in a String   332
12.2   Principle of Superposition   335
12.3   Periodic Waves   336
12.4   Types of Waves   341
12.5   Standing Waves   343
12.6   Sound   348
12.7   Interference of Sound Waves   352
12.8   Doppler Effect   357
12.9   Sound Intensity   360

13   Thermal Properties of Matter   370

13.1   Temperature   371
13.2   Thermal Expansion   373
13.3   Boyle's Law   377
13.4   Charles's Law   379
13.5   Atoms and Molecules   384
13.6   Kinetic Theory of Gases   386
13.7   Molecular Motion in Liquids   390

14   Heat   399

14.1   Internal Energy and Heat   400
14.2   Specific Heat Capacity   402
14.3   Changes of State   405
14.4   The Triple Point   410
14.5   Heat Conduction   413
14.6   Thermal Resistance   417
14.7   Convection   419
14.8   Radiation   421

15   Thermodynamics   433

15.1   First Law of Thermodynamics   434
15.2   Work Done by and on a Gas   436
15.3   Second Law of Thermodynamics   440
15.4   Carnot Engine   442
15.5   Engine Efficiency   445
15.6   Internal Combustion Engines   448
15.7   Refrigerators   454
15.8   Thermodynamics of Refrigeration   457
15.9   The Energy Problem   459

16   Electricity   469

16.1   Electric Charge   470
16.2   Coulomb's Law   473
16.3   Electricity and Matter   476
16.4   Electrical Conduction   476
16.5   What is a Field?   479
16.6   Electric Field   481
16.7   Electric Field Lines   485

17   Electric Energy   493

17.1   Electric Potential Energy   494
17.2   Potential Energy of Two Charges   496
17.3   Potential Difference   498
17.4   Cathode-Ray Tube   501
17.5   The Electron Volt   503
17.6   Electric Field Energy   504
17.7   Capacitance   506
17.8   Dielectric Constant   509
17.9   Capacitors in Combination   513

18 Electric Current 522

18.1	Electric Current	523
18.2	Ohm's Law	525
18.3	Resistivity	528
18.4	Electric Power	532
18.5	Resistors in Series	535
18.6	Resistors in Parallel	539
18.7	Electromotive Force	545
18.8	Impedance Matching	548
18.9	Kirchhoff's Rules	549
18.10	Capacitive Time Constant	553

19 Magnetism 565

19.1	The Nature of Magnetism	566
19.2	Magnetic Field	566
19.3	Magnetic Field of a Current	570
19.4	Magnetic Properties of Matter	574
19.5	Force on a Moving Charge	578
19.6	Force on a Current	583
19.7	Force Between Two Currents	587
19.8	Torque on a Current Loop	589
19.9	Magnetic Poles	593

20 Electromagnetic Induction 602

20.1	Electromagnetic Induction	603
20.2	Moving Wire in a Magnetic Field	605
20.3	Faraday's Law	607
20.4	The Generator	610
20.5	The Transformer	615
20.6	Inductance	619
20.7	Inductive Time Constant	622
20.8	Magnetic Energy	625
20.9	Electrical Oscillations	627

21 Alternating Current 636

21.1	Effective Current and Voltage	637
21.2	Phasors	639
21.3	Phase Relationships	640
21.4	Inductive Reactance	642
21.5	Capacitive Reactance	644
21.6	Impedance	646
21.7	Resonance	651
21.8	Power in AC Circuits	653
21.9	Impedance Matching	655
21.10	Parallel AC Circuits	656

22 Light 667

22.1	Electromagnetic Waves	668
22.2	Types of Em Waves	672
22.3	Huygens' Principle	676
22.4	Reflection	678
22.5	Refraction	681
22.6	Index of Refraction	684
22.7	Apparent Depth	688
22.8	Total Internal Reflection	690
22.9	Luminous Flux	693
22.10	Illumination	696

23 Lenses and Mirrors 705

23.1	Lenses	706
23.2	Image Formation	709
23.3	The Lens Equation	712
23.4	Magnification	715
23.5	The Camera	717
23.6	The Eye	720
23.7	Defects of Vision	722
23.8	The Microscope	725
23.9	The Telescope	726
23.10	Lens Aberrations	729
23.11	Spherical Mirrors	730

24 Physical Optics 743

24.1	Interference of Light	744
24.2	Double Slit	746
24.3	Diffraction Grating	748
24.4	X-Ray Diffraction	751
24.5	Thin Films	752
24.6	Diffraction	755
24.7	Polarization	758
24.8	Scattering	763

25 Atomic Physics 769

25.1	Quantum Theory of Light	770
25.2	X-Rays	774
25.3	Matter Waves	776
25.4	Mass and Energy	779
25.5	The Hydrogen Atom	782
25.6	Energy Levels	786
25.7	Atomic Spectra	788
25.8	Complex Atoms	791
25.9	Atomic Excitation	794
25.10	The Laser	795

26 Atoms in Combination   806

26.1 The Hydrogen Molecule   807
26.2 Covalent Bond   808
26.3 Solids and Liquids   810
26.4 Covalent and Ionic Solids   812
26.5 Van der Waals Bonds   814
26.6 Metallic Bond   816
26.7 Energy Bands   817
26.8 Impurity Semiconductors   820
26.9 Semiconductor Devices   822

27 The Nucleus   829

27.1 Nuclear Composition   830
27.2 Binding Energy   832
27.3 Radioactive Decay   835
27.4 Detecting Radiation   841
27.5 Half-Life   844

27.6   Radiometric Dating   846
27.7   Radiation Hazards   847
27.8   Fission   849
27.9   Nuclear Reactors   851
27.10 Nuclear Fusion   854

Appendix A   Useful Mathematics   861

Appendix B   Powers of Ten and Logarithms   869

Appendix C   Tables   876

Answers to Odd-Numbered Questions and Exercises   880

Photo Credits   903

Index   905

SIXTH EDITION

MODERN
TECHNICAL
PHYSICS

1

DESCRIBING MOTION

1.1 **Units**

1.2 **Speed**

1.3 **Velocity**

1.4 **Acceleration**

1.5 **Velocity and Acceleration**

1.6 **Displacement, Time, and Acceleration**

1.7 **Displacement, Velocity, and Acceleration**

1.8 **How to Solve Problems**

1.9 **Acceleration of Gravity**

1.10 **Free Fall**

Everything in the universe is in motion, from the electrons in atoms to the stars in the sky. We see people and cars moving on the ground, birds and airplanes in the sky, fish and ships in the sea. The most solid of buildings sway in the wind, and, though they do so at the speed of a growing fingernail, even the continents shift across the face of the earth. Understanding motion is thus a natural start toward understanding how the world around us works, which is the goal of physics. Historically, Galileo's experiments with moving bodies began the era of modern science. The subject of this chapter is motion in a straight line, the simplest motion of all.

1.1   UNITS

A few basic units are enough for all physical quantities.

The raw material of physics consists of measurements. Every measurement is a comparison. When we say a boat is 18 meters long, we mean that its length is 18 times a certain unit of length called the meter. The result of every measurement has two parts. One is a number (18 for the boat) to answer the question "How many?" The other is a unit (here the meter) to answer the question "Of what?"

All quantities in the physical world can be expressed in terms of only six basic units. For the first third of this book we will need only the units of length (the **meter,** m), time (the **second,** s), and mass (the **kilogram,** kg). Every other mechanical unit is some combination of two or all three of these. The unit of force, for instance, is the kg·m/s^2, which is called the **newton** (N) for convenience. The other basic units are the **kelvin** (K) for temperature, the **ampere** (A) for electric current, and the **candela** (cd) for luminous intensity (Table 1.1).

These units are part of the Système International, or SI, which is the current version of the metric system introduced in France two centuries ago to replace the hodgepodge of traditional units that were then making commerce and industry difficult. Today SI units are used by all scientists and in most of the world in everyday life as well. Although SI units have been legal in the United States for many years, they have not yet displaced such customary units as the foot and the pound outside the laboratory.

Whenever possible, basic units are defined in terms of quantities in nature that do not change. The second, for instance, is now defined in terms of the microwave radiation given off under certain circumstances by the ^{133}Cs atom: 1 s is equal to the time needed for 9,192,631,770 cycles of this radiation to be emitted. The meter in turn is defined in terms of the second and the speed of light: 1 m is the distance traveled in 1/299,792,458 s by light waves in a vacuum.

Because a basic unit in a system may not be convenient in size for a given measurement, other units have come into use within each system. Thus long distances are usually given in miles rather than in feet, or in kilometers rather than in meters.

Table 1.1   Basic SI units

Quantity	Unit
Length	meter (m)
Time	second (s)
Mass	kilogram (kg)
Temperature	kelvin (K)
Electric current	ampere (A)
Luminous intensity	candela (cd)

Table 1.2   Subdivisions and multiples of SI units are widely used. Each is designated by a prefix according to the corresponding power of ten.

Prefix	Power of ten	Abbre-viation	Pronunciation	Example
atto-	10^{-18}	a	at' toe	1 aC = 1 attocoulomb = 10^{-18} C
femto-	10^{-15}	f	fem' toe	1 fm = 1 femtometer = 10^{-15} m
pico-	10^{-12}	p	pee' koe	1 pf = 1 picofarad = 10^{-12} f
nano-	10^{-9}	n	nan' oe	1 ns = 1 nanosecond = 10^{-9} s
micro-	10^{-6}	μ	my' kroe	1 μA = 1 microampere = 10^{-6} A
milli-	10^{-3}	m	mil' i	1 mg = 1 milligram = 10^{-3} g
centi-	10^{-2}	c	sen' ti	1 cL = 1 centiliter = 10^{-2} L
kilo-	10^{3}	k	kil' oe	1 kN = 1 kilonewton = 10^{3} N
mega-	10^{6}	M	meg' a	1 MW = 1 megawatt = 10^{6} W
giga-	10^{9}	G	ji' ga	1 GeV = 1 gigaelectronvolt = 10^{9} eV
tera-	10^{12}	T	ter' a	1 Tm = 1 terameter = 10^{12} m
peta-	10^{15}	P	pe' ta	1 Ps = 1 petasecond = 10^{15} s
exa-	10^{18}	E	ex' a	1 EJ = 1 exajoule = 10^{18} J

The great advantage of the SI system is that it is a decimal system, which makes calculations easy, whereas the customary system is quite irregular in this respect. (1 km = 1000 m, for example, but 1 mi = 5280 ft.) The chief units in each system, together with their equivalents in the other system, are given inside the back cover of the book. The prefixes used with SI units of all kinds are listed in Table 1.2; powers-of-ten notation is reviewed in Appendix B.

Converting Units

Suppose we find from a European map that Amsterdam is 648 km from Berlin and want to know what this distance is in miles. Two rules apply to such a conversion:

1. **Units are treated in an equation in exactly the same way as algebraic quantities, and may be multiplied and divided by one another;**
2. **Multiplying or dividing a quantity by 1 does not affect its value.**

To convert 648 km to its equivalent in miles, we note from the table inside the back cover that

$$1 \text{ km} = 0.621 \text{ mi}$$

Therefore

$$0.621 \frac{\text{mi}}{\text{km}} = 1$$

and multiplying or dividing any quantity by 0.621 mi/km does not affect its value but only changes the units in which it is given. Hence we have

$$d = (648 \; \cancel{km}) \left(0.621 \, \frac{mi}{\cancel{km}} \right) = 402 \; mi$$

because km/km = 1 and so drops out. (When multiplied out completely, [648] [0.621] = 402.408. As discussed in Appendix B–4, we can keep only as many significant figures as there are in the least accurately known quantity used in a calculation. Since this means three significant figures here, the result must be expressed as 402 mi.)

◆   **E X A M P L E   1.1** _____

The piston displacement of the Volvo MD21A diesel engine is 2.11 liters, where 1 liter = 1000 cm³. Express this volume in cubic inches.

SOLUTION   The displacement of the engine is $V = 2110 \; cm^3$. Since 1 cm = 0.394 in, 1 cm³ = (0.394 in.)³ = 0.0612 in³, and

$$V = (2110 \; \cancel{cm^3}) \left(0.0612 \frac{in^3}{\cancel{cm^3}} \right) = 129 \; in^3$$ ◆

◆   **E X A M P L E   1.2** _____

Express a speed of 80 km/h in meters per second.

SOLUTION   Here two units are to be converted, which we can do in a single step:

$$v = \left(80 \frac{\cancel{km}}{\cancel{h}} \right) \left(1000 \frac{m}{\cancel{km}} \right) \left(\frac{1}{3600 s/\cancel{h}} \right) = 22 \; m/s$$ ◆

1.2 SPEED

Average and instantaneous speeds may differ.

As we all know, the speed of a moving object is the rate at which it covers distance. But there is more to the idea of speed than this simple statement.

The **average speed** $\bar{v}$ of an object that travels the distance x in the time interval t is

$$\bar{v} = \frac{x}{t} \qquad\qquad \textit{Average speed} \quad (1.1)$$

$$\text{Average speed} = \frac{\text{distance traveled}}{\text{time interval}}$$

Thus a car that went 180 km in 3.0 h had an average speed of

$$\bar{v} = \frac{x}{t} = \frac{180\,\text{km}}{3.0\,\text{h}} = 60\,\text{km/h}$$

The average speed of the car is only part of the story of its trip, however. Knowing $\bar{v}$ does not tell us whether the car had the same speed for the entire 3 h or sometimes went faster than 60 km/h and sometimes slower.

Suppose we are in a car starting from rest and note the time at which its odometer shows it has covered 100 m, 200 m, 300 m, and so on. The data might appear as in Table 1.3. When we plot these data on a graph, as in Fig. 1.1, we find that the line joining the various points is not a straight line but has an upward curve. In each successive equal time interval (as marked off at the bottom of the graph), the car covers a greater distance than before—it is going faster and faster.

Table 1.3

Total distance, m	0	100	200	300	400	500
Elapsed time, s	0	28	40	49	57	63

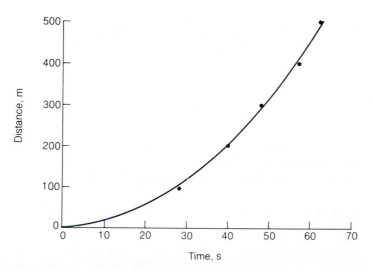

Figure 1.1

A graph of the data in Table 1.3. Each point represents the result of one measurement. The smooth curve is a reasonable fit to the data even though it does not quite go through every point.

Figure 1.2

The procedure for finding the instantaneous speed at $t = 40$ s from the data in Table 1.3. Because the graph of distance versus time is not a straight line, the instantaneous speed changes continuously.

Figure 1.3

A speedometer calibrated in m/s, km/h, and mi/h. The speedometer of a vehicle indicates its instantaneous speed.

Even though the car's speed is changing, at every moment it has a certain definite value (indicated by its speedometer). To find this **instantaneous speed** v at a particular time t, we draw a straight line tangent to the distance-time curve at that value of t. The length of the line does not matter. Then we find v from the tangent line by using the formula

$$v = \frac{\Delta x}{\Delta t}$$
           Instantaneous speed   (1.2)

where Δx is the distance interval between the ends of the tangent and Δt is the time interval between them. (Δ is the Greek capital letter *delta* and customarily means "change in.") From Fig. 1.2 the instantaneous speed of the car at $t = 40$ s is

$$v = \frac{\Delta x}{\Delta t} = \frac{100 \text{ m}}{10 \text{ s}} = 10 \text{ m/s}$$

Figure 1.3 is a speedometer dial calibrated in m/s, km/h, and mi/h. Evidently 10 m/s corresponds to 36 km/h or 22 mi/h. Table 1.4 shows the instantaneous speeds of the car at 10-s intervals as found from the graph of Fig. 1.2.

Table 1.4

Elapsed time, s	0	10	20	30	40	50	60
Instantaneous speed, m/s	0	2.5	5.0	7.5	10	12.5	15

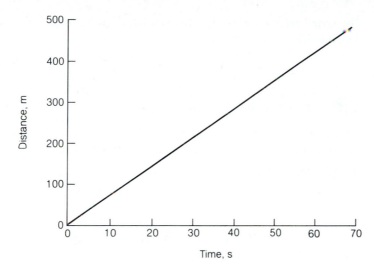

Figure 1.4

The distance-time graph of a car traveling at a constant speed of 7.5 m/s is a straight line.

Constant Speed

When the instantaneous speed of an object does not change, it is moving at **constant speed.** Figure 1.4 is a distance-time graph of a car that has a constant speed of 7.5 m/s: The curve is, of course, a straight line. For the case of constant speed, Eq. (1.2) can be rewritten to provide two useful formulas. The first gives the distance covered in a given period of time:

$$x = vt \qquad\qquad v = constant \quad (1.3)$$

Distance = (speed)(time)

The second formula gives the time needed to cover a given distance:

$$t = \frac{x}{v} \qquad\qquad v = constant \quad (1.4)$$

Time = $\dfrac{\text{distance}}{\text{speed}}$

If the speed of an object is constant, we can calculate how far it will go in a given period of time. Or, given the distance, we can calculate the time required.

◆   **E X A M P L E 1.3** _____

A woman standing in front of a cliff claps her hands, and 2.50 s later she hears an echo. How far away is the cliff? Assume that the speed of sound is 343 m/s.

SOLUTION   The total distance a sound travels in 2.50 s is

$$x = vt = (343 \text{ m/s})(2.50 \text{ s}) = 858 \text{ m}$$

The sound must travel to the cliff and back, so the cliff is half this distance away, which is $\frac{1}{2}$(858 m) = 429 m.   ◆

◆   E X A M P L E   1.4 _____

If the highway speed limit is raised from 55 mi/h to 65 mi/h, how much time is saved on a 100-mi trip made at the maximum permitted speed?

SOLUTION   At v_1 = 55 mi/h the time needed for the trip is t_1 = x/v_1, and at v_2 = 65 mi/h the time needed is t_2 = x/v_2. Hence the time saved is

$$\Delta t = t_1 - t_2 = \frac{x}{v_1} - \frac{x}{v_2} = \frac{100 \text{ mi}}{55 \text{ mi/h}} - \frac{100 \text{ mi}}{65 \text{ mi/h}} = 0.28 \text{ h}$$

Since 1 h = 60 min, the time saved in minutes is

$$\Delta t = (0.28 \text{ h})(60 \text{ min/h}) = 17 \text{ min}$$ ◆

◆   E X A M P L E   1.5 _____

A train leaves a station and travels at 80 mi/h for 3.5 h and then at 70 mi/h for 2.0 h to reach its destination. What was the train's average speed for the entire trip?

SOLUTION   To find the average speed of the train we must know the total distance x it has covered and the total time t involved. In the first part of the trip, which lasted 3.5 h, the trained covered

$$x_1 = v_1 t_1 = (80 \text{ mi/h})(3.5 \text{ h}) = 280 \text{ mi}$$

In the second part of the trip, which lasted 2.0 h, it covered

$$x_2 = v_2 t_2 = (70 \text{ mi/h})(2.0 \text{ h}) = 140 \text{ mi}$$

The total distance and total time of the trip are

$$x = x_1 + x_2 = (280 + 140) \text{ mi} = 420 \text{ mi}$$
$$t = t_1 + t_2 = (3.5 + 2.0) \text{ h} \quad = 5.5 \text{ h}$$

The train's average speed was therefore

$$v = \frac{x}{t} = \frac{420 \text{ mi}}{5.5 \text{ h}} = 76 \text{ mi/h}$$ ◆

1.3 VELOCITY

How fast in what direction.

The speed of a moving object specifies only how fast the object is going, regardless of its direction. If we are told that a car has a constant speed of 40 km/h, we do not know where it is headed even though it may be moving along a straight road. A complete description that includes both speed and direction is called **velocity.** If we are told that a car has a constant velocity of 40 km/h toward the west, we know all there is to know about its motion and can easily figure out where it will be in an hour, in two hours, or at any other time.

Because the rate of change of distance tells us only "how fast," the definition of velocity involves a different quantity, **displacement.** The displacement of an object that has moved is its change in position. Suppose an airplane leaves New York City and flies a distance of 640 km in an hour and then lands. From this information all we know is that the airplane must have landed somewhere inside a circle of radius 640 km whose center is New York. The airplane could have turned around and come back to New York and still flown 640 km. If the airplane was displaced by 640 km to the west, however, we know that it landed in Cleveland (Fig. 1.5).

An object in straight-line motion that is at the point x_1 at the time t_1 and at the point x_2 at the time t_2 has been displaced by $x_2 - x_1$. The **average velocity** of the object in the time interval $t_2 - t_1$ is

$$\bar{v}_{vel} = \frac{x_2 - x_1}{t_2 - t_1} \qquad \textit{Average velocity} \quad (1.5)$$

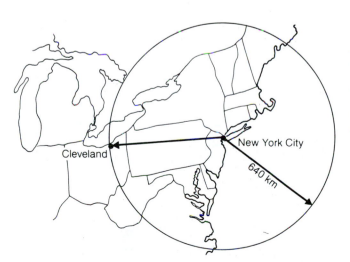

Figure 1.5

An airplane that leaves New York and flies for a distance of 640 km may land anywhere within the circle shown. If the airplane undergoes a displacement of 640 km to the west, it will land in Cleveland.

$$\text{Average velocity} = \frac{\text{displacement}}{\text{time interval}}$$

The average velocity and average speed of a moving object may not be numerically the same. The airplane just described had an average speed of 640 km/h. If it landed in Cleveland, its average velocity would be 640 km/h to the west. If instead the airplane returned to New York, its displacement would be 0 and its average velocity for the flight would also be 0.

On the other hand, the instantaneous speed of an object is always equal to the *magnitude* of its **instantaneous velocity.** An airplane whose instantaneous speed is 640 km/h might have an instantaneous velocity of 640 km/h to the west or 640 km/h to the north, or it might be flying in a circle at 640 km/h, but the magnitude of the airplane's instantaneous velocity is 640 km/h in each case.

1.4   ACCELERATION

Vroom!

In the real world few objects move at constant velocity for very long. An object whose velocity is increasing or decreasing, or whose direction is changed, is said to be **accelerated** (Fig. 1.6). In this chapter we are concerned only with straight-line motion; accelerations that involve changes in direction are discussed later.

Just as velocity is the rate of change of displacement with time, acceleration is the rate of change of velocity with time. If an object's velocity is v_0 to begin with and changes to v_f during a time interval t, its acceleration a is given by

$$a = \frac{v_f - v_0}{t}$$

Acceleration   (1.6)

$$\text{Acceleration} = \frac{\text{change in velocity}}{\text{time}}$$

Figure 1.6

The successive positions of three accelerated cars after equal periods of time. (a) The car is going faster and faster, so it travels a longer distance in each period of time. (b) The car is going slower and slower, so it travels a shorter distance in each period of time. (c) The car's speed is constant, but its direction changes.

(a) Increasing speed

(b) Decreasing speed

(c) Changing direction

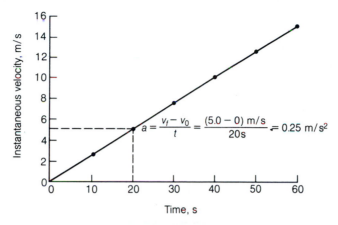

Figure 1.7

A graph of instantaneous velocity versus time for the data in Table 1.4. Although the car's velocity is not constant, it varies in a uniform way with time. This is an example of constant acceleration.

Let us return to the data on the car in Tables 1.3 and 1.4 and plot a graph of its instantaneous velocity v versus time, as in Fig. 1.7. All the points lie on a straight line, so v is directly proportional to t. Although the car's velocity is not constant, it varies in a uniform way with time; as time goes on, the velocity increases exactly in proportion. What is constant here is the car's acceleration.

From its definition, acceleration is expressed in terms of

$$\frac{\text{Speed}}{\text{time}} = \frac{\text{distance/time}}{\text{time}} = \frac{\text{distance}}{(\text{time})(\text{time})} = \frac{\text{distance}}{\text{time}^2}$$

An object whose speed increases by 10 m/s in each second would accordingly have its acceleration expressed as

$$a = 10\frac{\text{m/s}}{s} = 10 \text{ m/s}^2$$

The car whose motion we have been considering started out at $v_0 = 0$, and after $t = 20$ s its velocity was $v_f = 5.0$ m/s. Hence the car's acceleration is

$$a = \frac{v_f - v_0}{t} = \frac{(5.0 - 0)\text{ m/s}}{20\text{ s}} = 0.25 \text{ m/s}^2$$

If we make the same calculation for the later time $t = 40$ s when v_f is 10 m/s, we also get

$$a = \frac{(10 - 0)\text{ m/s}}{40\text{ s}} = 0.25 \text{ m/s}^2$$

Because the value of a is the same, the acceleration is constant. If the acceleration varied, we would get different values of a at different times.

Not all accelerations are constant, of course, but a great many are very nearly so. In what follows, all accelerations are assumed constant unless stated otherwise.

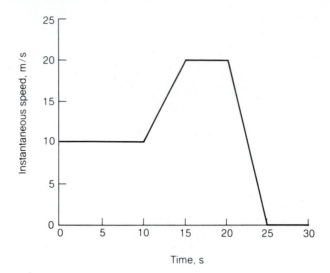

Figure 1.8

Instantaneous velocity versus
time for a certain car.

◆   E X A M P L E   1.6 _____

Discuss the motion of the car whose velocity-time graph is shown in Fig. 1.8.

SOLUTION   A horizontal line on a *v-t* graph means that *v* does not change, so the acceleration is 0 at first. A line sloping upward means a positive acceleration (*v* increasing), and a line sloping downward means a negative acceleration (*v* decreasing). Hence the car begins to move at the constant velocity of 10 m/s, then is accelerated at 2 m/s² to a velocity of 20 m/s and travels at 20 m/s for 5 s. The brakes are now applied and the car undergoes a negative acceleration of −4 m/s² until it comes to a stop at $t = 25$ s.   ◆

1.5   VELOCITY AND ACCELERATION

How fast after an acceleration?

Equation (1.6) can be solved for the final velocity v_f of an object whose initial velocity is v_0. The result is

$$v_f = v_0 + at \qquad\qquad \textit{Final velocity} \quad (1.7)$$

Final velocity = initial velocity + velocity change

See Fig. 1.9.

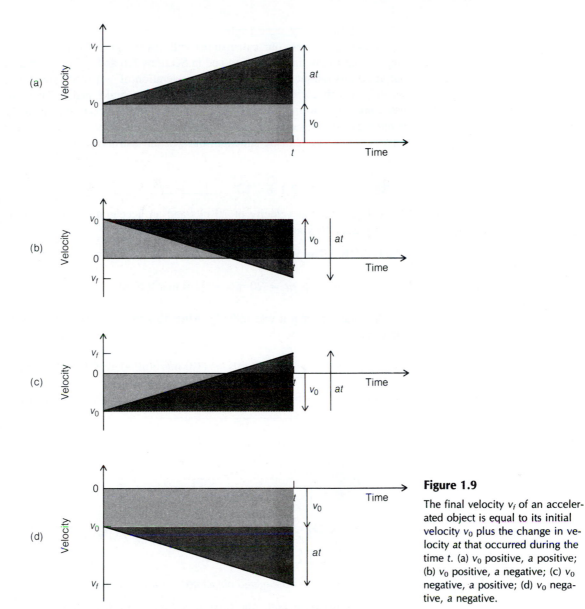

Figure 1.9

The final velocity v_f of an accelerated object is equal to its initial velocity v_0 plus the change in velocity at that occurred during the time t. (a) v_0 positive, a positive; (b) v_0 positive, a negative; (c) v_0 negative, a positive; (d) v_0 negative, a negative.

   In accelerated motion we have three quantities—displacement, velocity, and acceleration—that can be either positive or negative. In the case of a car, a negative velocity implies motion opposite to a specified direction, which we can call backward motion (though it does not matter which way the car actually faces). A negative displacement means the car ends up behind its starting point. A negative acceleration means the car is slowing down if v_0 is positive or that it is increasing in backward velocity if v_0 is negative. When v_0 is negative and a is positive, the car is going backward at a decreasing velocity.

All this may seem confusing, but if we are careful to put the right signs on the known quantities, our calculations will always give the right answers. The examples that follow in this section and in Sections 1.6 and 1.7 concern a car that starts out at 20 m/s (about 45 mi/h) with an acceleration of -1 m/s^2 (about -2.2 mi/h per second), which corresponds to a light foot on the brake pedal. Then, when the car has come to a stop, its driver takes her foot off the brake and instantly shifts into reverse to give the car the same backward acceleration of -1 m/s^2 it had before. This is not a very realistic situation, but it is an instructive one.

◆   **E X A M P L E   1.7** _____

A car has an initial velocity of 20 m/s and an acceleration of -1.0 m/s^2. Find its velocity after 10 s and after 50 s.

SOLUTION   The velocity after 10 s is

$$v_f = v_0 + at = 20 \text{ m/s} - (1.0 \text{ m/s}^2)(10 \text{ s}) = (20 - 10) \text{ m/s} = 10 \text{ m/s}$$

which is half of what it was initially. After 50 s the velocity of the car, if a stays the same, is

$$v_f = v_0 + at = 20 \text{ m/s} - (1.0 \text{ m/s}^2)(50 \text{ s}) = (20 - 50) \text{ m/s} = -30 \text{ m/s}$$

which is in the opposite direction and greater than the original velocity v_0.   ◆

1.6   DISPLACEMENT, TIME, AND ACCELERATION

How far an accelerated object goes in a given time.

As we have seen, the final velocity of an object that has been accelerated for the time t is given by $v_f = v_0 + at$. The next question to ask is, how far does the object go during the time interval t?

If we know the average velocity $\bar{v}$ during the time interval t, we can find $x = \bar{v}t$, the displacement of the object. Because a is constant, v is changing at a uniform rate, and the average velocity can be found in the same way as the average of two numbers:

$$\bar{v} = \frac{v_0 + v_f}{2}$$ *Average velocity under constant acceleration*   (1.8)

Figure 1.10

An object with the initial velocity v_0 at $t = 0$ that undergoes the constant acceleration a is displaced by $x = v_0 t + \frac{1}{2}at^2$ in the time t.

If a is not constant, this formula does not hold.

Here the initial velocity is v_0 and the final velocity is $v_0 + at$. Hence

$$\bar{v} = \frac{v_0 + v_0 + at}{2} = v_0 + \tfrac{1}{2}at$$

The displacement is accordingly

$$x = \bar{v}t = (v_0 + \tfrac{1}{2}at)\,t$$

which yields the very useful formula

$$x = v_0 t + \tfrac{1}{2}at^2 \qquad\qquad \textit{Displacement under constant acceleration} \quad (1.9)$$

Figure 1.10 illustrates this result. When the initial velocity is $v_0 = 0$, we have simply

$$x = \tfrac{1}{2}at^2 \qquad\qquad\qquad\qquad\qquad v_0 = 0 \quad (1.10)$$

◆   **E X A M P L E   1.8** _____

We return to the car with the initial velocity of 20 m/s and the acceleration of -1.0 m/s^2. Find its displacement after the first 10 s and after the first 50 s from the moment the acceleration beings.

SOLUTION   In the first 10 s the car travels

$$x = v_0 t + \tfrac{1}{2}at^2 = (20 \text{ m/s})(10\text{s}) + \tfrac{1}{2}(-1.0 \text{ m/s}^2)(10 \text{ s})^2$$
$$= (200 - 50)\text{ m} = 150\text{ m} = 0.15\text{ km}$$

and its displacement after 50 s is

$$x = v_0 t + \tfrac{1}{2}at^2 = (20 \text{ m/s})(50 \text{ s}) + \tfrac{1}{2}(-1.0 \text{ m/s}^2)(50 \text{ s})^2$$
$$= (1000 - 1250)\text{ m} = -250\text{ m} = -0.25\text{ km}$$

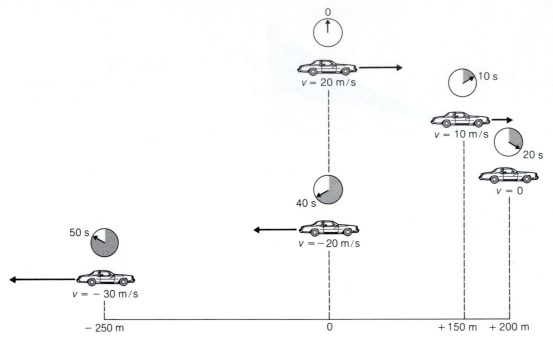

Figure 1.11

Position and velocity, at various times, of a car that has a constant acceleration of -1.0 m/s^2 and whose initial velocity is 20 m/s. At $t = 20$ s the car has come to a stop, and then it begins to move in the negative (backward) direction. At $t = 40$ s the car is back where it started, but moving in the direction opposite to its initial one. At $t = 50$ s the car is 250 m behind its starting point (Example 1.8).

This result means that the car is 0.25 km *behind* its starting point after 50 s have elapsed (Fig. 1.11).   ◆

1.7   DISPLACEMENT, VELOCITY, AND ACCELERATION

Additional formulas for accelerated motion.

It is not hard to find relationships among x, v_0, v_f, and a that do not directly involve the time t. The first step is to rewrite Eq. (1.6) for acceleration as

$$t = \frac{v_f - v_0}{a}$$

Now we substitute this expression for t into the formula for the displacement:

$$x = v_0 t + \tfrac{1}{2} a t^2$$

The result is

$$x = v_0 \frac{(v_f - v_0)}{a} + \tfrac{1}{2} a \frac{(v_f - v_0)^2}{a^2} = \frac{v_0 v_f}{a} - \frac{v_0^2}{a} + \frac{v_f^2}{2a} - \frac{v_0 v_f}{a} + \frac{v_0^2}{2a}$$

$$= \frac{v_f^2 - v_0^2}{2a} \qquad\qquad \textit{Displacement under constant acceleration} \quad (1.11)$$

Another, often useful, way to write this result is:

$$v_f^2 = v_0^2 + 2ax \qquad\qquad \textit{Velocity under constant acceleration} \quad (1.12)$$

◆   **E X A M P L E   1.9** _____

How far will the car in Examples 1.7 and 1.8 have gone when it comes to a stop?

SOLUTION   When the car is at rest, $v_f = 0$. Since $v_0 = 20$ m/s and $a = -1.0$ m/s^2, we have from Eq. (1.11)

$$x = \frac{v_f^2 - v_0^2}{2a} = \frac{0 - (20 \text{ m/s})^2}{(2)(-1.0 \text{ m/s}^2)} = \frac{400 \text{ m}^2/\text{s}^2}{2.0 \text{ m/s}^2} = 200 \text{ m} = 0.20 \text{ km}$$

The car comes to a stop after it has gone 0.20 km. Then, since the acceleration is negative and is assumed here to continue, it begins to move in the negative (backward) direction, as in Fig. 1.11.   ◆

Another useful formula that involves displacement and velocity follows from Eq. (1.8). Multiplying both sides by the time t gives

$$\bar{v}t = \left(\frac{v_0 + v_f}{2} \right) t$$

Since $\bar{v}t$ is the displacement x when the acceleration is constant,

$$x = \left(\frac{v_0 + v_f}{2} \right) t \qquad\qquad \textit{Displacement under constant acceleration} \quad (1.13)$$

This formula differs from Eq. (1.11) in that it contains the time t but not the acceleration a. Thus Eq. (1.13) is handy when the value of a neither is known nor has to be determined.

All problems that involve straight-line motion under constant acceleration can be solved by using the formulas developed thus far in this chapter. These formulas are listed in Table 1.5 for convenience. Depending on the problem, a particular formula may have to be rewritten to suit a particular combination of known and

Table 1.5   Formulas for motion under constant acceleration

$x = \left(\dfrac{v_0 + v_f}{2} \right) t$	(1.13)
$x = v_0 t + \tfrac{1}{2} a t^2$	(1.9)
$v_f = v_0 + at$	(1.7)
$v_f^2 = v_0^2 + 2ax$	(1.12)

unknown quantitities. For instance, Eq. (1.12) can be solved for four different quantities:

$$v_f = \sqrt{v_0^2 + 2ax} \qquad v_0 = \sqrt{v_f^2 - 2ax}$$

$$a = \frac{v_f^2 - v_0^2}{2x} \qquad x = \frac{v_f^2 - v_0^2}{2a}$$

1.8   HOW TO SOLVE PROBLEMS

Eight steps.

Until now the examples used to illustrate the ideas of the text were easy to understand and to solve. Unfortunately not all problems in physics are so simple. When faced with a seemingly difficult problem, newcomers to physics usually feel unsure of exactly how to proceed. This is a natural response. What is needed is an orderly approach, which can be summarized as follows:

1. **Draw a neat, reasonably accurate sketch where appropriate.**
2. **Write down all the known quantities. If necessary, convert them to the proper units. Put these quantities on the sketch.**
3. **Write down the unknown quantity (or quantities) and show it on the sketch as well.**
4. **Find the basic equation that connects known and unknown quantities.**
5. **Solve the basic equation for the unknown quantity. The apppendix may help: All the mathematics needed for this book is reviewed there.**
6. **Substitute the known quantities in the resulting formula, including all the units. If more information was given than needed to work out the problem, which sometimes happens, one or more of the known quantities will not have to be used.**
7. **Carry out the indicated operations to get the answer. Be sure the units are the same on both sides of the equals sign and that the correct number of significant figures appears in the answer.**
8. **Ask yourself if the answer is reasonable. Often a scale drawing can be made in Step 1 and the answer checked by a measurement on the drawing. If the units of the answer are strange to you, you might express the answer in more familiar units to see if it makes sense.**

◆   E X A M P L E   1.10 —————————————————————————

The driver of a train traveling at 30 m/s applies the brakes when he passes an amber signal. The next signal is 1.5 km down the track and the train reaches it 75 s later. Assuming a uniform acceleration, find the velocity of the train at the second signal.

SOLUTION   **Step 1.** *Sketch.* See Fig. 1.12.

Step 2. *Known Quantities.*

Initial velocity = v_0 = 30 m/s
Distance of acceleration = x = 1.5 km = 1500 m
Time of acceleration = t = 75 s

Step 3. *Unknown Quantity.*

Final velocity = v_f = ?

Step 4. *Basic Equation.* We do not need to know the actual value of the train's acceleration to find v_f if we use Eq. (1.13) which is

$$x = \left(\frac{v_0 + v_f}{2}\right)t$$

Step 5. *Working Equation.* We solve for v_f as follows:

$$x = \left(\frac{v_0 + v_f}{2}\right)t$$
$$2x = (v_0 + v_f)t$$
$$\frac{2x}{t} = v_0 + v_f$$
$$v_f = \frac{2x}{t} - v_0$$

Step 6. *Substitution.*

$$v_f = \frac{2x}{t} - v_0 = \frac{(2)(1500 \text{ m})}{75 \text{ s}} - 30 \text{ m/s}$$

Figure 1.12

Example 1.10.

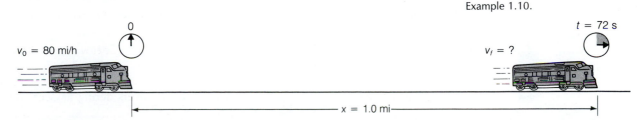

v_0 = 80 mi/h   0

t = 72 s

v_f = ?

———————————————— x = 1.0 mi ————————————————

Step 7. *Calculation.*

$$v_f = \frac{(2)(1500 \text{ m})}{75 \text{ s}} - 30 \text{ m/s} = 40 \text{ m/s} - 30 \text{ m/s} = 10 \text{ m/s}$$

Step 8. *Check.* The answer seems reasonable. One way to check further is to work out the train's acceleration. From Eq. (1.7) we have

$$v_f = v_0 + at$$
$$a = \frac{v_f - v_0}{t} = \frac{(10 - 30)\text{m/s}}{75 \text{ s}} = -0.27 \text{ m/s}^2$$

The train must slow down by 0.27 m/s in each second, which is plausible. With this value of a we obtain, as before,

$$v_f = v_0 + at = 30 \text{ m/s} + (-0.27 \text{ m/s}^2)(75 \text{ s})$$
$$= 30 \text{ m/s} - 20 \text{ m/s} = 10 \text{ m/s}$$

An elaborate check like this one is not always necessary, but it is a good idea if we are not sure of our algebra in Step 5 or of our arithmetic in Step 7.   ◆

Figure 1.13

Example 1.11.

Other problems may be more complicated than Example 1.10. In such cases, several intermediate calculations may be needed to find the final answer. Example 1.11 is such a problem. Although each step is not shown as completely as in Example 1.10, both the reasoning behind the method of attack and the algebra involved are given in enough detail to show how a complex problem can be broken up into a series of simpler ones.

◆   **E X A M P L E   1.11**

A car is stationary in front of a red traffic light. As the light turns green, a truck goes past at a constant velocity of 15 m/s. At the same moment, the car begins to accelerate at 1.25 m/s². When it reaches 25 m/s, the car continues at this velocity. When does the car pass the truck? How far have they gone from the traffic light at that time?

SOLUTION   (a) The car's final velocity is $v_1 = 25$ m/s, and the truck's constant velocity is $v_2 = 15$ m/s (Fig. 1.13). The car needs the time

$$t_1 = \frac{v_1}{a} = \frac{25 \text{ m/s}}{1.25 \text{ m/s}^2} = 20 \text{ s}$$

to reach the velocity v_1. To find the distance the car covers during its acceleration, we use Eq. (1.11) with $v_0 = 0$ and $v_f = v_1$. The result is

$$x_1 = \frac{v_1^2}{2a} = \frac{(25 \text{ m/s})^2}{(2)(1.25 \text{ m/s}^2)} = 250 \text{ m} = 0.25 \text{ km}$$

If t is the total time the car needs to catch up with the truck, the car travels the further distance

$$x_2 = v_1 (t - t_1)$$

for a total distance of

$$x = x_1 + x_2 = x_1 + v_1 (t - t_1)$$

The truck meanwhile covers the same distance in the same time at the constant velocity v_2, so

$$x = v_2 t$$

Setting equal these two formulas for x gives

$$v_2 t = x_1 + v_1(t - t_1)$$

$$t = \frac{x_1 - v_1 t_1}{v_2 - v_1} = \frac{250 \text{ m} - (25 \text{ m/s})(20 \text{ s})}{(15 \text{ m/s}) - (25 \text{ m/s})} = 25 \text{ s}$$

The car passes the truck after 25 s.

(b) We can use either formula for x. The second one gives

$$x = v_2 t = (15 \text{ m/s})(25 \text{ s}) = 375 \text{ m} = 0.38 \text{ km} \qquad \blacklozenge$$

According to legend, Galileo dropped a bullet and a cannon-ball from the Leaning Tower of Pisa to show that all falling objects have the same acceleration.

1.9   ACCELERATION OF GRAVITY

Experiment and observation are the foundations of science.

Drop a stone, and it falls. Does the stone fall at constant speed, or is it accelerated? Does the motion of the stone depend on its shape, its size, or its color?

Long ago such questions were answered by Greek philosophers, notably Aristotle, by reasoning based on what to them were self-evident principles. One such principle was that each kind of material had a natural place where it belonged and

Time   Speed

0   0

1 s   9.8 m/s

2 s   19.6 m/s

3 s   29.4 m/s

4 s   39.2 m/s

Figure 1.14

If air resistance is neglected, all freely falling objects near the earth's surface have the same acceleration of $g = 9.8$ m/s². The longer an object falls, the faster it falls.

toward which it tried to go. Thus fire rose naturally upward toward the sun and stars. Stones were "earthy" and so fell down toward their home in the earth. A big stone was more earthy than a small one and so should fall faster.

For over two thousand years almost nobody felt it necessary to perform experiments to seek information on the physical universe. Then Galileo (1564–1642) revolutionized science by doing just that: performing experiments. Galileo's universe turned out to be perhaps less charming than that of Aristotle, but it had the merit of fitting the facts. Modern science owes its success in understanding and utilizing natural phenomena to its reliance on experiment and observation.

What Galileo found, as the result of careful measurements, was that *all freely falling objects have the same acceleration* at the same place near the earth's surface. This acceleration, which is called the acceleration of gravity (symbol *g*), does not have the same value everywhere, for reasons given in Chapter 6. The worldwide average of *g* at sea level is 9.81 m/s², but *g* is 9.83 m/s² at the poles, 9.78 m/s² at the equator, and still less atop Mount Everest. In this book we use

$$g = 9.80 \text{ m/s}^2 \qquad \text{\textit{Acceleration of gravity}}$$

which is the average for the United States. In U.S. Customary Units $g = 32.2$ ft/s².

An object falling freely from rest thus has an instantaneous speed of 9.8 m/s (32 ft/s) after the first second, a speed of 19.6 m/s (64 ft/s) after the next second, and so on. The longer the time during which a stone falls after being dropped, the greater its speed when it hits the ground (Fig. 1.14). But the stone's *acceleration* is always the same.

Another aspect of Galileo's work deserves comment. His conclusion that all things fall with the same constant acceleration is an *idealization* of reality. The actual accelerations with which objects fall depend on many factors besides the location on the earth—for instance, the size and shape of the object and the density and state of the atmosphere. A stone, for example, falls faster than a feather does in air because of the effects of buoyancy and air resistance. Galileo saw that the basic phenomenon was a constant acceleration downward, with other factors acting merely to cause deviations from the constant value. In a vacuum the stone and feather fall with exactly the same acceleration (Fig. 1.15).

Why Raindrops Don't Kill

The drag force due to air resistance on an object of a given mass, size, and shape depends on the speed of the object—the faster it goes, the more the drag. In the case of a falling object, the drag force increases as the speed increases until finally it cannot go any faster. The object then continues to fall at a constant **terminal speed** (Fig. 1.16 and Table 1.6). The terminal speed of a person in free fall is about 54 m/s (120 mi/h), whereas it is only about 6.3 m/s (14 mi/h) with an open parachute. In the absence of air resistance, raindrops would reach the ground at speeds high enough to be dangerous.

A cloud consists of tiny water droplets or ice crystals whose terminal speeds are so small, often 1 cm/s or less, that very little updraft is needed to keep them

Figure 1.15

All objects that fall in a vacuum near the earth's surface have the same downward acceleration. In air, however, a stone falls faster than a feather because air resistance affects it less.

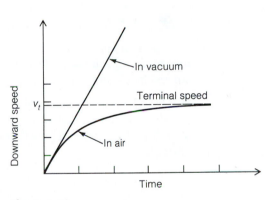

Figure 1.16

This graph compares how the speed of a falling body varies with time when it is in a vacuum and when it is in air. In a vacuum the speed is given by $v = gt$ and increases without limit until the body strikes the ground. In air the terminal speed of v_t is eventually reached, and the body continues to fall at this speed. The terminal speed of a person falling from an airplane is considerably reduced by a parachute, which allows a safe landing.

suspended indefinitely. When a cloud is rapidly cooled, however, the water droplets or ice crystals grow in size and weight until their terminal speeds become too great for updrafts to keep them aloft. The result is a fall of raindrops or snowflakes from the cloud.

Table 1.6 Some approximate terminal speeds (1 m/s = 2.2 mi/h)

Ball	Terminal speed	Ball	Terminal speed
16-lb shot	145 m/s	Tennis ball	30 m/s
Baseball	42	Basketball	20
Golf ball	40	Ping-Pong ball	9

1.10 FREE FALL

What goes up must come down.

We can apply the formulas for motion under constant acceleration to objects in free fall. It is important to keep in mind that the direction of the acceleration of gravity g is always downward, no matter whether we are dealing with a dropped object or with one that is initially thrown upward.

◆   **E X A M P L E   1.12** _____

A stone is dropped from the top of a building 1000 ft high. How long does it take the stone to reach the ground? (Neglect air resistance.) What is its velocity when it strikes the ground? (See Fig. 1.17.)

SOLUTION   In a problem like this one, where the motion is entirely downward, it is easiest to consider down as the positive direction, so that $y = 1000$ ft and $g = 32.2$ ft/s^2. The general formula for the distance traveled in the time t by an accelerated object is

$$y = v_0 t + \tfrac{1}{2} a t^2$$

Here $v_0 = 0$, since the stone is simply dropped from rest, and the acceleration is

$$a = g.$$

Hence we have
$$y = \tfrac{1}{2} g t^2$$

where y represents vertical distance from the starting point. Solving for t gives

$$t = \sqrt{\frac{2y}{g}} = \sqrt{\frac{(2)(1000 \text{ ft})}{32.2 \text{ ft/s}^2}} = 7.88 \text{ s}$$

Knowing the time of fall makes it simple to find the stone's final velocity:

$$v_f = v_0 + at = 0 + gt = (32.2 \text{ ft/s}^2)(7.88 \text{ s}) = 254 \text{ ft/s} \qquad ◆$$

The speed (that is, magnitude of velocity) that a dropped object has when it reaches the ground is the same as the speed with which it must be thrown upward from the

Figure 1.17

Example 1.17.

$v_o = 0$ $t = 0$

$y = 1000$ ft

$t = 7.9$ s

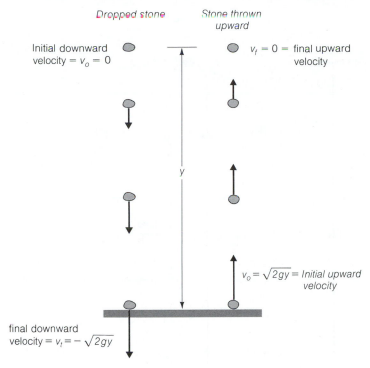

Figure 1.18

A stone dropped from a height y reaches the ground with the speed $\sqrt{2gy}$. In order to reach the height y, a stone thrown upward from the ground must have the minimum speed $\sqrt{2gy}$.

ground to rise to the same height (Fig. 1.18). To prove this statement, we refer to the formula $v_f^2 = v_0^2 + 2ax$ and replace the x with y to give

$$v_f^2 = v_0^2 + 2ay$$

When a stone is dropped, $a = g$ and $v_0 = 0$, so

$$v_f^2 = 0 + 2gy \qquad v_f = \sqrt{2gy}$$

When the stone is thrown upward, on the other hand, $a = -g$ (since the downward acceleration is opposite in direction to the upward initial speed), and at the top of its path $v = 0$. Hence

$$0 = v_0^2 - 2gy \qquad v_0 = \sqrt{2gy}$$

The speed is the same in both cases.

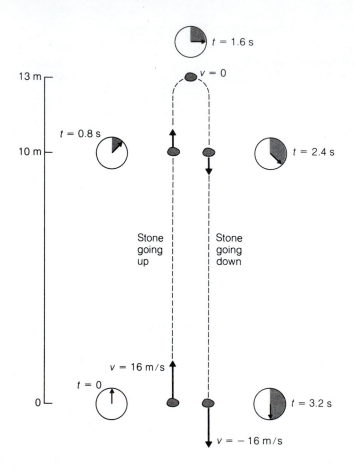

Figure 1.19

The path of a stone thrown upward with an initial speed of 16 m/s. Air resistance is neglected. (Example 1.13).

◆   **E X A M P L E   1.13** _____

A stone is thrown upward with an initial speed of 16 m/s (Fig. 1.19). (a) What will its maximum height be? (b) When will it return to the ground? (c) Where will it be in 0.8 s? (d) Where will it be in 2.4 s?

SOLUTION   (a) We make use of Eq. (1.11) to find the highest point the stone will reach, with up as positive ($+$) and down as negative ($-$). Here $v_0 = 16$ m/s and $a = -9.8$ m/s^2, and at the top of the path $x = y$ and $v_f = 0$. Hence

$$y = \frac{v_f^2 - v_0^2}{2a} = \frac{0 - (16\,\text{m/s})^2}{2(-9.8\,\text{m/s}^2)} = \frac{256\,(\text{m/s})^2}{19.6\,\text{m/s}^2} = 13\,\text{m}$$

(b) When will the stone strike the ground? An object takes precisely as long to fall from a certain height y as it does to rise that high (provided that y is its maximum height, as it is here), just as an object's final speed when dropped from a height y is the same as the initial speed required for it to get that high. From Eq. (1.10)

$$y = \tfrac{1}{2}gt^2$$

from which we find that

$$t = \sqrt{\frac{2y}{g}} = \sqrt{\frac{(2)(13\ \text{m})}{9.8\ \text{m/s}^2}} = 1.6\ \text{s}$$

Because the stone takes as long to rise as to fall, the total time it is in the air is twice 1.6 s, or 3.2 s. (We could just as well have started from $v_f = v_0 + at$ with $v_f = 0$ and $a = -g$ at the top of the path.)

(c) To find the height of the stone a given time after it was thrown upward, we make use of Eq. (1.9), $x = v_0 t + \frac{1}{2}at^2$, with $x = y$, $v_0 = 16$ m/s, and $a = -9.8$ m/s^2. For $t = 0.8$ s,

$$y = (16\ \text{m/s})(0.8\text{s}) - \tfrac{1}{2}(9.8\ \text{m/s}^2)(0.8\ \text{s})^2 = 10\ \text{m}$$

(d) When we substitute $t = 2.4$ s in the formula given above, the result is again

$$y = (16\ \text{m/s})(2.4\ \text{s}) - \tfrac{1}{2}(9.8\ \text{m/s}^2)(2.4\ \text{s})^2 = 10\ \text{m}$$

What this result means is that at 0.8 s the stone is at a height of 10 m on its way up, then it goes on further to its maximum height of 13 m, and at 2.4 s it is once more at a height of 10 m but now on the way down. ◆

Important Terms

To measure a quantity means to compare it with a standard quantity of the same kind called a **unit**. In a **system of units,** a set of basic units is specified from which all other units in the system are derived. **SI** units are used in everyday life in much of the world and universally by scientists.

The **average speed** of a moving object is the distance it covers in a time interval divided by the time interval. The object's **instantaneous speed** at a certain moment is the rate at which it is covering distance at that moment.

The **velocity** of an object describes both its speed and its direction of motion.

The **acceleration** of an object is the rate at which its velocity changes with time. Changes in direction as well as of speed are considered accelerations.

The **acceleration of gravity** is the acceleration of a freely falling body near the earth's surface. The symbol of the acceleration of gravity is g, and its value is 9.8 m/s^2 (32 ft/s^2).

Important Formulas

Speed: $\quad v = \dfrac{x}{t}$

Constant speed: $\quad x = vt$

$$t = \dfrac{x}{v}$$

Acceleration: $\quad a = \dfrac{v_f - v_0}{t}$

Motion under constant acceleration: $\quad x = \dfrac{v_0 + v_f}{2}t$

$$x = v_0 t + \tfrac{1}{2}at^2$$
$$v_f = v_0 + at$$
$$v_f^2 = v_0^2 + 2ax$$

Free fall from rest: $\quad y = \tfrac{1}{2}gt^2$

$$v_f = \sqrt{2gy}$$

(Air resistance is assumed negligible in the following exercises and problems.)

MULTIPLE CHOICE

1. The prefix micro represents
   a. 1/10
   b. 1/100
   c. 1/1000
   d. 1/1,000,000

2. A centimeter is
   a. 0.001 m
   b. 0.01 m
   c. 0.1 m
   d. 10 m

3. Of the following, the shortest is
   a. 1 mm
   b. 0.01 in
   c. 0.001 ft
   d. 0.00001 km

4. Of the following, the longest is
   a. 10^4 in
   b. 10^4 m
   c. 10^3 ft
   d. 0.1 mi

5. A height of 5 ft 8 in is equivalent to
   a. 173 cm
   b. 177 cm
   c. 207 cm
   d. 223 cm

6. The number of cubic centimeters in a cubic foot is approximately
   a. 1.7×10^3
   b. 1.7×10^4
   c. 2.8×10^4
   d. 1.7×10^5

7. The number of seconds in a month is approximately
   a. 2.6×10^6
   b. 2.6×10^7
   c. 2.6×10^8
   d. 2.6×10^9

8. On a distance-time graph, a horizontal straight line corresponds to motion at
   a. zero speed
   b. constant speed
   c. increasing speed
   d. decreasing speed

9. On a distance-time graph, a straight line sloping upward to the right corresponds to motion at
   a. zero speed
   b. constant speed
   c. increasing speed
   d. decreasing speed

10. On a velocity-time graph, the motion of a car traveling along a straight road with the uniform acceleration of 2 m/s^2 would appear as a
    a. horizontal straight line
    b. straight line sloping upward to the right
    c. straight line sloping downward to the right
    d. curved line whose downward slope to the right increases with time

11. The acceleration of a stone thrown upward is

a. greater than that of a stone thrown downward
b. the same as that of a stone thrown downward
c. smaller than that of a stone thrown downward
d. zero until it reaches the highest point in its motion

12. A stone is thrown upward from a roof at the same time as an identical stone is dropped from there. The two stones
    a. reach the ground at the same time
    b. have the same velocity when they reach the ground
    c. have the same acceleration when they reach the ground
    d. None of the above

13. A bicycle travels 12 km in 40 min. Its average speed is
    a. 0.3 km/h
    b. 8 km/h
    c. 18 km/h
    d. 48 km/h

14. A car that travels at 40 km/h for 2.0 h, at 50 km/h for 1.0 h, and at 20 km/h for 0.50 h, has an average speed of
    a. 31 km/h
    b. 40 km/h
    c. 45 km/h
    d. 55 km/h

15. A pitcher takes 0.10 s to throw a baseball, which leaves his hand with a velocity of 30 m/s. The ball's acceleration was
    a. 3 m/s^2
    b. 30 m/s^2
    c. 300 m/s^2
    d. 3000 m/s^2

16. How long does a car with an acceleration of 2.0 m/s^2 take to go from 10 m/s to 30 m/s?
    a. 10 s
    b. 20 s
    c. 40 s
    d. 400 s

17. A car undergoes a constant acceleration of 6.0 m/s^2 starting from rest. In the first second it travels
    a. 3 m
    b. 6 m
    c. 18 m
    d. 36 m

18. An airplane requires 20 s and 400 m of runway to become airborne, starting from rest. Its velocity when it leaves the ground is
    a. 20 m/s
    b. 32 m/s
    c. 40 m/s
    d. 80 m/s

19. A car has an initial velocity of 50 ft/s and an acceleration of 4 ft/s^2. In the first 10 s after the acceleration begins, the car travels
    a. 200 ft
    b. 500 ft
    c. 700 ft
    d. 900 ft

20. A car has an initial velocity of 15 m/s and an acceleration of -1.0 m/s^2. In the first 10 s after the acceleration begins,

the car travels

a. 25 m
b. 50 m
c. 100 m
d. 145 m

21. How far does the car in the previous question go before coming to a stop?

a. 113 m
b. 150 m
c. 225 m
d. 450 m

22. A wheel falls from an airplane flying horizontally at an altitude of 490 m. The wheel strikes the ground in

a. 10 s
b. 50 s
c. 80 s
d. 100 s

23. A stone is dropped from a cliff. After it has fallen 30 m, its velocity is

a. 17 m/s
b. 24 m/s
c. 44 m/s
d. 588 m/s

24. Two balls are thrown vertically upward, one with an initial velocity twice that of the other. The ball with the greater initial velocity will reach a height

a. $\sqrt{2}$ that of the other

b. twice that of the other
c. 4 times that of the other
d. 8 times that of the other

25. A ball thrown vertically upward at 80 ft/s continues to rise for approximately

a. 2.5 s
b. 5 s
c. 7.5 s
d. 10 s

26. In the previous question, how much time will elapse before the ball strikes the ground?

a. 2.5 s
b. 5 s
c. 7.5 s
d. 10 s

27. A ball is thrown downward at 5 m/s from a roof 10 m high. Its velocity when it reaches the ground is

a. 13 m/s
b. 14 m/s
c. 15 m/s
d. 20 m/s

28. Another ball is thrown upward at 5 m/s from the same roof. Its velocity when it reaches the ground is

a. 13 m/s
b. 14 m/s
c. 15 m/s
d. 20 m/s

QUESTIONS

1. In the following pairs of length units, which is the shortest? Inch, centimeter; yard, meter; mile, kilometer.

2. Figure 1.20 is a graph that shows the displacement of an object along a straight line plotted against time. Describe the motion of the object.

Figure 1.20

Question 2.

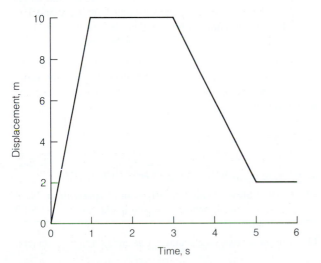

3. Can a rapidly moving object have the same acceleration as a slowly moving one? If so, give an example.

4. The acceleration of a certain moving object is constant in magnitude and direction. Must the path of the object be a straight line? If not, give an example.

5. Figure 1.21 shows how the velocity of a car varied during a period of time. (a) In what time interval was the car's acceleration highest? (b) In what time interval did the car cover the most distance?

Figure 1.21

Question 5.

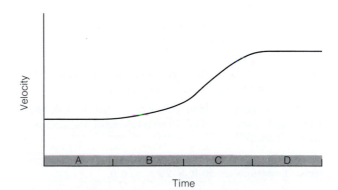

6. Figure 1.22 shows displacement-time graphs for nine cars.

 a. Which cars are or have been moving in the forward direction?
 b. Which cars are or have been moving in the backward direction?
 c. Which car has the highest constant velocity?
 d. Which car has the highest constant velocity in the forward direction?
 e. Which car has the highest constant velocity in the backward direction?
 f. Which cars have the same velocity?
 g. Which car has not moved at all?
 h. Which car has accelerated from rest to a constant velocity?
 i. Which car has been brought to a stop from an initial velocity in the forward direction?
 j. Which car has been brought to a stop from an initial velocity in the backward direction?

7. A movie is shown that appears to be of a ball falling through the air. From what appears on the screen, is there any way to determine whether the movie is actually of a ball being thrown upward but the film is being run backward in the projector?

8. The acceleration of gravity at the surface of Venus is 8.9 m/s². Would a ball thrown upward on Venus return to the ground sooner or later than a ball thrown upward with the same velocity on the earth?

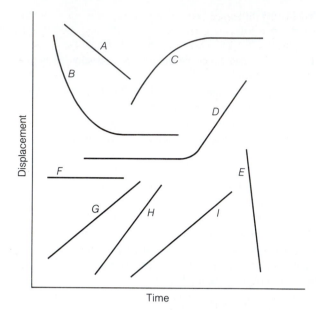

Figure 1.22

Question 6.

EXERCISES

1.1 Units

1. The tallest tree in the world is a Sequoia in California that is 368 ft high. How high is this in meters? In kilometers?

2. A moderately large tree gives off 300 gal of water daily to the atmosphere. How many cubic meters is this?

3. The speedometer of a European car is calibrated in kilometers per hour. What is the car's speed in miles per hour when the speedometer reads 50?

4. In 1968 the horse Dr. Fager ran a mile in 1 min 32.2 s. Find his average speed in meters per second.

5. An acre contains 4840 yd², where 1 yd = 3 ft. How many square meters is this? How many acres are there in a square kilometer?

6. Find the volume in cubic meters of a swimming pool whose dimensions are 50 ft × 20 ft × 6 ft.

7. A "board foot" is a unit of lumber measure that corresponds to the volume of a piece of wood 1 ft square and 1 in thick. How many cubic inches are there in a board foot? How many cubic feet? How many cubic centimeters?

8. Water emerges from the nozzle of a fountain in Arizona at 75 km/h and reaches a height of 170 m. Express the water speed in meters per second and in miles per hour and the height in feet.

1.2   Speed

9. A pitcher throws a baseball at 90 mi/h. How much time does it take the ball to reach the batter 60 ft away?

10. In 1977 Steve Weldon ate 91.44 m of spaghetti in 28.73 s. At the same speed, how long would it take him to eat 15.00 m of spaghetti?

11. The starter of a race stands at one end of a line of runners.

What is the difference between the arrival of the sound of his pistol at the nearest runner and at the most distant runner 10 m farther away? The speed of sound in air is 343 m/s. (In a sprint, 0.01 s can mean the difference between winning and coming in second.)

12. The speed of light is 3.0×10^8 m/s. How long does it take light to reach the earth from the sun, which is 1.5×10^{11} m away?

13. The average lifetime of many unstable elementary particles is about 10^{-23} s, and they move at nearly the speed of light (3.0×10^8 m/s) after being created in high-energy collisions between other elementary particles. As measured in the laboratory, how far does such a particle travel on the average before decaying?

14. A car travels 540 km in 4.5 h. How far will it go in 8.0 h at the same average speed? How long will it take to go 200 km at this speed?

15. An airplane takes off at 9:00 A.M. and flies at 300 km/h until 1:00 P.M. At 1:00 P.M. its speed is increased to 400 km/h, and it maintains this speed until it lands at 3:30 P.M. What is the airplane's average speed for the entire flight?

16. A car travels at 100 km/h for 2.0 h, at 60 km/h for the next 2.0 h, and finally at 80 km/h for 1.0 h. Find its average speed for the entire 5.0 h.

17. A police car moving at 140 km/h chases a speeding car moving at 110 km/h. How much closer to the speeding car does the police car get each minute?

18. A police car leaves in pursuit of a holdup car 0.50 h after the latter has left the scene of the crime at 75 mi/h. How fast must the police car go if it is to catch up with the holdup car in 1.00 h?

19. A Honda moving at 100 km/h is 80 km behind a Ford moving in the same direction at 60 km/h. How far does the Honda travel before it catches up with the Ford?

1.5 Velocity and Acceleration

20. A baseball moving at 20 m/s is struck by a bat and moves off in the opposite direction at 30 m/s. If the impact lasted 0.010 s, what was the baseball's acceleration?

21. (a) A car's velocity increases from 8 m/s to 20 m/s in 10.0 s. Find its acceleration. (b) The car's velocity then decreases from 20 m/s to 10 m/s in 5.0 s. Find its acceleration in this time interval.

22. The tires of a certain car begin to lose their grip on the pavement at an acceleration of 5.0 m/s². If the car has this acceleration, how many seconds does it require to reach a speed of 25 m/s starting from 10 m/s?

23. A car starts from a velocity of 10 m/s with an acceleration of 2.0 m/s². Find the time needed for the car to reach 20 m/s.

24. An airplane reaches its takeoff velocity of 60 m/s in 30 s starting from rest. Find the time it spends in going from 40 m/s to 60 m/s.

25. The velocity-time graph in Fig. 1.23 represents the motion of a golf cart. Find its acceleration at $t = 25$ s and at $t = 35$ s.

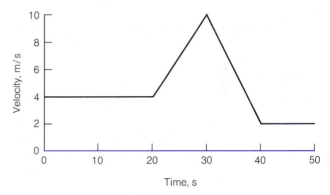

Figure 1.23

Exercise 25.

26. The brakes of a car moving at 14 m/s are suddenly applied and the car comes to a stop in 4.0 s. (a) What was its acceleration? (b) How long would the car take to come to a stop starting from 20 m/s with the same acceleration? (c) How long would the car take to slow down from 20 m/s to 10 m/s with the same acceleration?

27. A passenger in an airplane flying from New York to Los Angeles notes the time at which he passes over various cities and towns. With the help of a map he determines the distances between these landmarks, and compiles the table shown below. Plot the displacement of the airplane versus time from these data, and describe the airplane's motion with the help of the graph.

Time (P.M.):	4:00	5:12	5:41	6:14
Distance (km):	0	660	926	1267

Time (P.M.):	6:39	7:54	9:18	10:00
Distance (km):	1525	2300	3028	3392

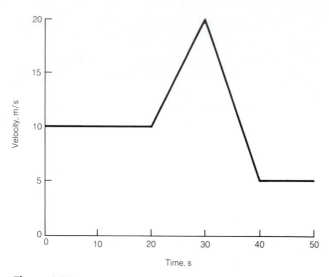

Figure 1.24

Exercise 28.

28. The velocity-time graph in Fig. 1.24 represents the motion of a certain car. (a) Find the car's acceleration at t = 15, 25, 35, and 45 s. (b) Find the distance covered by the car from t = 0 to t = 30 s and from t = 30 s to t = 50 s.

1.6   Displacement, Time, and Acceleration

29. A Porsche reaches a velocity of 42.0 km/h from a standing start in 15.5 s. What distance does it cover while doing so?

30. A Ferrari covers 300 ft from a standing start in 6.0 s at constant acceleration. Find its final velocity in mi/h.

31. A DC-8 airplane has a takeoff velocity of 80 m/s, which it reaches 35 s after starting from rest. (a) How much time does the airplane spend in going from 0 to 20 m/s? What distance does it cover in doing so? (b) How much time does the airplane spend in going from 60 to 80 m/s? What distance does it cover in doing so? (c) What is the minimum length of the runway?

32. A car starts from rest with an acceleration of 0.40 m/s². One minute later the acceleration stops, and the car continues at constant velocity for 5.0 min. The car's brakes are then applied, and the car comes to rest in 2.0 min with a uniform deceleration. (a) Show the car's motion on a velocity-time graph. (b) Find the total displacement of the car.

1.7   Displacement, Velocity, and Acceleration

33. A spacecraft takes 200 km after being launched to reach the escape speed from the earth of 11.2 km/s. Find its average acceleration in terms of g.

34. A golf cart has an acceleration of 0.4 m/s². What is its velocity after it has covered 10 m starting from rest?

35. A spacecraft has an acceleration of magnitude 5.0 g. What distance is needed for it to attain a velocity of 10 km/s?

36. A car moving at 20 m/s slows down at −1.5 m/s² to a velocity of 10 m/s. How far did the car go during the slow-down?

37. An express train passes a certain station at 20 m/s. The next station is 2 km away and the train reaches it 1.0 min later. (a) Did the train's velocity change? (b) If it did, what was its velocity at the second station? (Assume a constant acceleration.)

38. The time interval between a driver noticing a hazard ahead and her application of the brakes is typically 0.68 s. The maximum acceleration produced by the brakes of an average car might be 6.5 m/s². Find the distance required to stop such a car from the moment a hazard is seen when the initial velocity is (a) 25 km/h, (b) 50 km/h, and (c) 100 km/h.

1.10   Free Fall

39. Is it true that an object dropped from rest falls three times farther in the second second after being released than it does in the first second?

40. How fast must a ball be thrown upward to reach a height of 12 m?

41. Divers in Acapulco leap from a point 36 m above the sea. What is their velocity when they enter the water?

42. A stone is dropped from a cliff 490 m above its base. How long does the stone take to fall?

43. A ball dropped from the roof of a building takes 4.0 s to reach the street. How many feet high is the building?

44. A bullet is fired vertically upward and returns to the ground in 20 s. Find the height it reaches.

45. Find the initial and final velocities in ft/s of a ball thrown vertically upward that returns to the thrower 3.0 s later.

46. A stone is thrown vertically upward at 9.8 m/s. When will it reach the ground?

47. A ball is thrown vertically downward at 10 m/s. What is its velocity 1.0 s later? 2.0 s later?

48. A ball is thrown vertically upward at 10 m/s. What is its velocity 1.0 s later? 2.0 s later?

49. A lead pellet is propelled vertically upward by an air rifle with an initial velocity of 16 m/s. Find its maximum height.

50. At what velocity should a stone be thrown downward from a cliff 120 m high to reach the ground in 4.0 s?

51. An apple is thrown vertically downward from a cliff 48 m high and reaches the ground 2.0 s later. What was the apple's initial velocity?

52. From Fig. 1.16 we can see that the air resistance experienced by a falling object is not an important factor until a speed of about half its terminal speed is reached. The terminal speed of a golfball is 40 m/s. How much time is needed for a dropped golfball to reach a speed of half this? How far does it fall in this time?

53. The acceleration of gravity at the surface of Mars is 3.7 m/s^2. A stone thrown upward on Mars reaches a height of 15 m. (a) Find its initial velocity. (b) What is the total time of flight?

54. A movie scene in which a car is supposed to fall from a cliff 36 m high is filmed using a model car that falls from a model cliff 1.0 m high. The film will be shown at 24 frames/s. How fast should the camera be run for the result to be realistic?

55. In a novel a pirate forces a captured sailor to ''walk the plank'' (something unknown in the history of piracy), from which he falls into the sea 7.0 m below. The author claims the sailor's entire life flashes before him during his fall. (a) How much time is available for this? (b) With what velocity will the sailor enter the water?

56. A bagel is thrown vertically upward with a velocity of 60 ft/s from a bridge. On the way down the bagel just misses the bridge and falls into the water 5.0 s after having been thrown. Find the height of the bridge above the water.

57. A British parachutist bails out at an altitude of 150 m and accidentally drops his monocle. If he descends at the constant velocity of 6.0 m/s, how much time separates the arrival of the monocle on the ground from the arrival of the parachutist himself?

58. A Russian balloonist floating at an altitude of 150 m accidentally drops his samovar and starts to ascend at the constant velocity of 1.2 m/s. How high will the balloon be when the samovar reaches the ground?

59. An orangutan throws a coconut vertically upward at the foot of a cliff 40 m high while his mate simultaneously drops another coconut from the top of the cliff. The two coconuts collide at an altitude of 20 m. What was the initial velocity of the coconut that was thrown upward?

60. When a flea jumps, it accelerates through about 0.80 mm (a little less than the length of its legs) and is able to reach a height of as much as 10 cm. (a) Find the flea's acceleration (assumed constant) and its velocity at takeoff when this occurs. (b) When a person jumps, the acceleration distance is about 50 cm. If the acceleration of a person were the same as that of a flea, find the velocity at takeoff and the height that would be reached.

61. A helicopter is climbing at 8.0 m/s when it drops a pump near a leaking boat. The pump reaches the water 4.0 s afterward. How high was the helicopter when the pump was dropped? When the pump reached the water?

PROBLEMS

62. Figure 1.25 is a plot of displacement versus time for a beetle moving along a straight stick. (a) Plot the beetle's velocity versus time. (b) What is the beetle's average velocity during the period shown? (c) What is its average speed during the same period?

63. A snake is slithering toward you at 1.5 m/s. If you start walking when it is 5.0 m away, how fast must you go in order that the snake not overtake you when you have gone 40.0 m?

64. A car covers one-quarter of the distance to the next town at 40 km/h, another quarter at 50 km/h, and the rest at 90 km/h. Find the car's average speed for the entire distance.

65. A man drives one-third of the distance home at 25 mi/h. How fast must he drive the rest of the way in order to average 35 mi/h for the entire trip?

66. A woman on a bicycle climbs a hill at 4.0 m/s and then returns down the same hill at 8.0 m/s. What is her average speed for the entire trip?

Figure 1.25

Problem 62.

67. The odometer of a car is checked at 1.0-min intervals, and the readings below are obtained. Calculate the speed of the car in each time interval and plot the results on a graph. Describe the motion of the car with the help of this graph.

Time (min):	0	1.0	2.0	3.0
Distance (mi):	42.20	42.74	43.64	44.90

Time (min):	4.0	5.0	6.0	7.0
Distance (mi):	46.34	47.78	49.22	50.66

Time (min):	8.0	9.0	10.0
Distance (mi):	51.74	52.10	52.10

68. A European train passes successive kilometer posts at the times given below. Plot the data on a graph and determine whether the train's speed is constant over the entire distance. If it is not constant, plot the train's speed in each time interval versus time and find the acceleration. What are the train's initial and final speeds?

Distance (km):	0	1.0	2.0	3.0	4.0
Time (s):	0	64	114	156	193

Distance (km):	5.0	6.0	7.0	8.0
Time (s):	227	259	292	324

69. A car that is leaking oil at a constant rate is uniformly accelerated from rest starting out just when a drop falls. The first drop down the road is 0.80 m from the car's starting point. How far from the starting point is the fourth drop?

70. A sprinter accelerates from rest until he reaches a velocity of 40 ft/s and then continues running at this velocity. If he takes 11.0 s to cover 300 ft, what was his acceleration and how long did it last?

71. A bus travels 400 m between two stops. It starts from rest and accelerates at 1.5 m/s^2 until it reaches a velocity of 9.0 m/s. The bus continues at this velocity and then decelerates at 2.0 m/s^2 until it comes to a halt. Find the total time required for the journey.

72. A rocket is launched upward with an acceleration of 100 m/s^2. Eight seconds later the fuel is exhausted. Find (a) the highest velocity the rocket attains; (b) the maximum altitude; (c) the total time of flight; and (d) the velocity with which the rocket strikes the ground.

73. A girl throws a ball vertically upward at 10 m/s from the roof of a building 20 m high. (a) How long will it take the ball to reach the ground? (b) What will its velocity be when it strikes the ground?

74. A girl throws a ball vertically downward at 10 m/s from the roof of a building 20 m high. (a) How long will it take the ball to reach the ground? (b) What will its velocity be when it strikes the ground?

75. An elevator has a maximum acceleration of ± 1.5 m/s^2 and a maximum velocity of 6.0 m/s. Find the shortest period of time required for it to take a passenger to the tenth floor of a building from street level, a height of 50 m, with the elevator coming to a stop at this floor.

76. A person in an elevator drops an apple from a height 2.0 m above the elevator's floor and, with a stopwatch, times the fall of the apple to the floor. What is found (a) when the elevator is ascending with an acceleration of 1.0 m/s^2; (b) when it is descending with an acceleration of 1.0 m/s^2; (c) when it is ascending at a constant velocity of 3.0 m/s; (d) when it is descending at a constant velocity of 3.0 m/s; and (e) when the cable has broken and it is descending in free fall?

77. When a certain ball is dropped to the ground from a height of 2.0 m, it bounces back up to its original height. (a) Plot a graph of the height of the ball versus time for the first 3.0 s of its motion. (b) Plot a graph of the ball's velocity during the same period.

78. A champagne bottle is held upright 4 ft above the floor as the wire around its cork is removed. The cork then pops out, rises vertically, and falls to the floor 1.4 s later. (a) What height above the bottle did the cork reach? (b) What was the cork's initial velocity? (c) Its final velocity?

Answers to Multiple Choice

1. d	**8.** a	**15.** c	**22.** a
2. b	**9.** b	**16.** a	**23.** b
3. b	**10.** b	**17.** a	**24.** c
4. b	**11.** b	**18.** c	**25.** a
5. a	**12.** c	**19.** c	**26.** b
6. c	**13.** c	**20.** c	**27.** c
7. a	**14.** b	**21.** a	**28.** c

2

VECTORS

2.1 Vector Quantities

2.2 Vector Addition

2.3 Vector Subtraction

2.4 Resolving a Vector

2.5 Vector Addition by Components

2.6 Motion in a Vertical Plane

2.7 Projectile Flight

Some physical quantities are completely specified by just a number and a unit. If we say that a person's mass is 70 kg, that the area of a farm is 160 acres, or that the frequency of a sound wave is 660 cycles/s, there is nothing more to add. These are examples of **scalar quantities.**

Other quantities have directions as well as magnitudes associated with them. Your mass is the same whether you stand on your feet or on your head. However, if you drive north from Denver at 80 km/h you will end up in Canada, whereas if you drive south from Denver you will end up in Mexico. To say that your car's speed is 80 km/h, a scalar statement, is evidently not enough to describe its motion completely. A quantity whose direction is significant is called a **vector quantity.** Thus velocity, which incorporates direction as well as speed, is a vector quantity. Vector quantities occur often in physics, and their arithmetic is different from the ordinary arithmetic of scalar quantities.

Figure 2.1

The arrowed line is a vector that represents a 640-km westward displacement.

2.1 VECTOR QUANTITIES

Which way as well as how much.

Displacement, velocity, and acceleration, which were considered in Chapter 1, are vector quantities. Another vector quantity is **force.** A force is often spoken of as a "push" or a "pull." There is more to the concept of force than this, but it is all we need for the time being. Forces are responsible for all changes in motion: A force is needed to start a stationary object moving, to change its direction of motion, and to stop it. We have to know the direction as well as the magnitude of a force to know what its effects will be. Push this book down on the table and nothing happens. Push the book up and it will rise into the air. The SI unit of force (and of weight, since weight is a force) is the **newton** (N), which is equivalent to a little less than 0.225 lb. The weight of an apple is about 1 N.

We can represent a vector quantity by a straight line with an arrowhead at one end to show the direction of the quantity. The length of the line is proportional to the magnitude of the quantity. Such a line is called a **vector.**

Figure 2.1 shows how a 640-km westward displacement can be represented by a vector. The compass rose establishes the orientation of the displacement, and the distance scale establishes the relationship between length in the diagram and the actual length.

Other vector quantities besides displacements can also be represented by vectors. The length of the vector in each case is proportional to the magnitude of the quantity it represents, and its direction is the direction of the quantity. Thus a velocity whose magnitude is 25 m/s is represented on a scale of 1 mm = 1 m/s by an arrow 25 mm long, and a force whose magnitude is 1500 N is represented on a scale of 1 mm = 100 N by an arrow 15 mm long (Fig. 2.2). The directions of the arrows indicate the directions of the quantities.

Figure 2.2

(a) A vector that represents a velocity of 25 m/s. (b) A vector that represents a force of 1500 N.

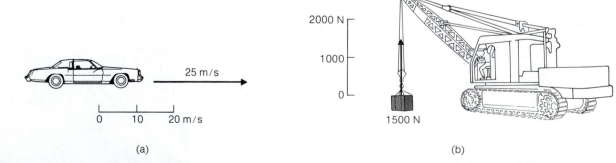

(a)

(b)

The symbols of vector quantities are customarily printed in boldface type (**F** for force), whereas italics are used both for scalar quantities (*m* for mass) and for the magnitudes of vectors. Thus if we denote the 640-km westward displacement by the symbol **A,** its magnitude is $A = 640$ km. In handwriting, vector quantities are indicated by placing arrows over their symbols—for instance, $\vec{A}$.

Figure 2.3

The displacements of 10.0 km north, **A**, and 4.0 km west, **B**, add to give a resultant displacement **R** of 10.8 km in a direction 22° west of north.

2.2 VECTOR ADDITION

The sum of two or more vectors is called their resultant.

Ordinary arithmetic is used to add two or more scalar quantities of the same kind together. Ten kg of potatoes plus 4 kg of potatoes equals 14 kg of potatoes. The same arithmetic is used for vector quantities of the same kind when the directions are the same. A total upward force of 100 N is required to lift a 100-N weight, regardless of whether the force is applied by one, two, or ten people. It is the total force on an object, not the individual forces, that affects the motion of the object.

What do we do when the directions are different? A person who walks 10.0 km north and then 4.0 km west does *not* undergo a displacement of 14.0 km from his starting point, even though he has walked a total distance of 14.0 km. A vector diagram provides a convenient method of determining his displacement. The procedure is to make a scale drawing of the successive displacements **A** and **B,** as in Fig. 2.3, and to join the starting point and the end point with a single vector **R.** The net displacement is **R,** whose length corresponds to 10.8 km and whose direction corresponds to 22° west of north. These values can be found by using a ruler and protractor.

The general rule for adding vectors is illustrated in Fig. 2.4. To add the vector **B** to the vector **A,** we follow these two steps:

---

1. **Shift B parallel to itself until its tail is at the head of A. In its new position B must still have its original length and direction.**

2. **Draw a new vector R (called the *resultant*) from the tail of A to the head of B. The vector R is equal to A + B.**

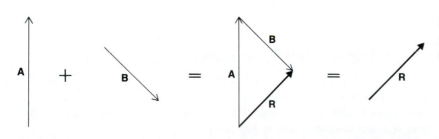

Figure 2.4

The addition of two vectors to obtain the resultant **R**.

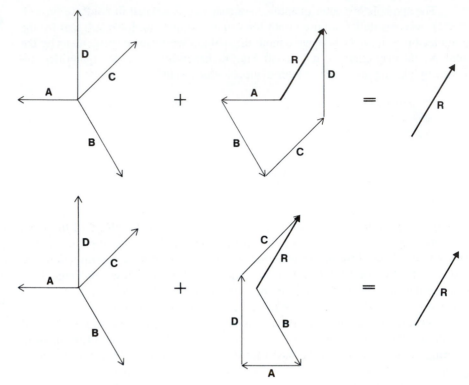

Figure 2.5

The addition of four vectors to obtain the resultant **R**. The order in which the vectors are added does not affect the magnitude or direction of **R**.

To find the magnitude R of the resultant **R,** we measure the length of **R** on the diagram and compare this length with the scale. The direction of **R** with respect to **A** or **B** may be found with a protractor. The order in which vectors are added does not matter, so that

$$\mathbf{A} + \mathbf{B} = \mathbf{B} + \mathbf{A}$$ *Vector addition*   (2.1)

Any number of vectors of the same kind can be added in this way. First place the tail of each vector at the head of the previous one, being sure to keep their lengths and original directions unchanged. Then draw a vector **R** from the tail of the first vector to the head of the last one. Figure 2.5 shows how four vectors are added together in two different ways to get the same resultant **R.**

Trigonometric Method

For an approximate result a ruler and protractor are sufficient. Trigonometry provides more accuracy and is especially easy to apply when the vectors to be added are perpendicular to each other. Basic trigonometry is reviewed in Appendixes A.7 and A.8 and is summarized in Fig. 2.6.

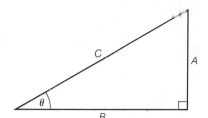

$\sin \theta = \dfrac{\text{opposite side}}{\text{hypotenuse}} = \dfrac{A}{C}$

$\cos \theta = \dfrac{\text{adjacent side}}{\text{hypotenuse}} = \dfrac{B}{C}$

$\tan \theta = \dfrac{\text{opposite side}}{\text{adjacent side}} = \dfrac{A}{B}$

Pythagorean theorem: $A^2 + B^2 = C^2$

Figure 2.6
Basic trigonometric formulas.

◆ **E X A M P L E** **2.1** _____

A person walks 10.0 km north and then 4.0 west, as in Fig. 2.3. Use trigonometry to find the person's displacement.

SOLUTION Because **A** is perpendicular to **B,** we can use the Pythagorean theorem to find the magnitude R of the resulting displacement **R.** According to this theorem, the square of the hypotenuse (the long side) of a right triangle equals the sum of the squares of the other two sides. Hence

$$R^2 = A^2 + B^2$$
$$R = \sqrt{A^2 + B^2} = \sqrt{(10.0\,\text{km})^2 + (4.0\,\text{km})^2} = 10.8\,\text{km}$$

The angle θ between **R** and **A** is specified by

$$\tan \theta = \frac{B}{A} = \frac{4.0\,\text{km}}{10.0\,\text{km}} = 0.40$$

When we look at a table of trigonometric functions, we find that the angle whose tangent is closest to 0.40 is 22°, since tan 22° = 0.404. To the nearest degree,

$$\theta = 22°$$

Alternatively, we can use an electronic calculator to find θ by entering 0.40 and pressing the $\tan^{-1}$ (sometimes arctan or inv tan) key. ◆

2.3 VECTOR SUBTRACTION

The negative of a vector points in the opposite direction.

It is sometimes necessary to subtract one vector from another. To subtract **A** from **B,** for example, we first form the negative of **A,** denoted **−A,** a vector that has

Figure 2.7

Vector subtraction.

$$\mathbf{B} - \mathbf{A} \nearrow = \mathbf{B} + \swarrow_{-\mathbf{A}} = \mathbf{B} = \mathbf{B} - \mathbf{A}$$

the same length as **A** but points in the opposite direction (Fig. 2.7). We then add −**A** to **B** in the usual manner. This procedure may be summarized as

$$\mathbf{B} - \mathbf{A} = \mathbf{B} + (-\mathbf{A})$$ *Vector subtraction*   (2.2)

Vector subtraction will be used in Chapter 6 to find the acceleration of an object that moves in a curved path.

2.4   RESOLVING A VECTOR

How to replace a vector with two or more other vectors called its components.

Just as we can add two or more vectors to give a single resultant vector, so we can break up a vector into two or more other vectors. The latter process is called **resolving** the vector. The new vectors are called the **components** of the original one (Fig. 2.8). As we will often find, the best way to analyze a physical situation may be to resolve a vector into components. Almost invariably the components are chosen to be perpendicular to one another. Usually two components, in directions labeled x and y, are enough, but sometimes three are needed, in x-, y-, and z-directions (Fig. 2.9).

Figure 2.10 shows a boy pulling a wagon with a rope at the angle θ above the ground. Only part of the force he exerts affects the horizontal motion, since the

Figure 2.8

(a) In vector addition, two or more vectors are combined to form a single vector. (b) In vector resolution, a single vector is replaced by two or more other vectors whose sum is the same as the original vector. Here **A** and **B** are the components of the vector **C**.

(a)   Vector addition $\mathbf{A} + \mathbf{B} = \mathbf{C}$

(b)   Vector resolution $\mathbf{C} = \mathbf{A} + \mathbf{B}$

$$F_x = F \cos \theta$$
$$F_y = F \sin \theta$$

Figure 2.9

How a vector **A** can be resolved into three mutually perpendicular components $\mathbf{A}_x$, $\mathbf{A}_y$, and $\mathbf{A}_z$.

Figure 2.10

The resolution of a force vector into horizontal and vertical components. The original vector is indicated by two short lines across it.

wagon moves horizontally while the force **F** is not a horizontal one. We can resolve **F** into two components, $\mathbf{F}_x$ and $\mathbf{F}_y$, where

$\mathbf{F}_x$ = horizontal component of **F**

$\mathbf{F}_y$ = vertical component of **F**

It is a good habit to distinguish between the original vector and its components when all are on the same diagram by drawing two short lines across the original vector, as in Fig. 2.10. This reminds us that the original vector can now be ignored since it has been replaced by its components.

The magnitudes of $\mathbf{F}_x$ and $\mathbf{F}_y$ are

$$F_x = F \cos \theta \tag{2.3}$$

$$F_y = F \sin \theta \tag{2.4}$$

The horizontal component $\mathbf{F}_x$ is responsible for the wagon's motion, while the vertical component $\mathbf{F}_y$ pulls upward on it. $\mathbf{F}_x$ is the projection of **F** in the horizontal direction, and $\mathbf{F}_y$ is the projection of **F** in the vertical direction.

◆ **E X A M P L E 2.2**

Figure 2.11

If the force the boy exerts on the wagon in Fig. 2.10 is 60.0 N and $\theta = 24°$, find F_x and F_y. Does their sum equal 60.0 N? Should it?

SOLUTION We have (Fig. 2.11)

$$F_x = F \cos \theta = (60.0 \text{ N})(\cos 24°) = 54.8 \text{ N}$$
$$F_y = F \sin \theta = (60.0 \text{ N})(\sin 24°) = 24.4 \text{ N}$$

The algebraic sum of the magnitudes F_x and F_y is 54.8 N + 24.4 N = 79.2 N, although the boy has exerted a force of only 60.0 N. Where is the mistake?

The answer is that there is no mistake. $\mathbf{F}_x$ and $\mathbf{F}_y$ are vectors whose directions are different, so they can *only* be added vectorially. The algebraic sum of the magnitudes F_x and F_y has no meaning. If we add $\mathbf{F}_x$ and $\mathbf{F}_y$ vectorially, with the help of the Pythagorean theorem we find that

$$F^2 = F_x^2 + F_y^2$$
$$F = \sqrt{F_x^2 + F_y^2} = \sqrt{(54.8 \text{ N})^2 + (24.4 \text{ N})^2} = 60.0 \text{ N}$$

which is equal to the force the boy exerts.   ◆

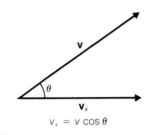

$v_x = v \cos \theta$

Figure 2.12

◆   **E X A M P L E   2.3**

A woman on the ground sees an airplane climbing at an angle of 35° above the horizontal. She gets into her car and by driving at 70 mi/h is able to stay directly below the airplane. What is the airplane's speed?

SOLUTION   The car's velocity is equal to the horizontal component $\mathbf{v}_x$ of the airplane's velocity $\mathbf{v}$ (Fig. 2.12). Since

$$v_x = v \cos \theta$$

we have for the airplane's speed

$$v = \frac{v_x}{\cos \theta} = \frac{70 \text{ mi/h}}{\cos 35°} = 85 \text{ mi/h}$$ ◆

The components of a vector do not have to be horizontal and vertical—they can have whatever orientation is convenient for the situation we are considering. A car on an inclined driveway is an example (Fig. 2.13). If its brakes are released, the force that causes the car to roll down the driveway is the component $\mathbf{F}_\parallel$ of the car's weight $\mathbf{w}$ along the driveway. The steeper the driveway, the greater the force $\mathbf{F}_\parallel$.

Figure 2.13

Weight is a force that acts vertically downward, but it can be resolved into components parallel and perpendicular to a surface of arbitrary orientation.

$F_\parallel = w \sin \theta$
$F_\perp = w \cos \theta$

◆ **E X A M P L E 2.4** _____

A car weighing 12.0 kN (2700 lb) is parked on a driveway that is at a 15° angle with the horizontal. Find the components of the car's weight parallel and perpendicular to the driveway.

SOLUTION The weight **w** of anything is the gravitational force the earth exerts on it, a force that acts vertically downward. Because **w** is vertical and $F_\perp$ is perpendicular to the road, the angle θ between **w** and $F_\perp$ is equal to the angle $\theta = 15°$ between the road and the horizontal. Hence

$$F_\parallel = w \sin \theta = (12.0 \text{ kN})(\sin 15°) = 3.1 \text{ kN}$$
$$F_\perp = w \cos \theta = (12.0 \text{ kN})(\cos 15°) = 11.6 \text{ kN}$$

◆

2.5 VECTOR ADDITION BY COMPONENTS

Divide and conquer.

As we have seen, we can add any number of vectors graphically just by drawing them head-to-tail. A line from the tail of the first vector to the head of the last is the resultant **R**. This is easy to do but not very accurate.

A problem of this kind can be solved by trigonometry in two ways. One is to begin by adding the first two vectors together, then adding a third to the resultant of the first two, a fourth to the resultant of the first three, and so on. This method usually means a lot of work with many chances for error.

A better way to add several vectors is to work in terms of their components. The procedure is as follows for vectors that lie in the same plane:

1. Resolve the initial vectors into their components in the x- and y-directions.
2. Add the components in the x-direction to give $\mathbf{R}_x$ and add the components in the y-direction to give $\mathbf{R}_y$. That is,

$\mathbf{R}_x = x$-component of **R**

$\quad = \mathbf{A}_x + \mathbf{B}_x + \mathbf{C}_x + \cdots = \text{sum of } x\text{-components}$ (2.5)

$\mathbf{R}_y = y$-component of **R**

$\quad = \mathbf{A}_y + \mathbf{B}_y + \mathbf{C}_y + \cdots = \text{sum of } y\text{-components}$ (2.6)

3. Find the magnitude and direction of the resultant **R** from the components $\mathbf{R}_x$ and $\mathbf{R}_y$. From the Pythagorean theorem,

$R = \sqrt{R_x^2 + R_y^2}$ (2.7)

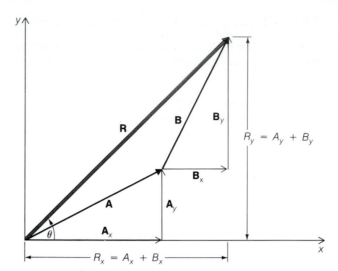

Figure 2.14

Vector addition by components.

The direction of **R** can be found from the values of the components by trigonometry. The best way to do this depends on the problem at hand.

Figure 2.14 shows the connection between head-to-tail vector addition and the component method for the case of two vectors.

◆   **E X A M P L E   2.5** _____

The sailboat *Ardent Spirit* is headed due north at a forward speed of 6.0 knots (kn). The pressure of the wind on its sails causes the boat to move sideways to the east at 0.5 kn. A tidal current is flowing to the southwest at 3.0 kn. What is the velocity of *Ardent Spirit* relative to the earth's surface?

Figure 2.15

A sailboat headed north at 6.0 kn is acted upon by a sideways force due to the wind that sets it to the east at 0.5 kn. A tidal current flows southeast at 3.0 kn. The resultant velocity of the boat relative to the earth's surface has a magnitude of 4.2 kn and is directed 22° west of north.

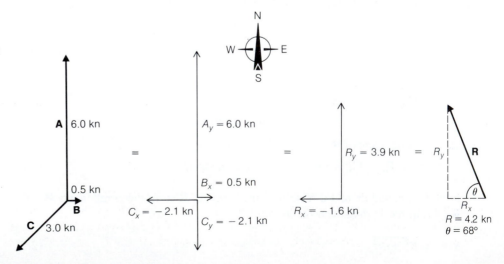

(A *knot* is a unit of speed equal to one nautical mile per hour. The nautical mile is widely used in air and sea navigation because it is the same in length as one minute [1'] of latitude, where $60' = 1°$. Since 1 nautical mile $= 1.852$ km $= 6076$ ft, 1 kn $= 1.852$ km/h $= 1.151$ mi/h.)

SOLUTION   For convenience, we shall call north the $+y$-direction, south the $-y$-direction, east the $+x$-direction, and west the $-x$-direction. With the help of Fig. 2.15 we see that the magnitudes of the components of the three velocity vectors are

$$A_x = 0 \qquad\qquad C_x = -(3.0 \text{ kn})(\cos 45°)$$
$$A_y = 6.0 \text{ kn} \qquad\quad = -2.1 \text{ kn}$$
$$B_x = 0.5 \text{ kn} \qquad C_y = -(3.0 \text{ kn})(\sin 45°)$$
$$B_y = 0 \qquad\qquad\quad = -2.1 \text{ kn}$$

We then add together the components in each direction:

$$R_x = A_x + B_x + C_x \qquad\qquad R_y = A_y + B_y + C_y$$
$$= 0 + 0.5 \text{ kn} - 2.1 \text{ kn} \qquad\quad = 6.0 \text{ kn} + 0 - 2.1 \text{ kn}$$
$$= -1.6 \text{ kn} \qquad\qquad\qquad = 3.9 \text{ kn}$$

The magnitude of the resultant velocity is

$$R = \sqrt{R_x^2 + R_y^2} = \sqrt{1.6^2 + 3.9^2} \text{ kn} = 4.2 \text{ kn}$$

Thus the speed of the boat relative to the earth is 4.2 kn. Its direction, as shown in the last part of the diagram, is west of north at an angle θ with west. We can find the value of θ by first finding the value of $\tan \theta$:

$$\tan \theta = \frac{R_y}{R_x} = \frac{3.9}{1.6} = 2.4 \quad \text{so} \quad \theta = 68° \qquad\qquad\qquad \blacklozenge$$

2.6   MOTION IN A VERTICAL PLANE

Only the vertical component of an object's motion is affected by gravity.

In Chapter 1 we considered two idealized kinds of straightline motion. One was the motion of something that could travel only horizontally—a car on a level road, for instance. Gravity could be ignored here. The other was vertical motion, and we had to take into account the downward acceleration of gravity g.

Figure 2.16

Ball *A* is dropped from the edge of a table while ball *B* is simultaneously rolled off the edge with the initial horizontal speed v_0.

Now we come to the general case of an object moving freely near the earth's surface. If the object's initial direction is not exactly up or down, gravity will cause its path to be curved rather than straight.

Our strategy in attacking problems of this kind is to resolve the object's acceleration **a** (assumed constant) and initial velocity $\mathbf{v}_0$ into their horizontal components $\mathbf{a}_x$ and $\mathbf{v}_{0x}$ and vertical components $\mathbf{a}_y$, $\mathbf{v}_{0y}$. With the help of the formulas of Chapter 1 we then examine separately the object's motion in each of these directions. Finally we combine $\mathbf{v}_x$ and $\mathbf{v}_y$ to find **v,** and x and y to find the displacement **s,** at any time t after the start of the motion by vector addition.

Suppose that we drop a ball *A* from the edge of a table while rolling an identical ball *B* off to the side (Fig. 2.16). At the moment the balls leave the table, *A* has zero velocity while *B* has the horizontal velocity $\mathbf{v}_0$. The velocity components of the balls therefore have the magnitudes

$$v_{Ax} = 0 \qquad v_{Bx} = v_0 \qquad v_{Ay} = 0 \qquad v_{By} = 0$$

Both balls reach the floor at the same time, even though *B* has traveled some distance x away from the table. The reason for this behavior is that the acceleration of gravity is the same for all bodies near the earth regardless of their state of motion. Both *A* and *B* started out with no vertical velocity, both had the same downward acceleration, and so both took the same period of time to fall.

If we consider vertically downward to be the $+y$-direction, an object in free fall is displaced by

$$y = \tfrac{1}{2} g t^2$$

in the time t when it starts with no vertical component of velocity. Hence both balls require the time

$$t = \sqrt{\frac{2y}{g}}$$

to reach the floor. If the table is 1.0 m high, then

$$t = \sqrt{\frac{(2)(1.0 \text{ m})}{9.8 \text{ m/s}^2}} = 0.45 \text{ s}$$

While it is falling, ball B is also moving horizontally with the speed v_0. When it strikes the floor, it will have been displaced horizontally by $x = v_0t$. Let us say that $v_0 = 5.0$ m/s. Therefore

$$x = v_0t = (5.0 \text{ m/s})(0.45 \text{ s}) = 2.3 \text{ m}$$

◆   **E X A M P L E   2.6** _____

Find the speeds with which balls A and B strike the floor.

SOLUTION   The final velocity of A has only the single component

$$v_{Ay} = gt$$

and so

$$v_A = v_{Ay} = (9.8 \text{ m/s}^2)(0.45 \text{ s}) = 4.4 \text{ m/s}$$

The final velocity of B, however, has both horizontal and vertical components, namely

$$v_{Bx} = v_0 \qquad v_{By} = gt$$

From Fig. 2.17 we see that since v_{Bx} is perpendicular to v_{By}, their vector sum $\mathbf{v}_B$ has the magnitude

$$v_B = \sqrt{v_{Bx}^2 + v_{By}^2} = \sqrt{v_0^2 + (gt)^2}$$

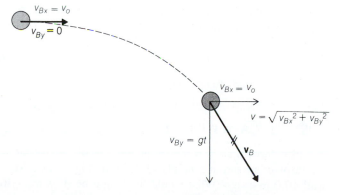

Figure 2.17

The horizontal and vertical components of the velocity of ball B must be added vectorially to determine the magnitude of its velocity.

Since $v_0 = 5.0$ m/s and $t = 0.45$ s,

$$v_B = \sqrt{(5.0 \text{ m/s})^2 + [(9.8 \text{ m/s}^2)(0.45 \text{ s})]^2} = 6.7 \text{ m/s} \qquad \blacklozenge$$

2.7   PROJECTILE FLIGHT

How to find the initial angle for maximum range.

In considering the flight of a projectile, let us ignore the curvature of the earth, the variation of g with altitude, and (for the moment) air resistance. If the initial velocity $\mathbf{v}_0$ of the projectile makes an angle of θ with level ground, its components have the magnitudes

$$v_{0x} = v_0\cos\theta \qquad \textit{Initial velocity components} \quad (2.8)$$

$$v_{0y} = v_0\sin\theta \qquad\qquad\qquad\qquad\qquad\qquad (2.9)$$

We can once again use the formulas for straight-line motion to examine the horizontal and vertical aspects of the projectile's flight separately, since these are independent of each other. Under the assumptions above, the horizontal velocity component v_x remains constant during the flight. The vertical component v_y, however, gradually drops to zero owing to the downward acceleration of gravity and then becomes more and more negative (meaning the projectile falls faster and faster) until the ground is reached. If we consider vertically upward to be the $+y$-direction, at the time t the projectile's velocity components become

$$v_x = v_{0x} \qquad \textit{Velocity components at later time} \quad (2.10)$$

$$v_y = v_{0y} - gt \qquad\qquad\qquad\qquad\qquad\qquad (2.11)$$

If the projectile starts from the point x_0, y_0, then at the time t its horizontal and vertical coordinates of position will be

$$x = x_0 + v_{0x}t \qquad \textit{Position of projectile} \quad (2.12)$$

$$y = y_0 + v_{0y}t - \tfrac{1}{2}gt^2 \qquad\qquad\qquad\qquad\qquad (2.13)$$

◆   **E X A M P L E   2.7** _____

A ball is thrown at 20.0 m/s at an angle of 65° above the horizontal. The ball leaves the thrower's hand at a height of 1.80 m (Fig. 2.18). At what height will it strike a wall 10.0 m away? What would the height be if v_0 were 30.0 m/s?

$v_0 = 20.0$ m/s

$\theta = 65°$

$y_0 = 1.80$ m/s

$y = ?$

$x = 10.0$ m

Figure 2.18

SOLUTION (a) First we find the time of flight t from the horizontal component of the ball's initial velocity, which is

$$v_{0x} = v_0\cos\theta = (20.0 \text{ m/s})\cos 65° = 8.45 \text{ m/s}$$

Since $x = 10.0$ m is the horizontal distance to the wall,

$$t = \frac{x}{v_{0x}} = \frac{10.0 \text{ m}}{8.45 \text{ m/s}} = 1.18 \text{ s}$$

The vertical component of the ball's initial velocity is

$$v_{0y} = v_0\sin\theta = (20.0 \text{ m/s})\sin 65° = 18.13 \text{ m/s}$$

The height at which the ball strikes the wall is therefore

$$y = y_0 + v_{0y}t - \tfrac{1}{2}gt^2$$
$$= 1.80 \text{ m} + (18.13 \text{ m/s})(1.18 \text{ s}) - \tfrac{1}{2}(9.80 \text{ m/s}^2)(1.18 \text{ s})^2$$
$$= 16.4 \text{ m}$$

(b) A similar calculation gives $y = 20.2$ m when $v_0 = 30.0$ m/s. The greater the initial velocity, the shorter the time of flight and the less the drop due to gravity during the flight.   ◆

Projectile Range

We can use the formulas derived above to find the range of a projectile launched from the ground, that is, how far from its starting point it will return to the ground. We first calculate the time of flight T (Fig. 2.19). The projectile will continue to rise until the vertical component of its velocity, given by

$$v_y = v_0 \sin \theta - gt$$

is zero. At this time t,

$$0 = v_0 \sin \theta - gt \quad \text{and} \quad t = \frac{v_0 \sin \theta}{g}$$

The projectile needs the same amount of time to return to the ground, and so its total time of flight is

Figure 2.19

In projectile flight, the horizontal component of velocity is constant in the absence of air resistance.

$$t = \frac{v_o \sin \theta}{g}$$

$$v_x = v_o \cos \theta$$

$$v_y = 0$$

$$v_y = +v_o \sin \theta$$

$$v_x = v_o \cos \theta$$

$$t = 0$$

$$v_x = v_o \cos \theta$$

$$v_y = -v_o \sin \theta$$

$$T = 2t = \frac{2v_o \sin \theta}{g}$$

Figure 2.20

In the absence of air resistance, the maximum range of a projectile occurs when it is sent off at an angle of 45°.

$$T = 2t = \frac{2v_0 \sin \theta}{g} \qquad \qquad \textit{Time of flight} \quad (2.14)$$

Since v_x is constant, the projectile's range R is

$$R = v_x T = (v_0 \cos \theta)\left(\frac{2v_0 \sin \theta}{g}\right) = \frac{2v_0^2}{g}\sin \theta \cos \theta$$

This formula can be simplified by making use of the trigonometric identity

$$\sin \theta \cos \theta = \tfrac{1}{2}\sin 2\theta$$

The range may therefore be written

$$R = \frac{v_0^2}{g}\sin 2\theta \qquad \qquad \textit{Range of projectile} \quad (2.15)$$

We note that R is a maximum when $\sin 2\theta = 1$, because 1 is the highest value the sine function can have. Since $\sin 90° = 1$, the maximum range R_{max} occurs when the initial angle θ is 45°:

$$R_{\text{max}} = \frac{v_0^2}{g} \qquad \qquad \textit{Maximum range} \quad (2.16)$$

Any other angle, greater or smaller, will result in a shorter range (Fig. 2.20). If θ_1 is an angle that leads to a range R, the other angle θ_2 for the same range is given by the formula

$$\theta_2 = 90° - \theta_1 \qquad \qquad \textit{Angles for same range} \quad (2.17)$$

◆ **E X A M P L E   2.8** _____

An arrow leaves a bow at 30 m/s. (a) What is its maximum range? (b) At what two angles could the archer point the arrow if it is to reach a target 70 m away?

Javelin about to be thrown at an
angle of 45°.

SOLUTION   (a) The maximum range is

$$R_{\max} = \frac{v_0^2}{g} = \frac{(30 \text{ m/s})^2}{9.8 \text{ m/s}^2} = 92 \text{ m}$$

(b) From Eq. (2.15) we obtain

$$\sin 2\theta_1 = \frac{Rg}{v_0^2} = \frac{(70 \text{ m})(9.8 \text{ m/s}^2)}{(30 \text{ m/s})^2} = 0.762$$

Thus

$$2\theta_1 = \sin^{-1} 0.762 = 50° \quad \text{and} \quad \theta_1 = 25°$$

The other angle for the same range is

$$\theta_2 = 90° - \theta_1 = 90° - 25° = 65° \qquad \qquad \blacklozenge$$

Effect of Air Resistance

Air resistance, of course, reduces the range of a projectile below the value given by Eq. (2.15). At a given speed the retardation is many times less for a ball heavy for its size, such as a baseball, than for a ball light for its size, such as a Ping-Pong ball. This is evident from the list of terminal speeds in Table 1.6. For a given

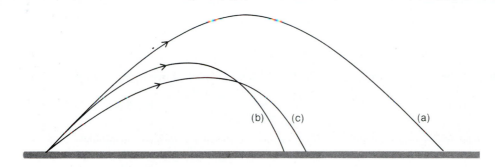

Figure 2.21

Paths of a baseball with the same initial speed after being struck by a bat. (a) Initial angle of 45° in vacuum. (b) Initial angle of 45° in air. (c) Initial angle of 40° in air.

projectile, the drag force increases with speed. A baseball for which $v_0 = 35$ m/s (about 80 mi/h) has a maximum range in air about 70% of R_{max} in vacuum, whereas when $v_0 = 50$ m/s (about 110 mi/h), the comparable ratio is down to 55%.

Because the horizontal component of a projectile's velocity decreases steadily as it moves through the air, its path is not symmetric about its highest point, as Fig. 2.21 shows. In this situation—that is, in real life—the maximum range occurs for an initial angle of less than 45°. The greater the drag force, the lower the angle for R_{max}. For a baseball struck hard by a bat, the optimum angle is about 40°.

Important Terms

A **scalar quantity** has magnitude only. A **vector quantity** has both magnitude and direction. Thus time is a scalar quantity, and force is a vector quantity.

A **vector** is an arrowed line whose length is proportional to the magnitude of some vector quantity and whose direction is that of the quantity. A **vector diagram** is a scale drawing of the various forces, velocities, or other vector quantities involved in the motion of a body.

In the graphical method of **vector addition,** the tail of each successive vector is placed at the head of the previous one, with their lengths and original directions kept unchanged. The **resultant** is a vector drawn from the tail of the first vector to the head of the last.

A vector can be **resolved** into two or more other vectors, called the **components** of the original vector. Usually the components of a vector are chosen to be in mutually perpendicular directions.

Important Formulas

Vector addition: $\mathbf{A} + \mathbf{B} = \mathbf{B} + \mathbf{A}$

Vector subtraction: $\mathbf{B} - \mathbf{A} = \mathbf{B} + (-\mathbf{A})$

Components of a vector: $\mathbf{A} = \mathbf{A}_x + \mathbf{A}_y$
$$A_x = A \cos \theta$$
$$A_y = A \sin \theta$$

Vector addition by components: $R = \sqrt{R_x^2 + R_y^2}$
where $R_x = A_x + B_x + C_x + \cdots$
$$R_y = A_y + B_y + C_y + \cdots$$

Angle θ between $\mathbf{R}$ and $\mathbf{R}_x$: $\tan \theta = \dfrac{R_y}{R_x}$

Projectile motion: $v_x = v_{0x}$
$$v_y = v_{0y} - gt$$
$$x = x_0 + v_{0x}t$$
$$y = y_0 + v_{0y}t - \tfrac{1}{2}gt^2$$

Range of projectile: $R = \dfrac{v_0^2}{g}\sin 2\theta$

Angles for same range: $\theta_2 = 90° - \theta_1$

(Air resistance is assumed negligible in the following exercises and problems.)

MULTIPLE CHOICE

1. Which of the following units could be associated with a vector quantity?

    a.   km/s^2 **b.**   kg/s
    c.   hours   **d.**   m^3

2. The minimum number of unequal forces whose vector sum can equal zero is

    a.   1   **b.**   2
    c.   3   **d.**   4

3. The magnitude of the resultant of two forces is a minimum when the angle between them is

    a.   0   **b.**   45°
    c.   90°   **d.**   180°

4. An engine block is supported by a rope hoist attached to an overhead beam. When the block is pulled to one side by a horizontal force exerted by another rope, the tension in the rope hoist

    a.   is less than before
    b.   is unchanged
    c.   is greater than before
    d.   may be any of the above, depending on the magnitude of the horizontal force

5. Which of the following pairs of displacements cannot be added to give a resultant displacement of 2 m?

    a.   1 m and 1 m   **b.**   1 m and 2 m
    c.   1 m and 3 m   **d.**   1 m and 4 m

6. Which of the following sets of forces cannot have a vector sum of zero?

    a.   10, 10, and 10 lb   **b.**   10, 10, and 20 lb
    c.   10, 20, and 20 lb   **d.**   10, 20, and 40 lb

7. Which of the following sets of displacements might be able to return a car to its starting point?

    a.   2, 8, 10, and 25 km
    b.   5, 20, 35, and 65 km
    c.   60, 120, 180, and 240 km
    d.   100, 100, 100, and 400 km

8. An airplane whose airspeed is 200 km/s is flying in a wind of 80 km/h. The speed of the airplane relative to the ground must be between

    a.   80 and 200 km/h   **b.**   80 and 280 km/h
    c.   120 and 200 km/h   **d.**   120 and 280 km/h

9. Fred walks 8 km north and then 6 km in a direction 30° north of east. His resultant displacement from his starting point is

    a.   11 km   **b.**   12 km
    c.   13 km   **d.**   14 km

10. The resultant of a 4-N force acting upward and a 3-N force acting horizontally is

    a.   1 N   **b.**   5 N
    c.   7 N   **d.**   12 N

11. The angle between the resultant in Multiple Choice 10 and the vertical is approximately

    a.   37°   **b.**   45°
    c.   53°   **d.**   60°

12. An airplane travels 100 mi to the north and then 200 mi to the east. The displacement of the airplane from its starting point is approximately

    a.   100 mi   **b.**   200 mi
    c.   220 mi   **d.**   300 mi

13. At what angle east of north should the airplane in Multiple Choice 12 have headed in order to reach its destination in a straight flight?

    a.   22°   **b.**   45°
    c.   50°   **d.**   63°

14. Two forces of 10 N each act on an object. The angle between the forces is 120°. The magnitude of their resultant is

    a.   10 N   **b.**   14 N
    c.   17 N   **d.**   20 N

15. A vector **A** lies in a plane and has the components A_x and A_y. The magnitude A_x of $\mathbf{A}_x$ is equal to

    a. $A - A_y$ **b.** $\sqrt{A} - \sqrt{A_y}$
    c. $\sqrt{A - A_y}$ **d.** $\sqrt{A^2 - A_y^2}$

16. An escalator has a velocity of 3.0 m/s at an angle of 60° above the horizontal. The vertical component of its velocity is

    a.   1.5 m/s   **b.**   1.8 m/s
    c.   2.6 m/s   **d.**   3.5 m/s

17. The following forces act on an object: 10 N to the north, 20 N to the southeast, and 5 N to the west. The magnitude of their resultant is

    a.   5 N   **b.**   10 N
    c.   13 N   **d.**   23 N

18. The direction of the resultant of the forces in Multiple Choice 17 is

    a.   east   **b.**   24° south of east
    c.   45° south of east   **d.**   63° south of east

19. Ball A is thrown horizontally, and ball B is dropped from the same height at the same moment.

    a.   Ball A reaches the ground first.
    b.   Ball B reaches the ground first.

c. Ball A has the greater speed when it reaches the ground.

d. Ball B has the greater speed when it reaches the ground.

20. A ball rolls off the edge of a horizontal roof at 10.0 m/s. Two seconds later the speed of the ball will be

a. 10.0 m/s
b. 19.6 m/s
c. 29.6 m/s
d. 22.0 m/s

21. A ball is thrown at a 30° angle above the horizontal with a speed of 10 ft/s. After 0.50 s the horizontal component of its velocity will be

a. 9 ft/s
b. 10 ft/s
c. 12 ft/s
d. 19 ft/s

22. A ball is thrown at a 40° angle below the horizontal at 8.0 m/s. After 0.40 s the horizontal component of its velocity will be

a. 5.1 m/s
b. 6.1 m/s
c. 6.7 m/s
d. 10.4 m/s

23. After 0.40 s the magnitude of the vertical component of the ball of Multiple Choice 22 is

a. 3.9 m/s
b. 5.1 m/s
c. 9.1 m/s
d. 10.0 m/s

24. Two identical projectiles are fired at the same angle. The initial speed of B is twice that of A. The range of B is

a. $R_B = R_A$
b. $R_B = \sqrt{2}\, R_A$
c. $R_B = 2R_A$
d. $R_B = 4R_A$

QUESTIONS

1. Is it correct to say that scalar quantities are abstract, idealized quantities with no precise counterparts in the physical world, whereas vector quantities can be said to properly represent reality?

2. A bird sits on an overhead power cable, which sags slightly under its weight. Is it possible to prevent any such sagging by applying enough tension to the cable?

3. The resultant of three vectors is zero. Must they all lie in a plane?

4. The resultant of two forces **X** and **Y** is **R**. **X** and **R** are shown in Fig. 2.22. Is **A** or **B** equal to **Y**?

5. Three forces, each of 10 N, act on the same object. What is the maximum total force they can exert on the object? The minimum total force?

6. Two boats with the same speed relative to the water set out to cross a river. Boat A heads directly across the river, and boat B heads upstream at an angle such that it arrives at a point on the other side opposite its starting point. (a) Which boat arrives at the other side first? (b) Which boat travels the greatest distance relative to the water? (c) Relative to the earth?

7. Three balls are thrown simultaneously at the same speed from a roof. Ball A is thrown vertically upward, ball B is thrown horizontally, and ball C is thrown vertically downward. (a) Do the balls reach the ground at the same time? If not, in which order do they reach the ground? (b) Do the balls have

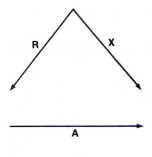

Figure 2.22
Question 4.

the same speeds when they reach the ground? If not, what is the order of their speeds?

8. A person at the masthead of a sailboat moving at constant velocity drops a wrench. The person is 20 m above the boat's deck at the time, and the stern of the boat is 20 m aft of the mast. Is there a minimum speed the sailboat can have such that the wrench does not land on the deck?

9. A hunter aims a rifle directly at a squirrel on a branch of a tree. The squirrel sees the flash of the rifle's firing. Should the squirrel stay where it is or drop from the branch in free fall at the instant the rifle is fired?

10. Does the speed of a projectile sent off at a 45° angle of elevation vary in its path? If so, where is the speed greatest and where is it least?

EXERCISES

2.2 Vector Addition

1. A 100-N sack of potatoes is suspended by a rope. A person pushes sideways on the sack with a force of 40 N. What is the tension in the rope?

2. A driver becomes lost and travels 12.0 km west, 5.0 km south, and 8.0 km east. Find the magnitude and direction of the car's displacement from the starting place.

3. A person walks 70 m to an elevator and then ascends 40 m. Find the magnitude and direction of the person's displacement from the starting place.

4. A woman jogs 4.2 km east and then 2.9 km south at a speed of 8.0 km/h. How much time would she have saved if she had jogged directly to her destination? In what direction should she have headed for this?

5. Two cars leave a crossroads at the same time, one headed north at 50 mi/h and the other headed east at 70 mi/h. How far apart are they after 0.5 h? After 2.0 h?

6. The resultant of two perpendicular forces has a magnitude of 40 N. If the magnitude of one of the forces is 25 N, what is the magnitude of the other force?

7. The ketch *Minots Light* is heading northwest at 7.0 kn through a tidal stream that is flowing southwest at 3.0 kn. Find the magnitude and direction of its velocity relative to the earth's surface.

8. A car traveled 16 km north and then 20 km east. The total time of travel was 0.75 h. Find the average speed and average velocity of the car for this trip.

9. A city block is bounded by north-south and east-west streets. The north-south sides of the block are 100 m long, and the east-west sides are 160 m long. Isabel starts from the southwest corner of the block and walks north, east, and south around the block to end up at the southeast corner 3.5 min later. Find her average speed and average velocity for the trip.

2.4 Resolving a Vector

10. A horizontal and a vertical force combine to give a resultant force of 10 lb that acts in a direction 40° above the horizontal. Find the magnitudes of the horizontal and vertical forces.

11. A woman pushes a 50-N lawn mower with a force of 25 N. If the handle of the lawn mower is 45° above the horizontal, how much downward force is the lawn mower exerting on the ground?

12. A sailboat cannot sail directly toward the wind but must "tack" back and forth at a certain angle with respect to the direction from which the wind is blowing. Which sailboat has the greater component of velocity to windward, the *Alpha*, whose velocity is 5.0 km/h at an angle of 40° away from the wind, or the *Beta*, whose velocity is 6.0 km/h at an angle of 50° away from the wind?

13. A horse is towing a barge along a canal with a rope that makes an angle of 25° with the canal. The horse exerts a force of 500 lb on the rope. How much force is exerted on the barge in the direction of its motion?

14. Two tugboats are towing a ship. Each exerts a horizontal force of 5.0 tons, and the angle between the two ropes is 30°. What is the resultant force exerted on the ship?

15. On a windless day, raindrops that fall on the side windows of a car moving at 10 m/s are found to make an angle of 50° with the vertical. Find the speed of the raindrops relative to the ground.

16. The shadow of an airplane taking off moves along the runway at 170 km/h. The sun is directly overhead and the airplane's air speed is 200 km/h. Find the angle at which the airplane is climbing.

17. An airplane is heading southeast when it takes off at an angle of 25° above the horizontal at 200 km/h. (a) What is the vertical component of its velocity? (b) What is the horizontal component of its velocity? (c) What is the component of the velocity of the plane toward the south?

18. An airplane whose speed is 150 km/h climbs from a runway at an angle of 20° above the horizontal. What is its altitude 1.00 min after takeoff? How many kilometers does it travel in a horizontal direction in this period of time?

2.5 Vector Addition by Components

19. Two billiard balls are rolling on a flat table. One has the velocity components $v_x = 1.0$ m/s, $v_y = 2.0$ m/s. The other has the velocity components $v_x = 2.0$ m/s, $v_y = 3.0$ m/s. If both balls started from the same point, what is the angle between their paths?

20. In going from one city to another, a car whose driver tends to get lost goes 30 mi north, 50 mi west, and 20 mi in a direction 30° south of east. How far apart are the cities? In what direction should the car have headed to travel directly from the first city to the second?

21. An airplane flies 200 km east from city *A* to city *B*, then 200 km south from city *B* to city *C*, and finally 100 km northwest to city *D*. How far is it from city *A* to city *D*? In what direction must the airplane head to return directly to city *A* from city *D*?

2.6 Motion in a Vertical Plane

22. An airplane is flying at an altitude of 8000 m at a speed of 900 km/h. At what distance ahead of a target must it drop a bomb?

23. A rifle is aimed directly at the bull's-eye of a target 50 m away. If the bullet's speed is 350 m/s, how far below the bull's-eye does the bullet strike the target?

24. An arrow shot horizontally from a cliff at 15 m/s lands 30 m away. How high is the cliff?

25. An airplane is in level flight at a speed of 400 km/h and an altitude of 1000 m when it drops a bomb. Find the bomb's speed when it strikes the ground.

26. A ball is rolled off the edge of a swimming pool and strikes the water 2.0 m away 0.30 s later. What was the initial speed of the ball? Its speed when it struck the water?

27. A rescue line is to be thrown horizontally from the bridge of a ship 30 m above sea level to a lifeboat 30 m away. What speed should the line have?

28. A horizontal stream of water comes out of a nozzle 1.2 m above the ground and strikes the ground 3.0 m away. At what speed does the water leave the nozzle? At what speed does it strike the ground?

29. An airplane is flying at an altitude of 500 m over the ocean at a speed of 60 m/s. A ship is moving along a path directly under that of the airplane but in the opposite direction at 10 m/s. At what horizontal distance from the ship should the pilot of the airplane drop a parcel meant to land on the ship's deck?

30. A ball is rolled off the edge of a table with a horizontal velocity of 3.0 ft/s. What will be the ball's velocity 0.10 s later?

31. A ball is thrown horizontally from the roof of a building 60 ft high at 80 ft/s. What will be the ball's velocity when it strikes the ground?

2.7 Projectile Flight

32. In April 1959 Miss Victoria Zacchini was fired 155 ft from a cannon in Madison Square Garden, New York City. What was the minimum muzzle speed of the cannon?

33. Find the minimum initial speed of a champagne cork that travels a horizontal distance of 11 m.

34. Find the range of an arrow that leaves a bow at 50 m/s at an angle of 50° above the horizontal.

35. A football leaves the toe of a punter at an angle of 40° above the horizontal. What is its minimum initial speed if it travels 120 ft?

36. A ball is thrown at 20.0 m/s at an angle of 60° above the horizontal. A wind blowing in the opposite direction reduces the ball's horizontal component of velocity by 5.0 m/s. How far away does the ball land?

37. A blunderbuss can fire a slug 100 m vertically upward. (a) What is its maximum horizontal range? (b) With what speeds will the slug strike the ground when fired upward and when fired so as to have maximum range?

38. A person can throw a ball a maximum distance of L. If the ball is thrown upward with the same initial speed, how high will it go?

PROBLEMS

39. The following forces act on an object resting on a level, frictionless surface: 10 N to the north, 20 N to the east, 10 N at an angle 40° south of east, and 20 N at an angle 50° west of south. Find the magnitude and direction of the resultant force acting on the object.

40. Find the magnitude and direction of the resultant of a 40-N force that acts at an angle of 63° clockwise from the $+y$-axis, a 15-N force that acts at an angle of 120° clockwise from the $+y$-axis, and a 30-N force that acts at an angle of 310° clockwise from the $+y$-axis.

41. The following horizontal forces act on an object: **A** has a magnitude of 6.0 lb and is in the $+y$-direction; **B** has a magnitude of 10.0 lb and is in the $-x$-direction; **C** has a magnitude of 8.0 lb and is at an angle of 45° clockwise from the $+x$-direction. Find the magnitude and direction of **A** + **B** − **C**.

42. The vector **A** has a magnitude of 20 cm and points 20° clockwise from the $+y$-direction. The vector **B** has a magnitude of 10 cm and points 60° clockwise from the $+y$-direction. Find the magnitude and direction **A** + **B**, **A** − **B**, and **B** − **A**.

43. The vector **A** has a magnitude of 10 cm and points 37° clockwise from the $+y$-direction. The vector **B** has a magnitude of 10 cm and points 37° clockwise from the $+x$-direction. Find the magnitude and direction of **A** + **B**, **A** − **B**, and **B** − **A**.

44. George walks 4.0 km in one direction and 2.0 km in another direction to end up 5.0 km from where he started. What is the angle between the two directions?

45. Nadia also walks 4.0 km in one direction and 2.0 km in another direction. However, she ends up only 3.0 km from where she started. What is the angle between the two directions?

46. A ball is thrown at 20 m/s from the roof of a building 25 m high at an angle of 30° below the horizontal. At what height above the ground will the ball strike the side of another building 20 m away from the first?

47. A ball is thrown horizontally toward the north from a rooftop at 8.0 m/s. A wind blowing from the east gives the ball a westward horizontal component of velocity of 10 m/s.

(a) What is the speed of the ball relative to the ground after 2.0 s? (b) What angle does its velocity make relative to the vertical at this time? (c) What angle does its velocity make relative to due north at this time?

48. What percentage increase in initial speed is required to increase the range of a javelin by 20%?

49. A baseball is struck by a batter and caught 100 m away 5.0 s later. (a) What was the baseball's initial speed? (b) What maximum height did it reach? (c) What was its speed 2.0 s after being struck? (d) What was its height at that time?

50. A ball is thrown at 12 m/s at an angle of 35° above the horizontal. (a) Find its velocity 1.0 s later. (b) At what time after it was thrown will the ball be headed at an angle of 20° above the horizontal? (c) At 20° below the horizontal?

51. A shell is fired at a velocity of 300 m/s at an angle of 30° above the horizontal. (a) How far does it go? What are its time of flight and maximum altitude? (b) At what other angle

could the shell have been fired to have the same range? What would its time of flight and maximum altitude have been in this case?

52. A golf ball leaves a tee at 180 ft/s and strikes the ground 600 ft away. At what two angles with the horizontal could it have begun its flight? Find the time of flight and maximum altitude in each case.

Answers to Multiple Choice

1. a	**7.** c	**13.** d	**19.** c
2. b	**8.** d	**14.** a	**20.** d
3. d	**9.** b	**15.** d	**21.** a
4. c	**10.** b	**16.** c	**22.** b
5. d	**11.** a	**17.** b	**23.** c
6. d	**12.** c	**18.** b	**24.** d

C H A P T E R

3

FORCE AND MOTION

3.1 First Law of Motion

3.2 Mass

3.3 Force

3.4 Second Law of Motion

3.5 Weight

3.6 Customary Units of Mass and Force

3.7 Third Law of Motion

3.8 Friction

3.9 Coefficient of Friction

T hus far we have managed to learn a good deal about motion while avoiding some very basic questions. To begin with, what makes anything move in the first place? Once moving, why do some things change direction, why do others speed up or slow down, and why do still others seem able to keep moving indefinitely at constant velocity? Questions such as these led Isaac Newton (1642–1727) to formulate three principles, based on observations he and others had made, that summarize so much of the behavior of moving bodies that they have become known as the *laws of motion*. These laws are the subject of this chapter.

(a) Sudden start

(b) Sudden stop

Figure 3.1

(a) When a car suddenly starts to move, the inertia of the passengers tends to keep them at rest relative to the earth, and so their heads move backward relative to the car. (b) When the car suddenly stops, inertia tends to keep the passengers moving, and so their heads move forward relative to the car. Both effects illustrate the first law of motion.

3.1   FIRST LAW OF MOTION

Constant velocity is as natural as being at rest.

Imagine a ball on a level floor. Left alone, it stays where it is. If we give it a push, the ball rolls for a while and then comes to a stop. The smoother the ball and the floor, the farther the ball rolls before stopping. With a perfectly round ball and a perfectly smooth and level floor, and with no air to impede the motion, would the ball ever stop?

There will never be a perfect ball and a perfect floor, of course. But we can come close. The result is that, as the resistance to its motion decreases, the ball does indeed go farther and farther for the same push. We can reasonably conclude that, under ideal conditions, the ball would keep rolling forever.

This conclusion was first reached by Galileo. Later it was stated by Newton, who was born in the year of Galileo's death, as his **first law of motion:**

An object at rest will remain at rest and an object in motion will continue in motion at constant velocity (constant speed in a straight line) in the absence of any interaction with something else.

The reluctance of an object to change its state of rest or of uniform motion is a property of matter known as **inertia.** When a car suddenly starts to move, its passengers feel themselves pushed backward (Fig. 3.1). What is actually happening is that inertia tends to keep their bodies where they initially were while the car begins to move. When the car suddenly stops, on the other hand, the passengers feel themselves pushed forward. What is actually happening is that inertia tends to keep their bodies moving while the car comes to a halt.

3.2   MASS

A measure of inertia.

A measure of the inertia of an object at rest is its **mass.** The more something resists being set in motion (assuming it is able to move freely), the greater its mass. The inertia of a bowling ball exceeds that of a basketball, as we can tell by kicking them in turn (Fig. 3.2), so the mass of the bowling ball exceeds that of the basketball.

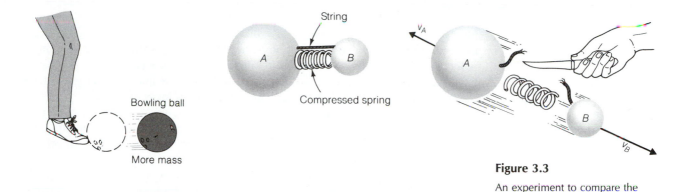

String

A B

Compressed spring

v_A

A

B

v_B

Figure 3.3

An experiment to compare the masses of two objects. When the string is cut, the object with the greater mass moves away at the lower speed.

Bowling ball

More mass

Basketball

Less mass

Figure 3.2

The mass of an object determines its inertia.

The greater the mass of an object, the greater its resistance to a change in its state of rest or of motion, as this shot-putter can testify.

We can compare the inertias—and hence the masses—of two objects, *A* and *B*, in a straightforward way. What we do in essence is put a small spring between them, push them together so the spring is compressed, and tie a string between them to hold the assembly in place (Fig. 3.3). Now we cut the string. The compressed spring pushes the objects apart, and *A* flies off to the left at the speed v_A while *B* flies off to the right at the speed v_B. Object *A* has a lower speed than object *B*, and we interpret this difference to mean that *A* exhibits more inertia than *B*.

We repeat the experiment a number of times using springs of different stiffness, so that the recoil speeds are different in each case. What we find each time is that *A* moves more slowly than *B*, and that, regardless of the exact values of v_A and v_B, their *speed ratio* v_B/v_A is always the same.

The fact that the speed ratio v_B/v_A is constant gives us a way to specify exactly what we mean by mass. If we denote the masses of *A* and *B* by the symbols m_A and m_B, respectively, we define the ratio of these masses to be

$$\frac{m_A}{m_B} = \frac{v_B}{v_A} \qquad (3.1)$$

The object with the greater mass has the lower speed, and vice versa. This procedure gives us an experimental method for finding the ratio between the masses of any two objects—it is an *operational definition*.

The next step is to choose an object to be the standard unit of mass. By international agreement this object is a platinum cylinder at Sèvres, France, called the *standard kilogram*. The mass of any other object in the world can be determined by a recoil experiment with the standard kilogram. (There are easier ways to measure mass, needless to say, but we are interested in basic principles for the present.) A liter of water, which is slightly more than a quart, has a mass of 1 kg.

Why bother to define mass in this roundabout way? After all, everybody knows that the mass of an object refers to the amount of matter it contains. The trouble is that "amount of matter" is a vague concept: It could refer to an object's volume, to the number of atoms it contains, or to yet other properties. It has proved most fruitful to choose the inertia of an object as a measure of the quantity of matter it contains and to define the object's mass in terms of this inertia as shown in an appropriate experiment.

In the customary system of units, the *slug* rather than the kilogram is the unit of mass. The slug and its relation to the pound (which is a unit of force, not of mass) will be discussed in Section 3.6.

3.3   FORCE

Any influence that can change the velocity of an object.

According to the first law of motion, the velocity of an object (which may be 0) remains constant as long as it does not interact with something else. When the object interacts with something else, its velocity changes.

We can interact with a football by kicking it, and the result is a change in the football's velocity from $v_1 = 0$ to some value v_2 (Fig. 3.4). Or the interaction can take the form of catching a moving football, in which case again the football's velocity changes. An interaction can lead to a change in the *direction* of **v** as well as to a change in its magnitude, as, for instance, when a football bounces off a tree. The concept of **force** can be used to put all this on a precise basis. In general:

A force is any influence that can change the velocity of an object.

This definition is in accord with the notion of a force as a "push" or a "pull," but it goes further since no direct contact is implied. No hand reaches up from the earth to pull a dropped stone downward, yet the increasing downward velocity of the falling stone testifies to the action of a force on it (Fig. 3.5).

Two or more forces may act on an object without affecting its state of motion if the forces cancel one another. What is needed for a velocity change is a **net force**, also called an **unbalanced force.** When an object is acted on by a set of forces whose

Figure 3.4

Three examples of how an interaction may give rise to a change in the velocity of an object.

Figure 3.5

A downward force exerted by the earth causes dropped objects to fall.

resultant is zero, the forces cancel, and the object is then in **equilibrium** (Fig. 3.6). But each of the forces acting by itself would accelerate the object.

In terms of net force the first law of motion becomes:

If no net force acts on it, an object at rest will remain at rest and an object in motion will remain in motion at constant velocity.

Figure 3.6

When several forces act on an object, they may cancel out one another to leave no net force.

3.4   SECOND LAW OF MOTION

How an object's acceleration is related to its mass and to the net force acting on it.

Newton's **second law of motion** gives us a quantitative definition of force:

The net force acting on an object equals the product of the mass and the acceleration of the object. The direction of the force is the same as that of the acceleration.

In equation form,

$$\mathbf{F} = m\mathbf{a}$$

Net force = (mass)(acceleration)

Second law of motion   (3.2)

According to the second law, a net force that gives an object twice the acceleration another force does must be twice as great as the other one (Fig. 3.7). An object moving to the right but going slower and slower is accelerated to the left.

Figure 3.7

The acceleration of an object is proportional to the net force applied to it. Successive positions of a block are shown at 1-s intervals while forces of *F*, *2F*, and *3F* are applied.

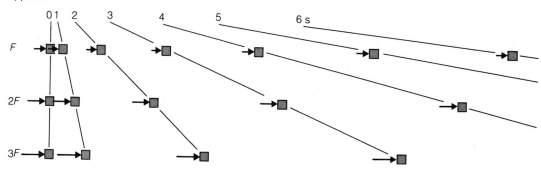

Figure 3.8

A force and the acceleration it produces are always in the same direction.

Applied force

Applied force

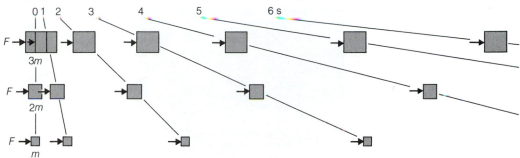

Figure 3.9

When the same force is applied to objects of different masses, the resulting accelerations are inversely proportional to the masses. Successive positions of blocks of mass m, $2m$, and $3m$ are shown at 1-s intervals while identical forces of F are applied.

Hence there must be a force toward the left acting on it (Fig. 3.8). The first law of motion is a special case of the second: When the net force on something is zero, its acceleration is also zero.

The definition of mass given earlier also fits nicely into the second law. The greater the mass of an object acted on by a given force, the smaller its acceleration and hence the smaller its final speed starting from rest (Fig. 3.9).

The relationship between force and acceleration means that the less acceleration an object has, the smaller the net force on it. If you drop to the ground from a height, you can reduce the force of the impact by bending your knees as you strike the ground so you come to a stop gradually instead of suddenly. The same reasoning can be applied to make cars safer. If a car's body is built to crumple progressively in a crash, the forces acting on the passengers will be smaller than if the car were rigid (Fig. 3.10).

In the SI system the unit of force is the **newton** (N):

A newton is that net force which, when applied to a 1-kg mass, gives it an acceleration of 1 m/s^2.

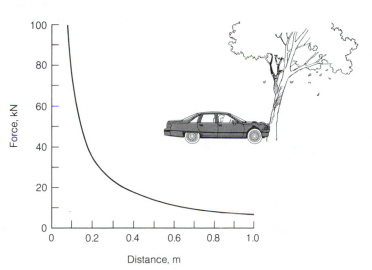

Figure 3.10

The farther something moves while being stopped, the less the force it experiences. This graph shows how the force on a 70-kg person in a car that crashes at 50 km/h (31 mi/h) varies with the distance the person moves during the impact. (10 kN = 10^4 N = 2248 lb)

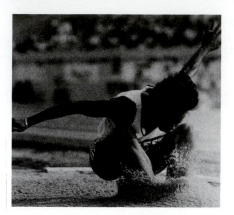

A gradual landing in sand reduces the acceleration of this long jumper when she strikes the ground and thereby reduces the force on her body.

The newton is not a basic unit and in calculations may have to be replaced by its equivalent in terms of the meter, the second, and the kilogram. This equivalent is easy to find:

$$F = ma$$
$$1\,\text{N} = (1\,\text{kg})(1\,\text{m/s}^2) = 1\,\text{kg} \cdot \text{m/s}^2$$

A newton is about 0.225 lb, a little less than ¼ lb. A pound is nearly 4½ N.

The second law of motion is a powerful tool for analyzing problems that involve force and motion. Let us examine a few such problems to become familiar with how the second law is applied.

◆   **E X A M P L E   3.1** _____

A 60-g tennis ball approaches a racket at 30 m/s, is in contact with the racket's strings for 5.0 ms (1 ms = 1 millisecond = 10^{-3} s), and then rebounds at 30 m/s (Fig. 3.11). What was the average force the racket exerted on the ball?

Figure 3.11

$v_f = -30\,\text{m/s}$

$v_o = 30\,\text{m/s}$

SOLUTION The tennis ball experienced a change in velocity of

$$\Delta v = v_f - v_0 = (-30 \text{ m/s}) - (30 \text{ m/s}) = -60 \text{ m/s}$$

so its acceleration was

$$a = \frac{\Delta v}{\Delta t} = \frac{-60 \text{ m/s}}{5.0 \times 10^{-3} \text{ s}} = -1.2 \times 10^4 \text{ m/s}^2$$

The corresponding force is, since 60 g = 0.060 kg,

$$F = ma = (0.060 \text{ kg})(-1.2 \times 10^4 \text{ m/s}^2) = -720 \text{ N} = -0.72 \text{ kN}$$

The minus sign means that the force was in the opposite direction to that of the ball when it approached the racket. The Customary equivalent of 0.72 kN is 162 lb. This force may seem high, but not if we remember that it was exerted for only 0.0050 s. ◆

◆ **E X A M P L E   3.2** _____

During performances of the Bouglione Circus in 1976, John Tailor was fired from a compressed-air cannon whose barrel was 20 m long. Tailor emerged from the cannon (twice daily, three times on Saturdays and Sundays) at 40 m/s. If Tailor's mass was 70 kg, find the average force on him during the firing of the cannon (Fig. 3.12).

40 m/s   **Figure 3.12**

70 kg

20 m

SOLUTION   We start by finding Tailor's acceleration with the help of Eq. (1.12),

$$v_f^2 = v_0^2 + 2ax$$

Here $v_0 = 0$, $v_f = 40$ m/s, and $x = 20$ m, so

$$v_f^2 = 0 + 2ax$$

$$a = \frac{v_f^2}{2x} = \frac{(40 \text{ m/s})^2}{(2)(20 \text{ m})} = 40 \text{ m/s}^2$$

The corresponding average force is

$$F = ma = (70 \text{ kg})(40 \text{ m/s}^2) = 2800 \text{ N} = 2.8 \text{ kN}$$

which is about 630 lb.   ◆

◈   **E X A M P L E   3.3** _____

A horizontal force of 10 N is applied to a 4.0-kg block that is at rest on a perfectly smooth, level surface. Find the speed of the block and how far it has gone after 6.0 s.

SOLUTION   We begin by finding the block's acceleration, which is

$$a = \frac{F}{m} = \frac{10 \text{ N}}{4.0 \text{ kg}} = \frac{10 \text{ kg} \cdot \text{m/s}^2}{4.0 \text{ kg}} = 2.5 \text{ m/s}^2$$

The speed of the block after 6.0 s is

$$v = at = (2.5 \text{ m/s}^2)(6.0 \text{ s}) = 15 \text{ m/s}$$

For the distance the block travels in $t = 6.0$ s at an acceleration of $a = 2.5$ m/s², we require Eq. (1.9), $x = v_0 t + \frac{1}{2}at^2$. The block started from rest, so $v_0 = 0$ and

$$x = \frac{1}{2}at^2 = \frac{1}{2}(2.5 \text{ m/s}^2)(6.0 \text{ s})^2 = 45 \text{ m}$$

Figure 3.13

Successive distances and speeds of a 4-kg mass acted on by a 10-N force.

After 6.0 s a 4.0-kg mass acted on by a net force of 10 N will have gone 45 m and have a speed of 15 m/s (Fig. 3.13). ◆

3.5 WEIGHT

Weight is a force.

The force with which the earth attracts an object is called its **weight.** If you weigh 700 N (157 lb), that means the earth is pulling you down with a force of 700 N. Weight (a vector quantity) is different from mass (a scalar quantity), which is a measure of inertia. Weight and mass, however, are closely related.

The weight of a stone is the force that gives it the acceleration of gravity g when it is dropped. If the stone's mass is m, then the magnitude of the downward force on it, which is its weight w, can be found from the second law of motion, $F = ma$. Since $F = w$ and $a = g$,

$$w = mg \qquad \text{Weight} \quad (3.3)$$

Weight = (mass)(acceleration of gravity)

The mass of an object is a more basic property than its weight because the pull of gravity on it is not the same everywhere. This pull is less on a mountaintop than at sea level, and less at the equator than near the poles because the earth bulges slightly at the equator. A person who weighs 700 N in Lima, Peru, would weigh about 703 N in Oslo, Norway. On the surface of Mars the same person would weigh only 266 N, so he or she would be able to jump much higher there than on the earth's surface. However, a ball could not be thrown any faster on Mars because $F = ma$ and the ball's mass m is the same.

In working out problems that concern the second law of motion, it is usually best to begin by sketching the forces that act on the object involved. This is called a **free-body diagram.** Forces that the object exerts on anything else should not be shown, because such forces do not affect the object's motion.

◆ **E X A M P L E   3.4** _____

A loaded elevator whose total mass is 800 kg is suspended by a cable whose maximum permissible tension is 20,000 N. What is the greatest upward acceleration possible for the elevator under these circumstances? What is the maximum possible downward acceleration? (Tension can be a confusing concept. See Fig. 3.14.)

Figure 3.14

The tension T in a cable is the magnitude of the force any part of the cable exerts on the adjoining part. The tension is the same in both directions in the cable, and, if the cable has no mass, it is the same along the entire cable. Since only cables of negligible mass will be considered here, we can think of T as the magnitude of the force either end of the cable exerts on whatever it is attached to.

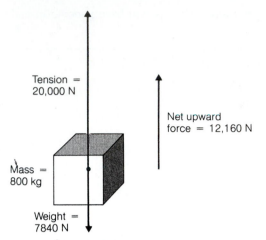

Figure 3.15

A free-body diagram showing that the net upward force on an elevator of mass 800 kg is 12,160 N when the tension in its supporting cable is 20,000 N.

Tension = 20,000 N

Net upward force = 12,160 N

Mass = 800 kg

Weight = 7840 N

SOLUTION   When the elevator is at rest or moving at constant velocity, the tension in the cable is just its weight of

$$w = mg = (800 \text{ kg})(9.8 \text{ m/s}^2) = 7840 \text{ N}$$

To accelerate the elevator upward, an additional tension F is required in order to provide a net upward force (Fig. 3.15). Since the total tension T cannot exceed 20,000 N, the greatest accelerating force available is

$$F = T - w = 20{,}000 \text{ N} - 7840 \text{ N} = 12{,}160 \text{ N}$$

The elevator's acceleration when this net force is applied is

$$a = \frac{F}{m} = \frac{12{,}160 \text{ N}}{800 \text{ kg}} = 15.2 \text{ m/s}^2$$

For the elevator to exceed the downward acceleration of gravity $g = 9.8$ m/s^2, a downward force besides its own weight is needed. Since this cannot be provided by a supporting cable, the maximum acceleration is 9.8 m/s^2.   ◆

◆   **E X A M P L E   3.5**

Figure 3.16 shows a 12-kg block, A, which hangs from a string that passes over a pulley and is connected at its other end to a 30-kg block, B, which rests on a frictionless table. Find the accelerations of the two blocks under the assumption that the string is massless and the pulley is massless and frictionless. What is the tension in the string?

Figure 3.16

Both blocks have the same acceleration.

SOLUTION Because the blocks are joined by the string, their accelerations have the same magnitude a even though they are different in direction. The net force on B is equal to the tension T in the string. From the second law of motion, taking the left as the $+$ direction so that a will come out positive,

$$F_B = T = m_B a$$

In the case of A, the net force is the difference between its weight $m_A g$, which acts downward, and the tension T in the string, which acts upward. If we consider the downward direction as positive, so that the two accelerations will have the same sign,

$$F_A = m_A g - T = m_A a$$

We can eliminate T, whose value we do not know at this point, by substituting $T = m_B a$ from the first equation in the second equation. This step permits us to find a:

$$m_A g - m_B a = m_A a$$
$$m_A g = (m_A + m_B)a$$
$$a = \frac{m_A g}{m_A + m_B} = \frac{(12 \text{ kg})(9.8 \text{ m/s}^2)}{12 \text{ kg} + 30 \text{ kg}} = 2.8 \text{ m/s}^2$$

The tension in the string is

$$T = m_B a = (30 \text{ kg})(2.8 \text{ m/s}^2) = 84 \text{ N}$$

If B were fixed in place, the tension would equal the weight of A, which is $m_A g = 118$ N. Here the tension is less because B moves in response to the pull of A's weight, but it is not zero because of B's inertia. ◆

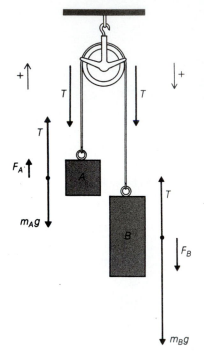

Figure 3.17

The force on each block is the difference between its weight and the tension in the string.

◆   E X A M P L E   3.6

Figure 3.17 shows the same two blocks, A and B, suspended by a string on either side of a massless, frictionless pulley. Find the accelerations of the two blocks and the tension in the string.

SOLUTION   Here A moves upward and B moves downward, both with accelerations having the same magnitude a. Applying the second law of motion to the two blocks gives

$$F_A = T - m_A g = m_A a$$
$$F_B = m_B g - T = m_B a$$

where we have considered up as $+$ for A and down as $+$ for B. From the first equation $T = m_A a + m_A g$. Using this value of T in the second equation permits us to find a:

$$m_B g - m_A a - m_A g = m_B a$$
$$(m_B - m_A)g = (m_A + m_B)a$$
$$a = \frac{(m_B - m_A)g}{m_A + m_B} = \frac{(30\,\text{kg} - 12\,\text{kg})(9.8\,\text{m/s}^2)}{12\,\text{kg} + 30\,\text{kg}}$$
$$= 4.2\,\text{m/s}^2$$

The tension in the string may be found from the first equation above:

$$T = m_A a + m_A g = m_A(a + g) = (12\,\text{kg})(4.2 + 9.8)\text{m/s}^2 = 168\,\text{N} \qquad ◆$$

◆   E X A M P L E   3.7

A rope whose working strength is 2000 N is used to tow a 1000-kg car up a $10°$ incline, as in Fig. 3.18. Find the maximum acceleration that can be given to the car.

SOLUTION   The component of the weight mg of the car that is parallel to the incline has the magnitude $mg \sin \theta$. If T is the maximum tension in the rope, the maximum net force along the incline that can be applied to the car is

$$F = T - mg \sin \theta$$

From the second law of motion,

$$T - mg \sin \theta = ma$$
$$a = \frac{T}{m} - g \sin \theta = \frac{2000\,\text{N}}{1000\,\text{kg}} - (9.8\,\text{m/s}^2)(\sin 10°) = 0.30\,\text{m/s}^2$$

On a level road, $\theta = 0$ and $\sin \theta = 0$, and the maximum acceleration would be 2.0 m/s². ◆

Figure 3.18

Example 3.7.

3.6 CUSTOMARY UNITS OF MASS AND FORCE

The slug and the pound.

In the U.S. Customary System the unit of mass is the **slug** and the unit of force is the **pound** (lb). An object whose mass is 1 slug experiences an acceleration of 1 ft/s² when a net force of 1 lb acts on it. Table 3.1 compares units of mass and weight in the SI and customary systems.

The slug is an unfamiliar unit because in everyday life weights rather than masses are specified in the customary system. We go shopping for 10 lb of apples,

Table 3.1   Units of mass and weight

System of units	Unit of mass	Unit of weight	Acceleration of gravity g	To find mass m given weight w	To find weight w given mass m
SI	kilogram (kg)	newton (N)	9.80 m/s²	$m \text{ (kg)} = \dfrac{w \text{ (N)}}{9.80 \text{ m/s}^2}$	$w \text{ (N)} = m \text{ (kg)} \times 9.80 \text{ m/s}^2$
Customary	slug	pound (lb)	32.2 ft/s²	$m \text{ (slugs)} = \dfrac{w \text{ (lb)}}{32.2 \text{ ft/s}^2}$	$w \text{ (lb)} = m \text{ (slugs)} \times 32.2 \text{ ft/s}^2$

Conversion of units: 1 slug = 14.6 kg   1 N = 0.225 lb
1 kg = 0.0685 slug   1 lb = 4.45 N

not 1/3 slug of apples. In the SI system, on the other hand, masses are normally specified. European grocery scales are calibrated in kilograms, not in newtons.

As a practical matter, especially when traveling outside the United States, it can be handy to know how to find the weight in pounds of a mass in kilograms, or the mass in kilograms of a weight in pounds. The relations between kilograms and pounds (for $g = 9.80$ m/s^2 = 32.2 ft/s^2 only!) are as follows:

> One kilogram corresponds to 2.21 pounds in the sense that the weight of 1 kilogram is 2.21 pounds.
>
> One pound corresponds to 0.454 kilogram in the sense that the mass of 1 pound is 0.454 kilogram.

Thus a 60.0-kg woman weighs (60.0 kg)(2.21 lb/kg) = 133 lb, and a 170-lb man has a mass of (170 lb)(0.454 kg/lb) = 77.2 kg.

◆   E X A M P L E   3.8

A net force of 15 lb acts on an 8.0-lb object. Find the object's acceleration.

SOLUTION   First we calculate the mass of the object, which is

$$m = \frac{w}{g} = \frac{8.0 \text{ lb}}{32 \text{ ft/s}^2} = 0.25 \text{ slug}$$

From the second law of motion $F = ma$ we find that the acceleration is

$$a = \frac{F}{m} = \frac{15 \text{ lb}}{0.25 \text{ slug}} = 60 \text{ ft/s}^2$$

We can work out problems like this in a single step by using w/g in place of m in the second law, so that $F = (w/g)a$ and

$$a = \frac{F}{w/g} = \frac{Fg}{w} = \frac{(15 \text{ lb})(32 \text{ ft/s}^2)}{8.0 \text{ lb}} = 60 \text{ ft/s}^2 \qquad ◆$$

◆   E X A M P L E   3.9

A 3000-lb car has an initial speed of 10 mi/h. How much force is required to accelerate the car to a speed of 50 mi/h in 9.0 s?

SOLUTION   We begin by converting the speeds from mi/h to ft/s. From the table inside the back cover of this book, 1 mi/h = 1.47 ft/s. To two significant figures,

$$v_0 = (10 \text{ mi/h})\left(1.47\frac{\text{ft/s}}{\text{mi/h}}\right) = 15 \text{ ft/s}$$

$$v_f = (50 \text{ mi/h})\left(1.47 \frac{\text{ft/s}}{\text{mi/h}}\right) = 73 \text{ ft/s}$$

The car's acceleration is

$$a = \frac{v_f - v_0}{t} = \frac{(73 - 15)\text{ft/s}}{9.0 \text{ s}} = 6.4 \text{ ft/s}^2$$

The force that must act on the car is therefore

$$F = ma = \left(\frac{w}{g}\right)a = \left(\frac{3000 \text{ lb}}{32 \text{ ft/s}^2}\right)(6.4 \text{ ft/s}^2) = 600 \text{ lb} \qquad \blacklozenge$$

3.7 THIRD LAW OF MOTION

Action and reaction.

We push against a heavy table and it does not move. Evidently the table is pushing back just as hard as we push against it. The table stays in place because our force on it is matched by the opposing force of friction between its base and the floor. We ourselves do not move because the force of the table on us is matched by a similar opposing force between our shoes and the floor.

Suppose now that we and the table are on a frozen lake whose surface is so slippery on a warm day that there is no friction. Again we push on the table, which this time moves away from us as a result (Fig. 3.19). But we can stick to the ice no

Force exerted by person on table

Force exerted by table on person

Figure 3.19

Pushing a table on a frozen lake results in person and table moving apart in opposite directions. There is no way to exert a force without an equal and opposite force coming into being. The two forces act on different objects.

Figure 3.20

According to the third law of motion, $\mathbf{F}_{AB} = -\mathbf{F}_{BA}$.

better than the table can, and we find ourselves sliding backward. No matter what we do, pushing on the table always means that the table pushes back on us.

Considerations of this kind led Newton to his **third law of motion:**

When an object exerts a force on another object, the second object exerts on the first a force of the same magnitude but in the opposite direction.

If we call one of the interacting objects *A* and the other *B*, then according to the third law of motion

$$\mathbf{F}_{AB} = -\mathbf{F}_{BA} \qquad\qquad \textit{Third law of motion} \quad (3.4)$$

Here $\mathbf{F}_{AB}$ is the force *A* exerts on *B* and $\mathbf{F}_{BA}$ is the force *B* exerts on *A* (Fig. 3.20).

Action and Reaction

The third law of motion always applies to two different forces on two different bodies—the **action force** that one body exerts on another, and the equal but opposite **reaction force** that the second exerts on the first.

This book weighs about 20 N. When it rests on a table, it presses down on the table with a force of 20 N (Fig. 3.21). The table pushes upward on the book with a reaction force of 20 N. Why doesn't the book fly upward into the air? The answer is that the upward force of 20 N on the book merely balances its weight of 20 N, which acts downward. If the table were not there to cancel out the latter 20-N force, the book would, of course, be accelerated downward.

The process of walking is another example of the third law at work. We push backward with one foot, and the ground pushes forward on us. The forward reaction force exerted by the ground causes us to move forward. At the same time, the backward force of our foot causes the earth to move backward (Fig. 3.22). Owing to the earth's enormously greater mass, its motion is too small to be detected, but it does move.

Figure 3.21

The action-reaction forces between a book and a table and between the table and the floor.

Figure 3.22

When we push backward on the earth with one foot, the opposite reaction force of the earth pushes forward on us. The latter force causes us to move forward.

Figure 3.23

If we throw a snowball while standing on a frozen lake, the reaction force pushes us backward.

Suppose we find ourselves on a frozen lake with a melted surface. Now we cannot walk because the lack of friction prevents us from exerting a backward force on the ice that would produce a forward force on us. But what we can do is exert a force on some object we may have with us, say a snowball. We throw the snowball forward by applying a force to it. At the same time, the snowball is pressing back on us with the identical force but in the opposite direction, and in consequence we find ourselves moving backward (Fig. 3.23).

It is sometimes arbitrary which force of an action-reaction pair to consider action and which reaction. For instance, we cannot really say that the gravitational pull of the earth on an apple is the action force and that the pull of the apple on the earth is the reaction force, or the other way around. When we push on the ground while walking, however, it is legitimate to call this force the action force and the force with which the ground pushes back on us the reaction force.

◆ **E X A M P L E 3.10** ————————————————————

A 1.0-kg block A and a 3.0-kg block B are in contact on a frictionless horizontal surface, as in Fig. 3.24. (a) A horizontal force of $F = 10$ N is applied to block A. Find the force this block exerts on block B. (b) The same force is now applied to block B. Find the force block A exerts on block B in this case.

$m_A = 1.0$ kg
$m_B = 3.0$ kg
$F = 10$ N

Figure 3.24

SOLUTION   (a) Because the blocks stick together, they have the same acceleration, which is

$$a = \frac{F}{m_A + m_B} = \frac{10\,\text{N}}{4.0\,\text{kg}} = 2.5\,\text{m/s}^2$$

To give block B this acceleration, the force block A exerts on it must be

$$F_{AB} = m_B a = (3.0\,\text{kg})(2.5\,\text{m/s}^2) = 7.5\,\text{N}$$

(b) The force block B exerts on block A has the magnitude

$$F_{BA} = m_A a = (1.0\,\text{kg})(2.5\,\text{m/s}^2) = 2.5\,\text{N}$$

The reaction force of block A on block B must have the same magnitude but the opposite direction. As a check, we note that the net force to the left on block B is $10\,\text{N} - 2.5\,\text{N} = 7.5\,\text{N}$, which is what is needed to give this block the acceleration a.   ◆

3.8   FRICTION

Sometimes useful, sometimes not.

Frictional forces act to impede relative motion between two surfaces in contact. Friction is quite different from inertia. Inertia refers to the tendency of an object to maintain its state of rest or motion at constant velocity when no net force acts on it. Even the smallest net force, however, can accelerate an object despite its inertia. **Friction,** on the other hand, refers to an actual force that arises to oppose relative motion between contacting surfaces.

Often friction is desirable. The fastening action of nails, screws, and bolts and the resistive action of brakes depend on it. Walking would be impossible without friction. In many situations, however, friction merely reduces efficiency, and great efforts are made in industry to minimize it through the use of lubricants—notably grease and oil—and special devices—notably the wheel. About half the power of a typical automobile engine is wasted in overcoming friction in the engine itself and in its drive train.

Let us consider what happens when we try to move a box across a level floor, as in Fig. 3.25. As we begin to push, the box stays in place because the floor exerts a force on the bottom of the box, which opposes the force we apply. This opposition force is friction, and it arises from the nature of the contact between the floor and the box. As we push harder, the frictional force also increases to match our efforts, until finally we are able to exceed the frictional force and begin to move the

(a)

(b)

Figure 3.25

(a) As force is applied to a box on a level floor, the frictional resistive force increases to a certain maximum, decreases somewhat as the box begins to move, and then remains constant. (b) The net force on the box equals the applied force minus the frictional force. The frictional force is always equal to or less than the applied force.

box. The frictional force is parallel to the contacting surfaces of the floor and the box.

As indicated in Fig. 3.25(a), the force between two stationary surfaces in contact that prevents motion between them is called **static friction.** In a given case, static friction has a certain maximum value called **starting friction.** When the applied force exceeds starting friction, one surface breaks away and begins to move relative to the other. The **kinetic friction** (or **sliding friction**) that occurs after that is less than the starting friction, so less force is needed to keep the box of Fig. 3.25 in motion than the force needed to set it moving in the first place. ("Kinetic" is a word of Greek origin that implies motion.)

Because starting friction is always greater than kinetic friction, a car's tires provide more traction when they are not slipping. This holds true whether the intention is to accelerate the car forward (spinning the wheels on snow or ice is not very effective) or to slow it down (locking the wheels by applying too much force to the brakes reduces the retarding force on the car).

Friction has two chief causes. One is the mechanical interlocking of irregularities in the two surfaces, as in Fig. 3.26, which prevents one surface from sliding evenly past the other. The second cause is the tendency for materials in very close contact to stick together because of attractive forces between their respective atoms; the origin of such forces is discussed later. The tiny "welds" that occur where projections on the surfaces are pressed together must be broken in order to move one

Figure 3.26

A lubricant reduces friction by providing a soft material able to flow readily to separate surfaces in contact.

surface over the other. As the motion continues, new welds form and must be broken in turn.

Lubricants reduce friction by separating two contacting surfaces with a layer of softer material. Instead of rubbing against each other, the surfaces rub against the lubricant (Fig. 3.26). Depending on the specific application, the best lubricant may be a gas, a liquid, or a solid. Most lubricants are oils derived from petroleum. Grease consists of oil to which a thickening agent has been added to prevent the oil from running out from between the surfaces involved. The joints of the limbs of the human body are lubricated by a substance called **synovial fluid,** which resembles blood plasma.

The Wheel

There is no relative motion between the rim of a wheel and a smooth surface over which it rolls if wheel and surface are both rigid and thus no frictional resistance must be overcome. By reducing friction so drastically, the wheel makes it possible to carry loads from one place to another without the enormous forces that dragging them would require. Most aspects of our technological civilization rely on the wheel in one way or another.

Figure 3.27

Both the flattening of a wheel and the indentation of the surface it presses on contribute to rolling friction.

Neither wheels nor the surfaces on which they travel can ever be perfectly rigid. Figure 3.27 shows the flattening of the wheel and the indentation of the surface, which both contribute to **rolling friction.** Because the wheel and the surface must be constantly deformed as the wheel rolls, a force is needed to keep the wheel rolling. However, this force is usually many times smaller than that needed to overcome sliding friction. Balls and rollers are widely used to reduce the friction on a rotating shaft by replacing sliding friction with rolling friction. Tire deformation accounts for most of the frictional force on a car at speeds of up to 50 to 80 km/h, when air resistance starts to dominate (see Fig. 10.28).

3.9   COEFFICIENT OF FRICTION

The greater the normal force, the greater the friction, regardless of the area in contact.

It is a matter of experience that the frictional force exerted by one surface on another depends on two factors:

1. The kinds of surfaces that are in contact;
2. The perpendicular force with which either surface is pressed against the other. This perpendicular force is usually referred to as the **normal force,** symbol F_N.

The more tightly two objects are pressed together, the greater the friction between them. For this reason an empty box is easier to push across a floor than a similar box loaded with something heavy (Fig. 3.28). Equally familiar is the effect of the nature of the contacting surfaces. For instance, we need more than three times as much force to push a wooden box across a wooden floor as we do to push a steel box of the same weight across a steel floor. Interestingly enough, the area in contact between the two surfaces is not important. It is just as hard to push a small 50-kg box over a given floor as it is to push a large 50-kg box of the same material over the same floor.

To a good degree of approximation, the following formulas relate the normal force F_N pressing one surface against another to the frictional force F_f that results:

$$F_f \leq \mu_s F_N \qquad \text{Static friction} \quad (3.5)$$

$$F_f = \mu_k F_N \qquad \text{Kinetic friction} \quad (3.6)$$

The quantity μ (Greek letter *mu*) is called the **coefficient of friction** and is a constant for a given pair of surfaces. Table 3.2 lists the coefficients of static and kinetic friction for a number of sets of surfaces.

When an object resting on a surface is pushed, the force of static friction increases with the applied force until the limiting value of $\mu_s F_N$ is reached. That is, when no motion occurs, $\mu_s F_N$ gives the starting frictional force, not the actual frictional force: The actual frictional force has the same magnitude as the applied force, but is in the opposite direction. When the applied force is greater than the starting frictional force of $\mu_s F_N$, motion begins and the coefficient of kinetic friction must then be used. In this case Eq. (3.6) gives the frictional force, which no longer depends on the applied force and is the same over a fairly wide range of speeds.

As mentioned earlier, the normal force between an object and a surface with which it is in contact can be taken either as the force the object exerts on the surface or as the reaction force the surface exerts on the object, since both have the same magnitude. If we use a free-body diagram to help solve problems, however, we are

Figure 3.28

The greater the normal force with which one surface is pressed against the other, the greater the force of friction between them.

Table 3.2 Approximate coefficients of static and kinetic friction for various materials in contact

Materials in contact	Coefficient of static friction, μ_s	Coefficient of kinetic friction, μ_k
Wood on wood	0.5	0.3
Wood on stone	0.5	0.4
Steel on steel (smooth)	0.15	0.09
Metal on metal (lubricated)	0.03	0.03
Leather on wood	0.5	0.4
Rubber tire on dry concrete	1.0	0.7
Rubber tire on wet concrete	0.7	0.5
Glass on glass	0.94	0.4
Steel on Teflon	0.04	0.04
Bone on bone (dry)		0.3
Bone on bone (lubricated)		0.003

Figure 3.29

Free-body diagram of a box on a level floor. The normal force F_N has the magnitude mg here. When the applied force F_A exceeds the frictional force F_f, the net force $F_A - F_f$ acts to accelerate the box.

limited to forces that act *on* the object. When friction is involved, this means we must use the reaction force on the object for the normal force F_N. In the examples that follow, this convention will be used.

◆   **EXAMPLE   3.11** ———————————————

A 100-kg wooden crate is at rest on a level stone floor (Fig. 3.29). (a) What is the minimum horizontal force needed to start the crate moving? (b) What is the minimum horizontal force needed to keep the crate in motion at constant speed? (c) What will happen if a horizontal force of 500 N is applied to the crate?

SOLUTION   (a) The starting frictional force is given by Eq. (3.5) using the coefficient of static friction for these surfaces from Table 3.2:

$$F_f(\text{starting}) = \mu_s F_N = \mu_s mg = (0.5)(100 \text{ kg})(9.8 \text{ m/s}^2) = 490 \text{ N}$$

The crate will begin to move only if a force of more than 490 N is applied to it.

(b) When the crate is moving, the coefficient of kinetic friction must be used to find the frictional force, which is

$$F_f(\text{kinetic}) = \mu_k mg = (0.4)(100 \text{ kg})(9.8 \text{ m/s}^2) = 392 \text{ N}$$

Although at least 490 N of force must be exerted to set the crate in motion, only 392 N of force is needed to continue the motion at constant speed.

(c) Since the applied force of $F_A = 500$ N exceeds the starting force of 490 N, the crate will begin to move. Since the applied force also exceeds the force of sliding friction $F_f = 392$ N, the crate will be accelerated. The net force on the moving crate is

$$F = F_A - F_f = 500 \text{ N} - 392 \text{ N} = 108 \text{ N}$$

The crate's acceleration is, from the second law of motion $F = ma$,

$$a = \frac{F}{m} = \frac{108 \text{ N}}{100 \text{ kg}} = 1.08 \text{ m/s}^2 \qquad\qquad ◆$$

◆   **EXAMPLE   3.12** ———————————————

The coefficient of kinetic friction between a rubber tire and a dry concrete road is 0.700. What is the distance in which a car will skid to a stop on such a road if its brakes are locked when it is moving at 50.0 mi/h?

SOLUTION   The first step is to calculate the acceleration of the car under these circumstances. The normal force F_N is the car's weight of mg, and the frictional force is

$$F = \mu_k F_N = \mu_k mg$$

From the second law of motion, $F = ma$, and

$$\mu_k mg = ma$$
$$a = \mu_k g = (0.700)(32.0 \text{ ft/s}^2) = 22.4 \text{ ft/s}^2$$

The car's weight does not matter. We can find the distance x from $v_f^2 = v_0^2 + 2ax$. Here $v_f = 0$ and

$$v_0 = (50.0 \text{ mi/h})\left(1.467\frac{\text{ft/s}}{\text{mi/h}}\right) = 73.4 \text{ ft/s}$$

Therefore, since $a = -22.4 \text{ ft/s}^2$ (the minus sign is needed because the acceleration is negative),

$$x = \frac{-v_0^2}{2a} = -\frac{(73.4 \text{ ft/s}^2)}{(2)(-22.4 \text{ ft/s}^2)} = 120 \text{ ft}$$ ◆

◆ **E X A M P L E 3.13** _____

A person whose shoes have leather heels is walking on a wooden floor. (a) Find the maximum angle the forward leg may make with the vertical in order that the heel not slip on the floor. (b) How is this angle affected if the floor is wet, which reduces the coefficient of static friction?

SOLUTION   (a) The geometry of the situation is shown in Fig. 3.30. The force **F** that the foot exerts on the floor, if there is no slip, has the reaction force $\mathbf{F}_R$, whose components parallel and perpendicular to the floor are, respectively,

$$F_x = F \sin \theta \qquad F_y = F \cos \theta$$

The condition for the heel not to slip is that the frictional force $\mathbf{F}_f$ have the same magnitude as $\mathbf{F}_x$. Since

$$F_f = \mu_s F_N = \mu_s F \cos \theta$$

we have for the limiting angle

$$F_f = F_x \qquad \mu_s F \cos \theta = F \sin \theta \quad \text{and} \quad \tan \theta = \mu_s$$

From Table 3.2 the coefficient of static friction for leather on wood is 0.5, so

$$\tan \theta = 0.5 \quad \text{and} \quad \theta = 27°$$

If θ is equal to or less than 27°, the heel will not slip.

Figure 3.30

If slipping is not to occur, the angle θ must be smaller than a certain critical angle that depends on the coefficient of static friction between the heel and the floor.

(b) Reducing μ_s reduces θ, so smaller steps have to be taken on a slippery surface. Slipping is a sudden process because, as it starts, the smaller coefficient of kinetic friction takes over, and the frictional force drops sharply. This is why it is hard to keep from falling once slipping begins.   ◆

◆   **E X A M P L E   3.14** _____

A wooden chute is being built, along which wooden crates of merchandise are to be slid down into the basement of a store. (a) What angle with the horizontal should the chute make if the crates are to slide down at constant speed? (b) With what force must a 100-kg crate be pushed in order to start it sliding down the chute if the angle of the chute is that found in (a)?

SOLUTION   (a) Figure 3.31 shows the various forces that act on the crate. The component **F** of the crate's weight mg parallel to the chute has the magnitude

$$F = mg \sin \theta$$

and the reaction force $\mathbf{F}_N$ of the chute on the crate has the magnitude

$$F_N = mg \cos \theta$$

When the crate slides down at constant speed, there is no net force acting on it, according to Newton's first law of motion. Hence the downward force along the chute must exactly balance the force of kinetic friction, which means that

$$F = \mu_k F_N$$
$$mg \sin \theta = \mu_k mg \cos \theta$$
$$\mu_k = \frac{\sin \theta}{\cos \theta} = \tan \theta$$

From Table 3.2 the value of μ_k for wood on wood is 0.3, and so $\theta = 17°$.

Figure 3.31

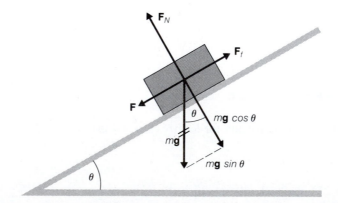

(b) We note that the coefficient of static friction here is $\mu_s = 0.5$. Hence the force of static friction to be overcome is

$$F_f = \mu_s F_N = \mu_s mg \cos \theta$$

This is greater than the force of kinetic friction, and so a crate will not begin to move without a push. The force component along the plane due to the crate's own weight is $F = mg \sin \theta$. Here F is less than F_f, and so, if we call F_A the applied force parallel to the plane required to move the crate,

$$\text{Applied force} + \text{forward component of weight} = \text{backward frictional force}$$
$$F_A + mg \sin \theta = \mu_s mg \cos \theta$$

$$F_A = mg(\mu_s \cos \theta - \sin \theta)$$
$$= (100 \text{ kg})(9.8 \text{ m/s}^2)(0.5 \cos 17° - \sin 17°) = 182 \text{ N} \qquad \blacklozenge$$

Important Terms

The **inertia** of an object refers to the apparent resistance it offers to changes in its state of motion. The property of matter that manifests itself as inertia is called **mass.** The unit of mass in the SI system is the **kilogram,** in the customary system it is the **slug.**

A **force** is any influence that can cause an object to be accelerated. The unit of force in the SI system is the **newton,** in the customary system it is the **pound.**

The **weight** of an object is the gravitational force exerted on it by the earth. The weight of an object is proportional to its mass.

Newton's **first law of motion** states that, in the absence of a net force acting on it, an object at rest will remain at rest and an object in motion will continue in motion at constant velocity. The **second law of motion** states that a net force acting on an object causes it to have an acceleration proportional to the magnitude of the force and inversely proportional to the object's mass. The acceleration is in the same direction as the force. The **third law of motion** states that when an object exerts a force on another object, the second object exerts on the first a force of the same magnitude but in the opposite direction.

The **free-body** diagram of an object shows only the forces that act *on* the object. Such a diagram is helpful in working out problems that involve the second law of motion.

The term **friction** refers to the resistive forces that arise to oppose the motion of an object past another with which it is in contact. **Kinetic friction** is the frictional resistance an object in motion experiences. **Static friction** is the frictional resistance that prevents a stationary object from being set in motion.

The **coefficient of static friction** is the constant of proportionality for a given pair of contacting surfaces not in relative motion that relates the maximum frictional force between them to the normal force with which one presses against the other. The **coefficient of kinetic friction** is this constant for surfaces in relative motion.

Important Formulas

Second law of motion: $\mathbf{F} = m\mathbf{a}$

Weight: $\mathbf{w} = m\mathbf{g}$

Third law of motion: $\mathbf{F}_{AB} = -\mathbf{F}_{BA}$

Frictional force: $F_f \leq \mu_s F_N$ (Static friction)
$F_f = \mu_k F_N$ (Kinetic friction)

MULTIPLE CHOICE

1. As an airplane climbs,
   a. its mass increases
   b. its mass decreases
   c. its weight increases
   d. its weight decreases

2. The weight of an object
   a. is the quantity of matter it contains
   b. refers to its inertia
   c. is basically the same quantity as its mass but expressed in different units
   d. is the force with which it is attracted to the earth

3. The acceleration of gravity on Mars is 3.7 m/s². Compared with her mass and weight on the earth, an astronaut on Mars has
   a. less mass and less weight
   b. less mass and more weight
   c. the same mass and less weight
   d. less mass and the same weight

4. An automobile that is towing a trailer is accelerating on a level road. The force that the automobile exerts on the trailer is
   a. equal to the force the trailer exerts on the automobile
   b. greater than the force the trailer exerts on the automobile
   c. equal to the force the trailer exerts on the road
   d. equal to the force the road exerts on the trailer

5. When a horse pulls a wagon, the force that causes the horse to move forward is the force
   a. he exerts on the wagon
   b. the wagon exerts on him
   c. he exerts on the ground
   d. the ground exerts on him

6. The action and reaction forces referred to in Newton's third law of motion
   a. act on the same object
   b. act on different objects
   c. need not be equal in magnitude and need not have the same line of action
   d. must be equal in magnitude but need not have the same line of action

7. A woman whose weight is 500 N is standing on the ground. The force the ground exerts on her is
   a. less than 500 N
   b. equal to 500 N
   c. more than 500 N
   d. any of the above, depending on her location on the earth's surface

8. A jumper of weight w presses down on the ground with a force of magnitude F and becomes airborne as a result. The magnitude of the force the ground exerted on the jumper must have been
   a. equal to w and less than F
   b. equal to w and equal to F
   c. more than w and equal to F
   d. more than w and more than F

9. When an object exerts a normal force on a surface, the force is
   a. equal to the object's weight
   b. frictional in origin
   c. parallel to the surface
   d. perpendicular to the surface

10. To set an object in motion on a surface usually requires
    a. less force than to keep it in motion
    b. the same force as that needed to keep it in motion
    c. more force than to keep it in motion
    d. only as much force as is needed to overcome inertia

11. The frictional force between two surfaces in contact does not depend on
    a. the normal force pressing one against the other
    b. the areas of the surfaces
    c. if the surfaces are stationary or in relative motion
    d. whether or not a lubricant is used

12. When a 1.0-N force acts on a 1.0-kg object that is able to move freely, the object receives
    a. a speed of 1.0 m/s
    b. an acceleration of 0.102 m/s²
    c. an acceleration of 1.0 m/s²
    d. an acceleration of 9.8 m/s²

13. When a 1.0-N force acts on a 1.0-N object that is able to move freely, the object receives
    a. a speed of 1.0 m/s
    b. an acceleration of 0.102 m/s²
    c. an acceleration of 1.0 m/s²
    d. an acceleration of 9.8 m/s²

14. When a 1.0-lb force acts on a 1.0-slug object that is able to move freely, the object receives
    a. a speed of 1.0 ft/s
    b. an acceleration of 0.031 ft/s²
    c. an acceleration of 1.0 ft/s²
    d. an acceleration of 32 ft/s²

15. When a 1.0-lb force acts on a 1.0-lb object that is able to move freely, the object receives
    a. a speed of 1.0 ft/s

b. an acceleration of 0.031 ft/s^2
c. an acceleration of 1.0 ft/s^2
d. an acceleration of 32 ft/s^2

16. The mass of 8 lb of pineapples is
    a. 0.25 slug   **b.** 0.5 slug
    c. 4 slugs   **d.** 256 slugs

17. The weight of 600 g of salami is
    a. 0.061 N   **b.** 5.9 N
    c. 61 N   **d.** 5.9 kN

18. A certain force gives a 5.0-kg object an acceleration of 2.0 m/s^2. The same force would give a 20-kg object an acceleration of
    a. 0.5 m/s^2 **b.** 2.0 m/s^2
    c. 4.9 m/s^2 **d.** 8.0 m/s^2

19. A force of 10 N gives an object an acceleration of 5.0 m/s^2. What force would be needed to give it an acceleration of 1.0 m/s^2?
    a. 1 N   **b.** 2 N
    c. 5 N   **d.** 50 N

20. An upward force of 600 N acts on a 50-kg dumbwaiter. The dumbwaiter's acceleration is
    a. 0.82 m/s^2 **b.** 2.2 m/s^2
    c. 11 m/s^2 **d.** 12 m/s^2

21. A force of 300 lb slows down a wagon from 60 ft/s to 20 ft/s in 10 s. The wagon's weight is
    a. 75 lb   **b.** 300 lb
    c. 1200 lb   **d.** 2400 lb

22. A net force of 1.0 N acts on a 2.0-kg object, initially at rest, for 2.0 s. The distance the object moves during that time is
    a. 0.5 m   **b.** 1.0 m
    c. 2.0 m   **d.** 4.0 m

23. A 500-g ball lying on the ground is kicked with a force of 250 N. If the kick lasted 0.020 s, the ball flew off with a speed of
    a. 0.01 m/s   **b.** 0.1 m/s
    c. 2.5 m/s   **d.** 10 m/s

24. An 80-kg firefighter slides down a brass pole 8.0 m high in a fire station while exerting a frictional force of 600 N on the pole. He reaches the bottom of the pole in the time of
    a. 1.5 s   **b.** 1.9 s
    c. 2.6 s   **d.** 7.0 s

25. A brick has the dimensions 8 cm $\times$ 16 cm $\times$ 32 cm. The force of starting friction between the brick and a wooden floor is
    a. greatest when the brick rests on the 8 cm $\times$ 16 cm face

b. greatest when the brick rests on the 8 cm $\times$ 32 cm face
c. greatest when the brick rests on the 16 cm $\times$ 32 cm face
d. the same regardless of which face it rests on

26. The coefficient of static friction between two wooden surfaces is
    a. 0.5   **b.** 0.5 N
    c. 0.5 kg/N   **d.** 0.5 N/kg

27. A force of 40 N is needed to set a 10-kg steel box moving across a wooden floor. The coefficient of static friction is
    a. 0.08   **b.** 0.25
    c. 0.4   **d.** 2.5

28. The coefficient of static friction for steel on ice is 0.1. The force needed to set a 70-kg skater in motion is about
    a. 0.1 N   **b.** 0.7 N
    c. 7 N   **d.** 70 N

29. A horizontal force of 150 N is applied to a 51-kg carton on a level floor. The coefficient of static friction is 0.5 and that of kinetic friction is 0.4. The frictional force acting on the carton is
    a. 150 N   **b.** 200 N
    c. 250 N   **d.** 500 N

30. A car is moving at 20 m/s on a road such that the coefficient of static friction between its tires and the road is 0.80. The minimum distance in which the car can be stopped is
    a. 26 m
    b. 33 m
    c. 51 m
    d. undeterminable unless the car's mass is known

31. A toboggan reaches the foot of a hill at a speed of 4.0 m/s and coasts on level snow for 15 m before coming to a stop. The coefficient of kinetic friction is
    a. 0.004   **b.** 0.05
    c. 0.16   **d.** 0.27

32. A force of 40 lb is needed to set a 100-lb steel box moving across a wooden floor. The coefficient of static friction is
    a. 0.08   **b.** 0.25
    c. 0.4   **d.** 2.5

33. The coefficients of static and sliding friction for wood on wood are respectively 0.50 and 0.30. If a 100-lb wooden box is pushed across a horizontal wooden floor with just enough force to overcome the force of static friction, its acceleration is
    a. 0.5 ft/s^2 **b.** 6.4 ft/s^2
    c. 9.6 ft/s^2 **d.** 16 ft/s^2

QUESTIONS

1. When a body is accelerated, a force is invariably acting on it. Does this mean that, when a force is applied to a body, it is invariably accelerated?

2. Compare the tension in the coupling between the first two cars in a train being pulled by a locomotive with the tension in the coupling between the last two cars when the train's speed is increasing.

3. It is less dangerous to jump from a high wall onto loose earth than onto a concrete pavement. Why?

4. Measurements are made of distance versus time for three moving objects. The distances are found to be directly proportional to t, t^2, and t^3, respectively. What can you say about the net force acting on each of the objects?

5. When a force equal to its weight is applied to a body free to move, what is its acceleration?

6. A person in an elevator suspends a 1.0-kg mass from a spring balance. What is the nature of the elevator's motion when the balance reads 9.0 N? 9.8 N? 10.0 N?

7. Can anything ever have a downward acceleration greater than g? If so, how can this be accomplished?

8. When you ''weigh'' an object on one pan of an equal-arm balance by putting standard masses on the other pan until the beam is horizontal, are you actually determining the object's weight or its mass? What if you use a spring balance?

9. Can we conclude from the third law of motion that a single force cannot act on a body?

10. Since the opposite forces of the third law of motion are equal in magnitude, how can anything ever be accelerated?

11. An engineer designs a propeller-driven spacecraft. Because there is no air in space, he incorporates a supply of oxygen as well as a supply of fuel for the motor. What do you think of the idea?

12. A car with its engine running and in gear goes up a hill and then down on the other side. What forces cause it to move upward? Are they the same as the forces that then cause it to move downward?

13. A bird is in a closed box that rests on a scale. Does the reading on the scale depend on whether the bird is standing on the bottom of the box or is flying around inside it? What would happen if the bird were in a cage instead?

14. Ships are often built on ways that slope down to a nearby body of water. Normally a ship is launched before most of its interior and superstructure have been installed, and is completed when afloat. Is this done because the additional weight would cause the ship to slide down the ways prematurely?

15. A person is pushing a box across a floor. Can the frictional force on the box ever exceed its own weight? Can the frictional force ever exceed the applied force?

EXERCISES

3.4   Second Law of Motion

1. A 25-g snail goes from rest to a speed of 3.0 mm/s in 5.0 s. How much horizontal force does the ground exert on it?

2. A 430-g soccer ball lying on the ground is kicked and flies off at 25 m/s. If the duration of the impact was 0.010 s, what was the average force on the ball?

3. An empty truck whose mass is 2000 kg has a maximum acceleration of 1.0 m/s². What is its maximum acceleration when it is carrying a 1000-kg load?

4. A car whose mass (including the driver) is 1600 kg has a maximum acceleration of 1.2 m/s². If three 80-kg passengers are also in the car, find its maximum acceleration then.

5. A net horizontal force of 4000 N is applied to a 1400-kg car. What will the car's speed be after 10 s if it started from rest?

6. A bicycle and its rider together have a mass of 80 kg. If the bicycle's speed is 6.0 m/s, how much force is needed to bring it to a stop in 4.0 s?

7. A 12,000-kg airplane launched by a catapult from an aircraft carrier is accelerated from 0 to 200 km/h in 3.0 s. (a) How many times the acceleration of gravity is the airplane's acceleration? (b) What is the average force the catapult exerts on the airplane?

8. A force of 20 N gives an object an acceleration of 5.0 m/s². (a) What force would be needed to give the same object an acceleration of 1.0 m/s²? (b) What force would be needed to give an acceleration of 10 m/s²?

9. A 2000-kg truck is braked to a stop in 15 m from an initial speed of 12 m/s. How much force was required?

10. A 100-kg motorcycle carrying a 70-kg rider comes to a stop in 40 m from a speed of 50 km/h when its brakes are applied. Find the force exerted by the brakes and the force experienced by the rider.

11. A driver shifts into neutral when her 1200-kg car is moving at 80 km/h and finds that its speed has dropped to 65 km/h 10 s later. What was the average drag force acting on the car?

12. The driver of the car in Exercise 11 presses down on the gas pedal and thereby applies a force of 1300 N to the car. How long does it take the car to go from 65 km/h back to 80 km/h?

13. A billy goat running at 5.0 m/s butts his head against a tree. If his head has a mass of 4.0 kg and it comes to a stop in 5.0 mm, find the average force of the impact. What do you think will happen to the goat?

14. A car moving at 10 m/s (22.4 mi/h) strikes a stone wall. (a) The car is very rigid and the 80-kg driver comes to a stop in a distance of 20 cm. What is his average acceleration and how does it compare with the acceleration of gravity *g*? How much force acted on him? Express this force in both newtons and pounds. (b) The car is so constructed that its front end gradually collapses on impact, and the driver comes to a stop in a distance of 100 cm. Answer the same questions for this situation.

3.5 Weight

15. Two objects whose masses are 5.0 kg and 10.0 kg are suspended by strings as in Fig. 3.32. Find the tension in each string.

Figure 3.32

Exercise 15.

16. A 100-kg man slides down a rope at constant speed. (a) What is the minimum breaking strength the rope must have? (b) If the rope has precisely this strength, will it support the man if he tries to climb back up?

17. A force of 20 N acts on a body whose weight is 8.0 N. (a) What is the mass of the body? (b) What is its acceleration?

18. A force of 20 N acts on a body whose mass is 4.0 kg. (a) What is the weight of the body? (b) What is its acceleration?

19. A mass of 8.0 kg and another of 12.0 kg are suspended by a string on either side of a frictionless pulley. Find the acceleration of each mass.

20. How much force must you supply to give a 1.0-kg object an upward acceleration of 2*g*? A downward acceleration of 2*g*?

21. (a) A 1000-kg elevator has a downward acceleration of 1.0 m/s². What is the tension in its supporting cable? (b) The same elevator has an upward acceleration of 1.0 m/s². What is the tension in its cable now?

22. (a) A 400-kg load of bricks on a 50-kg pallet is raised with an acceleration of 0.40 m/s². What is the tension in the supporting rope? (b) The empty pallet is lowered with a downward acceleration of 0.60 m/s². What is the tension in the rope now?

23. A parachutist whose total mass is 100 kg is falling at 50 m/s when her parachute opens. Her speed drops to 7.0 m/s in a vertical distance of 40 m. What total force did her harness have to withstand? How many times her weight is this force?

24. A balloon whose total mass is *M* is falling with the constant acceleration *a*. How much ballast should be thrown overboard to make the balloon float at constant altitude?

25. A person stands on a scale in an elevator. When the elevator is at rest, the scale reads 700 N. When the elevator starts to move, the scale reads 600 N. (a) Is the elevator going up or down? (b) Does it have a constant speed? If so, what is this speed? (c) Does it have a constant acceleration? If so, what is this acceleration?

26. A 60-kg person stands on a scale in an elevator. How many newtons does the scale read (a) when the elevator is ascending with an acceleration of 1.0 m/s²; (b) when it is descending with an acceleration of 1.0 m/s²; (c) when it is ascending at the constant speed of 3.0 m/s; (d) when it is descending at the constant speed of 3.0 m/s; (e) when the cable has broken and the elevator is descending in free fall?

27. A sprinter presses on the ground with a force equal to three times his own weight at a 50° angle with the horizontal at the start of a race. What is his forward acceleration?

28. A 10-kg crate and a 100-kg crate are both sliding without friction down a plane inclined at 20° with the horizontal. What is the acceleration of each crate?

3.6 Customary Units of Mass and Force

29. A force of 100 lb acts upon a body whose mass is 1.5 slugs. (a) What is the weight of the body? (b) What is its acceleration?

30. A force of 100 lb acts upon a body whose weight is 96 lb. (a) What is the mass of the body? (b) What is its acceleration?

31. How much force is needed to give a 6-lb box an upward acceleration of 4 m/s²?

32. Find the force needed to accelerate a 2200-lb car from a standing start to a speed of 25 mi/h in 15 s.

33. A weight of 9.0 lb and another of 6.0 lb are suspended by a string on either side of a frictionless pulley. Find the acceleration of each weight.

34. A 2400-lb elevator is supported by a cable that can safely withstand a tension of no more than 3000 lb. (a) What is the maximum upward acceleration the elevator can have? (b) The maximum downward acceleration?

3.9 Coefficient of Friction

35. A 100-kg wooden crate rests on a level wooden floor. What is the minimum force required to move it at constant speed across the floor?

36. A woman prevents a 2.0-kg brick from falling by pressing it against a vertical wall. The coefficient of static friction is 0.60. What force must she use? Is this more or less than the weight of the brick?

37. An eraser is pressed against a vertical blackboard with a horizontal force of 10 N. The coefficient of static friction between eraser and blackboard is 0.20. Find the force parallel to the blackboard required to move the eraser.

38. A very strong 100-kg man is having a tug-of-war with a 1500-kg elephant. The coefficient of static friction for both on the ground is 1.0. (a) Find the force needed to move the elephant. (b) What would happen if the man could actually exert such a force?

39. A tennis ball whose initial speed is 2.0 m/s rolls along a floor for 5.0 m before coming to a stop. Find the coefficient of rolling friction.

40. The coefficient of rolling friction of a baseball rolling on the ground is 0.080. If the ball comes to a stop in 5.0 s, what was its initial speed in feet per second?

41. A driver sees a cow in the road ahead and applies the brakes so hard that they lock and the car skids to a stop in 80 ft. If the road is concrete and it is a dry day, what was the car's speed in miles per hour when the brakes were applied?

42. In what distance will a car skid to a stop on a dry concrete road if its brakes are locked when it is moving at 80 km/h?

43. Find the maximum possible accelerations the engines of the following cars can provide on a level road when the coefficient of static friction is μ. In each case the front wheels of the car support 60% of its weight.

    a. a car with rear-wheel drive

    b. a car with front-wheel drive

    c. a car with four-wheel drive

44. A 1250-kg car reaches a speed of 2.0 m/s after being pushed by two people for 12 m starting from rest. If the coefficient of rolling friction is 0.0070, what force did the people exert on the car?

45. A 5.0-lb wooden block whose initial speed is 9.0 ft/s slides on a smooth floor for 6.0 ft before it comes to a stop. (a) Find the coefficient of kinetic friction. (b) How much force would be needed to keep the block moving at constant speed across the floor?

46. A truck moving at 15 m/s is carrying a 2000-kg steel girder that rests on the bed of the truck without any fastenings. The coefficient of static friction between the girder and the truck bed is 0.50. (a) Find the minimum distance in which the truck can come to a stop without having the girder move forward. (b) If the girder had a mass of 3000 kg, would this distance be different?

47. The coefficient of static friction between a tire and a certain road is 0.70. (a) What is the maximum possible acceleration of a car on this road? (b) What is the minimum distance in which the car can be stopped when it is moving at 100 km/h?

48. A railway boxcar is set in motion along a track at the same initial velocity as a truck on a parallel road. If the only horizontal forces acting on both vehicles are due to rolling friction with the respective coefficients of 0.0045 and 0.040, which vehicle will come to a stop first? How many times farther will the other vehicle travel?

49. A sled slides down a snow-covered hill at constant speed. If the hillside is 10° above the horizontal, what is the coefficient of kinetic friction between the runners of the sled and the snow?

50. A cyclist finds that she is able to coast at constant speed along a road that slopes downward at an angle of 1° with the horizontal. If she and her bicycle together have a mass of 70 kg, what is the force required to propel them at the same constant speed along a level road?

51. A 500-lb crate is being slid down a ramp that makes an angle of 20° with the horizontal. The coefficient of kinetic friction is 0.30. How much force parallel to the plane must be applied to the crate if it is to slide down at constant speed? In which direction must the force be applied?

PROBLEMS

52. A box slides down a frictionless plane that makes an angle of 35° with the horizontal. A horizontal force of half the box's weight presses the box against the plane. What is the box's acceleration?

53. A 40-kg kangaroo exerts a constant force on the ground in the first 60 cm of her jump, and rises 2.0 m higher. When she carries a baby kangaroo in her pouch, she can rise only 1.8 m higher. What is the mass of the baby kangaroo?

54. In the arrangement shown in Fig. 3.33 find the mass M that will result in the 2.0-kg block remaining at rest. Assume that the pulleys are massless and frictionless.

Figure 3.33

Problem 54.

Figure 3.34

Problem 55.

55. Three objects whose masses are 2.0 kg, 3.0 kg, and 4.0 kg are joined by strings, as in Fig. 3.34. The 2.0-kg object is pulled upward with a force of 120 N. Find the tensions in the strings between the objects.

56. Find the accelerations of the blocks shown in Fig. 3.35 and the tensions in the strings that join them. Neglect friction and the mass of the pulley.

Figure 3.35

Problem 56.

57. The drag force R on a falling object of mass m due to air resistance is proportional to the square of the object's speed v, so that $R = kv^2$ where k depends on the properties of the object. Find the terminal speed of the object in terms of m, k, and g.

58. If μ is the coefficient of static friction between the cart A and block B in Fig. 3.36, what acceleration must the cart have if B is not to fall?

Figure 3.36

Problem 58.

59. A 5.0-kg block resting on a horizontal surface is attached to a 5.0-kg block that hangs freely by a string passing over a pulley, an arrangement similar to that shown in Fig. 3.16. (a) If there is no friction between the first block and the surface, what is the block's acceleration? (b) If the coefficient of kinetic friction between the first block and the surface is 0.20, what is the block's acceleration?

60. A 10.0-kg block resting on a horizontal surface is attached to a 5.0-kg block that hangs freely by a string passing over a pulley, an arrangement similar to that shown in Fig. 3.16. (a) If there is no friction between the first block and the surface, what is the block's acceleration? What is the block's acceleration if the coefficient of kinetic friction between the first block and the surface is (b) 0.40? (c) 0.50? (d) 0.60?

61. Two wooden blocks rest on a wooden table, as in Fig. 3.37. The coefficient of kinetic friction is 0.30. Find the minimum force needed to move the 3.0-kg block at constant speed in each case.

62. A horse pulls a 300-kg sled at constant speed over level snow by a rope that is 35° above the horizontal. If the coefficient of kinetic friction is 0.10, find the force the horse exerts on the rope.

63. A ship whose mass is 2.0×10^7 kg rests on launching ways that slope down to the water at an angle of 6°. The ways are greased to reduce the coefficient of kinetic friction to 0.11. Will the ship slide down the ways into the water by itself? If not, find the force needed to winch the ship into the water.

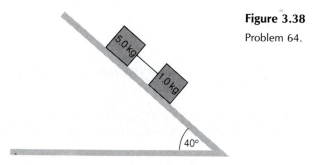

Figure 3.38

Problem 64.

64. Two blocks joined by a string are sliding down an inclined plane, as in Fig. 3.38. The coefficient of kinetic friction between the 5.0-kg block and the plane is 0.40 and that between the 1.0 kg block and the plane is 0.10. Find the tension in the string.

65. A block slides down an inclined plane 9.0 m long that makes an angle of 38° with the horizontal. The coefficient of kinetic friction is 0.25. If the block starts from rest, find the time required for it to reach the foot of the plane.

66. If the block in Problem 65 has a mass of 50 kg, find the force parallel to the plane required to move it upward at constant speed.

67. A skier starts from rest and slides 50 m down a slope that makes an angle of 40° with the horizontal. She then continues sliding on level snow. (a) If the coefficient of kinetic friction between skis and snow is 0.10 and air resistance is neglected, what is the speed of the skier at the foot of the slope? (b) How far away from the foot of the slope does she come to a stop?

68. A block takes twice as long to slide down an inclined plane that makes an angle of 35° with the horizontal as it does to fall freely through the same vertical distance. What is the coefficient of kinetic friction?

Answers to Multiple Choice

1. d	**10.** c	**19.** b	**28.** d
2. d	**11.** b	**20.** b	**29.** a
3. c	**12.** c	**21.** d	**30.** a
4. a	**13.** d	**22.** b	**31.** b
5. d	**14.** c	**23.** d	**32.** c
6. b	**15.** d	**24.** c	**33.** b
7. b	**16.** a	**25.** d	
8. c	**17.** b	**26.** a	
9. d	**18.** a	**27.** c	

4

ENERGY

4.1 Work

4.2 Work Done Against Gravity

4.3 Power

4.4 Energy

4.5 Kinetic Energy

4.6 Potential Energy

4.7 Conservation of Energy

4.8 Conservative and Nonconservative Forces

The word **energy** is part of everyday life. We say an active person is energetic. We hear a candy bar described as being full of energy. We complain about the cost of the electric energy that lights our lamps and turns our motors. We worry about someday running out of the energy stored in coal and oil. We argue about whether nuclear energy is a blessing or a curse. Exactly what is meant by energy?

In general, energy refers to an ability to accomplish change. All changes in the physical world involve energy in some way. But "change" is not a very precise concept, and we must be sure of exactly what we are talking about before going further. The best way to introduce the complicated and many-sided idea of energy is to begin with the simpler idea of work.

4.1 WORK

A measure of the change a force produces.

As we know, forces can produce changes. Forces set things in motion, change their paths, and bring them to a stop. Forces push things together and pull them apart. The physical quantity called **work** is a measure of the amount of change (in a broad sense) to which a force gives rise when it acts on something.

Suppose we push against a wall. When we stop, nothing has happened even though we exerted a force on the wall. But if we apply the same force to a ball, the ball flies through the air when we let it go (Fig. 4.1). The difference is that the ball moved during our push, whereas the wall did not.

Or we might try to lift a heavy barbell. If we fail, the world is exactly the same afterward. If we succeed, though, the barbell is now up in the air, which represents a change. As before, the difference is that in the second case an object moved while we exerted a force on it.

To make these ideas precise, work is defined as follows:

The work done by a constant force F acting on an object while it undergoes a displacement x is equal to the magnitude of the force component F_x in the direction of the displacement multiplied by the magnitude x of the displacement.

If the angle between the force **F** and the displacement **x** is θ, as in Fig. 4.2 (a), then

$$F_x = F \cos \theta$$

Figure 4.1

Work is done by a force when the object on which it acts is displaced while the force is applied.

No work done

Work done

(a) $W = Fx \cos \theta$ (b) $W = Fx \cos 0° = Fx$ (c) $W = Fx \cos 90° = 0$

Figure 4.2

The work done by a force depends on the angle θ between the force and its displacement. (a) In general, $W = Fx \cos \theta$. (b) When **F** is parallel to **x**, $W = Fx$. (c) When **F** is perpendicular to **x**, $W = 0$ since **F** then has no component in the direction of **x** and therefore does not affect its motion.

and so the work done is

$$W = F_x x = (F \cos \theta)\, x$$

We can therefore define work by the equation

$$W = Fx \cos \theta \qquad\qquad Work \quad (4.1)$$

Work is a scalar quantity. No direction is associated with work, even though it depends on two vector quantities, force and displacement. Figure 4.2 summarizes the meaning of Eq. (4.1). When **F** is in the same direction as **x**, $\theta = 0$ and $\cos \theta = 1$, so

$$W = Fx \quad (\textbf{F} \text{ parallel to } \textbf{x}) \qquad\qquad (4.2)$$

But if **F** is perpendicular to **x**, $\theta = 90°$, and $\cos \theta = 0$, so in this case $W = 0$.

The preceding definition is a great help in clarifying the effects of forces. Unless a force acts through a displacement, no work is done no matter how great the force. (Although we may get tired after pushing on a wall for a long time, we still have done no work on the wall if the wall remains in place. We get tired because muscle fibers constantly contract and release when exerting a steady force even if no motion occurs.) Even if an object moves, no work is done on it unless a force acting on it has a component in its direction of motion.

If the direction of $\textbf{F}_x$ is opposite to that of the displacement **x**, then $\cos \theta = \cos 180° = -1$. The work done by the force on the object in this case is negative. An example is the work done by a frictional force, since $\textbf{F}_f$ is always opposite to **x**. In such a situation we can regard the work as being done *by* the moving object instead of *on* it, as when $\textbf{F}_x$ is in the same direction as **x**.

Joule and Foot-pound

The SI unit of work is the **joule**, abbreviated J. One joule is the work done by a force of 1 N acting through a distance of 1 m. That is,

$$1 \text{ joule} = 1 \text{ J} = 1 \text{ N·m}$$

Work is done in lifting a barbell, but no work is done while it is being held in the air even though this may be tiring.

The joule is named after the English scientist James Joule (1818–1889) and is pronounced "jool." To raise an apple from your waist to your mouth takes about 1 J of work.

◆   **E X A M P L E   4.1**

A person pulls an 80-kg crate 20 m across a level floor using a rope that is 30° above the horizontal. The person exerts a force of 150 N on the rope (Fig. 4.3). How much work is done?

Figure 4.3

SOLUTION   The work done is

$$W = Fx \cos \theta = (150 \text{ N})(20 \text{ m})(\cos 30°)$$
$$= (150)(20)(0.866) \text{ J} = 2.6 \times 10^3 \text{ J} = 2.6 \text{ kJ}$$

Evidently the 80-kg mass of the crate has no significance here—the force exerted on the crate determines how much work is done.   ◆

In the U.S. Customary System the unit of work is the **foot-pound,** abbreviated ft · lb. One ft · lb is equal to the work done by a force of 1 lb acting through a distance of 1 ft. A ft · lb is a little more than $1\frac{1}{3}$ J.

◆   **E X A M P L E   4.2**

A 100-lb wooden box is pushed aross a horizontal floor with a force of 50 lb. The coefficient of kinetic friction is $\mu_k = 0.40$. (a) Find the work done in pushing the box 60 ft. (b) How much work went into overcoming friction and how much into accelerating the box?

SOLUTION   (a) The work done by the applied fore of 50 lb is

$$W = F_x = (50 \text{ lb})(60 \text{ ft}) = 3.0 \times 10^3 \text{ ft} \cdot \text{lb}$$

(b) Since the normal force here is the box's weight w, the frictional force is

$$F_f = \mu_k F_N = \mu_k w = (0.40)(100 \text{ lb}) = 40 \text{ lb}$$

The work done to overcome this force is

$$W_f = F_f x = (40 \text{ lb})(60 \text{ ft}) = 2400 \text{ ft} \cdot \text{lb}$$

Hence the work that went into accelerating the box was

$$3000 \text{ ft} \cdot \text{lb} - 2400 \text{ ft} \cdot \text{lb} = 6.0 \times 10^2 \text{ ft} \cdot \text{lb} \qquad \blacklozenge$$

4.2   WORK DONE AGAINST GRAVITY

This work depends only on the height, not on the path.

The work needed to lift an object of mass m against gravity is easy to figure out. The force of gravity on the object is simply its weight $w = mg$. Therefore the force that must be exerted to lift the object has the same magnitude of mg and is in the same direction as the height h to which the object is raised. Since $F_x = mg$, $x = h$, and $\cos \theta = \cos 0 = 1$, the work done is

$$W = mgh \qquad\qquad \textit{Work done against gravity} \quad (4.3)$$

To lift an object of mass m to the height h requires the work mgh (Fig. 4.4).

Figure 4.4

(a) The work mgh must be done to lift an object of mass m to a height h. (b) When an object of mass m falls from a height h, the force of gravity does the work mgh on it. (c) The force of gravity does no work on objects that move parallel to the earth's surface.

$W = $ work done by person $= mgh$

h

$\mathbf{F} = mg$

(a)

$W = $ work done by gravity $= mgh$

mg

h

(b)

x

mg

$W = 0$

(c)

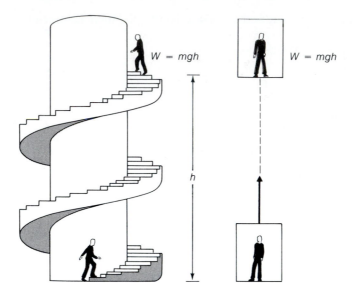

Figure 4.5

In the absence of friction, the work done in lifting a mass m to a height h is mgh regardless of the exact path taken.

It is important to note that only the height h is involved in work done against the force of gravity. The particular route taken by an object being raised is not significant (Fig. 4.5). Excluding any frictional effects, exactly as much work must be done to climb a flight of stairs as to go up in an elevator to the same floor (though not by the person involved!). Moving an object parallel to the earth's surface requires no work to be done against gravity because $h = 0$.

◆   **E X A M P L E   4.3** _____

How much work is done in lifting a 300-lb load of bricks to a height of 60 ft on a building under construction?

SOLUTION   Here $mg = w = 300$ lb and $h = 60$ ft, so

$$W = mgh = wh = (300 \text{ lb})(60 \text{ ft}) = 1.8 \times 10^4 \text{ ft} \cdot \text{lb}$$ ◆

◆   **E X A M P L E   4.4** _____

Eating a banana enables a person to perform about 4.0×10^4 J of work. To what height does eating a banana enable a 60-kg woman to climb?

SOLUTION   From $W = mgh$ we have

$$h = \frac{W}{mg} = \frac{4.0 \times 10^4 \text{ J}}{(60 \text{ kg})(9.8 \text{ m/s}^2)} = 68 \text{ m}$$ ◆

4.3 POWER

The rate of doing work.

Often the time needed to carry out a task is as important as the actual work needed. If we are willing to wait, even a tiny motor can raise an elevator as high as we like. If we are in a hurry, however, we want a motor whose work output is rapid in terms of the total work needed. Thus the rate at which work is being done is significant. This rate is called **power**. The more powerful something is, the faster it can do work.

Figure 4.6

If an amount of work W is done in a time interval t, the power involved is

$$P = \frac{W}{t} \qquad\qquad \textit{Power} \quad (4.4)$$

$$\text{Power} = \frac{\text{work done}}{\text{time interval}}$$

In the SI system, where work is measured in joules and time in seconds, the unit of power is the **watt**:

$$1 \text{ watt} = 1 \text{ W} = 1 \text{ J/s}$$

The **kilowatt-hour** (kWh) is often used as a unit of work. Since 1 kW = 1000 W = 1000 J/s and 1 h = 3600 s,

$$1 \text{ kilowatt-hour} = 1 \text{ kWh} = 3.60 \times 10^6 \text{ J} = 3.60 \text{ MJ}$$

◆ **E X A M P L E   4.5** _____

An electric motor with an output of 15 kW provides power for the elevator of a six-story building. If the total mass of the loaded elevator is 1000 kg, what is the minimum time needed for it to rise the 30 m from the ground floor to the top floor (Fig. 4.6)?

SOLUTION The work done in raising the elevator through a height h is

$$W = mgh$$

Since $P = W/t$, the time needed for the motor to raise the elevator by 30 m is

$$t = \frac{W}{P} = \frac{mgh}{P} = \frac{(1000 \text{ kg})(9.8 \text{ m/s}^2)(30 \text{ m})}{15 \times 10^3 \text{ W}} = 20 \text{ s} \qquad ◆$$

The U.S. customary unit of power is the **foot · pound/second** (ft · lb/s), which is commonly replaced by the **horsepower** (hp) in engineering. The horsepower is equal to 550 ft · lb/s, so that

$$1 \text{ hp} = 550 \text{ ft} \cdot \text{lb/s} = 746 \text{ W} = 0.746 \text{ kW}$$
$$1 \text{ kW} = 738 \text{ ft} \cdot \text{lb/s} = 1.34 \text{ hp}$$

Few actual horses can develop as much as 1 hp for very long.

Force, Speed, and Power

How much power is delivered when a constant force **F** does work on an object moving at the constant velocity **v**? If θ is the angle between **F** and **v**, then

$$P = \frac{W}{t} = \frac{Fx \cos \theta}{t}$$

Since $x/t = v$,

$$P = Fv \cos \theta \qquad\qquad\qquad\qquad Power \quad (4.5)$$

When **F** is in the same directon as **v**,

$$P = Fv \quad (\textbf{F parallel to v})$$
$$\text{Power} = (\text{force})(\text{speed})$$

◆   **E X A M P L E   4.6** _____

A swimmer develops an average power of 200 W as she covers 100 m in 80.0 s. What is the resistive force exerted by the water on her?

SOLUTION   The swimmer's speed is $v = 100$ m/80.0 s $= 1.25$ m/s. Since $P = Fv$ here,

$$F = \frac{P}{v} = \frac{200 \text{ W}}{1.25 \text{ m/s}} = 160 \text{ N}$$

which is about 36 lb.   ◆

Efficiency

The power delivered by a machine of any kind is always less than the power supplied to it because of the inevitable presence of friction. The **efficiency** of a machine is the ratio between these powers:

$$\text{Eff} = \frac{\text{power output}}{\text{power input}} = \frac{P_{\text{out}}}{P_{\text{in}}} \qquad\qquad Efficiency \quad (4.6)$$

◆   **E X A M P L E   4.7**

A trash compactor with a $\frac{1}{4}$-hp motor can apply a crushing force of 1 ton (2000 lb). How fast does its ram move, assuming 80% efficiency?

SOLUTION   The power developed by the motor is

$$P_{in} = (\tfrac{1}{4}\,hp)\left(550\,\frac{ft \cdot lb/s}{hp}\right) = 138\,ft \cdot lb/s$$

From Eq. (4.6) the power output of the compactor is

$$P_{out} = (Eff)(P_{in}) = (0.80)(138\,ft \cdot lb/s) = 110\,ft \cdot lb/s$$

Because $P = Fv$ the ram's speed is

$$v = \frac{P_{out}}{F} = \frac{110\,ft \cdot lb/s}{2000\,lb} = 0.055\,ft/s$$

which can be expressed in inches per second as

$$v = \left(0.055\frac{ft}{s}\right)\left(12\frac{in.}{ft}\right) = 0.66\,in./s$$

◆

4.4   ENERGY

Three broad categories.

Now that we know what work is, we can define energy:

Energy is the property that gives something the capacity to do work.

When we say that something has energy, we mean it is able (directly or indirectly) to exert a force on something else and do work on it. On the other hand, when we do work on something, we add to it an amount of energy equal to the work done. The unit of energy is the same as that of work, the joule.

There are three broad categories of energy:

1. **Kinetic energy**, the energy something possesses by virtue of its motion;
2. **Potential energy**, the energy something possesses by virtue of its position;
3. **Rest energy**, the energy something possesses by virtue of its mass.

In these descriptions the word *something* was used instead of *object* because, as we will see later, such nonmaterial entities as force fields and massless particles may also possess energy.

All modes of energy possession fit into one or another of these three categories. For instance, it is convenient for many purposes to think of heat as a separate form of energy, but what this term actually refers to is the sum of the kinetic energies of the randomly moving atoms and molecules in a body of matter. "Chemical energy" is actually the electric potential energy of electrons in the atoms and molecules of which matter is composed; this energy is given off in chemical reactions, such as the burning of gasoline in a car engine. Wave motion involves the interchange of kinetic and potential energies. Rest energy—the $E_0 = m_0 c^2$ that most of us have heard of, if perhaps not fully understood—will be considered in Chapter 27.

4.5   KINETIC ENERGY

The energy of motion.

When we do work on a ball by throwing it, what becomes of this work?

Let us suppose we apply the constant force **F** to the ball for a distance x before it leaves our hand, as in Fig. 4.7(a). The work done on the ball is therefore, since $\cos \theta = 1$,

$$W = Fx \tag{4.7}$$

The mass of the ball is m. As we throw it, its acceleration has the magnitude

$$a = \frac{F}{m}$$

according to the second law of motion, $\mathbf{F} = m\mathbf{a}$ (Fig. 4.7b).

We know from the formula $v_f^2 = v_0^2 + 2ax$ that when an object starting from rest ($v_0 = 0$) undergoes an acceleration of magnitude a through a distance x, its final speed v is related to a and x by

$$v_f^2 = 2ax \tag{4.8}$$

This relationship is pictured in Fig. 4.7(c).

If we now substitute F/m for a in Eq. (4.8), we find that

$$v_f^2 = 2ax = 2\left(\frac{F}{m}\right)x$$

Figure 4.7

Successive steps in deriving the formula KE = $\frac{1}{2}mv^2$ for the kinetic energy of a moving object.

which we can rewrite as

$$Fx = \tfrac{1}{2}mv_f^2 \qquad (4.9)$$

The quantity on the left-hand side, Fx, is the work our hand has done in throwing the ball, as in Eq. (4.7). The quantity on the right-hand side, $\frac{1}{2}mv_f^2$, must therefore be the energy acquired by the ball as a result of the work we did on it. This energy is **kinetic energy**, energy of motion. That is, we interpret the preceding equation as follows (Fig. 4.7d):

$$Fx = \tfrac{1}{2}mv_f^2$$

Work done on ball = kinetic energy of ball

The symbol for kinetic energy is KE. The kinetic energy of an object of mass m and velocity v is therefore

$$KE = \tfrac{1}{2}mv^2 \qquad\qquad \textit{Kinetic energy} \quad (4.10)$$

A moving object is able to perform an amount of work equal to $\tfrac{1}{2}mv^2$ in the course of being stopped. Although Eq. (4.10) was derived for a particular situation, it is a perfectly general result.

◆   **E X A M P L E   4.8** _____

A 1.0-lb hammer head strikes a nail at a speed of 10 ft/s and drives it exactly $\tfrac{1}{4}$ in. into a wooden board (Fig. 4.8). What is the average force on the nail?

Figure 4.8

$$\tfrac{1}{2}mv^2 = Fx$$

$$F = \frac{wv^2}{2gx} = \frac{(1.0\ \text{lb})(10\ \text{ft/s})^2}{(2)(32\ \text{ft/s}^2)(\tfrac{1}{48})\ \text{ft}} = 75\ \text{lb}$$

◆

_____

The v^2 factor means that KE goes up very rapidly with increasing speed. Consider a 1200-kg car at 25 km/h (16 mi/h) and at 100 km/h (62 mi/h). Because 1 km/h = 0.278 m/s, the respective speeds are 7.0 and 27.8 m/s, and so the corresponding KE values are (Fig. 4.9):

$$KE_1 = \tfrac{1}{2}mv_1^2 = \tfrac{1}{2}(1200\ \text{kg})(7.0\ \text{m/s})^2 = 2.9 \times 10^4\ \text{J}$$
$$KE_2 = \tfrac{1}{2}mv_2^2 = \tfrac{1}{2}(1200\ \text{kg})(27.8\ \text{m/s})^2 = 46.4 \times 10^4\ \text{J}$$

1200 kg

25 km/h

$KE_1 = 2.9 \times 10^4$ J

100 km/h

$KE_2 = 46.4 \times 10^4$ J

Figure 4.9

Kinetic energy is proportional to v^2. A car going 4 times as fast as another car of the same mass has 16 times as much KE.

At 100 km/h the car has 16 times as much kinetic energy as it does at 25 km/h. The fact that kinetic energy, and hence ability to do work (that is, damage), is proportional to the square of the speed is responsible for the severity of automobile accidents at high speeds.

◆ **E X A M P L E   4.9** _____

What is the power output of the engine of a 1200-kg car if the car can go from 25 km/h to 100 km/h in 12 s?

SOLUTION   The work needed to accelerate the car is

$$W = KE_2 - KE_1 = 46.4 \times 10^4 \text{ J} - 2.9 \times 10^4 \text{ J} = 43.5 \times 10^4 \text{ J}$$

The power needed to provide this amount of work in 12 s is

$$P = \frac{W}{t} = \frac{43.5 \times 10^4 \text{ J}}{12 \text{ s}} = 3.63 \times 10^4 \text{ W} = 36.3 \text{ kW}$$

which is equivalent to

$$P = (36.3 \text{ kW})\left(1.34\frac{\text{hp}}{\text{kW}}\right) = 49 \text{ hp}$$

◆

Figure 4.10

A raised stone has potential energy with respect to the ground.

Raised stone has PE = mgh

m

h

4.6   POTENTIAL ENERGY

The energy of position.

When we drop a stone from a height h, it falls faster and faster until it strikes the ground. If we lift the stone afterward, we see that it has done work by making a shallow hole in the ground (Fig. 4.10). At its original height the stone must have had the capacity to do work even though it was not moving at the time. The work the stone could do by falling to the ground is called its **potential energy,** symbol PE.

Work done by stone in making hole = PE

As we know, the work needed to raise a stone of mass m to a height h is

$$W = mgh \tag{4.3}$$

The same amount of work can be done *by* the stone after dropping through the same height h. Accordingly the potential energy of the stone is

$$PE = mgh \qquad \textit{Gravitational potential energy} \quad (4.11)$$

The gravitational potential energy of an object depends on the reference level from which its height h is measured. For example, the potential energy of a 1.0-kg book held 10 cm above a desk is

$$PE = mgh = (1.0 \text{ kg})(9.8 \text{ m/s}^2)(0.10 \text{ m}) = 0.98 \text{ J}$$

with respect to the desk (Fig. 4.11). However, if the book is 1.0 m above the floor of the room, its potential energy is

$$PE = mgh = (1.0 \text{ kg})(9.8 \text{ m/s}^2)(1.0 \text{ m}) = 9.8 \text{ J}$$

with respect to the floor. And the book may conceivably be 100 m above the ground, so its potential energy is

$$PE = mgh = (1.0 \text{ kg})(9.8 \text{ m/s}^2)(100 \text{ m}) = 980 \text{ J}$$

with respect to the ground. The height h in the formula $PE = mgh$ means nothing unless the base height $h = 0$ is specified.

PE of book	Relative to
0.98 J	Desk
9.8 J	Floor
980 J	Ground

Figure 4.11

The potential energy of an object depends on the reference level from which its height h is measured. The object here is a 1.0-kg book.

Figure 4.12

◆   **E X A M P L E   4.10**

Compare the potential energy of a 1200-kg car at the top of a hill 30 m high (Fig. 4.12) with its kinetic energy when moving at 100 km/h.

1200 kg

PE = 3.5 × 10⁵ J

h = 30 m

SOLUTION   The car's potential energy relative to the bottom of the hill is

$$PE = mgh = (1200 \text{ kg})(9.8 \text{ m/s}^2)(30 \text{ m}) = 3.5 \times 10^5 \text{ J}$$

Earlier we found that the kinetic energy of the car at 100 km/h is 4.6×10^5 J, which is greater than its potential energy at the top of the hill. This means that a crash at 100 km/h (62 mi/h) into a stationary obstacle will yield more work—that is, do more damage—than dropping the car 30 m (98 ft).   ◆

◆   **E X A M P L E   4.11** _____

Electricity demand varies a great deal throughout the day, and to provide generating capacity for peak periods that lies idle at other times is very expensive. It is more efficient to even out the variations by always running a generating plant at full capacity and in some way storing the surplus energy of off-peak periods to be tapped when needed. But how can this be done on the scale required? In one approach, at Dinorwig, in Wales, turbines connected to electric motor-generators pump water 568 m up to a mountain reservoir at night and at other times of low energy demand. When more electric power is needed, the water is allowed to fall back through the turbines to run the motor-generators in their generator mode. If 85% of the potential energy of the stored water is converted into electricity, how much water must pass through the turbines per second to produce the system's maximum power output of 1680 MW?

SOLUTION   The PE of a mass m of stored water is mgh. If 85% of this PE can be recovered as work W done to produce electricity in the generators,

$$W = 0.85 \text{ PE} = 0.85mgh$$

The corresponding power output P is

$$P = \frac{W}{t} = 0.85gh\,\frac{m}{t}$$

Since $P = 1680 \text{ MW} = 1680 \times 10^6$ W, the rate of water flow is

$$\frac{m}{t} = \frac{P}{0.85gh} = \frac{1680 \times 10^6 \text{ W}}{(0.85)(9.8 \text{ m/s}^2)(568 \text{ m})} = 3.55 \times 10^5 \text{ kg/s}$$

This is 355 tonnes per second, where 1 tonne = 1 "metric ton" = 10^3 kg. A tonne of water occupies 1 m^3, so 355 tonnes occupies 355 m^3, the volume of a small house.   ◆

What we have been speaking of as the potential energy of an object is actually not a property of the object itself. It is a property of the *system* of the object and whatever exerts on the object a force that depends on the object's position. Thus the PE of a stone held above the earth's surface is shared by both the stone and the earth. When the stone is let go, both it and the earth move toward each other. But the earth's motion is too small to detect because of its immense mass relative to that of the stone. It is therefore appropriate for us to assign the entire PE of the system to the stone. In the case of two interacting objects more nearly equal in mass, however, we must keep in mind that the PE belongs to the system, not to just one of the objects.

Other Kinds of PE

Until now we have considered only one kind of potential energy, namely, the gravitational PE of an object raised above some reference level near the earth's surface. The concept of potential energy is a much more general one, however, for it refers to the energy something has as a consequence of its position regardless of the nature of the force acting on it. The earth itself, for instance, has gravitational potential energy with respect to the sun, since if its orbital motion were to stop, it would fall toward the sun (Fig. 4.13). An iron nail has magnetic potential energy

Figure 4.13

Three examples of potential energy. In each case the PE is a property of the entire system.

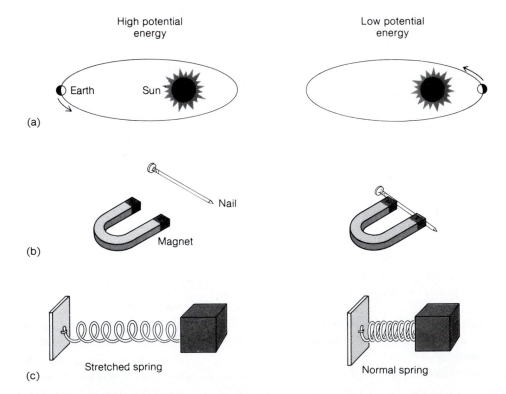

High potential energy   Low potential energy

(a)

(b)

Earth   Sun   Nail   Magnet

(c)   Stretched spring   Normal spring

with respect to a nearby magnet, since it will fly to the magnet if released. An object at the end of a stretched spring has elastic potential energy with respect to its position when the spring has its normal extension, since if let go the object will move as the spring contracts. In each of these cases the object in question has the potentiality of doing work in its original position.

4.7   CONSERVATION OF ENERGY

A fundamental law of nature.

The elastic potential energy of this bow becomes kinetic energy of the arrow when the bowstring is released.

A **conservation principle** states that, in a system of any kind isolated from the rest of the universe, a certain quantity keeps the same value it originally had no matter what changes the system undergoes.

As an example, the law of conservation of mass revolutionized chemistry by holding that the total mass of the products of a chemical reaction is the same as the total mass of the original substances. Thus the increase in mass of a piece of iron when it rusts indicates that the iron has combined with some other material, rather than having decomposed, as the early chemists believed. In fact, the gas oxygen was discovered in the course of seeking this other material.

Given one or more conservation principles that apply to a given system, we know at once which classes of events can take place in the system and which cannot. Thus when iron rusts, the gain in mass means that it has combined chemically with something else. In physics we can often draw conclusions about the behavior of the particles that make up a system without a detailed study by basing our analysis on the conservation of some particular quantities.

The first conservation principle we will study is that of **conservation of energy:**

The total amount of energy in a system isolated from the rest of the universe always remains constant, although energy transformations from one form to another may occur within the system.

This principle is perhaps the most fundamental generalization in all of science; no violation of it has ever been found.

A falling ball provides a simple example of conservation of energy. As it falls, its initial potential energy is converted into kinetic energy, so that the total energy of the stone remains the same. The potential energy of a 1-kg ball 50 m above the ground is $mgh = 490$ J, and its total energy is 490 J until it interacts with the ground and transfers energy to it (Fig. 4.14).

Height	1-kg ball	PE = mgh	KE = ½mv²	PE + KE
50 m	●	490 J	0 J	490 J
40		392	98	490
30		294	196	490
20		196	294	490
10		98	392	490
0	●	0	490	490

Figure 4.14

The total energy of a falling ball remains constant as its potential energy is transformed into kinetic energy.

Another example is the motion of a planet about the sun. Planetary orbits are elliptical, so that at different points in its orbit the planet is at different distances from the sun. When the planet is close to the sun, it has a low PE, just as a stone near the ground has a low PE; when the planet is far from the sun, it has a high PE (Fig. 4.13a). Since the sum of the planet's PE and KE must be constant, we conclude that the kinetic energy of the planet is at a maximum when it is nearest the sun and at a minimum when it is farthest from the sun. The earth's orbital speed varies between 29 and 30 km/s for this reason.

Newton's laws of motion enable us—in principle—to solve all problems that involve forces and moving objects. However, these laws are easy to use only in the simplest cases because in order to apply them we must take into account all the various forces acting on each object at every point in its path. This is usually a difficult and complicated procedure. The advantage of the principle of conservation of energy is that it tells us about the relationship between the initial and final states of motion of some object or system of objects without our having to investigate what happens in between.

  E X A M P L E   4.12 _____

A skier is sliding downhill at a constant speed of 8.0 m/s when she reaches an icy patch on which her skis move with negligible friction (Fig. 4.15). If the icy patch is 10 m high, what is the skier's speed at its bottom?

SOLUTION Because the path of the skier on the icy patch is complicated, to apply $\mathbf{F} = m\mathbf{a}$ here would be very difficult. Using conservation of energy, however, makes the problem quite easy. At the top of the icy patch, the skier's initial kinetic energy

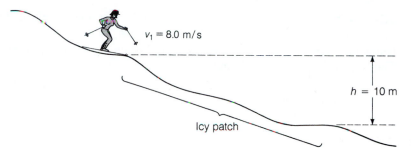

$v_1 = 8.0$ m/s

$h = 10$ m

Icy patch

Figure 4.15

is

$$KE_1 = \tfrac{1}{2}mv_1^2$$

where $v_1 = 8.0$ m/s. Her potential energy relative to the bottom of the patch is

$$PE = mgh$$

where $h = 10$ m. At the bottom the skier's kinetic energy is $KE_2 = KE_1 + PE$, and so

$$\tfrac{1}{2}mv_2^2 = \tfrac{1}{2}mv_1^2 + mgh$$
$$v_2^2 = v_1^2 + 2gh$$
$$v_2 = \sqrt{v_1^2 + 2gh} = \sqrt{(8.0 \text{ m/s})^2 + (2)(9.8 \text{ m/s}^2)(10 \text{ m})} = 16 \text{ m/s}$$

We did not have to know the skier's mass to find her final speed. ◆

Center of Mass

Sometimes we must take into account the size and shape of the object whose motion we are examining, which we have not had to do thus far. The first step is to identify the object's **center of mass** (CM), which is the point at which we can consider all its mass concentrated. Often the location of the CM is obvious; in the case of a uniform sphere, for instance, it is the geometric center. (In Chapter 8 we will learn how to find the CMs of more complicated bodies.) We now treat the object as a particle located at the CM. This procedure works provided the various forces on the object all act through its CM. If they don't, the object tends to turn under their influence, and what happens then is the subject of Chapter 7.

Figure 4.16

(a)

CM

$\dfrac{L}{2}$

L

(b)

◆ **E X A M P L E   4.13** _____

A chain L long lying on a frictionless table starts from rest to slide over the edge of the table, as in Fig. 4.16(a). What will the speed of the chain be as its last link passes over the edge?

SOLUTION   The center of mass is at the chain's midpoint. When the last link of the chain has just passed over the edge of the table, as in Fig. 4.16(b), its CM is $L/2$ below the edge, and the chain as a whole has lost $mgh = mg(L/2)$ of potential energy. The lost PE has become KE, and so

$$\tfrac{1}{2}mgL = \tfrac{1}{2}mv^2$$
$$v = \sqrt{gL} \qquad\qquad\qquad \blacklozenge$$

4.8   CONSERVATIVE AND NONCONSERVATIVE FORCES

Only a conservative force gives rise to potential energy.

As we have seen, we can do work on something and thereby give it kinetic or potential energy, which in turn can reappear as work. When we throw a stone upward into the air, the work we do appears first as kinetic energy. As the stone rises, the kinetic energy gradually becomes potential energy. At its highest point the stone has only potential energy, which, as the stone begins to fall, is converted back into kinetic energy. Finally, when the stone strikes the ground and makes a hole in it, the kinetic energy turns to work.

But what happens to the energy used in opposing frictional forces? To find the answer, all we need do is rub one piece of wood against another. After a short time, it is obvious that the rubbed surfaces of the wood are warmer than before. This is a quite general observation: Work done against frictional forces always produces a rise in the temperature of the objects involved. What is happening is that the energy that disappears from the motion of each object as a whole reappears as additional KE of random motion of the molecules of which the objects are composed. This additional molecular energy is manifested as a rise in temperature, as will be explained in Chapter 13.

This discussion is summarized in the **work-energy theorem:**

Work done *on* object = change in object's KE

+ change in object's PE

+ work done *by* object

The kinetic energy of the object may either increase or decrease; the same is true of its potential energy. It is the work done *by* the object that may be converted into heat.

◆ E X A M P L E 4.14

A 25-kg box is pulled up a ramp 20 m long and 3.0 m high by a constant force of 120 N (Fig. 4.17). If the box starts from rest and has a speed of 2.0 m/s at the top, what is the force of friction between box and ramp?

Figure 4.17

SOLUTION According to the work–energy theorem, the total work W done by the applied force of 120 N must equal the work W_f done against friction (which becomes heat) plus the change ΔKE in the box's kinetic energy and the change ΔPE in its potential energy. Hence

$$W = W_f + \Delta KE + \Delta PE$$
$$W_f = W - \Delta KE - \Delta PE$$

The total work done is

$$W = Fs = (120 \text{ N})(20 \text{ m}) = 2400 \text{ J}$$

Because the box starts from rest, its change in kinetic energy is equal to its kinetic energy at the top of the ramp, and so

$$\Delta KE = \tfrac{1}{2}mv^2 = \tfrac{1}{2}(25 \text{ kg})(2.0 \text{ m/s})^2 = 50 \text{ J}$$

The change in the box's potential energy is

$$\Delta PE = mgh = (25 \text{ kg})(9.8 \text{ m/s}^2)(3.0 \text{ m}) = 735 \text{ J}$$

Hence the work done against friction is

$$W_f = W - \Delta KE - \Delta PE = (2400 - 50 - 735) \text{ J} = 1615 \text{ J}$$

Because $W_f = F_f s$, the frictional force is

$$F_f = \frac{W_f}{s} = \frac{1615 \text{ J}}{20 \text{ m}} = 81 \text{ N} \qquad \qquad \blacklozenge$$

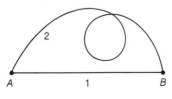

Figure 4.18

The work done by a conservative force (such as gravity) that acts on an object going from A to B is the same for either path. The work done by a nonconservative force (such as friction) will be different if the object takes path 2 rather than path 1.

Clearly we can divide forces into two categories. **Conservative forces** include the gravitational, magnetic, and elastic forces shown in Fig. 4.13. The work done by (or against) a force of this kind depends only on the endpoints of the motion of an object it acts on. The path taken between the endpoints does not matter. Such a force can give rise to a potential energy: If the path is reversed, the work done can reappear as kinetic energy.

The work done by (or against) a **nonconservative force,** on the other hand, depends on the path taken. Friction is an example. In Fig. 4.18 the work done to move something from A to B against a frictional force will be greater if path 2 is taken instead of path 1. If a conservative force were acting, the work would be the same. A nonconservative force does not give rise to potential energy: The work done cannot be returned by reversing the path. Nonconservative forces are also called **dissipative forces.**

Of course, the work done against a nonconservative force in going from A to B is not really lost. This work merely is changed into a form of energy—heat, light, sound, and so forth—that is not recoverable by bringing the object back to A. The term nonconservative forces is somewhat misleading because the forces actually do conserve energy. What they do not conserve is the KE + PE of an object they act on.

Important Terms

Work is a measure of the change (in a general sense) a force gives rise to when it acts on something. When an object is displaced while a force acts on it, the work done by the force is equal to the product of the displacement and the component of the force in the direction of the displacement. The SI unit of work is the **joule;** the customary unit is the **foot · pound.**

The rate at which work is done is called **power.** The SI unit of power is the **watt,** which is equal to 1 J/s; the customary unit is the **foot · pound/second.** The **horsepower** is equal to 550 ft · lb/s.

Energy is that which may be converted into work. When something has energy, it is able to perform work or, in a general sense, to change some aspect of the physical world. The unit of energy is also the joule.

The three broad categories of energy are **kinetic energy,** which is the energy something has by virtue of its motion; **potential energy,** which is the energy something has by virtue of its position; and **rest energy,** which is the energy something has by virtue of its mass.

The principle of **conservation of energy** states that the total amount of energy in a system isolated from the rest of the universe always remains constant, although energy transformations from one form to another may occur within the system.

Work done by a **conservative force** (such as gravity) is independent of the path taken; such a force can give rise to a potential energy. Work done by a **nonconservative force** (such as friction) varies with the path taken and is dissipated; such a force cannot give rise to a potential energy.

Important Formulas

Work: $W = Fx \cos \theta$

Work in lifting object: $W = wh = mgh$

Power: $P = \dfrac{W}{t} = Fv \cos \theta$

Kinetic energy: $KE = \frac{1}{2}mv^2$

Gravitational potential energy: $PE = wh = mgh$

MULTIPLE CHOICE

1. Which of the following is not a unit of power?
  a. joule-second
  b. watt
  c. newton-meter per second
  d. horsepower

2. To keep a vehicle moving at the speed v requires a force F. The power needed is

  a. Fv
  b. $\frac{1}{2}Fv^2$
  c. F/v
  d. F/v^2

3. A dropped ball loses PE as it falls. Gravity is an example of
  a. conservative force
  b. nonconservative force
  c. dissipative force
  d. any of the above, depending on the reference level

4. The work done in moving an object from A to B against a nonconservative force
  a. cannot be recovered by moving it from B to A
  b. does not depend on the path taken between A and B
  c. is always entirely converted into heat
  d. disappears forever

5. A 2.00-kg book is held 1.00 m above the floor for 50 s. The work done is
  a. 0
  b. 10.2 J
  c. 100 J
  d. 980 J

6. The work done in lifting 30 kg of bricks to a height of 20 m on a building under construction is
  a. 61 J
  b. 600 J
  c. 2940 J
  d. 5880 J

7. Six thousand foot-pounds of work is used to raise an object to a height of 100 ft. The weight of the object is
  a. 1.875 lb
  b. 18.75 lb
  c. 60 lb
  d. 1920 lb

8. The amount of work a 0.60-kW electric drill can do in 1.0 min is
  a. 0.6 J
  b. 36 J
  c. 0.6 kJ
  d. 36 kJ

9. A sedentary person requires about 6 million J of energy per day. This rate of energy consumption is equivalent to about
  a. 70 W
  b. 335 W
  c. 600 W
  d. 250 kW

10. A 100-lb boy runs up a staircase to a floor 15 ft higher in 5.45 s. His power output is
  a. 1/64 hp
  b. 1/2 hp
  c. 1 hp
  d. 275 hp

11. A 60-kg woman whose average power is 30 W can climb a mountain 2.0 km high in
  a. 1.0 h
  b. 1.1 h
  c. 5.4 h
  d. 11 h

12. A 200-kg bucket of cement is accelerated upward by a force of 2200 N. The acceleration continues for 5.00 s. The power needed at the end of this period is
  a. 1.44 kW
  b. 11.8 kW
  c. 13.2 kW
  d. 121 kW

13. An 800-kg white horse has a power output of 1.00 hp. The maximum force it can exert at a speed of 3.0 m/s is
  a. 0.25 kN
  b. 2.2 kN
  c. 2.6 kN
  d. 3.6 kN

14. A road slopes upward so that it climbs 1.0 m for each 12 m of distance along the road. A car whose weight is 10 kN moves up the road at a constant speed of 24 m/s. The minimum power the car's engine develops is
  a. 1.7 kW
  b. 2.0 kW
  c. 20 kW
  d. 0.24 MW

15. A 1.0-kg mass has a kinetic energy of 1.0 J when its speed is
  a. 0.45 m/s
  b. 1.0 m/s
  c. 1.4 m/s
  d. 4.4 m/s

16. A 1.0-N weight has a kinetic energy of 1.0 J when its speed is
  a. 0.45 m/s
  b. 1.0 m/s
  c. 1.4 m/s
  d. 4.4 m/s

17. A 1.0-slug mass has a kinetic energy of 1.0 ft · lb when its speed is
  a. 0.25 ft/s
  b. 1.0 ft/s
  c. 1.4 ft/s
  d. 8.0 ft/s

18. A 1.0-lb weight has a kinetic energy of 1.0 ft · lb when its speed is
  a. 0.25 ft/s
  b. 1.0 ft/s
  c. 1.4 ft/s
  d. 8.0 ft/s

19. Car A has a mass of 1000 kg and a speed of 60 km/h, and car B has a mass of 2000 kg and a speed of 30 km/h. The kinetic energy of car A is
  a. half that of car B
  b. equal to that of car B
  c. twice that of car B
  d. four times that of car B

20. A 1000-kg car whose speed is 80 km/h has a kinetic energy of
  a. 2.52×10^4 J
  b. 2.47×10^5 J
  c. 2.42×10^6 J
  d. 3.20×10^6 J

21. A 1000-kg car has a KE of 450 kJ. The car's speed is
a. 21 m/s   b. 30 m/s
c. 45 m/s   d. 900 m/s

22. A 3200-lb car whose speed is 60 mi/h has a kinetic energy of
a. 1.8×10^5 ft · lb
b. 3.9×10^5 ft · lb
c. 7.7×10^5 ft · lb
d. 1.4×10^7 ft · lb

23. A 1.0-kg mass has a potential energy of 1.0 J relative to the ground when it is at a height of
a. 0.12 m   b. 1.0 m
c. 9.8 m   d. 32 m

24. A 1.0-N weight has a potential energy of 1.0 J relative to the ground when it is at a height of
a. 0.10 m   b. 1 m
c. 9.8 m   d. 32 m

25. A 1.0-slug mass has a potential energy of 1.0 ft · lb relative to the ground when it is at a height of
a. 0.031 ft   b. 1.0 ft
c. 9.8 ft   d. 32 ft

26. A 1.0-lb weight has a potential energy of 1.0 ft · lb relative to the ground when it is at a height of
a. 0.031 ft   b. 1.0 ft
c. 9.8 ft   d. 32 ft

27. A total of 4900 J is expended in lifting a 50-kg mass. The mass was raised to a height of
a. 10 m   b. 98 m
c. 960 m   d. 245 km

28. A 16-slug mass is lifted to a height of 10 ft. Its potential energy is
a. 5 ft · lb   b. 20 ft · lb
c. 160 ft · lb   d. 5120 ft · lb

29. The height above the ground of a child on a swing varies from 50 cm at the lowest point to 200 cm at the highest point.

The maximum speed of the child is
a. about 5.4 m/s
b. about 7.7 m/s
c. about 29 m/s
d. dependent on the child's mass

30. The 2.0-kg head of an ax is moving at 60 m/s when it strikes a log. If the blade of the ax penetrates 20 mm into the log, the average force it exerts is
a. 3 kN   b. 90 kN
c. 72 kN   d. 180 kN

31. The car of a roller coaster starts from rest at A in Fig. 4.19. If there is no friction or air resistance, the car's speed at B is
a. 4.5 m/s   b. 14 m/s
c. 20 m/s   d. 0.39 km/s

32. The car's speed at C is
a. 13 m/s   b. 15 m/s
c. 20 m/s   d. 0.16 km/s

33. The car's speed at D is
a. 13 m/s   b. 15 m/s
c. 20 m/s   d. 0.24 km/s

34. The car's speed at E is
a. 0   b. 4.5 m/s
c. 13 m/s   d. 15 m/s

Figure 4.19

Multiple Choice 31–34.

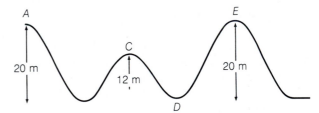

QUESTIONS

1. Under what circumstances (if any) is no work done on a moving object even though a net force acts on it?

2. The potential energy of a golf ball in a hole is negative with respect to the ground. Under what circumstances (if any) is its kinetic energy negative?

3. Does every moving body possess kinetic energy? Does every stationary body possess potential energy?

4. A golf ball and a Ping-Pong ball are dropped in a vacuum chamber. When they have fallen halfway to the bottom, how do their speeds compare? Their kinetic energies? Their potential energies?

5. Identical twins Ann and Betty set out to climb a mountain. Ann chooses a slope that averages 30° above the horizontal, and Betty chooses a slope that averages 40°. Compare the amounts of work the twins do.

6. (a) Where is the KE of the pendulum bob of Fig. 4.20 a minimum? A maximum? (b) Where is the PE of the bob a minimum? A maximum?

Figure 4.20

Question 6.

Figure 4.21

Question 7.

7. When the valve connecting the two tanks shown in Fig. 4.21 is opened, water from *A* flows into *B* until their levels are the same. What percentage of the original PE of the water in tank *A* is changed into KE and eventually into heat in this process?

8. How can the existence of nonconservative forces be reconciled with the principle of conservation of energy?

EXERCISES

4.1 Work

1. The sun exerts a force of 4.0×10^{28} N on the earth, and the earth travels 9.4×10^{11} m in its annual orbit around the sun. How much work is done by the sun on the earth in the course of a year?

2. A horse is towing a barge with a rope that makes an angle of 20° with the canal. If the horse exerts a force of 400 N, how much work does it do in moving the barge 1.0 km?

3. A 500-N box is pushed across a horizontal floor with a force of 250 N. The coefficient of kinetic friction is 0.40. (a) Find the work done in pushing the box 20 m. (b) How much work went into overcoming friction and how much into accelerating the box?

4. A 50-kg box is pushed across a horizontal floor by a horizontal force of 180 N. The coefficient of kinetic friction is 0.30. (a) Find the work done in pushing the box 20 m. (b) How much work went into overcoming friction and how much into accelerating the box?

5. A 60-lb wooden box is pushed a distance of 50 ft on a horizontal stone floor by a force just sufficient to overcome the friction between box and floor. The coefficient of kinetic friction is 0.40. (a) What is the required force? (b) How much work does the force do?

4.2 Work Done Against Gravity

6. The work done to lift a 30-kg mass is 4.0 kJ. If the mass is at rest before and after its elevation, how high does it go?

7. A uniform 500-lb lamppost 30 ft long is lying on the ground. How much work is needed to raise it to a vertical position?

8. A woman drinks a bottle of beer and proposes to work off its 460-kJ energy content by exercising with a 20-kg barbell. If each lift of the barbell from chest height to over her head is through 60 cm and the efficiency of her body is 10% under these circumstances, how many times must she lift the barbell?

9. (a) A force of 130 N is used to lift a 12-kg mass to a height of 8.0 m. How much work is done by the force? (b) A force of 130 N is used to push a 12-kg mass on a horizontal, frictionless surface for a distance of 8.0 m. How much work is done by the force?

10. An 80-kg person carrying a 10-kg backpack climbs to the summit of a 3000-m mountain starting from sea level along a route that averages 30° above the horizontal. Find the work the person does.

11. A horizontal force of 5.0 N is used to push a box up a ramp 5.0 m long that is at an angle of 15° above the horizontal. How much work is done by the force?

4.3 Power

12. A 15-kW motor is used to hoist an 800-kg bucket of concrete to the twentieth floor of a building under construction, a height of 90 m. If no energy is lost, how much time is required?

13. A weightlifter raises a 150-kg barbell from the floor to a height of 2.2 m in 0.80 s. What is his average power output during the lift?

14. In 1932 five members of the Polish Olympic ski team climbed from the 5th to the 102nd floor of the Empire State Building, a distance of approximately 350 m, in 21 min. If one of these men had a mass of 70 kg, how much power did he develop during the ascent?

15. The anchor windlass of a boat must be able to raise a total load (anchor plus chain) of 800 kg at a speed of 0.50 m/s. What should the minimum rating of the motor be, in kilowatts?

16. A car needs 13 kW to move along a level road at 60 km/h. What is the sum of the air resistance and rolling friction on the car at that speed?

17. Each of the four engines of a DC-8 airplane develops 7500 hp when the cruising speed is 240 m/s. How much thrust does each engine produce under these circumstances?

18. A boat is powered by three 2600-kW diesel engines. At the top speed of 27.4 m/s, the boat's propellers produce a total force of 250 kN. (This force equals in magnitude the resistance of the water to the boat's motion at that speed; hence the force does not accelerate the boat.) Find the efficiency of the boat's propulsion system.

19. A hoist whose efficiency is 80% is powered by an 8.0-kW motor. At what speed can the hoist raise an 800-N load?

20. A 75-kg man carrying 10-kg pack climbs a mountain 2800 m high in 10 h. (a) What is his average power output? (b) The efficiency with which his body utilizes food energy for climbing is 15%. How many joules of food energy were needed for the climb?

21. A crane whose motor has a power input of 5.0 kW raises a 1200-kg beam through a height of 30 m in 90 s. Find the efficiency.

22. A boy pulls a sled with a force of 10 lb for 100 ft. The rope attached to the sled is at an angle of 30° above the horizontal. (a) How much work is done? (b) If the boy moves the sled 100 ft in 45 s, find his power output in horsepower.

23. An escalator carries passengers from one floor of a building to another 35 ft higher. It is designed to have a capacity of 200 passengers per minute, assuming an average weight per passenger of 150 lb. Find the required horsepower of the motor if half the work it does is dissipated as heat.

24. The bilge pump of a boat is able to raise 200 L of water per minute through a height of 1.2 m. If the pump is 60% efficient, how much power must be supplied to the pump? The mass of 1 liter of water is 1 kg.

25. A person's metabolic processes can usually operate at a power of 6.0 W/kg of body mass for several hours at a time. If a 60-kg woman carrying a 12-kg pack is walking uphill with an energy-conversion efficiency of 20%, at what rate, in meters per hour, does she ascend?

4.5   Kinetic Energy

26. Find the kinetic energy of a 115-kg ostrich running at 15 m/s.

27. Find the speed of a 2.0-g insect whose kinetic energy is 0.010 J.

28. Find the kinetic energy of a 256-lb ostrich running at 50 ft/s.

29. Find the kinetic energy of a 3200-lb car moving at 40 mi/h.

30. A raindrop of mass 1.0 mg falls at 3.0 m/s when there is no wind. Find its Ke when a wind of 5.0 m/s is blowing.

31. Find the kinetic energy of a 70-kg runner who covers 400 m in 45 s at constant speed.

32. Is the work needed to bring a car's speed from 10 to 20 mi/h less than, equal to, or .more than the work needed to bring its speed form 90 to 100 mi/h? If the amounts of work are different, what is the ratio between them?

33. A 2.0-kg ball is at rest when a horizontal force of 5.0 N is applied. In the absence of friction, what is the speed of the ball after it has gone 10 m?

34. A 1.0-kg trout is hooked by a fisherman and swims off at 2.5 m/s. The fisherman stops the trout in 50 cm by braking his reel. How much tension is exerted on the line?

35. A certain 800-kg car has a motor whose power output is 30 kW. If the car is carrying two passengers whose total mass is 150 kg, how long will it need to accelerate from 70 to 110 km/h?

36. A 70-kg sprinter pushes on the starting blocks for 0.20 s and leaves them with a speed of 5.0 m/s. He then continues to accelerate for a further 5.0 s until his speed is 12 m/s. Find the reaction force exerted by the starting blocks on the sprinter and his average power output in each of the accelerations.

37. What is the power output of a 1200-kg car that can go from 25 km/h to 100 km/h in 12 s? Neglect air resistance and rolling friction.

4.6   Potential Energy

38. A 3.0-kg stone is dropped from a height of 100 m. Find its kinetic and potential energies when it is 50 m from the ground.

39. A ball is dropped from a height of 60 m. How high above the ground will it be when one-third of its total energy is KE?

40. A 120 lb woman jumps off a wall 4.0 ft high and lands with her knees stiff. If her body is compressed by 1.0 in. on impact, what is the average force exerted on her by the ground?

What would the force be if she had bent her knees on impact, so that she came to a stop in 10 in?

41. In the operation of a certain pile driver, a 500-kg hammer is dropped from a height of 5.0 m above the head of a pile. If the pile is driven 20 cm into the ground with each impact of the hammer, what is the average force on the pile when struck?

42. A bullet moving at 500 m/s has a kinetic energy of 2500 J and a potential energy of 0.50 J at a certain moment. What is the bullet's mass and how high above the ground is it at that moment?

43. A ball is dropped from a height of 1.0 m and loses 10% of its kinetic energy when it bounces on the ground. To what height does it rise?

44. A certain frog's hind legs produce a force of 2.5 times its weight through a vertical distance of 90 mm, at which point the frog becomes airborne. (a) What is the frog's speed at takeoff? (b) What height above the ground does the frog reach? Assume a vertical jump.

45. A waterfall is 30.0 m high, and 1.00×10^4 kg of water flows over it per second. (a) How much power does this flow represent? (b) If all this power could be converted to electricity, how many 100-W light bulbs could be supplied?

4.7 Conservation of Energy

46. A force of 50 lb is used to lift a 40-lb statue, initially at rest, to a height of 20 ft. (a) What is the change in the statue's PE? (b) In its KE?

47. A woman skis down a slope 300 ft high. Her speed at the foot of the slope is 60 ft/s. What percentage of her initial potential energy was dissipated?

48. At her highest point a 40-kg girl on a swing is 2.0 m from the ground, while at her lowest point she is 0.80 m from the ground. What is her maximum speed? On another swing a 50-kg boy undergoes exactly the same motion. What is his maximum speed?

49. Find the height of the bar a pole-vaulter can clear on the basis of the following assumptions: His running speed is 8.0 m/s; his center of mass is initially 1.1 m above the ground, and he pulls himself upward along the pole 0.60 m as the pole swings into a vertical position; and all his initial KE is converted into work done to raise his center of mass sufficiently to clear the bar.

50. A man uses a rope and system of pulleys to lift an 80-kg object to a height of 2.0 m. He exerts a force of 220 N on the rope and pulls a total of 8.0 m of rope through the pulleys in the course of raising the object, which is at rest afterward. (a) How much work does the man do? (b) What is the change in the object's PE? (c) If the answers to (a) and (b) are different, explain.

51. (a) A force of 8.0 N is used to push a 0.50-kg ball over a horizontal, frictionless table a distance of 3.0 m. If the ball starts from rest, what is its final KE? (b) The same force is used to lift the same ball a height of 3.0 m. If the ball starts from rest, what is its final KE?

52. A force of 200 N is used to lift a 15-kg object, initially at rest, to a height of 8.0 m. No friction is present. What is the speed of the object at this height?

53. An 800-kg car coasts down a hill 40 m high with its engine off and its driver's foot pressing on the brake pedal. At the top of the hill the car's speed is 6.0 m/s, and at the bottom it is 20.0 m/s. How much energy was converted to heat on the way down?

PROBLEMS

54. A person pulls a 150-lb crate at constant speed 80 ft across a level floor using a rope that is 30° above the horizontal. If the coefficient of kinetic friction between crate and floor is 0.30, how much work is done by the person? (Assume the rope is attached to the center of mass of the crate.)

55. A 10-kg block is pulled up a plane 20° above the horizontal at constant speed by a force parallel to the plane. The coefficient of kinetic friction between the block and the plane is 0.40. How much work is done by the force in moving the block 2.0 m along the plane?

56. A 50-kg box is pulled 10 m up a plane 15° above the horizontal at constant speed by a force acting 25° above the

plane, as in Fig. 4.22. The coefficient of kinetic friction between the box and the plane is 0.30. How much work is done by the applied force?

Figure 4.22

Problem 56.

57. A 70-kg block of balsa wood is 1.5 m high and has a base 60 cm × 60 cm. The block is tilted about one edge until it falls over on its side. How much work had to be done to tilt the block to the point where it began to fall over?

58. A well 50 cm in diameter is to be dug to a depth of 20 m. The earth taken out must be lifted 1.2 m above the ground into a truck that carries it away. If each m^3 of the earth has a mass of 1300 kg, what is the minimum work needed to remove the earth from the well?

59. A 15-lb shot is thrown 54 ft. What was its minimum initial KE?

60. An object of mass m is accelerated uniformly from rest for a time t during which it covers the distance x. Find the work done on the object.

61. A 2.0-kg block is pulled up a ramp by a constant force parallel to the ramp of 15 N. The ramp is 2.0 m long and inclined at 30° above the horizontal. If the coefficient of kinetic friction between the block and the ramp is 0.30, what is the speed of the block at the top of the ramp?

62. A brick slides down a wooden plank 2.0 m long tilted so that one end is at a height of 1.0 m. The brick's speed at the bottom is 2.5 m/s. The plank is then sanded smooth and waxed so that the coefficient of friction is half what it was before, and the brick is slid down it again. What is the new speed of the brick at the bottom?

63. The bob of a pendulum 5.0 ft long is pulled aside so that the string is 40° from the vertical. When the bob is released, with what speed will it pass through the bottom of its path?

64. Steam enters a 10-MW-output turbine at 800 m/s and emerges at 100 m/s. Assuming 90% mechanical efficiency, determine what mass of steam passes through the turbine per second.

65. An escalator 14 m long is carrying a 70-kg person from one floor to another floor 8.0 m higher. The linear speed of the escalator is 1.0 m/s. (a) How much work does the escalator do in carrying the person to the top? What is its power output while doing so? (b) The person is walking up the escalator at

0.8 m/s. Answer the same questions for this situation. (c) The person is walking down the escalator at 0.8 m/s. Answer the same questions for this situation.

66. A 20-kg crate is dropped from a height of 15 m and reaches the ground in 1.9 s. If the force of air resistance on the crate is constant, what is its magnitude? What percentage of the original potential energy of the crate is lost?

67. The power required to propel a 1200-kg car at 40 km/h on a level road is 30 kW. (a) How much resistance must the car overcome at this speed? (b) How much power is needed for the car to ascend an 8° hill at the same speed?

68. A 60-kg woman is riding a 10-kg bicycle at a constant speed of 20 km/h. Friction and air resistance total 25 N. Find her power output (a) when the road is horizontal and (b) when it goes up a 5° hill.

69. A 120-lb woman uses 0.15 hp of power to ride a 25-lb bicycle on a level road at 20 ft/s. If she can develop 0.35 hp, what is the angle of the steepest hill she can climb at the same speed?

70. A wooden block whose initial speed is 6.0 m/s starts to slide up a plane inclined at 25° above the horizontal. (a) If the coefficient of kinetic friction is 0.30, how far up the plane does the block go? (b) What will the block's speed be after it has slid back down the plane to its starting point?

Answers to Multiple Choice

1. a	**10.** b	**19.** c	**28.** d
2. a	**11.** d	**20.** b	**29.** a
3. a	**12.** c	**21.** b	**30.** d
4. a	**13.** a	**22.** b	**31.** c
5. a	**14.** c	**23.** a	**32.** a
6. d	**15.** c	**24.** b	**33.** c
7. c	**16.** d	**25.** a	**34.** a
8. d	**17.** c	**26.** b	
9. a	**18.** d	**27.** a	

5

MOMENTUM

5.1 Linear Momentum

5.2 Impulse

5.3 Conservation of Momentum

5.4 Rocket Propulsion

5.5 Inelastic Collisions

5.6 Elastic Collisions

5.7 Energy Transfer

5.8 Coefficient of Restitution

Because the universe is so complex, we need the help of many different quantities to describe its various aspects economically. Besides length, time, and mass we have already found velocity, acceleration, force, work, and energy to be useful, and more are to come. None of these quantities is sacred—we can certainly give up any of them, but only at the expense of making it harder to understand what is going on around us. The idea behind defining each physical quantity is to single out something that unifies a wide range of observations. Then we can boil down a large number of separate discoveries about nature into a brief, clear statement, for instance, the second law of motion or the law of conservation of energy. In this chapter we learn how the concepts of linear momentum and impulse can help us to analyze the behavior of moving bodies.

High momentum
Baseball

m_1 $m_1 = m_2$

Low momentum $v_1 > v_2$

m_2

High momentum
Iron shot

m_1 $m_1 > m_2$

Low momentum $v_1 = v_2$

m_2

Figure 5.1

The linear momentum $\mathbf{p} = m\mathbf{v}$ of a moving body is a measure of its tendency to continue in motion at constant velocity.

5.1   LINEAR MOMENTUM

The product of mass and velocity.

We all know that a baseball struck squarely by a bat is harder to stop than the same ball thrown gently. We also know that the heavy iron ball used for the shotput is harder to stop than a baseball whose speed is the same (Fig. 5.1). These observations suggest that a measure of the tendency of a moving body to continue moving at constant velocity is the product $m\mathbf{v}$ of its mass m and its velocity $\mathbf{v}$.

The quantity $m\mathbf{v}$ is called the **linear momentum** (symbol $\mathbf{p}$) of a moving body:

$$\mathbf{p} = m\mathbf{v} \qquad\qquad\qquad \textit{Linear momentum} \quad (5.1)$$

Linear momentum is a vector quantity whose direction is the direction of $\mathbf{v}$. The kinetic energy of a moving body, which also depends on its mass and velocity because $\mathrm{KE} = \frac{1}{2}mv^2$, is a scalar quantity with magnitude only. The different significances of linear momentum and kinetic energy will be discussed later in this chapter.

Because $m\mathbf{v}$ describes the tendency of a moving body to pursue a straight path at constant velocity, it is referred to as linear momentum. A different quantity, **angular momentum,** describes the tendency of a spinning body, such as a top, to continue to spin. When there is no question as to which is meant, linear momentum is usually referred to simply as momentum.

◆   **E X A M P L E   5.1** _____

What effect on its momentum does doubling the kinetic energy of a moving object have?

SOLUTION   The object's initial kinetic energy is $\mathrm{KE}_1 = \frac{1}{2}mv_1^2$, and its final kinetic energy is $\mathrm{KE}_2 = \frac{1}{2}mv_2^2$. Since $\mathrm{KE}_2 = 2\mathrm{KE}_1$,

$$\tfrac{1}{2}mv_2^2 = 2(\tfrac{1}{2}mv_1^2)$$
$$v_2^2 = 2v_1^2$$
$$v_2 = \sqrt{2}\,v_1$$

Hence

$$p_2 = mv_2 = \sqrt{2}\,mv_1 = \sqrt{2}\,p_1$$

The object's momentum has increased by a factor of $\sqrt{2}$.   ◆

5.2 IMPULSE

What is needed to cause a change in momentum.

To set something in motion from rest, a force must be applied for a period of time. We might expect that the greater the force and the longer the time, the more momentum the object will have. This expectation is correct, and the product $\mathbf{F}\,\Delta t$ of a constant force $\mathbf{F}$ and the time interval Δt during which it acts accordingly has the status of a physical quantity in its own right. This quantity is called **impulse:**

$$\text{Impulse} = \mathbf{F}\,\Delta t \qquad\qquad (5.2)$$

Impulse, like momentum, is a vector quantity. Let us see how the momentum of an object is affected when it receives a certain impulse.

The second law of motion states that the force $\mathbf{F}$ applied to an object of constant mass m that undergoes the acceleration $\mathbf{a}$ is given by $\mathbf{F} = m\mathbf{a}$. When a force $\mathbf{F}$ is applied at the time $t_1 = 0$ to an object whose initial velocity is $\mathbf{v}$, at the later time $t_2 = \Delta t$ its velocity will have changed to $\mathbf{v} + \Delta\mathbf{v}$ (Fig. 5.2). The object's acceleration in this time interval is

$$\mathbf{a} = \frac{\text{velocity change}}{\text{time interval}} = \frac{\Delta\mathbf{v}}{\Delta t}$$

and so

$$\mathbf{F} = m\mathbf{a} = m\frac{\Delta\mathbf{v}}{\Delta t}$$

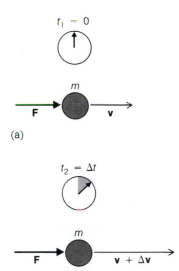

Figure 5.2

(a) At the time $t_1 = 0$, a force $\mathbf{F}$ is applied to a body whose velocity is $\mathbf{v}$. (b) At the later time $t_2 = \Delta t$, the velocity of the body is $\mathbf{v} = \mathbf{v} + \Delta\mathbf{v}$.

The greater the impulse given to a pitched ball by a bat, the greater the change in the ball's momentum.

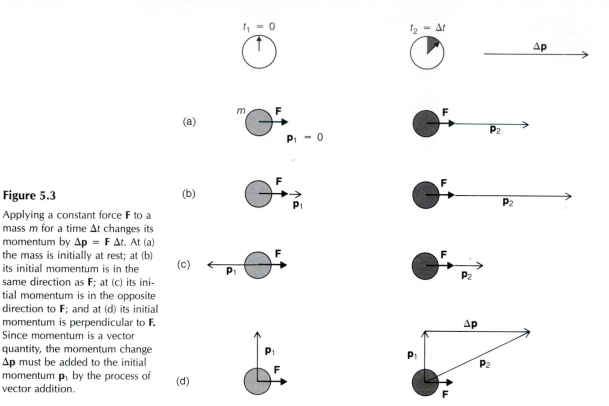

Figure 5.3

Applying a constant force **F** to a mass m for a time Δt changes its momentum by $\Delta \mathbf{p} = \mathbf{F}\,\Delta t$. At (a) the mass is initially at rest; at (b) its initial momentum is in the same direction as **F**; at (c) its initial momentum is in the opposite direction to **F**; and at (d) its initial momentum is perpendicular to **F**. Since momentum is a vector quantity, the momentum change $\Delta \mathbf{p}$ must be added to the initial momentum $\mathbf{p}_1$ by the process of vector addition.

which we can rewrite as

$$\mathbf{F}\,\Delta t = m\,\Delta \mathbf{v}$$

Evidently the impulse provided by the force equals the momentum change of the object:

$$\mathbf{F}\,\Delta t = \Delta(m\mathbf{v}) = \Delta \mathbf{p} \qquad (5.3)$$

Impulse = momentum change

Figure 5.3 shows the effect of applying the constant force **F** for the time Δt to several objects with different momenta $\mathbf{p}_1$. In each case the final momentum $\mathbf{p}_2$ is obtained by finding the vector sum $\mathbf{p}_2 = \mathbf{p}_1 + \mathbf{F}\,\Delta t$.

In the SI the unit of impulse is the **newton-second** (N·s), and the unit of momentum is the kg·m/s; they are actually the same, of course, but it is often convenient to distinguish between them in this way. In the U.S. Customary System, the corresponding units are the **pound·second** (lb·s) and the **slug·foot/second** (slug·ft/s).

◆ E X A M P L E 5.2 _____

The head of a golf club is in contact with a 46-g golf ball for 0.50 ms (1 ms = 1 millisecond = 10^{-3} s), and as a result the ball flies off at 70 m/s. Find the average force that was acting on the ball during the impact.

SOLUTION   The ball starts from rest, hence its momentum change is

$$\Delta mv = (0.046 \text{ kg})(70 \text{ m/s}) = 3.22 \text{ kg·m/s}$$

From Eq. (5.3) we therefore have

$$F = \frac{\Delta mv}{\Delta t} = \frac{3.22 \text{ kg·m/s}}{5.0 \times 10^{-4} \text{ s}} = 6.4 \times 10^3 \text{ N}$$

which is equivalent to 1450 lb. A force of the same magnitude but acting in the opposite direction (the **recoil force**) acts on the club's head during the impact, in accord with the third law of motion. No golf club could withstand such a static load, but the impact is so brief that its only effect on the shaft is to bend it temporarily by a few centimeters.   ◆

5.3   CONSERVATION OF MOMENTUM

A vector principle.

Energy and work are scalar quantities that have magnitude only. Despite the fundamental and all-inclusive character of the law of conservation of energy, it cannot by itself enable us to solve most problems that involve interacting bodies. A simple example is the firing of a rifle. To conserve energy here, the kinetic energies of the bullet and the recoiling rifle, plus the heat and sound energy that are liberated, must equal the chemical energy released by the explosive. However, this does not tell us how the total energy is divided among the rifle, the bullet, and the atmosphere. Indeed, because energy is a scalar quantity, conserving it does not even imply that the bullet and rifle must move in opposite directions. To solve a problem like this, when we have no detailed knowledge of the active forces, an additional principle of a vector nature is required.

Let us consider two particles A and B that collide and thereby exert forces on each other. At every instant during their interaction these forces, $\mathbf{F}_{AB}$ acting on B and $\mathbf{F}_{BA}$ acting on A, obey Newton's third law of motion,

$$\mathbf{F}_{AB} = -\mathbf{F}_{BA}$$

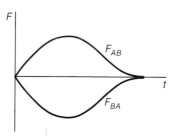

Figure 5.4

Particles A and B collide and exert the forces $\mathbf{F}_{AB}$ and $\mathbf{F}_{BA}$ on each other. As shown in the graph, these forces are equal in magnitude and opposite in direction at all times. Their impulses are also equal and opposite, so the total momentum of A and B is left unchanged by the collision, although it may be distributed differently between them.

This is shown in Fig. 5.4. Hence the impulses exchanged must be equal and opposite,

$$\mathbf{F}_{AB}\,\Delta t \;=\; -\mathbf{F}_{BA}\,\Delta t$$

and the *total* momentum of the system of A and B together is the same after the collision as it was before.

This result can be generalized to any system of two or more particles. If no net force from outside the system acts on it, the total linear momentum of the system, which is the sum

$$\Sigma\mathbf{p} \;=\; \mathbf{p}_1 + \mathbf{p}_2 + \mathbf{p}_3 + \cdots \tag{5.4}$$

of the individual momenta of its particles, cannot change. (The symbol Σ is the Greek capital letter sigma and stands for "sum of.") Thus we have the principle of **conservation of linear momentum:**

When the vector sum of the external forces acting on a system of particles equals zero, the total linear momentum of the system remains constant.

Of course, as we saw above, interactions within the system may change the *distribution* of the total momentum $\Sigma\mathbf{p}$ among the various particles in the system, but $\Sigma\mathbf{p}$ stays the same.

The total of the kinetic energies of the particles need not be constant, however. The KE involved in an explosion is zero to start with, because it all comes from the chemical energy stored in the explosive material, but it is not zero afterward. In a collision some or even all of the KE of relative motion of the colliding objects may disappear into heat and sound energy. In both cases the total momentum does not change.

Explosions

Let us consider a specific example easily treated with the help of conservation of momentum. Suppose we have an isolated particle of mass m, initially at rest, that suddenly explodes into two particles of masses m_1 and m_2, which fly apart (Fig. 5.5). The forces acting on the original particle that caused it to break up were internal ones, and no external force was present. Since m has the initial momentum of zero, the final momentum of m_1 and m_2, when added together, must also be zero. Hence

$$m\mathbf{v} \;=\; 0 \;=\; m_1\mathbf{v}_1 + m_2\mathbf{v}_2$$

and

$$m_1\mathbf{v}_1 \;=\; -m_2\mathbf{v}_2 \qquad \textit{Momentum conservation in explosion} \tag{5.5}$$

where $\mathbf{v}_1$ and $\mathbf{v}_2$ are the final velocities of the two fragments. We note immediately

Figure 5.5

The total momentum of a system of particles remains constant if no external forces act on the system.

that these velocities must be in opposite directions along the same line. This problem could *not* be solved starting from $\mathbf{F} = m\mathbf{a}$, since we do not know exactly what the forces were that acted during the explosion.

In the customary system, where the weight w of an object rather than its mass $m = w/g$ is usually given, we have $w_1\mathbf{v}_1/g = w_2\mathbf{v}_2/g$ and

$$w_1\mathbf{v}_1 = -w_2\mathbf{v}_2 \tag{5.6}$$

The g's have canceled out, so we need not find the masses of the objects involved if we know their weights.

◆   **E X A M P L E   5.3** ────────────────────────────

An astronaut in orbit outside an orbiting space station throws her 800-g camera away in disgust when it jams (Fig. 5.6). If she and her space suit together have a mass of 100 kg and the speed of the camera is 12 m/s, how far away from the space station will she be in 1.0 h?

Figure 5.6

The initial momentum of this system was 0. Hence the forward momentum of the thrown camera is equal in magnitude to the backward momentum of the astronaut who threw it.

SOLUTION   Here $m_1 = 0.80$ kg, $m_2 = 100$ kg, and $v_1 = 12$ m/s. From Eq. (5.5),

$$v_2 = -\frac{m_1}{m_2}v_1 = -\left(\frac{0.80 \text{ kg}}{100 \text{ kg}}\right)(12 \text{ m/s}) = -0.096 \text{ m/s}$$

After $t = 1.0$ h $= 3600$ s the astronaut will be

$$x = v_2t = (0.096 \text{ m/s})(3600 \text{ s}) = 346 \text{ m} = 0.35 \text{ km}$$

away from the space station. The total momentum of the system of astronaut plus camera remains zero until other forces act on them.   ◆

Figure 5.7

Conservation of momentum in rocket flight. The downward momentum of the exhaust gases is exactly balanced by the upward momentum of the rocket itself.

Figure 5.8

In a liquid-fueled rocket, the reacting substances are mixed and ignited in a combustion chamber and the resulting exhaust gases escape at high speed through a steerable nozzle. The liberated chemical energy becomes kinetic energy of the rocket and its exhaust. Liquid hydrogen and liquid oxygen are used to power many rockets, and their product is water vapor, H_2O. The combination of 1 kg of hydrogen and 8 kg of oxygen releases 245 MJ.

5.4   ROCKET PROPULSION

Momentum conservation is the basis.

The principle underlying rocket flight is conservation of momentum. The total momentum of a rocket on its launching pad is zero. When it is fired, the exhaust gases shoot downward at high speed, and the rocket moves upward to balance the momentum of the gases (Fig. 5.7). Rockets do not operate by "pushing" against their launching pads, the air, or anything else; in fact, they perform best in space, where there is no atmosphere to impede their motion. The energy of the rocket and its exhaust comes from chemical energy stored in the fuel (Fig. 5.8). The total momentum of the system of rocket plus exhaust, which is initially zero, does not remain constant after a launch from the earth, because of the impulses provided by air resistance and the earth's gravitational pull.

The upward force exerted on a rocket by the expulsion of exhaust gases is called **thrust**. If a mass Δm of exhaust gas comes out at the velocity $\mathbf{v}$ relative to the rocket in the time interval Δt, the momentum change of the gas relative to the rocket is $\Delta (m\mathbf{v})$. From Eq. (5.3) $\mathbf{F} \, \Delta t = \Delta (m\mathbf{v})$, so the associated force on the rocket is

$$\mathbf{F} = \frac{\Delta(m\mathbf{v})}{\Delta t} = \mathbf{v} \, \frac{\Delta m}{\Delta t} \qquad\qquad \textit{Thrust} \quad (5.7)$$

The thrust of a rocket is the product of the exhaust speed and the rate at which fuel is consumed. In the case of a rocket that burns 100 kg of fuel per second with an exhaust speed of 3 km/s, the thrust is

$$F = v \, \frac{\Delta m}{\Delta t} = (3000 \text{ m/s})(100 \text{ kg/s}) = 3 \times 10^5 \text{ N}$$

which is about 33 tons.

When the spacecraft is launched from the earth's surface, the thrust of its motor must exceed its initial weight $m_0 g$ for it to rise from the ground. If the thrust remains constant, the spacecraft acceleration increases as its fuel is burnt. A maximum acceleration of several times g in magnitude is usual.

Rocket propulsion is a gradual rather than an instantaneous process, with the fuel burned and ejected as exhaust gases at a certain rate instead of in one lump. As a result part of the momentum of the exhaust is wasted in pushing forward unburned fuel. When this factor is taken into account, the ultimate speed of a rocket (neglecting air resistance and gravity) turns out to be directly proportional to the speed of the exhaust gases and to the logarithm of the ratio of the rocket's initial mass to its final mass after the fuel has been consumed. A typical modern rocket might have an exhaust speed of 3 km/s with 75% of its initial mass consisting of fuel, which gives a final speed of 4.2 km/s. This is over 9000 mi/h, which is very fast by ordinary standards but still only about half the minimum speed needed to put a satellite in orbit around the earth (see Section 6.6).

To attain higher speeds than a single rocket is capable of, two or more rocket stages can be used. The first stage is a large rocket whose payload is another, smaller rocket. When the fuel of the first stage has been consumed, its fuel tanks and engine are cast loose. Then the second stage is fired starting from a high initial speed instead of from rest and without the burden of the fuel tanks and engine of the first stage. This process can be repeated a number of times, depending on the final speed required. The *Saturn V* launch vehicle that propelled the *Apollo 11* spacecraft to the moon in July 1969 used three stages, as shown in Fig. 5.9. At original ignition the entire assembly was 111 m long and had a mass of 2.9×10^6 kg (3240 tons).

$v = 0$

Apollo 11 spacecraft

Third stage

Second stage

Saturn V launch vehicle

First stage

First stage is ignited.

$v = 2760$ m/s

First stage drops off, second stage is ignited.
$\Delta v = 2760$ m/s

$v = 6935$ m/s

Second stage drops off, third stage is ignited.
$\Delta v = 4125$ m/s

$v = 10,835$ m/s

Third stage drops off, spacecraft continues to moon.
$\Delta v = 3900$ m/s

Figure 5.9

The *Saturn V* launch vehicle that propelled the *Apollo 11* spacecraft to the moon for the first manned landing used three rocket stages.

Apollo 11 lifts off its pad on July 16, 1969 to begin the first human visit to the moon.

5.5 INELASTIC COLLISIONS

KE is not conserved in such collisions.

Momentum is conserved in all collisions as well as in all explosions. When two or more objects collide, their final total momentum equals their initial total momentum. *The essential effect of a collision is to redistribute the total momentum of the colliding objects.*

All collisions conserve momentum, but not all of them conserve kinetic energy as well. Collisions fall into three categories (Fig. 5.10):

1. **Elastic collisions,** which conserve kinetic energy.
2. **Inelastic collisions,** which do not conserve kinetic energy. Some KE is lost to heat, sound energy, and so forth.
3. **Completely inelastic collisions,** in which the objects stick together afterward. In such collisions the KE loss is the maximum possible.

Given the initial masses and velocities of the objects involved, both elastic and completely inelastic collisions can be analyzed to give the final velocities. Although few actual collisions fit exactly into these extreme categories, many come close, so it is worth examining them.

Perfectly elastic; KE is conserved.

Inelastic; some KE is lost.

Perfectly inelastic; maximum KE is lost.

Figure 5.10

The three possible outcomes of a ball striking the ground illustrate the three categories of collisions.

 E X A M P L E 5.4 _____

A 5-kg lump of clay that is moving at 10 m/s to the left strikes a 6-kg lump of clay moving at 12 m/s to the right. The two lumps stick together after they collide. Find the final speed of the composite object and the kinetic energy dissipated in the collision.

SOLUTION   This is an example of a completely inelastic collision. If we call the mass of the final object M and its velocity V, conservation of linear momentum requires that

$$\text{Final momentum} = \text{initial momentum}$$
$$MV = m_1 v_1 + m_2 v_2$$

Adopting the convention that motion to the right is + and to the left is −, we have

$$m_1 = 5 \text{ kg} \qquad m_2 = 6 \text{ kg} \qquad M = m_1 + m_2 = 11 \text{ kg}$$
$$v_1 = -10 \text{ m/s} \qquad v_2 = +12 \text{ m/s} \qquad V = ?$$

Figure 5.11

In a completely inelastic collision, the colliding bodies stick together. Kinetic energy is not conserved in such an event. Linear momentum is conserved in all collisions.

Solving for V yields

$$V = \frac{m_1v_1 + m_2v_2}{M} = \frac{(5 \text{ kg})(-10 \text{ m/s}) + (6 \text{ kg})(12 \text{ m/s})}{11 \text{ kg}} = 2 \text{ m/s}$$

Since V is positive, the composite body moves off to the right (Fig. 5.11).

Energy and momentum are distinct concepts. The lumps of clay before the collision have the kinetic energies

$$KE_1 = \tfrac{1}{2}m_1v_1^2 = \tfrac{1}{2}(5 \text{ kg})(-10 \text{ m/s})^2 = 250 \text{ J}$$
$$KE_2 = \tfrac{1}{2}m_2v_2^2 = \tfrac{1}{2}(6 \text{ kg})(12 \text{ m/s})^2 = 432 \text{ J}$$

After the collision the new lump of clay has the kinetic energy

$$KE_3 = \tfrac{1}{2}MV^2 = \tfrac{1}{2}(11 \text{ kg})(2 \text{ m/s})^2 = 22 \text{ J}$$

The total kinetic energy prior to the collision was $432 + 250$ or 682 J, whereas afterward it is only 22 J. The difference of 660 J was dissipated largely into heat in the collision, with some probably being lost to sound energy as well.   ◆

It is important to keep in mind the directional character of linear momentum. Sometimes a problem involves bodies that move along the same line, as in the preceding example, but in general they may move in two or three dimensions, and we must then take this into account by a vector calculation.

◆   **E X A M P L E   5.5**

A 60-kg man is sliding east on the frictionless surface of a frozen pond at a velocity of 0.50 m/s. He is struck by a 1.00-kg snowball whose velocity is 20 m/s toward the north. If the snowball sticks to the man, what is his final velocity?

SOLUTION   Here linear momentum must be conserved separately in both the east-west and the north-south directions, which we will call the *x*- and *y*-axes, respectively.

Since

$$m_1 = 60 \text{ kg} \qquad m_2 = 1.00 \text{ kg} \qquad M = m_1 + m_2 = 61 \text{ kg}$$

$$v_{1x} = 0.50 \text{ m/s} \qquad v_{2x} = 0 \qquad V_x = ?$$

$$v_{1y} = 0 \qquad v_{2y} = 20 \text{ m/s} \qquad V_y = ?$$

we have (see Fig. 5.12)

Final momentum = initial momentum

x-direction: $MV_x = m_1 v_{1x} + m_2 v_{2x} = (60 \text{ kg})(0.50 \text{ m/s}) = 30 \text{ kg·m/s}$

y-direction: $MV_y = m_1 v_{1y} + m_2 v_{2y} = (1.00 \text{ kg})(20 \text{ m/s}) = 20 \text{ kg·m/s}$

Hence the magnitude MV of the momentum of man + snowball after the collision is

$$MV = \sqrt{(MV_x)^2 + (MV_y)^2} = \sqrt{(30 \text{ kg·m/s})^2 + (20 \text{ kg·m/s})^2} = 36 \text{ kg·m/s}$$

and the corresponding final speed is

$$V = \frac{MV}{M} = \frac{36 \text{ kg·m/s}}{61 \text{ kg}} = 0.59 \text{ m/s}$$

We can specify the direction in which the man + snowball combination moves after the collision in terms of the angle θ between the $+y$-direction (which is north) and MV:

$$\tan \theta = \frac{MV_x}{MV_y} = \frac{30 \text{ kg·m/s}}{20 \text{ kg·m/s}} = 1.5$$

$$\theta = 56°$$

Thus man + snowball move in a direction 56° to the east of north (Fig. 5.13).   ◆

Frozen
pond

Before collision

After collision

Figure 5.12

Figure 5.13

5.6   ELASTIC COLLISIONS

They conserve kinetic energy.

In an elastic collision no kinetic energy is lost. Suppose we have an object of initial speed v_1 that collides elastically with another object of initial speed v_2. Both objects move along the same straight line before and after the collision. If the speeds of the objects after the collision are, respectively, v_1' and v_2', then

$$v_1 - v_2 = -(v_1' - v_2') \qquad (5.8)$$

Initial relative velocity = $-$(final relative velocity)

This equation follows from the conservation of both KE and momentum (see Problem 57). What it means is that the effect of the collision is to reverse the direction of the relative velocity without changing its magnitude. Thus *the relative speed with which the objects approach each other equals the relative speed with which they move apart afterward.*

◆   **E X A M P L E   5.6** _____

A ball rolling on a level table strikes head-on an identical ball that is at rest, as in Fig. 5.14. What is the result of the collision?

SOLUTION   Here $v_2 = 0$, so from Eq. (5.8)

$$v_1 = v_2' - v_1'$$

From conservation of momentum, since the balls have the same mass m,

$$m_1 v_1 + m_2 v_2 = m_1 v_1' + m_2 v_2'$$
$$mv_1 + 0 = mv_1' + mv_2'$$
$$v_1 = v_1' + v_2'$$

Setting equal these two formulas for v_1 gives

$$v_2' - v_1' = v_1' + v_2'$$
$$-2v_1' = 0$$
$$v_1' = 0$$

The first ball has no speed after the collision, which means it comes to a stop. For the final speed of the second ball, which was originally at rest, we can use either of the formulas for v_1 with $v_1' = 0$. Thus

$$v_1 = v_2' - v_1' = v_2' - 0 = v_2'$$

The second ball begins to move with the speed v_1. The two balls have exchanged their initial velocities.

Figure 5.14

When a rolling ball makes a head-on collision with an identical stationary ball, the first ball stops and the second begins moving with the first's initial velocity.

◆

(a)

(b)

Figure 5.15

The behavior of this arrangement of suspended steel balls is determined by the conservation of both momentum and kinetic energy.

The results of the preceding problem help us to understand the operation of the toy shown in Fig. 5.15, which consists of a number of identical steel balls suspended by strings. When one ball at the left is pulled out and released, it swings to strike the row of stationary balls, and one ball at the right swings out in response. Why not two balls, or indeed all the others?

The answer is that both momentum and KE can be conserved *only* if a single ball swings out on the right. If two balls were to swing out, their joint mass would be twice that of the ball on the left, and to conserve momentum their initial speeds would have to be half that of the ball on the left. But the combined kinetic energies of the two balls would then be half that of the ball on the left:

	One ball on left		Two balls on right	
Momentum:	mv	$=$	$m\left(\dfrac{v}{2}\right) + m\left(\dfrac{v}{2}\right)$	$= mv$
Kinetic energy:	$\tfrac{1}{2}mv^2$	$\neq$	$\tfrac{1}{2}m\left(\dfrac{v}{2}\right)^2 + \tfrac{1}{2}m\left(\dfrac{v}{2}\right)^2$	$= \tfrac{1}{4}mv^2$

Only if half the initial KE is lost can two balls swing out. The same reasoning shows why, if two balls are pulled out at the left and released, as in Fig. 5.15(b), two balls at the right will swing out after the collision, and so on.

What about the case when more than half the balls are pulled out to the left? In that event the balls at the left cannot give up all their momentum and KE to the balls at the right, and so several of the pulled-out balls continue to move to the right after the collision.

5.7   ENERGY TRANSFER

It is a maximum when the colliding objects have the same mass.

A moving object strikes a stationary one, and as a result the second object is set in motion. What is the mass ratio between the two that will cause the struck object to have the highest possible speed after the impact? What mass ratio will lead to the greatest transfer of energy to the struck object? These are not questions of abstract interest only but are closely connected with a variety of actual problems that range from the design of golf clubs to the design of nuclear reactors.

For simplicity we will confine ourselves to elastic head-on collisions in which both objects move along the same straight line. Let us consider an object of mass m_1 and initial speed v_1 that strikes a stationary object of mass m_2, after which their respective speeds are v_1' and v_2'. From conservation of momentum,

$$m_1 v_1 = m_1 v_1' + m_2 v_2'$$
$$m_1 (v_1 - v_1') = m_2 v_2' \tag{5.9}$$

From Eq. (5.8), since $v_2 = 0$,

$$v_1 = v_2' - v_1' \tag{5.10}$$

Combining Eqs. (5.9) and (5.10) gives for the final speed of m_1

$$v_1' = \frac{m_1 - m_2}{m_1 + m_2} v_1 \tag{5.11}$$

and for the final speed v_2' of m_2

$$v_2' = \frac{2m_1}{m_1 + m_2} v_1 \tag{5.12}$$

Some interesting general conclusions follow from these formulas (Fig. 5.16):

1. When m_1 is less than m_2, v_1' is opposite to v_1: The lighter object bounces off the heavier one. A ball striking a wall is an extreme example. Here, since m_2 is virtually infinite compared with m_1, $v_1' = -v_1$.
2. When $m_1 = m_2$, $v_1' = 0$ and $v_2' = v_1$. The colliding object stops, and the target one moves off with the same speed. This is the case considered in the preceding section.
3. When m_1 is greater than m_2, as in the case of a tennis serve, the colliding object continues in the same direction after the impact but with reduced speed while

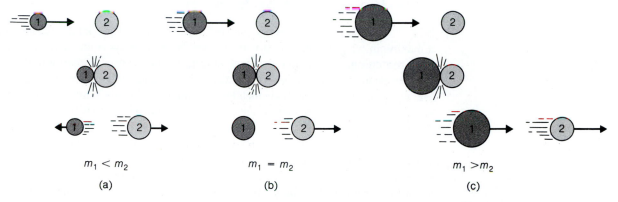

$m_1 < m_2$ (a) $m_1 = m_2$ (b) $m_1 > m_2$ (c)

Figure 5.16

How the effects of an elastic head-on collision with a stationary target depend on the relative masses of the two objects.

the target object moves ahead of it at a faster pace. When m_1 is much greater than m_2, the colliding object loses little speed while the target one is given a speed nearly twice v_1.

The ratio between the kinetic energy KE_2' transferred to the initially stationary object and the kinetic energy KE_1 of the colliding object can be found with the help of Eq. (5.12):

$$\frac{\text{Transferred KE}}{\text{Original KE}} = \frac{KE_2'}{KE_1} = \frac{\frac{1}{2}m_2 v_2'^2}{\frac{1}{2}m_1 v_1^2} = \frac{4m_1 m_2}{(m_1 + m_2)^2} = \frac{4(m_2/m_1)}{(1 + m_2/m_1)^2} \qquad (5.13)$$

This formula is plotted in Fig. 5.17. Evidently the transfer of energy is a maximum for $m_1 = m_2$, when *all* the energy of m_1 is given to m_2. This is the situation illustrated in Fig. 5.16(b). We note that, for a given mass ratio m_2/m_1, it does not matter which object was initially moving and which object was at rest.

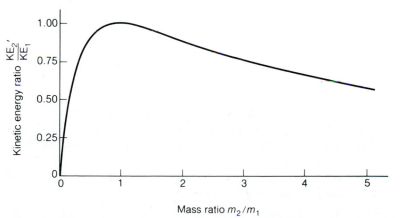

Figure 5.17

Energy transfer in an elastic head-on collision between a moving object and a stationary one.

Impedance Matching

The fact that the energy transfer in a collision of the kind described here is a maximum when the objects involved have the same mass is an example of **impedance matching. Impedance** is a general term for opposition to energy transfer. Here it corresponds to inertia, but the concept of impedance has a broader significance. We will encounter impedance matching again later in this book when we look at energy transfer in wave motion and in direct-current and alternating-current electric circuits.

From Eq. (5.13) it would seem that the best mass for the head of a golf club would be the same as that of a golf ball, in order that all the energy of the club be given to the ball (assuming an elastic collision). In this case $v_2' = v_1$. However, according to Eq. (5.12) the greater the mass m_1 of the clubhead, the more the ball's speed v_2' will exceed v_1, up to a limit of $2v_1$. The trouble with a heavy clubhead is twofold: It is hard to swing a heavy golf club as fast as a light one, and the more m_1 exceeds m_2, the smaller is the proportion of KE that is transferred to the ball—the extra effort does not provide a commensurate return. Experience has led golfers to use clubheads whose masses are typically about four times the 46-g mass of a golf ball where maximum distance is required.

5.8   COEFFICIENT OF RESTITUTION

How much bounce.

If a collision is not perfectly elastic, the loss of some of the original KE means that the relative speed of the objects involved after the collision will be less than their relative speed of approach. The ratio between the relative speeds is called the **coefficient of restitution,** symbol e:

$$e = \frac{v_2' - v_1'}{v_1 - v_2} \qquad \textit{Coefficient of restitution} \quad (5.14)$$

$$\text{Coefficient of restitution} = \frac{\text{relative speed after collision}}{\text{relative speed before collision}}$$

In a perfectly elastic collision, $e = 1$. In a perfectly inelastic collision, when the objects stick together and $v_2' = v_1'$, $e = 0$; the KE loss here is the maximum possible. A coefficient of restitution always applies jointly to the colliding objects, not to either one of them by itself.

When the coefficient of restitution is taken into account in a collision between a moving object and a stationary one,

$$v_2' = \frac{(1 + e)m_1}{m_1 + m_2} v_1$$ (5.15)

$$\frac{KE_2'}{KE_1 \cdot} = \frac{(1 + e)^2 m_1 m_2}{(m_1 + m_2)^2} = \frac{(1 + e)^2 (m_2/m_1)}{(1 + m_2/m_1)^2}$$ (5.16)

In the case of a golf club striking a golf ball, typical values might be $m_1 = 200$ g, $m_2 = 46$ g, and $e = 0.7$, from which we find that the ball moves off with a speed 38% greater than the speed of the clubhead and carries with it 44% of the clubhead's original energy. If the collision were perfectly elastic, $e = 1$ and these figures would be 63% and 61% respectively.

◆   **E X A M P L E   5.7** _____

A tennis ball dropped on the floor from a height of 100 cm rebounds to a height of 60 cm. Find the coefficient of restitution.

SOLUTION   Because the floor does not move, $v_2 = v_2' = 0$. Let us consider upward as $+$ and downward as $-$. The speed of the ball when dropped from the height $h = 100$ cm is $v_1 = -\sqrt{2gh}$. The ball's speed v_1' when it rebounds from the floor is given by $v_1' = +\sqrt{2gh'}$, where $h' = 60$ cm is the height it reaches. Hence, from Eq. (5.14),

$$e = \frac{v_2' - v_1'}{v_1 - v_2} = \frac{0 - v_1'}{v_1 - 0} = \frac{-\sqrt{2gh'}}{-\sqrt{2gh}} = \sqrt{\frac{h'}{h}}$$

This is a general conclusion for something dropped from a height h that rebounds to the new height h' from a level surface (Fig. 5.18). Here,

$$e = \sqrt{\frac{h'}{h}} = \sqrt{\frac{60 \text{ cm}}{100 \text{ cm}}} = \sqrt{0.6} = 0.77$$

◆

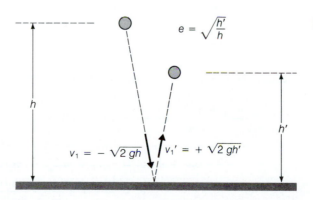

$$e = \sqrt{\frac{h'}{h}}$$

Figure 5.18

Figure 5.19

◆   **E X A M P L E**   **5.8**

A 2.0-kg ball moving at 2.0 m/s to the right collides head-on with a 1.0-kg ball moving at 3.0 m/s to the left (Fig. 5.19). The coefficient of restitution is 0.70. Find the speeds and directions of the balls after the collision.

SOLUTION   Let us call motion to the right $+$ and motion to the left $-$, so that

$$m_1 = 2.0\,\text{kg} \qquad m_2 = 1.0\,\text{kg}$$
$$v_1 = +2.0\,\text{m/s} \qquad v_2 = -3.0\,\text{m/s}$$
$$v_1' = ? \qquad v_2' = ?$$

From conservation of momentum,

$$\text{Momentum before} = \text{momentum after}$$
$$m_1v_1 + m_2v_2 = m_1v_1' + m_2v_2'$$

Since we have two unknown quantities, the final speeds v_1' and v_2', we need another equation relating them. From Eq. (5.14),

$$\text{Coefficient of restitution} = e = \frac{v_2' - v_1'}{v_1 - v_2}$$

Solving both equations for v_2' gives

$$v_2' = \frac{m_1v_1 + m_2v_2 - m_1v_1'}{m_2}$$
$$v_2' = e(v_1 - v_2) + v_1'$$

Because both of these formulas equals v_2', they must be equal to each other. Hence

$$\frac{m_1v_1 + m_2v_2 - m_1v_1'}{m_2} = e(v_1 - v_2) + v_1'$$
$$m_1v_1 + m_2v_2 - m_1v_1' = em_2\,(v_1 - v_2) + m_2v_1'$$
$$(m_1 + m_2)v_1' = m_1v_1 + m_2v_2 - em_2(v_1 - v_2)$$
$$v_1' = \frac{m_1v_1 + m_2v_2 - em_2(v_1 - v_2)}{m_1 + m_2} = -0.83\,\text{m/s}$$

With this value of v_1' we can find v_2':

$$v_2' = e(v_1 - v_2) + v_2' = 2.7 \text{ m/s}$$

The first ball reverses its direction and moves off to the left at 0.83 m/s, which is slower than its original speed. The second ball also reverses its direction and moves off to the right at 2.7 m/s, also slower than its original speed. As we expect when e is less than 1, some of the initial KE of the system has been lost in the collision.

To check these results, we can calculate the total momentum of the system before and after the collision:

$$\text{Before: } m_1v_1 + m_2v_2 = (2.0 \text{ kg})(2.0 \text{ m/s}) + (1.0 \text{ kg})(-3.0 \text{ m/s})$$
$$= +1.0 \text{ kg} \cdot \text{m/s}$$
$$\text{After: } m_1v_1' + m_2v_2' = (2.0 \text{ kg})(-0.83 \text{ m/s}) + (1.0 \text{ kg})(2.7 \text{ m/s})$$
$$= +1.0 \text{ m/s} \qquad \blacklozenge$$

Important Terms

The **linear momentum** of an object is the product of its mass and velocity. Linear momentum is a vector quantity having the direction of the object's velocity.

The **impulse** of a force is the product of the force and the time during which it acts. Impulse is a vector quantity having the direction of the force. When a force acts on an object that is free to move, its change in momentum equals the impulse given it by the force.

The law of **conservation of momentum** states that when the vector sum of the external forces acting on a system of particles equals zero, the total linear momentum of the system remains constant.

The **thrust** of a rocket is the force that results from the expulsion of exhaust gases.

In an **elastic collision** kinetic energy is conserved. In an **inelastic collision** KE is not conserved. A **completely inelastic collision** is one in which the objects stick together on impact, which results in the maximum possible KE loss. Linear momentum is conserved in all collisions.

The **coefficient of restitution** is the ratio between the relative speeds of two colliding objects after and before they collide. It equals 1 for an elastic collision and 0 for a completely inelastic collision.

Important Formulas

Linear momentum: $\mathbf{p} = m\mathbf{v}$

Impulse and momentum change: $\mathbf{F}\,\Delta t = \Delta(m\mathbf{v})$

Thrust: $\mathbf{F} = \mathbf{v}\dfrac{\Delta m}{\Delta t}$

Elastic collision: $v_1 - v_2 = -(v_1' - v_2')$

Coefficient of restitution: $e = \dfrac{v_2' - v_1'}{v_1 - v_2}$

MULTIPLE CHOICE

1. An object at rest may have
 a. velocity
 b. momentum
 c. kinetic energy
 d. potential energy

2. An object in motion need not have
 a. velocity
 b. momentum
 c. kinetic energy
 d. potential energy

3. An object that has momentum must also have
   a. acceleration   b. impulse
   c. kinetic energy   d. potential energy

4. Momentum is most closely related to
   a. kinetic energy   b. potential energy
   c. impulse   d. power

5. Momentum can be expressed in
   a. N/s   b. N·s
   c. N·m   d. N·m/s

6. When the velocity of a moving object is doubled,
   a. its acceleration is doubled
   b. its momentum is doubled
   c. its kinetic energy is doubled
   d. its potential energy is doubled

7. An iron sphere of mass 30 kg has the same diameter as an aluminum sphere of mass 10.5 kg. The spheres are simultaneously dropped from a cliff. When they are 10 m from the ground, they have identical
   a. accelerations   b. momenta
   c. potential energies   d. kinetic energies

8. If a shell fired from a cannon explodes in midair,
   a. its total momentum increases
   b. its total momentum decreases
   c. its total kinetic energy increases
   d. its total kinetic energy decreases

9. What never changes when two or more objects collide is
   a. the momentum of each one
   b. the kinetic energy of each one
   c. the total momentum of all the objects
   d. the total kinetic energy of all the objects

10. In an elastic collision
    a. momentum is conserved but not KE
    b. KE is conserved but not momentum
    c. momentum and KE are both conserved
    d. neither momentum nor KE is conserved

11. In an inelastic collision
    a. momentum is conserved but not KE
    b. KE is conserved but not momentum
    c. momentum and KE are both conserved
    d. neither momentum nor KE is conserved

12. In a perfectly elastic collision between two objects, their relative speed after the collision is
    a. 0
    b. less than their relative speed before the collision
    c. equal to their relative speed before the collision
    d. more than their relative speed before the collision

13. A ball whose momentum is **p** strikes a wall and bounces off. The change in the ball's momentum is
    a. 0   b. **p**
    c. 2**p**   d. −2**p**

14. Object *A* strikes object *B*, which is initially at rest. The maximum transfer of energy from *A* to *B* occurs when m_A is
    a. less than m_B b. equal to m_B
    c. greater than m_B
    d. any of the above, depending on the speed of *A*

15. In the collision in Multiple Choice 14, object *B* will acquire the most speed when m_A is
    a. less than m_B b. equal to m_B
    c. greater than m_B
    d. any of the above, depending on the speed of *A*

16. If a rocket of initial mass *m* is to rise from its launching pad, its initial thrust must exceed
    a. $\frac{1}{2}mg$ b. *mg*
    c. 2*mg*   d. $\frac{1}{2}mg^2$

17. The average momentum of a 70-kg runner who covers 400 m in 50 s is
    a. 8.75 kg·m/s   b. 57 kg·m/s
    c. 560 kg·m/s   d. 5488 kg·m/s

18. A 1280-lb white horse is cantering at 20 ft/s. Its linear momentum is
    a. 400 slug·ft/s   b. 800 slug·ft/s
    c. 8000 slug·ft/s   d. 25,600 slug·ft/s

19. A 60-lb girl and a 50-lb boy face each other on frictionless roller skates. The girl pushes the boy, who moves away at a speed of 4.0 ft/s. The girl's speed is
    a. 2.1 ft/s   b. 3.3 ft/s
    c. 4.0 ft/s   d. 4.8 ft/s

20. An astronaut whose total mass is 100 kg ejects 1.0 g of gas from his propulsion pistol at a speed of 50 m/s. His recoil speed is
    a. 0.5 mm/s   b. 5 mm/s
    c. 5 cm/s   d. 50 cm/s

21. A 200-g ball moving at 5.0 m/s strikes a wall perpendicularly and rebounds elastically at the same speed. The impulse given to the wall is
    a. 1 N·s   b. 2 N·s
    c. 2.5 N·s   d. 10 N·s

22. A stationary 60-g tennis ball is struck by a racket that exerts an average force of 300 N on it for a time of 0.0050 s. The KE of the ball after the impact is
    a. 0.019 J   b. 0.068 J
    c. 1.5 J   d. 19 J

23. An 800-kg car moving at 80 km/h overtakes a 1200-kg car moving at 40 km/h in the same direction. If the two cars

stick together, the wreckage has an initial speed of
- a   8 km/h
- b.   40 km/h
- c.   56 km/h
- d.   60 km/h

24. The cars in Multiple Choice 23 are moving in opposite directions and collide head-on. If they stick together, the wreckage now has an initial speed of
- a.   8 km/h
- b.   40 km/h
- c.   56 km/h
- d.   60 km/h

25. The total KE lost by the cars in Multiple Choice 24 is
- a.   0
- b.   less than that lost by the cars in Multiple Choice 23
- c.   the same as that lost by the cars in Multiple Choice 23
- d.   more than that lost by the cars in Multiple Choice 23

26. A 1200-kg car moving north at 50 km/h collides with a 1600-kg car moving east at 30 km/h. The cars stick together,

and the wreckage moves off at an initial speed of
- a.   27 km/h
- b.   38 km/h
- c.   39 km/h
- d.   58 km/h

27. The direction in which the wreckage of the preceding collision moves off is
- a.   39° E of N
- b.   45° E of N
- c.   51° E of N
- d.   54° E of N

28. The additional force needed to maintain the speed of a pickup truck at a constant 20 m/s when rain begins to fall and accumulates in its back at a rate of 3.0 kg/min is
- a.   0
- b.   1 N
- c.   10 N
- d.   60 N

29. A ball dropped from a height of 3.0 m bounces up to a height of 1.0 m. The coefficient of restitution is
- a.   0.33
- b.   0.58
- c.   1.7
- d.   3.0

QUESTIONS

1. Is it possible for an object to have more kinetic energy but less momentum than another object? Less kinetic energy but more momentum?

2. When the momentum of an object is doubled in magnitude, what happens to its kinetic energy?

3. When the kinetic energy of an object is halved in magnitude, what happens to its momentum?

4. When a rocket explodes in midair, how are its total momentum and total kinetic energy affected?

5. How is the principle of conservation of linear momentum related to the definition of mass given in Chapter 3 and to Newton's first law of motion? In what way does this principle go beyond the definition of mass and the first law of motion?

6. (a) When an object at rest breaks up into two parts that fly off, must they move in exactly opposite directions? (b) When

a moving object strikes a stationary one and the two do not stick together, must they move off in exactly opposite directions?

7. A railway car is at rest on a frictionless track. A man at one end of the car walks to the other end. (a) Does the car move while he is walking? (b) If so, in which direction? (c) What happens when the man comes to a stop?

8. An empty coal car coasts at a certain speed along a level railroad track without friction. (a) It begins to rain. What happens to the speed of the car? (b) The rain stops, and the collected water gradually leaks out. What happens to the speed of the car now?

9. Give an example of a collision in which all the KE of the participating objects is lost. What happens to their total momentum?

EXERCISES

5.1   Linear Momentum

1. Find the momentum of a 50-g bullet whose KE is 250 J.

2. Find the momentum of a giraffe galloping at 20 km/h whose KE is 6.0 kJ.

3. Find the momentum of a 3200-lb car moving at 30 mi/h.

4. Find the momentum of a 0.040-lb bullet whose kinetic energy is 1600 ft · lb.

5. Car A has a mass of 1000 kg and is moving at 60 km/h. Car B has a mass of 2000 kg and is moving at 30 km/h. Compare the kinetic energies and momenta of the two cars.

5.2   Impulse

6. A 1000-kg car strikes a tree at 30 km/h and comes to a stop in 0.15 s. Find its initial momentum and the average force on the car while it is being stopped.

7. A 160,000-kg DC-8 airplane is flying at 870 km/h. (a) Find its momentum. (b) If the thrust its engines develop is 340 kN, how much time is needed for the airplane to reach this speed starting from rest? Neglect air resistance, changes in altitude, and the fuel consumed by the engines.

8. A $\frac{1}{3}$-lb baseball reaches a batter with a speed of 75 ft/s.

After it has been struck, it leaves the bat at 100 ft/s in the opposite direction. If the ball was in contact with the bat for 1.0 ms, find the average force exerted on it during this period.

9. A 4.0-kg seagull flies at 15 m/s into the windshield of an airplane flying in the opposite direction at 180 m/s. If the impact lasts 1.0 ms, find the average force on the windshield.

10. Water emerging from a hose at a rate of 2.0 L/s and a speed of 8.0 m/s strikes a person. If the water loses all its momentum on impact, find the force on the person. (The mass of 1 L of water is 1 kg.)

11. A 2.0-kg pail rests on a scale. Water is poured into the pail from a height of 1.5 m at a rate of 100 g/s. (a) What is the reading on the scale (in kilograms) 1.0 min after the pouring began? (b) The pouring stops just after 1.0 min. What is the reading on the scale later?

12. Wheat falls vertically at 50 kg/s on a horizontal conveyor belt whose speed is 4.0 m/s. How much force is needed to keep the speed of the belt constant? How much power is needed to operate the belt?

5.3   Conservation of Momentum

13. A hunter has a rifle that can fire 60-g bullets with a speed of 900 m/s. A 40-kg leopard springs at him at 10 m/s. How many bullets must the hunter fire into the leopard in order to stop him in his tracks?

14. A hunter in a rowboat loses the oars and decides to set the boat in motion by firing his rifle astern five times. The total weight of the boat and its contents is 300 lb, the weight of each bullet is 0.040 lb, and the speed of the bullets is 2000 ft/s. Assume water resistance is negligible. How far does the boat go in 10 min?

15. A 50-kg girl throws a 5.0-kg pumpkin at 10 m/s to a 50-kg boy, who catches it. If both are on a frictionless frozen lake, how fast does each of them move backward?

16. A 150-lb astronaut is drifting forward in an orbiting space shuttle at 1.00 in/s. A fellow astronaut throws her a 0.50-lb orange, which she catches. If she then starts to move backward at 0.50 in/s, what was the speed of the orange?

17. An astronaut outside an orbiting spacecraft uses a pistol that ejects a gas in order to maneuver in space. Suppose the astronaut and her space suit have a total mass of 100 kg and the pistol ejects 12 g of gas per second at a speed of 650 m/s. How long should the astronaut operate the pistol in order to have a speed of 1.0 m/s?

18. A spacecraft moving at 10.0 km/s breaks apart into two pieces of equal mass that continue moving in the original direction. If the speed of one of the pieces is 4.0 km/s, what is the speed of the other one?

19. A spacecraft moving at 10.0 km/s breaks apart into two pieces of equal mass, one of which moves off at 4.0 km/s in a direction opposite to the original direction. Find the speed and direction of the other piece.

20. An object at rest breaks up into two fragments that fly apart. Express the ratio of their kinetic energies in terms of their masses.

21. A 70-kg man and a 50-kg woman are in a 60-kg rubber dinghy. The man dives into the water with a horizontal speed of 3.0 m/s. If his swimming speed is 1.0 m/s, can he return to the dinghy? If not, can the woman change the dinghy's motion enough by diving off it at 3.0 m/s in the opposite direction? Could she then return to the dinghy herself if her swimming speed were also 1.0 m/s?

5.4   Rocket Propulsion

22. The motor of a 2000-lb rocket standing vertically on the ground is being tested. Fuel is being burned at the rate of 10 lb/s with an exhaust speed of 10,000 ft/s. Find the force needed to hold the rocket down.

23. The motors of a spacecraft provide a total thrust of 1.8 MN at takeoff. If the exhaust speed is 2.5 km/s, at what rate is fuel being consumed?

24. A rocket fired from the earth's surface ejects 2% of its mass at a speed of 2.0 km/s in the first second of its flight. Find the initial acceleration of the rocket.

25. A certain rocket consumes 25 kg of fuel per second, and its exhaust has a speed of 3000 m/s. If the initial mass of the rocket is 5000 kg of which 3500 kg is fuel, what are its initial and final accelerations? Ignore the decrease in g with altitude.

26. A rocket launched vertically from the earth has a mass of 2500 kg at takeoff. (a) If the rocket's initial acceleration is 0.50 g and it is using fuel at the rate of 15 kg/s, what is the speed of the exhaust gases? (b) What is the acceleration of the rocket 1 min later? Ignore the decrease in g with altitude.

5.5   Inelastic Collisions

27. A neutron of mass 1.67×10^{-27} kg and speed 1.00×10^5 km/s collides with a stationary deuteron of mass 3.34×10^{-27} kg. The two particles stick together. What is the speed of the composite particle (called a triton)?

28. A 30-kg girl who is running at 3.0 m/s jumps on a stationary 10-kg sled on a frozen lake. How fast does the sled then move?

29. A 3.0-g bullet moving at 4.0 km/s strikes an 8.0-kg wooden block resting on a frictionless table. The bullet passes through the block and comes out with a speed of 0.20 km/s. What is the speed of the block?

30. A moving stone strikes a stationary lump of clay whose mass is three times that of the stone and becomes embedded in it. What proportion of the stone's initial KE is lost?

31. An 0.50-kg stone moving at 4.0 m/s overtakes a 4.0-kg lump of clay moving at 1.0 m/s. The stone becomes embedded in the clay. (a) What is the speed of the composite body after the collision? (b) How much KE is lost in the collision?

32. A 1000-kg car moving east at 80 km/h collides head-on with a 1500-kg car moving west at 40 km/h, and the two cars stick together. (a) Which way does the wreckage move and with what initial speed? (b) How much KE is lost in the collision?

33. A 1000-kg car moving east at 80 km/h overtakes a 1500-kg car moving east at 40 km/h and collides with it. The two cars stick together. (a) What is the initial speed of the wreckage? (b) How much KE is lost in the collision?

5.6 Elastic Collisions

34. A neutron of mass 1.67×10^{-27} kg and speed 1.00×10^5 m/s collides head-on with a stationary deuteron of mass 3.34×10^{-27} kg. The particles do not stick together, and the deuteron moves off at 6.67×10^4 m/s. What is the speed of the neutron? Is the collision elastic?

35. The 176-g head of a golf club is moving at 45 m/s when it strikes a 46-g golf ball and sends it off at 65 m/s. Find the final speed of the clubhead after the impact, assuming that the mass of the club's shaft can be neglected.

36. A billiard ball moving at 4.00 m/s strikes another billiard ball at rest and moves off at 3.46 m/s in a direction 30° from its original line of motion. What is the velocity of the target ball?

5.7 Energy Transfer

37. A 10-kg iron ball rolling at 2.0 m/s on a horizontal surface strikes a 1.0-kg wooden ball of the same size that is at rest. What is the speed of each ball after the collision? What proportion of the iron ball's original KE was transferred to the wooden ball?

38. A 46-g golf ball flies off at 1.5 times the speed of the clubhead that struck it. What was the mass of the clubhead? What proportion of the clubhead's initial KE was transferred to the ball?

5.8 Coefficient of Restitution

39. A steel ball bearing is dropped on a steel plate from a height of 2.0 m. If the coefficient of restitution is 0.97, find the height to which the ball rebounds.

40. A rubber ball is dropped on the ground from a height of 150 cm and on its second rebound reaches a height of 50 cm. Find the coefficient of restitution.

41. A 5.0-kg ball moving at 6.0 m/s strikes a stationary 3.0-kg ball and continues in the same direction at 2.0 m/s. (a) Find the speed and direction of the 3.0-kg ball after the collision. (b) What is the coefficient of restitution?

42. A 3.0-kg ball moving at 6.0 m/s strikes a stationary 5.0-kg ball and rebounds in the opposite direction at 1.0 m/s. (a) Find the speed and direction of the 5.0-kg ball after the collision. (b) What is the coefficient of restitution?

43. A 12-lb ball moving at 20 ft/s strikes a stationary 30-lb ball that then moves off at 8.0 ft/s. (a) Find the speed and direction of the 12-lb ball after the collision. (b) What is the coefficient of restitution?

PROBLEMS

44. Some years ago the record distance for the shotput was 22 m and that for the discus throw was 68 m. The mass of the shot was 7.25 kg and that of the discus was 2.00 kg. Neglecting air resistance and assuming that each athlete hurled his projectile at the optimal angle, compare the initial KEs and momenta of the projectiles. What does this comparison suggest?

45. A certain howitzer has a barrel 7.6 m long and fires a 65-kg projectile 20 cm in diameter at a muzzle speed of 400 m/s. Use three different ways, each based on an important physical principle, to find the average force on the projectile while it is in the howitzer barrel.

46. The elastic potential energy of the compressed spring of Fig. 3.3 is 12 J,. and the masses of the two objects are, respectively, $m_A = 2.0$ kg and $m_B = 1.0$ kg. What speed does each object have when the string is cut?

47. A certain rocket's motor consumes 100 lb of fuel per second, which it exhausts at 9000 ft/s. If the initial weight of the rocket is 10.00 tons of which 7.00 tons is fuel, find its initial and final accelerations.

48. Coal falls vertically at 100 kg/s on a horizontal conveyor belt whose speed is 4.0 m/s. Compare the power needed to keep the speed of the belt constant with the rate at which the coal gains KE. If there is any difference, account for it.

49. An object at rest explodes into three fragments of equal mass. Two of the fragments fly off in directions 60° apart with equal speeds of 10 m/s. What are the speed and direction of the third fragment?

50. A 1200-kg car traveling east at 30 km/h collides with an 1800-kg car traveling north at 20 km/h. The cars stick together after the collision. What is the velocity (magnitude and direction) of the wreckage?

51. An 0.50-kg stone moving north at 4.0 m/s collides with a 4.0-kg lump of clay moving west at 1.0 m/s. The stone becomes embedded in the clay. What is the velocity (magnitude and direction) of the composite body after the collision?

52. A 1.0-kg ball moving at 2.0 m/s in the $+x$-direction collides head-on with a 2.0-kg ball moving at 3.0 m/s in the $-x$-direction. The coefficient of restitution is 0.80. Find the velocities of the balls after the collision.

53. A 1.0-kg ball moving at 5.0 m/s in the $+x$-direction collides head-on with a 2.0-kg ball moving at 5.0 m/s in the $-x$-direction. The coefficient of restitution is 0.70. Find the velocities of the balls after the collision.

54. A 500-kg cannon on a wheeled carriage is at the foot of a 25° slope with its barrel horizontal. The cannon fires a 30-kg shell at 300 m/s in a direction opposite to the slope. (a) What is the cannon's initial speed up the slope? (b) How far along the slope does the cannon travel before it comes to a stop and starts to roll back down? Disregard friction.

55. A ballistic pendulum consists of a wooden block of mass M suspended by long strings from the ceiling. A bullet of mass m and speed v is fired horizontally into the block, which swings away until its height is the amount h above its original height. (a) Find a formula that gives v in terms of g and the readily measurable quantities m, M, and h. (b) A 5.0-g bullet is fired into the 2.0-kg block of a ballistic pendulum, which rises by 10 cm. Find the bullet's speed.

Figure 5.20

Problem 56.

56. As shown in Fig. 5.20, a magnet A attached to a string R long is pulled out to one side so the string is horizontal. When the magnet is let go, it swings downward and strikes an iron cube B of the same mass that is resting on a frictionless surface. The two stick together and swing upward on the other side. What is the maximum value of θ, the angle between the string and the vertical?

57. Verify Eq. (5.8) in the following way. Set equal the initial and final KEs of the two objects and use the algebraic relationship $a^2 - b^2 = (a + b)(a - b)$ to obtain $m_1(v_1 + v_1')$ $(v_1 - v_1') = m_2(v_2' + v_2)(v_2' - v_2)$. Then set equal the initial and final momenta of the objects to obtain $m_1(v_1 - v_1') = m_2(v_2' - v_2)$. From these equations obtain Eq. (5.8).

58. A billiard ball moving at 4.0 m/s strikes a pair of billiard balls at rest. One of the latter moves off at 2.0 m/s in a direction 60° from the line of motion of the first ball, and the other moves off at 3.0 m/s in a direction 30° from the line of motion of the first ball on the other side. What is the velocity of the first ball after the collision?

59. A billiard ball at rest is struck by another ball of the same mass whose speed is 5.0 m/s. After an elastic collision the striking ball goes off at an angle of 40° with respect to its original direction of motion and the struck ball goes off at an angle of 50° with respect to this direction. Find the final speeds of both balls.

60. A billiard ball at rest is struck by another billiard ball of the same mass whose speed is 4.0 m/s. After an elastic collision the striking ball goes off at an angle of 25° with respect to its original direction of motion. Find the angle the struck ball makes with this direction and the final speeds of both balls.

Answers to Multiple Choice

1. d	9. c	17. c	25. d
2. d	10. c	18. b	26. a
3. c	11. a	19. b	27. a
4. c	12. c	20. a	28. b
5. b	13. d	21. b	29. b
6. b	14. b	22. d	
7. a	15. c	23. c	
8. c	16. b	24. a	

CIRCULAR MOTION AND GRAVITATION

6.1   **Centripetal Force**

6.2   **Centripetal Acceleration**

6.3   **Magnitude of Centripetal Force**

6.4   **Motion in a Vertical Circle**

6.5   **Gravitation**

6.6   **Earth Satellites**

6.7   **Weightlessness**

Almost everything in the natural world travels in a curved path. Often these paths are circles or very nearly circles. For instance, the orbits of the earth and the other planets about the sun are almost circular, as is the orbit of the moon around the earth. On a smaller scale a handy way to visualize an atom is to imagine its electrons circling around its central nucleus. Of course, circular motion is no novelty in our own experience—it is hard to think of any machine in which circular motion of some kind is not involved.

6.1 CENTRIPETAL FORCE

A curved path requires an inward pull.

Tie a ball to the end of a string and whirl it around your head, as in Fig. 6.1. What you will find is that your hand must pull on the string to keep the ball going in a circle. Let go of the string, and the ball flies off tangentially to the circle.

The force needed to make an object follow a curved path is called **centripetal force,** which means ''center-seeking force.'' Without it, circular motion cannot occur. In general:

Centripetal force is the force perpendicular to the velocity of an object moving along a curved path. The centripetal force is directed toward the center of curvature of the path.

An object that moves in a circle at constant speed is said to undergo **uniform circular motion.** As shown in Fig. 6.2, the centripetal force $\mathbf{F}_c$ on such an object always points to the center of the circle. This means that $\mathbf{F}_c$ is perpendicular to the direction of $\mathbf{v}$ at all times, so the centripetal force does no work on the object, and the object's speed v remains constant.

A centripetal force acts whenever rotational motion occurs, since such a force is needed to change the path of a particle from the straight line that it would normally follow to a curved path. Gravitation provides the centripetal forces that keep the planets moving around the sun and the moon around the earth. Friction between

Figure 6.1

When a ball is whirled at the end of a string, the tension in the string provides the centripetal force that keeps the ball moving in a circle.

Centripetal force

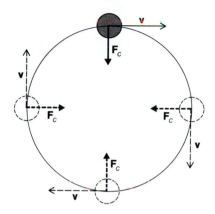

Figure 6.2

Even though the velocity **v** of an object
traveling in a circle at constant speed
has the same magnitude along its path,
the direction of **v** changes constantly.
The inward force that causes this change
in direction is called centripetal force, **F_c**.

Centripetal forces

Figure 6.3

The centripetal force exerted when a car rounds a curve on a
level road is provided by friction between its tires and the
road.

tires and road provides the centripetal force needed by a car in rounding a curve in a
level road (Fig. 6.3). If the tires are worn and the road is wet or icy, the frictional
force may be too small to permit the car to turn.

6.2 CENTRIPETAL ACCELERATION

It produces a change in direction without a change in speed.

How much centripetal force do we need to keep a given object moving in a
circle at a certain speed? To find out, we begin by asking what the object's acceleration
is. In the following derivation the object will be a particle, but the same arguments
and conclusion hold for the circular motion of the center of mass of an object of any
size.

In Fig. 6.4(a) a particle is shown traveling along a circular path of radius r
at the constant speed v. At $t = 0$ the particle is at the point A, where its velocity is
$\mathbf{v}_A$, and at $t = \Delta t$ the particle is at the point B, where its velocity is $\mathbf{v}_B$. The change
$\Delta \mathbf{v}$ in the particle's velocity in the time interval Δt is $\Delta \mathbf{v} = \mathbf{v}_B - \mathbf{v}_{A'}$ and its
acceleration is $\mathbf{a} = \Delta \mathbf{v}/\Delta t$.

The vector triangle whose sides are $-\mathbf{v}_A$, $\mathbf{v}_B$, and $\Delta \mathbf{v}$ is similar to the space
triangle whose sides are OA, OB, and s, as we can see from Fig. 6.4(c). Since v is
the constant speed of the particle, the magnitudes of $-\mathbf{v}_A$ and $\mathbf{v}_B$ are both v. Also,

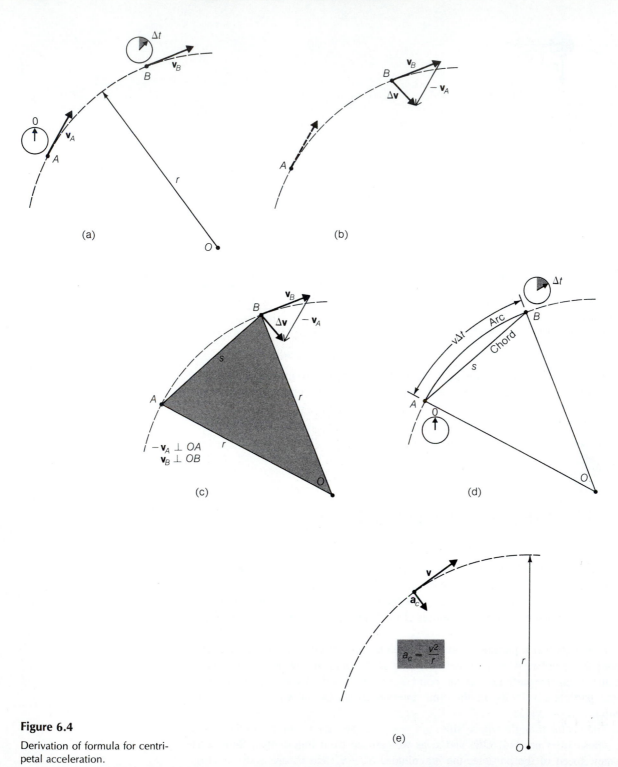

Figure 6.4

Derivation of formula for centripetal acceleration.

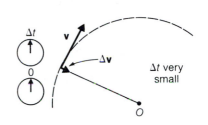

Δt large

Δt small

Δt very small

Figure 6.5

The centripetal acceleration of a particle in uniform circular motion points toward the center of the circle.

OA and OB are radii of the circle, so their lengths are both r. Corresponding sides of similar triangles are proportional, hence

$$\frac{\Delta v}{v} = \frac{s}{r} \quad \text{and} \quad \Delta v = \frac{vs}{r}$$

The distance the particle actually covers in going from A to B is the arc joining these points, the length of which is $v\,\Delta t$. The distance s, however, is the chord joining A and B, as in Fig. 6.4(d). We are finding the *instantaneous* acceleration of the particle, which means that A and B are very close together. In this case the chord and arc are equal. Hence

$$s = v\,\Delta t$$

and we have

$$\Delta v = \frac{v^2 \Delta t}{r}$$

Since the magnitude of the particle's acceleration is $a_c = \Delta v / \Delta t$,

$$a_c = \frac{v^2}{r} \qquad\qquad \text{Centripetal acceleration} \quad (6.1)$$

The inward **centripetal acceleration** of a particle in uniform circular motion is proportional to the square of its speed and inversely proportional to the radius of its path. See Fig. 6.4(e).

In Fig. 6.4(c) $\Delta\mathbf{v}$ does not quite point toward O, the center of the particle's circular path. When A and B are very close together, however, $\Delta\mathbf{v}$ *does* point toward O (Fig. 6.5). Because the acceleration $\mathbf{a}_c$ is an instantaneous acceleration, we only care about the case when Δt and hence s are extremely small. The direction of $\mathbf{a}_c$ is accordingly radially inward.

Period of Circular Motion

The time T needed by an object in uniform circular motion to complete an orbit is called its **period.** If r is the radius of the circle, its circumference is $2\pi r$ (Fig.

Figure 6.6

An object in uniform circular motion whose period is T has a speed of $2\pi r/T$.

6.6). The speed of the object is the distance it covers (here $2\pi r$) divided by the time needed (here T), so

$$v = \frac{2\pi r}{T} \qquad\qquad\qquad \textit{Speed in orbit of period } T \quad (6.2)$$

This formula is often useful in working out problems in uniform circular motion.

◆   **E X A M P L E   6.1** _____

The ball in Fig. 6.1 is whirled in a horizontal circle 80 cm in radius. If the ball makes two revolutions per second, what is its centripetal acceleration?

SOLUTION   The ball takes $T = 0.5$ s per orbit. Using Eq. (6.2) for the ball's speed gives, from Eq. (6.1), the centripetal acceleration

$$a_c = \frac{v^2}{r} = \left(\frac{2\pi r}{T}\right)^2 \left(\frac{1}{r}\right) = \frac{4\pi^2 r}{T^2} = \frac{(4\pi^2)\,(0.8 \text{ m})}{(0.5 \text{ s})^2} = 126 \text{ m/s}^2$$

which is nearly 13 times the acceleration of gravity g.   ◆

6.3   MAGNITUDE OF CENTRIPETAL FORCE

How much of an inward pull?

From the second law of motion we can see that the centripetal force $\mathbf{F}_c$ on an object of mass m in uniform circular motion is $\mathbf{F}_c = m\mathbf{a}_c$, which has the magnitude

$$F_c = \frac{mv^2}{r} \qquad\qquad\qquad \textit{Centripetal force} \quad (6.3)$$

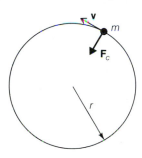

The centripetal force on an object in uniform circular motion is equal in magnitude to mv^2/r.

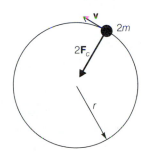

Doubling the mass doubles the required centripetal force.

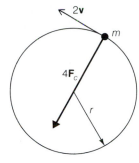

Doubling the speed, however, quadruples the required centripetal force.

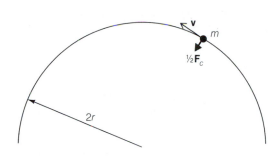

Doubling the radius of the circle halves the required centripetal force.

Figure 6.7

Centripetal force.

Figure 6.7 illustrates how F_c varies with m, v, and r. Now we know why a car rounding a curve is hard to control when it is heavy (large m), when it is going fast (large v means very large v^2), and when the curve is sharp (small r).

◆ **EXAMPLE 6.2** _____

(a) Find the centripetal force needed by a 1200-kg car to make a turn of radius 40 m at a speed of 25 km/h (16 mi/h). (b) Assuming the road is level, find the minimum coefficient of static friction between the car's tires and the road that will permit the turn to be made (Fig. 6.8).

SOLUTION   (a) The car's speed is

$$v = (25 \text{ km/h}) \left(0.278 \, \frac{\text{m/s}}{\text{km/h}} \right) = 7.0 \text{ m/s}$$

Accordingly the centripetal force is

$$F_c = \frac{mv^2}{r} = \frac{(1200 \text{ kg})(7.0 \text{ m/s})^2}{40 \text{ m}} = 1470 \text{ N} = 1.5 \text{ kN}$$

Figure 6.8

A centripetal force of 1470 N (330 lb) is needed by this car to make the turn shown. The corresponding coefficient of static friction between its tires and the road is $\mu_s = 0.125$.

(b) The frictional force is given by

$$F_f = \mu_s F_N$$

To find the coefficient of static friction μ_s, we substitute 1470 N for F_f and the car's weight of $w = mg$ for the normal force F_N. Thus

$$\mu_s = \frac{F_f}{F_N} = \frac{F_c}{mg} = \frac{1470 \text{ N}}{(1200 \text{ kg})(9.8 \text{ m/s}^2)} = 0.13$$

which is available under most driving conditions.

It is worth noting that the coefficient of friction required here does not depend on the car's mass, since

$$\mu_s = \frac{F_c}{mg} = \frac{mv^2/r}{mg} = \frac{v^2}{gr}$$

Because the required value of μ_s depends on v^2, high-speed turns on a level road can be dangerous. In this example, increasing the speed to 60 km/h (37 mi/h) increases the needed μ_s to 0.72, which is too much for a wet road, and the car would skid. ◆

Braking during a fast turn may be risky. Suppose a car is rounding a curve on a flat road at the highest speed possible without the tires slipping on the road. This means that the centripetal force on the car is the most that can be transferred to its tires from the road by static friction. If the driver now applies the brakes, an additional force is exerted on the tires at right angles to the centripetal force on them. The resultant of these two forces is now more than static friction between the tires and the road can provide. The tires lose their grip on the road, the lower coefficient of kinetic friction takes over, and the car skids toward the outside of the curve. As Table 3.2 shows, the coefficient of static friction between a rubber tire and a concrete road is about 50% greater than the corresponding coefficient of sliding friction.

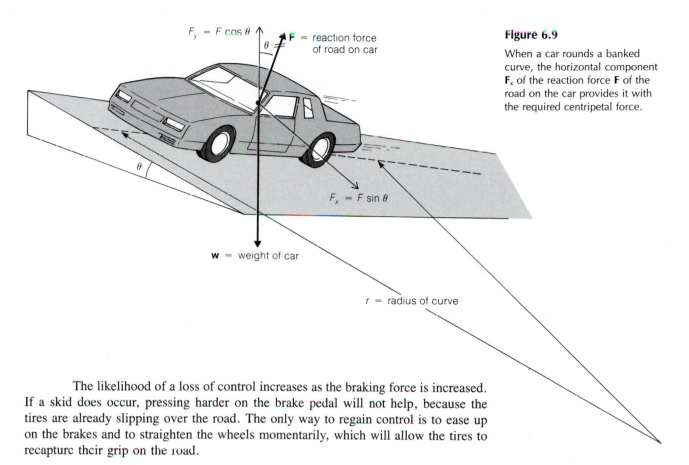

Figure 6.9

When a car rounds a banked curve, the horizontal component F_x of the reaction force F of the road on the car provides it with the required centripetal force.

The likelihood of a loss of control increases as the braking force is increased. If a skid does occur, pressing harder on the brake pedal will not help, because the tires are already slipping over the road. The only way to regain control is to ease up on the brakes and to straighten the wheels momentarily, which will allow the tires to recapture their grip on the road.

◆   E X A M P L E   6.3 _____

Usually the friction between its tires and the road is enough to provide a car with the centripetal force it needs to make a turn. If the car's speed is high or the road surface is slippery, however, the available frictional force may not be enough and the car will skid. To reduce the chance of skids, highway curves are often **banked** so that the roadbed tilts inward. The horizontal component of the reaction force of the road on the car (the action force is the car pressing on the road) then furnishes the required centripetal force. Find the proper banking angle for a car making a turn of radius r at the speed v.

SOLUTION   The reaction force F of the road on the car is perpendicular to the roadbed, as in Fig. 6.9, since friction is not involved here. This force can be resolved into two components, F_y, which supports the weight w of the car, and F_x, which is available to provide centripetal force. From the diagram,

$F_x = F \sin\theta$ = horizontal component of reaction force

$F_y = F \cos\theta$ = vertical component of reaction force

At high speeds friction between its tires and the road may not provide enough centripetal force for a vehicle to round a turn on a level road. If the turn is banked, additional centripetal force is provided by the reaction force of the road on the vehicle.

Figure 6.10

(a) In level flight at constant velocity, the lift produced by an airplane's wings equals its weight to give no resultant force perpendicular to the airplane's direction of motion. (b) When the airplane banks, a resultant force comes into being whose horizontal component provides the centripetal force needed for a turn. The vertical component of the resultant force is downward, so the airplane loses altitude as it turns. (c) If the airplane's speed is increased as it banks, the greater lift can balance its weight to permit a horizontal turn.

where θ is the angle between the roadbed and the horizontal. Since F_x furnishes the centripetal force F_c,

$$F \sin \theta = \frac{mv^2}{r}$$

The vertical component F_y of the reaction force equals the car's weight mg, and so

$$F \cos \theta = mg$$

We divide the first of these equations by the second to obtain

$$\frac{F \sin \theta}{F \cos \theta} = \frac{mv^2}{mgr}$$

$$\tan \theta = \frac{v^2}{gr} \qquad\qquad \textit{Banking angle} \quad (6.4)$$

The tangent of the proper banking angle θ varies directly with the square of the car's speed and inversely with the radius of the curve. The mass of the car does not matter. When a car goes around a curve at precisely the design speed, the reaction force of the road provides the centripetal force. If the car goes more slowly than this, friction tends to keep it from sliding down the inclined roadway; if the car goes faster, friction tends to keep it from skidding outward.

    The same considerations apply to an airplane making a turn, in which case Eq. (6.4) specifies the angle its wings should make with the horizontal (Fig. 6.10). ◆

6.4   MOTION IN A VERTICAL CIRCLE

Total energy is constant but speed is not.

    Let us now look at what happens when an object moves in a vertical rather than a horizontal circle. The object's total energy remains constant, and all of it is kinetic energy at the bottom of the circle (Fig. 6.11). At the top of the circle, however, the total energy is divided between KE and the maximum potential energy relative to the bottom, so the object's speed is least there. The motion is circular but not uniform. The object goes faster as it goes downward and slower as it goes upward.

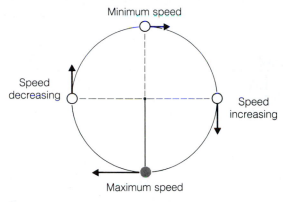

Figure 6.11

The speed of an object moving freely in a vertical circle is not constant. Shown are velocity vectors at four different positions on the circle.

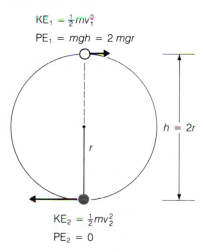

Figure 6.12

◆ **E X A M P L E 6.4**

A ball is being swung at constant energy at the end of an 80-cm string, as in Fig. 6.12. If its speed at the top of the circle is 3.5 m/s, what is its speed at the bottom?

SOLUTION At the top of the circle the ball has the potential energy

$$PE_1 = mgh = mg(2r) = 2mgr$$

and at the bottom $PE_2 = 0$. Since the total energy is constant,

$$KE_2 + PE_2 = KE_1 + PE_1$$
$$\tfrac{1}{2}mv_2^2 + 0 = \tfrac{1}{2}mv_1^2 + 2mgr$$
$$v_2^2 = v_1^2 + 4gr$$
$$v_2 = \sqrt{v_1^2 + 4gr}$$

Here $v_1 = 3.5$ m/s and $r = 0.80$ m, so the speed of the ball at the bottom of its path is

$$v_2 = \sqrt{(3.5 \text{ m/s})^2 + (4)(9.8 \text{ m/s}^2)(0.80 \text{ m})} = 6.6 \text{ m/s} \qquad ◆$$

When an object is whirled in a vertical circle at the end of a string, how does the tension in the string vary with the object's position?

The centripetal force $\mathbf{F}_c$ on the object at any point is the vector sum of the tension $\mathbf{T}$ in the string and the component of the object's weight $m\mathbf{g}$ toward the center

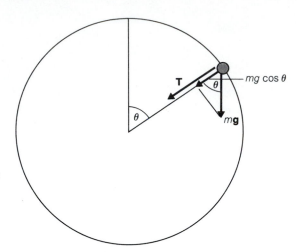

Figure 6.13

The centripetal force on an object moving in a vertical circle is the sum of the tension T in the string and the component $mg \cos \theta$ of the object's weight toward the center of the circle.

of the circle. As we can see from Fig. 6.13, when the string is at the angle θ with the vertical,

$$F_c = T + mg \cos \theta$$

and so

$$T = F_c - mg \cos \theta \qquad\qquad \textit{String tension} \quad (6.5)$$

At the top of the circle, $\theta = 0$ and $\cos \theta = 1$, so

$$T_1 = \frac{mv_1^2}{r} - mg \qquad\qquad \textit{Top of circle} \quad (6.6)$$

At the bottom of the circle, $\theta = 180°$ and $\cos \theta = -1$, so

Figure 6.14

For a given speed v, the tension in the string is least at the top of the circle, most at the bottom.

$$T_2 = \frac{mv_2^2}{r} + mg \qquad\qquad \textit{Bottom of circle} \quad (6.7)$$

For a constant speed v, the tension in the string is least at the top of the circle and most at the bottom (Fig. 6.14).

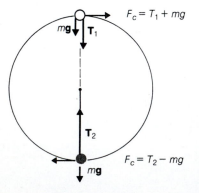

◆ **E X A M P L E   6.5**

An airplane pulls out of a dive in a circular arc whose radius is 3000 ft. The speed of the airplane is a constant 600 ft/s (409 mi/h). Find the force with which the 80-kg pilot presses down on his seat at the bottom of the arc.

SOLUTION   The downward force **F** the pilot exerts on his seat is the reaction to the upward force of the seat on him that both supports his weight and keeps him in his circular path (Fig. 6.15). The force of the seat on the pilot plays the same role

Upward force *on* pilot exerted by airplane's seat = force needed to support pilot's weight + centripetal force needed for pilot to move in circular path.

Downward force on seat exerted *by* pilot = pilot's weight + reaction force to centripetal force.

Figure 6.15

Forces exerted by and on the pilot of an airplane making a vertical circle at constant speed.

the tension in the string does for an object being whirled in a vertical circle. Hence Eq. (6.7) applies here with T_2 corresponding to F, and we have

$$F = \frac{mv^2}{r} + mg = \left(\frac{w}{g}\right)\left(\frac{v^2}{r}\right) + w$$

$$= \frac{(180 \text{ lb})(600 \text{ ft/s})^2}{(32.2 \text{ ft/s}^2)(3000 \text{ ft})} + 180 \text{ lb} = (671 + 180) \text{ lb} = 851 \text{ lb}$$

The pilot presses down on his seat with a force of 851 lb, nearly five times his actual weight.

Because there has been no compensating increase in his muscular strength, the pilot may be unable to move his arms and legs in order to control the airplane. A further complication is the tendency of the pilot's blood to leave his head because of inertia, leaving him with impaired vision ("blacked out") and perhaps unconscious. Special pressure suits have been devised that prevent disturbances in blood supply and in the positions of internal organs during severe accelerations in flight.

The forces exerted by and on a pilot or astronaut vary with the individual's mass. Hence it is often convenient to speak instead of the acceleration, by custom in units of g. Here the pilot's acceleration is

$$a = \frac{F}{m} = \frac{F}{w/g} = \left(\frac{F}{w}\right)g = \left(\frac{851 \text{ lb}}{180 \text{ lb}}\right)g = 4.73g$$

◆

Critical Speed

Anyone who has whirled a ball on a string knows that, if the ball's speed at the top is less than a certain critical value v_0, the string goes slack and the upper part of the ball's path flattens out. To find this critical speed we set the string tension $T_1 = 0$ in Eq. (6.6), with the result

$$T_1 = \frac{mv_0^2}{r} - mg = 0$$

$$v_0 = \sqrt{rg} \qquad\qquad\qquad \textit{Critical speed} \quad (6.8)$$

For a ball with an 80-cm string, as in the first example in this section,

$$v_0 = \sqrt{(0.8.0 \text{ m})(9.8 \text{ m/s}^2)} = 2.8 \text{ m/s}$$

◆   **E X A M P L E   6.6**

A sled starts from rest and slides down the frictionless track shown in Fig. 6.16 to loop the loop without falling off. If $r = 10$ m, what is the minimum value of h?

SOLUTION   In the loop the force exerted on the sled by the track takes the place of tension in a string. Equation (6.8) therefore applies here for the minimum speed at the top of the loop, so $v_0 = rg$, and the sled's kinetic energy there is

$$\text{KE} = \tfrac{1}{2}mv_0^2 = \tfrac{1}{2}mrg$$

At the top of the track the sled is $h - 2r$ above the top of the loop, and its potential energy there relative to the top of the loop is

$$\text{PE} = mg(h - 2r)$$

If there is no friction or air resistance, a sled that starts from a height 2.5 times the radius of a vertical loop above the bottom of the loop will not leave the track at the top of the loop.

Figure 6.16

Because there is no friction,

$$PE = KE$$
$$mg(h - 2r) = \tfrac{1}{2}mrg$$
$$h = \frac{r}{2} + 2r = \frac{5}{2}r$$

With $r = 10$ m, $h = 25$ m. ◆

6.5 GRAVITATION

A fundamental force.

The earth and the other planets have approximately circular orbits around the sun. We conclude that the planets are being acted on by centripetal forces that act toward the sun. This much was generally understood by the middle of the seventeenth century, when Newton turned his mind to the question of exactly what the nature of the centripetal forces was.

Newton proposed that the inward force exerted by the sun that is responsible for the planetary orbits is one example of a universal interaction, called **gravitation,** that occurs between all objects in the universe by virtue of their possession of mass. Another example of gravitation, according to Newton, is the attraction of the earth for nearby objects. Thus the centripetal acceleration of the moon and the downward acceleration g of objects dropped near the earth's surface are both due to the gravitational pull of the earth.

Newton was able to arrive at the form of the **law of universal gravitation** from an analysis of the motions of the planets about the sun:

Every object in the universe attracts every other object with a force directly proportional to the product of their masses and inversely proportional to the square of the distance separating them.

The law of gravitation in equation form is

$$F_{\text{grav}} = G\frac{m_A m_B}{r^2}$$

Gravitational force (6.9)

Figure 6.17

The gravitational forces between two spherical objects.

where m_A and m_B are the masses of any two objects and r is the distance between them. The quantity G is a universal constant whose value is

$$G = 6.67 \times 10^{-11} \frac{\text{N} \cdot \text{m}^2}{\text{kg}^2} \qquad \textit{Gravitational constant}$$

The direction of the gravitational force is always along a line joining the two objects A and B. The force on A exerted by B is equal in magnitude to that on B exerted by A but is in the opposite direction (Fig. 6.17). A homogeneous spherical object, or one composed of homogeneous spherical shells, behaves gravitationally as if all its mass were concentrated at its center.

◆ **E X A M P L E   6.7** _____

A grocer puts a 100-kg lead block under the pan of his scale. By how much does this increase the reading of the scale when 1.0 kg of onions are on the pan, if the centers of mass of the lead and of the onions are 30 cm apart?

SOLUTION   The gravitational force of the lead on the onions is

$$F = G\frac{m_A m_B}{r^2} = \left(6.67 \times 10^{-11} \frac{\text{N} \cdot \text{m}^2}{\text{kg}^2}\right) \frac{(100 \text{ kg})(1.0 \text{ kg})}{(0.30 \text{ m})^2} = 7.4 \times 10^{-8} \text{ N}$$

The increase in the scale reading is therefore

$$m = \frac{F}{g} = \frac{7.4 \times 10^{-8} \text{ N}}{9.8 \text{ m/s}^2} = 7.6 \times 10^{-9} \text{ kg} = 0.0000076 \text{ g}$$

so it is hardly worth the effort. Blowing gently on the onions will increase the reading over a million times more.   ◆

_____

Astronauts in the *Apollo 11* spacecraft saw this view of the earth as they orbited the moon, part of whose bleak landscape appears in the foreground. Gravity is responsible for the roundness of the earth and other large astronomical bodies.

The earth is round because gravitational forces squeeze it into this shape. If a part of the earth were to stick out very far, the pull of the rest would lead to strong pressures on the material underneath. This material would then flow out sideways until the bump became level or nearly so. The downward forces around the rim of a large hole would similarly cause the underlying material to flow into it. The earth's mountains and ocean basins are actually very small-scale irregularities—the total range from the Pacific depths to the summit of Mount Everest is less than 20 km, not much compared with the earth's radius of 6400 km.

Because the earth is rotating, its equatorial region bulges outward, just as a ball on a string swings outward when whirled around (Fig. 6.18). The earth is about 0.34% away from being a perfect sphere (apart from surface irregularities). Venus, whose "day" is 243 of our days, turns so slowly that its distortion is negligible. Saturn, at the other extreme, spins so rapidly that it is out of round by 9.6%.

Acceleration of Gravity

The gravitational force of the earth on an object varies as $1/r^2$, where r is the distance of the object from the center of the earth. This force is the object's weight of $w = mg$. Because m is constant, we conclude that the acceleration of gravity g varies as $1/r^2$. If we let g_0 be the value of g at the earth's surface, where $r = r_e$, then the value of g at any radius r is

Perfect sphere

Actual shape of earth (exaggerated)

Figure 6.18

The earth is an oblate spheroid in shape because of its rotation, with a difference of 43 km between its polar and equatorial diameters.

$$g = \left(\frac{r_e}{r}\right)^2 g_0 \qquad \text{Acceleration of gravity at radius } r \quad (6.10)$$

The earth's radius is

$$r_e = 6.37 \times 10^6 \text{ m} = 2.09 \times 10^7 \text{ ft} = 3960 \text{ mi}$$

◆ **E X A M P L E 6.8** —————————————————————

A woman weighs 128 lb on the earth's surface. At what height above the surface would she weigh 80 lb?

SOLUTION Because $w = mg$, we have from Eq. (6.10)

$$mg = \left(\frac{r_e}{r}\right)^2 mg_0$$

$$w = \left(\frac{r_e}{r}\right)^2 w_0$$

where w_0 is the woman's weight at the earth's surface. Solving for r gives

$$\frac{r_e}{r} = \sqrt{\frac{w}{w_e}}$$

$$r = r_e \sqrt{\frac{w_e}{w}} = (3960 \text{ mi}) \sqrt{\frac{128 \text{ lb}}{80 \text{ lb}}} = 5009 \text{ mi}$$

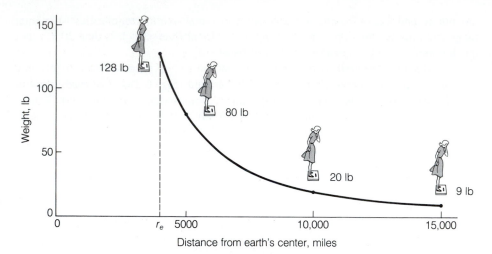

Figure 6.19

How the weight of a 4-slug person varies with distance from the earth's center. The person's weight on the earth's surface is 128 lb.

Hence her height h above the earth's surface is

$$h = r - r_e = (5009 - 3960) \text{ mi} = 1049 \text{ mi}$$

Figure 6.19 shows how the woman's weight varies with distance from the earth's center. ◆

Chapter 1 mentioned that g varies around the earth, from 9.78 m/s² at the equator to 9.83 m/s² at the poles. Because the earth bulges at the equator (see Fig. 6.18), the closer we get to it, the farther we are from the earth's center. This accounts for 0.02 m/s² of the maximum difference of 0.05 m/s². The rest follows from the earth's turning around its axis. The centripetal acceleration of a point on the earth's surface increases from 0 at the poles to 0.03 m/s² at the equator. The earth's gravitational pull provides this acceleration, and the acceleration of a falling body relative to the surface is reduced accordingly.

6.6 EARTH SATELLITES

Why they don't fall down.

The first artificial satellite, *Sputnik I,* was launched in 1957 by the Soviet Union. Since then thousands of others have been put into orbits around the earth, most of them by the United States and the Soviet Union. Men and women have been in orbit regularly since 1961, when a Russian cosmonaut circled the globe at an altitude of about 160 km. The first American went into orbit the following year.

What keeps an artificial earth satellite from falling down? The answer, of course, is that it *is* falling down, but, like the moon, at just such a rate as to circle the earth in a stable orbit. Let us use what we know about gravitation and circular motion to investigate the orbits of earth satellites. In the following discussion we will neglect the frictional resistance of the atmosphere, which ultimately brings down all artificial satellites.

Near the earth the gravitational force on an object is its weight *mg*. For uniform circular motion around the earth, this force must provide the object with the centripetal force mv^2/r. The condition for a stable orbit is therefore $mg = mv^2/r$, so that

$$v = \sqrt{rg} \qquad\qquad \textit{Satellite speed} \quad (6.11)$$

From Eqs. (6.10) and (6.11) the satellite speed at the radius *r* is

$$v = \sqrt{rg} = \sqrt{\frac{r_e^2 g_0}{r}} \qquad \textit{Orbit of earth satellite} \quad (6.12)$$

For a circular orbit just above the earth's surface,

$$v_0 = \sqrt{r_e g_0} = \sqrt{(6.4 \times 10^6\,\text{m})(9.8\,\text{m/s}^2)} = 7.9 \times 10^3\,\text{m/s}$$

which is nearly 18,000 mi/h. Anything sent off tangentially to the earth's surface at this speed will become a satellite of the earth. If it is sent off at a higher speed, its orbit will be elipitical rather than circular (Fig. 6.20).

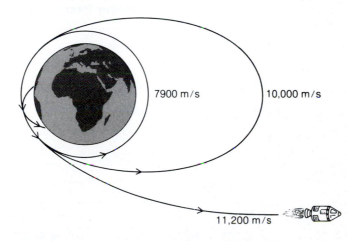

7900 m/s   10,000 m/s

11,200 m/s

Figure 6.20

Depending on its speed, an object projected horizontally above the earth's surface may fall back to the earth, revolve around the earth in a circular orbit, revolve around the earth in an elliptical orbit, or escape permanently from the earth into space.

An object sent off at a sufficiently high speed can escape permanently from the earth. The required speed, called the **escape speed,** is given by

$$v_{\text{escape}} = \sqrt{2}v_0 = \sqrt{2r_e g_0} \qquad\qquad \textit{Escape speed} \quad (6.13)$$

The escape speed from the earth is 11.2×10^3 m/s, which is about 25,000 mi/h.

A satellite can be placed in a circular orbit of any radius if it is provided with a small rocket motor to give it the required impulse at the right altitude. More than 3000 satellites are now in orbits that range from about 160 km to 35,900 km above the earth. The closer satellites are used to keep watch on the earth's surface, both for military purposes and for such scientific ones as keeping track of the weather and studying earth resources. Five satellites at an altitude of 1230 km are used in the highly accurate Transit system of marine navigation.

Some satellites circle the equator with a period of exactly 1 day so that they remain indefinitely over a particular location. A satellite in such a **geostationary orbit** can "see" about a third of the earth's surface. Nearly 200 of the satellites now in geostationary orbits are used to relay radio communications from one point to another, notably for intercontinental telephone calls and television broadcasts. Such satellites, as well as most of the others, obtain their electrical power from sunlight with the help of photoelectric cells.

◆   **E X A M P L E   6.9** _____

Find the altitude of a geostationary orbit.

SOLUTION   According to Eq. (6.2) the speed of an object in a circular orbit of period T is

$$v = \frac{2\pi r}{T}$$

where here $T = 1$ day $= 86,400$ s. Another formula for the satellite speed is given by Eq. (6.12),

$$v = \sqrt{\frac{r_e^2 g_0}{r}}$$

Setting the two formulas equal enables us to eliminate v:

$$\frac{2\pi r}{T} = \sqrt{\frac{r_e^2 g_0}{r}}$$

$$r = \sqrt[3]{\frac{r_e^2 g_0 T^2}{4\pi^2}} = \sqrt[3]{\frac{(6.4 \times 10^6 \text{ m})^2 (9.8 \text{ m/s}^2)(8.64 \times 10^4 \text{s})^2}{4\pi^2}}$$

$$= \sqrt[3]{759 \times 10^{20}} \text{ m} = \sqrt[3]{75.9 \times 10^{21}} \text{ m} = 4.23 \times 10^7 \text{ m}$$

Figure 6.21
(Not to scale.)

The corresponding altitude h (Fig. 6.21) above the earth's surface is

$$h = r - r_e = 42.3 \times 10^6 \text{ m} - 6.4 \times 10^6 \text{ m} = 35.9 \times 10^6 \text{ m}$$

which is about 22,000 miles. ◆

6.7 WEIGHTLESSNESS

How to lose weight without dieting.

Because an earth satellite is always falling toward the earth, an astronaut inside one feels "weightless." In reality, there *is* a gravitational force acting on him. What is missing to his senses is the upward reaction force provided by a stationary platform underneath him—the seat of a chair, the floor of a room, the ground itself. Instead of pushing back, the floor of the satellite falls just as fast as he does toward the earth.

It is useful to distinguish between the *actual weight* of an object, which is the gravitational force acting on it, and its *apparent weight,* which is the force it exerts on whatever it rests upon. We can think of the apparent weight of a person as the reading on a bathroom spring scale the person is standing on. Astronauts in an earth satellite have no apparent weight because they do not press down on the floor of the satellite.

A person who jumps off a diving board is just as "weightless" in her fall as an astronaut, since nothing restricts her acceleration toward the earth either. But a person standing on the ground is acted on by *both* the downward force of gravity and the upward reaction force of the ground. The latter force is what prevents her from simply dropping all the way down to the center of the earth. The human body (indeed, all living things on the earth) evolved in the presence of both these forces, and various body functions, such as blood circulation, do not seem to take place efficiently in a "weightless" state. Future spacecraft designed for long journeys may be set in rotation so that inertia will cause astronauts to press against the cabin sides, which will then press back (action-reaction again) and so bring about a situation corresponding to that on the earth's surface.

The earth exerts a gravitational force on an astronaut in orbit. The astronaut feels "weightless" because no reaction force acts on him or her.

Important Terms

An object traveling in a circle at constant speed is said to be undergoing **uniform circular motion.**

The velocity of an object in circular motion continually changes in direction, although its magnitude may remain constant. The acceleration that causes the object's velocity to change is called **centripetal acceleration,** and it points toward the center of the object's circular path.

The inward force that provides an object in circular motion with its centripetal acceleration is called **centripetal force.**

Newton's **law of universal gravitation** states that every object in the universe attracts every other object with a force directly proportional to both their masses and inversely proportional to the square of the distance separating them.

Important Formulas

Period of orbit: $T = \dfrac{2\pi r}{v}$

Centripetal acceleration: $a_c = \dfrac{v^2}{r}$

Centripetal force: $F_c = \dfrac{mv^2}{r}$

Law of gravitation: $F_{\text{grav}} = G\dfrac{m_A m_B}{r^2}$

Acceleration of gravity: $g = \left(\dfrac{r_e}{r}\right)^2 g_0$

Satellite orbit: $v = \sqrt{rg} = \sqrt{\dfrac{r_c^2 g_0}{r}}$

MULTIPLE CHOICE

1. The acceleration of an object undergoing uniform circular motion is constant in
   a. magnitude only
   b. direction only
   c. both magnitude and direction
   d. neither magnitude nor direction

2. The centripetal force on a car rounding a curve on a level road is provided by
   a. gravity
   b. friction between its tires and the road
   c. the torque applied to its steering wheel
   d. its brakes

3. The radius of the path of an object in uniform circular motion is doubled. The centripetal force needed if its speed remains the same is
   a. half as great as before
   b. the same as before
   c. twice as great as before
   d. four times as great as before

4. The centripetal force needed to keep the earth in orbit is provided by
   a. inertia
   b. its rotation on its axis
   c. the gravitational pull of the sun
   d. the gravitational pull of the moon

5. The shape of the earth is closest to that of
   a. a perfect sphere   **b.** an egg
   c. a football   **d.** a grapefruit

6. The gravitational acceleration of an object
   a. has the same value everywhere in space
   b. has the same value everywhere on the earth's surface
   c. varies somewhat over the earth's surface
   d. is greater on the moon because of its smaller diameter

7. A hole is drilled to the center of the earth and a stone is dropped into it. When the stone is at the earth's center, compared with the values at the earth's surface
   a. its mass and weight are both unchanged
   b. its mass and weight are both zero
   c. its mass is unchanged and its weight is zero
   d. its mass is zero and its weight is unchanged

8. If the earth were three times farther from the sun than it is, the gravitational force exerted on it by the sun would be
   a. three times as large as it is now
   b. nine times as large as it is now
   c. one-third as large as it is now
   d. one-ninth as large as it is now

9. The moon's mass is 1.2% of the earth's mass. Relative to the gravitational force the earth exerts on the moon, the gravitational force the moon exerts on the earth

   a. is smaller
   b. is the same
   c. is greater
   d. depends on the phase of the moon

10. The speed needed to put a satellite in orbit does not depend on

   a. the radius of the orbit
   b. the shape of the orbit
   c. the value of g at the orbit
   d. the mass of the satellite

11. An astronaut is "weightless" in a spacecraft

   a. as it takes off from the earth
   b. in certain orbits only
   c. in all orbits
   d. only when the escape speed is exceeded

12. A 0.50-kg ball moves in a circle 40 cm in radius at a speed of 4.0 m/s. Its centripetal acceleration is

   a. 10 m/s^2 **b.** 20 m/s^2
   c. 40 m/s^2 **d.** 80 m/s^2

13. The centripetal force on the ball in Multiple Choice 12 is

   a. 10 N   **b.** 20 N
   c. 40 N   **d.** 80 N

14. A toy cart at the end of a string 0.70 m long moves in a circle on a table. The cart has a mass of 2.0 kg, and the string has a breaking strength of 40 N. The maximum speed of the cart is approximately

   a. 1.9 m/s   **b.** 3.7 m/s
   c. 11.7 m/s   **d.** 16.7 m/s

15. On a rainy day the coefficient of friction between a car's tires and a certain level road surface is reduced to half its usual value. The maximum safe speed for rounding the curve is

   a. unchanged
   b. reduced to 25% of its usual value
   c. reduced to 50% of its usual value
   d. reduced to 71% of its usual value

16. A car is traveling at 50 km/h on a road such that the coefficient of friction between its tires and the road is 0.50. The minimum turning radius of the car is

   a. 9.9 m   **b.** 39 m
   c. 0.95 km   **d.** 5.1 km

17. A 2.0-kg stone at the end of a string 1.0 m long is whirled in a vertical circle. The tension in the string is 52 N when the stone is at the bottom of the circle. The stone's speed then is

   a. 4 m/s   **b.** 5 m/s
   c. 6 m/s   **d.** 7 m/s

18. A 3.2-lb stone at the end of a string 2.0 ft long is whirled in a vertical circle. The tension in the string is 5.0 lb when the stone is at the top of the circle. The stone's velocity then is

   a. 1.8 ft/s   **b.** 6.0 ft/s
   c. 10 ft/s   **d.** 12.8 ft/s

19. A car traveling at 20 m/s just leaves the road at the top of a hump. The radius of curvature of the hump is

   a. 20 m   **b.** 41 m
   c. 3.9 km
   d. impossible to determine without knowing the car's mass

20. Mars is about 1.5 times as far from the sun as the earth and its mass is about 0.10 times the earth's mass. Relative to the gravitational force the sun exerts on the earth, the force it exerts on Mars is about

   a. 0.0044 as much   **b.** 0.0067 as much
   c. 0.044 as much   **d.** 0.067 as much

21. A woman whose mass is 60 kg on the earth's surface is in a spacecraft at a height of twice the earth's radius (that is, 2 earth radii) above the earth's surface. Her mass there is

   a. 6.7 kg   **b.** 15 kg
   c. 20 kg   **d.** 60 kg

22. A man whose weight is 160 lb on the earth's surface is also in the spacecraft of Multiple Choice 21. His weight there is

   a. 40 lb   **b.** 80 lb
   c. 160 lb   **d.** 320 lb

23. A 200-kg satellite circles the earth in an orbit 7.0 × 10^6 m in radius. At this altitude $g = 8.2$ m/s^2. The speed of the satellite is

   a. 38 m/s   **b.** 0.85 km/s
   c. 7.6 km/s   **d.** 7.9 km/s

24. Earth satellite A has an orbit four times greater in radius than the orbit of satellite B. The orbital speed of A is

   a. $v_B/4$ **b.** $v_B/2$
   c. $2v_B$ **d.** $4v_B$

QUESTIONS

1. Under what circumstances, if any, can an object move in a circular path without being accelerated?

2. A car makes a clockwise turn on a level road at too high

a speed and overturns. Do its left wheels leave the ground first, or its right wheels?

3. A person swings an iron ball in a vertical circle at the end of a string. At what point in the circle is the string most likely to break? Why?

4. Where should you stand on the earth's surface to experience the most centripetal acceleration? The least?

5. A track team on the moon could set new records for the high jump (if they did not need space suits, of course) because of the smaller gravitational pull there. Could sprinters also improve their times for the 100-m dash?

6. Compare the meanings of g and G.

7. Why is Eq. (6.11), which gives the speed a satellite must have for a stable orbit around the earth, the same as Eq. (6.8), which gives the critical speed of a ball being whirled in a vertical circle at the end of a string?

8. For the moon to have the same orbit it has now, what would its speed have to be if (a) the moon's mass were double

its present mass, and (b) the earth's mass were double its present mass?

9. An earth satellite is placed in an orbit whose radius is half that of the moon's orbit. Is its time of revolution longer or shorter than that of the moon?

10. Two satellites are launched from a certain station with the same initial speeds relative to the earth's surface. One is launched toward the west, the other toward the east. Will there by any difference between their orbits? If so, what will the difference be and why?

11. Must a space vehicle in continuously powered flight reach the escape speed in order to leave the earth permanently?

12. An airplane makes a vertical circle in which it is upside down at the top of the loop. Will the pilot fall out of his seat if he has no belt to hold him in place?

EXERCISES

6.2   Centripetal Acceleration

1. A phonograph record 12 inches in diameter rotates $33\frac{1}{3}$ times per minute. (a) What is the linear speed of a point on its rim? (b) What is the centripetal acceleration of a point on its rim?

2. The minute hand of a large clock is 0.50 m long. (a) What is the linear speed of its tip in meters per second? (b) What is the centripetal acceleration of the tip of the hand?

3. What is the minimum radius at which an airplane flying at 300 m/s can make a U-turn if its centripetal acceleration is not to exceed $4.0g$?

4. If the earth were to spin so fast that a person at the equator were weightless, what would the length of the day be?

5. An astronaut in training is seated at the end of a horizontal arm 7.0 m long. How many revolutions per second must the arm make for the astronaut to experience a horizontal acceleration of $4.0g$?

6. A car starts from rest to go around a circular track 200 m in radius. If the car's speed increases at a rate of 4.0 m/s², how long will it take for the car's centripetal acceleration to equal its linear acceleration?

6.3   Magnitude of Centripetal Force

7. What is the centripetal force needed to keep a 3.0-kg mass moving in a circle of radius 0.50 m at a speed of 8.0 m/s?

8. A string 1.0 m long breaks when its tension is 100 N.

What is the greatest speed at which it can be used to whirl a 1.0-kg stone? (Neglect the gravitational pull of the earth on the stone.)

9. What is the centripetal force needed to keep a 130-lb skater moving in a circle of radius 10 ft at a speed of 5.0 ft/s?

10. A 2000-kg car is rounding a curve of radius 200 m on a level road. The maximum frictional force the road can exert on the tires of the car is 4000 N. What is the highest speed at which the car can round the curve?

11. A 10-kg counterweight is bolted to the rim of a flywheel 1.8 m in diameter that is rotating 300 times per minute. How much force must the bolts withstand?

12. A car is traveling at 30 mi/h on a level road where the coefficient of static friction between its tires and the road is 0.70. Find the minimum turning radius of the car.

13. A 2000-kg car is rounding a curve of radius 200 m on a level road. The maximum frictional force the road can exert on the tires of the car is 4000 N. What is the highest speed at which the car can round the curve?

14. A box is resting on the flat floor in the rear of a station wagon moving at 15 m/s. What is the minimum radius of a turn the station wagon can make if the box is not to slip? Assume that $\mu_s = 0.40$.

15. A dime is placed 10 cm from the center of a record. The coefficient of static friction between coin and record is 0.30. Will the coin remain where it is or will it fly off when the record turns at $33\frac{1}{3}$ rev/min? At 78 rev/min?

16. British railway engineers refer to the centripetal acceleration of a train rounding a curve in terms of "cant deficiency," the banking angle that would be needed for a passenger to feel no sideways force relative to the train. What is the centripetal acceleration that corresponds to the maximum permitted cant deficiency of 4.25°?

17. An airplane traveling at 500 km/h banks at an angle of 45° as it makes a turn. What is the radius of the turn in kilometers? Assume that the rudder is not used to make the turn.

18. A highway curve has a radius of 300 m. (a) At what angle should it be banked for a traffic speed of 100 km/h? (b) If the curve is not banked, what is the minumum coefficient of friction required between tires and road?

19. A curve in a road 8.0 m wide has a radius of 60 m. How much higher than its inner edge should the outer edge of the road be if it is to be banked properly for cars traveling at 30 km/h?

6.4   Motion in a Vertical Circle

20. A string 1.0 m long is used to whirl a 0.50-kg stone in a vertical circle. What is the tension in the string when the stone is at the top of the circle moving at 5.0 m/s?

21. A string 0.80 m long is used to whirl a 2.0-kg stone in a vertical circle. What must be the speed of the stone at the top of the circle if the string is to be just taut? How does this speed compare with that required for a 1.0-kg stone in the same situation?

22. The 200-g head of a golf club moves at 45 m/s in a circular arc of 1.0 m radius. How much force must the player exert on the handle of the club to prevent it from flying out of his hands at the bottom of the swing? Assume that the shaft of the club has negligible mass.

23. A road has a hump 12 m in radius. What is the minimum speed at which a car will leave the road at the top of the hump?

24. A pendulum has a string 1.0 m long. If the tension in the string is twice the bob's weight when the string is vertical, what is the speed of the bob at that point in its path?

25. A physics teacher swings a pail of water in a vertical circle 4.0 ft in radius at constant speed. What is the maximum time per revolution if the water is not to spill?

26. An airplane whose speed is a constant 600 km/h pulls out of a dive in a circular arc. If the 80-kg pilot presses down on his seat with a force of 4.0 kN, what is the radius of the arc?

27. A ball of mass m is swung in a vertical circle on a string just fast enough for the string to be barely taut at the top of

the circle. Find the tension in the string at the bottom of the circle.

6.5   Gravitation

28. Two identical lead spheres whose centers are 2.0 m apart attract each other with a force of 1.0×10^{-5} N. Find the mass of each sphere.

29. A 2.0-kg mass is 1.0 m away from a 5.0-kg mass. What is the gravitational force (a) that the 5.0-kg mass exerts on the 2.0-kg mass, and (b) that the 2.0-kg mass exerts on the 5.0-kg mass? (c) If both masses are free to move, what are their respective accelerations in the absence of other forces?

30. A bull and a cow elephant, each of mass 2000 kg, attract each other gravitationally with a force of 1.0×10^{-5} N. How far apart are they?

31. The earth's average orbital radius is 1.5×10^{11} m and its average orbital speed is 3.0×10^4 m/s. From these figures, together with the value of G, find the mass of the sun.

32. The mass of the planet Saturn is 5.7×10^{26} kg and that of the sun is 2.0×10^{30} kg. The average distance between them is 1.4×10^{12} m. (a) What is the gravitational force the sun exerts on Saturn? (b) Assuming that Saturn has a circular orbit, find its orbital speed.

33. An object dropped near the earth's surface falls 4.9 m in the first second. How far does the moon fall toward the earth in each second? Why doesn't the moon ever reach the earth?

34. A woman has a mass of 55 kg on the earth's surface. (a) What would she weigh at a distance from the earth's center of three times the earth's radius? (b) What would her mass be there?

35. A man weighs 180 lb on the earth's surface. (a) What would he weigh at a distance from the earth's center of three times the earth's radius? (b) What would his mass be there?

6.6   Earth Satellites

36. A spacecraft with the kinetic energy KE circles the earth in an orbit just above its surface. How much kinetic energy would it need to escape permanently from the earth?

37. In terms of the earth's radius, how far is it from the center of the earth to a point outside the earth where $g = 0.98$ m/s^2?

38. What is the acceleration of a meteor when it is 1 earth's radius above the surface of the earth? Two earth's radii?

39. The radius of the earth is 6.4×10^6 m. What is the acceleration of a meteor when it is 8.0×10^6 m from the center of the earth?

40. A satellite is to be put into orbit around the moon just above its surface. What should its speed be? Assume that the moon's radius is half that of the earth and that the acceleration of gravity at its surface is $g/6$.

41. Find the speed of an earth satellite whose orbit is 400 km above the earth's surface. What is the period of the orbit?

42. Find the altitude of an earth satellite whose period is 40 h.

43. The new Global Positioning System (GPS) developed by the United States employs 18 satellites that circle the earth in 12 h. Find the orbit radius that corresponds to this period.

PROBLEMS

47. A 500-g model airplane flies in a horizontal circle while attached to the ground with a wire 10 m long that is at an angle of 40° above the ground. If the airplane takes 5.0 s to complete each circle, what is the tension in the wire?

48. A girl on a merry-go-round who is sitting 5.0 m from its axis is holding a string with a ball at its end. If the string makes an angle of 3° with the vertical, how long does it take the merry-go-round to make a complete rotation?

49. A bicycle and rider of total mass 80 kg go around a banked circular track 40 m in radius at a speed of 16 m/s. If the reaction force of the track on the bicycle is perpendicular to the track, find its magnitude.

50. A car whose speed is 90 km/h rounds a curve 180 m in radius that is properly banked for a speed of 45 km/h. Find the minimum coefficient of friction between tires and road that will permit the car to make the turn.

6.7   Weightlessness

44. If the apparent weight of a person in an airplane is not to exceed 2.0 mg, what is the maximum radius of a circular arc in which the airplane can pull out of a dive at a constant speed of 300 km/h?

45. If the apparent weight of a person in an airplane is not to exceed 2.0 mg, what is the maximum horizontal acceleration the airplane can have?

46. A 50-kg skier goes over a bump 10 m in radius at a speed of 9.0 m/s. What is her apparent weight at the top of the bump? How fast would she have to go to fly off the top?

51. The moon's mass is 7.3×10^{22} kg, and the average radius of its orbit is 3.8×10^{8} m. At what point could an object be placed between the earth and the moon where it would experience no resultant force? (Neglect the gravitational attractions of the sun and the other planets.)

Answers to Multiple Choice

1. a	**7.** c	**13.** b	**19.** b
2. b	**8.** d	**14.** b	**20.** c
3. a	**9.** b	**15.** d	**21.** d
4. c	**10.** d	**16.** b	**22.** a
5. d	**11.** c	**17.** a	**23.** c
6. c	**12.** c	**18.** d	**24.** b

C H A P T E R

7

ROTATIONAL MOTION

7.1 **Angular Measure**

7.2 **Angular Speed**

7.3 **Rotational Kinetic Energy**

7.4 **Moment of Inertia**

7.5 **Combined Translation and Rotation**

7.6 **Angular Acceleration**

7.7 **Torque**

7.8 **Work and Power**

7.9 **Angular Momentum**

Until now we have been looking at translational motion, motion in which something shifts in position from one moment to the next. But rotational motion is just as common. The sun, the moon, planets, stars, and galaxies of stars all rotate as they sweep through space. Down on the earth, wheels, pulleys, propellers, drills, and audio disks rotate in order to do their jobs. In the atomic world, protons, neutrons, and electrons all rotate, and their rotations affect how they form atoms and how atoms form molecules, liquids, and solids. Here our concern is mainly with the rotational motion of a rigid body, one whose shape does not change as its spins. Strictly speaking, there is no such thing, but in practice many rotating bodies distort too little to matter. As we will find, all the formulas that describe rotational motion are exact analogs of the formulas we have been using to describe translational motion.

7.1 ANGULAR MEASURE

The radian is the most convenient unit.

We are accustomed to measuring angles in degrees, where 1° is 1/360 of a full rotation. That is, a complete turn represents 360°.

A better unit for our present purposes is the **radian** (rad). The radian is defined with the help of a circle drawn with its center at the vertex of the angle in question. If the circle's radius is r and the arc cut by the angle is s as in Fig. 7.1, then the angle in radians is given by

$$\theta = \frac{s}{r} = \frac{\text{arc length}}{\text{radius}} \qquad\qquad \textit{Radian measure} \quad (7.1)$$

The angle θ between two radii of a circle, in radian measure, is the ratio of the arc s to the radius r. An angle of 1 rad has an arc length that is the same as the radius.

It is easy to find the conversion factor between degrees and radians, and vice versa. There are 360° in a complete circle, and the number of radians in a complete circle is

$$\theta = \frac{s}{r} = \frac{2\pi r}{r} = 2\pi$$

because the circumference of a circle of radius r is $2\pi r$ (Fig. 7.2). Hence

$$360° = 2\pi \text{ rad}$$

from which we find that

$$1° = 0.01745 \text{ rad} \quad \text{and} \quad 1 \text{ rad} = 57.30° \qquad \textit{Radians and degrees}$$

The conversions $(2\pi \text{ rad}/360°)$ and $(360°/2\pi \text{ rad})$ are easier to remember.

Figure 7.1

The ratio of arc to radius gives the magnitude of an angle in radians.

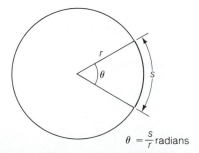

$\theta = \frac{s}{r}$ radians

The radian is an odd kind of unit because it has no dimensions—an angle in radians is a ratio of lengths, so it is really a pure number. In calculations the unit "rad" is always dropped at the end unless the result is an angular quantity.

◆ **E X A M P L E 7.1** _____

A phonograph record 30.0 cm in diameter turns through an angle of 120°. How far does a point on its rim travel?

SOLUTION First the angle is converted from degrees to radians:

$$\theta = (120°)\left(\frac{2\pi \text{ rad}}{360°}\right) = 2.09 \text{ rad}$$

The radius of the record is 15.0 cm, and so, from Eq. (7.1),

$$s = r\theta = (15.0 \text{ cm})(2.09 \text{ rad}) = 31.4 \text{ cm} \qquad ◆$$

◆ **E X A M P L E 7.2** _____

A television image with 525 horizontal lines is being shown on a picture tube whose screen is 50 cm high. If a viewer's eyes can resolve detail to 0.0003 rad—about 1′ (one minute of arc), where 60′ = 1°—how far away should the eyes be from the screen in order to just be able to see the separate lines?

SOLUTION The distance between adjacent lines is $s = 0.50$ m/525, hence

$$r = \frac{s}{\theta} = \frac{0.50 \text{ m}}{(525)(0.0003 \text{ rad})} = 3.2 \text{ m}$$

Because θ is so small here, we can ignore the difference between the arc s and the chord d (Fig. 7.3). Even when $\theta = 10°$, this difference is only 0.11%. At large angles, however, it is appreciable: When $\theta = 90°$, for instance, the difference between s and d is 10%. ◆

_____

Sometimes it is useful to express angles in radian measure in terms of π itself. For example, an angle of 90° is $\frac{1}{4}$ of a complete circle, and so

$$90° = (\tfrac{1}{4} \text{ circle})\left(2\pi \frac{\text{rad}}{\text{circle}}\right) = \frac{\pi}{2} \text{ rad}$$

Of course, this has the same numerical value as

$$(90°)\left(\frac{2\pi \text{ rad}}{360°}\right) = 1.571 \text{ rad}$$

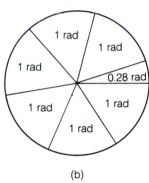

Figure 7.2

(a) 360° = 2π rad, so 1 rad = 57.30°. (b) A circle contains 2π rad, which is about 6.28 rad.

Figure 7.3

For small angles the difference between the arc s and the chord d is usually negligible.

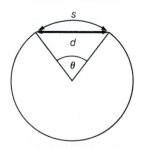

Table 7.1   Degrees to radians

0° =	0 rad	
30° =	$\pi/6$ rad =	0.524 rad
45° =	$\pi/4$ rad =	0.785 rad
60° =	$\pi/3$ rad =	1.047 rad
90° =	$\pi/2$ rad =	1.571 rad
180° =	π rad =	3.142 rad
270° =	$3\pi/2$ rad =	4.712 rad
360° =	2π rad =	6.283 rad
720° =	4π rad =	12.566 rad

because $\pi/2 = 1.571$. Table 7.1 gives some other examples of radian measure in terms of π and as decimals.

7.2   ANGULAR SPEED

Round and round it goes.

If a rotating body turns through the angle θ in the time t, its average **angular speed** ω (Greek letter *omega*) is

$$\omega = \frac{\theta}{t} \qquad \text{Angular speed} \quad (7.2)$$

When θ is in radians and t in seconds, which are the usual units for these quantities, the unit of ω is the rad/s.

Two other common units of angular speed are the revolution per second (rps) and the revolution per minute (rpm), where

$$1\,\frac{\text{rev}}{\text{s}} = \left(1\,\frac{\text{rev}}{\text{s}}\right)\left(2\pi\,\frac{\text{rad}}{\text{rev}}\right) = 2\pi\,\text{rad/s} = 6.28\,\text{rad/s}$$

$$1\,\frac{\text{rev}}{\text{min}} = \left(1\,\frac{\text{rev}}{\text{min}}\right)\left(2\pi\,\frac{\text{rad}}{\text{rev}}\right)\left(\frac{1\,\text{min}}{60\,\text{s}}\right) = \frac{2\pi}{60}\,\text{rad/s} = 0.105\,\text{rad/s}$$

Suppose we have a particle moving with the uniform speed v in a circle of radius r, as in Fig. 7.4. This particle travels the distance $s = vt$ in the time t. The angular distance through which it moves in that time is

$$\theta = \frac{s}{r} = \frac{vt}{r}$$

Figure 7.4

The angular speed of a particle in uniform circular motion is $\omega = v/r$.

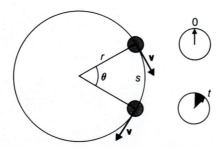

so that its angular speed is

$$\omega = \frac{\theta}{t} = \frac{vt}{rt}$$

or

$$\omega = \frac{v}{r} \qquad\qquad \text{Angular speed} \quad (7.3)$$

$$\text{Angular speed} = \frac{\text{linear speed}}{\text{path radius}}$$

This relationship can be written another way:

$$v = \omega r \qquad\qquad\qquad (7.4)$$

Linear speed = (angular speed)(path radius)

The formulas of this section are valid only when ω is expressed in radian measure.

Axis of Rotation

The **axis of rotation** of a rigid body turning in place is that line of particles which does not move (Fig. 7.5). Sometimes the axis of rotation is a line in space. All other particles of the rotating body move in circles about the axis. Since $v = \omega r$, the farther a particle is from the axis, the greater its linear speed, although all the particles of the body (except those on the axis) have the same angular speed.

◆   **E X A M P L E   7.3** _____

Find the linear speeds of points 2.0 cm and 15.0 cm from the axis of a phonograph record rotating at $33\frac{1}{3}$ rpm (Fig. 7.6).

$v = 53$ cm/s

7.0 cm/s

$\omega = 33\frac{1}{3}$ rpm

Figure 7.6

Figure 7.5

The axis of rotation is that line of particles which does not move in a rotating body. (It may be a line in space.)

A drill bit must have a particular linear speed for cutting into a given material without damage. The larger the radius of the drill bit, the slower it needs to turn to achieve the correct linear speed.

SOLUTION   The angular speed of the record is

$$\omega = (33\tfrac{1}{3}\text{ rpm}) \left(\frac{2\pi \text{ rad/s}}{60 \text{ rpm}} \right) = 3.5 \text{ rad/s}$$

Hence a point 2.0 cm from the axis has a linear speed of

$$v = \omega r = \left(3.5 \frac{\text{rad}}{\text{s}} \right) (2.0 \text{ cm}) = 7.0 \text{ cm/s}$$

and a point on the record's rim, where $r = 15.0$ cm, has a linear speed of

$$v = \left(3.5 \frac{\text{rad}}{\text{s}} \right) (15.0 \text{ cm}) = 53 \text{ cm/s}$$

$\blacklozenge$

7.3   ROTATIONAL KINETIC ENERGY

A spinning body has KE even though it may be going nowhere.

The correspondences between the linear and angular versions of displacement and speed are rather obvious. Less obvious—in fact, not obvious at all—is the rotational equivalent of mass.

Mass is a measure of the reluctance of a body at rest to be set in translational motion or, once in motion, to have its velocity change. Certainly the same body is also reluctant to be set in rotation, and if already rotating, it tends to continue to do so. We need a measure of this rotational inertia. Perhaps the easiest way to find a suitable one is by considering the kinetic energy of a rotating body.

A rotating body has KE because its particles are moving even though the body as a whole remains in place. The speed of a particle that is the distance r from the axis of a rigid body rotating with the angular velocity ω is, as we know, $v = \omega r$ (Fig. 7.7). If the particle's mass is m, its kinetic energy is therefore

$$\text{KE} = \tfrac{1}{2}mv^2 = \tfrac{1}{2}m\omega^2 r^2 \tag{7.5}$$

The body consists of many particles that need not have the same mass or be the same distance from the axis. However, all the particles have the common angular speed ω. This means that the total kinetic energy of all the particles may be written

$$\text{KE} = \Sigma\tfrac{1}{2}mv^2 = \tfrac{1}{2}(\Sigma mr^2)\omega^2 \tag{7.6}$$

Figure 7.7

The kinetic energy of each particle of a rotating body depends on the square of its distance r from the axis of rotation.

where the symbol Σ means, as before, "sum of." Equation (7.6) states that the kinetic energy of a rotating rigid body is equal to half the sum of the mr^2 values of all its particles multiplied by the square of its angular speed ω.

The quantity

$$I = \Sigma mr^2 \qquad\qquad Moment\ of\ inertia \quad (7.7)$$

is called the **moment of inertia** of the body. It has the same value regardless of the body's state of motion. The farther a given particle is from the axis of rotation, the faster it moves and the greater is its contribution to the KE of the body. The moment of inertia of a body depends on the way in which its mass is distributed relative to its axis of rotation. It is perfectly possible for one body to have a greater moment of inertia than another even though its mass may be much the smaller of the two.

The KE of a body of moment of inertia I rotating with the angular velocity ω is therefore

$$KE = \tfrac{1}{2}I\omega^2 \qquad\qquad Rotational\ kinetic\ energy \quad (7.8)$$

Evidently the rotational analog of mass is moment of inertia, just as the rotational analog of linear speed is angular speed. We will find further support for the correspondence of mass and moment of inertia in the rest of this chapter.

7.4 MOMENT OF INERTIA

The rotational analog of mass.

A rigid body is made up of a large number of separate particles whose masses are m_1, m_2, m_3, and so on (Fig. 7.8). To find the moment of inertia of such a body about a specified axis, we first multiply the mass of each of these particles by the square of its distance from the axis (r_1^2, r_2^2, r_3^2, and so on). Then we add all the mr^2 values. That is,

$$I = \Sigma mr^2 = m_1 r_1^2 + m_2 r_2^2 + m_3 r_3^2 + \cdots \qquad (7.9)$$

The units are of I the kg $\cdot$ m^2 and the slug $\cdot$ ft^2.

 E X A M P L E   7.4 _____

Find the moment of inertia of a thin ring of mass M and average radius R about an axis passing through its center and perpendicular to the plane in which it lies.

Figure 7.8

A rigid body consists of a large number of particles each of which has a certain mass m and distance r from the axis.

Axis

R

m_n

m_3

m_1 m_2

Figure 7.9

SOLUTION   As in Fig. 7.9, we proceed by subdividing the ring into n segments, each of which is at a distance R from the axis. Hence

$$I = m_1R^2 + m_2R^2 + m_3R^2 + \cdots + m_nR^2$$
$$= (m_1 + m_2 + m_3 + \cdots + m_n)R^2$$

But the sum of the masses of the segments is the same as the total mass M of the ring, and so

$$I = MR^2$$ ◆

When a body consists of a continuous distribution of matter, the more particles we imagine it to contain, the more accurate will be our value of its moment of inertia. While I can be calculated for a few simple bodies without difficulty by Eq. (7.9), in general either considerable labor or the use of integral calculus is required. Figure 7.10 gives the moments of inertia of several regularly shaped bodies in terms of the total mass M and dimensions of each.

Figure 7.10

Moments of inertia of various bodies each of mass M, about indicated axes.

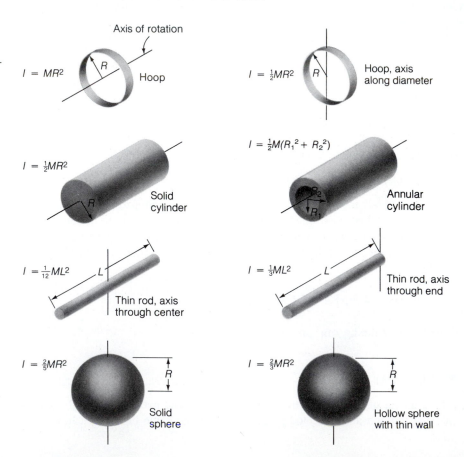

◆   **E X A M P L E   7.5**

Rotating flywheels have been proposed for energy storage in electric power plants. The flywheels would be set in motion during off-peak periods by electric motors, which would act as generators to return the stored energy during times of heavy demand. (a) Find the kinetic energy of a 1.0×10^5-kg cylindrical flywheel whose radius is 2.0 m that rotates at 400 rad/s. (b) For how many hours could the flywheel supply energy at the rate of 1.0 MW?

SOLUTION   (a) From Fig. 7.10 the moment of inertia of a solid cylinder is $I = \frac{1}{2}MR^2$. The kinetic energy of the cylinder is therefore

$$KE = \tfrac{1}{2}I\omega^2 = \tfrac{1}{2}(\tfrac{1}{2}MR^2)\omega^2 = \tfrac{1}{4}MR^2\omega^2$$
$$= (\tfrac{1}{4})(1.0 \times 10^5 \text{kg})(2.0 \text{ m})^2(400 \text{ rad/s})^2 = 1.6 \times 10^{10} \text{ J}$$

(b) Because power = energy/time, 1 MW = 10^6 W, and 1 h = 3600 s, we have

$$t = \frac{KE}{P} = \frac{1.6 \times 10^{10} \text{ J}}{1.0 \times 10^6 \text{ W}} = 1.6 \times 10^4 \text{ s} = 4.4 \text{ h}$$

◆

7.5   COMBINED TRANSLATION AND ROTATION

The total KE is the sum of translational KE and rotational KE.

When a rigid body is both moving through space and undergoing rotation, its total kinetic energy is the sum of its translational and rotational kinetic energies. The translational KE is calculated on the basis that the body is a particle whose linear speed is the same as that of the body's center of mass. The rotational KE is calculated on the basis that the body is rotating about an axis that passes through the center of mass. Thus

$$KE = KE_{\text{translation}} + KE_{\text{rotation}} = \tfrac{1}{2}mv^2 + \tfrac{1}{2}I\omega^2 \qquad (7.10)$$

where m is the body's mass, v is the velocity of its center of mass, I is its moment of inertia about an axis through the center of mass, and ω is its angular speed about that axis.

◆   E X A M P L E   7.6 _____

A 16-lb wooden ball 0.50 ft in radius is rolling at a speed of 10 ft/s. Find its total kinetic energy.

SOLUTION   The ball's moment of inertia is, from Fig. 7.10,

$$I = \tfrac{2}{5}mR^2 = \tfrac{2}{5}\left(\frac{w}{g}\right)R^2 = \tfrac{2}{5}\left(\frac{16\text{ lb}}{32\text{ ft/s}^2}\right)(0.50\text{ ft})^2 = 0.050\text{ slug} \cdot \text{ft}^2$$

and its angular speed is

$$\omega = \frac{v}{r} = \frac{10\text{ ft/s}}{0.50\text{ ft}} = 20\text{ rad/s}$$

The ball's total kinetic energy is

$$\text{KE} = \tfrac{1}{2}mv^2 + \tfrac{1}{2}I\omega^2 = \tfrac{1}{2}\frac{w}{g}v^2 + \tfrac{1}{2}I\omega^2$$

$$= \tfrac{1}{2}\left(\frac{16\text{ lb}}{32\text{ ft/s}^2}\right)(10\text{ ft/s})^2 + \tfrac{1}{2}(0.050\text{ slug} \cdot \text{ft}^2)(20\text{ rad/s})^2$$

$$= 25\text{ ft} \cdot \text{lb} + 10\text{ ft} \cdot \text{lb} = 35\text{ ft} \cdot \text{lb}$$ ◆

◆   E X A M P L E   7.7 _____

Consider a uniform cylinder of radius R and mass m that is poised at the top of an inclined plane (Fig. 7.11). Will it have a greater speed at the bottom if it slides down without friction or if it rolls down without sliding?

SOLUTION   In the case of sliding, we set the cylinder's initial PE of mgh equal to its final KE of $\tfrac{1}{2}mv^2$, and find as usual that

$$mgh = \tfrac{1}{2}mv^2$$
$$v = \sqrt{2gh}$$

Figure 7.11

Sliding cylinder
(no friction)

Rolling
cylinder

In the case of rolling, the cylinder has both translational and rotational KE at the bottom, so that

$$mgh = \tfrac{1}{2}mv^2 + \tfrac{1}{2}I\omega^2$$

The moment of inertia of a cylinder is $I = \tfrac{1}{2}mR^2$. If it rolls without slipping, its linear and angular velocities are related by the formula $\omega = v/R$, so

$$mgh = \tfrac{1}{2}mv^2 + \tfrac{1}{2}(\tfrac{1}{2}mR^2)\left(\frac{v^2}{R^2}\right) = \tfrac{1}{2}mv^2 + \tfrac{1}{4}mv^2 = \tfrac{3}{4}mv^2$$

$$v = \sqrt{\tfrac{4}{3}gh}$$

The cylinder moves more slowly when it rolls down the plane than when it slides without friction because some of the available energy is absorbed by its rotation. The total energy of the cylinder at the bottom of the plane is the same in both cases. ◆

7.6   ANGULAR ACCELERATION

The formulas for accelerated rotation have the same forms as those for accelerated linear motion.

A rotating body need not have a uniform angular speed ω, just as a moving particle need not have a uniform linear speed v. If the angular speed of a body changes by an amount $\Delta\omega$ in the time interval Δt, its average **angular acceleration** α (Greek letter *alpha*) is

$$\alpha = \frac{\Delta\omega}{\Delta t} \qquad\qquad \textit{Angular acceleration}\quad (7.11)$$

The unit of angular acceleration is the rad/s^2.

A particle moving in a circle of radius r that has an angular acceleration α also has a linear acceleration a_T tangential to its path, because its orbital speed is changing (Fig. 7.12). From the definition $\omega = v/r$, if r is constant

$$\Delta\omega = \frac{\Delta v}{r} \quad \text{and} \quad \alpha = \frac{\Delta\omega}{\Delta t} = \frac{\Delta v}{r\Delta t}$$

Figure 7.12

The angular and tangential accelerations of a particle moving in a circle are proportional to each other.

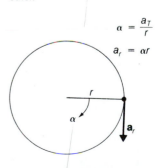

$$\alpha = \frac{a_T}{r}$$

$$a_r = \alpha r$$

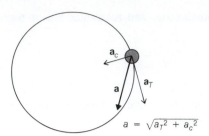

$$a = \sqrt{a_T{}^2 + a_c{}^2}$$

Figure 7.13

The tangential and centripetal accelerations of a particle in circular motion are always perpendicular.

Because the component of the particle's linear acceleration along the direction of its velocity **v** has the magnitude $a_T = \Delta v/\Delta t$, we see that

$$\alpha = \frac{a_T}{r} \qquad (7.12)$$

$$\text{Angular acceleration} = \frac{\text{tangential acceleration}}{\text{path radius}}$$

Conversely,

$$a_T = \alpha r \qquad (7.13)$$

Tangential acceleration = (angular acceleration)(path radius)

We must be careful to distinguish between the **tangential acceleration** a_T of a particle, which represents a change in its speed, and its **centripetal acceleration** a_c, which represents a change in its direction of motion. A particle in circular motion with the speed v has, as we know, the centripetal acceleration

$$a_c = \frac{v^2}{r}$$

directed toward the center of its circular path. The accelerations $\mathbf{a}_T$ and $\mathbf{a}_c$ are therefore always perpendicular (Fig. 7.13). *All* particles in circular motion have centripetal accelerations. Only those particles whose speed changes, however, have tangential accelerations. The centripetal acceleration of a particle moving in a circle of radius r can be expressed in terms of its angular velocity ω as

$$a_c = \omega^2 r \qquad \qquad \textit{Centripetal acceleration} \quad (7.14)$$

because $v = \omega r$.

◆ **E X A M P L E   7.8** _____

Find the magnitude of the total linear acceleration of a particle moving in a circle of radius 0.40 m at the instant when the angular speed is 2.0 rad/s and the angular acceleration is 5.0 rad/s².

SOLUTION   From Eqs. (7.13) and (7.14), for the particle's tangential and centripetal accelerations, we have

$$a_T = \alpha r = \left(5.0\,\frac{\text{rad}}{\text{s}^2}\right)(0.40\text{ m}) = 2.0\text{ m/s}^2$$

$$a_c = \omega^2 r = \left(2.0\,\frac{\text{rad}}{\text{s}}\right)^2(0.40\text{ m}) = 1.6\text{ m/s}^2$$

We must add $\mathbf{a}_T$ and $\mathbf{a}_c$ vectorially, because they are vector quantities and are in different directions. As shown in Fig. 7.13, the magnitude a of the vector sum $\mathbf{a}$ of $\mathbf{a}_T$ and $\mathbf{a}_c$ is

$$a = \sqrt{a_T^2 + a_c^2} = \sqrt{2.0^2 + 1.6^2}\text{ m/s}^2 = 2.6\text{ m/s}^2 \qquad \blacklozenge$$

Comparison with Linear Motion

The formulas we obtained in Chapter 1 for the linear motion of a particle under constant acceleration all have counterparts in angular motion. Because the derivations are the same, they are simply listed in Table 7.2.

Table 7.2

Linear motion	Angular motion	
$x = v_0 t + \frac{1}{2}at^2$	$\theta = \omega_0 t + \frac{1}{2}\alpha t^2$	(7.15)
$x = \left(\dfrac{v_0 + v_f}{2}\right)t$	$\theta = \left(\dfrac{\omega_0 + \omega_f}{2}\right)t$	(7.16)
$v_f = v_0 + at$	$\omega_f = \omega_0 + \alpha t$	(7.17)
$v_f^2 = v_0^2 + 2ax$	$\omega_f^2 = \omega_0^2 + 2\alpha\theta$	(7.18)

◆ **E X A M P L E   7.9** _____

A motor starts rotating from rest with an angular acceleration of 12.0 rad/s^2. (a) What is the motor's angular speed 4.0 s later? (b) How many revolutions does it make in this period of time?

SOLUTION   (a) From Eq. (7.17) we find

$$\omega_f = \omega_0 + \alpha t = 0 + (12.0\text{ rad/s}^2)(4.0\text{ s}) = 48\text{ rad/s}$$

(b) Because we know ω_0, ω_f, α, and t, we can use Eq. (7.15), (7.16), or (7.18) to find θ. Eq. (7.15) has the advantage that it involves only the quantities we are given, so any errors made or rounding done in finding ω_f cannot affect the value of θ. From Eq. (7.15)

$$\theta = \omega_0 t + \tfrac{1}{2}\alpha t^2 = 0 + \tfrac{1}{2}(12.0\,\text{rad/s}^2)(4.0\,\text{s})^2 = 96\,\text{rad}$$

Because 1 rev $= 2\pi$ rad,

$$\theta = \frac{96\,\text{rad}}{2\pi\,\text{rad/rev}} = 15.3\,\text{rev}$$

$\blacklozenge$

7.7   TORQUE

The rotational analog of force.

According to Newton's second law of motion, a net force applied to a body gives it a linear acceleration. What can give a body an angular acceleration?

Everyday experience can guide us here. When we want to turn a bolt, we exert a force **F** on a wrench whose jaws are on the bolt's head (Fig. 7.14). The stronger the force, the greater the turning effect. Also, the longer the perpendicular distance L from the line of action of the force to the pivot point O, which is the bolt, the greater the turning effect. The distance L is called the **moment arm** of the force.

Figure 7.14

A measure of the turning effect of a force about a pivot point O is its torque τ, which is the product FL of the magnitude F of the force and the moment arm L. In (a) the moment arm L is longest, and so τ is a maximum. In (d) the line of action of the force passes through O, and so $L=0$ and $\tau = 0$.

The product of the magnitude F of the force and its moment arm L is called the **torque** of the force about O. To twist something is to apply a torque to it. The symbol for torque is τ (Greek letter *tau*), so that

$$\tau = FL \qquad\qquad\qquad Torque \quad (7.19)$$

Torque = (force)(moment arm)

In the SI system the unit of torque is the newton · meter. In the U.S. Customary System it is the pound · foot (lb · ft).

◆ **E X A M P L E   7.10** _____

The cylinder-head bolts on the diesel engine of a truck are supposed to be tightened to a torque of 20 lb · ft. If the center of the handle on the wrench used is 11 in from the axis of the bolt, what force perpendicular to the handle should be applied?

SOLUTION The moment arm here is $L = $ (11 in)/(12 in/ft) $= 0.92$ ft. From Eq. (7.19) the required force is

$$F = \frac{\tau}{L} = \frac{20 \text{ lb} \cdot \text{ft}}{0.92 \text{ ft}} = 22 \text{ lb}$$

◆

Torque and Angular Acceleration

Let us now see how torque is related to angular acceleration. Figure 7.15 shows a particle of mass m that moves in a circle of radius r. A force **F** that acts on the particle tangentially to the particle's path gives it the acceleration a_T according to the formula

$$F = ma_T$$

Because the tangential acceleration a_T here is equal to αr,

$$F = mr\alpha$$

Multiplying both sides of this equation by r gives

$$Fr = mr^2\alpha$$

We recognize Fr as the torque τ of the force F about the axis of rotation of the particle and mr^2 as the particle's moment of inertia I. The latter equation therefore states that

$$\tau = I\alpha \qquad\qquad\qquad (7.20)$$

Torque = (moment of inertia)(angular acceleration)

Figure 7.15

A force **F** that acts on a particle moving in a circle of radius r exerts a torque of magnitude $\tau = Fr$ when **F** is perpendicular to the particle's path.

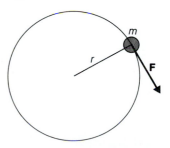

Although Eq. (7.20) was derived for the case of a single particle, it is also valid for any rotating body, provided that the torque and moment of inertia are both calculated about the same axis.

The formula $\tau = I\alpha$ is the fundamental law of motion for rotating bodies in the same sense that $F = ma$ is the fundamental law of motion for bodies moving through space. In rotational motion, torque plays the same role that force does in translational motion. The angular acceleration experienced by a body when the torque τ acts on it is

$$\alpha = \frac{\tau}{I}$$

Thus α is directly proportional to the torque and inversely proportional to the body's moment of inertia.

◆ **E X A M P L E   7.11**

0

$t = 6$ s

F = 2 N

$\omega_0 = 120$ rad/s

$\omega_f = 0$

Figure 7.16

A 2.0-kg grindstone 10 cm in radius is turning at 120 rad/s. The motor is switched off, and a chisel is pressed against the grindstone with a force whose tangential component is 2.0 N (Fig. 7.16). How long will it take the grindstone to come to a stop?

SOLUTION   The grindstone is a solid cylinder and so its moment of inertia is

$$I = \tfrac{1}{2}MR^2 = \tfrac{1}{2}(2.0\,\text{kg})(0.10\,\text{m})^2 = 0.010\,\text{kg} \cdot \text{m}^2$$

The torque the chisel exerts on the grindstone is

$$\tau = -Fr = -(2.0\,\text{N})(0.10\,\text{m}) = -0.20\,\text{N} \cdot \text{m}$$

where the minus sign means that the torque acts to slow down the grindstone. The angular acceleration of the grindstone is

$$\alpha = \frac{\tau}{I} = \frac{-0.20\,\text{N} \cdot \text{m}}{0.010\,\text{kg} \cdot \text{m}^2} = -20\,\text{rad/s}^2$$

We now call on Eq. (7.17), $\omega_f = \omega_0 + \alpha t$, and set $\omega_f = 0$ to correspond to the grindstone coming to a stop. The result is

$$\alpha t = -\omega_0$$

$$t = \frac{-\omega_0}{\alpha} = \frac{-120\,\text{rad/s}}{-20\,\text{rad/s}^2} = 6.0\,\text{s}$$

◆

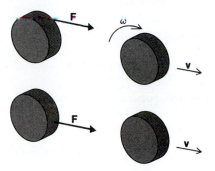

Figure 7.17

A force whose line of action passes through the center of gravity of a body cannot cause it to rotate.

When a net force acts on a body able to move freely, the body will have both linear and angular accelerations unless the line of action of the force passes through the body's center of mass (Fig. 7.17). In the latter case, if the body was not rotating initially, it will keep its original orientation during its motion.

7.8 WORK AND POWER

No surprises here.

Mechanical energy is usually transmitted by rotary motion. The power output of almost every modern engine emerges via a rotating shaft, and more often than not this power is expended in some form of rotation as well. The tires of a car, the propeller of a ship, the bit of a drill, the vanes of a centrifugal pump, and the rotor of a dynamo all function by turning.

Let us consider a shaft of radius r on which a tangential force **F** acts, as in Fig. 7.18. After a time t the shaft has turned through the angle θ, the point at which

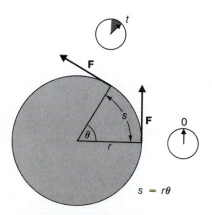

Figure 7.18

The power transmitted by a rotating shaft is the product of the torque τ on the shaft and its angular speed ω, hence $P = \tau\omega$.

$s = r\theta$

the force is applied has moved through the distance s, and the force has done the amount of work

$$W = Fs = Fr\theta$$

We recognize that Fr is the torque τ applied to the shaft, so

$$W = \tau\theta \qquad\qquad\qquad\qquad \textit{Rotational work} \quad (7.21)$$

Work = (torque)(angular displacement)

Power is the rate at which work is done, which gives us here

$$P = \frac{W}{t} = \frac{Fr\theta}{t} = Fr\frac{\theta}{t}$$

Because $Fr = \tau$ and $\theta/t = \omega$, the angular speed of the shaft, we have

$$P = \tau\omega \qquad\qquad\qquad\qquad \textit{Rotational power} \quad (7.22)$$

Power = (torque)(angular speed)

When energy is to be transmitted at a certain rate P by means of a rotating shaft, the higher the angular speed, the lower the torque needed, and vice versa. Equation (7.22) is the rotational analog of the formula

$$P = Fv$$

Power = (force)(speed)

◆   **E X A M P L E   7.12** _____

A truck engine develops 230 hp at 5000 rpm. How much torque is exerted on the crankshaft?

SOLUTION   We begin by converting 230 hp and 5000 rpm to their equivalents in lb · ft/s and rad/s:

$$P = (230\ \text{hp})\left(550\frac{\text{ft} \cdot \text{lb/s}}{\text{hp}}\right) = 1.27 \times 10^5\ \text{ft} \cdot \text{lb/s}$$

$$\omega = (5000\ \text{rpm})\left(0.1047\frac{\text{rad/s}}{\text{rpm}}\right) = 524\ \text{rad/s}$$

From Eq. (7.22) we find that the torque is

$$\tau = \frac{P}{\omega} = \frac{1.27 \times 10^5\ \text{ft} \cdot \text{lb/s}}{524\ \text{rad/s}} = 242\ \text{lb} \cdot \text{ft}$$

◆

◆ E X A M P L E 7.13 _____

The alternator on a truck engine produces 250 W of electric power when it turns at 3600 rpm, which is 377 rad/s. The alternator's pulley has a radius of 40.0 mm. If the alternator is 95.0% efficient, find the difference between the tensions in the taut and slack parts of the V-belt connecting it to the engine.

SOLUTION   The efficiency of the alternator is Eff $= P_{out}/P_{in}$, and so the input power to its pulley is

$$P_{in} = \frac{P_{out}}{Eff} = \frac{250 \text{ W}}{0.950} = 263 \text{ W}$$

The torque on the alternator pulley is

$$\tau = \frac{P_{in}}{\omega}$$

Therefore the net force on the rim of the pulley must be

$$F = \frac{\tau}{r} = \frac{P_{in}/\omega}{r} = \frac{P_{in}}{\omega r} = \frac{263 \text{ W}}{(377 \text{ rad/s})(0.0400 \text{ m})} = 17.4 \text{ N}$$

which is about 4 lb. This is the difference between the tensions in the two parts of the V-belt (Fig. 7.19).   ◆

Figure 7.19

The torque on a V-belt pulley is due to the difference in the tensions in the taut and slack parts of the V-belt.

7.9   ANGULAR MOMENTUM

Its conservation is an important physical principle.

We have all noticed that rotating objects tend to keep spinning. A top would spin forever if friction did not slow it down. The rotational equivalent of linear momentum is called **angular momentum,** and **conservation of angular momentum** is the formal way to describe the tendency of a spinning object to continue to spin.

L

Thumb in direction
of angular
momentum vector

Fingers of right hand
in direction of
rotational motion

Figure 7.20

Angular momentum is a vector
quantity whose direction is given
by the right-hand rule shown
here.

When a skater brings her arms
and legs close to her axis of rotation after starting a spin, her moment of inertia is decreased and
she spins faster as a result.

The magnitude L of the angular momentum of a body depends on its moment of inertia I and angular speed ω in the same way its linear momentum depends on its mass m and linear speed v:

$$L = I\omega \qquad\qquad \textit{Angular momentum} \quad (7.23)$$

Angular momentum = (moment of inertia) (angular speed)

Like linear momentum, angular momentum is a vector quantity. The direction associated with the angular momentum $\mathbf{L}$ of a spinning object is given by the right-hand rule shown in Fig. 7.20.

We recall from Chapter 5 that, if no net force acts on a body, its linear momentum is constant. Similarly, if no net torque acts on a rigid body, its angular momentum is constant. In fact, the angular momentum of a body does not change when $\Sigma\tau = 0$ even if it is *not* a rigid body. If a body is so altered during its motion that its moment of inertia changes, ω also changes to keep $\mathbf{L}$ the same. Thus we have the useful principle of **conservation of angular momentum**:

When the sum of the external torques acting on a system of particles is zero, the total angular momentum of the system remains constant.

In equation form, if $I_1\omega_1$ is the initial angular momentum of a system of particles and $I_2\omega_2$ is its final angular momentum, then if no torque has acted on the system

$$I_1\omega_1 = I_2\omega_2 \qquad\qquad \textit{Conservation of angular momentum} \quad (7.24)$$

A skater or ballet dancer doing a spin capitalizes on conservation of angular momentum. In Fig. 7.21 a skater is shown starting her spin with her arms and one leg outstretched. By bringing her arms and extended leg inward, she reduces her moment of inertia considerably and consequently spins faster.

I large, ω small   I small, ω large

Figure 7.21

Angular momentum is conserved when a skater executes a spin.

◆ **E X A M P L E 7.14** _____

A skater has the moment of inertia 3.0 kg · m² when her arms are outstretched and 1.0 kg · m² when her arms are brought to her sides. She starts to spin at the rate of 1.0 rev/s when her arms are outstretched, and then pulls her arms to her sides. (a) What is her final angular speed? (b) How much (if any) work did she have to do?

SOLUTION (a) Because the conversion factors would only cancel anyway, we can use the rev/s as the unit of angular speed. Here $I_1 = 3.0$ kg · m², $I_2 = 1.0$ kg · m², and $\omega_1 = 1.0$ rev/s. From conservation of angular momentum

$$I_1\omega_1 = I_2\omega_2$$
$$\omega_2 = \frac{I_1}{I_2}\omega_1 = \left(\frac{3.0\,\text{kg} \cdot \text{m}^2}{1.0\,\text{kg} \cdot \text{m}^2}\right)(1.0\,\text{rev/s}) = 3.0\,\text{rev/s}$$

The skater spins three times faster than she did with her arms outstretched.

(b) The work the skater does must equal the change (if any) in her kinetic energy. To calculate KE here, we must use the rad/s as the unit of angular velocity. Because 1 rev = 2π rad, $\omega_1 = 2\pi$ rad/s and $\omega_2 = 6\pi$ rad/s. The skater's initial and final kinetic energies are therefore

$$\text{KE}_1 = \tfrac{1}{2}I_1\omega_1^2 = \tfrac{1}{2}(3.0\,\text{kg} \cdot \text{m}^2)(2\pi\,\text{rad/s})^2 = 59\,\text{J}$$
$$\text{KE}_2 = \tfrac{1}{2}I_2\omega_2^2 = \tfrac{1}{2}(1.0\,\text{kg} \cdot \text{m}^2)(6\pi\,\text{rad/s})^2 = 178\,\text{J}$$

Although the skater's angular momentum stayed the same, her kinetic energy did not. The new KE is three times the old KE, and the work she had to do is $W = \text{KE}_1 - \text{KE}_2 = 119\,\text{J}$.   ◆

As we have seen, there are many points of correspondence between angular and linear motion at constant speed and under constant acceleration. Table 7.3 lists

Table 7.3 Comparison of linear and angular quantities

	Linear quantity		Angular quantity	
Distance	$x = v_0 t + \frac{1}{2}at^2$	Angle	$\theta = \omega_0 t + \frac{1}{2}\alpha t^2$	
	$x = \left(\dfrac{v_0 + v_f}{2}\right)t$		$\theta = \left(\dfrac{\omega_0 + \omega_f}{2}\right)t$	
Speed	$v = v_0 + at$	Angular speed	$\omega = \omega_0 + \alpha t$	
	$v^2 = v_0^2 + 2ax$		$\omega^2 = \omega_0^2 + 2\alpha\theta$	
Acceleration	$a = \Delta v/\Delta t$	Angular acceleration	$\alpha = \Delta\omega/\Delta t$	
Mass	m	Moment of inertia	I	
Force	$F = ma$	Torque	$\tau = I\alpha$	
Momentum	$p = mv$	Angular momentum	$L = I\omega$	
Work	$W = Fx$	Work	$W = \tau\theta$	
Power	$P = Fv$	Power	$P = \tau\omega$	
Kinetic energy	$\text{KE} = \frac{1}{2}mv^2$	Kinetic energy	$\text{KE} = \frac{1}{2}I\omega^2$	

Figure 7.22

The faster a top spins, the more stable it is. When most of its angular momentum has been lost through friction, the top falls over.

Figure 7.23

A gyroscope.

the principal ones. Although the symbols are different, the formulas are the same because the basic concepts involved are so closely related.

Spin Stabilization

Because angular momentum is a vector quantity, a torque must be applied to change the orientation of the axis of rotation of a spinning body as well as to change the magnitude of its angular velocity. The greater the magnitude of **L,** the more torque is needed for it to deviate from its original direction. This is the principle behind the spin stabilization of projectiles such as footballs and rockets. Such projectiles are set spinning about axes in their directions of motion so that they do not tumble and thereby increase air resistance. A top is another illustration of the vector nature of angular momentum. A stationary top set on its tip falls over at once, but a rotating top stays upright until most of its angular momentum is lost by friction between its tip and the ground (Fig. 7.22).

A **gyroscope** is a disk with a high moment of inertia whose axis is mounted on gimbals, as in Fig. 7.23. When it is set in motion, conservation of angular momentum keeps the disk in a fixed orientation in space regardless of how its support is moved. Thus a gyroscope can be used on an airplane as an artificial horizon when the true horizon is not visible and as the basis of a compass that operates independently of the earth's magnetic field.

Conservation of angular momentum provides an elegant way to reduce the angular speed of a spin-stabilized object in space where friction is not available. The spacecraft that carried the European satellite *ESRO I* into orbit was set rotating at about 100 rpm to stabilize it before the last rocket stage was separated. Afterward the rotation rate was slowed to about 20 rpm by allowing two strings with masses at their ends to unwind from the spinning satellite, thereby increasing I and lowering ω. The strings were then let go, and the masses flew off into space together with the unwanted angular momentum.

Important Terms

The **radian** is a unit of angular measure equal to 57.30°. If a circle is drawn whose center is at the vertex of an angle, the angle in radian measure is equal to the ratio between the arc of the circle cut by the angle and the radius of the circle. A full circle contains 2π radians.

The **angular speed** ω of a rotating body is the angle through which it turns per unit time. The **angular accelerations** α of a rotating body is the rate of change of its angular speed with respect to time.

The **axis of rotation** of a rigid body turning in place is that line of particles which does not move.

All particles in circular motion experience centripetal accelerations, but only those particles whose angular speed changes have **tangential accelerations.**

The **moment of inertia** I of a body about a given axis is the rotational analog of mass of the body is distributed about the axis.

The **torque** of a force about a particular axis is the product of the magnitude of the force and the perpendicular distance from the line of action of the force to the axis. The latter distance is called the **moment arm** of the force.

The **angular momentum** L of a rotating body is the product $I\omega$ of its momentof inertia and angular speed. The principle of **conservation of angular momentum** states that the total angular momentum of a system of particles remains constant when no net external torque acts on it.

Important Formulas

Angle in radian measure: $\quad \theta = \dfrac{s}{r}$

$$\theta = \left(\dfrac{\omega_0 + \omega_f}{2}\right) t$$

Angular speed: $\quad \omega = \dfrac{\theta}{t}$

Linear speed: $\quad v = \omega r$

Moment of inertia: $\quad I = \Sigma m r^2$

Rotational kinetic energy: $\quad \text{KE} = \frac{1}{2}I\omega^2$

Motion under constant angular acceleration:

$$\theta = \omega_0 t + \tfrac{1}{2}\alpha t^2$$
$$\omega_f = \omega_0 + \alpha t$$
$$\omega_f^2 = \omega_0^2 + 2\alpha\theta$$

Angular acceleration: $\quad \alpha = \dfrac{\Delta\omega}{\Delta t} = \dfrac{a_T}{r}$

Torque: $\quad \tau = FL = I\alpha$

Work and power: $\quad W = \tau\theta$
$$P = \tau\omega$$

Angular momentum: $\quad L = I\omega$

MULTIPLE CHOICE

1. In a rigid object undergoing uniform circular motion, a particle that is a distance R from the axis of rotation
    a. has an angular speed proportional to R
    b. has an angular speed inversely proportional to R
    c. has a linear speed proportional to R
    d. has a linear speed inversely proportional to R

2. The centripetal acceleration of a particle in circular motion
    a. is less than its tangential acceleration
    b. is equal to its tangential acceleration
    c. is more than its tangential acceleration
    d. may be more or less than its tangential acceleration

3. The rotational analog of force in linear motion is
    a. moment of inertia
    b. angular momentum
    c. torque
    d. weight

4. The rotational analog of mass in linear motion is
    a. moment of inertia   b. angular momentum
    c. torque   d. angular speed

5. A quantity not directly involved in the rotational motion of an object is
    a. mass   b. moment of inertia

    c. torque   **d.** angular speed

6. All rotating objects at sea level that have the same mass and angular velocity also have the same
    a. angular momentum
    b. moment of inertia
    c. kinetic energy
    d. gravitational potential energy

7. The moment of inertia of an object does not depend on
    a. its mass
    b. its size and shape
    c. its angular speed
    d. the location of the axis of rotation

8. Of the following properties of a yo-yo moving in a circle, the one that does not depend on the radius of the circle is the yo-yo's
    a. angular speed
    b. angular momentum
    c. linear speed
    d. centripetal acceleration

9. A yo-yo being swung in a circle need not possess
    a. angular speed
    b. angular momentum
    c. angular acceleration
    d. centripetal acceleration

10. The total angular momentum of a system of particles
    a. remains constant under all circumstances
    b. changes when a net external force acts on the system
    c. changes when a net external torque acts on the system
    d. may or may not change under the influence of a net external torque, depending on the direction of the torque

11. A hoop and a disk of the same mass and radius roll down an inclined plane. At the bottom they have the same
    a. angular speed   **b.** angular momentum
    c. KE of rotation   **d.** gravitational PE

12. A solid disk rolls from rest down an inclined plane. A hole is then cut in the center of the disk, and the resulting doughnut-shaped object is allowed to roll down the same inclined plane. Its speed at the bottom of the plane is
    a. less than that of the solid disk
    b. the same as that of the solid disk
    c. greater than that of the solid disk
    d. any of the above, depending on the size of the hole

13. A solid iron sphere A rolls down an inclined plane, while an identical sphere B slides down the plane in a frictionless manner.
    a. Sphere A reaches the bottom first.

    b. Sphere B reaches the bottom first.
    c. They reach the bottom together.
    d. Which one reaches the bottom first depends on the angle of the plane.

14. At the bottom of the plane, the total KE of sphere A in Multiple Choice 13 is
    a. less than that of sphere B
    b. equal to that of sphere B
    c. more than that of sphere B
    d. more or less than that of sphere B, depending on the angle of the plane

15. The number of radians in half a circle is
    a. $\pi/2$ **b.** π
    c. 2π **d.** 4π

16. An angle of 20° is equivalent to
    a. 0.17 rad   **b.** 0.20 rad
    c. 0.35 rad   **d.** 3.18 rad

17. An angle of 540° is equivalent to
    a. $2\pi/3$ rad   **b.** π rad
    c. $3\pi/2$ rad   **d.** 3π rad

18. An angle of $\pi/6$ rad is equivalent to
    a. 15°   **b.** 30°
    c. 60°   **d.** 90°

19. A wheel is 1 m in diameter. When it makes 30 rev/min, the linear speed of a point on its circumference is
    a. $\pi/2$ m/s   **b.** π m/s
    c. 30π m/s   **d.** 60π m/s

20. The shaft of a motor turns at a constant 3000 rpm. In 40 s it turns through
    a. 2000 rad   **b.** 6283 rad
    c. 12,000 rad   **d.** 12,566 rad

21. An object undergoes a uniform angular acceleration. In the time t since the object started rotating from rest, the number of turns it makes is proportional to
    a. $\sqrt{t}$ **b.** t
    c. t^2 **d.** t^3

22. A wheel that starts from rest has an angular speed of 20 rad/s after being uniformly accelerated for 10 s. The total angle through which it has turned in these 10 s is
    a. 2π rad   **b.** 40π rad
    c. 100 rad   **d.** 200 rad

23. A motor takes 8.00 s to go from 60 rad/s to 140 rad/s at constant angular acceleration. The total angle through which its shaft turns during this time is
    a. 320 rad   **b.** 640 rad
    c. 800 rad   **d.** 1600 rad

24. A flywheel rotating at 10 rev/s is brought to rest by a constant torque in 15 s. In coming to a stop, the flywheel makes

a. 75 rev   b. 150 rev
c. 472 rev   d. 600 rev

25. A constant torque of 50 lb · ft acts on a 40-lb flywheel whose moment of inertia is 10 slug · ft². The time needed for the flywheel to reach 75 rad/s starting from rest is

a. 5.5 s   b. 15 s
c. 19 s   d. 600 s

26. A torque of 30 N · m acts on a 5.0-kg wheel of moment of inertia 2.0 kg · m². If the wheel starts from rest, after 10 s it will have turned through

a. 750 rad   b. 1500 rad
c. 3000 rad   d. 6000 rad

27. The final KE of the wheel in Multiple Choice 26 will be

a. 9 kJ   b. 23 kJ
c. 45 kJ   d. 56 kJ

28. A tangential force of 10 N acts on the rim of a wheel 80 cm in diameter. The wheel takes 1.5 s to complete its first revolution starting from rest. The wheel's moment of inertia is

a. 0.72 kg · m²   b. 0.96 kg · m²
c. 1.8 kg · m²   d. 4.5 kg · m²

29. A motor develops 56 kW when its shaft turns at 300 rad/s. The torque on the shaft

a. is 30 N · m
b. is 187 N · m
c. is 17 kN · m
d. depends on the shaft diameter

QUESTIONS

1. Many flywheels have most of their mass concentrated around their rims. What is the advantage of this?

2. The density of an object is its mass divided by its volume. The density of iron is about three times the density of aluminum. If an iron cylinder and an aluminum cylinder have the same length and the same mass, which has the greater moment of inertia?

3. A square and a rectangle of the same mass are cut from a sheet of metal. Which has the greater moment of inertia about a perpendicular axis through its center?

4. Will a car coast downhill faster if it has heavy tires or light tires?

5. A hollow cylinder and a solid cylinder having the same mass and diameter are released from rest simultaneously at the top of an inclined plane. Which reaches the bottom first?

6. Which will roll to the bottom of an inclined plane first, a can that contains ice or an identical can that contains water?

7. An aluminum cylinder of radius R, a lead cylinder of radius R, and a lead cylinder of radius $2R$ all roll from rest

down an inclined plane. In what order will they reach the bottom?

8. A person on a bicycle presses down on each pedal through the front half of its circle. In which position is the torque a maximum? In which positions is it zero?

9. What determines the moment arm of a force?

10. The lower the center of gravity of an object, the more stable it is. However, it is easier to balance a billiard cue vertically when its tip is on one's finger with its heavy handle up in the air rather than with its handle on the finger. Why?

11. Why does a diver tuck her knees to her chest when turning somersaults during a dive?

12. All helicopters have two propellers. Some have both propellers on vertical axes but rotating in opposite directions, and others have one on a vertical axis and one on a horizontal axis perpendicular to the helicopter body at the tail. Why is a single propeller never used?

13. If the polar ice caps melt, how will the length of the day be affected?

EXERCISES

7.1 Angular Measure

1. An apple pie is cut into nine equal pieces. What angle (in radians) is included between the sides of each piece?

2. How many radians does the second hand of a clock turn through in 30 s? In 90 s? In 105 s? Express the answers in terms of π.

3. The resolution of the human eye is about 1′, where 60′ = 1°. (a) Express an angle of 1′ in radians. (b) What is the length of the smallest detail that can be discerned when an object 25 cm away is being examined? (This is the distance of most distinct vision.)

4. In a good light a golden eagle can detect a 16-in hare at a distance of 1.0 mi. What angle does this represent, in radians?

7.2   Angular Speed

5. What is the angular speed in radians per second of the hour, minute, and second hands of a clock?

6. A car makes a U-turn in 5.0 s. What is its average angular speed?

7. In 1976 Kazuya Shiozaki made 49,299 turns in 5 h 37 min of skipping rope. Find the average angular speed of the rope.

8. The shaft of a motor rotates at the constant angular speed of 3000 rpm. How many radians will it have turned through in 10 s?

9. The blades of a rotary lawnmower are 30 cm long and rotate at 315 rad/s. Find the linear speed of the blade tips and their angular speed in rpm.

10. A drill bit 0.25 in in diameter is turning at 1200 rpm. Find the linear speed of a point on its circumference in ft/s.

11. In 1941 a wind-driven electric power station was set up at Grandpa's Knob, Vermont. The propeller was 40 m in diameter. When the propeller was turning at 30 rev/min, what was the linear speed of one of the blade tips in kilometers per hour?

12. A steel cylinder 40 mm in radius is to be machined in a lathe. At how many revolutions per second should it rotate in order that the linear speed of the cylinder's surface be 70 cm/s?

13. A rotating platform is to be used to test aircraft equipment under accelerations of 6g. If the equipment is 50 cm from the axis, what angular speed is needed?

14. The propeller of a boat rotates at 100 rad/s when the speed of the boat is 6.0 m/s. The diameter of the propeller is 40 cm. What is the speed of the tip of the propeller?

7.3   Rotational Kinetic Energy

7.4   Moment of Inertia

15. The earth's mass is 6.0×10^{24} kg, and its radius is 6.4×10^6 m. Find its rotational KE under the assumption that it is a uniform sphere.

16. The baton of a drum majorette is a 0.70-lb uniform rod 2.0 ft long. What is its KE when it is twirled at an angular speed of 10 rad/s?

17. A 7.0-kg bowling ball 30 cm in diameter rolls at a speed of 5.0 m/s. What is its total KE?

18. The wheel shown in Fig. 7.24 has a diameter of 1.2 m. Its rim has a mass of 12 kg, and each spoke has a mass of 1.0 kg. Find the moment of inertia of the wheel.

Figure 7.24
Exercise 18.

19. The seat of an adjustable piano stool has the mass m and the moment of inertia I. If the seat is given an initial angular speed of ω in such a direction that the seat rises and there is no friction, how much higher is the seat when it stops turning?

7.5   Combined Translation and Rotation

20. A hoop rolls down an inclined plane. What fraction of its total KE is associated with its rotation?

21. A solid sphere of mass M_1 and radius R_1 and a hollow sphere of mass M_2 and radius R_2 start rolling down an inclined plane at the same moment. Find the ratio of their linear speeds at the bottom of the plane.

7.6   Angular Acceleration

22. A wheel starts from rest at a constant angular acceleration and turns through 500 radians in 10 s. (a) What is this angular acceleration? (b) What is the final angular speed?

23. A truck undergoes an acceleration of 0.25 m/s². If its wheels are 1.0 m in diameter, what is their angular acceleration?

24. A circular saw blade rotating at 15 rev/s is brought to a stop in 125 revolutions. How much time did this take?

25. An engine idling at 10 rev/s is accelerated at 2.5 rev/s² to 20 rev/s. How many revolutions does it make during this acceleration?

26. A phonograph turntable slows down to a stop in 20 s from an initial angular speed of 3.5 rad/s. (a) What is its angular acceleration? (b) How many turns does it make while slowing down?

27. A record is dropped on a turntable rotating at $33\frac{1}{3}$ rpm. The record slips briefly before reaching the same speed as the turntable. If the turntable rotates through 20° while the record is slipping, find the record's angular acceleration.

7.7 Torque

28. The torque required to fracture a person's tibia (the large bone in the lower leg) is about 100 N·m. If the distance between the pivoting heel of the safety binding on a ski and the toe release is 35 cm, what is the maximum horizontal force at which the release should be set to open?

29. The torque used to tighten a $\frac{1}{4}$-in bolt should not exceed 80 lb · in. Express this torque in newton · meters.

30. A **couple** consists of two forces of equal magnitude that act in opposite directions along parallel lines of action a distance d apart. If the magnitude of each force is F, what is the torque exerted by a couple?

31. A 60-lb cylindrical flywheel 1.0 ft in radius is acted upon by a torque of 5.0 lb · ft. (a) If it starts from rest, how much time is required to accelerate it to an angular speed of 30 rpm? (b) What is its KE at this angular speed?

32. A rope 1.0 m long is wound around the rim of a drum of radius 12 cm and moment of inertia 0.020 kg · m². The rope is pulled with a force of 2.5 N. (a) Assuming that the drum is free to rotate without friction, find its final angular speed. (b) What is its final KE? (c) How much work is done by the force?

33. A torque of 500 N · m is applied to a turbine rotating at 200 rad/s. After 40 s its speed has doubled. What is the turbine's moment of inertia?

34. A cylinder whose axis is fixed has a string wrapped around it that is pulled with a force equal to the cylinder's weight. Show that the acceleration of the string is equal to $2g$.

35. A flywheel whose moment of inertia is 8 slug · ft² is rotating at 50 rad/s. (a) What constant torque is needed to slow it down to 10 rad/s in 20 s? (b) What is the angular displacement of the flywheel while it is being slowed down? (c) How much KE does the flywheel lose?

7.8 Work and Power

36. The cutting tool of a lathe exerts a torque of 15 lb · ft on a steel shaft. Find the power needed to turn the shaft at 8.0 rev/s.

37. A 1.0-hp motor rotates at 1200 rpm. How much torque can it exert?

38. A diesel engine can exert 140 N · m of torque when it is turning at 250 rad/s. Find the corresponding power output.

39. A high-speed elevator is being planned to lift a total load of 4000 kg at 6.0 m/s. The winding drum is to be 1.8 m in diameter. Neglecting losses, how much power is required from the driving motor? At what angular speed should the power be developed?

40. A lathe whose motor has a power output of 2.0 kW is being used to cut a round bar 12 cm in diameter that is turning at 200 rpm. If 10% of the motor's power is lost in friction, what is the maximum force the cutting tool can exert on the bar?

41. An electric motor develops 5.0 kW at 2000 rpm. The motor delivers its power through a spur gear 20 cm in diameter. If two teeth of the gear transmit torque to another gear at the same instant, find the force exerted by each gear tooth.

42. A V-8 engine that develops 200 kW at 90 rev/s is coupled through a frictionless 5:1 reduction gear to a windlass drum 80 cm in diameter. (a) What is the heaviest mass the windlass can raise? (b) At what linear speed will it raise such a load?

43. A 30-rev/s motor operates a pump through a V-belt drive. The motor pulley is 20 cm in diameter, and the tension in the belt is 180 N on one side and 70 N on the other. What is the power output of the motor?

7.9 Angular Momentum

44. What are the initial and final angular momenta of the flywheel in Exercise 35?

45. A boy with his hands in his lap is spinning on a swivel chair at 1.00 rev/s. When he extends his arms, he slows down to 0.40 rev/s. (a) What is the ratio of the moments of inertia of the boy + chair combination before and after he extended his arms? (b) What is the ratio of the kinetic energies before and after?

46. A phonograph record of mass m and radius R is dropped on a turntable with the same radius and the mass M whose initial angular speed is ω. The turntable is rotating freely with no torque acting on it. What is the final angular speed of the system? Assume that the turntable is a uniform disk.

47. (a) What is the angular momentum of a particle of mass m that moves in a circle of radius r at the speed v? (b) An earth satellite follows an elliptical orbit in which its maximum altitude above the earth's surface is 2000 km and its minimum altitude is 400 km. Use the result of (a) to find the ratio between the maximum and minimum speeds of the satellite.

48. A model train whose mass is 100 g runs at 0.80 m/s (relative to the track) around a circular track mounted on the rim of a uniform 4.0-kg disk whose radius is 1.2 m. The disk can turn freely about a vertical axis through its center. If the disk was at rest when the train started to move, what is the angular speed of the disk?

PROBLEMS

49. A neutron speed selector consists of two disks of a neutron-absorbing material mounted 1.0 m apart on a shaft that turns at 1000 rpm, as in Fig. 7.25. Each disk has a radial slot, and the angular separation of the slots is 3″ (1° = 60′ and 1′ = 60″). What is the speed of the neutrons that can pass through this device?

Figure 7.25

Problem 49 (3″ = 3 sec of arc).

50. The tires of a car on a muddy road are 35 cm in radius and are rotating at 40 rad/s. What is the maximum distance a lump of mud can be thrown behind a tire?

51. A uniform thin rod of length L is pivoted about a horizontal axis at one end. If it is released from a horizontal position, what will its maximum angular velocity be?

52. A uniform knitting needle 20 cm long is balanced on its point on a table. After a moment it falls over. Find the speed with which the upper end of the needle strikes the table.

53. A bottle of mass m is tied to the free end of a string wrapped around a uniform cylinder of mass M and radius R whose axis is horizontal. If the cylinder can turn freely, what is the linear speed of the bottle after it has fallen through a height h?

54. A 3.0-kg hoop 1.0 m in diameter rolls down an inclined plane 10 m long that is at an angle of 20° with the horizontal. (a) What is the angular speed of the hoop at the bottom of the plane? (b) What is its linear speed? (c) What is its rotational kinetic energy? (d) What is its total kinetic energy?

55. The uniform door to a bank vault is 2.0 m high and 1.0 m wide and is supported by frictionless hinges. To determine the door's mass, a person pushes against its free end with a constant force of 250 N always directed perpendicularly to its plane and finds that 1.5 s is needed to turn the door through 90°. What is the door's mass? (*Note:* The formula for the moment of inertia of a uniform rectangular plate about one of its edges is the same as one of the formulas given in Fig. 7.10.)

56. An object of mass m on a frictionless table is attached to the end of a string that passes through a frictionless hole in

Figure 7.26

Problem 56.

the table, as shown in Fig. 7.26. The object is moving in a circle of radius R with the angular speed ω when the string is pulled through the hole until the object's orbital radius is $R/2$. (a) What is the new angular speed of the object? (b) How much work had to be done to pull in the string?

57. Two 0.40-kg balls are joined by a 1.0-m string and set whirling through the air at 5.0 rev/s about a vertical axis through the center of the string. After a while the string stretches to 1.2 m. (a) What is the new angular speed? (b) What are the initial and final kinetic energies of each ball? If these are different, account for the difference.

58. A woman is sitting on a piano stool with 3.0-kg dumbbells in her outstretched arms 60 cm from the axis of the stool. She is given a push and starts to turn at 1.0 rev/s. Her arms become tired, and she performs 70 J of work to pull the dumbbells to her lap, where they are 10 cm from the axis. Find the woman's moment of inertia about the axis of the stool. Neglect the contribution of her arms to the moment of inertia of the system.

59. A disk of moment of inertia 1.0 slug · ft^2 that is rotating at 100 rad/s is pressed against a similar disk that is initially at rest but is able to rotate freely. The two disks stick together and rotate as a unit. (a) What is the final angular speed of the combination? (b) If any kinetic energy was lost, where did it go? (c) The two disks are now separated. What are their new angular speeds? (d) If any kinetic energy was lost in the separation, where did it go?

Answers to Multiple Choice

1. c	9. c	17. d	25. b
2. d	10. c	18. b	26. a
3. c	11. d	19. a	27. b
4. a	12. a	20. d	28. a
5. a	13. b	21. c	29. b
6. d	14. b	22. c	
7. c	15. b	23. c	
8. a	16. c	24. a	

8

EQUILIBRIUM AND MECHANICAL ADVANTAGE

8.1   Translational Equilibrium

8.2   Rotational Equilibrium

8.3   Center of Gravity

8.4   Finding a Center of Gravity

8.5   Mechanical Advantage

8.6   Some Simple Machines

T hus far we have been concerned with moving objects, the study of which is called dynamics. Now, in a change of pace, we look at statics, the study of objects at rest. Our interest centers on the conditions under which something acted on by two or more forces is nevertheless not accelerated. It is not enough that such forces should balance out to give no net force: If the lines of action of the forces do not meet at a common point, they will produce a torque that starts the object spinning. Evidently an object at rest cannot have either a net force or a net torque acting on it. This is a simple enough idea, but one that is not always so simple to apply. At the end of the chapter we will examine mechanical advantage, another topic of practical importance.

8.1   TRANSLATIONAL EQUILIBRIUM

It occurs when force components in each direction add up to zero.

We begin by considering only forces that act through the same point on an object. This allows us to ignore torques. Later we consider situations in which torques are important.

When the forces that act on an object have a vector sum of zero, the object is said to be in **translational equilibrium.** With no net force there is no linear acceleration. Of course, such an object need not be at rest but may be moving along a straight path at constant speed.

The condition for translational equilibrium is easy to express:

$$\Sigma \mathbf{F} = 0 \qquad\qquad\qquad \textit{Translational equilibrium} \quad (8.1)$$

As before, Σ means "sum of," and here $\mathbf{F}$ refers to the various forces that act *on* the object under study.

In many equilibrium situations all the forces lie in the same plane. When this is the case, we can establish a set of *x-y* coordinate axes wherever convenient in the plane and then resolve each force $\mathbf{F}$ into the components $\mathbf{F}_x$ and $\mathbf{F}_y$. Thus we can replace Eq. (8.1), which is a vector equation, with the two scalar equations

$$\text{Sum of } x \text{ force components} = \Sigma F_x = 0 \qquad\qquad (8.2)$$

$$\text{Sum of } y \text{ force components} = \Sigma F_y = 0 \qquad\qquad (8.3)$$

Is is usually much easier to calculate the components of each force present and to make use of Eqs. (8.2) and (8.3) than it is to work with the forces themselves in Eq. (8.1).

There are five steps to follow in working out problems concerning the translational equilibrium of an object:

---

1. **Draw a sketch of the forces that act on the object. (As we know, this is called a *free-body* diagram.) Do not show the forces that the object exerts on anything else, because such forces do not affect the equilibrium of the object itself.**

2. **Choose a convenient set of coordinate axes and resolve the various forces acting on the object into components along these axes. Be sure to use + and − signs consistently. If the *y*-axis is vertical and the *x*-axis is horizontal, for instance, an upward force would be considered as +, a downward force as −, a force to the right as +, and a force to the left as − (Fig. 8.1).**

Figure 8.1

Sign conventions for the x- and y-components of a force.

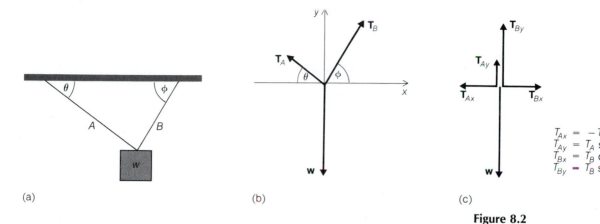

$$T_{Ax} = -T_A \cos \theta$$
$$T_{Ay} = T_A \sin \theta$$
$$T_{Bx} = T_B \cos \phi$$
$$T_{By} = T_B \sin \phi$$

(a) (b) (c)

Figure 8.2

(a) A box of weight w is suspended by two ropes. (b) A free-body diagram of the forces acting on the box. (c) At equilibrium the sum ΣF_x of the horizontal force components and the sum ΣF_y of the vertical force components each equals zero.

3. **Set the sum of the force components along each axis equal to 0, as specified in Eqs. (8.2) and (8.3). This is the condition for equilibrium.**

4. **Solve the resulting equations for the unknown quantity or quantities.**

5. **Substitute numerical values of the known quantities to find the answer.**

An example will make clear this procedure. Figure 8.2 shows a box of weight w that is supported by two weightless ropes, A and B. The ropes are at the respective angles θ and ϕ with the horizontal. We note that the tension T in a rope is a force along its length in a direction away from its point of attachment at either end, as we saw in Fig. 3.14.

We begin by resolving the tension in each rope into components in the x- and y-directions, so that we have T_{Ay} and T_{By} upward, T_{Ax} to the left, and T_{Bx} to the right. From Fig. 8.2(b) we have

$$T_{Ax} = -T_A \cos \theta \qquad T_{Bx} = T_B \cos \phi \qquad \textit{Horizontal components}$$
$$T_{Ay} = T_A \sin \theta \qquad T_{By} = T_B \sin \phi \qquad \textit{Vertical components}$$

We may now replace the actual tensions T_A and T_B by their components, as in Fig. 8.2(c). For the forces in the vertical and horizontal directions to cancel separately,

$$\Sigma F_x = T_{Bx} + T_{Ax} = 0 \qquad \textit{Horizontal direction} \quad (8.4)$$
$$\Sigma F_y = T_{Ay} + T_{By} - w = 0 \qquad \textit{Vertical direction} \quad (8.5)$$

◆ **EXAMPLE 8.1** _____

In the preceding situation $w = 100$ N, $\theta = 37°$, and $\phi = 60°$. Find the tension in each rope.

SOLUTION   From Eq. (8.4) we find that

$$T_{Bx} + T_{Ax} = 0$$

$$T_B \cos \phi - T_A \cos \theta = 0$$

$$T_B = T_A \frac{\cos \theta}{\cos \phi} = T_A \frac{\cos 37°}{\cos 60°} = 1.6 T_A$$

From Eq. (8.5) we find that

$$T_{Ay} + T_{By} - w = 0$$

$$T_A \sin \theta + T_B \sin \phi - 100 \, \text{N} = 0$$

Substituting $1.6 T_A$ for T_B in the last equation, we obtain

$$T_A \sin \theta + 1.6 T_A \sin \phi - 100 \, \text{N} = 0$$

$$T_A(\sin 37° + 1.6 \sin 60°) = 100 \, \text{N}$$

$$T_A = 50 \, \text{N}$$

The tension in rope A is 50 N. Because $T_B = 1.6 T_A$ here,

$$T_B = (1.6)(50 \, \text{N}) = 80 \, \text{N}$$

The tension in rope B is 80 N. The algebraic sum of the tensions in the two ropes is 130 N, which is more than the weight being supported, but this sum means nothing. The vector sum of $\mathbf{T}_A$ and $\mathbf{T}_B$ is 100 N acting upward, and this cancels the downward force $\mathbf{w} = 100$ N. ◆

◈   **E X A M P L E   8.2** _____

A 100-N box is suspended from the end of a horizontal strut, as in Fig. 8.3. Find the tension in the cable supporting the strut under the assumption that the strut's weight is negligible.

Figure 8.3

(a) A box of weight w is suspended from the end of a horizontal strut held in place by a cable attached to the wall. (b) A free-body diagram of the forces acting on the end of the strut. (c) At equilibrium $\Sigma F_x = 0$ and $\Sigma F_y = 0$.

(a)

(b)

(c)

SOLUTION   It is easiest to consider the equilibrium of the end of the strut. The three forces that act on the end of the strut are the tension **T** in the cable, the outward force **F**$_s$ exerted by the strut itself, and the weight **w** of the box. The horizontal and vertical components of the tension T are, from the diagram,

$$T_x = -T \cos 30° \qquad T_y = T \sin 30°$$

The end of the strut is in equilibrium when

$$\Sigma F_x = T_x + F_s = 0 \quad \text{and} \quad \Sigma F_y = T_y - w = 0$$

All we need is the second of these equations to find T:

$$T_y - w = 0$$
$$T \sin 30° = w$$
$$T = \frac{w}{\sin 30°} = \frac{100 \text{ N}}{0.500} = 200 \text{ N}$$

◆ **EXAMPLE 8.3** _____

Find the inward force the strut of the previous example exerts on the wall.

SOLUTION   Because this force is horizontal, we need the equation that expresses the equilibrium of the strut in the x-direction, which is

$$\Sigma F_x = T_x + F_s = 0$$

With $T = 200$ N and $T_x = -T \cos 30°$,

$$\Sigma F_x = -T \cos 30° + F_s = 0$$
$$F_s = T \cos 30° = (200 \text{ N})(\cos 30°) = 173 \text{ N} \qquad ◆$$

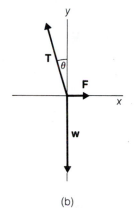

◆ **EXAMPLE 8.4** _____

A 200-kg motor is suspended from the ceiling by a cable 6.0 m long, as in Fig. 8.4(a). If the coefficient of static friction between his shoes and the floor is $\mu_s = 0.6$, the maximum horizontal force a 70-kg person can exert on the motor before he starts to slip is $F = \mu_s \, mg = 412$ N. Find the horizontal displacement of the motor when a horizontal force of 412 N is applied to it.

SOLUTION   Since we know that L in Fig. 8.4(a) is 6.0 m, we must find the angle θ in order to determine the displacement s. We begin by resolving the tension **T** in the cable into its horizontal and vertical components. These are

$$T_x = -T \sin \theta \qquad T_y = T \cos \theta$$

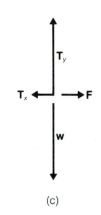

Figure 8.4

Next we substitute into Eqs. (8.2) and (8.3) to obtain

$$\Sigma F_x = T_x + F = -T\sin\theta + F = 0$$
$$\Sigma F_y = T_y - w = T\cos\theta - w = 0$$

We now solve the first equation for $\sin\theta$ and the second for $\cos\theta$:

$$T\sin\theta = F \quad \text{or} \quad \sin\theta = \frac{F}{T} \qquad T\cos\theta = w \quad \text{or} \quad \cos\theta = \frac{w}{T}$$

Because $\sin\theta/\cos\theta = \tan\theta$,

$$\tan\theta = \frac{\sin\theta}{\cos\theta} = \frac{F/T}{w/T} = \frac{F}{w} = \frac{F}{mg} = \frac{412\,\text{N}}{(200\,\text{kg})(9.8\,\text{m/s}^2)} = 0.210$$

and

$$\theta = \tan^{-1} 0.210 = 12°$$

It is not necessary to know the tension T in order to find θ. From Fig. 8.4(a) we have for the horizontal displacement s of the motor

$$s = L\sin\theta = (6.0\,\text{m})(\sin 12°) = 1.2\,\text{m} \qquad \blacklozenge$$

◆   **E X A M P L E   8.5** _____

A boom hinged at the base of a vertical mast is used to lift a 2000-lb load, as in Fig. 8.5. Find the tension in the cable from the top of the mast to the end of the boom. The weight of the boom itself can be neglected.

SOLUTION   We shall consider the equilibrium of the end of the boom. As shown in Fig. 8.5, three forces act here: the weight **w** of the boat, the tension **T** in the cable,

Figure 8.5

The vectors in (b) are not drawn to scale.

and the outward reaction force **R** of the boom. Because **R** is along the same line as the boom,

$$\theta_3 = \theta_1 = 40°$$

The sum of the acute angles of a right triangle is always 90° (see Appendix A–8), hence

$$\theta_2 + \theta_4 = 90°$$
$$\theta_4 = 90° - \theta_2 = 90° - 60° = 30°$$

The x- and y-components of **T** and **R** are as follows:

$$T_x = -T \cos \theta_4 \qquad R_x = R \cos \theta_3$$
$$T_y = T \sin \theta_4 \qquad R_y = R \sin \theta_3$$

Hence the conditions for equilibrium in the x- and y-directions are

$$\Sigma F_x = -T \cos \theta_4 + R \cos \theta_3 = 0$$
$$\Sigma F_k\ T \sin \theta_4 + R \sin = \theta_3 - w = 0$$

We are not interested in the value of R. To eliminate R we begin by solving the first equilibrium equation for it:

$$R \cos \theta_3 = T \cos \theta_4$$
$$R = T \left(\frac{\cos \theta_4}{\cos \theta_3} \right)$$

Substituting this value of R in the second equilibrium equation gives

$$T \sin \theta_4 + T \left(\frac{\cos \theta_4}{\cos \theta_3} \right) \sin \theta_3 - w = 0$$

Since $\sin \theta_3 / \cos \theta_3 = \tan \theta_3$,

$$T \sin \theta_4 + T \cos \theta_4 \tan \theta_3 - w = 0$$
$$T (\sin \theta_4 + \cos \theta_4 \tan \theta_3) = w$$
$$T = \frac{w}{\sin \theta_4 + \cos \theta_4 \tan \theta_3}$$
$$= \frac{2000\ \text{lb}}{\sin 30° + (\cos 30°)(\tan 40°)}$$
$$= \frac{2000\ \text{lb}}{1.227} = 1630\ \text{lb}$$

Another way to eliminate R is to align the x- and y-axes so that the x-axis is along the boom. Then we need consider only the equilibrium of the boom end in the new y-direction, which will involve only components of **w** and **T.** The advantage of proceeding as we did here is that, although there is a little more algebra, the geometry is less confusing when the $x =$ axis is horizontal and the y-axis is vertical.   ◆

8.2   ROTATIONAL EQUILIBRIUM

The sum of the torques about any point must be zero.

When the lines of action of the various forces that act on an object intersect at a common point, they do not tend to set the object in rotation. Such forces are said to be **concurrent.**

If the lines of action of the various forces do *not* intersect, the forces are **nonconcurrent,** and the object may be set into rotation even though the vector sum of the forces may equal zero. In Fig. 8.6(a) the three applied forces are concurrent and, if their vector sum is zero, the object is in equilibrium. In Fig. 8.6(b) the same forces are nonconcurrent, and they combine to produce a counterclockwise torque.

By convention a torque that tends to produce a counterclockwise rotation is considered positive and a torque that tends to produce a clockwise rotation is considered negative (Fig. 8.7). Thus the condition for an object to be in **rotational equilibrium** is that the sum of the torques acting on it about any point, using the convention given above for plus and minus signs, be zero:

$$\Sigma \tau = 0 \qquad\qquad\qquad \textit{Rotational equilibrium} \quad (8.6)$$

Of course, if the various forces that act do not all lie in the same plane, the sum of the torques in each of three mutually perpendicular planes must be zero.

It is possible to prove that if the sum of the torques on an object in translational equilibrium is zero about any point, it is also zero about all other points. Hence

Figure 8.6

(a) The lines of action of concurrent forces intersect at a common point. (b) When the lines of action do not intersect at a common point, the forces are nonconcurrent and the object cannot be in rotational equilibrium.

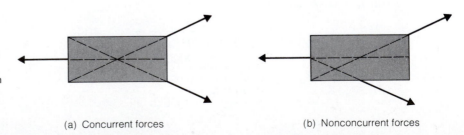

(a) Concurrent forces   (b) Nonconcurrent forces

(a) $\tau = FL$ (b) $\tau = -FL$

Figure 8.7

(a) A torque that tends to produce a counterclockwise rotation is considered positive. (b) A torque that tends to produce a clockwise rotation is considered negative.

the location of the point about which torques are calculated in an equilibrium problem is completely arbitrary; *any* point will do. (Of course, *all* torques must be calculated about this point.) Let us verify this statement with an example.

◆ **E X A M P L E 8.6**

Figure 8.8 shows a rod 2.0 m long that has weights of 10 N and 30 N at its ends. We assume that the weight of the rod is negligible. At what point should the rod be picked up if it is to have no tendency to rotate? In other words, where is the balance point of the rod?

SOLUTION 1   We first compute torques about the unknown balance point. If x is the distance of the 30-N weight from this point, the 10-N weight is $(2.0 \text{ m} - x)$ from it on the other side. The torques these weights exert are

$$\tau_1 = w_1 L_1 = +30x \text{ N}$$
$$\tau_2 = -w_2 L_2 = -10(2.0 \text{ m} - x) \text{ N}$$

The torque τ_2 is negative because it acts clockwise. Equilibrium will result when

$$\Sigma\tau = \tau_1 + \tau_2 = 30x \text{ N} - 10(2.0 \text{ m} - x) \text{ N} = 0$$
$$30x = 10(2.0 \text{ m} - x)$$
$$40x = 20 \text{ m}$$
$$x = 0.5 \text{ m}$$

Figure 8.8

Torques are computed about the unknown balance point of the rod in Solution 1.

Figure 8.9

Torques are computed about the center of the rod in Solution 2.

When the rod is picked up 0.5 m from the 30-N weight, the two weights exert opposite torques of the same magnitude (15 N · m) about this point, so the rod is in balance. ◆

SOLUTION 2   Let us solve the same problem by calculating torques about the middle of the rod, as shown in Fig. 8.9. Here y represents the distance between the balance point and the middle of the rod. We now have three torques to consider:

$$\tau_1 = w_1L_1 = + (30\,\text{N})(1.0\,\text{m}) = + 30\,\text{N} \cdot \text{m}$$
$$\tau_2 = FL_2 = - 40y\,\text{N}$$
$$\tau_3 = w_2L_3 = - (10\,\text{N})(1.0\,\text{m}) = - 10\,\text{N} \cdot \text{m}$$

The condition for equilibrium is

$$\Sigma\tau = \tau_1 + \tau_2 + \tau_3 = 30\,\text{N} \cdot \text{m} - 40y\,\text{N} - 10\,\text{N} \cdot \text{m} = 0$$

from which we find

$$y = 0.5\,\text{m}$$

The location of the balance point is the same regardless of the point about which we calculate torques. It is usually wise to calculate torques about the point of application of one of the forces that act on an object, because this makes it unnecessary to consider the torque produced by that force.   ◆

Not all equilibrium situations are necessarily stable. For instance, a cone balanced on its apex is in equilibrium, but it will fall over when disturbed even slightly (Fig. 8.10). This is an example of an **unstable equilibrium.** The same cone on its base will return to its original position if tipped over a little; hence it is in **stable equilibrium** on its base. There is a third possibility as well, illustrated by a cone lying on its side. If such a cone is displaced, it stays in equilibrium in its new position without tending either to move further or to return to where it was before. A cone on its side is said to be in **neutral equilibrium.**

Stable
equilibrium

Unstable
equilibrium

Neutral
equilibrium

Figure 8.10

These cones are all in equilibrium, but only one of them is in a
stable position.

8.3 CENTER OF GRAVITY

The balance point of an object.

The **center of gravity** (CG) of an object is that point from which it can be
suspended without tending to rotate (Fig. 8.11). Each particle in an object has a
certain weight and therefore exerts a torque about whatever point the object is
suspended from. There is only one point for an object about which all these torques
cancel out no matter how it is oriented; this is its center of gravity. For equilibrium
purposes we can therefore regard the entire weight of an object as a downward force
acting from its CG.

We can now recognize the distinctions between the different kinds of equilibrium shown in Fig. 8.10. The left-hand cone is in stable equilibrium because its
CG has to be raised to tip it over; the center cone is in unstable equilibrium because
any change in its orientation lowers its CG; and the right-hand cone is in neutral
equilibrium because the height of its CG does not change when it is rolled along on
its side.

If the rod in Figs. 8.8 and 8.9 had the weight w instead of being weightless,
we could take into account its effect on the location of the balance point by including
the torque due to a force of magnitude w acting downward at the center of the rod.
The CG of a uniform object of regular shape is located at its geometrical center. The
CG of an irregular object need not even be located within the object itself. The CG
of a seated person, for example, is a few inches in front of the person's abdomen.

Center of
gravity

Figure 8.11

A body suspended from its center
of gravity is in equilibrium in any
orientation.

Figure 8.12

◆   E X A M P L E   8.7

A beam that projects beyond its supports is called a **cantilever.** A diving board is an example. Find the forces exerted by the two supports of the 4.0-m, 50-kg uniform diving board shown in Fig. 8.12 when a 60-kg woman stands at its end.

SOLUTION   The weight of the diving board is $w_1 = m_1 g_1 = 490$ N, and that of the woman is $w_2 = m_2 g = 588$ N. It is obvious that the force F_1 exerted by the left-hand support must be downward. (If we nevertheless were to consider F_1 as upward, the result would be a negative value for F_1, signifying the opposite direction.) We can calculate F_1 without knowing F_2 by computing torques about the point where F_2 is applied. The CG of the board is at its middle, which is 1.2 m from the pivot point. Hence the three torques that act about this point are

$$\tau_1 = +F_1 x_1 = +(F_1)(0.8 \text{ m}) = 0.8 F_1 \text{ m}$$
$$\tau_2 = -w_1 x_2 = -(490 \text{ N})(1.2 \text{ m}) = -588 \text{ N} \cdot \text{m}$$
$$\tau_3 = -w_2 x_3 = -(588 \text{ N})(3.2 \text{ m}) = -1882 \text{ N} \cdot \text{m}$$

Adding the torques and setting their sum equal to zero gives

$$\Sigma \tau = \tau_1 + \tau_2 + \tau_3 = 0.8 F_1 \text{ m} - 588 \text{ N} \cdot \text{m} - 1882 \text{ N} \cdot \text{m} = 0$$
$$F_1 = 3088 \text{ N}$$

We can find F_2 by considering the translational equilibrium in the y-direction of the loaded board:

$$F_2 = F_1 + w_1 + w_2 = 3088 \text{ N} + 490 \text{ N} + 588 \text{ N} = 4166 \text{ N} \qquad ◆$$

◆   E X A M P L E   8.8

A horizontal beam 12.0 ft long is supported at one end by a vertical post and at the other by a cable that makes an angle of 40° with the beam, as in Fig. 8.13. A load of 3000 lb is suspended from the outer end of the beam, which itself weighs 430 lb. Find the tension in the cable.

SOLUTION   Four forces act on the beam in this problem: the load w_1, the beam's own weight w_2, which may be regarded as acting at its center, the tension **T** in the cable, and the force **F** exerted by the post on the inner end of the beam. By calculating torques about the inner end of the beam we eliminate F, a convenient simplication. The torques τ_1 exerted by the load, τ_2 by the beam's weight, and τ_3 by the cable on the beam about its inner end are respectively

$$\tau_1 = w_1 L_1 = +(3000 \text{ lb})(12.0 \text{ ft}) = 36{,}000 \text{ lb} \cdot \text{ft}$$
$$\tau_2 = w_2 L_2 = +(430 \text{ lb})(6.0 \text{ ft}) = 2580 \text{ lb} \cdot \text{ft}$$
$$\tau_3 = T_y L_1 = -(T \sin \theta) L_1 = -(12.0T)(\sin 40°)\text{ft} = -7.71T \text{ ft}$$

(In finding τ_3 the vertical component $\mathbf{T}_y$ of the tension in the rope was used as the force; the length L_1 of the beam is the moment arm because the beam is horizontal. The sign of τ_3 is negative, since it acts clockwise; the other torques are positive, since they act counterclockwise.) The condition for the equilibrium of the beam is

$$\Sigma\tau = \tau_1 + \tau_2 + \tau_3 = 36{,}000 \text{ lb} \cdot \text{ft} + 2580 \text{ lb} \cdot \text{ft} - 7.71T \text{ ft} = 0$$

Figure 8.13

By calculating torques about the inner end of the beam, the problem of finding the value of T is simplified. The horizontal component T_x of the tension in the rope is not shown.

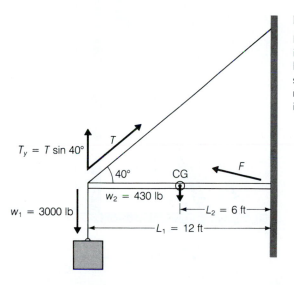

Solving for the magnitude of the tension T gives

$$7.71T = 36,000 \text{ lb} + 2580 \text{ lb} = 38,580 \text{ lb}$$

$$T = \frac{38,580 \text{ lb}}{7.71} = 5.00 \times 10^3$$

◆

◆   **E X A M P L E   8.9** _____

A ladder 4.0 m long is leaning against a frictionless wall with its lower end 1.6 m away from the wall, as in Fig. 8.14. If the ladder weighs 150 N, what forces does it exert on the wall and on the ground?

SOLUTION   Because the wall is frictionless, the only force it can exert on the ladder is the horizontal reaction force $\mathbf{F}_1$. The other forces on the ladder are its weight $\mathbf{w}$, which acts downward from its midpoint, and the reaction force $\mathbf{F}_2$ exerted by the ground. It is worth mentioning that the direction of $\mathbf{F}_2$ need not lie along the line of the ladder.

Because F_{2y}, the upward component of $\mathbf{F}_2$, and $\mathbf{w}$ are the only vertical forces and $\mathbf{w}$ is in the $-y$-direction,

$$\Sigma F_y = F_{2y} - w = 0$$
$$F_{2y} = w = 150 \text{ N}$$

Figure 8.14

To simplify the calculation, the pivot point about which torques are calculated is chosen as the bottom of the ladder.

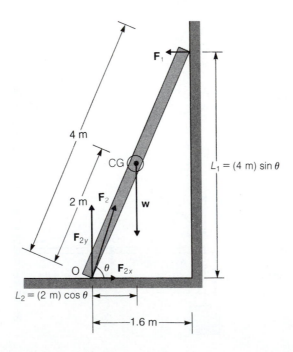

Similarly the horizontal force $\mathbf{F}_1$ acting to the left must have the same magnitude as $\mathbf{F}_{2x}$ acting to the right:

$$\Sigma F_x = F_{2x} - F_1 = 0$$
$$F_{2x} = F_1$$

Before going further we need the value of the angle θ. From Fig. 8.14

$$\cos \theta = \frac{1.6 \text{ m}}{4.0 \text{ m}} = 0.40$$
$$\theta = \cos^{-1} 0.40 = 66°$$

Let us find F_1 first. The easiest way is to calculate torques about the lower end of the ladder, because then we will have only the two torques produced by $\mathbf{w}$ and by $\mathbf{F}_1$ to consider. We have

$$\tau_1 = F_1 L_1 = (F_1)(4.0 \text{ m})\sin 66° = 3.65 F_1 \text{ m}$$
$$\tau_2 = -w L_2 = -(150 \text{ N})(2.0 \text{ m})\cos 66° = -122 \text{ N} \cdot \text{m}$$

The sum of these torques must be zero for the ladder to be in equilibrium, so

$$\tau = \tau_1 + \tau_2 = 0$$
$$\tau_1 = -\tau_2$$
$$3.65 F_1 \text{ m} = -(-122 \text{ N} \cdot \text{m}) = 122 \text{ N} \cdot \text{m}$$
$$F_1 = \frac{122 \text{ N} \cdot \text{m}}{3.65 \text{ m}} = 33.4 \text{ N}$$

The force the wall exerts on the ladder is horizontal with the magnitude 33.4 N, and this is equal and opposite to the force the ladder exerts on the wall.

Now we must find $\mathbf{F}_2$. Since $F_{2x} = F_1$ and $F_{2y} = w$, the total force the ground exerts on the ladder is

$$F_2 = \sqrt{F_{2x}^2 + F_{2y}^2} = \sqrt{F_1^2 + w^2}$$
$$= \sqrt{(33.4 \text{ N})^2 + (150 \text{ N})^2} = 154 \text{ N}$$

The force the ladder exerts on the ground has the same magnitude. The angle θ between $\mathbf{F}_2$ and the vertical is

$$\theta = \tan^{-1} \frac{F_{2x}}{F_{2y}} = \tan^{-1} \frac{33.4 \text{ N}}{150 \text{ N}} = \tan^{-1} 0.223 = 13° \qquad \blacklozenge$$

8.4   FINDING A CENTER OF GRAVITY

A straightforward calculation.

An experimental way to find the CG of an irregular object is shown in Fig. 8.15. An analytical procedure follows from the definition of CG. What we do is imagine the object divided into two or more separate elements whose CGs are known, then consider these elements as particles joined by weightless rods, and finally calculate the balance point of the assembly. If the object is complex—for instance, the hull of a ship—a great many elements must be considered for an accurate result, but the method is still straightforward.

To see how such a calculation can be made, let us consider a system of three particles whose masses are m_1, m_2, and m_3. They are located x_1, x_2, and x_3, respectively, from one end of a weightless rod, as in Fig. 8.16. The CG of the system is located at some distance X from the end of the rod. This means that the torque exerted about this end by a single particle of mass $M = m_1 + m_2 + m_3$ located at X will be exactly the same as the torque exerted by the actual system of the three particles. Hence

$$\Sigma\tau = m_1 g x_1 + m_2 g x_2 + m_3 g x_3 = MgX = (m_1 + m_2 + m_3)gX$$

so that

$$X = \frac{m_1 x_1 + m_2 x_2 + m_3 x_3}{m_1 + m_2 + m_3}$$

Figure 8.15

To find the center of gravity of a flat body, suspend it and a plumb bob successively from two different points on its edge. The center of gravity is located at the intersection of the two lines of action of the plumb bob.

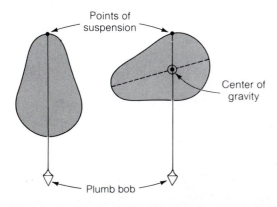

Figure 8.16

The center of gravity of a complex object can be found by considering it as a system of particles and applying Eq. (8.7).

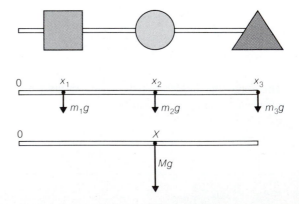

This formula can be generalized to a system of any number of particles (or objects that can be represented by particles located at their CGs) in a straight line, giving

$$X = \frac{m_1 x_1 + m_2 x_2 + m_3 x_3 + \cdots}{m_1 + m_2 + m_3 + \cdots}$$

Center of gravity of   (8.7)
composite object

◆   **E X A M P L E   8.10** _____

The hand, forearm, and upper arm of a woman have the respective masses 0.4 kg, 1.2 kg, and 1.9 kg, and their CGs are, respectively, 0.60 m, 0.40 m, and 0.15 m from her shoulder joint (Fig. 8.17). Find the distance of the CG of her entire unbent arm from the shoulder joint.

Figure 8.17

SOLUTION   From Eq. (8.7)

$$X = \frac{(0.4\,\text{kg})(0.60\,\text{m}) + (1.2\,\text{kg})(0.40\,\text{m}) + (1.9\,\text{kg})(0.15\,\text{m})}{(0.4 + 1.2 + 1.9)\,\text{kg}}$$

$$= 0.29\,\text{m}$$

The CG of the entire arm is 0.29 m from the shoulder joint.   ◆

_____

If the irregular object whose CG is to be found lies in a plane rather than along a straight line (or is three-dimensional), the same procedure is applied along two (or three) coordinate axes. An example will make this clear.

◆   **E X A M P L E   8.11** _____

Find the CG of the L-shaped steel plate shown in Fig. 8.18.

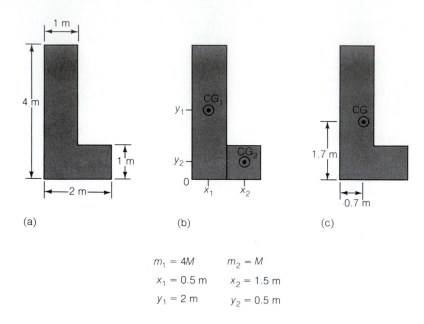

$$m_1 = 4M \qquad m_2 = M$$
$$x_1 = 0.5 \text{ m} \qquad x_2 = 1.5 \text{ m}$$
$$y_1 = 2 \text{ m} \qquad y_2 = 0.5 \text{ m}$$

Figure 8.18

SOLUTION   We imagine the plate to consist of two sections, one a rectangle 4 m long by 1 m wide and the other a square 1 m on a side. The CGs of these sections are at their geometric centers, as shown in Fig. 8.18(b). Now we replace each section by a particle at its CG. If the steel plate has a mass of M per square meter, particle 1 has a mass of $4M$ and particle 2 has a mass of M, because their areas are, respectively, 4 m² and 1 m². The x- and y-coordinates of the CG of the entire plate are as follows:

$$X = \frac{m_1 x_1 + m_2 x_2}{m_1 + m_2} = \frac{(4M)(0.5 \text{ m}) + (M)(1.5 \text{ m})}{4M + M} = 0.7 \text{ m}$$

$$Y = \frac{m_1 y_1 + m_2 y_2}{m_1 + m_2} = \frac{(4M)(2 \text{ m}) + (M)(0.5 \text{ m})}{4M + M} = 1.7 \text{ m}$$

The location of this point is shown in Fig. 8.18(c). We did not need to know the value of M in this case, because the steel plate is uniform.   ◆

Net Force and Center of Gravity

The response of an object to a net force acting on it depends on whether or not the line of action of the force passes through the object's CG. In the case where the line of action passes through the CG, the force has no effect on the object's rotational motion (Fig. 8.19). If the object is not rotating to begin with, such a force will simply accelerate it in the direction of the force. If the line of action of the force does not pass through the CG, however, the object will start to spin as well as to move.

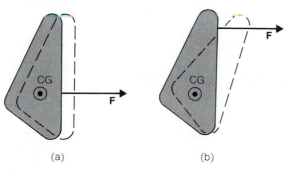

Figure 8.19

(a) A body remains in rotational equilibrium when the line of action of an applied force passes through its CG. (b) When the line of action of an applied force does not pass through its CG, the body is given a rotational as well as a translational acceleration.

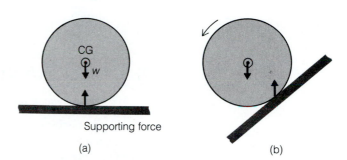

Figure 8.20

A wheel rolls downhill on an inclined surface because the line of action of the supporting force does not pass through its CG. This leads to a net torque on the wheel.

A wheel on a level surface has its CG directly above its point of contact with the surface, as in Fig. 8.20(a). The line of action of the reaction force of the surface on the wheel, which is what supports the wheel, passes through the wheel's CG. Unless it is pushed, the wheel has no acceleration. If the surface is tilted, though, the line of action of the supporting force does not intersect the wheel's CG. The result is a torque on the wheel that starts it rolling (Fig. 8.20b).

We have all noticed that a car tilts backward when it is accelerated, and that it tilts forward when the brakes are applied. These effects occur because the forces applied to the car in each case are horizontal ones that act where the tires touch the road, and are not in line with the car's CG. The car rotates when the resulting torques act on it until the springs of the suspension system restore rotational equilibrium (Fig. 8.21).

Figure 8.21

A car tilts backward when accelerated (a) and forward when braked (b) because the lines of action of the applied forces in each case do not pass through the car's CG. In (a) the car is assumed to have rear-wheel drive. If the car has front-wheel drive, the reaction forces act on its front tires, but the situation is otherwise the same.

Reaction forces of road on tires when accelerated

(a)

Reaction forces of road on tires when braking

(b)

8.5   MECHANICAL ADVANTAGE

Multiplying a force.

A **machine** is a device that transmits force or torque for a definite purpose. All machines, however complicated, are actually combinations of only three basic machines: the lever, the inclined plane, and the hydraulic press. Thus the train of gears that carries power from the engine of a car to its wheels is a development of the lever; the screw jack that can raise one end of the car from the ground is a development of the inclined plane; and the brake system that permits a touch of the foot to stop the car is a development of the hydraulic press.

The **actual mechanical advantage** (AMA) of a machine is the ratio between the output force F_{out} it exerts and the input force F_{in} that is furnished to it:

$$\text{AMA} = \frac{F_{out}}{F_{in}} \quad \textit{Actual mechanical advantage} \quad (8.8)$$

$$\text{Actual mechanical advantage} = \frac{\text{output force}}{\text{input force}}$$

An actual mechanical advantage greater than 1 means that the output force exceeds the input force, whereas an actual mechanical advantage less than 1 means that the output force is smaller than the input force. Usually the AMA is greater than 1, which makes it possible for a relatively small applied force to accomplish a task ordinarily beyond its capacity. However, sometimes an AMA less than 1 is useful. An example is a pair of scissors, where the range of motion is increased at the expense of a reduced force.

Ideal Mechanical Advantage

The **ideal mechanical advantage** (IMA) of a machine is its MA under ideal conditions. The actual MA, on the other hand, takes into account friction and any other dissipative factors that may be acting. Because these factors can never be entirely eliminated, the AMA of every machine is less than its IMA.

To obtain a general formula for IMA that can be applied to any machine, let us look at the imaginary frictionless machine of Fig. 8.22. When we apply a force F_{in} to the rope at the right of the machine, a force of F_{out} is exerted on the load at the left. How the machine works does not matter—it could be a system of pulleys, a system of gears, a hydraulic system, or something else.

Because there is no friction, the work W_{in} done *on* the machine by the input force must equal the work W_{out} done *by* the machine on the load. If the load moves through the distance s_{out} when the agent (our hand, perhaps) that applies the input

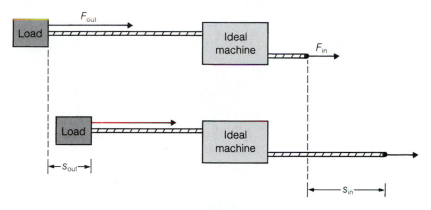

Figure 8.22

An ideal machine operates without friction to transform an input force F_{in} to an output force F_{out} that may be greater or smaller than F_{in}. If F_{out} is greater than F_{in}, as shown here, the distance s_{out} through which the load is moved is less than the distance s_{in} through which the input force acts; the reverse is true if F_{out} is smaller than F_{in}.

force moves through the distance s_{in}, then since $W = Fs$ we have

$$W_{out} = W_{in}$$
$$F_{out}s_{out} = F_{in}s_{in}$$
$$\frac{F_{out}}{F_{in}} = \frac{s_{in}}{s_{out}} \qquad \qquad \textit{Ideal machine} \quad (8.9)$$

The ratio of the distances is the inverse of the ratio of the forces. The effect of friction is always to increase the input force F_{in} needed to produce a given output force F_{out}, but the distance ratio s_{in}/s_{out} does not change. Hence we can use s_{in}/s_{out} to define IMA:

$$\text{IMA} = \frac{s_{in}}{s_{out}} \qquad \textit{Ideal mechanical advantage} \quad (8.10)$$

$$\text{Ideal mechanical advantage} = \frac{\text{input distance}}{\text{output distance}}$$

◆   **E X A M P L E   8.12** _____

Find the ideal mechanical advantage of the block and tackle shown in Fig. 8.23.

SOLUTION   When the free end of the rope is pulled with the force F_{in} through a distance d, the movable block is raised through a height of $\frac{1}{4}d$ because there are four strands that must be shortened. Hence

$$\text{IMA} = \frac{s_{in}}{s_{out}} = \frac{d}{\frac{1}{4}d} = 4$$

In general, the IMA of a block and tackle is equal to the number of strands of rope that support the movable block and thereby the load. The strand to which the tension F_{in} is applied in Fig. 8.23 does not contribute to supporting the movable block; the upper pulley merely serves to change the direction of the applied force. ◆

Figure 8.23

The theoretical mechanical advantage of a block and tackle is equal to the number of strands of rope that support the movable block, in this case 4.

Efficiency

In an actual machine, the presence of friction means that F_{in} must be greater than that specified in Eq. (8.9), and W_{in} must be greater than W_{out}. The lost work is work done to overcome friction and ends up as heat. Since the efficiency of any machine is

$$\text{Efficiency} = \frac{\text{work output}}{\text{work input}}$$

we have here

$$\text{Eff} = \frac{W_{out}}{W_{in}} = \frac{F_{out}s_{out}}{F_{in}s_{in}}$$

$$= \frac{\text{AMA}}{\text{IMA}} \qquad\qquad \textit{Efficiency} \quad (8.11)$$

The efficiency of a machine is equal to the ratio of its actual and ideal mechanical advantages. If we know the ideal mechanical advantage and the efficiency of a machine,

$$F_{out} = (\text{Eff})(\text{IMA})(F_{in}) \qquad\qquad\qquad (8.12)$$

In some machines, such as the lever, the efficiency may be very close to 100%, whereas in others, such as the screw jack, it may be 10% or less.

◆   E X A M P L E   8.13

A woman uses a chain hoist to lift a 240-lb motor. She exerts a force of 27 lb on the chain and pulls a total of 24 ft of chain through the pulleys in order to raise the motor 2.5 ft. (a) Find the ideal mechanical advantage, actual mechanical advantage, and efficiency of the hoist. (b) How much force is needed to lift a 400-lb motor with the same hoist?

SOLUTION   (a) We do not have to know how a chain hoist works in order to find these quantities. Here $F_{out} = 240$ lb, $F_{in} = 27$ lb, $s_{out} = 2.5$ ft, and $s_{in} = 24$ ft, and so

$$\text{IMA} = \frac{s_{in}}{s_{out}} = \frac{24 \text{ ft}}{2.5 \text{ ft}} = 9.6$$

$$\text{AMA} = \frac{F_{out}}{F_{in}} = \frac{240 \text{ lb}}{27 \text{ lb}} = 8.9$$

$$\text{Eff} = \frac{\text{AMA}}{\text{IMA}} = \frac{8.9}{9.6} = 0.93 = 93\%$$

(b) We must use the actual mechanical advantage to find F_{in}. From Eq. (8.8)

$$F_{in} = \frac{F_{out}}{\text{AMA}} = \frac{400 \text{ lb}}{8.9} = 45 \text{ lb}$$ ◆

When two or more machines are coupled together so that the output of one is the input of the next, the overall mechanical advantage (IMA or AMA) of the combination is the product of the mechanical advantages of each one:

$$\text{MA} = (\text{MA}_1)(\text{MA}_2)(\text{MA}_3) \ldots \qquad \textit{Overall MA} \quad (8.13)$$

The overall efficiency is similarly given by the product of the individual efficiencies:

$$\text{Eff} = (\text{Eff}_1)(\text{Eff}_2)(\text{Eff}_3) \ldots \qquad \textit{Overall efficiency} \quad (8.14)$$

Thus the overall IMA of two machines connected with one feeding the other that each have an IMA of 8 is $(8)(8) = 64$. If their efficiencies are each 75%, the overall efficiency is $(0.75)(0.75) = 0.56 = 56\%$.

◆   E X A M P L E   8.14

Figure 8.24 shows a pulley system that has a higher mechanical advantage than a block and tackle using the same number of pulleys. Find its ideal mechanical advantage.

Figure 8.24

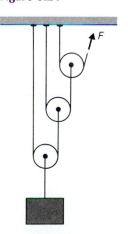

SOLUTION   In this arrangement, each pulley has an IMA of 2 and its output force acts as the input force to the one below it. From Eq. (8.13) we have

$$IMA = (IMA_1)(IMA_2)(IMA_3) = (2)(2)(2) = 8$$ ◆

8.6   SOME SIMPLE MACHINES

Using mechanical advantage.

Let us now look at two of the three basic machines, the lever and the inclined plane, and some of their developments. The hydraulic press, the third basic machine, is considered in Chapter 10.

The Lever

A **lever** is a rigid rod pivoted at a point called the **fulcrum.** Figure 8.25 shows the most familiar type of lever, with a load at one end being raised by a force applied at the other.

Figure 8.25

The lever. The ratio of the input and output distances equals the ratio of the corresponding lever arms.

$$IMA = \frac{s_{in}}{s_{out}} = \frac{L_{in}}{L_{out}}$$

From Fig. 8.25 we see that, because the angles through which the ends of the lever move are the same, the arcs s_{in} and s_{out} are respectively proportional to the lever arms L_{in} and L_{out}. The ideal mechanical advantage of the lever is therefore

$$IMA = \frac{s_{in}}{s_{out}} = \frac{L_{in}}{L_{out}} \qquad \qquad \textit{The lever} \quad (8.15)$$

The IMA of the lever equals the inverse ratio of the lever arms. A 4:1 ratio of lever arms, for instance, means an IMA of 4 when the input force is applied to the longer arm, an IMA of $\frac{1}{4}$ when it is applied to the shorter arm.

◆   **E X A M P L E   8.15** _____

A steel rod 2.00 m long is to be used to pry up one end of a 500-kg crate. If a workman can exert a downward force of 350 N on one end of the rod, where should he place a block of wood to act as the fulcrum?

SOLUTION   The weight of the crate is $w = mg = 4900$ N. A force of half this is needed to lift one end, so the IMA must be

$$IMA = \frac{F_{out}}{F_{in}} = \frac{w/2}{F_{in}} = \frac{2450 \text{ N}}{350 \text{ N}} = 7.00$$

Hence the ratio of the lever arms should be 7.00:

$$L_{in} = 7L_{out}$$

Since the lever is 2.00 m long,

$$L_{in} + L_{out} = 7.00L_{out} + L_{out} = 2.00 \text{ m}$$
$$L_{out} = \frac{2.00 \text{ m}}{8.00} = 0.25 \text{ m} = 25 \text{ cm}$$

The fulcrum should be placed 25 cm from the crate.   ◆

The lever of Fig. 8.25, where the fulcrum is between the load and the applied force, is called a Class I lever. In a Class II lever, of which the wheelbarrow is an example, the load is between the fulcrum and the applied force, as in Fig. 8.26. In a Class III lever, of which the human forearm is an example, the load and fulcrum are at the ends with the applied force between them.

Figure 8.26

The three classes of lever.

Torque Transmission

Levers are handicapped by the limited angles through which they can operate. Such developments of the lever as belt and gear drives, however, can be used on a continuous basis, and one or another of them is an important element in nearly every motor-driven machine.

Figure 8.27 shows two pulleys joined by a V-belt. As we know, torque τ and angular displacement θ in rotational motion are the analogs of force F and displacement s in linear motion. Accordingly we have for the definitions of actual mechanical advantage and ideal mechanical advantage in a belt drive system

$$\text{AMA} = \frac{\tau_{\text{out}}}{\tau_{\text{in}}} \qquad\qquad\qquad \textit{Belt drive} \quad (8.16)$$

$$\text{IMA} = \frac{\theta_{\text{in}}}{\theta_{\text{out}}} \qquad\qquad\qquad \textit{Belt drive} \quad (8.17)$$

The output torque is given in terms of the efficiency and the ideal mechanical advantage by

$$\tau_{\text{out}} = (\text{Eff})(\text{IMA})(\tau_{\text{in}}) \qquad\qquad (8.18)$$

As in any machine, mechanical advantage in a belt drive is achieved at the expense of motion: The higher the output torque relative to the input torque, the more turns of the driving pulley are needed per turn of the driven one. The difference

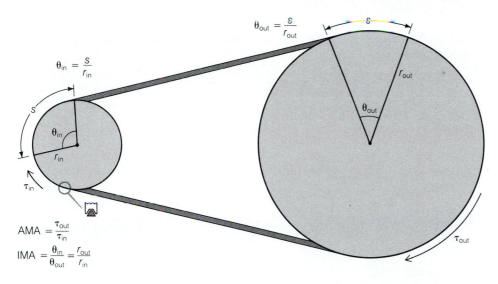

$\theta_{out} = \dfrac{s}{r_{out}}$

$\theta_{in} = \dfrac{s}{r_{in}}$

r_{out}

θ_{out}

θ_{in}

r_{in}

τ_{in}

τ_{out}

$AMA = \dfrac{\tau_{out}}{\tau_{in}}$

$IMA = \dfrac{\theta_{in}}{\theta_{out}} = \dfrac{r_{out}}{r_{in}}$

Figure 8.27

A V-belt drive.

between the IMA and the AMA of a belt system is due to friction both in the belt itself and in the shaft bearings. The higher the tension in the belt, the more the friction from both sources, so the belt should only be tight enough to prevent it from slipping on the pulleys.

If the belt does not slip, the distance through which a point on the rim of the driving (input) pulley moves in any time interval is equal to the distance through which a point on the rim of the driven (output) pulley moves; both distances are marked s in Fig. 8.27. In this time the pulleys turn through the respective angles θ_{in} and θ_{out}. From Eq. (7.1)

$$\theta_{in} = \frac{s}{r_{in}} \qquad \theta_{out} = \frac{s}{r_{out}}$$

where r_{in} is the radius of the driving pulley and r_{out} is the radius of the driven pulley. Since

$$\frac{\theta_{in}}{\theta_{out}} = \frac{s/r_{in}}{s/r_{out}} = \frac{r_{out}}{r_{in}}$$

we have another formula for the ideal mechanical advantage of a belt drive system:

$$IMA = \frac{r_{out}}{r_{in}} = \frac{d_{out}}{d_{in}} \qquad\qquad \textit{Belt drive} \quad (8.19)$$

Here d_{in} and d_{out} are the pulley diameters. A driving pulley that is smaller than the driven one, as in Fig. 8.27, means a mechanical advantage greater than 1 and so an output torque greater than the input torque.

According to Eq. 7.2, angular speed ω is given by $\Delta\theta/\Delta t$. In a given time interval Δt the driving pulley turns through $\Delta\theta = \theta_{in}$ and the driven pulley turns through $\Delta\theta = \theta_{out}$. Hence

$$\omega_{in} = \frac{\theta_{in}}{\Delta t} \qquad \omega_{out} = \frac{\theta_{out}}{\Delta t}$$

and

$$\text{IMA} = \frac{\theta in}{\theta_{out}} = \frac{\omega_{in}}{\omega_{out}} \tag{8.20}$$

Comparing this result with Eq. (8.19) leads to the conclusion that

$$\frac{\omega_{in}}{\omega_{out}} = \frac{d_{out}}{d_{in}} \qquad\qquad \textit{Belt drive} \quad (8.21)$$

The ratio of angular speeds is the inverse of the ratio of pulley diameters. When the motor (input) pulley is smaller than the machine (output) pulley, the machine turns at a smaller number of rpm than the motor but the torque provided to the machine shaft is higher than that of the motor; when the motor pulley is larger than the machine pulley, the machine speed is higher but the available torque is lower.

◆   **E X A M P L E   8.16** _____

A V-belt is used to transmit power from a 1750-rpm electric motor to a band saw, as in Fig. 8.28. The motor and saw pulleys have diameters of 3.000 in and 5.000 in respectively, and the diameter of the wheels over which the saw blade runs is 14.000 in. Find the linear speed of the blade.

SOLUTION   The angular speed of the saw shaft is, from Eq. (8.21),

$$\omega_{out} = \left(\frac{d_{in}}{d_{out}}\right)\omega_{in} = \left(\frac{3.000 \text{ in}}{5.000 \text{ in}}\right) 1750 \text{ rpm} = 1050 \text{ rpm}$$

In radian measure this is

$$\omega_{out} = \left(2\pi\frac{\text{rad}}{\text{rev}}\right)\left(1050\frac{\text{rev}}{\text{min}}\right) = 6600 \text{ rad/min}$$

Because the wheel diameter is 14.000 in, its radius is $R = 7.000 \text{ in} = \frac{7}{12}$ ft $= 0.5833$ ft. The linear speed of the saw blade is therefore

$$v = \omega_{out}R = (6600 \text{ rad/min})(0.5833 \text{ ft}) = 3850 \text{ ft/min}$$

◆

Figure 8.28

A band saw. (Example 8.16)

(a) Spur gears

(b) Helical gears

Figure 8.29

(a) Spur gears have their teeth cut parallel to the axes of rotation. (b) Helical gears have curved teeth cut in a spiral pattern at an angle to their axes; several teeth of meshing helical gears are in contact at all times, which makes a set of these gears smoother in operation and able to bear greater torque.

Gear drives (Fig. 8.29) have the advantage over belt drives of being able to transmit very high torques with no chance of slippage. They have the disadvantages of greater cost and weight and require more precise installation.

Because the number of teeth N on the rim of a gear is proportional to its diameter, we can replace the ratio of diameters in Eqs. (8.19) and (8.20) by the ratio of gear teeth in a pair of meshing gears. Accordingly

$$\text{IMA} = \frac{N_{\text{out}}}{N_{\text{in}}} = \frac{\omega_{\text{in}}}{\omega_{\text{out}}} \qquad\qquad \textit{Gear drive} \quad (8.22)$$

$$\text{IMA} = \frac{\text{output gear teeth}}{\text{input gear teeth}} = \frac{\text{input angular speed}}{\text{output angular speed}}$$

A chain drive combines features of both belt and gear drives. Its MA equals the ratio of the diameters of the driven and driving sprocket wheels; the greater the MA, the greater the multiplication of the driving torque.

◆   **E X A M P L E   8.17**

A gasoline engine develops a power of 11.6 kW at an angular speed of 240 rad/s. What gear ratio is needed to provide an output torque of 200 N · m, assuming 92% efficiency in the gear system?

SOLUTION   The engine produces the torque

$$\tau_{in} = \frac{P}{\omega} = \frac{11{,}600 \text{ W}}{240 \text{ rad/s}} = 48.3 \text{ N} \cdot \text{m}$$

and so the required AMA is, as in the case of belt drive,

$$\text{AMA} = \frac{\tau_{out}}{\tau_{in}} = \frac{200 \text{ N} \cdot \text{m}}{48.3 \text{ N} \cdot \text{m}} = 4.14$$

The corresponding IMA is, from Eq. (8.22), equal to the gear ratio:

$$\text{IMA} = \frac{N_{out}}{N_{in}} = \frac{\text{AMA}}{\text{Eff}} = \frac{4.14}{0.92} = 4.5$$

The output gear should have 4.5 times as many teeth as the gear on the engine shaft.   ◆

The Inclined Plane

We are so accustomed to using the inclined plane in everyday life that it may be hard to think of it as a "machine." So instinctive an act as choosing a gradual slope of a hill to walk up instead of a steep slope is based on the principle of the inclined plane. Figure 8.30 shows an inclined plane along which a crate of weight w is being pushed. The plane is L long and h high. From Eq. (8.10) we have

$$\text{IMA} = \frac{s_{in}}{s_{out}} = \frac{L}{h} \qquad\qquad \textit{Inclined plane}\quad (8.23)$$

Figure 8.30

The inclined plane.

$$\text{MA} = \frac{w}{F} = \frac{L}{h}$$

◆   E X A M P L E   8.18

A 150-kg safe on frictionless casters is to be raised 1.20 m off the ground to the bed of a truck. Planks 4.00 m long are available for the safe to be rolled along. How much force is needed to push the safe up to the truck?

SOLUTION   Here IMA = AMA = F_{out}/F_{in}, so from Eq. (8.23) we have

$$F_{in} = \frac{F_{out}}{IMA} = \frac{mg}{L/h} = \frac{mgh}{L} = \frac{(150\,\text{kg})(9.80\,\text{m/s}^2)(1.20\,\text{m})}{4.00\,\text{m}} = 441\,\text{N} \qquad ◆$$

The **screw** and the **wedge** are both developments of the inclined plane. As we can see from Fig. 8.31, a screw is an inclined plane wrapped around a cylinder to form a continuous helix. Aside from its use in splitting logs, leveling objects, and holding doors open, the wedge (Fig. 8.32) provides the operating principle of all cutting tools. A knife or a chisel is obviously a wedge, but so are the teeth of a saw and the abrasive chips of a grindstone.

The **pitch** p of a screw is the distance between adjacent threads (Fig. 8.33), and the screw travels this far in each complete rotation. (In the Customary System, the number of threads per inch is specified, which is the reciprocal of p. Thus a screw that has 16 threads per inch has a pitch of $\frac{1}{16}$ in = 0.0625 in.) The IMA of a screw depends upon the lever arm L used to turn it. In a screwdriver L is the handle radius, while in a wrench it is the distance from where the wrench is grasped to the center of its jaws. When a screw of pitch p is turned through one rotation, the applied force moves through a circle of circumference $2\pi L$ and the screw advances the distance p. Hence

$$IMA = \frac{s_{in}}{s_{out}} = \frac{2\pi L}{p} \qquad\qquad \textit{The screw} \quad (8.24)$$

Owing to the considerable friction that occurs between threads in contact, a screw is a relatively inefficient machine, but this friction is normally an advantage inasmuch as it prevents the screw from backing out under load.

Figure 8.31

A screw is an inclined plane wrapped around a cylinder.

Figure 8.32

The wedge.

Figure 8.33

A bolt is an example of the screw. The smaller its pitch, the greater the mechanical advantage.

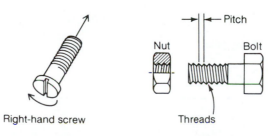

Right-hand screw

Nut   Bolt

Pitch

Threads

◆   E X A M P L E   8.19 _____

A wrench whose effective lever arm is 20 cm is used to tighten a nut whose pitch is 1.5 mm. The efficiency is 5.0%, and a force of 50 N (about 11 lb) is applied to the wrench. Find the force exerted by the nut as it is screwed down.

SOLUTION   Since 1 cm = 10 mm, L = 200 mm here, and so

$$F_{out} = (Eff)(IMA)(F_{in}) = (Eff)\left(\frac{2\pi L}{p}\right)(F_{in})$$

$$= (0.050)\left[\frac{(2\pi)(200 \text{ mm})}{1.5 \text{ mm}}\right](50 \text{ N}) = 2094 \text{ N} = 2.1 \text{ kN}$$

which is about 740 lb—42 times the force applied to the wrench. Evidently the screw is a very powerful device.   ◆

Important Terms

When the net force acting on an object is zero, the object is in **translational equilibrium.** When the net torque acting on an object is zero, the object is in **rotational equilibrium.**

If the lines of action of the forces that act on an object meet at a common point, the forces are said to be **concurrent.**

The **center of gravity** of an object is that point from which it can be suspended in any orientation without tending to rotate. The weight of an object can be considered a downward force acting on its center of gravity.

If the lines of action of the forces that act on an object meet at a common point, the forces are said to be **concurrent.**

A **machine** is a device that transmits force or torque. The three basic machines are the **lever,** the **inclined plane,** and the **hydraulic press.**

The **actual mechanical advantage** (AMA) of a machine is the ratio between the output force (or torque) it exerts and the input force (or torque) that is furnished to it. The **ideal mechanical advantage** (AMA) is its value when friction is absent.

The **efficiency** of a machine is the ratio between its actual and ideal mechanical advantages; it is always less than 100%.

Important Formulas

Translational equilibrium: $\Sigma F_x = 0$
$\Sigma F_y = 0$
$\Sigma F_z = 0$

Rotational equilibrium: $\Sigma \tau = 0$ about any point

Actual echanical advantage: $AMA = \dfrac{F_{out}}{F_{in}} = \dfrac{\tau_{out}}{\tau_{in}}$

Ideal mechanical advantage: $IMA = \dfrac{s_{in}}{s_{out}} = \dfrac{\theta_{in}}{\theta_{out}} = \dfrac{\omega_{in}}{\omega_{out}}$

Efficiency: $Eff = \dfrac{AMA}{IMA}$

MULTIPLE CHOICE

1. Which of the following sets of horizontal forces could leave an object in equilibrium?
- **a.** 25, 50, and 100 N
- **b.** 5, 10, 20, and 50 N
- **c.** 8, 16, and 32 N
- **d.** 20, 20, and 20 N

2. Which of the following sets of horizontal forces could not leave an object in equilibrium?
- **a.** 6, 8, and 10 lb
- **b.** 10, 10, and 10 lb
- **c.** 10, 20, and 30 lb
- **d.** 20, 40, and 80 lb

3. If an object is free to move in a plane, the number of scalar equations that must be satisfied for it to be in equilibrium is
- **a.** 2
- **b.** 3
- **c.** 4
- **d.** 6

4. If an object is free to move in three dimensions, the number of scalar equations that must be satisfied for it to be in equilibrium is
- **a.** 2
- **b.** 3
- **c.** 4
- **d.** 6

5. An object in equilibrium may not have
- **a.** any forces acting on it
- **b.** any torques acting on it
- **c.** velocity
- **d.** acceleration

6. Two ropes are used to support a stationary weight W. The tensions in the ropes must
- **a.** each be $W/2$
- **b.** each be W
- **c.** have a vector sum of magnitude W
- **d.** have a vector sum of magnitude greater than W

7. A weight is suspended from the middle of a rope whose ends are at the same level. In order for the rope to be perfectly horizontal, the forces applied to the ends of the rope
- **a.** must be equal to the weight
- **b.** must be greater than the weight
- **c.** might be so great as to break the rope
- **d.** must be infinite in magnitude

8. If the sum of the torques on an object in equilibrium is zero about a certain point, it is
- **a.** zero about no other point
- **b.** zero about some other points
- **c.** zero about all other points

d. any of the above, depending on the situation

9. In an equilibrium problem the point about which torques are computed
- **a.** must pass through one end of the object
- **b.** must pass through the object's center of gravity
- **c.** must intersect the line of action of at least one force acting on the object
- **d.** may be located anywhere

10. The center of gravity of an object
- **a.** is always at its geometrical center
- **b.** is always in the interior of the object
- **c.** may be outside the object
- **d.** is sometimes arbitrary

11. An object of mass m is moving at the speed v toward another object, also of mass m, that is at rest. The speed of the center of gravity of the system is
- **a.** $v/2$
- **b.** v
- **c.** $2v$
- **d.** v^2

12. Which of the following cannot be increased by using a machine of some kind?
- **a.** Force
- **b.** Torque
- **c.** Speed
- **d.** Work

13. The minimum number of pulleys in a block and tackle needed to achieve an IMA of 6 is
- **a.** 3
- **b.** 4
- **c.** 5
- **d.** 6

14. The highest IMA that can be obtained with a system of two pulleys is
- **a.** 1
- **b.** 2
- **c.** 3
- **d.** 4

15. A 100-lb box is suspended by a rope from an overhead support. If a horizontal force of 58 lb is applied to the box, the rope will make an angle with the vertical of
- **a.** 30°
- **b.** 45°
- **c.** 60°
- **d.** 75°

16. A 5.0-N picture is supported by two strings that run from its upper corners to a nail on the wall. If each string makes a 40° angle with the vertical, the tension in each is
- **a.** 3.3 N
- **b.** 3.9 N
- **c.** 5.0 N
- **d.** 10.0 N

17. The system shown in Fig. 8.34 is in equilibrium. The mass of A is
- **a.** 0.5 kg
- **b.** 1 kg
- **c.** $\sqrt{2}$ kg
- **d.** 2 kg

Figure 8.34

Multiple choice 17.

18. A weight of 1.0 kN is suspended from the end of a horizontal strut 1.5 m long that projects from a wall. A supporting cable 2.5 m long goes from the end of the strut to a point on the wall 2.0 m above the strut. The tension in the cable is

a. 0.8 kN

b. 1.0 kN

c. 1.25 kN

d. 1.67 kN

19. The inward force the strut in Multiple Choice 18 exerts on the wall is

a. 0.75 kN

b. 0.94 kN

c. 1.25 kN

d. 1.34 kN

20. A 60-kg object is attached to one end of a 40-kg steel tube 2.4 m long. The distance from the loaded end to the balance point is

a. 48 cm

b. 60 cm

c. 80 cm

d. 160 cm

21. A uniform wooden plank 2.0 m long that weighs 200 N is supported by a sawhorse 400 mm from one end. What downward force on the end nearest the sawhorse is needed to keep the plank level?

a. 200 N

b. 300 N

c. 500 N

d. 800 N

22. The entire 200-lb weight of a door 2.00 m high and 0.80 m wide is supported by a hinge at a lower corner. The force the door exerts on the hinge at the top corner is

a. 40 lb

b. 80 lb

c. 200 lb

d. 400 lb

23. The force the door of Multiple Choice 22 exerts on the lower hinger is

a. 200 lb

b. 204 lb

c. 215 lb

d. 240 lb

24. A person pries up one end of a 400-lb crate with a steel pipe 6.0 ft long. If the force exerted is 50 lb, the distance of the fulcrum from the crate is

a. 4 in

b. 8 in

c. 9 in

d. 16 in

25. One end of a plank 4.0 m long is supported 1.0 m above the ground. A 10-kg box on frictionless wheels is placed on the plank. The force parallel to the plank needed to keep the box stationary is

a. 25 N

b. 95 N

c. 98 N

d. 0.39 kN

26. A machine with an IMA of 5 and an AMA of 4 is used to raise a 10-kg load by 4.0 m. The work input to the machine is

a. 40 J

b. 0.16 kJ

c. 0.39 kJ

d. 0.49 kJ

QUESTIONS

1. What forces should be included in a free-body diagram of an object? What forces should not be included? Should the object's weight be included? If the object rests on a surface, should the reaction force of the surface be included?

2. If the earth followed a circular orbit around the sun, its speed would be constant. In this case would the earth be in equilibrium?

3. A ladder rests against a frictionless wall. (a) In what direction must be the force the ladder exerts on the wall? (b) How is the force the ladder exerts on the ground related to the weight of the ladder? Why?

4. Would you expect any difference in how soon you get tired between standing with your feet together and standing with them some distance apart? Why?

5. The sign convention for angular momentum is the same as that for torque. Is the angular momentum of the minute hand of a clock positive or negative when you look at the clock's face? When you turn the clock over?

6. In an equilibrium problem, under what circumstances is it necessary to consider the torques exerted by the various forces?

7. About which point should the torques exerted by the various forces that act on a body be calculated when this is necessary?

8. Why is a person with his feet apart more stable than if his feet were together?

9. What must be true of the AMA of a machine meant to increase force? Of a machine meant to increase speed?

10. What effect does lubrication have on the IMA of a machine? On its AMA?

EXERCISES

8.1 Translational Equilibrium

1. A horizontal beam 6.0 m long projects from the wall of a building. A guy wire that makes a 40° angle with the horizontal is attached to the outer end of the beam. When a weight of 100 N is attached to the end of the beam, what is the tension in the guy wire? (Neglect the weight of the beam.)

2. A 40,000-kg airplane is provided with 130 kN of thrust by its engines as it takes off at an angle of 25° above the horizontal. If the velocity of the airplane is constant, how much upward force on it is provided by the vertical component of the lift produced by the flow of air over its wings?

3. A 50-kg object is suspended from two ropes that each make an angle of 30° with the vertical. What is the tension in each rope?

4. The arrangement shown in Fig. 8.35 is in equilibrium. Find the tensions in the strings *A* and *B*.

Figure 8.35

Exercise 4.

5. The arrangement shown in Fig. 8.36 is in equilibrium. Find the mass of *A*.

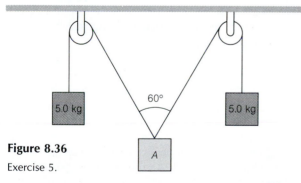

Figure 8.36

Exercise 5.

6. A 60-kg girl sitting in a canvas sling is being hoisted up a mast by a rope that passes through a pulley at the masthead and then down to a winch on the deck. She presses her feet against the mast to hold herself clear. How much force must her legs exert if her CG is 60 cm from the mast when she is 4.0 m below the masthead?

7. A horizontal force of 80 lb acts on a 50-lb object suspended by a string. What are the direction and magnitude of the force the string must provide to keep the object at rest?

8. A 500-kg load of bricks is lifted by a crane alongside a building under construction. A horizontal rope is used to pull the load to where it is required, and the supporting cable is then at an angle of 10° from the vertical. What is the tension in the rope? In the supporting cable?

9. A 600-lb crate is suspended by two ropes, one that makes an angle of 20° with the vertical and the other that makes an angle of 30° with the vertical. Find the tension in each rope.

10. A gong is suspended by two ropes. One makes an angle of 25° with the vertical and is under a tension of 70 N. The other rope makes an angle of 40° with the vertical. Find the mass of the gong and the tension in the second rope.

11. Find the maximum load that can be supported by the arrangement shown in Fig. 8.37 if the force on the strut is not to exceed 3000 N.

Figure 8.37

Exercise 11.

12. The femur is the large bone in the thigh. When a fractured femur is set, the thigh muscles tend to pull the segments together so tightly that the healed bone would be shorter than its original length. To prevent this, a traction apparatus like the one shown in Fig. 8.38 can be attached to the lower leg

Figure 8.38

Exercise 12.

to exert a force that opposes the muscular contraction. The advantage of this particular arrangement is that the injured leg can be moved around a fair amount without much effect on the applied force. Find the magnitude of the resultant force on the thigh produced by the apparatus and the lower leg and its direction relative to the horizontal for the situation shown. Assume a mass of 3.5 kg for the lower leg.

13. A bird sits on a telephone wire midway between two poles 20 m apart. The wire, assumed weightless, sags by 50 cm. If the tension in the wire is 90 N, what is the mass of the bird?

14. A car is stuck in the mud. To get it out, the driver ties one end of a rope to the car and the other to a tree 30 ft away. He then pulls sideways on the rope at its midpoint. If he exerts a force of 100 lb, how much force is applied to the car when he has pulled the rope 1.5 ft to one side?

15. Find the compressive force on the boom of the crane shown in Fig. 8.39 and the tension in the horizontal supporting cable. Assume both cables are fixed at the end of the boom.

3500 kg

Figure 8.39

Exercise 15.

8.2   Rotational Equilibrium

16. When a person stands on one foot with his heel raised, the entire reaction force of the floor, which is equal to his weight w, acts upward on the ball of his foot, as in Fig. 8.40.

Figure 8.40

Exercise 16.

In order to raise his heel, he must apply the upward force F_1 via his Achilles tendon, so the downward force F_2 on his ankle is greater than his weight w. Find the values of F_1 and F_2 for a 75-kg man for whom $L_1 = 50$ mm and $L_2 = 150$ mm. If the distance L_1 were greater, would this mean an increase or a decrease in the muscular force F_1?

17. A 15-kg child and a 25-kg child sit at opposite ends of a 4.0-m seesaw pivoted at its center. Where should a third child whose mass is 20 kg sit in order to balance the seesaw?

18. The driver of a car applies a force of 200 N to the brake pedal shown in Fig. 8.41. What is the resulting tension T in the brake cable?

Figure 8.41

Exercise 18.

19. One end of a 50-lb plank 16 ft long rests on the floor and the other end is held up by a carpenter so that the plank makes an angle of 25° with the floor. If the force the carpenter exerts is perpendicular to the plank, what is its magnitude?

20. A weightless ladder 6.0 m long rests against a frictionless wall at an angle of 60° above the horizontal. A 75-kg person is 1.2 m from the top of the ladder. What horizontal force at the bottom of the ladder is needed to keep it from slipping?

21. A fishing rod 3.0 m long is attached to a pivoting holder at its lower end. The rod is 50° above the horizontal. A fisherman grasps the rod 1.0 m from its lower end. What horizontal force must the fisherman exert to keep the rod at this angle when reeling in a 10-kg fish from directly below the rod's upper end?

22. The lower end of a weightless rod 2.0 m long is hinged to a wall, and a load of 200 N is suspended from its upper end, as in Fig. 8.42. A horizontal rope 60 cm long joins the middle of the rod to the wall. Find the tension in the rope.

Figure 8.42

Exercise 22.

Figure 8.43

Problem 28.

8.3 Center of Gravity

23. A uniform plank 5.0 m long that weighs 300 N rests on two sawhorses 1.0 m from each end. How near either end of the plank can a 700-N painter stand without tipping the plank?

24. A square table has legs at three of its corners. Will the table fall over? If not, what is the maximum mass that can be placed on the unsupported corner before it falls over?

25. The wheels of a certain truck are 1.8 m apart, and the truck falls over when tilted sideways at 30° from the horizontal. How high above the road is its CG?

26. A 60-kg woman stands upright with her feet 20 cm apart and her arms outstretched on both sides. In her right hand she holds a dumbbell 70 cm from the centerline of her body. Find the maximum mass the dumbbell can have if she is not to topple over.

27. A crate whose total mass is 80 kg rests on a floor such that the coefficient of static friction is 0.40. The crate is 1.5 m high and 1.0 m square, and its CG is at its geometric center. (a) Find the minimum horizontal force needed to start the crate sliding across the floor. (b) What is the maximum height above the floor at which this force can be applied without tipping over the crate? (*Hint:* Assume the crate has just begun to tip over so only the far edge of its base is touching the floor, and calculate moments about this edge.)

28. A uniform beam that weighs 10 lb is supported by two vertical wires *A* and *B*. A box that weighs 35 lb is placed on the beam, as shown in Fig. 8.43. Find the tensions in the two wires.

29. Three people are carrying a horizontal ladder 4.0 m long. One of them holds the front end of the ladder, and the other two hold opposite sides of the ladder the same distance from

its far end. What is the distance of the latter two people from the far end of the ladder if each person supports one-third of the ladder's weight?

30. A person holds a 10-kg board that is 2.0 m long in a horizontal position by supporting it with one hand underneath it 40 cm from one end while the other hand presses down at that end, as shown in Fig. 8.44. Find the force each hand exerts. What percentage of the board's weight is each force?

Figure 8.44

Problem 30.

31. The front and rear axles of a 24-kN truck are 4.0 m apart. The CG of the truck is located 2.5 m behind its front axle. Find the weight supported by the front wheels of the truck.

32. A 250-N bag of cement is placed on a 4.0-m-long plank 1.8 m from one end. The plank itself weighs 80 N. Two men pick up the plank, one at each end. How much weight must each support?

33. The CG of a 300-lb polar bear standing on all fours is 40 in from her hind paws and 32 in from her forepaws. Find the force the ground exerts on each of her paws.

34. A butcher with his upper arm vertically at his side and his forearm horizontal presses down on a scale with his hand, which is 40 cm from his elbow joint. His forearm and hand together weigh 30 N, and their CG is 20 cm from his elbow joint. The triceps muscle on the back of his upper arm is attached 2.5 cm behind the center of the elbow joint and it is exerting an upward force of 1.0 kN. What is the reading on the scale in kilograms?

35. A 40-kg boom 3.0 m long is hinged to a vertical mast and held in position by a rope at its end that is attached to the mast 1.0 m above the hinge pin. If the boom is horizontal and uniform, find the tension in the rope.

36. A 20-kg object is suspended from one end of a 30-kg wooden beam 3.0 m long. Where should the beam be picked up so that it remains horizontal?

37. The front wheels of a certain car are found to support 600 kg and the rear wheels 400 kg. The car's wheelbase (distance between axles) is 2.2 m. How far from the forward axle is the CG of the car?

8.4   Finding a Center of Gravity

38. The moon does not actually revolve around the earth: Both revolve around the center of mass of the earth-moon system. Find the location of this center of mass in terms of r_e, the earth's radius, given that the earth-moon distance is $60r_e$ and that the masses of the two bodies are 6.0×10^{24} kg and 7.3×10^{22} kg, respectively.

39. The asymmetric dumbbell shown in Fig. 8.45 is balanced on a frictionless surface. Eventually the dumbbell falls over with the 2-kg ball moving to the right. What is the final position of the 1-kg ball? (*Hint:* Consider the motion of the CG of the dumbbell.)

Figure 8.45
Problem 39.

40. A 60.0-kg woman is sitting with her upper body and lower legs vertical and her thighs horizontal. The mass of her upper body is 40.0 kg and its CG is 32 cm above her hips; the mass of her thighs is 13.0 kg and their CG is 16 cm in front of her hips; and the mass of her lower legs and feet is 7.0 kg and their CG is 23 cm below her knees and 37 cm in front of her hips. Find the location of the CG of the seated woman.

41. Find the location of the CG of the flat object of uniform thickness shown in Fig. 8.46.

Figure 8.46
Problem 41.

8.5   Mechanical Advantage

42. In the frictionless pulley system shown in Fig. 8.47, a force of 40 N is needed to support the 10-kg object. What is the mass of the lower pulley?

Figure 8.47
Problem 42.

43. A block and tackle is used to pull a car out of the mud. A block with two pulleys is fastened to the car's bumper, and another block with three pulleys is tied to a tree. One end of a rope is made fast to the bumper and threaded through the various pulleys. Neglecting friction, determine how much force is applied to the car when two people exert a total force of 900 N on the free end of the rope.

44. A person exerts a force of 32 lb to lift a 350-lb compressor using a pulley system that is 91% efficient. When 30 ft of rope have been pulled through the system, through what height was the compressor raised?

8.6 Some Simple Machines

45. The human forearm is a Class III lever. The elbow is the fulcrum, the applied muscular force is exerted at a point perhaps 25 mm in front of it, and the load is at the hand, perhaps 33 cm from the elbow. How much force must the muscles of the arm develop to lift a 20-kg object held horizontally by the hand? How many pounds is this?

46. A vertical-axis capstan drum 60 cm in diameter was used to pull in the anchor chain of a whaler a century ago. Six horizontal arms 1.2 m long protruded from the top of the drum, and a man on each arm exerted a force of 300 N on it. The efficiency of the system was 85%. (a) Find the force applied to the anchor chain. (b) How far must each man have walked to raise 100 m of chain?

47. In a certain winch a spur gear with 10 teeth is used to drive another spur gear with 40 teeth. The efficiency of the winch is 80%. (a) If the input gear is turned at 8 revolutions/min, how fast does the output gear turn? (b) If a torque of 30 lb · ft is applied to the input gear, what is the output torque?

48. Find the force exerted on the free end of the rope by the geared windlass shown in Fig. 8.48 when a force of 150 N is applied to the crank handle. Neglect friction.

49. A gasoline engine that develops 30 kW at 60 rps is coupled to a 4:1 reduction gearbox of 95% efficiency. Find its output torque.

50. A ramp 25 m long slopes down 1.2 m to the edge of a lake. What force is needed to pull a 300-kg boat on an 80-kg trailer up along the ramp if friction is negligible?

Figure 8.48

Exercise 48.

51. A block and tackle with one fixed and one movable pulley is used to drag a 40-kg dolly carrying a 200-kg load up a ramp 8.0 m long and 2.0 m high. If friction is negligible, what is the force needed on the end of the rope?

52. A machinist's vise has a screw of 3.0-mm pitch and a handle 13 cm long. In the absence of friction, how much force is developed between its jaws when a force of 40 N is applied to the end of the handle?

53. A wood screw with a pitch of 2.0 mm is being turned by a screwdriver whose handle is 40 mm in diameter. A pilot hole has been drilled for the screw, and grease is used to increase the efficiency to 10%. If a force of 5.0 N is applied to the handle of the screwdriver, with how much force does the screw advance?

54. A screw jack whose pitch is 0.50 in has a handle 3.0 ft long. A force of 17 lb must be applied to the handle in order to raise a load of 2500 lb. Find the efficiency of the jack.

PROBLEMS

55. Two uniform rods pivoted together at one end stand like an inverted V on a table. If the angle between each rod and the table is θ, find the minimum coefficient of static friction between each rod and the table.

56. A 500-kg load is supported by a pair of hinged legs, as

shown in Fig. 8.49. There is no friction between the legs and the floor. Find the tension in the rope that joins the midpoints of the legs.

57. A 100-kg load is supported by two hinged legs, as shown in Fig. 8.50. Find the force each leg exerts on the ground.

Figure 8.49

Problem 56.

Figure 8.50

Problem 57.

58. A 50-kg load is supported by two hinged legs, as shown in Fig. 8.51. Find the force each leg exerts on the ground.

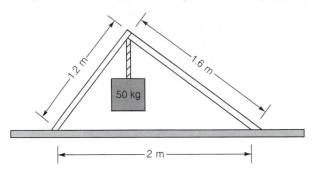

Figure 8.51

Problem 58.

59. A boom hinged at the base of a vertical mast is used to lift a 1500-kg load, as shown in Fig. 8.52. Find the tension in the cable from the top of the mast to the end of the boom. The weight of the boom itself can be neglected.

Figure 8.52

Problem 59.

60. (a) Find the tension T in the wire and the force F the horizontal beam exerts when the structure shown in Fig. 8.53 is at rest. What is the direction of **F**? (b) Find T and F when the structure is rotating twice per second. What is the direction of **F** now? Neglect the masses of the beam and the wire.

Figure 8.53

Problem 60.

61. A 200-lb uniform horizontal beam 12.0 ft long is supported at one end by a rigid post and at the other by a rope that makes an angle of 40° with the beam. A load of 1000 lb is suspended from the outer end of the beam. Find the tension in the rope.

62. The flat, uniform metal plate shown in Fig. 8.54 has a circular hole 10 cm in radius. Find the location of the CG of the plate.

Figure 8.54

Problem 62.

63. The arm of a certain person has a mass of 3.0 kg and its CG is 28 cm from the shoulder joint. When the arm is outstretched so that it is horizontal, the shoulder muscle supporting it has a line of action 15° above the horizontal. (a) If this muscle is attached to the upper arm 13 cm from the shoulder joint, what is the force developed by the muscle? (b) Find the force when the outstretched hand 60 cm from the shoulder joint is used to support a 1.0-kg load.

64. A uniform 10-kg ladder 2.5 m long is placed against a frictionless wall with its base on the ground 80 cm from the wall. Find the magnitudes of the forces exerted on the wall and on the ground.

65. A uniform 15-kg ladder 3.0 m long rests against a frictionless wall at a point 2.4 m off the floor. Find the vertical and horizontal components of the force exerted by the ladder on the floor.

66. A uniform 12-kg ladder 2.8 m long rests against a vertical frictionless wall with its lower end 1.1 m from the wall. If the ladder is to stay in place, what must be the minimum coefficient of friction between the bottom of the ladder and the ground?

67. A door 2.40 m high and 0.80 m wide has hinges at the top and bottom of one edge. The entire 200-N weight of the door is supported by the upper hinge. Find the magnitude and direction of (a) the force the door exerts on the upper hinge; (b) the force the lower hinge exerts on the door.

68. A door 3.0 m high and 1.2 m wide has hinges on one edge that are 30 cm above the bottom and 30 cm below the top. The entire 500-N weight of the door is supported by the lower hinge. Find the magnitude and direction of (a) the force the door exerts on the lower hinge; (b) the force the upper hinge exerts on the door.

69. A 0.50-hp 1750-rpm electric motor is coupled to a 5.0-in diameter circular saw blade with a V-belt. The motor pulley is 6.0 in. in diameter and the saw pulley is 2.0 in. in diameter. With how much force do the teeth of the saw strike a piece of wood?

70. On a certain bicycle, the sprocket wheel to which the pedals are attached is 20 cm in diameter and the sprocket wheel on the rear axle is 10 cm in diameter. The wheels of the bicycle are 65 cm in diameter. How many times must the pedals be turned per minute in order that a rider travel at 6.0 m/s?

Answers to Multiple Choice

1. d	8. c	15. a	22. b
2. d	9. d	16. a	23. b
3. b	10. c	17. c	24. b
4. d	11. a	18. c	25. a
5. d	12. d	19. a	26. d
6. c	13. c	20. a	
7. d	14. d	21. b	

C H A P T E R

MECHANICAL PROPERTIES OF MATTER

9.1   Density

9.2   Elasticity

9.3   Structure of Solids

9.4   Young's Modulus

9.5   Shear

9.6   Bulk Modulus

9.7   Building over Space

The usual way a physicist approaches a complex situation is first to set up a simple model that represents the main features of the situation. Then, if predictions based on this model agree reasonably well with observation and experiment, the model is further refined until agreement is even better. Our work in physics thus far is an illustration of this process. We began by treating objects as though they were just particles, and later we went on to consider them as rigid bodies. Actually, of course, there is no such thing as a rigid body—we can stretch, compress, or twist the strongest piece of steel if we apply suitable forces. In this chapter we come even closer to reality by looking at some of the mechanical properties of materials.

9.1 DENSITY

Mass per unit volume.

A characteristic property of every substance is its **density,** which is its mass per unit volume. When we speak of lead as a ''heavy'' metal and of aluminum as a ''light'' one, what we really mean is that lead has a greater density than aluminum: A cubic meter of lead has a mass of 11,300 kg, whereas a cubic meter of aluminum has a mass of only 2700 kg.

The symbol for density is ρ (Greek letter *rho*). If the volume V of a substance has the mass m,

$$\rho = \frac{m}{V} \qquad\qquad\qquad \textit{Density} \quad (9.1)$$

$$\text{Density} = \frac{\text{mass}}{\text{volume}}$$

Although in the SI system the proper unit of density is the kg/m^3, densities are frequently given in g/cm^3. Because there are 10^3 g in a kilogram and 100 cm in a meter, hence 10^6 cm^3 in a cubic meter,

$$1 \text{ g/cm}^3 = 10^3 \text{ kg/m}^3$$

Densities in the conventional system are properly expressed in $slugs/ft^3$. However, **weight density** $D = w/V$, the weight of a substance per unit volume, is often used; its unit is the $pound/ft^3$. The difference between ρ and D is a factor of g. Thus the density of pure water is $\rho = 1.94$ $slugs/ft^3$ and its weight density is $D = \rho g = 62$ lb/ft^3. Because g varies with location, ρ is a more fundamental quantity than D. The densities and weight densities of some common substances are given in Table 9.1.

◆ **E X A M P L E   9.1** _____

A mooring anchor is being made by pouring concrete into a cylindrical mold whose radius is 18 in. How thick should the concrete be in order that the anchor weigh 3000 lb?

SOLUTON The area of a circle of radius r is πr^2. Hence the volume of a cylinder of radius r and height h is $V = Ah = \pi r^2 h$ and

$$h = \frac{V}{\pi r^2}.$$

Table 9.1   Densities of various substances at atmospheric pressure and room temperature

Substance	Mass density, d			Weight density, D
	kg/m³	g/cm³	slugs/ft³	lb/ft³
Air	1.3	1.3×10^{-3}	2.5×10^{-3}	8×10^{-2}
Alcohol (ethyl)	7.9×10^2	0.79	1.5	48
Aluminum	2.7×10^3	2.7	5.3	1.7×10^2
Balsa wood	1.3×10^2	0.13	0.25	8
Blood (37°C)	1.06×10^3	1.06	2.05	66
Bone	1.6×10^3	1.6	3.1	9.9×10^2
Brass	8.4×10^3	8.4	16	5.2×10^2
Carbon dioxide	2.0	2.0×10^{-3}	3.8×10^{-3}	0.12
Concrete	2.3×10^3	2.3	4.5	1.4×10^2
Copper	8.9×10^3	8.9	17	5.5×10^2
Gasoline	6.8×10^2	0.68	1.3	42
Gold	1.9×10^4	19	38	1.2×10^3
Helium	0.18	1.8×10^{-4}	3.5×10^{-4}	1.1×10^{-2}
Hydrogen	0.09	9×10^{-5}	1.8×10^{-4}	5.4×10^{-3}
Ice	9.2×10^2	0.92	1.8	58
Iron and steel	7.8×10^3	7.8	15	4.8×10^2
Lead	1.1×10^4	11	22	7×10^2
Mercury	1.4×10^4	14	26	8.3×10^2
Nitrogen	1.3	1.3×10^{-3}	2.4×10^{-3}	7.7×10^{-2}
Oak	7.2×10^2	0.72	1.4	45
Oxygen	1.4	1.4×10^{-3}	2.8×10^{-3}	9×10^{-2}
Pine	3.7×10^2	0.37	0.73	23
Silver	1.05×10^4	10.5	20	6.6×10^2
Water, pure	1.00×10^3	1.00	1.94	62
Water, sea	1.03×10^3	1.03	2.00	64

Because weight density is defined as $D = w/V$, the volume of the concrete is $V = w/D$. Here $r = 18$ in $= 1.5$ ft and, from Table 9.1, $D = 1.4 \times 10^2$ lb/ft³, so

$$h = \frac{V}{\pi r^2} = \frac{w}{\pi r^2 D} = \frac{3000 \text{ lb}}{\pi (1.5 \text{ ft})^2 (1.4 \times 10^2 \text{ lb/ft}^3)} = 3.0 \text{ ft} \qquad \blacklozenge$$

 EXAMPLE 9.2 _____

A 50-g bracelet is suspected of being gold-plated lead instead of pure gold. When it is dropped into a full glass of water, 4.0 cm³ of water overflows (Fig. 9.1). Is the bracelet pure gold? If not, what proportion of its mass is gold?

SOLUTION The density of the bracelet is

$$\rho = \frac{m}{V} = \frac{50 \text{ g}}{4.0 \text{ cm}^3} = 12.5 \text{ g/cm}^3$$

This is less than the density of gold, which is 19 g/cm³, so the bracelet is not pure gold. To find the proportion of gold it contains, we note that

$$m_{\text{gold}} + m_{\text{lead}} = 50 \text{ g}$$

$$V_{\text{gold}} + V_{\text{lead}} = \frac{m_{\text{gold}}}{\rho_{\text{gold}}} + \frac{m_{\text{lead}}}{\rho_{\text{lead}}} = 4.0 \text{ cm}^3$$

From the first equation, $m_{\text{lead}} = 50 \text{ g} - m_{\text{gold}}$. Substituting this expression for m_{lead} and the values $\rho_{\text{gold}} = 19$ g/cm³ and $\rho_{\text{lead}} = 11$ g/cm³ in the second equation yields

$$\frac{m_{\text{gold}}}{19 \text{ g/cm}^3} + \frac{50 \text{ g}}{11 \text{ g/cm}^3} - \frac{m_{\text{gold}}}{11 \text{ g/cm}^3} = 4.0 \text{ cm}^3$$

$$m_{\text{gold}} = 14 \text{ g}$$

Hence the proportion of gold in the bracelet is 14 g/50 g $= 0.28 = 28\%$. ◆

Figure 9.1

The density of this bracelet is $\rho = m/V = (50 \text{ g})/(4.0 \text{ cm}^3) = 12.5$ g/cm³.

◆ **E X A M P L E   9.3** _____

We want to pump water from a well 20 m deep at the rate of 500 liters/min. Assuming 60% efficiency, how powerful should the pump motor be?

SOLUTION The required power output of the pump is

$$P_{\text{out}} = \frac{\text{work}}{\text{time}} = \frac{W}{t} = \frac{mgh}{t} = \frac{\rho \, Vgh}{t}$$

Because 1 liter $= 1 \text{ L} = 10^{-3} \text{ m}^3$ and 1 min $= 60$ s, the flow rate of 500 liters/min is equivalent to

$$\frac{V}{t} = \left(500 \frac{\text{L}}{\text{min}}\right)\left(10^{-3}\frac{\text{m}^3}{\text{L}}\right)\left(\frac{1}{60 \text{ s/min}}\right) = 8.33 \times 10^{-3} \text{ m}^3/\text{s}$$

The power output is therefore

$$P_{\text{out}} = \rho g h \left(\frac{V}{t}\right) = \left(1.00 \times 10^3 \frac{\text{kg}}{\text{m}^3}\right)\left(9.8 \frac{\text{m}}{\text{s}^2}\right)(20 \text{ m})\left(8.33 \times 10^{-3}\frac{\text{m}^3}{\text{s}}\right)$$

$$= 1633 \text{ W}$$

In general, efficiency = power output/power input, so the pump motor must have a power of

$$P_{\text{in}} = \frac{P_{\text{out}}}{\text{Eff}} = \frac{1633 \text{ W}}{0.60} = 2722 \text{ W} = 2.7 \text{ kW}$$

$\blacklozenge$

The **specific gravity** of a substance is its density relative to that of water and is a pure number. Because the density of water is almost exactly 1 g/cm^3, the specific gravity of a substance is very nearly equal to the numerical value of its density when expressed in g/cm^3. Because the density of aluminum is 2.7 g/cm^3, its specific gravity is 2.7. Specific gravity is also called **relative density.**

9.2   ELASTICITY

Tension, compression, and shear.

Although most solid bodies seem to be solid and unyielding, we can nevertheless deform them either temporarily or permanently by applying **stresses.** (When equal and opposite forces act on a body, the stress on the body is the magnitude of either force divided by the area over which it is exerted; see Section 9.4.) Stresses fall into three categories: tensions, compressions, and shears. These are illustrated in Fig. 9.2.

A **tensile stress** is applied to an object when equal and opposite forces act away from each other on its ends. Such a stress tends to lengthen the object. A **compressive stress** is applied when such forces act toward each other. Such a stress tends to shorten the object. A **shear stress** is applied when the equal and opposite forces have different lines of action. Such a stress tends to alter the shape of the object without changing its volume.

Thus tension stretches an object, compression shrinks it, and shear twists it.

Figure 9.2

The three types of stress.

Figure 9.3

The increases in length of three wires 1 m long and 1 mm in diameter when a tensile force of 100 N is applied.

The response of an object to a given stress depends on its composition, shape, temperature, and so on. Figure 9.3 shows the increases in length of identical steel, copper, and aluminum wires fixed in place at one end when a force of 100 N is applied to the other end.

The amount by which a particular object is deformed is directly proportional to the applied stress, provided the stress is not too much. In the case of tension, for example, we might find that supporting a 20-kg mass with a certain thin wire stretches the wire by 1 mm (Fig. 9.4). Doubling the mass to 40 kg will stretch it by 2 mm, tripling the mass to 60 kg will stretch it by 3 mm, and so on. When the mass is removed, the wire returns to its original length.

This proportionality is called **Hooke's law** and may be written

$$F = kx \qquad\qquad \textit{Hooke's law} \quad (9.2)$$

where F is the applied tension, x the resulting elongation, and k a constant whose value depends on the nature and dimensions of the object under stress. The force constant k is higher for materials such as steel than it is for materials such as lead. It is proportional to the cross-sectional area of the object, so a thick wire of a given material has a higher value of k than a thin wire of the same material. Relationships like Eq. (9.2) also apply to solids under shear and to all states of matter under compression.

Figure 9.4

The elongation of a wire is proportional to the stress applied to it, provided the elastic limit is not exceeded. When the stress is removed, the wire returns to its original length.

Figure 9.5

Graph of the elongation of an iron rod as more and more tension is applied to it. When a tension below the elastic limit is applied and then removed, the rod returns to its original length, which is elastic behavior. When a tension exceeding the elastic limit is applied and removed, the rod does not contract fully but remains permanently longer, which is plastic behavior.

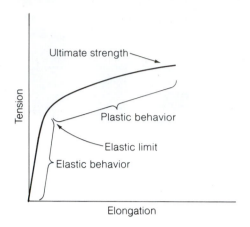

Elastic Limit and Ultimate Strength

The term **elastic limit** refers to the maximum stress that can be applied to an object without its being permanently deformed as a result. When its elastic limit is exceeded, the object may or may not be far from breaking. Brittle substances like glass or cast iron break at or near their elastic limits. Bone is another material that remains elastic until its breaking point is reached, and so it cannot be permanently changed in size or shape by applying a force. The elastic limits of many materials decrease with temperature, which is why metals are often heated before being formed into shape by bending or hammering.

Most metals (cast iron is one exception) can be deformed a great deal beyond their elastic limits, a property known as **ductility.** Copper is a very ductile metal; although a copper wire whose cross-sectional area is 1 mm^2 reaches its elastic limit when a force of about 150 N is applied, it will not rupture until the force has more than doubled.

A graph of the elongation of an iron rod versus the tension applied to it is shown in Fig. 9.5. At first the graph is a straight line, which corresponds to Hooke's

Forces in the earth's crust acting over thousands or millions of years can deform even the strongest rock.

Figure 9.6

Four methods of shaping metals by plastic deformation. Most metals deform considerably before breaking.

Figure 9.7

A force applied to an object with a crack causes a concentration of stress at the end of the crack. The smaller the diameter of the end, the greater the stress concentration and the more likely the crack is to spread.

Diameter of crack end

law. Past the elastic limit the graph flattens out, which means that each increase in tension by a given amount produces a greater increase in length than it did below the elastic limit: The rod stretches more readily. If the tension is removed after having exceeded the elastic limit, the rod remains longer than it was originally; it has undergone **plastic deformation.** The **ultimate strength** of the rod is the greatest tension it can withstand, and it corresponds to the highest point on the curve.

Many metals (cast iron is a notable exception) can be deformed considerably beyond their elastic limits. A **malleable** metal is one that has a large plastic range in compression. A piece of such a metal can be formed into a desired shape by rolling, forgoing, or extrusion (Fig. 9.6). A **ductile** metal is one that has a large plastic range in tension and can be formed into rod or wire by being drawn through a die. The two terms are sometimes used interchangeably.

An object may fail through **fatigue** after repeated applications of stresses that are well under its original breaking strength. Tiny defects in the internal structure of the material grow a little each time a stress acts, and eventually cracks appear that lead to rupture (Fig. 9.7). The chief factors are the level of stress and the total number of stress cycles. The number of cycles needed for failure decreases rapidly with increasing stress. A metal bar that can tolerate a million applications of a force equal to 20% of its ultimate strength may break after only 10,000 or 20,000 applications of a force twice as great. The problem is clearly most severe in machinery—an engine operating at 3000 rpm undergoes 180,000 stress cycles per hour. Most failures in aircraft components are due to fatigue.

Stress concentrations occur wherever there is a sharp corner or notch, even a scratch, so parts designed for high-speed machines are always highly polished. Imperfections in a weld produce stress concentrations that promote fatigue failure, and welded parts therefore need generous margins of safety. Corrosion can accelerate fatigue by producing pits in a surface from which cracks can develop, and also by causing the cracks to grow faster.

9.3   STRUCTURE OF SOLIDS

How elasticity and plasticity originate.

Most solids are **crystalline,** with their atoms arranged in regular patterns. Usually a sample of such a solid does not consist of a single perfect crystal but is an assembly of a great many tiny crystals. A few solids, such as pitch, wax, and glass, have irregular structures much like those of liquids. They are said to be **amorphous** (''without form'') solids.

The atoms in a crystal have a certain normal spacing and resist being pulled farther apart or being squeezed closer together. For a simple mental picture of a crystal, we can imagine little springs joining the atoms, as in Fig. 9.8. Each spring represents a **bond** between a pair of atoms. (Such bonds are electrical in nature, as we will learn in Chapter 27.) In terms of this model, it is easy to see why even the smallest stress can deform a crystal and why it returns to its original form when a stress below the elastic limit is removed.

Few crystals have completely regular structures. A common crystal defect is a missing line of atoms, which is called a **dislocation**. Dislocations make it possible to see how a solid can be permanently deformed without breaking. We might think the sliding of layers of atoms over one another is responsible. However, calculations show that sliding of this kind—which involves the simultaneous rupture of millions of bonds between atoms—would need forces about a thousand times stronger than those actually required.

Figure 9.9(a) shows a crystal with a missing row of atoms. When equal and opposite forces are applied to the crystal along different lines of action, as in Fig. 9.9(b), the dislocation shifts to the right as the atoms in the layer with the missing row shift their bonds, one row at a time, with the atoms in the layer above. In Fig. 9.9(c) the dislocation has reached the end of the crystal, which is now permanently deformed. Because the bonds were shifted one row at a time instead of all at once, much less force was needed than would have been required to deform a perfect solid.

To increase the strength of a solid, it is necessary to impede the motion of dislocations in its structure. There are two chief ways to do this in a metal. One is to increase the number of dislocations by hammering the metal or squeezing it between rollers. The dislocations then become so numerous and tangled together that they interfere with each other's motion. This effect is called **work hardening.**

Figure 9.8

A simple model of a crystalline solid. Each spring represents the bond between adjacent atoms, which resist being pulled farther apart or pressed closer together than their normal spacing. No such springs actually exist, but they help us to visualize the origins of elastic behavior in solids.

(a)   (b)   (c)

Figure 9.9

The motion of a dislocation in crystal under stress results in a permanent deformation.

Another approach is to add foreign atoms to the metal that act as roadblocks to the progress of dislocations. Thus adding small amounts of carbon, chromium, manganese, and other elements to iron turns it into the much stronger steel by hindering the ability of dislocations to move through the metal. The most suitable composition for steel depends on how it is going to be used. For example, a carbon content of more than about 0.75% plus appropriate treatment gives a very hard steel (the steel used in files and razor blades typically contains 1.25% carbon), but such steels are difficult to weld. When welding is required, low-carbon steels are preferred.

9.4   YOUNG'S MODULUS

Strain is proportional to stress below the elastic limit.

Suppose we apply equal and opposite forces F to the ends of a bar so that it is under tension or compression, as in Fig. 9.10. If the cross-sectional area of the bar is A, the **stress** on it is defined as the ratio between F and A:

$$\text{Stress} = \frac{\text{force}}{\text{area}} = \frac{F}{A} \tag{9.3}$$

Figure 9.10

The stress on a bar in tension or compression is given by F/A.

The SI unit of stress is the newton per square meter, which is called the **pascal** (Pa) in honor of the French scientist and philosopher Blaise Pascal (1623–1662). Thus

$$1 \text{ pascal} = 1 \text{ Pa} = 1 \text{ N/m}^2$$

The pascal is a very small unit: The stress you exert by pressing hard with your thumb on a table is about a million pascals. In the Customary System the unit of stress is the pound per square inch, or **psi**:

$$1 \text{ psi} = 1 \text{ lb/in}^2 = 6.89 \times 10^3 \text{ Pa} = 6.89 \text{ kPa}$$

◆   E X A M P L E   9.4 _____

A 70-kg swami lies on a bed of nails 1.0 cm apart whose points each have an area of 1.0 mm^2. If the area of the swami's body in contact with the bed is 0.50 m^2 and the threshold stress for pain is 1 MPa (10^6 Pa), how disagreeable is the experience for the swami?

SOLUTION   The bed has 1 nail/cm^2. Because the area in contact with the nails is $0.50 \text{ m}^2 = 5000 \text{ cm}^2$, the swami's weight of $mg = 686$ N is supported by 5000 nails. Each nail's area is $1.0 \text{ mm}^2 = 1.0 \times 10^{-6} \text{ m}^2$, so the stress each nail exerts on the swami is

$$p = \frac{F}{A} = \frac{686 \text{ N}}{(5000)(1.0 \times 10^{-6} \text{ m}^2)} = 1.37 \times 10^5 \text{ Pa} = 0.137 \text{ MPa}$$

This is well below the 1-MPa threshold for pain.   ◆

The relative change in length of something under tension or compression is called **strain.** Strain is equal to the ratio between the change in length ΔL and the original length L_0:

$$\text{Strain} = \frac{\text{change in length}}{\text{original length}} = \frac{\Delta L}{L_0} \qquad (9.4)$$

Because strain is a ratio of lengths, it has no units.

Experience shows that, below the elastic limit, *strain is proportional to stress*. Under a given force a thin rod will stretch more than a thick one, and a long rod will stretch more than a short one. The **modulus of elasticity** of a material subjected to a certain kind of stress is defined as the ratio between the stress and the strain that occurs because of it:

$$\text{Modulus of elasticity} = \frac{\text{stress}}{\text{strain}}$$

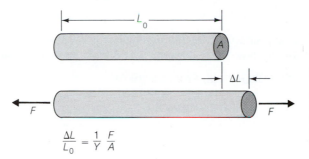

$$\frac{\Delta L}{L_0} = \frac{1}{Y}\frac{F}{A}$$

Figure 9.11

An object under tension stretches by an amount that depends on the value of Y for the material of which it is composed. Y is called Young's modulus.

For tensile or compressive stresses, the modulus of elasticity is called **Young's modulus**, symbol Y. In the case of a rod of initial length L_0 and cross-sectional area A in which a tensile or compressive force F produces a change ΔL in its length (Fig. 9.11),

$$Y = \frac{F/A}{\Delta L/L_0} \qquad\qquad \textit{Young's modulus} \quad (9.5)$$

Hence we have

$$\frac{\Delta L}{L_0} = \frac{1}{Y}\frac{F}{A} \qquad\qquad \textit{Tension or compression} \quad (9.6)$$

The rod increases in length by ΔL if it is in tension and decreases by that amount if it is in compression. The value of Y depends on the composition of the rod. Young's moduli for a number of common substances are given in Table 9.2.

Table 9.2   Typical elastic moduli (1 Pa = 1 N/m²)

Material	Young's modulus, Y ($\times 10^{10}$ Pa)	($\times 10^6$ lb/in²)	Shear modulus, S ($\times 10^{10}$ Pa)	($\times 10^6$ lb/in²)	Bulk modulus, B ($\times 10^{10}$ Pa)	($\times 10^6$ lb/in²)
Aluminum	7.0	10	2.4	3.4	7.0	10
Brass	9.1	13	3.6	5.1	6.1	8.5
Concrete	2.0	2.9				
Copper	11	16	4.2	6.0	14	20
Glass	5.5	7.8	2.3	3.3	3.7	5.2
Granite	4.5	6.5			4.5	6.5
Iron	19	28	7.0	10	10	14
Lead	1.6	2.3	0.56	0.8	0.77	1.1
Pine wood (parallel to grain)	1.0	1.5				
Steel	20	29	8.4	12	16	23

◆   **E X A M P L E   9.4** _____

A copper wire 1.00 mm in diameter and 2.0 m long is used to support a mass of 5.0 kg. By how much does the wire stretch under this load?

SOLUTION   (a) From Eq. (9.6)

$$\Delta L = \frac{L_0 F}{Y A}$$

Here, because the radius of a wire 1.00 mm in diameter is 5.0×10^{-4} m, we have

$$L_0 = 2.0\,\text{m}$$
$$F = mg = (5.0\,\text{kg})(9.8\,\text{m/s}^2) = 49\,\text{N}$$
$$Y = 1.1 \times 10^{11}\,\text{N/m}^2$$
$$A = \pi r^2 = \pi (5.0 \times 10^{-4}\,\text{m})^2 = 7.85 \times 10^{-7}\,\text{m}^2$$

and so

$$\Delta L = \frac{(2.0\,\text{m})(49\,\text{N})}{(1.1 \times 10^{11}\,\text{N/m}^2)(7.85 \times 10^{-7}\,\text{m}^2)} = 1.1 \times 10^{-3}\,\text{m} = 1.1\,\text{mm} \quad ◆$$

◆   **E X A M P L E   9.6** _____

A sagging floor is jacked up and a steel girder 10.0 ft long whose cross-sectional area is 6.0 in^2 is put in place underneath. When the jack is removed, a sensitive strain gauge shows that the girder has been compressed by 0.0080 in. Find the load the girder is supporting.

SOLUTION   In order to use Eq. (9.6), both L_0 and ΔL must be in the same units. Here it is convenient to express both in inches, with $L_0 = (10.0\,\text{ft})(12.0\,\text{in./ft}) = 120$ in. From Eq. (9.6) we find for the load on the girder

$$F = YA\frac{\Delta L}{L_0} = \left(29 \times 10^6 \frac{\text{lb}}{\text{in}^2}\right)(6.0\,\text{in}^2)\left(\frac{0.0080\,\text{in}}{120\,\text{in}}\right) = 11{,}600\,\text{lb} = 5.8\,\text{tons} \quad ◆$$

The proportionality between strain and stress is valid only when the elastic limit is not exceeded. Table 9.3 is a list of the elastic limits and ultimate strengths under tension of several materials. The elastic limit is the same in both tension and compression, but the ultimate strength may be much greater in compression. The compressive strength of cast iron, for instance, is more than three times its tensile strength, and in the case of concrete the difference is a factor of ten. A material otherwise ideal for a certain purpose may have to be reinforced with something else to provide the missing tensile strength. Thus steel rods are commonly embedded in the concrete used for buildings and bridges, and glass fibers are embedded in the polyester resin used to make boat hulls.

Table 9.3   Typical elastic limits and ultimate strengths

Material	Elastic limit		Ultimate strength			
			Tension		Compression	
	$(\times\ 10^8\ \text{Pa})$	$(\times\ 10^4\ \text{lb/in}^2)$	$(\times\ 10^8\ \text{Pa})$	$(\times\ 10^4\ \text{lb/in}^2)$	$(\times\ 10^8\ \text{Pa})$	$(\times\ 10^4\ \text{lb/in}^2)$
Aluminum	1.8	2.6	2.0	2.9	2.0	2.9
Concrete	a		0.03	0.04	0.14	0.20
Copper	1.5	2.2	3.4	4.9	3.4	4.9
Granite	a		0.05	0.07	1.3	1.8
Iron (cast)	1.6	2.3	1.7	2.5	5.5	8.0
Pine wood (parallel to grain)	0.3	0.4	0.4	0.6	0.35	0.5
Steel	2.5	3.6	5.0	7.3	5.0	7.3

a Elastic limits in tension and compression are the same as the corresponding ultimate strengths.

In general, elastic limit and ultimate strength depend upon the history of an object as well as upon its composition: Hot-rolled and cold-rolled steel have different properties, as do annealed and tempered steel. Further, repeated cycles of stress tend to weaken an object through fatigue, as mentioned earlier. For these reasons the values in Table 9.3 are only approximate, although they are typical of each material.

◆   **E X A M P L E   9.7** _____

What is the minimum diameter of a copper wire able to support a 5.0-kg mass without exceeding its elastic limit?

SOLUTION   The applied force is

$$F = mg = (5.0\ \text{kg})(9.8\ \text{m/s}^2) = 49\ \text{N}$$

We note from Table 9.3 that the elastic limit of copper is 1.5×18^8 Pa. Hence

$$\frac{F}{A} = 1.5 \times 10^8\ \text{Pa} = 1.5 \times 10^8\ \text{N/m}^2$$

$$A = \frac{49\ \text{N}}{1.5 \times 10^8\ \text{N/m}^2} = 3.27 \times 10^{-7}\ \text{m}^2$$

Because the cross-sectional area of a wire of radius r is $A = \pi r^2$,

$$r = \sqrt{\frac{A}{\pi}} = \sqrt{\frac{3.27 \times 10^{-7}\ \text{m}^2}{\pi}}$$

$$= \sqrt{1.04 \times 10^{-7}\ \text{m}^2} = \sqrt{10.4 \times 10^{-8}\ \text{m}^2} = 3.2 \times 10^{-4}\ \text{m}$$

and the corresponding diameter is 6.4×10^{-4} m, which is 0.64 mm.   ◆

◆ E X A M P L E 9.8 _____

A nylon rope 0.75 in. in diameter has a breaking strength of 15,000 lb. Find the breaking strength of similar ropes (a) 0.50 in and (b) 1.00 in. in diameter.

SOLUTION (a) Since the breaking stress F/A is the same for all the ropes and $A = \pi d^2/4$,

$$\frac{F_2}{A_2} = \frac{F_1}{A_1}$$

$$F_2 = F_1\left(\frac{A_2}{A_1}\right) = F_1\left(\frac{\pi d_2^2/4}{\pi d_1^2/4}\right) = F_1\left(\frac{d_2}{d_1}\right)^2$$

The ratio of breaking strengths is proportional to the square of the ratio of diameters. With the 0.75-in rope as 1 and the 0.50-in rope as 2, the breaking strength of the 0.50-in rope is

$$F_2 = F_1\left(\frac{d_2}{d_1}\right)^2 = 15{,}000\,\text{lb}\left(\frac{0.50\,\text{in}}{0.75\,\text{in}}\right)^2 = 6667\,\text{lb}$$

To two significant figures, this is 6.7×10^3 lb.
(b) Here $d_2 = 1.00$ in, so

$$F_2 = 15{,}000\,\text{lb}\left(\frac{1.00\,\text{in}}{0.75\,\text{in}}\right)^2 = 26{,}667\,\text{lb}$$

To two significant figures, this is 2.7×10^4 lb. ◆

Figure 9.12

In shear there is a change in shape without a change in volume. The angle ϕ is the angle of shear.

9.5 SHEAR

The shear modulus of a material is a measure of its rigidity.

Shear stresses change the shape of an object on which they act but not its volume. The situation is much like that of a book whose covers are pushed out of alignment, as in Fig. 9.12. The layers of atoms, which are analogous to the pages of the book, are displaced sideways, but the spacing of the layers, which corresponds to the thickness of the pages, remains the same.

Let us consider a block of thickness d whose lower face is fixed in place and on whose upper face the force F acts (Fig. 9.13). A measure of the relative distortion of the block caused by the shear stress is the angle ϕ, called the **angle of shear**.

Figure 9.13

The angle of shear ϕ (in radians) is equal to s/d. The greater the shear modulus S, the more rigid the material.

Because this angle is always small, its value in radians is equal to the ratio s/d between the displacement s of the block's faces and the distance d between them. The greater the area A of these faces, the less they will be displaced by the shear force F. Therefore

$$\text{Shear stress} = \frac{\text{force}}{\text{area}} = \frac{F}{A}$$

$$\text{Shear strain} = \text{angle of shear} = \phi = \frac{s}{d}$$

and the stress-strain equation for shear is

$$\text{Shear modulus} = \frac{\text{shear stress}}{\text{shear strain}}$$

$$S = \frac{F/A}{\phi} = \frac{F/A}{s/d}$$

Hence we have

$$\phi = \frac{s}{d} = \frac{1}{S}\frac{F}{A} \qquad\qquad \textit{Shear} \quad (9.7)$$

Here the applied forces are *parallel* to the faces on which they act and not perpendicular as in the case of tension and compression.

The quality S is called the **shear modulus,** and the values of S for various substances are given in Table 9.2. The higher the value of S, the more rigid the material; S is sometimes referred to as the **modulus of rigidity** for this reason. It is interesting to note that the shear modulus of any material is usually a good deal smaller than its Young's modulus. This means that it is easier to slide the atoms of a solid past one another than it is to pull them apart or squeeze them together.

The shear strength of a material is the greatest shear stress an object of that material can withstand before breaking. The greater the shear strength of a material, the more force must be applied to cut a sheet of it with a pair of scissors (or their industrial equivalent) or to punch a hole in the sheet.

Figure 9.14

◆ **E X A M P L E   9.9**

Ordinary mild steel ruptures when a shear stress of about 3.5×10^8 Pa is applied. Find the force needed to punch a 10-mm diameter hole in a steel sheet 3.0 mm thick.

SOLUTION The area across which the shear stress is exerted here is the cylindrical inner surface of the hole (Fig. 9.14), so that

$$A = \pi dh = \pi(1.0 \times 10^{-2}\ \text{m})(3.0 \times 10^{-3}\ \text{m}) = 9.4 \times 10^{-5}\ \text{m}^2$$

Since we are given that $(F/A)_{max} = 3.5 \times 10^8$ Pa, we have

$$F_{max} = (3.5 \times 10^8\ \text{Pa})(9.4 \times 10^{-5}\ \text{m}^2) = 3.3 \times 10^4\ \text{N}$$

which is nearly 4 tons.   ◆

9.6 BULK MODULUS

What happens when an object is squeezed uniformly.

If we squeeze something on all sides, it shrinks in volume by some amount ΔV from its original volume of V_0. Only force components perpendicular to an object's surface act to compress it; parallel components lead to shearing strains. If the compressive force per unit area F/A is the same over the entire surface of the object, as in Fig. 9.15, then

$$\text{Volume stress} = \frac{\text{force}}{\text{area}} = \frac{F}{A}$$

Figure 9.15

Bulk compression.

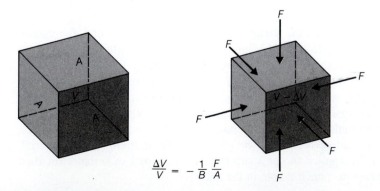

$$\frac{\Delta V}{V} = -\frac{1}{B}\frac{F}{A}$$

$$\text{Volume strain} = \frac{\text{change in volume}}{\text{original volume}} = \frac{\Delta V}{V_0}$$

and

$$\text{Bulk modulus} = -\frac{\text{volume stress}}{\text{volume strain}}$$

$$B = -\frac{F/A}{\Delta V/V_0}$$

Thus we have for the relative change in volume

$$\frac{\Delta V}{V_0} = -\frac{1}{B}\frac{F}{A} \qquad\qquad \textit{Uniform compression} \quad (9.8)$$

The minus sign corresponds to the fact that an increase in force leads to a decrease in volume. The quantity B is called the **bulk modulus.** Typical values of B are given in Table 9.2.

The perpendicular stress F/A is usually called **pressure,** symbol p, as discussed in Chapter 10. Hence we can also write

$$\frac{\Delta V}{V_0} = -\frac{p}{B} \qquad\qquad (9.9)$$

Liquids can support neither tensions nor shears, but they do tend to resist compression. Bulk moduli for several liquids are given in Table 9.4. Interatomic forces within a liquid are weaker than those within a solid, which is why bulk moduli of liquids are smaller than those of solids. A lot of pressure is nevertheless needed to compress a liquid by more than a slight amount. For instance, to compress a volume of water by 1% requires an inward force per unit area of 2.3×10^7 Pa, which is equivalent to 3300 lb/in^2.

Table 9.4 Bulk moduli of liquids at room temperature

Liquid	Bulk modulus, B ($\times 10^9$ Pa)	Bulk modulus, B ($\times 10^5$ lb/in^2)
Alcohol, ethyl	0.90	1.3
Benzene	1.05	1.5
Kerosene	1.3	1.9
Mercury	26	38
Oil, lubricating	1.7	2.5
Water	2.3	3.3

◆   E X A M P L E   9.10 _____

Verify the preceding statement.

SOLUTION   The bulk modulus of water is 2.3×10^9 Pa and, by hypothesis, $\Delta V/V_0 = -0.01$. Hence

$$\frac{F}{A} = -B\frac{\Delta V}{V_0} = -(2.3 \times 10^9 \, \text{Pa})(-0.01) = 2.3 \times 10^7 \, \text{Pa}$$ ◆

9.7   BUILDING OVER SPACE

How materials govern design.

Three successive inventions for building over space have shaped the course of architecture. The earliest, and still the most widely used, is the **post-and-beam** arrangement shown in Fig. 9.16(a), in which two vertical posts hold up a horizontal beam. Before steel came into general use in the last century, the width that could be spanned by a beam was severely limited by the nature of the materials available, which were mainly wood and stone. Wooden beams cannot support really large loads unless the span is narrow, and obtaining timber of adequate size and quality has always been a problem in much of the world. Stone can be quite strong in compression but is very much less so in tension, and the lower part of a beam is under tension.

Figure 9.16

Three ways to build over space. (a) A horizontal beam experiences tensile as well as compressive stresses. (b) The principal stresses in an arch are compressive. An arch made of stones or bricks, which are weak in tension but strong in compression, can support a considerable load. (c) Modern structural materials such as steel and reinforced concrete are strong in tension as well as compression and can be used as cantilevers supported only at one end.

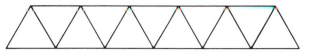

Figure 9.17

The members of a truss are arranged in triangles.

Narrow doorways and numerous interior supports were accordingly needed in buildings of even modest size that used post-and-beam construction until steel beams came into use in the nineteenth century.

Figure 9.18

The structural design of a proposed 135-story building uses trusses to cope with wind loads.

The great advantage of the **arch** is that the principal stresses imposed on it by a load are compressive (Fig. 9.16(b)). Stone and brick are sufficiently strong in compression for quite large arches, and also for domes, which are the three-dimensional equivalent of arches. The arch first came into wide use in Roman times, although it had been known thousands of years earlier. Innumerable ancient structures based on the arch and the dome still stand, testimony to the sound engineering principle behind them.

The third, and most recent, development in building over space is the **cantilever,** a beam held in place at one end only (Fig. 9.16(c)). Reinforced concrete as well as steel can be used in cantilever construction. The tall buildings of today do not have the load-bearing walls of the past but have internal skeletons instead. The outer parts of their floors and their thin ''curtain'' walls, often almost entirely glass, are supported by cantilever beams that extend outward from central frames.

Solid beams are often replaced by **trusses.** A truss is a structure made up of straight members of wood or metal arranged in triangles (Fig. 9.17). A truss using the same amount of material as a solid beam is much more rigid; or, for the same rigidity, a truss is much lighter than the corresponding solid beam. The members of a truss are in tension or in compression, not in shear, and hence can be relatively thin. The longest truss in the world is a 375-m bridge span.

Tall buildings are themselves vertical cantilevers because of the enormous forces winds can exert on them. The higher a skyscraper, the more severe the bending loads it must be able to withstand. For this reason a 135-story building proposed for New York City's Columbus Circle was designed as a trussed tube to maximize stiffness (Fig. 9.18).

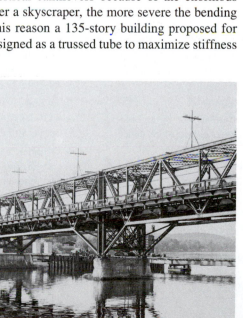

Truss bridge over the Amoskeag River in New Hampshire.

Important Terms

The **density** of a substance is its mass per unit volume. Its **weight density** is its weight per unit volume.

The **specific gravity** of a substance is its density relative to that of water. Specific gravity is also called **relative density.**

The three categories of stress forces are **tension,** in which equal and opposite forces that act away from each other are applied to a body; **compression,** in which equal and opposite forces that act toward each other are applied to a body; and **shear,** in which equal and opposite forces that do not act along the same line of action are applied to a body. A tensile stress tends to elongate a body, a compressive stress to shorten it, and a shearing stress to change its shape without changing its volume.

Hooke's law states that the amount of deformation experienced by a body under stress is proportional to the magnitude of the

stress. Thus the elongation of a wire is proportional to the tension applied to it.

The **elastic limit** is the maximum stress to which a solid can be subjected without being permanently altered. Hooke's law is valid only when the elastic limit is not exceeded.

The **stress** on an object is the applied force per unit area; its SI unit is the **pascal,** equal to 1 N/m^2, and its customary unit is the **psi,** equal to 1 lb/in^2. **Strain** is the resulting change in a dimension of the object relative to its original value.

A **modulus of elasticity** of a material is the ratio between a particular kind of applied stress and the resulting strain, provided the elastic limit is not exceeded. **Young's modulus** applies to tension and compression; the **shear modulus** applies to shear; and the **bulk modulus** applies to uniform volume compression.

Important Formulas

Density: $\rho = \dfrac{m}{V}$

Weight density: $D = \dfrac{w}{v}$

Hooke's law: $F = kx$

Tension or linear compression: $\dfrac{\Delta L}{L_0} = \dfrac{1}{Y}\dfrac{F}{A}$

Shear: $\phi = \dfrac{s}{d} = \dfrac{1}{S}\dfrac{F}{A}$

Uniform compression: $\dfrac{\Delta V}{V_0} = -\dfrac{1}{B}\dfrac{F}{A} = -\dfrac{p}{B}$

MULTIPLE CHOICE

1. Which of the following quantities is independent of the size and shape of an object composed of a given material?
- **a.** Volume
- **b.** Mass
- **c.** Weight
- **d.** Density

2. When equal and opposite forces are exerted on an object along different lines of action, the object is said to be under
- **a.** tension
- **b.** compression
- **c.** shear
- **d.** elasticity

3. Ductility refers to the ability of a metal to
- **a.** be deformed temporarily
- **b.** be deformed permanently
- **c.** shrink under compression
- **d.** break under tension

4. When a force acts on an object, the stress on it is equal to
- **a.** the relative change in its dimensions

- **b.** the applied force per unit area
- **c.** Young's modulus
- **d.** the elastic limit

5. A shear stress that acts on an object affects its
- **a.** length
- **b.** width
- **c.** volume
- **d.** shape

6. Another name for the shear modulus of a material is
- **a.** Young's modulus
- **b.** modulus of rigidity
- **c.** bulk modulus
- **d.** ductility

7. The only elastic modulus that applies to liquids is
- **a.** Young's modulus
- **b.** shear modulus
- **c.** modulus of rigidity
- **d.** bulk modulus

8. According to Hooke's law, the force needed to elongate an elastic object by an amount x is proportional to
a. x
b. $1/x$
c. x^2
d. $1/x^2$

9. The stress on a wire supporting a load does not depend on
a. the wire's length
b. the wire's diameter
c. the mass of the load
d. the acceleration of gravity

10. The density of brass is 8.4 g/cm^3. The volume of a 200-g brass monkey is approximately
a. 0.042 cm^3
b. 0.41 cm^3
c. 24 cm^3
d. 1.7 L

11. Gasoline has a specific gravity of 0.68. One liter of gasoline weighs
a. 0.68 N
b. 1.4 N
c. 6.7 N
d. 6.8 N

12. A quart of water weighs 2.1 lb and a quart of antifreeze weighs 2.3 lb. The specific gravity of the antifreeze is
a. 0.2
b. 0.9
c. 1.1
d. 2.3

13. A cable stretches by the amount a under a certain load. If it is replaced by a cable of the same material but half as long and half the diameter, the same load will stretch it by
a. $a/4$
b. $a/2$
c. a
d. $2a$

14. The original cable in Multiple Choice 13 could support a maximum load of w without exceeding the elastic limit. The elastic limit will not be exceeded by the new cable up to a load of
a. $w/4$
b. $w/2$
c. w
d. $2w$

15. An iron wire 1.0 m long with a square cross section 2.0 mm on a side is used to support a 100-kg load. Its elongation is
a. 0.0027 mm
b. 0.27 mm
c. 1.3 mm
d. 3.7 mm

16. A brass rod 5.0 ft long with a square cross section 0.50 in on a side is used to support a 520-lb load. Its elongation is
a. 0.0008 in
b. 0.0048 in
c. 0.0096 in
d. 0.048 in

17. Young's modulus for aluminum is 7.00×10^{10} Pa. The force needed to stretch by 1.0 mm an aluminum wire 2.0 mm in diameter and 800 mm long is
a. 2.75 N
b. 275 N
c. 1.10 kN
d. 2.75 kN

18. The elastic limit of aluminum is 1.80×10^8 Pa. The maximum mass the wire in Multiple Choice 17 can support without exceeding the elastic limit is
a. 58 kg
b. 116 kg
c. 232 kg
d. 565 kg

19. A wire 10 m long with a cross-sectional area of 0.10 cm^2 stretches by 13 mm when a load of 100 kg is suspended from it. The Young's modulus for this wire is
a. 0.77×10^{10} Pa
b. 7.5×10^{10} Pa
c. 7.7×10^{10} Pa
d. 9.3×10^{10} Pa

20. The ultimate strength in compression of aluminum is 2×10^8 Pa. The maximum mass an aluminum cube 1 cm on each edge can support is about
a. 2×10^2 kg
b. 2×10^3 kg
c. 2×10^4 kg
d. 2×10^5 kg

21. The force needed to punch a hole 8.0 mm square in a steel sheet 3.0 mm thick whose shear strength is 2.5×10^8 Pa is
a. 2.7 kN
b. 6 kN
c. 24 kN
d. 48 kN

22. When a pressure of 2.0 MPa is applied to a sample of kerosene, it contracts by 0.15%. The bulk modulus of kerosene is
a. 1.7 MPa
b. 1.997 MPa
c. 2.003 MPa
d. 1.3 GPa

QUESTIONS

1. Two boys wish to break a string. Are they more likely to do this if each takes one end of the string and they pull against each other, or if they tie one end of the string to a tree and both pull on the free end? Why?

2. A rubber band is easy to stretch at first, but after it has been extended by a certain amount, additional stretching be-comes more difficult at a greater rate. Make a rough stress-strain graph for rubber in tension.

3. A three-legged stool has one leg of aluminum, one of copper, and one of steel. The legs have the same dimensions. If the load on the stool is on its exact center, which leg is under the greatest stress and which under the least stress?

Which leg experiences the greatest strain and which the least strain?

4. A cable is replaced by another one of the same length and same material but of twice the diameter. How does this affect the maximum load that can be supported by the cable? What would be the result of using a cable of the same diameter and material but twice the length of the original one?

5. Two wires are made of the same material, but wire A is twice as long and has twice the diameter of wire B. Find the elongation of wire B relative to that of wire A when both are subjected to the same load.

6. The elastic moduli of a particular material are related to one another, and each of them can be expressed in terms of the other two. For instance, the theory of elasticity shows that Young's modulus Y can be expressed in terms of the shear modulus S and the bulk modulus B by the formula

$$Y = \frac{9SB}{(3B + S)}$$

EXERCISES

9.1   Density

1. A room is 5.0 m long, 4.0 m wide, and 3.0 m high. What is the mass of the air it contains?

2. A coil of sheet steel 0.0300 in thick and 20.0 in wide weighs 240 lb. Find the length of the steel in the coil.

3. Mammals have approximately the same density as fresh water. Find the volume in liters of a 55-kg woman and the volume in cubic meters of a 140,000-kg blue whale.

4. A 1200-kg concrete slab that measures 2.0 m × 1.0 m × 20 cm is delivered to a building under construction. Does it contain steel reinforcing rods or is it plain concrete?

5. The radius of the earth is 6.37×10^6 m, and its mass is 5.98×10^{24} kg. (a) Find the average density of the earth. (b) The average density of rocks at the earth's surface is 2.7×10^3 kg/m³. What must be true of the matter of which the earth's interior is composed? Is it likely that the earth is hollow and peopled by another species, as the ancients believed?

6. If gold costs $300/oz, how many millimeters on a side does a $10,000 cube of gold measure?

7. Water and oil are poured into the ends of a glass tube bent into a U-shape, as in Fig. 9.20. Find the density of the oil in grams per cubic centimeter.

8. A 200-g bottle has a mass of 340 g when filled with water and 344 g when filled with blood plasma. What is the density of the plasma?

Check this formula for three of the materials listed in Table 9.2. Give several reasons not to expect perfect agreement.

Figure 9.19

Question 7.

7. Figure 9.19 shows the concrete floor of a building that is continued outside to form a cantilevered balcony. Because of their expense, steel rods are only used to reinforce concrete where tensile stresses occur. Show where reinforcing rods are most needed in the floor and balcony of the diagram.

Figure 9.20

Exercise 7.

9.2   Elasticity

9. When a coil spring is used to support a 12-kg object, the spring stretches by 40 mm. What is the force constant of the spring?

10. A coil spring has a force constant of 1000 N/m. How much will it stretch when it is used to support an object whose mass is 8.0 kg?

11. A coil spring has a force constant of 20 lb/ft. How much will it stretch when it is used to support an object whose mass is 0.50 slug?

12. The length of a spring is measured when a series of forces is applied to it. The results are as follows:

Applied force, N	0	5	10	15	20	25	30
Spring length, cm	10.0	10.8	11.7	12.5	13.3	14.2	16.4

From a graph of applied force versus elongation for the spring, estimate its elastic limit.

9.4 Young's Modulus

13. A nylon rope 10.0 mm in diameter breaks when a load of 25 kN is applied to it. What would you estimate for the breaking strength of a nylon rope 6.0 mm in diameter? 14.0 mm in diameter?

14. The actual dimensions of a "2 × 4" timber are 1.5 × 3.5 in. Find the maximum weight in tons that a pine 2 × 4 standing on end can support.

15. How safe is it for a 75-kg circus acrobat to balance on the index finger of his right hand, whose bones have a minimum cross-sectional area of 0.50 cm²?

16. Find the maximum force a hammer can apply to a steel nail 2.5 mm in diameter if the nail's elastic limit is not to be exceeded.

17. A human hair 60 μm in diameter can just support a mass of 55 g (equivalent to about 2 oz of weight). How does the ultimate strength in tension of hair compare with that of aluminum?

18. In a test firing, the three main engines of the space shuttle *Columbia* developed a total thrust of 4.5×10^6 N. Eight steel bolts 90 mm in diameter were used to hold the *Columbia* to its launch pad during the firing. What was the ratio between the ultimate strength of the bolts and the applied stress?

19. The table below shows by how much a metal rod 120 cm long whose cross-sectional area is 16 mm² stretches when various loads are applied. From a stress-strain graph for the metal, find its Young's modulus.

Load, kN	0	1	2	3	4	5
ΔL, mm	0	0.63	1.25	1.88	2.50	3.13

20. A cube of pine wood 50 mm on an edge is held in the jaws of a vise by a force of 2000 N. By how much is the wood compressed?

21. A steel wire 1.0 m long and 1.0 mm square in cross section supports a mass of 6.0 kg. By how much does it stretch?

22. An 80-kg man has femurs (the femur is the bone of the thigh) 42 cm long and 11 cm² in average cross-sectional area. Find the approximate change in length of each femur as the man walks. Young's modulus for bone is 1.5×10^{10} Pa.

23. A steel column 12 ft long and of 4.3 in² cross-sectional area supports a load of 3.0 tons. By how much is it shortened?

24. A wire 10 ft long with a cross-sectional area of 0.12 in² stretches by 0.02 in when a load of 600 lb is suspended from it. Find the value of Young's modulus for this wire.

25. A concrete post 20 ft high is not supposed to be compressed by more than 0.10 in when it supports a load of 40 tons. What should the diameter of the post be?

26. By how much can a copper wire 2.0 m long be stretched before its elastic limit is exceeded?

27. An iron pipe 3.0 m long is used to support a sagging floor. The inside diameter of the pipe is 10 cm and its outside diameter is 11 cm. When the force on it is 15 kN, by how much is it compressed?

28. A certain kind of brick has a density of 2400 kg/m³ and an ultimate strength in compression of 9.6×10^7 Pa. Find the maximum height to which a wall of this brick can be built before the lowest brick disintegrates.

29. A wall of lead bricks exactly 1.0 m high is used to shield a sample of radium. Each brick was originally a cube exactly 10 cm on an edge. What is the change in height of the lowest brick when the wall has been erected?

30. A typical sample of compact bone has a tensile strength of 1.3×10^8 Pa and a compressive strength of 1.7×10^8 Pa. Suppose you have cylinders of the same length and diameter of bone, aluminum, and steel. For each cylinder find the ratio between the maximum tensile and compressive forces it can withstand and its mass. How does bone compare with aluminum and steel as a structural material?

31. A 30-mm cube of raspberry gelatin on a table is subjected to a shearing force of 0.50 N. The upper surface is displaced by 5.0 mm. What is the shear modulus of the gelatin?

32. Two steel plates are riveted together with 10 rivets each 5.0 mm in diameter. If the maximum shear stress the rivets can withstand is 3.5×10^8 Pa, how much force applied parallel to the plates is needed to shear off the rivets?

33. A punch press that exerts a force of 20 kN is employed to punch 1.0-cm-square holes in sheet aluminum. If the shear strength of aluminum is 70 MPa, what is the maximum thickness of aluminum sheet that can be used?

34. A slot 8.00 in long and 0.50 in wide is to be punched in a steel sheet 0.060 in thick whose shear strength is 5.0×10^4 lb/in². Find the force needed.

9.6 Bulk Modulus

35. The pressure at a depth of 1.0 km in the ocean exceeds sea-level atmospheric pressure by about 10 MPa. If an iron anchor whose volume at the surface is 400 cm³ is lowered to a depth of 1.0 km, by how much does its volume decrease?

36. The deepest known point in the oceans, 11 km below the surface, is found in the Marianas Trench southwest of Guam in the Pacific. At that depth the water pressure exceeds sea-level pressure by 0.11 GPa. The density of sea water at sea level is 1.03×10^3 kg/m³ and its bulk modulus is 2.3 GPa. What is the density of sea water at a depth of 11 km?

PROBLEMS

38. One gram of gold can be beaten out into a foil 1.00 m² in area. (Thus an ounce of gold can yield 300 ft² of foil.) How many atoms thick is such a foil? The mass of a gold atom is 3.27×10^{-25} kg.

39. A 1.0-kW motor drives a pump that raises gasoline from an underground tank through a height of 3.2 m. If the pump is 65% efficient, find the rate of flow of gasoline in L/min.

40. A vise is used to hold a 2.0-in cube of pine wood while it is being worked on. The cube is in contact with the jaws of the vise, whose screw has a pitch of $\frac{3}{16}$ in and whose handle is grasped 4.0 in from the screw axis. If there is no friction and a force of 15 lb is applied to the handle, find the amount by which the cube is compressed.

41. A helicopter whose mass is 900 kg has rotor blades 5.0 m in radius. Find the speed of the air forced downward by its blades when the helicopter hovers at rest above the ground.

42. The parachute of a 60-kg woman fails to open, and she falls into a snowbank at a terminal speed of 60 m/s, coming to a stop 1.5 m below the surface of the snow. If the area of the woman's body that strikes the snow is 0.20 m² and the stress required for serious injury to body tissues is 5.0×10^5 Pa, is she likely to survive the impact?

43. A steel cable whose cross-sectional area is 2.5 cm² supports a 1000-kg elevator. The elastic limit of the cable is 3.0×10^8 Pa. What is the maximum upward acceleration that can be given the elevator if the tension in the cable is to be no more than 20% of the elastic limit?

44. A certain material has a density of ρ and a tensile strength of U. How long a rod of this material can be suspended from one end without breaking under its own weight? What is this length in the case of aluminum? Steel? (Assume the density of steel to be the same as that of iron.)

45. A copper wire and a steel wire are being used side by side to support a load. (a) If they have the same diameter and each stretches by the same amount, what proportion of the load does each one support? (b) What should the ratio of their diameters be if they are each to support half the load?

46. A bar 2.0 m long is suspended horizontally from its ends by steel wires 1.0 m long. The wires have radii of 1.0 mm.

A mass of 20 kg is then attached 40 cm from one end of the bar. What is the difference in length between the two wires?

47. A wire 1.0 m long of negligible mass with a cross-sectional area of 1.0 mm² is supported at both ends so that it is very nearly horizontal. When a mass of 1.0 kg is attached to the midpoint of the wire, the wire stretches so that the midpoint sags by 20 mm. Find the approximate value of Young's modulus for the wire's material.

Figure 9.21

Problem 48.

48. When a torque τ is applied to a cylinder, the angle θ (in radians) through which it twists depends on the cylinder's length L, radius r, and shear modulus S according to the formula

$$\theta = \frac{2\tau L}{\pi S r^4}$$

The geometry of the situation is shown in Fig. 9.21. If the angle of twist of a solid steel drive shaft 50 mm in diameter and 2.5 m long is not to exceed 2°, find the maximum power the shaft can transmit at 900 rpm.

Answers to Multiple Choice

1. d	7. d	13. d	19. c
2. c	8. a	14. a	20. b
3. b	9. a	15. c	21. c
4. b	10. c	16. c	22. d
5. d	11. c	17. b	
6. b	12. c	18. a	

CHAPTER 10

10

FLUIDS

10.1   Pressure

10.2   Hydraulic Press

10.3   Pressure and Depth

10.4   Buoyancy

10.5   Fluid Flow

10.6   Bernoulli's Equation

10.7   Applications of Bernoulli's Equation

10.8   Viscosity

10.9   Surface Tension

As the name implies, a fluid is a substance that flows readily. Gases and liquids are fluids. Its ability to flow enables a fluid to exert an upward force on an immersed object such as a ship, to multiply an applied force in a hydraulic press, and to provide "lift" to an airplane wing. In what follows we will find that there is nothing mysterious about these properties of fluids, properties that follow from the laws of physics we already know.

10.1 PRESSURE

Normal force per unit area.

When a force **F** acts perpendicular to a surface of area A, the **pressure** p exerted on the surface is the ratio between the magnitude F of the force and the area. Because a perpendicular force is also described as a **normal force,** we can say that pressure is the magnitude of normal force per unit area:

$$p = \frac{F}{A} \qquad\qquad\qquad Pressure \quad (10.1)$$

$$\text{Pressure} = \frac{\text{normal force}}{\text{area}}$$

As we learned in Chapter 9, the SI unit of pressure is the **pascal** (Pa), equal to 1 N/m^2. The U.S. customary unit of pressure is the lb/in^2 or **psi**.

Because the pascal is very small, the **bar** is widely used instead, where

$$1 \text{ bar} = 10^5 \text{ Pa}$$

The average pressure of the earth's atmosphere at sea level is 1.013 bar (14.7 lb/in^2). Meteorologists commonly use the **millibar** (mb) for convenience, where 1 mb = 10^{-3} bar, so that 1013 mb corresponds to average atmospheric pressure. Actual atmospheric pressures usually range between 970 and 1040 mb.

◆   **E X A M P L E   10.1** _____

The flat roof of a house is 10.0 m long and 8.0 m wide and has a mass of 7500 kg. Before a severe storm the people in the house closed its doors and windows so tightly that the air pressure inside remained at 1013 mb even when the outside pressure fell to 980 mb. Compare the upward force on the roof with its weight.

SOLUTION   The area of the roof is (10.0 m) (8.0 m) = 80 m^2. The difference Δp between the pressures on the inside and outside of the roof is

$$\Delta p = (1013 \text{ mb} - 980 \text{ mb})(100 \text{ Pa/mb}) = 3.3 \text{ kPa}$$

Because the pressure on the inside of the roof is greater, the net force on it is upward with the magnitude

$$F = A\Delta p = (80 \text{ m}^2)(3.3 \text{ kPa}) = 264 \text{ kN}$$

This is 59,400 lb! The roof's weight is

$$w = mg = (7500 \text{ kg})(9.8 \text{ m/s}^2) = 74 \text{ kN}$$

Thus the upward force on the roof is nearly four times its weight. If the roof is not well attached to the walls of the house, and if the windows do not break first, the roof will be lifted off during the storm. Clearly it is unwise to seal a building when a large drop in air pressure is coming.   ◆

Three ways to measure pressure are shown in Fig. 10.1. As a rule what is found is the difference between the unknown pressure and atmospheric pressure. This difference is **gauge pressure,** whereas the true pressure, called **absolute pressure,**

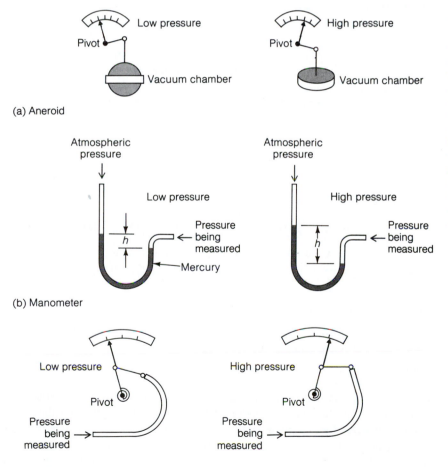

(a) Aneroid

(b) Manometer

(c) Bourdon tube

Figure 10.1

Three types of pressure gauge. (a) An aneroid measures pressure in terms of the amount by which the thin, flexible ends of an evacuated metal chamber are pushed in or out by the external pressure. (b) A manometer measures pressure in terms of the difference in height h of two mercury columns, one open to the atmosphere and the other connected to the source of the unknown pressure. (c) A Bourdon tube straightens out when the internal pressure exceeds the external pressure.

Figure 10.2

On an absolute pressure scale, 0 corresponds to a perfect vacuum. On a gauge pressure scale, 0 corresponds to atmospheric pressure, which averages 1.013 bars (1.013 × 10^5 Pa). Shown is an absolute pressure of 3 bar, which is equivalent to a gauge pressure of 2 bar.

includes atmospheric pressure. That is,

$$p = p_{\text{gauge}} + p_{\text{atm}} \tag{10.2}$$

Absolute pressure = gauge pressure + atmospheric pressure

A tire inflated to a gauge pressure of 2 bar contains air at an absolute pressure of about 3 bar because sea-level atmospheric pressure is 1.013 bar (Fig. 10.2).

◆   **E X A M P L E   10.2** _____

The weight of a 2400-lb car is supported equally by its four tires. The gauge pressure of the air in the tires is 28.0 lb/in^2. Find the area of each tire that is in contact with the ground.

SOLUTION   The load on each tire consists of the portion of the car's weight it supports plus the weight of a column of the atmosphere whose cross-sectional area equals the area of the tire in contact with the ground, since this part of the tire has no air at atmospheric pressure under it to provide an equal upward force. Hence only the gauge pressure of the air in the tires, which is the excess over atmospheric pressure, is effective in supporting the car's weight. Each tire must support 600 lb. The area of each tire in contact with the ground is therefore

$$A = \frac{F}{P} = \frac{600 \text{ lb}}{28.0 \text{ lb/in}^2} = 21.4 \text{ in}^2 \qquad\qquad ◆$$

Three Properties of Pressure in a Fluid

Pressure is a useful quantity because fluids flow under stress instead of being deformed elastically as solids are. The absence of rigidity in fluids has three significant consequences:

_____

1. The forces a fluid at rest exerts on the walls of its container, and vice versa, always act perpendicular to the walls.

_____

If this were not so, any sideways force by a fluid on a wall would be accompanied, according to the third law of motion, by a sideways force back on the fluid, which would cause the fluid to move parallel to the wall. But the fluid is at rest, so the force must be perpendicular to the container walls. A *moving* fluid is another matter. As we will learn in Section 10.8, frictional forces act between a moving fluid and, for instance, the walls of a pipe or the bank of a river. The difference between a body of liquid in contact with a solid and two solids in contact is that in the former case there is no static friction between them.

---

2. An external pressure exerted on a fluid is transmitted uniformly throughout the volume of the fluid.

If this were not so, the fluid would flow from a region of high pressure to one of low pressure, which would equalize the pressure. We must keep in mind, however, that this statement, which is known as **Pascal's principle,** refers to a pressure imposed from outside the fluid. The fluid at the bottom of a container is always under more pressure than that at the top owing to the weight of the overlying fluid. A notable example is the earth's atmosphere (Fig. 10.3), although such pressure differences are ordinarily significant only for liquids.

---

3. The pressure on a small surface in a fluid is the same regardless of the orientation of the surface.

If this were not so, again, the fluid would flow in such a way as to equalize the pressure (Fig. 10.4).

Figure 10.3

Variation of atmospheric pressure with height above sea level. At an altitude of 50 km, the pressure is down to about 1/1000 of its sea-level value. (1 mb = 100 Pa.)

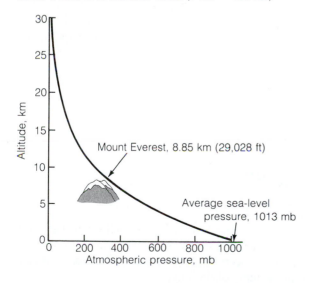

Figure 10.4

At a given depth, the force exerted by the pressure in a fluid is the same in all directions.

Figure 10.5

A piston pump. During the suction stroke, the inlet valve opens and the fluid is drawn into the cylinder. When the piston is pushed down, the inlet valve closes, the outlet valve opens, and the fluid is forced out.

Suction stroke   Pressure stroke

Figure 10.6

Two types of rotary pump. (a) Impeller pump. The fluid is compressed when the rubber fins of the impeller are bent over at the flattened top of the chamber. (b) Centrifugal pump. The rotating blades set the fluid in motion.

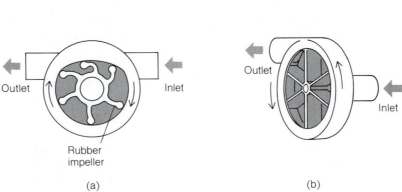

(a)   (b)

A **pump** is a device for creating a pressure difference. The operation of the familiar piston pump is illustrated in Fig. 10.5. A pump of this kind can be used to provide high pressure (to inflate a tire or a football, for instance) or low pressure (to pump water from a well or the bilge of a boat, for instance). The heart circulates blood through the lungs and through the body by means of pumps of this kind, with the contraction and relaxation of the muscular walls of the heart chambers taking the place of piston strokes. Rotary pumps are more suited to motor drive; two examples are shown in Fig. 10.6.

10.2 HYDRAULIC PRESS

The pressure is the same on both pistons, but the force is not.

Two of the three basic machines, the lever and the inclined plane, were considered in Chapter 8. The third, the **hydraulic press,** is based on Pascal's principle, which is property 2 given in the previous section: An external pressure on a fluid is transmitted uniformly throughout the volume of the fluid.

$$MA = \frac{F_{out}}{F_{in}} = \frac{A_{out}}{A_{in}} = \frac{L_{in}}{L_{out}}$$

Figure 10.7

The hydraulic press makes use of the fact that pressure exerted on a fluid is transmitted equally throughout the fluid. Valve 1 is closed and valve 2 open on the downstroke of the input piston. Valve 1 is open and valve 2 closed on the upstroke, when additional fluid is drawn into the input cylinder to enable its piston to make another stroke.

Figure 10.7 shows a hydraulic press being used in a garage to lift a car for servicing. We will assume that the fluid is incompressible and that there are no other sources of inefficiency. If we exert a force F_{in} on the input piston, whose area is A_{in}, the pressure on the fluid is $p = F_{in}/A_{in}$. This pressure is transmitted by the fluid to the output piston, where $p = F_{out}/A_{out}$. The pressure is the same on both pistons, so that

$$p = \frac{F_{in}}{A_{in}} = \frac{F_{out}}{A_{out}}$$

and

$$\frac{F_{out}}{F_{in}} = \frac{A_{out}}{A_{in}}$$

A small input force can be considerably increased simply by having the output piston much larger in area than the input piston.

Because the fluid is incompressible, the fluid shifted from one cylinder to the other has the same volume $V = AL$, where L is the distance through which each piston has moved. Therefore

$$V = A_{in}L_{in} = A_{out}L_{out}$$

and the ideal mechanical advantage of the hydraulic press is

$$IMA = \frac{F_{out}}{F_{in}} = \frac{A_{out}}{A_{in}} = \frac{L_{in}}{L_{out}} \qquad \qquad \textit{Hydraulic press} \quad (10.3)$$

Figure 10.8

A hydraulic vehicle brake. When the brake pedal is pushed, the pressure on the brake fluid in the master cylinder is transmitted to each brake cylinder where the force on the pistons pushes the brake shoe against the brake drum. The friction between the brake linings and the drums is what slows the car down. In a disk brake, the pistons of the brake cylinder press pads against the sides of a metal disk mounted on the inside of each wheel.

Thus the input piston must move through a large range in order to move the output piston through a small one. If the piston areas are in the ratio 100:1, a force of 1 N can lift a weight of 100 N, but for every centimeter the weight is raised, the input piston must move down 100 cm. The purpose of the valves shown in Fig. 10.7 is to permit the output piston to be raised by a series of short strokes of the input piston.

For a hydraulic press that is not 100% efficient,

$$F_{\text{out}} = (\text{Eff}) \frac{A_{\text{out}}}{A_{\text{in}}} F_{\text{in}}$$

The principle of the hydraulic press can be applied in a variety of ways. Common industrial uses include presses of various kinds, garage lifts, control systems in airplanes, and vehicle brakes (Fig. 10.8).

The hydraulic rams on this backhoe convert liquid pressure into applied forces. The pressure is provided by an engine-driven pump.

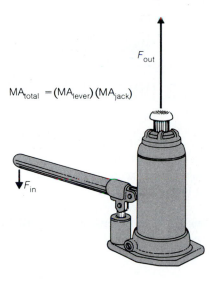

$$MA_{total} = (MA_{lever})(MA_{jack})$$

Figure 10.9

Hydraulic jack.

◆ **E X A M P L E 10.3**

The input and output pistons of a hydraulic jack are, respectively, 10 mm and 40 mm in diameter. A lever with a mechanical advantage of 6.0 is used to apply force to the input piston (Fig. 10.9). How much mass can the jack lift if a force of 180 N (about 40 lb) is applied to the lever and the system is 100% efficient?

SOLUTION   The mechanical advantage of the jack is

$$IMA = \frac{A_{out}}{A_{in}} = \frac{(\pi r_{out})^2}{(\pi r_{in})^2} = \left(\frac{d_{out}}{d_{in}}\right)^2 = \left(\frac{40 \text{ mm}}{10 \text{ mm}}\right)^2 = 16$$

The total mechanical advantage of the lever + jack system is

$$IMA_{total} = (IMA_{lever})(IMA_{jack}) = (6.0)(16) = 96$$

Hence the output force of the system when the input force is 180 N is

$$F_{out} = (IMA)(F_{in}) = (96)(180 \text{ N}) = 1.73 \times 10^4 \text{ N}$$

If the force equals a weight that the jack is lifting, the corresponding mass is

$$m = \frac{w}{g} = \frac{1.73 \times 10^4 \text{ N}}{9.8 \text{ m/s}^2} = 1.8 \times 10^3 \text{ kg}$$

which is nearly 2 tons.   ◆

Figure 10.10

(a) The gauge pressure a fluid in a tank exerts on the bottom is equal to ρgh. (b) The same formula holds for the fluid pressure at any depth h below the surface.

Figure 10.11

The total pressure at a depth h in a fluid is the sum of the fluid pressure and the external pressure exerted on the fluid surface.

10.3 PRESSURE AND DEPTH

Pressure in a fluid increases with depth.

The pressure in a fluid becomes greater with increasing depth because of the weight of the overlying material. In a liquid the pressure change is marked: Most submarines cannot descend more than a few hundred meters without the danger of collapsing.

Suppose we have a tank of height h and cross-sectional area A that is filled with a fluid of density ρ (Fig. 10.10). The volume of the tank is $V = Ah$, and the weight of the fluid it contains is

$$w = mg = (\rho V)g = \rho gAh$$

The pressure p_{fluid} the fluid exerts on the bottom of the tank is its weight divided by the area of the bottom, with the result that

$$p_{\text{fluid}} = \frac{F}{A} = \frac{w}{A} = \rho gh \qquad (10.4)$$

The pressure difference between the top and the bottom of the tank is proportional to the height of the fluid column and to the fluid density.

This result also applies to *any* depth h in a fluid, whether at the bottom or not, because the fluid below that depth does not add to the weight pressing down there.

The *total* pressure within a fluid, of course, also depends on the pressure p_{external} exerted on its surface by the atmosphere or, perhaps, by a piston (Fig. 10.11). Thus, in general, the total pressure at a depth h in a fluid of density ρ is $p = p_{\text{external}} + p_{\text{fluid}}$:

$$p = p_{\text{external}} + \rho gh \qquad \textit{Pressure at depth } \text{h} \textit{ in a fluid} \quad (10.5)$$

◆   **E X A M P L E   10.4**

Find the pressure on a scuba diver when she is 15.0 m below the surface of the sea. (*Scuba* comes from the first letters of "*self-contained underwater breathing apparatus.*")

SOLUTION   The density of seawater is 1.03×10^3 kg/m³. Hence the pressure on the diver due to the sea alone is

$$\rho gh = \left(1.03 \times 10^3 \frac{\text{kg}}{\text{m}^3}\right)\left(9.80 \frac{\text{m}}{\text{s}^2}\right)(15.0 \text{ m}) = 1.51 \times 10^5 \text{ Pa} = 1.51 \text{ bar}$$

High pressure air from a scuba diver's tank passes through a regulator valve that reduces the pressure to that of water at the depth of the diver. The deeper the diver goes, the greater the water pressure, and the faster the air in the tank is used up.

Because atmospheric pressure is 1.01 bar, the total pressure on the diver is

$$p = p_{\text{external}} + pgh = (1.01 + 1.51) \text{ bar} = 2.52 \text{ bar}$$

which is about $2\frac{1}{2}$ times atmospheric pressure. She is not crushed, because the pressure inside her body increases as she descends to match the pressure exerted on it.

As the diver returns to the surface, it is essential that she breathe out continuously to allow the air pressure in her lungs to decrease at the same rate as the pressure on her decreases. If this is not done, the pressure difference will cause her lungs to expand and rupture. Similarly a face mask that covers the nose as well as the eyes should be used instead of just goggles when scuba diving. Air exhaled through the nose can then keep the external pressure on the eyeballs the same as the internal pressure, which prevents damage to the blood vessels of the eyes. (See Section 24.6 for the reason why goggles or a face mask are needed for proper vision underwater.) ♦

Figure 10.12

Arterial blood pressures are measured with the help of an inflatable cuff that is wrapped around the upper arm. A pump is used to inflate the cuff until the flow of blood stops, and air is then let out by means of the valve until the flow begins again. The stethoscope is used to monitor the blood flow.

Blood Pressure

A person's arterial blood pressures are usually measured with the help of an inflatable cuff wrapped around the upper arm at the level of the heart (Fig. 10.12). A stethoscope is used to monitor the sound of the blood flowing through an artery below the cuff. The cuff is first inflated until the flow of blood stops. Then the pressure of the cuff is gradually reduced until the blood just begins to flow, which is recognized by a gurgling sound in the stethoscope. This pressure, called **systolic,** represents the maximum pressure the heart produces in the artery. The pressure in the cuff is then further reduced until the gurgling stops, which corresponds to the restoration of normal blood flow. The pressure at this time, called **diastolic,** represents the pressure in the artery between the contractions of the heart.

Physicians express blood pressures in **torr,** where 1 torr is the pressure exerted by a column of mercury 1 mm high; it is equivalent to 133 Pa. The torr was formerly referred to as the "millimeter of mercury," abbreviated mm Hg. The unit

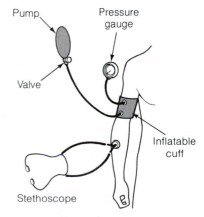

Pump

Pressure gauge

Valve

Inflatable cuff

Stethoscope

Figure 10.13

Arterial blood pressure rises and falls as the heart contracts and relaxes. The pressures shown are typical for a healthy person. The time scale corresponds to 70 heartbeats per minute.

is named after Evangelista Torricelli (1608–1647), the Italian physicist who invented the barometer, which measures atmospheric pressure. Average atmospheric pressure is 760 torr. In a healthy person the systolic and diastolic blood pressures are, respectively, about 120 and 80 torr (Fig. 10.13).

 E X A M P L E   10.5 _____

The average pressure of the blood in a person's arteries is 100 torr at the same elevation as the heart. Find the average pressure in an artery in the head of a standing person, say 40 cm above the heart, and in an artery in the foot, say 120 cm below the heart. The density of blood is 1.06×10^3 kg/m³.

SOLUTION   The difference in pressure in each case is $\Delta p = \rho g h$. Here

$$\Delta p_1 = \rho g h_1 = (1.06 \times 10^3 \text{ kg/m}^3)(9.8 \text{ m/s}^2)(0.4 \text{ m}) = 4.16 \times 10^3 \text{ Pa}$$
$$\Delta p_2 = \rho g h_2 = (1.06 \times 10^3 \text{ kg/m}^3)(9.8 \text{ m/s}^2)(1.2 \text{ m}) = 12.5 \times 10^3 \text{ Pa}$$

Because 1 torr = 133 Pa, $\Delta p_1 = 31$ torr and $\Delta p_2 = 94$ torr. Hence the pressure in the artery in the head is $(100 - 31)$ torr = 69 torr and that in the artery in the foot is $(100 + 94)$ torr = 194 torr. The arteries that lead to the head expand and contract as needed to keep the flow of blood to the brain constant despite changes in the elevation of the head relative to the heart. Such expansions and contractions require a few seconds to be completed, which explains why sitting up suddenly from a horizontal position may lead to a momentary dizzy sensation.   ◆

10.4 BUOYANCY

Sink or swim.

When an object is placed in water (or any other fluid), an upward force acts on it. This effect, called **buoyancy,** makes it possible for balloons to float in the air and ships to float in the sea.

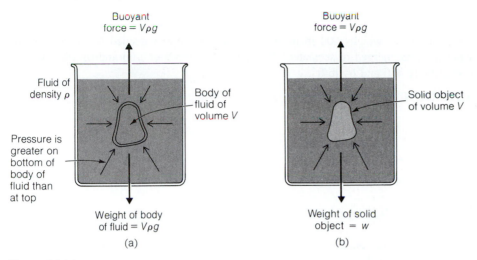

Figure 10.14

The buoyant force on a submerged object is equal to the weight of the fluid it displaces.

Imagine a solid object of volume V submerged in a fluid of density ρ. A body of fluid of the same size and shape in the same place is acted on by a buoyant force that is the vector sum of all the forces the rest of the fluid exerts on it, as in Fig. 10.14(a). The buoyant force must be upward because the pressure (and hence the upward force) on the bottom of the body is greater than the pressure (and hence the downward force) on its top. The horizontal forces on the sides of the body cancel one another out. Because the body of fluid is in equilibrium, the buoyant force on it equals its weight, which is $V\rho g$.

Now we replace the body of fluid by the solid object, as in Fig. 10.14(b). The various pressures remain the same, so the buoyant force of $V\rho g$ also remains the same. We conclude that

$$F_{\text{buoyant}} = V\rho g \qquad\qquad \textit{Archimedes' principle} \quad (10.6)$$

Buoyant force = weight of displaced fluid

This result is known as **Archimedes' principle,** after its discoverer.

The buoyant force on an object in a fluid is equal to the weight of the fluid the object displaces.

Archimedes' principle holds whether the object floats or sinks. If the object's weight is greater than the buoyant force, it sinks. If its weight is less than the buoyant force, it floats, in which case the volume V refers only to the submerged part.

◆   E X A M P L E   10.6 _____

An iron anchor weighs 600 lb in air. How much force is needed to lift the anchor when it is submerged in seawater? The weight densities of iron and of seawater are respectively 480 and 64 lb/ft³.

SOLUTION   Since $D = w/V$, the anchor's volume is

$$V = \frac{w_i}{D_i} = \frac{600 \text{ lb}}{480 \text{ lb/ft}^3} = 1.25 \text{ ft}^3$$

The weight of seawater displaced by the anchor is

$$w_s = D_s V = (64 \text{ lb/ft}^3)(1.25 \text{ ft}^3) = 80 \text{ lb}$$

The buoyant force on the anchor is therefore 80 lb and the net force needed to lift the submerged anchor is

$$F = w_i - w_s = 600 \text{ lb} - 80 \text{ lb} = 520 \text{ lb}$$ ◆

◆   E X A M P L E   10.7 _____

An iceberg is a chunk of freshwater ice that has broken off from an ice cap, such as those that cover Greenland and Antarctica, or from a glacier at the edge of the sea. (More than three-quarters of the world's fresh water is in the form of ice.) Find the proportion of the volume of an iceberg that is submerged.

SOLUTION   If V_{ice} is the iceberg's total volume and V_{sub} is its submerged volume, then

Weight of iceberg = weight of displaced water

$$V_{ice}\rho_{ice}g = V_{sub}\rho_{seawater}g$$

$$\frac{V_{sub}}{V_{ice}} = \frac{\rho_{ice}}{\rho_{seawater}} = \frac{9.2 \times 10^2 \text{ kg/m}^3}{1.03 \times 10^3 \text{ kg/m}^3} = 0.89$$

Thus 89% of the volume of an iceberg is below sea level (Fig. 10.15).

Figure 10.15

Only 11% of the volume of an iceberg is above sea level.

◆

The **hydrometer** is a simple instrument based on Archimedes' principle that is used to measure the density of a liquid (Fig. 10.16). By squeezing and then releasing the rubber bulb, the liquid is drawn into a glass tube that contains a smaller sealed tube with a weight at its bottom. The higher the sealed tube floats, the denser the liquid.

The density of the acid in a car's storage battery depends on its state of charge, which means that a hydrometer can be used to establish the condition of the battery. A fully charged battery typically has a specific gravity of about 1.28 (that is, its density is 1.28 g/cm^3), which drops to about 1.15 when the battery is completely discharged. A hydrometer can also be used to determine the freezing point of the ethylene glycol antifreeze solution in the cooling system of a car, because the freezing point depends on the proportion of ethylene glycol it contains. Ethylene glycol has a specific gravity of 1.11, and so the greater the specific gravity of the solution, the higher the proportion of antifreeze and the lower the freezing point. For example, a specific gravity of 1.04 corresponds to a 28% solution and to a freezing point of $-14°$C.

Figure 10.16

A hydrometer.

10.5 FLUID FLOW

A simple model for a complex subject.

Fluids in motion often behave in complex and unpredictable ways. However, we can understand many aspects of fluid flow on the basis of a simple model that in many cases is reasonably realistic. The liquids in this model are supposed to be incompressible and to have no viscosity. (*Viscosity* is the term used to describe internal frictional resistances in a fluid; see Section 10.8.) With no viscosity, layers of fluid slide freely past one another and past other surfaces, so that we can apply the model to such liquids as water but not to such liquids as honey.

Another approximation we will make is that the fluid undergoes **laminar** (or **streamline**) **flow** only. In streamline flow, which is illustrated in Fig. 10.17, every particle of liquid passing a particular point follows the same path (called a *streamline*) as the particles that passed that point before. Furthermore, the direction in which the individual fluid particles move is always the same as the direction in which the fluid as a whole moves.

Figure 10.17

Laminar and turbulent flows around an obstacle.

Laminar and turbulent flows.

At the other extreme is **turbulent flow,** in which whirls and eddies occur, such as those in a cloud of cigarette smoke or at the foot of a waterfall. Turbulence generally develops at high speeds and when there are obstructions or sharp bends in the path of the fluid.

The volume of liquid that flows through a pipe per unit time is easy to figure out. If the average speed of the liquid in the pipe of Fig. 10.18 is v, each part of the stream travels the distance vt in the time interval t. The volume of liquid that moves the distance vt in the time t is vt multiplied by the pipe's cross-sectional area A, or vtA. Therefore the rate of flow R of liquid through the pipe is

$$R = \frac{vtA}{t} = vA \qquad\qquad\qquad \textit{Rate of flow} \quad (10.7)$$

The rate of flow is the product of the liquid speed and the cross-sectional area of the pipe.

Figure 10.18

The rate of flow of liquid through a pipe is equal to the product of the cross-sectional area of the pipe and the speed of the liquid.

Rate of flow = $R = vA$

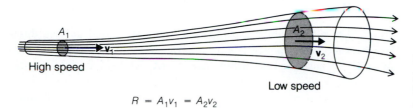

$$R = A_1v_1 = A_2v_2$$

Figure 10.19

In laminar flow, liquid speed is inversely proportional to the cross-sectional area of the pipe.

If the pipe size varies, the speed of the liquid also varies so as to keep R constant, so that

$$v_1A_1 = v_2A_2 \qquad\qquad \textit{Equation of continuity} \quad (10.8)$$

Hence a liquid flows faster through a constriction in a pipe and slower through a dilation. As in Fig. 10.19, streamlines drawn close together signify rapid motion, whereas streamlines far apart signify slow motion.

Everyone who has watered a lawn is familiar with Eq. (10.8). Partly closing the nozzle (or obstructing the end of the hose with a thumb, as in Fig. 10.20) means a smaller A and hence a faster stream of water.

◆   **E X A M P L E   10.8** _____

Water flows at 2.5 m/s through a garden hose whose inside radius is 6.0 mm. (a) What should the nozzle radius be for water to leave it at 10.0 m/s? (b) What is the rate of flow of water through the hose in liters per minute?

SOLUTION   (a) From Eq. (10.8), with $A = \pi r^2$, we have

$$v_1\,(\pi r_1^2) = v_2\,(\pi r_2^2)$$

$$r_2 = r_1\sqrt{\frac{v_1}{v_2}} = (6.0\text{ mm})\sqrt{\frac{2.5\text{ m/s}}{10.0\text{ m/s}}} = 3.0\text{ mm}$$

(b) From Eq. (10.7) the rate of flow is

$$R = v_1A_1 = v_1\,(\pi r_1^2) = (2.5\text{ m/s})(\pi)(0.0060\text{ m})^2 = 5.65 \times 10^{-4}\text{ m}^3\text{/s}$$

The same result would have been found from $R = v_2A_2$, of course. Because 1 m^3 = 10^3 L and 1 min = 60 s,

$$R = \left(5.65 \times 10^{-4}\,\frac{\text{m}^3}{\text{s}}\right)\left(10^3\,\frac{\text{L}}{\text{m}^3}\right)\left(60\,\frac{\text{s}}{\text{min}}\right) = 34\text{ L/min}$$

◆

Figure 10.20

The equation of continuity in action. Reducing the cross-sectional area of the flow increases the fluid speed.

Power

The rate at which a pump does work on a fluid passing through it is $P = Fv$, where the force applied is $F = pA$. Hence the power output of the pump is $P = pAv$, and, because the rate of flow is $R = vA$, we have useful formula

$$P = pR \hspace{4cm} \textit{Power in fluid flow} \quad (10.9)$$
Power = (pressure)(rate of flow)

◆ **E X A M P L E   10.9**

Oxygenated blood from the lungs is pumped by the left ventricle of the heart into a large artery called the aorta, whose diameter is typically 20 mm (Fig. 10.21). When a person is resting, the rate of flow of blood might be 6.0 L/min. (a) Find the average speed of blood in the aorta under these circumstances. (b) Find the power output of the left ventricle of a person at rest, assuming an average pressure of 100 torr at the aorta. (The right ventricle, which pumps blood at the same rate to the lungs where it gives up carbon dioxide and absorbs oxygen, has a smaller power output because there is less resistance to the flow of blood through the lungs than through the rest of the body.)

Figure 10.21

Schematic diagram of the human circulatory system. The heart consists of two pumps, one that pushes blood through the lungs and another that pushes blood through the rest of the body.

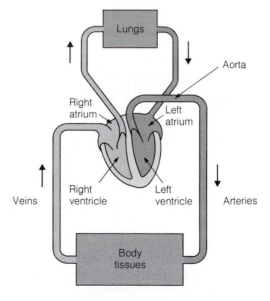

SOLUTION (a) Because $1\,\text{L} = 10^{-3}\,\text{m}^3$ and $1\,\text{min} = 60\,\text{s}$, the rate of blood flow here is $R = 6.0\,\text{L/min} = 0.10\,\text{L/s} = 1.0 \times 10^{-4}\,\text{m}^3$. The speed of the blood is therefore

$$v = \frac{R}{A} = \frac{R}{\pi r^2} = \frac{1.0 \times 10^{-4}\,\text{m}^3}{\pi (1.0 \times 10^{-2}\,\text{m})^2} = 0.32\,\text{m/s}$$

(b) From Eq. (10.9) the power output of the left ventricle is

$$P = pR = (100 \text{ torr})(133 \text{ Pa/torr})(1.0 \times 10^{-4} \text{ m}^3/\text{s}) = 1.3 \text{ W}$$

Both the pressure and the rate of flow increase during physical activity. Because the power output of the heart increases faster than the power output of the other muscles during exercise, a normally sedentary person should not attempt vigorous exercise without a gradual conditioning program first. ◆

10.6 BERNOULLI'S EQUATION

Conservation of energy in liquid flow.

When a liquid flowing in a pipe enters a region where the pipe diameter is smaller, its speed increases. A change in speed is an acceleration, which means that a net force must be acting on the liquid. This force can only come from a difference in pressure. Evidently the pressure must be higher in the large-diameter part of the pipe, because the liquid increases in speed on its way to the constriction.

The liquid speed is also affected by changes in the height of the pipe. If the liquid rises, it slows down, and if the liquid falls, it speeds up. Thus we expect the pressure, the speed, and the height of a moving liquid to be related in some way. The relationship turns out to be

$$p_1 + \rho g h_1 + \tfrac{1}{2}\rho v_1^2 = p_2 + \rho g h_2 + \tfrac{1}{2}\rho v_2^2 \qquad \textit{Bernoulli's equation} \quad (10.10)$$

Here p_1, h_1, and v_1 are, respectively, the pressure, the height above some reference level, and the speed of a liquid of density ρ at the point 1 in a body of the liquid, and p_2, h_2, and v_2 are the values of these quantities at another point, 2 (Fig. 10.21).

Equation (10.10) is known as **Bernoulli's equation,** after Daniel Bernoulli (1700–1782), who first derived it. According to Bernoulli's equation the quantity $p + \rho g h + \tfrac{1}{2}\rho v^2$ has the same value at all points in an incompressible liquid with negligible viscosity that undergoes laminar flow. The effect of viscosity is to dissipate mechanical energy into heat. If the viscosity of the liquid cannot be ignored, the quantity $p + \rho g h + \tfrac{1}{2}\rho v^2$ decreases in the direction of flow.

To obtain Bernoulli's equation, we consider a curved pipe of nonuniform cross section through which a liquid flows, as in Fig. 10.22. Let us apply the work-energy principle to a parcel of the liquid of volume $v_1 t A_1$ as it enters at the left in the time t and to the same parcel as it leaves at the right. The mass of the parcel is

$$m = \rho V = \rho v_1 t A_1 = \rho v_2 t A_2$$

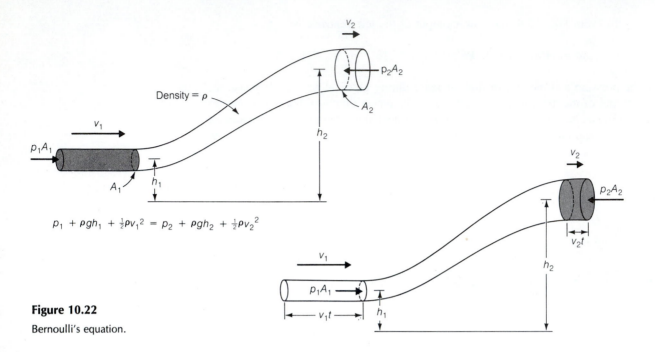

$$p_1 + \rho g h_1 + \tfrac{1}{2}\rho v_1{}^2 = p_2 + \rho g h_2 + \tfrac{1}{2}\rho v_2{}^2$$

Figure 10.22

Bernoulli's equation.

because vA has the same value at 1 and 2. The net amount of work ΔW done on the liquid parcel as it passes from 1 to 2 must equal the net change in its potential energy ΔPE as its height goes from h_1 to h_2, plus the net change in its kinetic energy ΔKE as its speed goes from v_1 to v_2. That is,

$$\Delta W = \Delta\text{PE} + \Delta\text{KE}$$

The work done *on* the parcel at 1 is the force $p_1 A_1$ acting on it multiplied by the distance $v_1 t$ through which the force acts. The work done *by* the parcel at 2 is the force $p_2 A_2$ multiplied by the distance $v_2 t$ through which the force acts. The *net* work done on the parcel is therefore

$$\Delta W = p_1 A_1 v_1 t - p_2 A_2 v_2 t = \frac{p_1 m}{\rho} - \frac{p_2 m}{\rho}$$

The change in the potential energy of the parcel in going from 1 to 2 is

$$\Delta\text{PE} = mgh_2 - mgh_1$$

and the change in its kinetic energy is

$$\Delta\text{KE} = \tfrac{1}{2}mv_2^2 - \tfrac{1}{2}mv_1^2$$

Therefore

$$\Delta W = \Delta PE + \Delta KE$$

$$\frac{p_1 m}{\rho} - \frac{p_2 m}{\rho} = mgh_2 - mgh_1 + \tfrac{1}{2}mv_2^2 - \tfrac{1}{2}mv_1^2$$

When we divide through by the common factor m, multiply by ρ, and rearrange terms, the result is Bernoulli's equation,

$$p_1 + \rho gh_1 + \tfrac{1}{2}\rho v_1^2 = p_2 + \rho gh_2 + \tfrac{1}{2}\rho v_2^2$$

◆ **E X A M P L E   10.10**

Water flows upward through the pipe shown in Fig. 10.23 at 15 L/s. If water enters the lower end of the pipe at 3.0 m/s, what is the difference in pressure between the two ends?

SOLUTION The first step is to find the water speed v_2 at the upper end of the pipe. From Eq. (10.7), $R = vA$, this is

$$v_2 = \frac{R}{A_2} = \frac{(15 \text{ L/s})(10^{-3} \text{ m}^3\text{/L})}{(20 \text{ cm}^2)(10^{-4} \text{ m}^2\text{/cm}^2)} = 7.5 \text{ m/s}$$

We now substitute the known quantities v_1, v_2, and $h_2 - h_1$ into Bernoulli's equation. The density of water is $\rho = 1.0 \times 10^3$ kg/m^3, hence

$$\Delta p = p_1 - p_2 = \rho g(h_2 - h_1) + \tfrac{1}{2}\rho (v_2^2 - v_1^2)$$

$$= \left(1.0 \times 10^3 \frac{\text{kg}}{\text{m}^3}\right)\left(9.8 \frac{\text{m}}{\text{s}^2}\right)(1.5 \text{ m})$$

$$+ \tfrac{1}{2}\left(1.0 \times 10^3 \frac{\text{kg}}{\text{m}^3}\right)\left[\left(7.5 \frac{\text{m}}{\text{s}}\right)^2 - \left(3.0 \frac{\text{m}}{\text{s}}\right)^2\right]$$

$$= 38 \text{ kPa}$$

Figure 10.23

v_2

$A_2 = 20$ cm²

1.5 m

$v_1 = 3.0$ m/s

◆

Figure 10.24

Torricelli's theorem.

$$v = \sqrt{2gh}$$

10.7 APPLICATIONS OF BERNOULLI'S EQUATION

Heavier-than-air flight is one of them.

In many situations the speed, pressure, or height of a liquid is constant, and simplified forms of Bernoulli's equation hold. Thus when a liquid column is stationary, we see that the pressure difference between two depths in it is

$$p_2 - p_1 = \rho g(h_1 - h_2) \qquad\qquad \textit{Liquid at rest} \quad (10.11)$$

which is just what Eq. (10.4) states. Evidently the latter formula is included in Bernoulli's equation.

Another straightforward result occurs when $p_1 = p_2$. As an example, Fig. 10.24 shows a liquid emerging from an opening in a tank. The liquid pressure equals atmospheric pressure both at the top of the tank and at the opening. If the opening is small compared with the cross section of the tank, the liquid level in the tank will fall slowly enough for the liquid speed at the top of the tank to be assumed zero. If the speed of the liquid as it leaves the opening is v and the difference in height between the top of the liquid and the opening is h, Bernoulli's equation reduces to

$$\tfrac{1}{2}\rho v^2 = \rho g h$$

$$v = \sqrt{2gh} \qquad\qquad \textit{Torricelli's theorem} \quad (10.12)$$

The speed with which the liquid comes out is the same as the speed of a body falling from rest from the height h. This result is called **Torricelli's theorem,** and, like the relationship between pressure and depth, it is a special case of Bernoulli's equation. The rate at which liquid flows through the opening may be found from Eq. (10.7) if its area A is known. The volume of liquid per unit time is

$$R = vA = A\sqrt{2gh} \qquad\qquad \textit{Flow from orifice} \quad (10.13)$$

◆ **E X A M P L E   10.11**

A boat strikes an underwater rock and opens a pencil-sized crack 7.0 mm wide and 150 mm long in its hull 65 cm below the waterline. The crew takes 5.00 min to locate the crack and plug it up. How much water entered the boat in the meantime?

SOLUTION   The area of the crack is (7.0 mm)(150 mm) = 1050 mm^2 = 1.05 × 10^{-3} m^2, and so, from Eq. (10.13),

$$R = (1.05 \times 10^{-3} \, \text{m}^2) \sqrt{(2)(9.8 \, \text{m/s}^2)(0.65 \, \text{m})} = 3.75 \times 10^{-3} \, \text{m}^3/\text{s}$$

In 5.00 min = 300 s the volume of water that came in was

$$V = Rt = (3.75 \times 10^{-3} \, \text{m}^3/\text{s})(300 \, \text{s}) = 1.1 \, \text{m}^3$$

This is over a ton, a substantial amount.   ◆

Here is an illustration of a further special case of Bernoulli's equation.

◆ **E X A M P L E   10.12**

Geysers in Yellowstone National Park and elsewhere shoot columns of water as high as 50 m into the air at more or less regular intervals. If the water at the mouth of a geyser is assumed to be initially at rest, what subsurface pressure is needed for the water to reach such a height?

SOLUTION   If we let 1 refer to the mouth of the geyser and 2 to the top of the water column, $h_1 = 0$, $h_2 = 50$ m, $p_2 = p_{\text{atm}} = 1.0 \times 10^5$ Pa, and $v_1 = v_2 = 0$. From Bernoulli's equation

$$p_1 - p_2 = \rho g(h_2 - h_1)$$
$$p_1 = p_{\text{atm}} + \rho g h_2$$
$$= 1.0 \times 10^5 \, \text{Pa} + (1.0 \times 10^3 \, \text{kg/m}^3)(9.8 \, \text{m/s}^2)(50 \, \text{m})$$
$$= 1.0 \times 10^5 \, \text{Pa} + 4.9 \times 10^5 \, \text{Pa} = 5.9 \times 10^5 \, \text{Pa}$$

The absolute pressure at the mouth of the geyser is 5.9 bar, and the gauge pressure there is 4.9 bar.   ◆

Lift

The most interesting special case of Bernoulli's equation occurs when there is no change in height during the motion of the liquid (Fig. 10.25). Here

$$p_1 + \tfrac{1}{2}\rho v_1^2 = p_2 + \tfrac{1}{2}\rho v_2^2 \qquad\qquad \textit{Pressure and speed} \quad (10.14)$$

Figure 10.25

In a horizontal pipe the pressure is greatest when the velocity is least, and vice versa.

$$p_1 + \tfrac{1}{2}\rho v_1{}^2 = p_2 + \tfrac{1}{2}\rho v_2{}^2$$

which means that the pressure in the liquid is least where the speed is greatest, and vice versa.

A familiar application of Eq. (10.14) is the lifting force produced by the flow of air past the wing of an airplane, as in Fig. 10.26. (Air is, of course, a compressible fluid and so does not fit our model exactly, but the behavior predicted by Bernoulli's equation is not a bad approximation for gases at moderate speeds.) Air moving past the upper surface of the wing must travel faster than air moving past the lower surface. This is indicated by the closeness of the streamlines near the upper surface. The difference in speed gives a reduced pressure over the top of the wing, which leads to an upward force on the wing. The greater the difference in air speeds around the upper and lower surfaces, the greater the lift that is produced, provided the wing shape is not so extreme that turbulence results (Fig. 10.26c). An additional source of lift is the deflection of air from the angled bottom of the wing, which leads to an upward force on it.

Air flows around the tip of each wing from the high-pressure region under it to the low-pressure region above it (Fig. 10.27). The result is a trail of whirling air (called a **vortex**) behind each wingtip that can be dangerous for small aircraft flying too closely behind large ones.

The lift force F_{lift} on a wing of area A due to the difference in air speeds between its surfaces is, from Eq. (10.14),

$$F_{\text{lift}} = (p_1 - p_2)A = \tfrac{1}{2}\rho A(v_2^2 - v_1^2)$$

Here v_1 is the air speed below the wing and v_2 is its speed above the wing. How these speeds are related to the speed v of the airplane depends on the shape of the wing and on the angle the wing makes with the direction of the airflow ahead of it. In general $(v_2^2 - v_1^2)$ is proportional to v^2, so F_{lift} is proportional to Av^2. Thus the faster an airplane moves, the smaller the wing area needed to provide it with lift. Because a small wing also exhibits less resistance to motion through the air (less "drag") than a large one, it is clear why high-speed airplanes have relatively small

Figure 10.26

Air flow past a wing. (a) The air speed is greater over the upper surface, as indicated by the closer streamlines, and so the pressure is lower there and a net upward force called lift acts on the wing. (b) When the wing is angled upward, air deflected from its lower surface provides additional lift. (c) If the angle of attack is too great, turbulence reduces lift and increases drag.

(a)   (b)   (c)

Figure 10.27

Wingtip vortexes occur because air flows from the high-pressure region under a wing to the low-pressure region above it. They may extend for long distances behind a large aircraft.

Figure 10.28

The power needed to overcome the aerodynamic drag of a typical car increases with speed more rapidly than the rolling resistance of its tires does.

wings. The penalty such airplanes must pay is that their takeoff and landing speeds need to be correspondingly high.

At low speeds most of the resistance a car offers to motion comes from the flexing of its tires. As the speed increases, aerodynamic drag becomes more and more important, until finally the car's engine cannot provide enough power to overcome it and the car can go no faster (Fig. 10.28). Drag depends on the shape of a car as well as on its speed, so much attention is paid nowadays to streamlining. A streamlined car, however, is much like a wing section (Fig. 10.29), and the resulting lift at high speeds reduces the grip of the tires on the road. Because both steering and braking depend on this grip, the result can be dangerous. Ridges (''spoilers'') are often placed on the rear ends of fast cars to reduce lift by interfering with laminar flow, but an increase in drag is the inevitable result.

Air flow over a car being checked in a wind tunnel. Laminar flow gives least resistance.

Figure 10.29

A car streamlined to reduce drag experiences aerodynamic lift and may be hard to control at high speeds unless its shape is modified with ''spoilers'' that interfere with the laminar flow over the car.

Flame

Air

Gas

Figure 10.30

Blowtorch nozzle.

Figure 10.31

The Venturi meter. The rate of flow of a fluid through the meter can be found from A_1, A_2, and H.

The nozzle of a blowtorch (Fig. 10.30) provides another application of the reduced pressure that accompanies an increase in fluid speed. The gas leaves the supply tube through a small orifice, which means that its speed is high as it escapes. The resulting low pressure sucks in air through holes in a sleeve around the end of the tube to mix the oxygen needed for burning with the gas. In this arrangement the flame is hotter than if the gas simply burns at the orifice without being mixed with air first. The laboratory Bunsen burner has a similar construction.

Still another application of Eq. (10.14) is the **Venturi meter,** which provides a convenient method for determining the rate of flow of a liquid through a pipe. As shown in Fig. 10.31, a Venturi meter consists of a constriction in a pipe with manometers connected to the wide and narrow sections. The liquid height in the manometer at the constriction is lower because the pressure there is lower, and the height difference H is proportional to the pressure difference, which in turn depends on the speed difference according to Eq. (10.14).

10.8 VISCOSITY

Internal friction in a fluid.

The viscosity of a fluid is a kind of internal friction that prevents neighboring layers of the fluid from sliding freely past one another. Figure 10.32 shows how the speed of a fluid in a pipe varies with distance from the axis. The fluid in contact with the pipe's wall is stationary and, as we would expect, the speed increases to a maximum along the pipe's axis. The smaller the viscosity of the fluid, the greater the various speeds will be, but the characteristic parabolic shape of the speed profile will persist as long as the flow remains laminar.

Viscosity is given a quantitative definition in terms of the experiment shown in Fig. 10.33, in which a plate of area A is being pulled across a layer of fluid of thickness s. A hollow cylinder having a smaller rotating cylinder inside it, with a fluid layer between them, is a better arrangement for actual measurements.

Figure 10.32

Because of viscosity, the velocity of fluid in a pipe varies from zero at the pipe wall to a maximum along the axis.

Viscosity of fluid $= \eta = \dfrac{Fs}{Av}$

Figure 10.33

The viscosity of a fluid is defined in terms of the force needed to pull a flat plate at constant speed across a layer of the fluid.

For most fluids the force F required to pull the plate at the constant speed v turns out to be proportional to A and v and inversely proportional to s. The faster the motion and the thinner the layer of fluid, the more force is needed for a given plate area. Different fluids offer different degrees of resistance to the motion, but the force needed varies as Av/s for most of them provided the speed is not so great that turbulence occurs. Thus we can write

$$F = \frac{\eta A v}{s} \tag{10.15}$$

where η, the constant of proportionality, is called the **viscosity** of the fluid (η is the Greek letter *eta*). The SI unit of viscosity is the N $\cdot$ s/m^2, which is known as the **poiseuille** (Pl). An older unit, the *poise*, remains in common use, where 10 poise = 1 N·s/m^2; the *centipoise*, 0.01 poise, is equal to 10^{-3} N $\cdot$ s/m^2.

The viscosities of some common fluids are listed in Table 10.1. The viscosity of a liquid decreases with temperature as its molecules becomes less and less tightly bound to one another (see Chapter 13), a phenomenon familiar to anyone who has heated honey or molasses. One of the reasons a drop in body temperature is so dangerous is that the viscosity of the blood then increases, which impedes its flow. The viscosity of a gas, in contrast to that of a liquid, increases with temperature because the higher the temperature, the faster the gas molecules move and the more often they collide with one another (again, see Chapter 13).

The viscosity of motor oil is usually given in terms of SAE (Society of Automotive Engineers) index numbers. The higher the number, the greater the viscosity, so that SAE 30 oil is "thicker" than SAE 10 oil at a given temperature. Because viscosities of liquids decrease with temperature, SAE 30 oil has about the same viscosity at 20°C (68°F) as SAE 10 oil at 0°C (32°F). If SAE 30 oil were used in very cold weather, its resistance to flow would make an engine hard to start and it might not circulate adequately afterward; SAE 10 oil is better in such conditions. On the other hand, SAE 10 oil would be too thin on a hot summer day to provide proper lubrication, and SAE 30 oil would be appropriate then. "Multigrade" oils are formulated to be usable over a wide range of temperatures.

Table 10.1   Viscosities of various fluids at atmospheric pressure and the indicated temperature

Substance	Temperature (°C)	Viscosity (Pl)
Gases		
Air	0	1.7×10^{-5}
	20	1.8×10^{-5}
	100	2.2×10^{-5}
Water vapor	100	1.3×10^{-5}
Liquids		
Alcohol (ethyl)	20	1.2×10^{-3}
Blood plasma	37	1.3×10^{-3}
Blood, whole (varies with speed)	37	$\sim 2 \times 10^{-3}$
Glycerin	20	0.83
Water	0	1.8×10^{-3}
	20	1.0×10^{-3}
	40	0.66×10^{-3}
	60	0.47×10^{-3}
	80	0.36×10^{-3}
	100	0.28×10^{-3}

Flow through a Pipe

Let us return to a fluid moving through a pipe. If the pipe is cylindrical with the length L and inside radius r, and a fluid of viscosity η is in laminar flow through it under the influence of a pressure difference $\Delta p = p_1 - p_2$, as in Fig. 10.34, the rate of flow is

$$R = \frac{\pi r^4 \, \Delta p}{8 \eta L} \qquad \qquad \textit{Poiseuille's law} \quad (10.16)$$

Equation (10.16) is known as **Poiseuille's law**. The dependence of the rate of flow on the viscosity η and on the pressure gradient $\Delta p/L$ are both about what we might expect, but the variation with r^4 is remarkable: Halving the radius of a pipe reduces R by a factor of 16 if Δp stays the same. The radius of the pipe plays a far more important role than its length does with respect to viscous resistance.

Figure 10.34

Poiseuille's law describes fluid flow through a pipe.

$\Delta p = p_1 - p_2$

◆   E X A M P L E   10.13

Atherosclerosis is the medical term for a common condition in which arteries are narrowed by deposits of tissue called plaque. If the flow of blood is to continue at its usual rate, a higher pressure is required. A consequence of atherosclerosis is therefore an elevated blood pressure, which has many undesirable effects on the body, one of them being that the heart must work harder to circulate the blood. (a) Find the percentage increase in pressure needed to maintain R constant when the radius of an artery is decreased by 10%. (b) What is the corresponding increase in the power needed to maintain the flow of blood through that artery?

SOLUTION   (a) For a constant rate of flow R, the product $r^4\Delta p$ must be constant. Here $r_2 = 0.9r_1$, so

$$r_1^4\,\Delta p_1 = r_2^4\,\Delta p_2$$
$$\frac{\Delta p_2}{\Delta p_1} = \left(\frac{r_1}{r_2}\right)^4 = \left(\frac{r_1}{0.9r_1}\right)^4 = 1.52$$

The blood pressure must increase by 52% if R is to be unchanged.
(b) From Eq. (10.9) $P = \Delta pR$. Here R is the same, so

$$\frac{P_2}{P_1} = \frac{\Delta p_2}{\Delta p_1} = 1.52$$

The power output of the heart must increase by 52% also. Narrowing of the arteries usually means a reduced flow of blood as well as an increase in blood pressure and in the work done by the heart. ◆

10.9 SURFACE TENSION

A liquid surface is like a membrane under tension.

The surface of a liquid behaves remarkably like a membrane under tension. Thus a steel sewing needle placed horizontally on the surface of some water in a dish does not sink even though its density is nearly eight times that of water (Fig. 10.35). The needle rests in a depression in the water surface just as if that surface were a sheet of rubber stretched across the dish. The term **surface tension** for this effect is quite appropriate. Another example of surface tension is the tendency of a liquid drop to assume a spherical shape, just as an inflated balloon does.

The origin of surface tension lies in the fact that molecules on the surface of a body of liquid are acted on by a net inward force, whereas those in the interior are acted on by equal forces in all directions (Fig. 10.36). Because the surface molecules are all being pulled inward, the surface tends to contract to the minimum possible area. A sphere has the least area relative to its volume of any object, and so a liquid sample not acted on by any external forces (such as gravitation or air resistance if it is moving) will take on a spherical form. A liquid surface can support a small object such as a needle because the weight of the object is not enough to rupture the surface, which would involve first stretching it to a greater degree than that shown in Fig. 10.35.

Figure 10.35

The surface tension of a water surface supplements buoyancy in supporting a steel needle. The elastic character of a liquid surface resembles that of a membrane under tension.

Forces due to surface tension

Liquid surface

Weight of needle

Figure 10.36

Because a molecule on the surface of a body of liquid has no molecules above it to interact with, the net force on it is toward the body of the liquid.

Capillary tube

(a)   Adhesion > cohesion   (b)   Cohesion > adhesion

Figure 10.37

When the angle of contact θ is less than 90°, the liquid in a capillary tube rises; when θ is more than 90°, the liquid falls.

Water has an exceptionally high surface tension, which is sometimes a handicap. A **surfactant** is a substance whose addition to a liquid reduces its surface tension. Detergents are efficient surfactants, which enables even an extremely dilute solution of detergent in water to penetrate small crevices and to be absorbed by tightly woven textiles. Surfactants are also called **wetting agents** because they allow a water solution to spread uniformly over a surface instead of forming into separate droplets.

Capillarity

A familiar phenomenon is the rise of most liquids in a capillary tube. Capillarity is responsible for many familiar effects, such as the ability of paper and cloth fibers to absorb water. Two factors are involved: the **cohesion** of the liquid, which refers to the attractive forces its molecules exert on one another, and the **adhesion** of the liquid to the surface of a solid, which refers to the attractive forces the solid exerts on the liquid molecules.

If the adhesive forces exceed the cohesive ones, as they do in the case of water and glass, then the liquid tends to stick to the solid. It will rise in a capillary tube of that material since the attraction of a liquid molecule to the wall of the tube is greater than the attraction of this molecule to the others. In such an event the liquid surface is concave, as in Fig. 10.37(a), with an angle of contact θ that is characteristic of the liquid-solid combination. For water and clean glass, $\theta \approx 0$, as it is for any liquid that "wets" a particular solid, because adhesion must be much greater than cohesion for this to occur. For kerosene and glass, $\theta = 26°$, which reflects the smaller amount by which adhesion exceeds cohesion for these substances.

Figure 10.38

The narrower a tube, the higher a liquid rises in it (or falls if $\theta > 90°$).

When cohesion is greater than adhesion, as it is for mercury and glass, the liquid level in a capillary tube is lower than the level of the surrounding liquid. Now the liquid surface in the tube is convex, with an angle of contact greater than 90° (Fig. 10.37(b)). For mercury and glass, $\theta = 140°$. Water spreads out into a thin film on a clean horizontal glass surface, but mercury pulls itself into tiny droplets. The narrower a tube, the higher a given liquid rises in it (if $\theta < 90°$) or falls (if $\theta > 90°$), as in Fig. 10.38.

Important Terms

The **pressure** on a surface is the perpendicular force per unit area that acts on it. **Gauge pressure** is the difference between absolute pressure and atmospheric pressure.

Pascal's principle states that an external pressure exerted on a confined fluid is transmitted uniformly throughout its volume.

The **hydraulic press** is a machine consisting of two fluid-filled cylinders of different diameters connected by a tube. The input force is applied to a piston in one of the cylinders, and the output force is exerted by a piston in the other cylinder. The IMA of a hydraulic press is equal to the inverse ratio of the cylinder cross-sectional areas.

Archimedes' principle states that the buoyant force on a submerged object is equal to the weight of fluid it displaces.

In **laminar** (or **streamline**) **flow** every particle of fluid passing a particular point follows the same path, whereas in **turbulent flow** irregular whirls and eddies occur. The greater the speed of a fluid in streamline flow, the lower its pressure.

The **viscosity** of a fluid is a measure of its internal friction.

The **surface tension** of a liquid refers to the tendency of its surface to contract to the minimum possible area in any situation.

Important Formulas

Pressure: $\quad p = F/A = p_{gauge} + p_{atm}$

Pressure at depth h in a fluid: $\quad p = p_{external} + \rho g h$

Archimedes' principle: $\quad F_{buoyant} = V\rho g$

Equation of continuity: $\quad R = v_1 A_1 = v_2 A_2$

Bernoulli's equation: $\quad p + \rho g h + \frac{1}{2}\rho v^2 = $ constant

Poiseuille's law: $\quad R = \dfrac{\pi r^4 \, \Delta p}{8 \eta L}$

MULTIPLE CHOICE

1. Atmospheric pressure does not correspond to approximately
- **a.** 14.7 lb/in^2
- **b.** 98 N/m^2
- **c.** 1013 mb
- **d.** $1.013 \times 10^5 \text{ Pa}$

2. The hydraulic press is able to produce a mechanical advantage because
- **a.** the force a fluid exerts on a piston is always parallel to its surface
- **b.** an external pressure exerted on a fluid is transmitted uniformly throughout its volume
- **c.** at any depth in a fluid the pressure is the same in all directions
- **d.** the pressure in a fluid varies with its speed

3. In the operation of a hydraulic press, it is impossible for the output piston to exceed the input piston's
- **a.** displacement
- **b.** speed
- **c.** force
- **d.** work

4. Fluid at rest at the bottom of a container is
- **a.** under less pressure than the fluid at the top
- **b.** under the same pressure as the fluid at the top
- **c.** under more pressure than the fluid at the top
- **d.** any of the above, depending on the circumstances

5. The pressure at the bottom of a vessel filled with liquid does not depend on the
- **a.** acceleration of gravity
- **b.** liquid density
- **c.** height of the liquid
- **d.** area of the liquid surface

6. A person stands on a very sensitive scale and inhales deeply. The reading on the scale
- **a.** does not change
- **b.** increases
- **c.** decreases
- **d.** depends on the expansion of the person's chest relative to the volume of air inhaled

7. Buoyancy occurs because, with increasing depth in a fluid,
- **a.** the pressure increases
- **b.** the pressure decreases
- **c.** the density increases
- **d.** the density decreases

8. In order for an object to sink when placed in water, its average specific gravity (relative density) must be
- **a.** less than 1

b. equal to 1
c. more than 1
d. any of the above, depending on its shape

9. A cake of soap placed in a bathtub of water sinks. The buoyant force on the soap is
a. zero
b. less than its weight
c. equal to its weight
d. more than its weight

10. The density of fresh water is 1.00 g/cm^3 and that of seawater is 1.03 g/cm^3. A ship will float
a. higher in fresh water than in seawater
b. at the same level in fresh water and in seawater
c. lower in fresh water than in seawater
d. any of the above, depending on the shape of its hull

11. Bernoulli's equation is based on
a. the second law of motion
b. the third law of motion
c. the conservation of momentum
d. the conservation of energy

12. An express train goes past a station platform at high speed. A person standing at the edge of the platform tends to be
a. attracted to the train
b. repelled from the train
c. attracted or repelled, depending on the ratio between the speed of the train and the speed of sound
d. unaffected by the train's passage

13. The volume of liquid flowing per second out of an orifice at the bottom of a tank does not depend on
a. the area of the orifice
b. the height of liquid above the orifice
c. the density of the liquid
d. the value of the acceleration of gravity

14. Adding detergent to water
a. decreases its surface tension
b. does not change its surface tension
c. increases its surface tension
d. any of the above, depending on the detergent

15. Water neither rises nor falls in a silver capillary. This suggests that the contact angle between water and silver is
a. 0   b. 45°
c. 90°   d. 180°

16. A 10-lb brick has the dimensions 4 in × 6 in × 12 in. The minimum pressure it can exert when one of its faces rests on a horizontal surface is
a. 10 lb/ft^2 b. 20 lb/ft^2
c. 30 lb/ft^2 d. 60 lb/ft^2

17. A 20-kg piston rests on a sample of gas in a cylinder 10 cm in radius. The absolute pressure on the gas is
a. 0.0064 bar   b. 0.062 bar
c. 1.019 bars   d. 1.075 bars

18. The input piston of a hydraulic press is 20 mm in diameter, and the output piston is 10 mm in diameter. An input force of 1.0 N will produce an output force of
a. 0.25 N   b. 0.50 N
c. 2.0 N   d. 4.0 N

19. When the input piston of the press in Multiple Choice 18 is pushed in through 40 mm, the output piston is pushed out through
a. 10 mm   b. 20 mm
c. 80 mm   d. 160 mm

20. A manometer (Fig. 10.1b) whose upper end is evacuated and sealed can be used to measure atmospheric pressure. Such an instrument is called a mercury barometer, and the average height of the mercury column in it is 760 mm. If water were used in a barometer instead of mercury, the height of the column of water would be about
a. 56 mm   b. 760 mm
c. 10 m   d. 20 m

21. A viewing window 30 cm in diameter is installed with its center 3.0 m below the surface of an aquarium tank filled with seawater. The force the window must withstand is approximately
a. 22 N   b. 0.22 kN
c. 2.1 kN   d. 8.6 kN

22. An aquarium tank 25 cm long and 15 cm wide contains water 10 cm deep but no fish. The gauge pressure on its bottom is
a. 0.98 Pa   b. 98 Pa
c. 0.98 kPa   d. 9.8 kPa

23. If the tank in Multiple Choice 22 contained fish as well as water, with the water depth the same 10 cm, the pressure at the bottom would be
a. greater
b. the same
c. smaller
d. any of the above, depending on the total mass of the fish

24. An object suspended from a spring scale is lowered into a pail filled to the brim with a liquid, and 4 N of the liquid overflows. The scale indicates that the object weighs 6 N in the liquid. The weight in air of the object is
a. 2 N   b. 4 N
c. 6 N   d. 10 N

25. A wooden board 2.0 m long, 30 cm wide, and 40 mm

thick floats in water with 10 mm of its thickness above the surface. The mass of the board is
- a. 1.8 kg
- b. 18 kg
- c. 24 kg
- d. 176 kg

26. A force of 1000 lb is required to raise a concrete block to the surface of a freshwater lake. The force required to lift it out of the water is approximately
- a. 700 lb
- b. 1062 lb
- c. 1140 lb
- d. 1800 lb

27. The depth in fresh water at which the water density is 1% greater than its value at the surface is approximately
- a. 2.3×10^2 m
- b. 2.3×10^3 m
- c. 2.3×10^4 m
- d. 2.3×10^5 m

28. The total cross-sectional area of all the capillaries of a certain person's circulatory system is 0.25 m². If blood flows through the system at the rate of 100 cm³/s, the average speed of blood in the capillaries is
- a. 0.4 mm/s
- b. 4 mm/s
- c. 25 mm/s
- d. 400 mm/s

29. A certain person's heart beats 1.2 times per second and pumps 1.0×10^{-4} m³ of blood per beat against an average pressure of 14 kPa. The power output of the heart is
- a. 1.2 W
- b. 1.4 W

- c. 1.7 W
- d. 12 W

30. Water leaves the safety valve of a boiler at a speed of 30 m/s. The gauge pressure inside the boiler is
- a. 4.5 millibars
- b. 1.5 bars
- c. 4.5 bars
- d. 450 bars

31. A water tank springs a leak from which water emerges at 15 ft/s. The height of the water surface above the leak is
- a. 0.23 ft
- b. 1.9 ft
- c. 3.5 ft
- d. 7.0 ft

32. A horizontal pipe of 10 cm² cross-sectional area is joined to another horizontal pipe of 50 cm² cross-sectional area. The speed of water in the small pipe is 6.0 m/s, and the pressure there is 200 kPa. The speed of water in the large pipe is
- a. 0.24 m/s
- b. 1.2 m/s
- c. 3 m/s
- d. 13 m/s

33. The pressure in the large pipe of Multiple Choice 32 is
- a. 183 kPa
- b. 202 kPa
- c. 217 kPa
- d. 235 kPa

34. The rate of flow of water through the large pipe in Multiple Choice 32 is
- a. 0.6 L/s
- b. 1.2 L/s
- c. 3 L/s
- d. 6 L/s

QUESTIONS

1. A little water is boiled for a few minutes in a tin can, and the can is sealed while it is still hot. Why does the can collapse as it cools?

2. When a person drinks a soda through a straw, where does the force come from that causes the soda to move upward?

3. What is the absolute pressure of the air in a flat tire?

4. A jar is filled to the top with water, and a piece of cardboard is slid over the opening so that there is only water in the jar. If the jar is now turned over, will the cardboard fall off? What will happen if there is any air in the jar?

5. Why are the windows on commercial airliners much smaller than those on light single-engine aircraft?

6. The rolling resistance of a tire is due to its distortion as it rolls. Can you think of two ways to reduce rolling resistance?

7. The three containers shown in Fig. 10.39 are filled with water to the same height. Compare the pressures at the bottoms of the containers.

8. An ice cube floats in a glass of water filled to the brim. What will happen when the ice melts?

9. A wooden block is in such perfect contact with the bottom

A   B   C

Figure 10.39

Question 7.

of a water tank that there is no water beneath it. Is there a buoyant force on the block?

10. Two spheres of the same diameter but of different mass are dropped from a tower. If air resistance is the same for both, which will reach the ground first? Why?

11. In central Sweden a bridge carries the Göta Canal over a highway. What, if anything, happens to the load on the bridge when a boat passes across it?

12. A ship catches fire, and its steel hull expands as a result of the increase in temperature. What happens to the volume of water the ship displaces? What happens to the ship's freeboard, which is the height of its deck above sea level?

13. An aluminum canoe is floating in a swimming pool. After a while it begins to leak and sinks to the bottom of the pool. What happens to the water level in the pool?

14. A helium-filled balloon rises to a certain altitude in the atmosphere and floats there instead of rising indefinitely. Why?

15. Why does a steel ship float when the density of steel is nearly eight times that of water? Why does it sink if it springs a leak and water enters its hull?

16. Will there be any difference between the rates of flow from pipes A and B in Fig. 10.40? If so, from which pipe will the liquid flow faster?

17. Why is it more dangerous when ice forms on an airplane's wings than when it forms on its fuselage?

18. Because its surface is not perfectly smooth, a spinning baseball drags nearby air around with it. Use this fact to explain why a baseball thrown with a spin moves in a curved path.

Figure 10.40

Question 16.

19. Does a fluid whose density is greater than that of another fluid necessarily have a greater viscosity as well?

EXERCISES

10.1 Pressure

1. A 2.0-kg brick has the dimensions 7.5 cm × 15 cm × 30 cm. Find the pressures exerted by the brick on a table when it is resting on its various faces.

2. The two upper front teeth of a person exert a total force of 30 N on a piece of steak. If the biting edge of each tooth measures 10 mm × 1.0 mm, what is the pressure on the steak?

3. A tire inflated to a gauge pressure of 2.0 bars has a nail 3.0 mm in diameter embedded in it. What is the force tending to push the nail out of the tire?

4. The weight of a car is equally supported by its four tires, each of which has an area of 150 cm² in contact with the ground. If the gauge pressure of the air in the tires is 1.8 bars, find the mass of the car.

5. A DC-9 airplane weighing 90,000 lb is in level flight. The area of its wings is 990 ft². What is the average difference in pressure between the upper and lower surfaces of its wings?

6. A "plumber's helper" is a rubber cup with a wooden handle. If the diameter of the cup is 5.00 in, what is the maximum suction it can exert on a clogged pipe? How much force is needed on the handle to produce this suction?

7. The force on a phonograph needle whose point is 0.10 mm in radius is 0.20 N. What is the pressure in atmospheres that it exerts on the record?

8. A piston weighing 12 N rests on a sample of gas in a cylinder 5.0 cm in diameter. (a) What is the gauge pressure in the gas? (b) What is the absolute pressure in the gas?

9. A hypodermic syringe whose cylinder is 10 mm in diameter and whose needle is 1.0 mm in diameter is used to inject a liquid into a vein in which the blood pressure is 2.0 kPa. Find the minimum force needed on the plunger of the syringe.

10. A 7.0-g bullet leaves the 60-cm barrel of a 0.30-caliber rifle (whose bore diameter is 0.30 in. = 7.62 mm) at 600 m/s. Find the average pressure of the gases in the rifle barrel while the bullet is being accelerated.

10.2 Hydraulic Press

11. A force of 50 N is applied to the input piston of a hydraulic system. The area of the input piston is 60 cm², and that of the output piston is 15 cm². (a) How much force does the output piston exert? (b) If the input piston moves 2.0 cm per stroke, how far does the output piston move?

12. A lever with a mechanical advantage of 5.0 is attached to the pump piston of a hydraulic press. The area of the pump piston is 10 cm², and that of the output piston is 125 cm². (a) Find the force the output piston exerts when a force of 120 N is applied to the pump lever. (b) If each stroke of the lever moves the pump piston 3.0 cm, how many strokes are needed to move the output piston 30 cm?

13. The input and output pistons of a hydraulic jack are respectively 0.30 in and 1.60 in. in diameter. A lever with an IMA of 8.0 is used to apply force to the input piston. How much force must be exerted on the lever to lift 4000 lb if the efficiency of the system is 97%?

10.3 Pressure and Depth

14. The height of water at two identical dams is the same, but dam *A* holds back a lake containing 2 km³ of water, whereas dam *B* holds back a lake containing 1 km³ of water. What is the ratio of the total force exerted on dam *A* to that exerted on dam *B*?

15. In 1960 the U.S. Navy bathyscaphe *Trieste* descended to a depth of 10,920 m in the Pacific Ocean near Guam. Neglecting the increase in water density with depth, find the pressure on the *Trieste* at the bottom of its dive.

16. In a barometer on a particular day, a mercury column 766 mm high balanced the pressure of the atmosphere. What was that pressure in pascals? ($\rho_{Hg} = 1.36 \times 10^3$ kg/m³)

17. A person sucking hard on a thin tube can reduce the pressure in it to 90% of atmospheric pressure. How high can water rise up the tube?

18. The pressure of the blood in a patient's vein is 12 torr. Find the minimum height above the vein that a container of plasma must be held for the plasma to enter the vein. (The actual height is always much greater than this figure in order to achieve a reasonable flow.)

19. A standing woman whose head is 35 cm above her heart bends over so that her head is 35 cm below her heart. What is the change (in torr) in the blood pressure in her head?

20. If the density of the atmosphere were uniform at sea-level density instead of decreasing with altitude, how thick a layer would it form?

21. A submarine is at a depth of 100 ft in seawater. The interior of the submarine is maintained at average atmospheric pressure. Find the force that must be withstood by a square hatch 2.0 ft on a side.

22. Water and olive oil rise to the indicated heights in the device shown in Fig. 10.41. Find the density of olive oil.

Figure 10.41

Exercise 22.

10.4 Buoyancy

23. A wooden board 10.0 cm thick floats in water with its upper surface parallel to and 3.0 cm above the water surface. What is the density of the wood?

24. A cube of oak (relative density 0.72) 10 cm on each edge is floating in fresh water with one face parallel to the water surface. Find the height of this face above the water.

25. The density of a person is slightly less than that of water. Assuming that these densities are the same, find the buoyant force of the atmosphere on a 70-kg person.

26. A barge 35 m long and 6.0 m wide has a mass of 2.0×10^5 kg. What depth of seawater is needed to float it?

27. A woman whose mass is 60 kg stands on a rectangular swimming raft 3.0 m long and 2.0 m wide that is floating in fresh water. By how much does the raft rise after she dives off?

28. Six empty 200-L drums whose mass is 20 kg each are used as floats for a boat dock in a freshwater lake. The dock platform is 1.8×9.0 m and is made of pine planks 40 mm thick. What is the maximum number of 75-kg people who can stand on the platform without getting their feet wet? The density of pine is 370 kg/m³.

29. A 30-kg balloon is filled with 100 m³ of hydrogen. How much force is needed to hold it down?

30. A design has been proposed for a modern helium-filled airship that is to have a useful lift of 5.0×10^5 kg at sea level. (a) How many m³ of helium must the airship contain? (b) The length of the airship would be 400 m. If it were a cylinder, what would its diameter be?

31. The largest rigid airship ever built was the German *Hindenburg*, which was 245 m long and had a capacity of 2.0×10^5 m³ of hydrogen. It was destroyed in a fire at Lakehurst, N.J., in 1937 after crossing the Atlantic. What was the total payload of the *Hindenburg* including its structure but not including the hydrogen it contained?

32. A steel tank whose capacity is 200 L has a mass of 36 kg. (a) Will it float in seawater when empty? (b) When filled with fresh water? (c) When filled with gasoline?

33. A tank in the form of a cube 10.00 in on each edge weighs 8.00 lb. The tank is floating in fresh water with its bottom side parallel to the water surface. A liquid whose relative density is 1.20 is poured into the tank until the top of the tank is just at the water surface. How high is the liquid in the tank? Disregard the thickness of the tank's sides.

34. A brass bar weighs 40.0 lb in air and 32.2 lb in water. What is the weight density of brass?

35. A bracelet that appears to be gold is suspended from a spring scale that reads 50.0 g in air and 46.0 g when the bracelet is immersed in a glass of water. Is the bracelet pure gold? Justify your answer.

36. When a 500-g statue of a falcon suspended from a spring scale is immersed in water, the scale reads 400 g, and when it is suspended in benzene, the scale reads 412 g. What is the density of benzene?

37. A paperweight has an apparent mass of 70.0 g when immersed in water and 76.0 g when immersed in oil of density 0.70 g/cm^3. Find the mass and volume of the paperweight.

10.5   Fluid Flow

38. A pipe whose inside diameter is 30 mm is connected to three smaller pipes whose inside diameters are 15 mm each. If the liquid speed in the larger pipe is 1.0 m/s, what is its speed in the smaller pipes?

39. Water flows at a speed of 1.0 m/s through a hose whose internal diameter is 10 mm. What should the diameter of the nozzle be if the water is to emerge at 5.0 m/s?

40. Half a liter of gasoline emerges per second from a fuel hose at a speed of 50 cm/s. What is the diameter of the nozzle?

41. The nozzle of a garden hose has an inside diameter of 0.30 in. If water flows through the hose at 10 gal/min, with what speed does it leave the nozzle?

42. Water enters the hose of Exercise 41 from a pump at a pressure of 40 lb/in^2. If the pump is 60% efficient, what must be the minimum horsepower rating of the pump motor?

43. The left ventricle of a certain running man pumps 20 L of blood per minute into his aorta at an average pressure of 140 torr. Find the total power output of his heart under the assumption that his right ventricle has an output 20% as great as that of his left ventricle.

44. The power output of an athlete's heart during strenuous activity might be 8.0 W. If the blood pressure increases by 150 torr when it passes through the athlete's heart, find the corresponding rate of flow of blood in liters per minute.

10.7   Applications of Bernoulli's Equation

45. A water tank has a hole 3.0 cm^2 in area located 2.0 m below the water surface. How many kilograms of water emerge from the hole per second?

46. A tank contains gasoline under a gauge pressure of 3.0 bars. Neglecting the difference in height between the liquid

surface and the valve, determine at what speed the gasoline emerges when the valve is opened.

47. During the pumping phase of the heart's action, the gauge pressure of blood in the major arteries of a normal person at the level of the heart is about 120 torr. If one of these arteries is cut and blood spurts out vertically, how high will it go?

48. Figure 10.42 shows a hose whose internal cross-sectional area is 2.0 cm^2 being used to siphon water out of a tank. Find the speed with which the water emerges and the rate of flow in L/s.

Figure 10.42

Exercise 48.

49. Kerosene leaks out of a hole 0.50 in. in diameter at the bottom of a tank at a rate of 10 gal/min. How high is the kerosene in the tank?

10.8   Viscosity

50. A grease nipple on a car has a hole 0.50 mm in diameter and 5.0 mm long. If the viscosity of the grease used is 80 N · s/m^2, what is the pressure needed to force 1.0 cm^3 of grease into the nipple in 10 s? How many times atmospheric pressure is this?

51. Find the pressure gradient needed to pump 16 L of water at 20°C per minute through a pipe 6.0 mm in radius. Take the viscosity of the water into account.

52. A hypodermic syringe has a cylinder 8.0 mm in diameter and a needle 0.40 mm in diameter and 30 mm long. How much force is needed on the plunger to send water at 20°C through the needle at 0.10 cm^3/s?

53. A typical capillary blood vessel is 1.0 mm long and has a radius of 2.0 μm. (a) If the pressure difference between its ends is 20 torr, what is the average speed of the blood that flows through such a capillary? (b) If the heart pumps 80 cm^3 of blood through a person's body per second, how many capillaries does the body contain?

PROBLEMS

54. A person's head is 40 cm above his heart. The average arterial blood pressure at the level of the heart is 100 torr. Find the average arterial pressure in the person's head when standing and when lying down in an elevator accelerated upward at 3.0 m/s^2.

55. A lead fishing sinker suspended from a spring scale is immersed in oil whose density is 820 kg/m^3. The reading on the scale is 0.2 N. Find the mass of the sinker.

56. A helium-filled weather balloon 2.0 m in radius whose total mass including instruments is 20 kg is tethered to the ground by a rope. A wind causes the rope to make an angle of 25° with the vertical. Find the tension in the rope.

57. A sailboat has 8000 kg of lead ballast attached to the bottom of its keel, the center of gravity of which is 1.5 m below the water surface when the boat is level. When the boat is tilted by 20° from the vertical, what is the torque exerted by this ballast about an axis along the length of the boat in the plane of the water surface? Assume seawater.

58. Water emerges from a fire hose at a speed of 20 m/s and a rate of flow of 50 L/s. (a) Find the force with which the nozzle must be held. (b) Find the required power of the pump motor, assuming 50% overall efficiency.

59. A tank truck filled with gasoline springs a leak 1.5 m below the liquid surface. If the hole has an area of 100 cm^2 and is at the back of the tank where its wall is vertical, what is the force acting on the truck?

60. A 1360 hp pump throws a jet of water 130 m into the air in Geneva, Switzerland. (a) With what speed does the water leave the mouth of the fountain? (b) If the overall efficiency is 60%, how many kilograms of water per minute are thrown into the air?

61. A horizontal stream of water leaves an orifice 1.0 m above the ground and strikes the ground 2.0 m away. What is the gauge pressure behind the orifice?

62. An open tank 60 ft high that is filled with water has a hole 25 ft below the top. How far from the tank does the water strike the ground?

63. A horizontal pipe 2.0 mm in radius at one end gradually increases in size so that it is 5.0 mm in radius at the other end. The pipe is 4.0 m long. Water is pumped into small end of the pipe at a speed of 8.0 m/s and a pressure of 2.0 bars. (a) Find the speed and pressure of the water at the pipe's large end. (b) The pipe is turned so as to be vertical with the small end underneath, so the water flows upward. Find the speed and pressure of the water at the large end of the pipe now.

64. A horizontal pipe of 7.00 in^2 cross-sectional area is joined to another horizontal pipe of 2.50 in^2 cross-sectional area; their total length is 6.00 ft. Water flows through the pipe at $0.600 \text{ ft}^3/\text{s}$, and the pressure in the large pipe is 30.0 lb/in^2. (a) What is the speed of the water in each pipe? (b) What is the pressure in the small pipe? (c) The pipes are turned so as to be vertical with the large end down, and the water flows upward at the same rate. What is the pressure at the upper end of the small pipe now?

65. Water flows upward through the pipe shown in Fig. 10.43 at 40 L/s. If the pressure at the lower end is 80 kPa, find the water speed at both ends and the pressure at the upper end.

66. Gasoline enters the pipe shown in Fig. 10.44 from the left at 3 m/s and 150 kPa. Find the speed and pressure at which gasoline leaves the pipe and the rate of flow in kilograms per second.

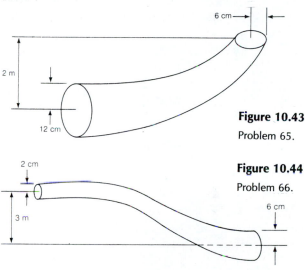

Figure 10.43

Problem 65.

Figure 10.44

Problem 66.

HARMONIC MOTION

11.1   Elastic Potential Energy

11.2   Simple Harmonic Motion

11.3   A Model of Simple Harmonic Motion

11.4   Position, Speed, and Acceleration

11.5   Trigonometric Notation

11.6   The Simple Pendulum

11.7   The Physical Pendulum

11.8   Damped Harmonic Oscillator

Many events in both nature and technology are periodic, with a certain motion repeating itself over and over again. As its name suggests, **simple harmonic motion** is the most basic kind of oscillation. A tree swaying in the wind, a child on a swing, the pistons of a car engine, the vibrating atoms in a solid—all are undergoing simple harmonic motion, or very nearly so. In fact, *every* periodic event is either simple harmonic or else the result of several such motions mixed together.

In harmonic motion the energy of the vibrations goes back and forth between kinetic and potential forms. The potential energy may be elastic rather than gravitational; we will find later that the notion of potential energy has even wider scope. The electrical equivalent of potential energy, in particular, can lead to electrical oscillations that closely resemble harmonic motion. It is these oscillations that make possible the electronic recording and transmission of sight and sound.

11.1 ELASTIC POTENTIAL ENERGY

The work done to deform an elastic object.

Open the lid of a jack-in-the-box and out pops Jack. How fast does he move? How high does he go? Not a very dignified problem for a serious science like physics, one might think, but the principles behind a jack-in-the-box apply equally well to any other system that stores elastic potential energy.

When we pull out a spring, it resists being stretched, and if we then let go, the spring returns to its original length. As we know, this is typical of **elastic** behavior. On the other hand, when we pull out a piece of taffy, it also resists being stretched, but if we then let go, nothing happens: The deformation is permanent. This is typical of **plastic** behavior.

In the case of the stretched spring, a **restoring force** comes into being that tries to return the spring to its normal length. The more we stretch the spring, the more the restoring force we must overcome (Fig. 11.1). Exactly the same thing happens when we compress the spring: It resists being shortened, and if we let go, the spring goes back to its normal length. Again a restoring force arises, and again the more the compression, the more the restoring force to be overcome.

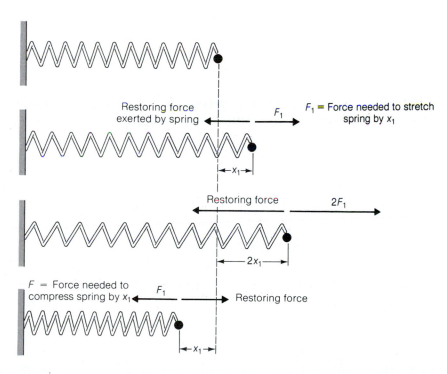

Figure 11.1

When a spring (or other elastic body) is stretched or compressed, a restoring force comes into being that tries to return the spring to its normal length.

The amount x by which an elastic solid is stretched or compressed by a force is directly proportional to the magnitude F of the force, provided the elastic limit is not exceeded. This proportionality is called Hooke's law, as mentioned in Chapter 9. Thus we can write.

$$F = kx \qquad\qquad\qquad \textit{Hooke's law} \quad (11.1)$$

where k is a constant whose value depends on the nature and dimensions of the object. A stiff spring has a higher value of k than a weak one.

◆ **E X A M P L E   11.1** _____

A spring of force constant k_1 is connected end-to-end to a spring of force constant k_2, as in Fig. 11.2. (a) What is the force constant k of the set of springs? (b) If $k_1 = 10.0$ N/m and $k_2 = 30.0$ N/m, what is k?

Figure 11.2

SOLUTION   (a) When a force F is applied to the set, each spring is acted on by F. Hence the elongations of the springs are, respectively,

$$x_1 = \frac{F}{k_1} \qquad x_2 = \frac{F}{k_2}$$

and the total elongation of the set is

$$x = x_1 + x_2 = \frac{F}{k_1} + \frac{F}{k_2} = F\frac{(k_1 + k_2)}{k_1 k_2}$$

Because $F = kx$ for the set, its force constant is

$$k = \frac{F}{x} = \frac{k_1 k_2}{k_1 + k_2}$$

(b) $k = \dfrac{(10.0 \text{ N/m})(30.0 \text{ N/m})}{10.0 \text{ N/m} + 30.0 \text{ N/m}} = 7.5 \text{ N/m}$

The force constant of the set is smaller than either k_1 or k_2.   ◆

The work needed to stretch (or compress) an object that obeys Hooke's law is easy to figure out. The work done by a force is the product of the magnitude of the force and the distance through which it acts. Here the force used in stretching the object is not constant but is proportional to the elongation x at each point in the stretching process. Because F is proportional to x, the *average* force $\overline{F}$ applied while the body is stretched from its normal length by an amount x to its final length is

$$\overline{F} = \frac{F_{\text{initial}} + F_{\text{final}}}{2} = \frac{0 + kx}{2} = \tfrac{1}{2}kx$$

since the initial force is 0 and final force is kx (Fig. 11.3). The work done in stretching the spring is the product of the average force $\overline{F} = \tfrac{1}{2}kx$ and the total elongation x, so that

$$W = \text{PE} = \tfrac{1}{2}kx^2 \qquad \textit{Elastic potential energy} \quad (11.2)$$

This formula is most often used in connection with springs: To stretch (or compress) a spring whose force constant is k by an amount x from its normal length requires $\tfrac{1}{2}kx^2$ of work to be done. This work goes into **elastic potential energy.** When the spring is released, its potential energy of $\tfrac{1}{2}kx^2$ is transformed into kinetic energy or into work done on something else (Fig. 11.4). Work done against frictional forces within the spring itself always absorbs some fraction of the available potential energy.

Figure 11.3

To compute the work done in stretching a body that obeys Hooke's law, the varying force that actually acts during the expansion may be replaced by the average force.

Figure 11.4

Some devices that make use of elastic potential energy in their operation.

◆ **E X A M P L E 11.2**

The horizontal spring shown in Fig. 11.5 has a force constant k of 90 N/m. Attached to the free end of the spring is a 1.4-kg block. If the spring is pulled out 50 cm from its equilibrium position and then released, what will the block's speed be when it returns to the equilibrium position?

Figure 11.5

A 1.4-kg block attached to a spring whose force constant is 90 N/m is pulled 50 cm from its equilibrium position. When the spring is released, its elastic potential energy of $\frac{1}{2}kx^2$ is converted into kinetic energy $\frac{1}{2}mv^2$. As the block's momentum compresses the spring on the other side of the equilibrium position, the kinetic energy is converted back into elastic potential energy.

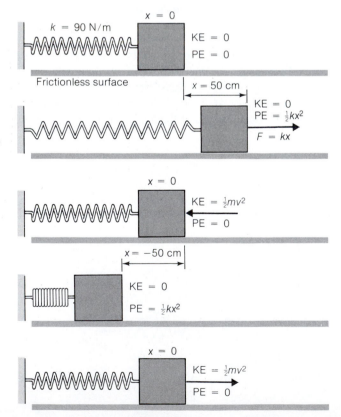

SOLUTION   When the spring is released, its elastic potential energy starts to be converted into kinetic energy of the block. We will assume that the spring's mass is small compared with that of the block and that its internal friction may be neglected. At the equilibrium position of the spring, $x = 0$, and all the initial potential energy of $\frac{1}{2}kx^2$ is now kinetic energy $\frac{1}{2}mv^2$. Hence

$$\tfrac{1}{2}mv^2 = \tfrac{1}{2}kx^2$$

$$v = \sqrt{\frac{k}{m}}\,x$$

$$= \sqrt{\frac{90\ \text{N/m}}{1.4\ \text{kg}}}\,(0.50\ \text{m}) = 4.0\ \text{m/s} \qquad\qquad \blacklozenge$$

11.2 SIMPLE HARMONIC MOTION

The energy of an oscillator shifts back and forth between PE and KE.

When a spring with something attached to it is pulled out and released, it does not merely go back to its equilibrium position and stop there. What happens is that the elastic potential energy $\frac{1}{2}kx^2$ of the spring is first turned into kinetic energy $\frac{1}{2}mv^2$ of the moving object. As the object's momentum compresses the spring on the other side of the equilibrium position, this KE is turned back into elastic PE (see Fig. 11.5). The compression $-x$ will have the same magnitude as the original extension x because the PE in both cases is the same and

$$\tfrac{1}{2}kx^2 = \tfrac{1}{2}k(-x)^2$$

Left to itself without friction, the spring-object system will keep oscillating back and forth indefinitely. The behavior of such a system is called **simple harmonic motion** (SHM).

Simple harmonic motion occurs whenever a force proportional to the displacement acts on a body in the opposite direction to its displacement from its normal position. The elastic restoring force of a stretched or compressed spring always tends to return the spring to its normal length, but the momentum of the moving mass compels it to overshoot in one direction and then in the other and thus to oscillate.

The **period** T of a system undergoing SHM is the time required for it to make one complete oscillation. (A complete oscillation is often called a **cycle**.) In the case of a spring, the period is the time the spring spends in going from its maximum extension, say, through its maximum compression and back to its maximum extension once more, as in Fig. 11.6.

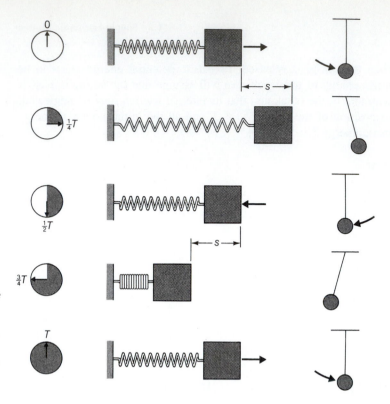

Figure 11.6

The period T of a body undergoing simple harmonic motion is the time required for it to make one complete oscillation. The maximum displacement on either side of the equilibrium position is called the *amplitude A* of the motion.

Figure 11.7

A pen attached to an oscillating object can be used to trace out a graph of the position of the object versus time on a sheet of paper pulled past the pen at a constant speed. The resulting graph has the same form as that of $\sin \theta$ plotted against θ, and accordingly the curve is said to be *sinusoidal*.

To find out how the position of an oscillating object varies with time, we can attach a pen to the object and pull a sheet of paper past it at a constant speed (Fig. 11.7). The graph has the same form as that of $\sin \theta$ plotted against θ, and accordingly the curve is said to be **sinusoidal.** Sinusoidal variations are found elsewhere in physics also—for example, in wave motion and in connection with alternating electric currents.

The maximum displacement A of an object undergoing harmonic motion on either side of its equilibrium position is called the **amplitude** of the motion:

$$A = x_{max}$$

Amplitude = maximum displacement from equilibrium position

Period of SHM

The period of all types of simple harmonic motion is given by the same formula,

$$T = 2\pi \sqrt{-\frac{x}{a}} \qquad \textit{Simple harmonic motion} \quad (11.3)$$

$$\text{Period} = 2\pi \sqrt{-\frac{\text{displacement}}{\text{acceleration}}}$$

The acceleration a is that experienced by the body when it is at the displacement x from its equilibrium position. This formula is derived in Section 11.3.

To calculate the acceleration of a mass attached to a stretched spring, we start with the second law of motion, $F = ma$, and substitute the **restoring force** $F_r = -kx$ because it is this force that causes the body to be accelerated. Thus

$$F_r = ma$$

$$a = \frac{F_r}{m} = \frac{-kx}{m} \qquad (11.4)$$

With this result we find that the period of a body of mass m attached to a spring of force constant k is

$$T = 2\pi \sqrt{-\frac{x}{a}} = 2\pi \sqrt{-\frac{x}{-kx/m}}$$

$$T = 2\pi \sqrt{\frac{m}{k}} \qquad \textit{Oscillating system of mass and spring} \quad (11.5)$$

It is worth noting that the period T does *not* depend on the amplitude. No matter how much or how little the spring is pulled out, exactly the same time is needed for each cycle. If A is small, the maximum acceleration is also small and the body moves back and forth slowly. If A is large, the maximum acceleration is also large and the body moves faster through the larger range to give the same period.

This feature of SHM is the basis of mechanical clocks and watches, which use the rotational oscillations of a coil spring or the swings of a pendulum—both simple harmonic motions—to keep a constant rate regardless of fluctuations in amplitude.

A quantity often used in describing harmonic motion is **frequency.** The frequency is the number of cycles that occur per unit time. Hence frequency, whose symbol is f, is the reciprocal of period T,

$$f = \frac{1}{T} \qquad\qquad \textit{Frequency} \quad (11.6)$$

The unit of frequency, the **hertz,** is named in honor of Heinrich Hertz (1857–1894), a pioneer in the study of electromagnetic waves: 1 hertz (Hz) = 1 cycle/s.

◆ **E X A M P L E 11.3**

When a ball is suspended from a spring, the spring stretches by 70 mm (Fig. 11.8). If the ball then oscillates up and down, what is its period? Its frequency?

SOLUTION The force exerted on the spring is the ball's weight of mg. Because $F = kx$, the force constant of the spring is

$$k = \frac{F}{x} = \frac{mg}{x}$$

Figure 11.8

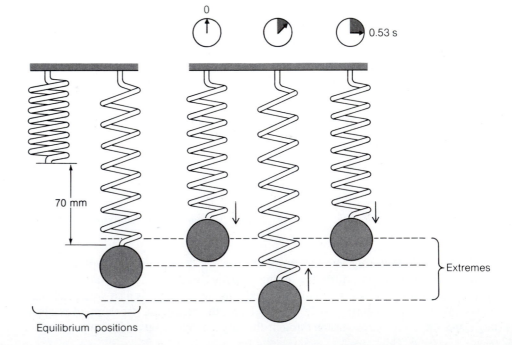

70 mm

0   0.53 s

Extremes

Equilibrium positions

The period of the oscillations is therefore

$$T = 2\pi \sqrt{\frac{m}{k}} = 2\pi \sqrt{\frac{mx}{mg}} = 2\pi \sqrt{\frac{x}{g}} = 2\pi \sqrt{\frac{0.070 \text{ m}}{9.8 \text{ m/s}^2}} = 0.53 \text{ s}$$

We did not have to know the ball's mass to find the period of its motion, because $x/g = m/k$ here. The frequency that corresponds to a period of 0.53 s is

$$f = \frac{1}{T} = \frac{1}{0.53 \text{ s}} = 1.9 \text{ cycles/s} = 1.9 \text{ Hz} \qquad \blacklozenge$$

◆   **E X A M P L E   11.4** _____

A 2000-lb car has a period of vertical oscillation of 2.00 s. What is the effective force constant of its springs?

SOLUTION   The first step is to solve Eq. (11.5) for the force constant k, which is done as follows:

$$T = 2\pi \sqrt{\frac{m}{k}} \qquad T^2 = \frac{4\pi^2 m}{k} \qquad k = \frac{4\pi^2 m}{T^2}$$

Here $m = w/g$, so

$$k = \frac{4\pi^2 w}{gT^2} = \frac{4\pi^2 (2000 \text{ lb})}{(32.2 \text{ ft/s}^2)(2.00 \text{ s})^2} = 613 \text{ lb/ft} \qquad \blacklozenge$$

11.3 A MODEL OF SIMPLE HARMONIC MOTION

How to derive the formula for the period of SHM.

To go directly from the formula for restoring force $F_r = -kx$ to Eq. (11.3) for the period T of SHM requires the use of calculus. However, simpler math can make the same link if we are clever enough to find a suitable model. Such a model is shown in Fig. 11.9.

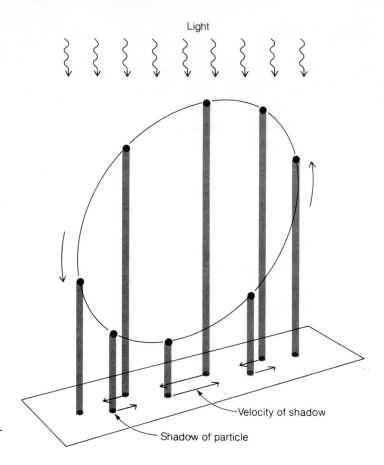

Figure 11.9

The shadow of a particle undergoing uniform circular motion executes simple harmonic motion.

The model makes use of a particle moving in a vertical circle at constant speed. The particle is lit from above, and it casts a shadow on a horizontal screen below its orbit. As the particle goes around the circle, its shadow oscillates back and forth. The shadow moves fastest at the center, slows down as it approaches each end of the path, comes to a stop, and then reverses its direction. We might suspect that the shadow is executing SHM. To verify this suspicion, we must show that the acceleration a of the shadow at any time is proportional to its displacement x from the center of its path and opposite in direction.

The acceleration of the shadow at any point is the horizontal component of the particle's acceleration a_c. As we know, a particle in uniform circular motion at the speed V in a circle of radius R experiences the centripetal acceleration

$$a_c = -\frac{V^2}{R}$$

where the minus sign indicates that the acceleration is inward toward the center of the circle. The horizontal component of this acceleration is the acceleration a of the shadow. Because corresponding sides of similar triangles are proportional, we see

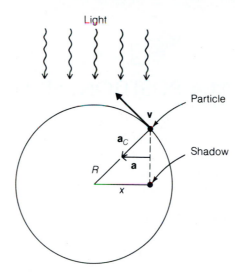

Figure 11.10

The acceleration of the shadow is the horizontal component of the particle's centripetal acceleration. The shadow's acceleration is proportional to its displacement x and is in the opposite direction.

from Fig. 11.10 that

$$\frac{a}{a_c} = \frac{x}{R} \quad \text{and so} \quad a = \frac{x}{R}a_c$$

Because $a_c = -V^2/R$, the acceleration of the shadow is

$$a = -\frac{x}{R}\frac{V^2}{R} = -\frac{V^2}{R^2}x \tag{11.7}$$

Thus the shadow's acceleration is proportional to its displacement x and in the opposite direction. This means that the shadow is indeed executing SHM.

Now let us use this model to find the period of an object in SHM. The circumference of a circle of radius R is $2\pi R$, and a particle moving around the circle with the constant speed V covers this distance in the time

$$T = \frac{2\pi R}{V}$$

From Eq. (11.7) we find that

$$\frac{R}{V} = \sqrt{-\frac{x}{a}}$$

and so

$$T = 2\pi \sqrt{-\frac{x}{a}} \tag{11.3}$$

This is the general formula for the period of SHM given earlier.

11.4 POSITION, SPEED, AND ACCELERATION

Formulas that apply to any harmonic oscillator.

The principle of conservation of energy gives us a way to express the speed of an object in SHM in terms of its frequency f, amplitude A, and displacement x. The total energy of the oscillator (object plus spring) is the sum of its kinetic and potential energies at any time, which are, respectively, $\frac{1}{2}mv^2$ and $\frac{1}{2}kx^2$. At either end of the motion, when $x = +A$ or $x = -A$, the object is stationary and has only the potential energy $\frac{1}{2}kA^2$. Hence

$$\text{Total energy} = \text{KE} + \text{PE}$$
$$\tfrac{1}{2}kA^2 = \tfrac{1}{2}mv^2 + \tfrac{1}{2}kx^2$$
$$mv^2 = k(A^2 - x^2)$$
$$v = \sqrt{k/m}\,\sqrt{A^2 - x^2}$$

From Eqs. (11.5) and (11.6) we know that

$$f = \frac{1}{T} = \frac{1}{2\pi}\sqrt{\frac{k}{m}}$$

which can be rewritten as

$$\sqrt{k/m} = 2\pi f \qquad\qquad (11.8)$$

The speed of the object when it has the displacement x is accordingly

$$v = 2\pi f\sqrt{A^2 - x^2} \qquad\qquad \textit{Speed at given displacement} \quad (11.9)$$

This formula gives only the absolute value of v; whether the sign of v is $+$ or $-$ depends on whether the body is at $+x$ or $-x$ and on whether it is on its way toward or away from the equilibrium position from there.

From Eq. (11.9) we see that the maximum speed $v_{\max}$ of the object, which occurs at the equilibrium position when $x = 0$, is

$$v_{\max} = 2\pi fA \qquad\qquad \textit{Maximum speed} \quad (11.10)$$

The maximum speed is proportional to both the frequency and the amplitude of the motion.

◆   E X A M P L E   11.5 _____

An object is in SHM with a frequency of 20 Hz and a maximum speed of 2.5 m/s. What is the object's displacement when its speed is 1.5 m/s?

SOLUTION   We begin by finding the amplitude A of the object's motion. From Eq. (11.10), $v_{max} = 2\pi fA$, we have

$$A = \frac{v_{max}}{2\pi f} = \frac{2.5 \text{ m/s}}{2\pi(20 \text{ Hz})} = 0.020 \text{ m}$$

which is 20 mm. To find the value of x that corresponds to a given speed v we must solve Eq. (11.9) for x:

$$v = 2\pi f\sqrt{A^2 - x^2}$$
$$v^2 = (2\pi f)^2(A^2 - x^2)$$
$$\left(\frac{v}{2\pi f}\right)^2 = A^2 - x^2$$
$$x^2 = A^2 - \left(\frac{v}{2\pi f}\right)^2$$
$$x = \sqrt{A^2 - \left(\frac{v}{2\pi f}\right)^2}$$

Finally we susbstitute $A = 0.020$ m, $v = 1.5$ m/s, and $f = 20$ Hz to obtain

$$x = \sqrt{(0.020 \text{ m})^2 - \left(\frac{1.5 \text{ m/s}}{2\pi(20 \text{ Hz})}\right)^2} = 0.016 \text{ m}$$

which is 16 mm.   ◆

_____

The energy of an oscillating object shifts back and forth between kinetic and potential forms. To find the total energy, we can calculate either KE_{max} or PE_{max}, with the help, respectively, of Eq. (11.10) or (11.8):

$$KE_{max} = \tfrac{1}{2}mv^2_{max} = 2\pi^2 mf^2A^2 \qquad\qquad \textit{Total energy}\quad (11.11)$$

$$PE_{max} = \tfrac{1}{2}kA^2 = 2\pi^2 mf^2A^2 \qquad\qquad\qquad\qquad\qquad (11.12)$$

The total energy depends on the square of the frequency and the square of the amplitude. Note that the total energy is not the sum of Eqs. (11.11) and (11.12), because when KE or PE reaches its maximum, the other is zero.

To find the acceleration of an object in SHM, we refer to Eq. (11.4), which states that $a = -kx/m$. On the basis of Eq. (11.8) this formula becomes

$$a = -4\pi^2 f^2 x \qquad \textit{Acceleration at given displacement} \quad (11.13)$$

The acceleration is always opposite in direction to the displacement, which, of course, is one of the conditions for SHM to occur. The maximum acceleration occurs at either end, when $x = \pm A$, and has the magnitude

$$a_{max} = 4\pi^2 f^2 A \qquad \textit{Maximum acceleration} \quad (11.14)$$

Figure 11.11

When the downward acceleration of the piston exceeds g, the coin will be left behind by the piston on its downstroke.

The maximum acceleration is proportional to the square of the frequency and to the amplitude.

Although these formulas were derived for the case of a vibrating spring, their validity is perfectly general and they apply to any type of harmonic oscillator, from the bob of a pendulum to an atom in a molecule.

◆ **E X A M P L E 11.6**

In a car engine, each piston moves up and down in an approximation of SHM. Given that a certain piston has a mass of 0.50 kg, a total travel ("stroke") of 120 mm, and a frequency of oscillation of 60 Hz (corresponding to 3600 rpm), find the maximum force it experiences, assuming SHM.

SOLUTION The maximum force is given by $F_{max} = ma_{max}$. Here, because the amplitude is half the total travel, $A = 60$ mm $= 0.060$ m, and

$$F_{max} = ma_{max} = 4\pi^2 mf^2 A = 4\pi^2(0.50 \text{ kg})(60 \text{ Hz})^2(0.060 \text{ m})$$
$$= 4264 \text{ N} = 4.3 \text{ kN}$$

This is nearly 1000 lb. ◆

◆ **E X A M P L E 11.7**

A piston undergoes SHM in a vertical direction with an amplitude of 2.5 in. A coin is placed on top of the piston (Fig. 11.11). What is the lowest frequency at which the coin will be left behind by the piston on its downstroke?

SOLUTION The coin will leave the piston when the piston's downward acceleration exceeds the acceleration of gravity g. The maximum downward acceleration of the piston occurs at the highest point of its motion, when $a_{max} = 4\pi^2 f^2 A$ according to Eq. (11.14). Hence we set a_{max} equal to g and solve for the frequency f, with

$A = 2.5$ in $= 0.21$ ft:

$$a_{max} = g = 4\pi^2 f^2 A$$

$$f = \frac{1}{2\pi}\sqrt{\frac{g}{A}}$$

$$= \frac{1}{2\pi}\sqrt{\frac{32 \text{ ft/s}^2}{0.21 \text{ ft}}} = 2.0 \text{ Hz}$$

◆

The variations of the displacement, velocity, and acceleration of a body undergoing SHM are plotted versus time in Fig. 11.12. The graphs are plotted on the assumption that the body is at $x = +A$ when $t = 0$. This corresponds to pulling out the body and letting it go at $t = 0$. At this instant the body's acceleration is a maximum and is opposite in direction to x, while the velocity is zero because the particle has not started to move.

When the particle is at the origin, $x = 0$ and the spring is at its normal length. Because the spring exerts no force on the body at this time, its acceleration is zero. The speed of the body is now a maximum because the system has no PE at this point.

When $x = -A$, the body is at the other extreme of its range, and its acceleration, again a maximum, is positive. The acceleration is once more in the direction of the origin, though now from the other side. All the energy of oscillation is potential, and so the body is stationary at this instant.

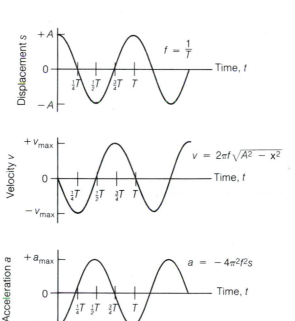

Figure 11.12

Displacement, velocity, and acceleration in simple harmonic motion.

11.5 TRIGONOMETRIC NOTATION

How to find x, v, and a at any time.

There are situations in which we need to know the position, the velocity, and the acceleration of a harmonic oscillator as functions of time. Figure 11.13 shows the same model used in Fig. 11.9, with a particle moving in a circle having a shadow that undergoes SHM. The particle has an angular speed of $\omega = 2\pi f$, and ω is called the **angular frequency** of the harmonic motion of its shadow. The unit of ω is, as usual, the radian/s.

The angular displacement of the particle at any time t is $\theta = \omega t$, its linear speed in the circle is ωA, and its centripetal acceleration is $\omega^2 A$, where A is the radius of the circle. From Fig. 11.13 we see that

$$x = A \cos \theta = A \cos \omega t \tag{11.15}$$

$$v = -\omega A \sin \theta = -\omega A \sin \omega t \tag{11.16}$$

$$a = -\omega^2 A \cos \theta = -\omega^2 A \cos \omega t \tag{11.17}$$

The sign convention used is that a quantity directed to the right is considered positive and one directed to the left is considered negative. Graphs of x, v, and a versus time were given in Fig. 11.12.

Figure 11.13

(a)−(c). The displacement, velocity, and acceleration of the shadow of the particle of Fig. 11.9 can be expressed in terms of the time t as shown here. The angular speed of the particle is ω, the magnitude of its linear velocity is ωA, and the magnitude of its centripetal acceleration is $\omega^2 A$.

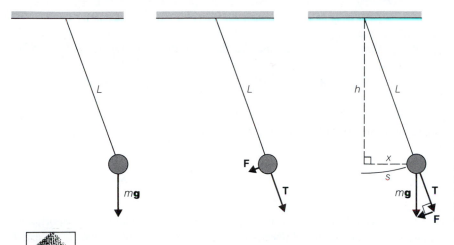

Figure 11.14

A pendulum executes simple harmonic motion when its oscillations are so small in amplitude that the chord x is very nearly equal in length to the arc s.

11.6 THE SIMPLE PENDULUM

Its period is independent of its mass.

A pendulum undergoes SHM as it swings back and forth, provided the arc through which it moves is not too large. We can see why this limitation arises by using Eq. (11.3) to find the period of a pendulum. We assume that the entire mass of the pendulum is concentrated in its bob and that the bob's dimensions are small compared with the pendulum's length. In the next section we will look into more realistic pendulums.

Figure 11.14 shows a pendulum of length L whose bob has the mass m. The weight $m\mathbf{g}$ of the bob, which acts vertically downward, may be resolved into two forces, $\mathbf{T}$ and $\mathbf{F}$, which act, respectively, parallel to and perpendicular to the supporting string L. That is,

$$\mathbf{T} + \mathbf{F} = m\mathbf{g}$$

The force $\mathbf{F}$ is the restoring force that acts to return the bob to the midpoint of its motion. The space triangle hLx and the vector triangle $\mathbf{T}m\mathbf{g}\mathbf{F}$ are similar, with each having a right angle and two sides of one being parallel to the two corresponding sides of the other. Hence

$$\frac{F}{x} = \frac{mg}{L}$$

The restoring force acting on the bob is therefore

$$F = -\frac{mgx}{L}$$

where the minus sign indicates that $\mathbf{F}$ points in the direction of decreasing x.

If the string is not far from the vertical, the horizontal distance x is nearly the same as the actual path length s. If they are equal, then F is proportional to $-s$ and the motion is simple harmonic. In this case the acceleration of the bob is

$$a = \frac{F}{m} = -\frac{g}{L}x$$

Substituting in Eq. (11.3), $T = 2\pi \sqrt{-x/a}$, gives

$$T = 2\pi \sqrt{\frac{L}{g}} \qquad\qquad\qquad \textit{Simple pendulum} \quad (11.18)$$

Provided that x is close to s, the period of a pendulum is proportional to the square root of its length and does not depend on the mass of its bob. If the arc through which the pendulum swings on either side of the vertical is 5°, its actual period exceeds that given by Eq. (11.18) by only 0.05%. Even if the arc is 20°, the discrepancy is less than 1%.

◆   E X A M P L E   11.8

How long should a pendulum be for it to have a period of 1.00 s?

SOLUTION   We first solve Eq. (11.18) for L:

$$T = 2\pi \sqrt{\frac{L}{g}} \qquad T^2 = \frac{4\pi^2 L}{g} \qquad L = \frac{gT^2}{4\pi^2}$$

Inserting the values $g = 9.8$ m/s^2 and $T = 1.00$ s gives

$$L = \frac{(9.8 \text{ m/s}^2)(1.00\text{s})^2}{4\pi^2} = 0.25 \text{ m} \qquad\qquad\qquad ◆$$

11.7 THE PHYSICAL PENDULUM

How to find the period of any swinging object.

An object of any shape will oscillate back and forth when it is pivoted at some point other than its center of gravity and given an initial push. Such an object is called a **physical pendulum.** We will find that the formula for the period of a simple pendulum also holds for a physical pendulum provided the length L is properly interpreted.

Figure 11.15 shows a physical pendulum pivoted at the point O. The pendulum has been pushed to one side so that the line from O to its center of gravity is at the angle θ from the vertical. The pendulum's weight mg acts from the center of gravity and produces the restoring torque

$$\tau = -mgx$$

where x is the horizontal distance between O and the center of gravity. The minus sign reflects the fact that the restoring torque is opposite in direction to the angular displacement θ of the pendulum.

When θ is small, the chord x is very nearly equal to the arc s. Assuming them to be equal is the same approximation made in analyzing the simple pendulum (see Fig. 11.14). Because h is the distance between the pivot point O and the center of gravity CG in Fig. 11.15, what we have for small θ is

$$x = s = h\theta$$

This means that

$$\tau = -mgh\theta$$

We know that a restoring force of the form $-kx$ leads to SHM, so we might suspect that a restoring torque of the form $-K\theta$ leads to a rotational version of SHM. If I is the moment of inertia of the pendulum about O and α is the angular acceleration produced by the restoring torque, then

$$\alpha = \frac{\tau}{I} = -\frac{mhg}{I}\theta \tag{11.19}$$

Comparing this formula with the equivalent one for a harmonic oscillator,

$$a = -\frac{k}{m}x \tag{11.4}$$

suggests that the period of a physical pendulum can be given by the general formula for the period of a harmonic oscillator,

$$T = 2\pi\sqrt{-\frac{x}{a}} \tag{11.5}$$

with θ/α replacing x/a. This idea turns out to be correct, and we have

$$T = 2\pi\sqrt{-\frac{\theta}{\alpha}} = 2\pi\sqrt{\frac{I}{mgh}} \qquad \textit{Physical pendulum} \quad (11.20)$$

$$\tau = -(mg)(h\sin\theta)$$

Figure 11.15

A physical pendulum pivoted at O. The center of gravity is marked CG.

◆   **E X A M P L E   11.9**

A uniform steel girder 15 ft long is suspended from one end by a crane. What is the period of its oscillations?

SOLUTION   From Fig. 7.10 the moment of inertia of a thin rod pivoted at one end is $I = \frac{1}{3}mL^2$, where L is the rod's length and m is its mass. The distance h from the pivot to the center of gravity of the girder is $L/2$. Hence the period of oscillation is

$$T = 2\pi\sqrt{\frac{I}{mgh}} = 2\pi\sqrt{\frac{mL^2/3}{mgL/2}} = 2\pi\sqrt{\frac{2L}{3g}}$$

$$= 2\pi\sqrt{\frac{(2)(15\text{ ft})}{(3)(32\text{ ft/s}^2)}} = 3.5\text{ s}$$

The period does not depend on the girder's mass, only on its length.   ◆

Figure 11.16

A simple pendulum of length $L = I/mh$ has the same period as that of a physical pendulum. The point C is the center of oscillation.

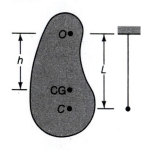

Center of Oscillation

The simple pendulum whose period is the same as that of a given physical pendulum may be found by setting equal the formulas for their respective periods of oscillation:

$$2\pi\sqrt{\frac{L}{g}} = 2\pi\sqrt{\frac{I}{mgh}}$$

$$L = \frac{I}{mh} \tag{11.21}$$

Thus the mass of a physical pendulum can be regarded as concentrated at a point C the distance $L = I/mh$ from the point O. This point is called the **center of oscillation** (Fig. 11.16) and has two interesting properties:

1. If the pendulum is pivoted at C instead of at O, it will oscillate with the same period as before and O will be the new center of oscillation.
2. If the pendulum is struck along a line of action through C, there will be no reaction force on the pivot at O. A baseball that strikes a bat at the latter's center of oscillation does not produce a sting in the batter's hands, for example (Fig. 11.17). The center of oscillation is often called the **center of percussion** for this reason. The concept of center of percussion plays an important part in the design of many mechanical devices.

Figure 11.17

When a physical pendulum is struck at its center of percussion, its pivot experiences no reaction force.

11.8 DAMPED HARMONIC OSCILLATOR

The greater the damping, the slower the oscillations and the sooner they die out.

Once set in motion, an ideal harmonic oscillator should continue oscillating forever. Actual oscillators do not behave this way. Although in some cases they may continue for a long time, if no further energy is fed into the oscillations their amplitude will decrease until they come to a stop. This phenomenon is called **damping.** If the amount of damping is very large, as it is for a car whose springs are damped with good shock absorbers, no oscillation at all occurs. After being displaced, the object gradually returns to its equilibrium position and goes no further.

In general, damping is caused by frictional forces. The potential energy of a stretched spring is not completely converted into kinetic energy at $x = 0$; some of it goes into work done against the frictional forces that are acting. At the end of each vibration the loss in energy leads to a shorter extension of the spring, until finally the amplitude becomes 0. Damping also acts to reduce the frequency of the motion—the greater the damping, the slower the oscillations.

Figure 11.18 shows the effects of different amounts of damping on a harmonic oscillator. In (a) the damping is small, and the oscillations steadily decrease in amplitude. In (b) the damping is so great that the displaced object never oscillates but returns to its equilibrium position very slowly; such an oscillator is said to be **overdamped**. A **critically damped** oscillator, as in (c), falls right on the line between the other two situations; it returns to its equilibrium position as rapidly as possible without overshooting, which would mean oscillating. Critical damping is desirable in many situations, such as that of the suspension of a car.

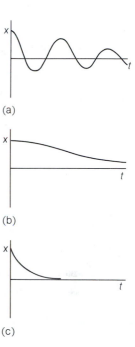

Figure 11.18

(a) Damped harmonic oscillator. (b) An overdamped oscillator returns to its equilibrium position very slowly. (c) A critically damped oscillator returns to its equilibrium position rapidly, but not so rapidly that it overshoots and begins to oscillate.

Important Terms

A body under stress has **elastic potential energy,** which is equal to the work done in deforming it.

Simple harmonic motion (SHM) is an oscillatory motion that occurs whenever a restoring force acts on a body in the opposite direction to its displacement from its equilibrium position, with the magnitude of the restoring force proportional to the magnitude of the displacement.

The **period** T of a system undergoing SHM is the time needed for it to make one complete oscillation. The **frequency** f of such a system is the number of complete oscillations it makes per unit time.

The **amplitude** of a body undergoing SHM is its maximum displacement on either side of its equilibrium position. The period of the motion is independent of the amplitude.

The **center of oscillation** of a pivoted object is that point at which it can be struck without producing a reaction force on its pivot.

In a **damped harmonic oscillator,** friction progressively reduces the amplitude of the vibrations.

Important Formulas

Elastic potential energy: $PE = \frac{1}{2}kx^2$

Harmonic oscillator: $T = \dfrac{1}{f} = 2\pi\sqrt{-\dfrac{x}{a}} = 2\pi\sqrt{\dfrac{m}{k}}$

Simple pendulum: $T = 2\pi\sqrt{\dfrac{L}{g}}$

Physical pendulum: $T = 2\pi\sqrt{\dfrac{I}{mgh}}$

MULTIPLE CHOICE

1. The product of the period and the frequency of a harmonic oscillator is always equal to

   a. 1
   b. π
   c. 2π
   d. the amplitude of the motion

2. The period of a simple harmonic oscillator is independent of its

   a. frequency   **b.** amplitude
   c. force constant   **d.** mass

3. The amplitude of the motion of an object undergoing SHM is

   a. its total range of motion
   b. its maximum displacement on either side of the equilibrium position
   c. its minimum displacement on either side of the equilibrium position
   d. the number of cycles per second it undergoes

4. The amplitude of a simple harmonic oscillator does not affect its

   a. frequency
   b. maximum speed
   c. maximum acceleration
   d. maximum KE

5. The amplitude of a simple harmonic oscillator is doubled. Which one or more of the following is also doubled?

   a. its frequency
   b. its period
   c. its maximum speed
   d. its total energy

6. Which one or more of the following statements about an object in SHM in a straight line is not true?

   a. Its acceleration can be 0 at times.
   b. Its acceleration can be greater than g at times.
   c. Its amplitude must be small.
   d. Its KE is constant.

7. An object undergoes simple harmonic motion. Its maximum speed occurs when its displacement from its equilibrium position is

   a. zero
   b. a maximum
   c. half its maximum value
   d. none of the above

8. In simple harmonic motion there is always a constant ratio between the displacement of the mass and its

   a. speed   **b.** acceleration
   c. period   **d.** mass

9. An object attached to a horizontal spring undergoes SHM on a frictionless surface. Its PE

   a. is 0
   b. equals its KE
   c. does not equal its KE
   d. may be any of the above, depending on its displacement

10. A pendulum comes very close to executing simple harmonic motion provided that

   a. its bob is not too heavy
   b. the supporting string is not too long
   c. the arc through which it swings is not too small
   d. the arc through which it swings is not too large

11. The period of a simple pendulum depends on its

   a. mass   **b.** length
   c. total energy   **d.** maximum speed

12. A pendulum clock is in an elevator. The clock will run fast when the elevator is

   a. rising at constant speed
   b. falling at constant speed
   c. accelerating upward
   d. accelerating downward

13. An object pivoted at an arbitrary point swings back and forth with the period T. The number of other points in the object at which it can be pivoted and have the same period is

   a. 0   **b.** at least 1
   c. at least 2   **d.** unlimited

14. An object pivoted at an arbitrary point is struck at its center of oscillation.
   a.   The object oscillates with a shorter period than if struck anywhere else.
   b.   The object oscillates with a longer period than if struck anywhere else.
   c.   There is no reaction force on the pivot.
   d.   The reaction force on the pivot is a maximum.

15. A spring whose force constant is k is cut in half. Each of the new springs has a force constant of
   a. $\frac{1}{2}k$
   b. k
   c. $2k$
   d. $4k$

16. A spring of force constant 1.0 N/m is joined end-to-end to a spring of force constant 2.0 N/m. The force constant of the combination is
   a.   0.67 N/m
   b.   1.0 N/m
   c.   1.5 N/m
   d.   3.0 N/m

17. A force of 0.2 N is needed to compress a certain spring by 2 cm. Its potential energy when compressed is
   a. 2×10^{-3} J
   b. 2×10^{-5} J
   c. 4×10^{-5} J
   d. 8×10^{-5} J

18. A 0.40-lb toy car is pressed against a horizontal spring of force constant 12.0 lb/ft. The spring is compressed 1.5 in and then the car is let go. If there is no friction, the car leaves the spring with a speed of
   a.   0.68 ft/s
   b.   3.9 ft/s
   c.   11 ft/s
   d.   46 ft/s

19. When a 1.00-kg mass is suspended from a spring, the spring stretches by 50.0 mm. The force constant of the spring is
   a.   0.20 N/m
   b.   1.96 N/m
   c.   49 N/m
   d.   196 N/m

20. If the suspended mass of Multiple Choice 19 oscillates up and down, its period will be approximately
   a.   0.032 s
   b.   0.071 s
   c.   0.45 s
   d.   4.5 s

21. The maximum speed of a particle that undergoes simple harmonic motion with a period of 0.50 s and an amplitude of 20 mm is
   a. π cm/s
   b. 2π cm/s
   c. 4π cm/s
   d. 8π cm/s

22. In order to oscillate at 10 Hz, a mass of 20 g should be suspended from a spring whose force constant is
   a.   2.5 N/m
   b.   8.9 N/m
   c.   12.6 N/m
   d.   79 N/m

23. A woman walks to the end of a diving board, whose end moves down by 35 cm as a result. If the woman bounces on the end of the board, the period of oscillation will be
   a.   0.19 s
   b.   1.2 s
   c.   1.4 s
   d.   dependent on her mass

24. A harmonic oscillator of mass 40 g has a period of 10 s. To reduce the period to 5 s, the mass should be changed to
   a.   10 g
   b.   20 g
   c.   80 g
   d.   160 g

25. The 2.00-lb piston in a compressor undergoes 20.0 cycles per second in which its total travel is 6.00 in. The maximum force on the piston
   a.   is 247 lb
   b.   is 494 lb
   c.   is 3.95×10^3 lb
   d.   cannot be calculated from the data given

26. A boy swings from a rope 4.9 m long. His approximate period of oscillation is
   a.   0.5 s
   b.   3.1 s
   c.   4.4 s
   d.   12 s

QUESTIONS

1. At what point or points in its motion is the energy of a harmonic oscillator entirely potential? At what point or points is its energy entirely kinetic?

2. Must a spring obey Hooke's law in order to oscillate?

3. A wooden object is floating in a bathtub. It is pressed down and then released. Under what circumstances will its oscillations be simple harmonic in nature?

4. The total energy of a harmonic oscillator is doubled. How does this affect the frequency of the oscillation? Its amplitude? The maximum speed?

5. On what, if anything, does the ratio between the maximum kinetic energy and maximum potential energy of a harmonic oscillator depend?

6. The Empire State Building in New York City oscillates in the wind with a period of about 8 s. Do you think the period is different on weekdays, when the building is full of people, from what it is on weekends, when the building is empty?

7. In what part of a pendulum's swing is the tension in its supporting string a maximum?

8. A body pivoted at some point is given an initial displacement and then released. Under what circumstances will it oscillate back and forth? Under what circumstances will the oscillations be simple harmonic in character? Under what circumstances will it behave like a simple pendulum?

9. What is the frequency of a pendulum whose normal period is T when it is in an elevator in free fall? When it is in an elevator descending at constant velocity? When it is in an elevator ascending at constant velocity?

10. Verify that T is expressed in units of time in Eqs. (11.5), (11.18), and (11.20).

11. Where should the center of percussion of an ax be located?

12. A loudspeaker generates sound waves that correspond in frequency and amplitude to the electrical oscillations that reach it from the output of an audio amplifier. Why is it desirable for the cone of a loudspeaker to be damped?

EXERCISES

11.1   Elastic Potential Energy

1. The work needed to compress a spring by 40 mm is 0.80 J. What is the force constant of the spring?

2. A 5.0-kg object is dropped on a vertical spring from a height of 2.0 m. If the maximum compression of the spring is 40 cm, what is its force constant?

3. A force of 2.0 N is needed to push a 100-g jack-in-the-box into its box, an operation in which the spring is compressed 10 cm. What will be the maximum speed of the jack-in-the-box when it pops out?

4. A toy rifle employs a spring whose force constant is 200 N/m. In use, the spring is compressed 50 mm, and when released, it propels a 5.0-g rubber ball. What is the ball's speed when it leaves the rifle?

5. A force of 5.00 lb compresses a spring by 3.50 in. (a) What is the elastic PE of the compressed spring? (b) A 0.500-lb ball is placed against the compressed spring. When the spring is let go, what is the ball's initial speed?

6. You have two springs whose force constants are, respectively, 50 N/m and 100 N/m. (a) The springs are connected end-to-end, and the combination is pulled out by a force of 10 N. What is the PE of the stretched springs? (b) The springs are then connected side-by-side and again pulled out by the same force. What is the PE of the stretched springs now?

11.2   Simple Harmonic Motion

7. A 100-g mass is suspended from a spring whose force constant is 50 N/m. The mass is then pulled down 10 mm and released. What is the amplitude of the resulting oscillations? What is their frequency?

8. A 25-kg portable gasoline-powered generating set is mounted on four springs, which are depressed by 4.0 mm when the set is put down. Find the natural frequency of vibration of the system.

9. A spring whose force constant is 15 lb/ft oscillates with a period of 0.60 s when a bag of onions is suspended from it. What is the weight of the onions?

Figure 11.19

Exercise 10.

10. Two springs with the same force constant k are connected to an object of mass m, as in Fig. 11.19. Find the period of oscillation in each case in the absence of friction.

11. A 50-g teacup suspended from a spring oscillates with a period of 1.5 s. When a 120-g mug is suspended from the same spring, what is its period of oscillation?

12. A spring has a 1.0-s period of oscillation when a 20-N weight is suspended from it. Find the elongation of the spring when a 50-N weight is suspended from it.

13. A 70-kg gymnast drops on a trampoline from a height of 60 cm. The trampoline sags 40 cm when the gymnast strikes it, and he then bounces up and down. If the motion is simple harmonic, find its period.

14. An object of unknown mass is suspended from a spring, and the system is set in oscillation. If the frequency of the oscillations is 2.0 Hz, by how much did the spring stretch when the object was suspended from it?

11.4   Position, Speed, and Acceleration

15. A harmonic oscillator has a period of 0.20 s and an amplitude of 10 cm. Find the speed of the moving object when it passes through the equilibrium position.

16. Atoms in a crystalline solid are in constant vibration at room temperature, where their motion has amplitudes in the neighborhood of 10^{-11} m. If the frequency of oscillation of one of the atoms in an iron bar is 2.5×10^{12} Hz, what are its maximum speed and acceleration?

17. A rack that undergoes SHM with an amplitude of 20 mm is used to test the ability of electronic equipment to survive accelerations. What frequency is needed to produce accelerations of $10g$?

18. The blade of a portable saber saw makes 3000 strokes per minute. Each stroke involves a total blade travel of 0.75 in. Find the maximum speed of the blade.

19. The prongs of a tuning fork vibrate in SHM at a frequency of 660 Hz and with an amplitude of 1.0 mm at their tips. Find the maximum speed and acceleration of the prong tips. Express the acceleration in terms of g.

20. A 3.0-kg object is suspended from a spring and oscillates at 5.0 Hz with an amplitude of 15 cm. (a) What is the total energy of the motion? (b) What is the object's maximum acceleration?

21. When a 2.00-lb wrench is suspended from a spring, the spring stretches by 3.00 in. The wrench is pulled down by a further 1.50 in and released. Find the frequency and total energy of the resulting oscillations.

22. A 16.0-lb object suspended form a spring oscillates at 3.00 Hz with a total energy of 22.0 ft · lb. What is the object's maximum acceleration?

23. A 1.0-lb piston moves up and down in SHM with an amplitude of 2.5 in. If the maximum force the piston can safely experience is 600 lb, what is the maximum frequency with which it can move?

24. An object is in SHM with an amplitude of 20 mm and a period of 0.20 s. (a) What is its speed when its displacement from its equilibrium position is 5.0 mm? (b) What is its displacement when its speed is 0.50 m/s?

25. An object whose mass is 1.0 kg hangs from a spring. When the object is pulled down 50 mm from its equilibrium position and released, it oscillates once per second. (a) What is the force constant of the spring? (b) What is the object's speed when it passes through its equilibrium position? (c) What is the maximum acceleration of the object?

26. An object whose mass is 0.0050 kg is in simple harmonic motion with a period of 0.040 s and an amplitude of 10.0 mm. (a) What is its maximum acceleration? (b) What is the maximum force on the object? (c) What is its acceleration when it is 5.0 mm from its equilibrium position? (d) What is the force on it at that point?

11.6 The Simple Pendulum

27. A chandelier is suspended from a high ceiling with a cable 6.0 m long. What is its period of oscillation?

28. A chandelier suspended from a high ceiling has a period of oscillation of 5.00 s. How many feet long is its cable?

29. Find the length of a pendulum whose period is 1.0 s on the surface of Mars, where the acceleration of gravity is 38% of its value on the earth's surface.

30. A pendulum whose length is 1.53 m oscillates 24.0 times per minute in a particular location. What is the acceleration of gravity there?

31. Find the period of a pendulum 1.00 ft long when it is suspended in (a) a stationary elevator; (b) an elevator falling at the constant speed of 20.00 ft/s; (c) an elevator falling at the constant acceleration of 7.00 ft/s²; (d) an elevator rising at the constant speed of 20.00 ft/s; (e) an elevator rising at the constant acceleration of 7.00 ft/s².

32. Find the period of a pendulum 50 cm long when it is suspended in (a) a stationary elevator; (b) an elevator falling at the constant speed of 5.0 m/s; (c) an elevator falling at the constant acceleration of 2.0 m/s²; (d) an elevator rising at the constant speed of 5.0 m/s; (e) an elevator rising at the constant acceleration of 2.0 m/s².

33. The bob of a simple pendulum of length L is pulled to one side through an angle θ with the vertical and then let go. What is the maximum speed of the bob?

11.7 The Physical Pendulum

34. A 200-g brass hoop 400 mm in diameter is suspended on a knife edge on which it rocks back and forth. The period of the oscillations is 1.27 s. Find the moment of inertia of the hoop about an axis through its circumference perpendicular to its plane.

35. An iron bar 80 cm long is suspended from one end. What is the period of its oscillations? What would be the length of a simple pendulum with the same period?

36. A broomstick suspended from one end oscillates with the same period as a simple pendulum 80 cm long. How long is the broomstick?

37. A leg may be approximated by a thin rod hinged at one end. (a) If its center of gravity is at its center, find the natural period of oscillation of a leg 1.0 m long on this basis. (b) With a stride of 0.8 m, what walking speed does this correspond to?

PROBLEMS

38. A steel wire 5.2 m long whose mass is 4.8 g hangs from an overhead beam. A 12-kg brass monkey is then attached to the lower end of the wire. How much additional energy is stored in the wire?

39. A spring 10 cm long whose force constant is 150 N/m is connected to a spring 20 cm long whose force constant is 100 N/m. The combination is then pulled out so that its ends are 50 cm apart. (a) Find the amount by which each spring is stretched. (b) A 400-g block on a frictionless surface is attached between the springs. Find its period of oscillation.

40. A hole is bored through the earth along a diameter. Inside the hole the acceleration of gravity varies as rg/R, where r is the distance from the center of the earth, R is the earth's radius of 6.4×10^6 m, and g is the acceleration of gravity at the earth's surface. A stone is dropped into the hole and executes simple harmonic motion about the center of the earth. Why? Find the period of this motion.

41. A simple pendulum has a period of 2.56 s at a certain place. When the length of the pendulum is increased by 800 mm, its period increases to 3.14 s. Find the original length of the pendulum and the value of g at that place.

42. A **torsion pendulum** consists of an object of moment of inertia I suspended by a wire or thin rod, as in Fig. 11.20.

Figure 11.20

Problem 42.

When the object is turned through an angle θ, a restoring torque of $\tau = -K\theta$ arises. If the object is then released, it will oscillate back and forth in SHM. Find the period of the oscillations in terms of I and K.

Figure 11.21

Problem 43.

43. Figure 11.21 shows a rotating wheel that drives a piston by means of a long connecting rod pivoted at both ends. The wheel's radius is 20 cm, and it is turning at 4.0 rev/s. (a) Where must the rod's attachment to the wheel be located at $t = 0$ in order that the equation $x = A \cos \omega t$ describe the piston's position? (b) What are the values of A and ω? (c) What are the maximum values of the piston's speed and acceleration?

Answers to Multiple Choice

1. a	8. b	15. c	22. d
2. b	9. d	16. a	23. b
3. b	10. d	17. a	24. a
4. a	11. b	18. b	25. a
5. c	12. c	19. d	26. c
6. c, d	13. b	20. c	
7. a	14. c	21. d	

12

WAVES

12.1 Pulses in a String

12.2 Principle of Superposition

12.3 Periodic Waves

12.4 Types of Waves

12.5 Standing Waves

12.6 Sound

12.7 Interference of Sound Waves

12.8 Doppler Effect

12.9 Sound Intensity

Throw a stone into a lake, and water waves spread out from the splash. Clap your hands, and sound waves carry the noise all around you. Switch on a lamp, and light waves flood the room. Water, sound, and light waves differ in important ways, but all share the basic properties of wave motion. A wave is a periodic disturbance—a back-and-forth change of water height in water waves, of air pressure in sound waves, of electric and magnetic field strength in light waves—that moves away from a source and carries energy with it. Information, too, can be carried by waves, which is the case for everything we see and hear.

A vibrating system and a wave going through a medium have much in common. In both, a certain characteristic motion recurs at regular intervals; in both, potential energy is continuously changed into kinetic energy and back; in both, the properties of matter provide restoring forces. On the other hand, the energy in a vibrating system is confined to the system, whereas waves bring energy from one place to another.

12.1 PULSES IN A STRING

An introduction to some important aspects of wave behavior.

If we give one end of a stretched string a quick shake, a kink or **pulse** travels down the string at some speed v (Fig. 12.1). If the string is uniform and completely flexible, the pulse keeps the same shape as it moves. It is worth examining the behavior of pulses in a string both because a wave can be regarded as a series of pulses and because it is easy to visualize what is going on.

The speed v of a pulse depends on the properties of the string—how heavy it is and how tightly it is stretched—and not on the shape of the pulse or on exactly how it is produced. Pulses move slowly down a slack, heavy rope, rapidly down a taut, light string. ("Heavy" and "light" here refer to the mass per unit length of a string, not its total weight.) When the mass per unit length of a string is large, the pulse speed is low because the inertia of each segment of the string is large and it therefore responds slowly to the forces acting on it. When the string is tightly stretched, the pulse speed is high because the tendency of the string to straighten out is greater.

These observations are reflected in the formula

$$v = \sqrt{\frac{T}{m/L}} \qquad\qquad \textit{Waves in a string} \quad (12.1)$$

for the speed of waves in a stretched string of mass m and length L that is under the tension T.

Figure 12.1

A pulse moves along a stretched string with a constant speed v.

◆ E X A M P L E   **12.1**

The stainless steel forestay of a racing sailboat is 20 m long, and its mass is 12 kg. To find its tension, the stay is struck by a hammer at the lower end and the return of the pulse is timed. If the time interval is 0.20 s, what is the tension in the stay?

SOLUTION   If L is the length of the stay, the pulse travels the distance $2L$ in the time t, so its speed is $v = 2L/t$. From Eq. (12.1), with $L = 20$ m, $t = 0.20$ s, and $m = 12$ kg, the tension in the stay is

$$T = \frac{mv^2}{L} = \frac{m}{L}\left(\frac{2L}{t}\right)^2 = \frac{4mL}{t^2} = \frac{(4)(12\text{ kg})(20\text{ m})}{(0.20\text{ s})^2} = 2.4 \times 10^4\text{ N} = 24\text{ kN}$$

which is 5400 lb.   ◆

Figure 12.2

The forward part of this traveling pulse is moving upward, and the rear part is moving downward.

The energy content of a moving pulse is partly kinetic and partly potential. As the pulse travels, its forward part is moving upward and its rear part is moving downward. Because the string has mass, there is a certain amount of KE associated with these up-and-down motions (Fig. 12.2). The PE is due to the tension in the string. Work had to be done in order to produce the pulse by pulling against the tension, and the deformed string accordingly possesses elastic PE.

Figure 12.3

A pulse reaching a fixed end of the string is inverted on reflection.

Reflection at a Boundary

When a pulse reaches the end of a string, it may be reflected back toward its starting point. Depending on how the end of the string is held in place, the reflected pulse may be inverted (upside down) or erect (right side up). Under just the right conditions, of course, the energy of the pulse may all be absorbed by the support, and the pulse will then disappear.

Suppose the end of the string is held firmly in place. When the pulse arrives there, the string exerts an upward force on the support (assuming the pulse is upward, as in Fig. 12.3). By the third law of motion, the support then exerts an equal and opposite reaction force on the string. This reaction force produces a pulse whose displacement is opposite to that of the original pulse but which otherwise has the same shape. The new inverted pulse then goes back along the string. Thus an upward pulse becomes a downward pulse upon reflection, and vice versa.

String exerts upward force on support when pulse arrives there.

Support exerts downward reaction force on string to produce inverted pulse moving in opposite direction.

If the end of the string is not held firmly in place, however, the reflected pulse is not inverted. Figure 12.4 shows the end of a string attached to a ring free to move up and down a frictionless rod. When the pulse arrives at this end, the string moves upward until its KE is completely converted into elastic PE, whereupon the end of the string moves downward again to send out a pulse reversed in direction but otherwise the same as the original one.

If the end of the string is held in just the right manner between complete rigidity and complete freedom, the pulse will not be reflected at all but will disappear when it reaches the end. A little experimentation with pulses in an actual string will

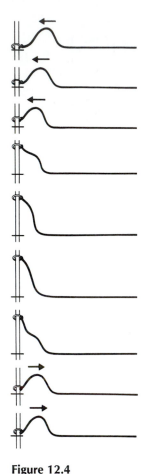

Figure 12.4

A pulse reaching a free end of the string is not inverted on reflection.

show that, whereas it is very easy to hold the far end of the string so that the reflected pulse is smaller than the original one, it is not easy at all to keep some reflection from taking place. It is equally hard to make perfect absorbers of sound, light, and water waves.

Reflection and Transmission at a Junction

So far we have been looking at pulses in a uniform string. Now let us connect two different strings together, one of them light (that is, with a low mass per unit length) and the other heavy (high mass per unit length). One end of the combination is fastened to something, and the other is given a shake to produce a pulse. Not surprisingly, the pulse passes from the first string to the second at the junction between them. But the transmission is not complete, since a reflected pulse also appears at the junction and proceeds in the opposite direction.

If the first string is the lighter one, the reflected pulse is inverted (Fig. 12.5). The greater inertia of the heavy string does not let it respond to the pulse as rapidly as the light string does, and an opposite reaction force occurs that causes the reflected pulse to be inverted, as though the junction were a rigid support. The energy of the original pulse is then split between the reflected and transmitted pulses. Since both strings have the same tension, the pulse travels more slowly in the heavy string. The length of the reflected pulse is the same as that of the original one, though its height is smaller since it has less energy. The length of the transmitted pulse, however, is shorter, because it moves more slowly than the original pulse while the time interval in which it comes into being is the same as that of the original pulse.

On the other hand, if the first string is the heavy one, the reflected pulse is erect (Fig. 12.6). The smaller inertia of the light string lets it follow the movements of the heavy one readily, and the situation is like that of a string whose end is able to move up and down freely. The light string does have some inertia, however, and so a reflected pulse again comes into being as well as a transmitted pulse. The pulse speed is higher in the light string, hence the pulse length is longer there than in the heavy one.

All types of waves, not just pulses in a stretched string, exhibit reflection and transmission at junctions between different media. For example, light waves are partially reflected and partially transmitted when they pass from air to glass. This is

Figure 12.5

When a pulse passes from a light string to a heavy string, reflection occurs with the reflected pulse being inverted. The transmitted pulse is right side up. Because both strings have the same tension, the pulse travels more slowly in the heavy string.

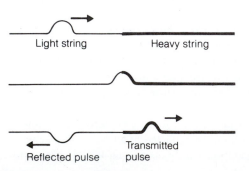

Light string   Heavy string

Reflected pulse

Transmitted pulse

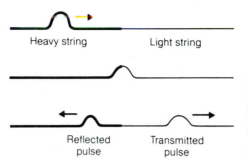

Figure 12.6

When a pulse passes from a heavy string to a light string, the reflected pulse stays right side up. The pulse speed is again less in the heavy string.

why we can see our images in a clear pane of glass (such as a shop window) even though the glass is transparent to light. The inversion of a pulse when it is reflected at a junction with a medium in which its speed is smaller also has a counterpart in the behavior of light waves, as we will find in Chapter 24.

12.2 PRINCIPLE OF SUPERPOSITION

What happens when two pulses meet.

Opposite ends of a stretched string are given upward shakes, and the pulses thus produced move along the string toward each other. What happens when they meet? The result is a larger pulse at the moment the pulses come together, and then the separate pulses reappear and continue unchanged in their original directions of motion. Each pulse proceeds as though the other did not exist (Fig. 12.7).

The **principle of superposition** is a statement of this behavior:

When two pulses travel past a point in a string at the same time, the displacement of the string at that point is the sum of the displacements each pulse would produce there by itself.

Figure 12.7

Two pulses moving in opposite directions along a stretched string. The pulses are unaffected by their crossing.

Figure 12.8

Complete cancellation occurs when two identical pulses with opposite displacements meet. At the instant of complete cancellation, the total energy of both pulses resides in the KE of the string segment where the cancellation occurs.

What if one of the pulses is inverted relative to the other? According to the superposition principle, if the pulses have the same sizes and shapes, their displacements ought to cancel out when they meet, only to reappear later on after they have passed the crossing point. Such behavior is indeed observed in practice. At the instant of complete cancellation, the total energy of both pulses resides in the KE of the string segment where the cancellation occurs (Fig. 12.8).

12.3 PERIODIC WAVES

Wave speed equals the product of frequency and wavelength.

In a periodic wave, one pulse follows another in regular succession. Sound waves, water waves, and light waves are almost always periodic, although in each case a different quantity varies as the wave passes.

In periodic waves a certain waveform—the shape of the individual waves—is repeated at regular intervals. Periodic waves of all kinds usually have **sinusoidal** waveforms. A stretched string down which such waves move has the same appearance as a graph of sin x (or cos x) versus x that is moved along the x-axis with the wave speed v (Fig. 12.9).

Figure 12.9

Most periodic waves have sinusoidal waveforms.

Figure 12.10

Each particle in the path of a sinusoidal wave executes SHM perpendicular to the wave direction.

Sinusoidal waves are common because the particles of matter in a medium that waves can travel through undergo simple harmonic motion when momentarily displaced from their equilibrium positions. The passage of a wave sets up coupled harmonic oscillations in the medium, with each particle behaving like a harmonic oscillator that has begun its cycle just a trifle later than the particle behind it (Fig. 12.10). The result in the case of waves in a stretched string is a waveform that is in essence a graph of how the position of a harmonic oscillator varies with time. As we know, this graph is a sine curve. (Light waves are also sinusoidal in character even though their existence does not involve the motion of material particles and they can travel through empty space.)

Three related quantities are useful in describing periodic waves:

1. The **wave speed** v, which is the distance through which each wave moves per second;
2. The **wavelength** λ (Greek letter *lambda*), which is the distance between adjacent crests or troughs;
3. The **frequency** f, which is the number of waves that pass a given point per second.

The unit of frequency is the **hertz** (Hz). As in the case of oscillations, 1 Hz = 1 cycle/s. Multiples of the hertz are used for high frequencies:

$$1 \text{ kilohertz} = 1 \text{ kHz} = 10^3 \text{ Hz}$$
$$1 \text{ megahertz} = 1 \text{ MHz} = 10^6 \text{ Hz}$$
$$1 \text{ gigahertz} = 1 \text{ GHz} = 10^9 \text{ Hz}$$

The wave speed, wavelength, and frequency of a train of waves are not independent of one another. In every second, f waves (by definition) go past a particular point, with each wave occupying a distance of λ (Fig. 12.11). Therefore a wave travels a total distance of $f\lambda$ per second, which is the wave speed v. Thus

$$v = f\lambda \qquad\qquad \textit{Wave speed} \quad (12.2)$$

Wave speed = (frequency)(wavelength)

which is a basic formula that applies to all periodic waves, sinusoidal or not.

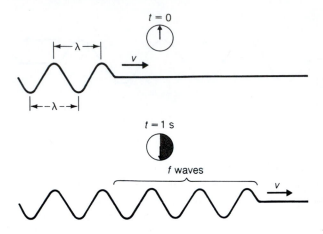

Figure 12.11

The speed of a wave is equal to the product of its frequency and wavelength.

v = wave speed
 = distance traveled per second
 = (number of waves passing a point per second) × (length of each wave)
 = $f\lambda$

◆   E X A M P L E   12.2 _____

A marine radar operating at a frequency of 9400 MHz emits groups of radio waves 0.0800 μs in duration. (The time needed for reflections of these groups to return indicates the distance of the target; see Section 22.2.) Radio waves, like light waves, are electromagnetic in nature and travel at 3.00×10^8 m/s. Find (a) the wavelength of these waves; (b) the length of each wave group, which is indicative of the precision with which the radar can measure distance; and (c) the number of waves in the group.

SOLUTION   (a) From Eq. (12.2) the wavelength is

$$\lambda = \frac{v}{f} = \frac{3.00 \times 10^8 \text{ m/s}}{9.40 \times 10^9 \text{ Hz}} = 3.19 \times 10^{-2}\,\text{m} = 3.19\,\text{cm}$$

(b) The length x of each wave group is simply

$$x = vt = (3.00 \times 10^8 \text{ m/s})(8.00 \times 10^{-8} \text{ s}) = 24.0 \text{ m}$$

(c) We can find the number n of waves in each group in either of these ways:

$$n = ft = (9.40 \times 10^9 \text{ Hz})(8.00 \times 10^{-8} \text{ s}) = 752 \text{ cycles} = 752 \text{ waves}$$

$$n = \frac{x}{\lambda} = \frac{24.0 \text{ m}}{3.19 \times 10^{-2}\,\text{m}} = 752 \text{ wavelengths} = 752 \text{ waves} \qquad ◆$$

_____

Sometimes it is more useful to consider the **period** T of a wave, which is the time required for one complete wave to pass a given point (Fig. 12.12). Because

Figure 12.12

The period of a wave is the time required for one complete wave to pass by a given point.

f waves pass by per second, the period of each wave is

$$T = \frac{1}{f}$$ *Wave period*   (12.3)

If five waves per second are passing by, for example, each wave has a period of $\frac{1}{5}$ s. In terms of period T, the formula for wave speed is

$$v = \frac{\lambda}{T}$$ (12.4)

◆   **E X A M P L E   12.3** _____

An anchored boat is observed to rise and fall once every 4.0 s as waves whose crests are 25 m apart pass by it. Find the frequency and speed of the waves.

SOLUTION   The frequency of the waves is

$$f = \frac{1}{T} = \frac{1}{4.0 \text{ s}} = 0.25 \text{ Hz}$$

and their speed is

$$v = \frac{\lambda}{T} = \frac{25 \text{ m}}{4.0 \text{ s}} = 6.25 \text{ m/s}$$ ◆

_____

The **amplitude** A of a wave refers to the maximum displacement from their normal positions of the particles that oscillate back and forth as the wave travels by (Fig. 12.13). The amplitude of a wave in a stretched string is the height of the crests above the original line of the string (or the depth of the troughs below the original line). The speed, frequency, and wavelength of a wave are independent of its amplitude, just as the period of a harmonic oscillator is independent of its amplitude.

Figure 12.13

The quantity A is the amplitude of the wave.

The correspondence between harmonic and wave motion leads to an interesting result. As we found in Section 11.4, a particle of mass m that undergoes simple harmonic motion of frequency f and amplitude A has a total energy of

$$E = 2\pi^2 m f^2 A^2$$

This dependence of energy on f^2 and on A^2 is also true for mechanical waves of all kinds. (A mechanical wave is one that involves moving matter, in contrast to, say, an electromagnetic wave.) Waves in a string are an example: Their energy per unit length is $2\pi^2 (m/L) f^2 A^2$, where m/L is the mass per unit length of the string.

Fourier's Theorem

We have been considering sinusoidal waves thus far. However, in a medium whose properties do not vary with wave frequency, everything that is true for a sinusoidal wave is also true for all other periodic waves. An interesting and important theorem by Jean Fourier (1768–1830) shows why this should be so.

What Fourier proved is that *any* periodic wave of frequency f, regardless of its waveform, can be thought of as a superposition of sinusoidal waves whose frequencies are f, $2f$, $3f$, and so on. The amplitudes of the various component waves depend on the precise character of the composite wave, and a mathematical procedure exists for finding these amplitudes. Figure 12.14 shows how just three waves of appropriate frequency and amplitude add up to give an approximation of a square wave. The more the waves that are included, the better the approximation becomes. Even an isolated pulse can be represented by a superposition of sinusoidal waves. Here, however, the frequencies must be very close to each other instead of being multiples of f, and a great many waves are needed.

Figure 12.14

Fourier synthesis of a square wave.

Figure 12.15

Longitudinal waves in a coil spring.

12.4 TYPES OF WAVES

Waves may be transverse, longitudinal, or a combination of both.

Waves in a stretched string are **transverse waves** because the individual segments of the string vibrate perpendicular to the direction in which the waves travel. **Longitudinal waves** occur when the individual particles of a medium vibrate back and forth parallel to the direction in which the waves travel (Fig. 12.15). Longitudinal waves are easy to produce in a long coil spring; each portion of the spring is alternately compressed and extended as the waves pass by. Longitudinal waves, then, are essentially density fluctuations. Sound waves are longitudinal.

Waves on the surface of a body of water (or other liquid) are a combination of longitudinal and transverse waves. If we were somehow to tag individual water molecules and follow them when a train of waves passes by, we would find that their paths are like those shown in Fig. 12.16. Each molecule moves in a circle with a period equal to the period of the wave, and does not undergo a permanent shift. Because successive molecules reach the tops of their orbits at slightly different times, the water surface takes the form of a series of crests and troughs. At the crest of a wave the molecules move in the direction the wave is traveling, whereas in a trough the molecules move in the opposite direction.

Water molecules move in circular orbits when wave passes by

Figure 12.16

Water molecules move in circular orbits about their original positions when a typical deep-water wave passes by. At the crest of a wave the molecules are moving in the direction the wave is traveling, whereas in the trough the molecules are moving in the opposite direction. There is no net motion of water involved in the motion of such a wave.

Water molecules move in circular orbits about their original positions when a typical deep-water wave passes by. At the crest of a wave the molecules are moving in the direction the wave is traveling, whereas in the trough the molecules are moving in the opposite direction. There is no net motion of water involved in the motion of such a wave.

The passage of a wave across the surface of a body of water, like the passage of a wave through any medium, involves the motion of a pattern. Energy is transported by virtue of the moving pattern, but there is no net transport of matter.

◆   E X A M P L E   **12.4** _____

The water waves in Example 12.3 have an amplitude of 60 cm. Find the speed of an individual molecule of water on the surface.

SOLUTION   The water molecules on the surface are moving in circles of radius 60 cm, so that as each wave passes by, the molecules travel a distance s equal to the circumference $2\pi r$ of the circle (Fig. 12.17). Hence

$$s = 2\pi r = 2\pi(0.60 \text{ m}) = 3.8 \text{ m}$$

Each wave takes $T = 4.0$ s to go past a given point, which means that the molecules must cover the 3.8-m circumference of their orbits in 4.0 s. The speed of each molecule is therefore

$$V = \frac{s}{T} = \frac{3.8 \text{ m}}{4.0 \text{ s}} = 0.95 \text{ m/s}$$

Figure 12.17

The speed of the *wave*, however, is 6.25 m/s, more than six times greater. Thus the motion of the pattern that constitutes a wave in a medium can be much more rapid than the motions of the individual particles of the medium, and energy can be transported by wave motion faster than might be possible through the net transport of matter. ◆

The speed of water waves in deep water depends on their wavelength: The longer the waves, the faster they move. Thus waves 15 m long travel at 5 m/s, but waves 150 m long travel at 15 m/s. In shallow water the orbits of the water molecules touch the bottom, which slows down the progress of the waves: The shallower the water, the more slowly they move, regardless of wavelength. The wave speed in water 1 m deep is only half the wave speed in water 4 m deep. (The slowing of the lower parts of the orbits is what causes wave crests to fall forward, or "break," where the bottom of a body of water rises near a shore.)

12.5 STANDING WAVES

In a given system only certain wavelengths are possible.

When we pluck a string whose ends are fixed, the string starts to vibrate in one or more loops (Fig. 12.18). These **standing waves** may be thought of as the result of waves that travel down the string in both directions, are reflected at the ends, proceed across to the opposite ends and are again reflected, and so on.

To appreciate how standing waves come into being, we draw upon our knowledge of how pulses in a string are reflected and of what happens when two pulses traveling in opposite directions meet. When a pulse in a string is reflected at a rigid support, the reflected pulse is inverted. A similar inversion occurs for periodic waves. Hence, although waveform and wavelength stay the same, the wave train is in effect shifted by $\frac{1}{2}\lambda$ so that a crest arriving at the end of the string is reflected as a trough, and vice versa.

Figure 12.18

Standing waves in a stretched string.

(a)

(b)

Figure 12.19

(a) Constructive interference. (b) Destructive interference.

Now let us look into how two waves in the same string interact. Applied to waves, the principle of superposition states:

When two or more waves of the same nature travel past a point at the same time, the displacement at that point is the sum of the instantaneous displacements of the individual waves.

The principle of superposition holds for all types of waves, including waves in a stretched string, sound waves, water waves, and light waves.

Thus every wave train proceeds independently of any others that may also be present. Should two waves with the same wavelength come together in such a way that crest meets crest and trough meets trough, the resulting composite wave will have an amplitude greater than that of either of the original waves. In this case the waves are said to **interfere constructively** with each other. Should the waves come together in such a way that crest meets trough and trough meets crest, the composite wave will have an amplitude less than that of the larger of the original waves. In this case the waves are said to **interfere destructively** with each other (Fig. 12.19).

Let us apply these ideas to a stretched string whose ends are fixed. When the string is plucked, waves move back and forth between its ends, inverting each time they are reflected. These waves interfere with each other in such a way that destructive interference always occurs at the ends. If the string has length L, destructive interference will occur at the ends when the waves have wavelengths of $\lambda = 2L$, L, $2L/3$, $L/2$, and so on (see Fig. 12.18). In the case of the shorter wavelengths, other points, called **nodes**, are present at intermediate positions where no motion of the string takes place.

Figure 12.20 shows how a standing wave pattern comes into being. The dotted curve represents a wave moving to the right, and the dashed curve a wave of the same wavelength and amplitude moving to the left. The sum of these waves is found by adding together their displacements at each point on the string, and is shown as a solid line. The addition is performed in Fig. 12.20 for four successive instants one-eighth of a period apart.

We note that the dotted and dashed curves do not always have displacements equal to 0 at the ends of the string. There is no contradiction here because it is their resultant, the solid curve, that corresponds to reality, and this curve behaves as it should.

Any type of wave can occur as a standing wave between suitable reflectors. The vibrating air columns in wind instruments and organ pipes—and in the throat, mouth, and nose of a person speaking—are standing sound waves, for instance. Standing light waves play an important role in the operation of lasers, as we will find later. And standing waves of a rather remarkable kind are the key to understanding the structure of the atom.

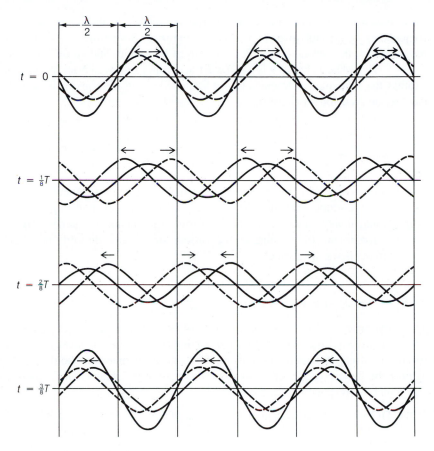

Figure 12.20

The origin of a standing wave. Waves (broken lines) that move to the left and to the right add up to give a stationary wave pattern (solid line).

Fundamental Frequency, Overtones, and Harmonics

The condition that nodes occur at each end of the string restricts the possible wavelengths of standing waves to

$$\lambda = \frac{2L}{n} \qquad n = 1, 2, 3, \ldots \qquad \text{Standing waves} \quad (12.5)$$

The lowest possible frequency of oscillation f_1 of a stretched string corresponds to the longest wavelength, $\lambda = 2L$. Thus

$$f_1 = \frac{v}{\lambda} = \frac{v}{2L} \tag{12.6}$$

Higher frequencies correspond to shorter wavelengths. From Eq. (12.5) we see that these higher frequencies can be represented in terms of f_1 by

$$f_n = nf_1 \quad \cdot \quad n = 2, 3, 4, \ldots \tag{12.7}$$

The frequency f_1 is called the **fundamental frequency** of the string, and the higher frequencies f_2, f_3, and so on, are called **overtones**. Sometimes f_1 is called the first **harmonic**, f_2 the second harmonic, f_3 the third harmonic, and so on.

The wave velocity v is $\sqrt{T/(m/L)}$ according to Eq. (12.1). We can therefore express the fundamental frequency of a stretched string in terms of its length, linear density, and tension by the formula

$$f_1 = \frac{1}{2L}\sqrt{\frac{T}{m/L}} \qquad \textit{Fundamental frequency of stretched string} \quad (12.8)$$

This formula is the basis for the design of stringed musical instruments such as pianos and violins. A short, light, taut string means a high fundamental frequency of vibration; a long, heavy, slack string means a low fundamental frequency. To "tune" a stringed instrument, the tensions in the various strings are adjusted until their fundamental frequencies are correct.

◆   **E X A M P L E   12.5** _____

The A string of a violin has a linear density of 0.60 g/m and an effective length of 330 mm. (a) Find the tension required for its fundamental frequency to be 440 Hz. (b) If the string is under this tension, how far from one end should it be pressed against the fingerboard in order to have it vibrate at a fundamental frequency of 495 Hz, which corresponds to the note B?

SOLUTION   (a) First we square both sides of Eq. (12.8) and then solve for T:

$$f_1^2 = \frac{T}{4L^2(m/L)}$$
$$T = 4L^2 f_1^2 (m/L) = 4(0.330\,\text{m})^2(440\,\text{Hz})^2(6.0 \times 10^{-4}\,\text{kg/m}) = 51\,\text{N}$$

This is about 11 lb.

(b) According to Eq. (12.8), $f_A/f_B = L_B/L_A$, so here

$$L_B = L_A\left(\frac{f_A}{f_B}\right) = (330\,\text{mm})\left(\frac{440\,\text{Hz}}{495\,\text{Hz}}\right) = 293\,\text{mm}$$

Hence the string should be pressed $(330 - 293)$ mm $= 37$ mm from one end.   ◆

The 88 "strings" of a piano are metal wires whose fundamental frequencies range from 27 Hz for the lowest bass note to 4186 Hz for the highest treble note. This span is so great that the density as well as the length and tension of the wires must be varied in order that the instrument be of reasonable size: The treble strings are thin steel wires, and the bass strings are wound with copper wire to increase their linear density m/L. The total tension of all the strings in a piano is about 20 tons and is borne by a heavy cast iron frame.

Resonance

The fundamental frequency and overtones of a stretched string are its natural frequencies of oscillation: If the string is plucked, one or more of these frequencies will be excited. Eventually internal friction in the string will cause the various vibrations to die out. However, if we apply a periodic force to the string whose frequency is the same as one of the string's natural frequencies, the standing waves continue as long as the periodic force supplies energy to the string. If the energy provided exceeds that lost to internal friction, the amplitude of the standing wave will increase until the string ruptures. (A column of soldiers can destroy a flimsy bridge by marching across it in step with one of its natural frequencies, although the bridge may be able to support the static load of the soldiers.)

The addition of energy to a system of some kind by a periodic force that varies with a frequency equal to one of the system's natural frequencies of vibration is called **resonance**. All rigid structures have natural frequencies of oscillation, even though their vibrations may be more complicated than those of a stretched string. These vibrations can be excited by a periodic force of the right frequency. The rattling of a car at certain speeds is a familiar example of resonance.

Earthquakes send out waves that cause the earth's surface to vibrate. The waves from a severe earthquake can be detected thousands of miles away. The Mexican earthquake of 1985 was especially destructive in Mexico City because the frequency of the waves happened to match the natural frequencies of many buildings there, notably slender ones of 8 to 15 stories. The acceleration of the ground was about $0.2g$, but resonance amplified the building vibrations to produce accelerations of over $1g$, well above what they could withstand.

Strong winds set up standing waves in the Tacoma Narrows Bridge in Washington State soon after its completion in 1940. The bridge collapsed as a result. Today bridges are stiffened to prevent such disaster.

Figure 12.21

Sound consists of longitudinal waves, representing condensations and rarefactions in the air in its path.

12.6 SOUND

Sound waves are pressure fluctuations in a solid, a liquid, or a gas.

Sound waves are longitudinal and consist of pressure fluctuations. The air (or other medium) in the path of a sound wave becomes alternately denser and rarer, and the resulting changes in pressure cause our eardrums to vibrate with the same frequency. This produces the physiological sensation of sound.

Most sounds are produced by vibrating objects. An example is the diaphragm of a loudspeaker (Fig. 12.21). When it moves outward, it pushes the air molecules in front of it closer together to form a region of high pressure that spreads out in front of the loudspeaker. The diaphragm then moves backward, expanding the volume available to nearby air molecules. Air molecules now flow toward the diaphragm, and a region of low pressure spreads out behind the high-pressure region. The continued vibrations of the diaphragm thus send out successive layers of condensation and rarefaction.

The speed of sound in air at sea level and at 20°C is 343 m/s, which is 1125 ft/s. This speed increases at about 0.6 m/s per °C because the random speeds of air molecules increase with temperature and so make the passage of pressure fluctuations more rapid.

The musical note A, the frequency of which is 440 Hz, represents a wavelength λ in air of

$$\lambda = \frac{v}{f} = \frac{343 \text{ m/s}}{440 \text{ Hz}} = 0.78 \text{ m}$$

This is about 31 in. A person singing this note produces pressure waves whose amplitude might be 1 Pa, about 0.001% of sea-level atmospheric pressure (Fig. 12.22).

A normal ear responds to sound waves with frequencies from about 20 Hz to about 20,000 Hz, which correspond to wavelengths in air from 17 m (54 ft) to 17 mm (0.65 in). The fundamental frequency of the speaking voice averages about 145 Hz in men and about 230 Hz in women. Even considering the overtones that are present, the frequencies contained in ordinary speech are, for the most part, below 1000 Hz.

Figure 12.22

The pressure variations that constitute the musical note A in air at an undisturbed pressure of 1 bar. The amplitude is approximately that which a singer would produce.

Table 12.1 Speed of sound

Medium	Speed, m/s	Speed, ft/s
Gases (0°C)		
Carbon dioxide	259	850
Air	331	1,086
Air, 20°C	343	1,125
Helium	965	3,165
Liquids (25°C)		
Ethyl alcohol	1,207	3,960
Water, pure	1,498	4,913
Water, sea	1,531	5,022
Solids		
Lead	1,200	4,000
Wood, mahogany	~4,300	~14,100
Iron and steel	~5,000	~16,400
Aluminum	5,100	16,730
Glass, pyrex	5,170	17,000
Granite	~6,000	~19,700

Sound waves are transmitted by solids, liquids, and gases. In general, the stiffer the material, the faster the waves travel. This is reasonable when we reflect that stiffness implies particles that are tightly coupled together and so are more immediately responsive to one another's motions. As in the case of waves in a stretched string, the less dense the material, the less the inertia and the higher the speed of the waves. The speed of sound in a fluid medium is given by $v = \sqrt{B/\rho}$ where B is the bulk modulus of the fluid and ρ its density. In a solid, $v = \sqrt{Y/\rho}$ where Y is the Young's modulus of the material. Table 12.1 lists the speed of sound in various materials.

The wavelength of the musical note A in seawater is

$$\lambda = \frac{v}{f} = \frac{1531 \text{ m/s}}{440 \text{ Hz}} = 3.5 \text{ m}$$

This is nearly five times the wavelength of the same note in air. When a sound wave produced in one medium enters another in which its speed is different, the frequency of the wave remains the same while the wavelength changes. This is in accord with the behavior of pulses that pass from one stretched string to another, as discussed earlier.

Ultrasound

Sound waves whose frequencies are about 20,000 Hz are called **ultrasonic** and can be detected by appropriate electromechanical devices. Ultrasonic waves are audible to many animals whose hearing organs are smaller than those of human beings

Figure 12.23

Echo sounding permits water depth to be rapidly and continuously determined from a boat. (a) A pulse of high-frequency sound waves is sent out by a suitable device on the boat's hull. (b) The time t needed for the pulse to return after being reflected by the sea floor is a measure of the water depth h.

and so are better able to resonate at higher frequencies. Figure 12.23 shows how pulses of ultrasound are used to determine the depth of the water under a boat or ship.

Ultrasound is widely used in medicine to produce images of internal parts of the body, including those of an unborn baby. Besides not being so harmful to living tissue as X-rays, ultrasound is much better able than X-rays to distinguish between liquids and soft tissues. The procedure is basically the same as that used in finding the depth of water under a boat: A pulse of ultrasound waves in a narrow beam is sent into a patient's body through the skin, and the reflections from the interfaces between different materials return at different times. By moving the beam a picture of the internal structure of the body can be built.

Shock Waves

As an object moves through a fluid—for instance, an airplane through the atmosphere—it disturbs the fluid, and the resulting pressure waves spread out in spherical shells at the speed of sound. If the airplane itself is moving at the speed of sound, the waves pile up in front of it, as in Fig. 12.24(b), to create a wall of high pressure—the ''sound barrier.'' For the airplane to go faster than sound, a great deal of force is needed to push it through this barrier, more than the force needed to give a comparable acceleration at speeds either less than (subsonic) or more than (supersonic) that of sound.

At supersonic speeds, an airplane outdistances the waves it produces, which gives rise to the pattern of crests shown in Fig. 12.24(c). Where successive crests overlap, constructive interference takes place, and the result is a conical shell of high pressure with the airplane at its apex. This shell, which moves with the speed of sound, is responsible for the sonic boom heard after a supersonic airplane has passed

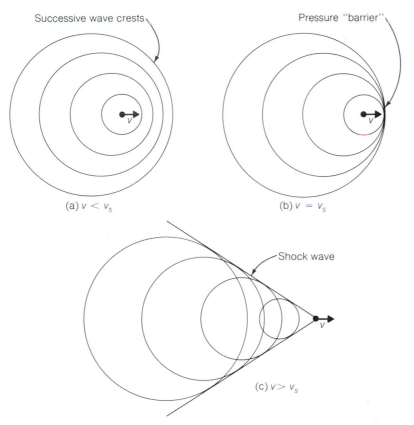

Successive wave crests

Pressure "barrier"

(a) $v < v_s$

(b) $v = v_s$

Shock wave

(c) $v > v_s$

Figure 12.24

Waves produced by an object moving at (a) subsonic, (b) sonic, and (c) supersonic speeds.

overhead. Because the rise in pressure is so sudden, such a **shock wave** can do physical damage to structures in its path even though the pressure change itself may not be very great. On a smaller scale, the crack of a whip is a sonic boom that occurs when the tip of the whip moves faster than sound.

Shock waves produced by a bullet moving faster than the speed of sound.

 E X A M P L E 12.6

The ratio between the speed of an airplane (or any other moving object) and the speed of sound is called the **Mach number** after the Austrian physicist Ernst Mach. Find the angle between the shock wave created by an airplane traveling at Mach 1.3 and the airplane's direction.

SOLUTION As shown in Fig. 12.25, in the time t the airplane travels the distance vt and the shock wave travels the distance $v_s t$. Hence the angle between the directions of the airplane and of the shock wave is specified by

$$\sin \theta = \frac{v_s t}{vt} = \frac{v_s}{v}$$

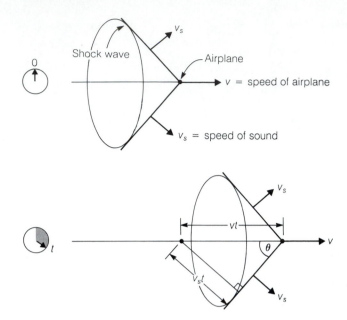

Figure 12.25

Geometry of a shock wave produced by a supersonic airplane.

Since a speed of Mach 1.3 means that $v = 1.3\,v_s$,

$$\sin \theta = \frac{v_s}{1.3\,v_s} = \frac{1}{1.3} = 0.769$$

$$\theta = \sin^{-1} 0.769 = 50°$$ ◆

12.7 INTERFERENCE OF SOUND WAVES

Musical sounds have complex waveforms.

Longitudinal as well as transverse waves exhibit interference. The standing waves that occur in an air column illustrate this, because they are produced by the interference of sound waves going back and forth between its ends. The analysis is the same as in the case of standing waves in a stretched string, except that nodes (places where there is no oscillation) occur only at closed ends of the tube. At an open end, air molecules can move in and out freely, so open ends are **antinodes** where the maximum amplitude is always found.

Figure 12.26 shows the possible modes of vibration in tubes open at one end and open at both ends. The curves in each drawing indicate how the displacement

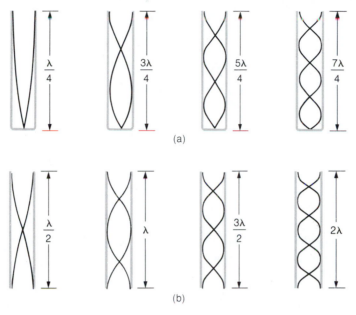

Figure 12.26

(a) Standing sound waves in a tube open at one end. The displacement maxima of the fundamental and the first three overtones are shown. A node occurs at the closed end and an antinode at the open end. (b) Standing sound waves in a tube open at both ends. An antinode occurs at both ends.

extremes vary along the length of each tube. In practice, the vibrations at the open end of a tube extend a little beyond it, so the effective length of the tube is slightly more than its actual length.

Wind instruments produce musical sounds by means of vibrating air columns. An organ has a separate pipe for each note. A woodwind, such as a flute or a clarinet, uses a single tube with holes whose opening or closing controls the effective length of the air column. Most brass instruments, such as the trumpet and the French horn, have valves connected to extra loops of tubing. Opening a valve adds to the length of the air column and thus produces a note of lower pitch. In a slide trombone the length of the air column is varied by sliding in or out a telescoping U-shaped tube.

Sound waves can be produced in a variety of ways. Shown here are vibrating strings (guitar and bass), vibrating air columns (saxophone and trumpet), and vibrating membranes (cymbal and drum).

Flute

Violin

Clarinet

Human voice

Figure 12.27

Waveforms of some musical sounds.

Vibrations of the players' lips as air is blown past them set up the standing waves in flutes, piccolos, and brass instruments. Vibrating reeds do this in clarinets, oboes, and saxophones.

The human voice is produced when the vocal cords in a person's throat set in vibration an air column that extends from the throat to the mouth and the nasal cavity above it. The shape of this column, which we adjust while speaking or singing by manipulating the mouth and tongue, determines the different vowel sounds by emphasizing some overtones and suppressing others. This shape also gives rise to the subtle differences that distinguish one person's voice from another's.

Natural sounds are never as regular as that shown in Fig. 12.22 but consist of mixtures of different frequencies that combine to give complex waveforms. The waveforms of sounds can be observed with the help of an oscilloscope, a device that displays electrical signals, such as those produced by a microphone in response to a sound, on the screen of a tube like a television picture tube. Figure 12.27 shows some waveforms of musical notes, which are mixtures of a fundamental frequency and some of its overtones. The precise mixture determines the quality, or timbre, of the note, which is different for each instrument and for different human voices. In the case of a flute, only the fundamental and first overtone are strong, giving a relatively simple waveform. In a clarinet, another woodwind, other overtones are also major contributors, leading to a "richer" sound.

Certain mixtures of frequencies are pleasing to the ear, and sounds that incorporate them are considered "musical." For instance, a tone combined with its first overtone, whose frequency is twice as great, appears harmonious to a listener. In music such an interval is called an octave because it includes eight notes. Another harmonious combination is a tone together with another whose frequency is 50% higher, so the frequencies are in the ratio 2:3. An example is C (262 Hz) plus G (392 Hz). Such an interval includes five notes (here C, D, E, F, G) and so is called a fifth. Somewhat less agreeable are two notes whose frequencies are in the ratio 4:5, for instance C and E (330 Hz), whose interval spans three notes and is called a third. The larger the numbers needed to express the ratio of frequencies, the less attractive the resulting sound is. Thus C and D are in the ratio 8:9 and seem discordant, and E and F, whose ratio is 15:16, are more discordant still. Ordinary sounds are mixtures of frequencies that have no special relationships with one another, and if the mixture seems particularly harsh, it qualifies as noise.

Beats

Let us look into what happens when two sounds of different frequency are combined. If two tuning forks (or other sources of single-frequency sound waves) whose frequencies are slightly different are struck at the same time, the sound that we hear fluctuates in intensity. At one instant we hear a loud tone, then virtual silence, then the loud tone again, then virtual silence, and so on. Why this happens is shown schematically in Fig. 12.28. The loud periods occur when the waves from the two forks interfere constructively, thus reinforcing one another, and the quiet periods occur when the waves interfere destructively, thus partially or wholly canceling out one another. These regular loudness pulsations are called **beats**.

A piano tuner can tell whether a piano string is in tune by listening for beats between the sound of that string and the sound of a tuning fork. If beats occur, the tension of the string is adjusted until they disappear. If you have access to a piano, you can exhibit beats by attaching a paper clip to one of the pair of strings used for each of the low notes. This changes the frequency of that string slightly. Striking the key for that note will set both strings vibrating, and you will hear beats.

Let us consider two waves with the frequencies f_1 and f_2, where $f_2 > f_1$. When the maxima of both waves occur together, there will be a loudness maximum, which is a beat. After a time T, their maxima will again coincide, and another beat will occur. In this time T, one of the waves has undergone one cycle more than the other, so that

$$f_2 T - f_1 T = 1 \quad \text{and} \quad f_2 - f_1 = \frac{1}{T}$$

The frequency of the beats is therefore

$$f_{\text{beat}} = \frac{1}{T} = f_2 - f_1 \qquad\qquad \textit{Beat frequency} \quad (12.9)$$

The sound we hear when beats are produced has a frequency that is the average of the two original frequencies, and the number of beats per second equals the difference between the two original frequencies. Thus the simultaneous vibrations of a 440-Hz tuning fork and an out-of-tune A string that vibrates at 444 Hz will produce a 442-Hz tone whose amplitude rises and falls four times per second.

Beats are hard to distinguish as such when the two frequencies differ by more than perhaps 10 Hz. When the frequencies are far enough apart, however, they give rise to an audible "difference tone" responsible for the lack of harmony of certain combinations of musical notes that was mentioned above. For instance, the notes E and F have the respective frequencies 330 Hz and 349 Hz, and when they are emitted together, the result is an additional 19-Hz difference tone that makes the combination dissonant.

(a)

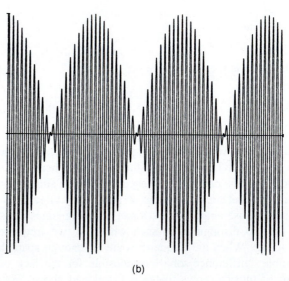

(b)

Figure 12.28

(a) The origin of beats in sound waves. (b) Beats produced by waves whose frequencies differ by 5%.

12.8 DOPPLER EFFECT

Relative motion between an observer and a wave source changes the perceived frequency.

If we stand beside a road when a police car goes by with its siren blowing, we cannot help but notice that the pitch of the siren drops suddenly as the car passes by. What happens is that the pitch of the siren as the car approaches is *higher* than the pitch when the car is stationary, and the pitch when the car recedes is *lower* than its normal one. We also find a change of pitch if the siren is at rest and we go past it in a rapidly moving car: The pitch is higher than usual as we approach the siren, and lower as we recede. The change in frequency of a sound brought about by relative motion between source and listener is called the **Doppler effect.**

The origin of the Doppler effect is straightforward. As a moving source emits sound waves, it tends to overtake those traveling in the same direction (Fig. 12.29a). Hence the distance between successive waves is smaller than usual. This distance is the wavelength of the sound, so the corresponding frequency is higher. At the same time the source is moving away from those of its waves that travel in the opposite direction. This increases the distance between successive waves behind it and thereby reduces their frequency.

Figure 12.29

The Doppler effect occurs when there is relative motion between a source of sound and a listener. The sign convention for Eq. (12.10) is shown.

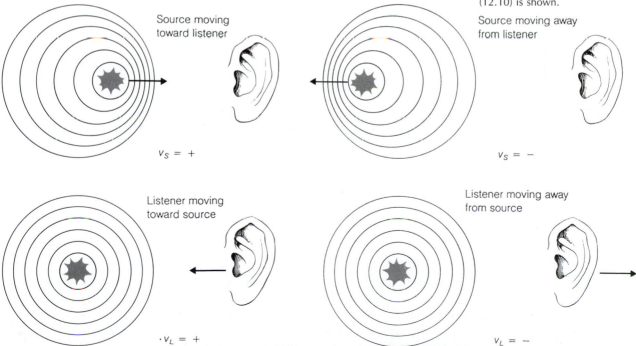

Source moving toward listener

$v_S = +$

Source moving away from listener

$v_S = -$

Listener moving toward source

$v_L = +$

Listener moving away from source

$v_L = -$

If the source is stationary, a listener moving toward it intercepts more sound waves per unit time than if she were at rest, and accordingly she hears a higher frequency (Fig. 12.29b). When the observer moves away from the source, fewer of the waves catch up with her per unit time, and she hears a lower frequency.

The relationship between the frequency f_L the listener hears and the frequency f_S produced by the source is

$$f_L = f_S \left(\frac{v + v_L}{v - v_S} \right) \qquad\qquad \textit{Doppler effect in sound} \quad (12.10)$$

Here v is the speed of sound in air, v_L is the speed of the listener (reckoned as $+$ for motion toward the source, as $-$ for motion away from the source), and v_S is the speed of the source (reckoned as $+$ for motion toward the listener, as $-$ for motion away from the listener). If the listener is stationary, $v_L = 0$, whereas if the source is stationary, $v_S = 0$.

◆   **E X A M P L E   12.7** _____

The frequency of a train's whistle is 1000 Hz. (a) The train is approaching a stationary man at 40 m/s. What frequency does the man hear? (b) The train is stationary, and the man is driving toward it in a car whose speed is 40 m/s. What frequency does the man hear now? (Unless otherwise specified, a temperature of 20°C is to be assumed.)

SOLUTION   (a) Here $f_s = 1000$ Hz, $v = 343$ m/s, $v_S = 40$ m/s, and $v_L = 0$. Hence the apparent frequency of the whistle is

$$f_L = f_S \left(\frac{v + v_L}{v - v_S} \right) = (1000 \text{ Hz}) \frac{343 \text{ m/s}}{(343 - 40) \text{ m/s}} = 1132 \text{ Hz}$$

(b) Again $f_S = 1000$ Hz and $v = 343$ m/s, but now $v_S = 0$ and $v_L = 40$ m/s. The apparent frequency of the whistle is therefore

$$f_L = f_S \left(\frac{v + v_L}{v - v_S} \right) = (1000 \text{ Hz}) \frac{(343 + 40) \text{ m/s}}{343 \text{ m/s}} = 1117 \text{ Hz}$$

We note that the answers to (a) and (b) are different.   ◆

Frequency measurements can be made so accurately that relatively small speeds can be detected by means of the Doppler effect. For example, the speed of blood in an artery does not exceed about 0.4 m/s even in the aorta. When an ultrasound beam is directed at an artery, the waves reflected from the moving blood cells exhibit a Doppler shift because the cells then act as moving wave sources. From this shift the speed of the blood can be found (see Problem 49).

Doppler Effect in Light

The Doppler effect in light differs from that in sound because light, which does not depend on a material medium for its transmission, has the same relative speed c to all observers regardless of their state of motion. The Doppler effect in light is represented by the formula

$$f = f_S \sqrt{\frac{1 + v/c}{1 - v/c}} \qquad \text{Doppler effect in light} \quad (12.11)$$

where f is the observed frequency, f_S is the frequency of the source, and v is the relative speed between source and observer. If source and observer are approaching each other, v is reckoned as $+$, and if they are receding from each other, v is reckoned as $-$. In a vacuum the speed of light is $c = 3.00 \times 10^8$ m/s; its value in air is very close to this. Unlike the case of sound, the Doppler effect in light does not distinguish between motion of a source and motion of an observer.

Visible light consists of electromagnetic waves in a certain frequency interval to which the eye responds. Other electromagnetic waves, such as those used in radar and in radio communication, also exhibit the Doppler effect in accord with Eq. (12.11). Doppler shifts in radar waves are widely used by police to determine vehicle speeds, and such shifts in radio waves emitted by a set of earth satellites form the basis of the highly accurate Transit system of marine navigation.

◆ **E X A M P L E   12.8** _____

A driver goes through a red light and, when he is arrested, claims the color he actually saw was green ($\lambda = 5.40 \times 10^{-7}$ m) and not red ($\lambda_S = 6.20 \times 10^{-7}$ m) because of the Doppler effect. The judge accepts this explanation and instead fines him for speeding at the rate of \$1 for each km/h he exceeded the speed limit of 80 km/h. What was the fine?

SOLUTION   To solve Eq. (12.11) for v, we first square both sides and then use routine algebra (see Appendix A) to obtain

$$v = c \left(\frac{f^2 - f_S^2}{f^2 + f_S^2} \right)$$

Here f, the observed frequency, is $c/\lambda = 5.56 \times 10^{14}$ Hz, and f_S, the source frequency, is $c/\lambda_S = 4.84 \times 10^{14}$ Hz. Hence

$$v = (3.00 \times 10^8 \text{ m/s}) \left[\frac{(5.56)^2 - (4.84)^2}{(5.56)^2 + (4.84)^2} \right]$$

$$= 4.14 \times 10^7 \text{ m/s} = 1.49 \times 10^8 \text{ km/h}$$

The fine is therefore \$$(1.49 \times 10^8 - 80) = $148,999,920$!   ◆

12.9 SOUND INTENSITY

The decibel is the unit of sound intensity level.

The rate at which a wave of any kind carries energy per unit cross-sectional area is called its **intensity** I. The unit of intensity is the watt/m^2. One joule of energy per second flows through a 1-m^2 surface perpendicular to the path of a wave whose intensity is 1 W/m^2 (Fig. 12.30). The minimum intensity a 1000-Hz sound wave must have in order to be audible is about 10^{-12} W/m^2. This corresponds to pressure variations of less than a billionth of sea-level atmospheric pressure—the human ear is a very sensitive instrument. At the other extreme, sound waves with intensities over about 1 W/m^2 damage the ear.

The human ear does not respond linearly to sound intensity. Doubling the intensity of a particular sound produces the sensation of a somewhat louder sound, but one that seems far less than twice as loud. For this reason the scale customarily used to measure the intensity level of a sound is logarithmic, which is a reasonable approximation of the actual response of the human ear. The unit is the **decibel** (dB), named after Alexander Graham Bell (1847–1922), who invented the telephone. A sound that is barely audible ($I_0 = 10^{-12}$ W/m^2) is given the value 0 dB. A 10-dB sound is, by definition, 10 times more intense than a 0-dB sound; a 20-dB sound is 10^2 (or 100) times more intense; a 30-dB sound is 10^3 (or 1000) times more intense; and so on. Formally, the **sound intensity level** β, in decibels, of a sound wave whose intensity in W/m^2 is I is given by

$$\beta = 10 \log \frac{I}{I_0} \qquad \text{\textit{Sound intensity level}} \quad (12.12)$$

(If $a = 10^b$, then b is the logarithm of a. Thus 3 is the logarithm of 1000 to the base 10 because $1000 = 10^3$. Logarithms are discussed in Appendix B.5.)

Ordinary conversation is usually about 60 dB, a sound intensity a million times greater than the faintest sound that can be heard. City traffic noise can be 80

Figure 12.30

The intensity of a wave is a measure of the rate at which it transports energy.

Wave intensity = $\dfrac{\text{Energy/s}}{\text{area}}$

dB, a rock band using amplifiers may reach 125 dB, and the noise of a jet airplane is about 140 dB at a distance of 30 m (Fig. 12.31). An extended exposure to sound intensity levels of over 90 dB usually leads to permanent hearing damage.

◆ **E X A M P L E 12.9**

Five trumpets are being played, each at an average sound intensity level of 70 dB. What is the resulting sound intensity level?

SOLUTION The intensity of one trumpet is I_1, for a sound intensity level of

$$\beta_1 = 10 \log \frac{I_1}{I_0} = 70 \text{ dB}$$

The intensity of five trumpets is $5I_1$. Because $\log xy = \log x + \log y$ (Appendix B.5) and $\log 5 = 0.70$, the sound intensity level of the trumpets is

$$\beta_5 = 10 \log \frac{5I_1}{I_0} = 10\left(\log 5 + \log \frac{I_1}{I_0}\right) = 10 \log 5 + 10 \log \frac{I_1}{I_0}$$
$$= 10(0.70) \text{ dB} + \beta_1 = 7 \text{ dB} + 70 \text{ dB} = 77 \text{ dB}$$

Hence increasing the intensity fivefold leads to only a 10% increase in the sound intensity level here. This is the reason why a solo instrument can be heard in a concerto even though a full orchestra is playing at the same time, and why one can carry on a conversation at a party even though many others are talking at the same time. ◆

Because sound waves spread out as they move away from their source, their intensity decreases with distance. Let us consider the sound waves from a source whose power output is P; that is, P joules of energy flow from the source per second. At the distance r from the source, the total power P is distributed over the $4\pi r^2$ area of a sphere of radius r. Hence the intensity I of the sound at this distance is

$$I = \frac{P}{4\pi r^2} \qquad \textit{Intensity and distance} \quad (12.13)$$

The sound intensity is inversely proportional to the square of the distance from the source. If we go from 5 m away from a source to 10 m away, the intensity drops to $\frac{1}{4}$ its former value.

An inverse-square law holds for the intensity of all waves that spread out freely in three dimensions—the intensity of the light from a lamp also varies as $1/r^2$, for example. The purpose of the concave mirror in a searchlight is to avoid the $1/r^2$ decrease in intensity by concentrating the light waves from a lamp into as nearly parallel a beam as possible. Cupping one's hands around one's mouth similarly helps one's voice to carry farther.

Figure 12.31

Decibel scale.

◆ **EXAMPLE 12.10**

The minimum sound intensity level needed for reasonable audibility is 20 dB. If a certain person speaking normally produces an intensity level of 40 dB at a distance of 1.0 m, what is the maximum distance at which the sound can be heard?

SOLUTION A difference of 20 dB is equivalent to an intensity ratio of $10^2 = 100$. Because $I_2/I_1 = r_1^2/r_2^2$,

$$r_2 = r_1 \sqrt{I_1/I_2} = (1.0 \text{ m})\sqrt{100} = 10 \text{ m} \qquad ◆$$

The ear is not equally sensitive to sounds of different frequencies. Maximum response occurs for sounds between 3000 and 4000 Hz, when the threshold for hearing is somewhat less than 0 dB. A 100-Hz sound, however, must have an intensity level of at least 40 dB to be heard. Sounds whose frequencies are below about 20 Hz **(infrasound)** and above about 20,000 Hz **(ultrasound)** are inaudible to almost everybody regardless of intensity. The intensity level at which a sound produces a feeling of discomfort in the ear is relatively constant at approximately 120 dB for all frequencies. Figure 12.32 shows how the thresholds of hearing and of discomfort vary with frequency. Hearing deteriorates with age, most notably at the high-frequency end of the spectrum. A typical person 60 years old has a threshold of hearing about 10 dB higher than the lowest curve of Fig. 12.32 for 2000-Hz tones, nearly 30 dB higher for 8000-Hz tones, and nearly 70 dB higher for 12,000-Hz tones.

Figure 12.32

The response of the ear to sound varies with frequency. Only 1% of the U.S. population can hear sounds with intensity levels that fall below the lower curve; 50% can hear sound with intensity levels that fall below the curve above it. Hearing acuity decreases with age.

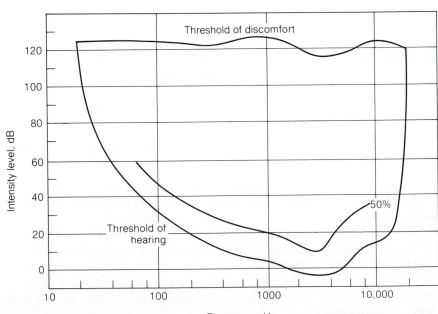

Power Gain

The decibel scale is widely used in electronics as well as in acoustics. If the power input to an amplifier or other signal-processing device is P_{in} and the power output is P_{out}, the **power gain** G in decibels is defined as

$$G = 10 \log \frac{P_{out}}{P_{in}} \qquad \text{Power gain} \quad (12.14)$$

Thus the power gain of an amplifier whose power input is 0.1 W and whose power output is 50 W is

$$G = 10 \log \frac{P_{out}}{P_{in}} = 10 \log \frac{50 \text{ W}}{0.1 \text{ W}} = 10 \log 500 = (10)(2.7) = 27 \text{ dB}$$

As mentioned earlier, the logarithm of a product is equal to the sum of the logarithms of the factors:

$$\log xyz = \log x + \log y + \log z$$

Because power gains are logarithmic quantities, the overall gain in decibels of a system of several devices is just the sum of their separate gains expressed in decibels:

$$G \text{ (overall)} = G_1 + G_2 + G_3 + \cdots \qquad (12.15)$$

For example, an audio system might consist of a 35-dB preamplifier, a -10-dB attenuator (a negative dB rating means a negative power gain, so the output is less than the input), and a 70-dB amplifier. The overall gain of the system is

$$G \text{ (overall)} = +35 \text{ dB} - 10 \text{ dB} + 70 \text{ dB} = +95 \text{ dB}$$

Important Terms

Wave motion involves the propagation of a change in a medium. Waves transport energy from one place to another.

The **frequency** of a series of periodic waves is the number of waves that pass a particular point per unit time, while their **wavelength** is the distance between adjacent crests or troughs. The **period** is the time required for one complete wave to pass a particular point. The **amplitude** of a wave is the maximum displacement of a particle of the medium on either side of its normal position when the wave passes.

Longitudinal waves occur when the individual particles of a medium vibrate back and forth in the direction in which the waves travel. **Transverse waves** occur when the individual particles of a medium vibrate from side to side perpendicular to the direction in which the waves travel. The vibrations of a stretched string are transverse waves.

The **principle of superposition** states that when two or more waves of the same nature travel past a given point at the same time, the amplitude at the point is the sum of the amplitudes of the individual waves. The interaction of different wave trains is called **interference. Constructive interference** occurs when the resulting composite wave has an amplitude greater than that of either of the original waves, and **destructive interference** occurs when the resulting composite wave has an amplitude less than that of either of the original waves.

Resonance occurs when periodic impulses are applied to a system at a frequency equal to one of its natural frequencies of oscillation.

Sound is a longitudinal wave phenomenon that results in periodic pressure variations. Sound intensity level is measured in **decibels** (dB).

A **shock wave** is a shell of high pressure produced by the motion of an object whose speed exceeds that of sound.

The **Doppler effect** refers to the change in frequency of a wave when there is relative motion between its source and an observer.

Important Formulas

Waves in a string: $\quad v = \sqrt{\dfrac{T}{m/L}}$

Wave motion: $\quad v = f\lambda = \dfrac{\lambda}{T}$

Doppler effect in sound: $\quad f_L = f_S\left(\dfrac{v + v_L}{v - v_S}\right)$

Doppler effect in light: $\quad f = f_S\sqrt{\dfrac{1 + v/c}{1 - v/c}}$

Sound intensity level: $\quad \beta(\text{dB}) = 10\log\dfrac{I}{I_0}$

MULTIPLE CHOICE

1. The speed of waves in a stretched string depends on
   a. the tension in the string
   b. the amplitude of the waves
   c. the wavelength of the waves
   d. the acceleration of gravity

2. The higher the frequency of a wave,
   a. the lower its speed
   b. the shorter its wavelength
   c. the greater its amplitude
   d. the longer its period

3. Of the following properties of a wave, the one that is independent of the others is its
   a. amplitude   b. speed
   c. wavelength   d. frequency

4. In a transverse wave the individual particles of the medium
   a. move in circles
   b. move in ellipses
   c. move parallel to the direction of travel
   d. move perpendicular to the direction of travel

5. An example of a purely longitudinal wave is
   a. a sound wave
   b. an electromagnetic wave
   c. a water wave
   d. a wave in a stretched string

6. An example of a wave that has both longitudinal and transverse characteristics is
   a. a sound wave

   b. an electromagnetic wave
   c. a water wave
   d. a wave in a stretched string

7. Two waves meet at a time when one has the instantaneous amplitude A and the other has the instantaneous amplitude B. Their combined amplitude at this time is
   a. $A + B$
   b. $A - B$
   c. between $A + B$ and $A - B$
   d. indeterminate

8. A wave of amplitude A interferes with another wave of the same kind whose frequency is different but whose amplitude is also A. The resulting wave
   a. has an amplitude of $2A$
   b. varies in amplitude between 0 and A
   c. varies in amplitude between A and $2A$
   d. varies in amplitude between 0 and $2A$

9. A string 180 cm long has a fundamental frequency of vibration of 300 Hz. The length of a similar string under the same tension whose fundamental frequency is 200 Hz is
   a. 120 cm   b. 147 cm
   c. 220 cm   d. 270 cm

10. Of the following procedures, which one or ones will lower the pitch of a guitar string?
   a. Lengthen the string.
   b. Use a lighter string.
   c. Increase the tension in the string.
   d. Decrease the tension in the string.

11. Sound travels fastest in
   a. air
   b. water
   c. iron
   d. a vacuum

12. Sound waves do not travel through
   a. solids
   b. liquids
   c. gases
   d. a vacuum

13. The amplitude of a sound wave determines its
   a. pitch
   b. loudness
   c. overtones
   d. resonance

14. What we recognize as the quality, or timbre, of a musical sound is due to
   a. its fundamental frequency
   b. its overtone structure
   c. its amplitude
   d. the presence of beats

15. A pure musical tone causes a thin wooden panel to vibrate. This is an example of
   a. an overtone
   b. harmonics
   c. resonance
   d. interference

16. When a sound wave goes from air into water, the quantity that remains unchanged is its
   a. speed
   b. amplitude
   c. frequency
   d. wavelength

17. Beats are the result of
   a. diffraction
   b. constructive interference
   c. destructive interference
   d. both constructive and destructive interference

18. A sonic boom is heard after an airplane has passed overhead. This means that the airplane
   a. is accelerating
   b. is climbing
   c. was just then passing through the sound barrier
   d. is traveling faster than sound

19. Relative to the radio signals sent out by a spacecraft headed away from the earth, the signals that are received on the earth
   a. have a lower speed
   b. have a lower frequency
   c. have a shorter wavelength
   d. have all of the above characteristics

20. Sound waves whose frequency is 300 Hz have a wavelength relative to sound waves in the same medium whose frequency is 600 Hz that is
   a. half as great
   b. the same
   c. twice as great
   d. four times as great

21. Radio amateurs are permitted to communicate on the ''10-meter band.'' What frequency of radio waves corresponds to a wavelength of 10 m? The speed of radio waves is 3.0×10^8 m/s.
   a. 3.3×10^{-8} Hz
   b. 3.0×10^7 Hz
   c. 3.3×10^7 Hz
   d. 3.0×10^9 Hz

22. A radio station broadcasts at a frequency of 660 kHz. The wavelength of these waves is
   a. 2.2 mm
   b. 455 m
   c. 4.55 km
   d. 1.98×10^{14} m

23. Waves in a lake are 5.00 m in length and pass an anchored boat 1.25 s apart. The speed of the waves is
   a. 0.25 m/s
   b. 4.00 m/s
   c. 6.25 m/s
   d. impossible to find from the information given

24. A boat at anchor is rocked by waves whose crests are 100 ft apart and whose speed is 25 ft/s. These waves reach the boat once every
   a. 0.25 s
   b. 2.0 s
   c. 4.0 s
   d. 2500 s

25. Two tuning forks of frequencies 310 and 316 Hz vibrate simultaneously. The number of times the resulting sound pulsates per second is
   a. 0
   b. 6
   c. 313
   d. 626

26. A steel wire has a fundamental frequency of vibration of 400 Hz. The fundamental frequency of another steel wire of the same length and under the same tension but whose diameter is twice as great is
   a. 100 Hz
   b. 200 Hz
   c. 400 Hz
   d. 800 Hz

27. In order to double the fundamental frequency of a wire under the tension T, the tension must be changed to
   a. $0.5T$
   b. $\sqrt{2T}$
   c. $2T$
   d. $4T$

28. How many times more intense is a 90-dB sound than a 40-dB sound?
   a. 5
   b. 50
   c. 500
   d. 10^5

29. Ten oboes produce a sound intensity level of 50 dB in a concert hall. The number of oboes needed to produce a level of 60 dB there is
   a. 15
   b. 20
   c. 100
   d. 200

QUESTIONS

1. A pulse sent down a long string eventually dies away and disappears. What happens to its energy?

2. When a wave passes a point in a stretched string, the energy of the string there changes back and forth between KE and elastic PE. What are the corresponding forms of energy for a water wave?

3. The amplitude of a wave is doubled. If nothing else is changed, how is the flow of energy affected?

4. A wave of frequency f and wavelength λ passes from a medium in which its speed is v to another medium in which its speed is $2v$. What are the frequency and wavelength of the wave in the second medium?

5. In general, in what state of matter does sound travel fastest? Why?

6. As a clarinet is played, the temperature of the air column inside it rises. What effect does this have on the frequencies of the notes it produces?

7. Verify that v has the dimensions of length/time in the formulas $v = \sqrt{T/(m/L)}$, $v = \sqrt{B/\rho}$, and $v = \sqrt{Y/\rho}$, which give the speed of waves, respectively, in a stretched string, in a fluid, and in a solid.

8. What is the relationship, if any, between the speed of sound in a given medium and the speed of a shock wave in that medium?

9. What property of a sound wave governs its pitch? Its loudness? Its quality?

10. How can constructive and destructive interference be reconciled with the principle of energy conservation?

11. A person has two tuning forks, one marked "256 Hz" and the other of unknown frequency. She strikes them simultaneously and hears 10 beats per second. "Aha," she says, "the other tuning fork has a frequency of 266 Hz." What is wrong with her conclusion?

12. A "double star" consists of two stars that revolve around their mutual center of mass. Telescopes cannot resolve a double star into its members, but an astronomer can recognize a double star from the characteristic frequencies in the light that reaches us from it. What property of these frequencies do you think enables the astronomer to identify their source as a double star?

13. The amplitude and wavelength of the wiggles in the grooves of a phonograph record correspond to the same quantities in the sound waves they represent. Inspection of a certain record reveals grooves whose wiggles have the same amplitude, but one has a wavelength three times shorter than the other. (a) If the audio system is linear, what will be the difference in the sound intensity the two sets of wiggles produce? (b) Will the difference in apparent loudness correspond to the difference in intensity?

EXERCISES

12.1   Pulses in a String

12.3   Periodic Waves

12.4   Types of Waves

1. A heavy rope hanging from the ceiling is given a transverse shake at its lower end. Show that the speed with which the resulting pulse travels up the rope increases as the square root of the height.

2. A certain groove in a phonograph record moves past the needle at a speed of 40 cm/s. The sound produced has a frequency of 3000 Hz. What is the wavelength of the wiggles in the groove?

3. A certain groove in a phonograph record moves past the needle at a speed of 30 cm/s. If the wiggles in the groove are 0.10 mm apart, what is the frequency of the sound that is produced?

4. Radio waves of very long wavelength can penetrate farther into seawater than those of shorter wavelength. The U.S. Navy communicates with submerged submarines by using radio waves whose frequency is only 76 Hz. What is the wavelength in air of those waves, in kilometers?

5. Water waves are observed approaching a lighthouse at a speed of 6.0 m/s. There is a distance of 7.0 m between adjacent crests. (a) What is the frequency of the waves? (b) What is their period?

6. In a ripple tank 90 cm across, a 6.0-Hz oscillator produces waves whose wavelength is 50 mm. Find the time the waves need to cross the tank.

7. An anchored boat rises by 1.2 m from its normal level and sinks by the same amount every 5.7 s as waves whose crests are 50 m apart pass it. Find the speed of the waves and the speed of the water molecules on the surface.

8. One end of a horizontal rope 12.0 m long is moved up and down seven times per second with a maximum displacement of 60 mm on each side of its equilibrium position. The rope has a mass of 1.00 kg and is under a tension of 40 N.

Find the amplitude, speed, frequency, and wavelength of the resulting waves.

12.5   Standing Waves

9. The G (196 Hz) string of a guitar is 650 mm long. What is the speed of sound in the string?

10. The vibrating part of a violin string whose linear density is 4.7 g/m is 30 cm long. What tension should the string be under if its fundamental frequency is to be 440 Hz, the musical note A?

11. A stretched wire 1.00 m long has a fundamental frequency of 300 Hz. (a) What is the speed of the waves in the wire? (b) What are the frequencies of the first three overtones?

12. A steel wire 1.00 m long whose mass is 10.0 g is under a tension of 400 N. (a) What is the wavelength of its fundamental mode of vibration? (b) What is the frequency of this mode? (c) What is the wavelength of the sound waves produced when the string vibrates in this mode?

13. The G and A strings of a violin have the respective fundamental frequencies of 196 Hz and 440 Hz. If the G string has a linear density of 3.0 g/m, what should the linear density of the A string be if both are to be under the same tension?

14. The vibrating part of the G string of a certain violin is 330 mm long and has a fundamental frequency of 196 Hz when under a tension of 50 N. (a) Find the linear density of the string. (b) Where should the string be pressed in order for it to vibrate at 220 Hz?

15. The vibrating part of the E string of a violin is 330 mm long and has a fundamental frequency of 659 Hz. What is its fundamental frequency when the string is pressed against the fingerboard at a point 60 mm from its end? What are the first and second overtones of the string under these circumstances?

12.6   Sound

(Assume that the speed of sound in sea-level air is 343 m/s = 1125 ft/s, which corresponds to a temperature of 20°C.)

16. The speed of sound is nearly the same in aluminum and steel, but the density of steel is almost three times that of aluminum. From this information find the approximate ratio of Young's moduli for the two metals.

17. There are two mechanisms by which a person determines the direction from which a sound comes. One compares the loudness of the sound at the left ear with that at the right ear, which is most effective at low frequencies. The other compares the phases of the waves that arrive at the two ears, which is most effective at high frequencies. (The **phase** of a wave is the part of its cycle it is in at a particular time and place.) The crossover point of equal effectiveness occurs at about 1200

Hz, and as a result sound with frequencies in the neighborhood of 1200 Hz are difficult to locate. How does the wavelength of a 1200-Hz sound compare with the distance between your ears?

18. An ultrasonic beam used in scanning body tissues has a frequency of 1.2 MHz. If the speed of sound in a particular tissue is 1540 m/s and the limit of resolution is equal to one wavelength, what is the size of the smallest detail that can be resolved?

19. A workman strikes a steel rail with a hammer, and the sound reaches an observer 1500 ft away both through the air and through the rail. How much time separates the two sounds?

20. A mine explodes at sea, and there is an interval of 5.0 s between the arrival of the sound through the water and its arrival through the air at a nearby ship. How far away is the mine from the ship?

21. An airplane is flying at 800 km/h at an altitude of 2.5 km. When the sound of the airplane's engines appears to a person on the ground to come from directly overhead, how far away is a point directly under the airplane?

22. A violin string is set in vibration at a frequency of 440 Hz. How many vibrations does it make while its sound travels 200 m in air?

23. A tuning fork vibrating at 440 Hz is placed in distilled water. (a) What are the frequency and wavelength of the waves produced within the water? (b) What are the frequency and wavelength of the waves produced in the surrounding air when the water waves reach the surface?

24. Find the angle between the shock wave created by an airplane traveling at 500 m/s just above sea level and the direction of the airplane.

25. The angle between the shock wave created by an airplane and the direction of the airplane is 60°. Find the Mach number corresponding to the speed of the airplane.

26. The speed of surface waves in shallow water of depth h is $\sqrt{gh}$. How fast is a boat moving in water 1.5 m deep if the total angular width of its wake is 50°?

12.7   Interference of Sound Waves

27. When a clarinet produces the note E (330 Hz), how far is its reed from the first open hole? (The reed acts as a closed end to the air column.)

28. A 440-Hz tuning fork is held over the open top of a glass test tube held vertically. Water is added to the tube until the air column inside it is in resonance with the vibrations of the tuning fork. When this occurs, the air column is 190 mm long. What is the speed of sound in the laboratory?

29. A standing sound wave of 50 Hz is set up between the walls of a room. How far apart are the walls?

30. (a) Express the frequencies of the standing waves in a tube open at one end and closed at the other in terms of the tube's length and the speed of sound. (See Eqs. 12.6 and 12.7 for how this is done for standing waves in a stretched string.) (b) Do the same for the standing waves in a tube open at both ends.

31. Determine whether the difference tone produced by the frequencies 264 Hz and 396 Hz will be consonant (pleasing to the ear) or dissonant.

32. Two identical steel wires have fundamental frequencies of vibration of 400 Hz. The tension in one of the wires is increased by 2%, and both wires are plucked. How many beats per second occur?

33. A wire under a tension of 50 N is vibrating at 200 Hz. If an identical wire under a tension of 55 N is also set vibrating, how many beats per second will be heard?

34. When two organ pipes open at both ends are sounded, 4 beats/s are heard. If the longer pipe is 1000 mm long, how long is the other pipe?

35. A whistle consists of two tubes, each closed at one end, that are 30 and 35 mm long. The chief frequency in the sound that is heard when the whistle is blown is that of the beats between the fundamental frequencies of the tubes. What is this frequency?

12.8   Doppler Effect

36. A person in a car is driving at 60 km/h toward a ferry whose whistle is blowing at 400 Hz. (a) What frequency does she hear? (b) The ferry leaves the dock and heads directly away from the driver at 15 km/h, still blowing its whistle. What frequency does she hear now?

PROBLEMS

47. A steel wire originally 1.00 m long is stretched so that the resulting strain is 1.00%. Find the fundamental frequency of vibration of the wire.

48. The smallest frequency change a person with normal hearing can detect is about 1% for frequencies over 500 Hz. What is the speed of a car with a siren if a person at rest can just detect a change in the siren's pitch as it passes by? (*Hint:* v_S is much smaller than v, so $v^2 - v_S^2 \approx v^2$.)

49. The speed with which blood flows through an artery can be determined from the Doppler shift in high-frequency sound waves sent into the artery and detected after reflection from

37. A fire engine has a siren whose frequency is 500 Hz. What frequency is heard by a stationary observer when the engine moves toward him at 12 m/s? When it moves away from him at 12 m/s?

38. A police car moving at 80 mi/h is chasing a truck moving in the same direction at 60 mi/h. The police car has its siren on. If the truck driver finds the siren's frequency to be 600 Hz, what frequency does the police-car driver hear?

39. A latecomer to a concert hurries down the aisle toward his seat so fast that the note middle C (262 Hz) appears 1 Hz higher in frequency. How fast is he going?

40. A spacecraft moving away from the earth at 97% of the speed of light transmits data at the rate of 1.00×10^4 pulses/s. At what rate are they received?

41. A galaxy in the constellation Ursa Major is receding from the earth at 15,000 km/s. If one of the characteristic wavelengths of the light the galaxy emits is 5.50×10^{-7} m, what is the corresponding wavelength measured by astronomers on the earth?

12.9   Sound Intensity

42. How many times more intense is a 60-dB sound than a 50-dB sound? Than a 40-dB sound? Than a 20-dB sound?

43. What is the intensity level in decibels of a sound whose intensity is 5.0×10^{-6} W/m²?

44. A siren produces a sound whose intensity is 1.0 W/m². What is the corresponding sound intensity level?

45. Sound from a busy street enters a room through an open window 80 cm square at a power of 6.4×10^{-6} W. What is the sound intensity level outside the window?

46. The acoustic power output of a certain purring cat is 4.0×10^{-10} W. How far away can a person of normal hearing just detect the purr?

the moving blood cells. What is done is to aim the source and receiver of the sound so that the reflections occur from cells moving away from them. Thus the sound waves striking a cell are Doppler shifted because of its motion away from the source, and the waves that reach the receiver are further shifted because they come from a reflector moving away from it. (a) Derive a formula for the frequency f_r of the waves that reach the receiver in terms of the source frequency f_S, the speed v of sound in blood (which is 1570 m/s), and the speed v_b of the blood. (b) Solve this formula for v_b. (c) Find the blood speed when the source frequency is 10^6 Hz and the Doppler shift is 40 Hz.

50. Train A is heading east at the speed v_A and train B is heading west at the speed v_B on an adjacent track. The locomotive on train A is blowing its whistle continuously. A passenger on train B observes the frequency of the sound from the whistle to be 307 Hz when the trains approach each other, 256 Hz when they are abreast, and 213 Hz when they recede from each other. From these figures find the speed of each train relative to the ground.

51. A chorus of 20 voices is singing at a sound intensity level of 70 dB. If the voices all have the same sound intensity level, what is it?

52. A riveting gun in a shipyard is producing 95 dB of noise. What is the sound intensity level when a second riveting gun begins to operate?

53. The average power output of a piano when it is playing a certain piece of music is 0.4 W. If the piano is assumed to radiate sound equally in all directions, what is the sound intensity level 10 m away?

54. A woman 50 m from a brass band in a park walks toward the band until the music she hears is 10 dB louder. How far from the band is she now?

Answers to Multiple Choice

1. a	9. d	17. d	25. b
2. b	10. a, d	18. d	26. b
3. a	11. c	19. b	27. d
4. d	12. d	20. c	28. d
5. a	13. b	21. b	29. c
6. c	14. b	22. b	
7. a	15. c	23. b	
8. d	16. c	24. c	

13

THERMAL PROPERTIES
OF MATTER

13.1   **Temperature**

13.2   **Thermal Expansion**

13.3   **Boyle's Law**

13.4   **Charles's Law**

13.5   **Atoms and Molecules**

13.6   **Kinetic Theory of Gases**

13.7   **Molecular Motion in Liquids**

S uppose we had a microscope with unlimited power. What would we find if we were to look at a drop of water under higher and higher magnifications? Would we continue to see a clear, structureless liquid?

No such microscope exists, but there is plenty of evidence obtained in other ways to show that, on a very small scale, water consists of individual particles. In fact, so does all other matter, whether solid, liquid, or gas. The ancient Greeks anticipated this finding over two thousand years ago. What they did not even suspect, however, is that these particles are in constant, random motion, and that what we perceive as temperature is closely related to the average kinetic energy of the particles. This relationship is the key to understanding a great many physical phenomena, such as why gases become hot when compressed and why evaporation cools liquids.

13.1 TEMPERATURE

Assigning numbers to hot and cold.

All of us know what temperature signifies in terms of our sense impressions, and we do not have to understand what temperature means in terms of the structure of matter to make use of the concept of temperature in a variety of ways. For the moment it is sufficient for us simply to regard temperature as that which gives rise to sensations of hot and cold.

A **thermometer** is a device used to measure temperature. Most substances expand when heated and contract when cooled, and the thermometers of everyday life are based on this property of matter. More precisely, they are based on the different rates of expansion of different materials (Fig. 13.1). Thus the length of the

Figure 13.1

Three types of thermometer based on thermal expansion.

(a) Two strips of different metals that are joined together bend to one side or the other with a change in temperature owing to different rates of expansion in the two metals, which is the basis of household oven thermometers. The higher the temperature, the greater the deflection. When cooled, such a bimetallic strip bends in the opposite direction. A thermostat uses a bimetallic strip to operate a switch that turns a heating or cooling system on and off at preset temperatures.

(b) In a constant-volume gas thermometer, which is a very sensitive laboratory instrument, the height of the mercury column at the left is adjusted until the mercury column at the right is at a fixed level. The difference in heights of the two mercury columns is a measure of the pressure needed to maintain the gas in a fixed volume, and hence a measure of the temperature.

(c) Mercury (or colored alcohol) expands more when heated than glass does, and so the length of the liquid column in a liquid-in-glass thermometer is a measure of the temperature of the thermometer bulb.

mercury column in the familiar mercury-in-glass thermometer indicates the temperature of the bulb because mercury expands more than glass when heated and contracts more than glass when cooled.

Thermal expansion is not the only property of matter that can be used to make a thermometer. The electrical resistance of most materials, for instance, varies with temperature. Another example is the color of an object heated until it glows. A poker thrust in a fire at first becomes a dull red, then bright red, orange, and yellow. At a high enough temperature it becomes "white hot." The color of a really hot object is thus a measure of its temperature, a fact used by astronomers when they study stars.

Before we can make a thermometer of any kind, we need a temperature scale. One way to establish such a scale is to note that water freezes into a solid (ice) and vaporizes into a gas (steam), both at definite temperatures at a given pressure. In the **Celsius scale** the freezing point of water at 1 atm pressure (or, more exactly, the point at which a mixture of ice and water is in equilibrium, with as much ice melting as water freezing) is called 0°, and the boiling point of water at 1 atm pressure (or, more exactly, the point at which a mixture of steam and water is in equilibrium) is called 100° (Fig. 13.2). Temperatures in the Celsius scale are written, for example, "40°C." In the United States the Celsius scale is sometimes called the **centigrade scale.**

Although the Celsius scale is used in most of the world, a different temperature scale called the **Fahrenheit scale** is commonly used for nonscientific purposes in some English-speaking countries. In the Fahrenheit scale the freezing point of

Figure 13.2

Calibrating a thermometer on the Celsius scale. A mixture of ice and water at atmospheric pressure is, by definition, at 0°C, and a mixture of steam and water at atmospheric pressure is, again by definition, at 100°C.

Mixture
of ice
and water

Mixture of
steam and water

water is 32°F and the boiling point of water is 212°F (Fig. 13.3). This means that 180°F separates the freezing and boiling points of water, whereas 100°C separates them in the Celsius scale. Therefore Fahrenheit degrees are 100/180, or 5/9, as large as Celsius degrees. We can convert temperatures from one scale to the other with the help of the formulas

$$T_F = \tfrac{9}{5}T_C + 32° \qquad \text{Celsius to Fahrenheit} \quad (13.1a)$$
$$T_C = \tfrac{5}{9}(T_F - 32°) \qquad \text{Fahrenheit to Celsius} \quad (13.1b)$$

EXAMPLE 13.1 _____

(a) The air temperature on a hot summer day is 95°F. What is the Celsius equivalent of 95°F? (b) Nitrogen freezes at −210°C. What is the Fahrenheit equivalent of −210°C?

SOLUTION (a) $T_C = \tfrac{5}{9}(T_F - 32°) = \tfrac{5}{9}(95° - 32°) = \tfrac{5}{9}(63°) = 35°C$
(b) $T_F = \tfrac{9}{5}T_C + 32° = \tfrac{9}{5}(-210°) + 32° = -378° + 32° = -346°F$ ♦

13.2 THERMAL EXPANSION

It is proportional to temperature change in nearly all cases.

A change in temperature causes most solids to change in length by an amount proportional to their original lengths and to the temperature change. A long steel bridge may vary in length by over a meter between summer and winter, which must be allowed for in its design.

If the original length of a rod of a certain material is L_0, its change in length ΔL after its temperature changes by ΔT is

$$\Delta L = aL_0 \Delta T \qquad \text{Thermal expansion} \quad (13.2)$$

Change in length = (a)(original length) (temperature change)

The quantity a, called the **coefficient of linear expansion,** is a constant whose value depends on the nature of the material. Different substances expand (and contract) to different extents. A lead rod, for example, changes in length by 60 times as much as a quartz rod of the same initial length when both are heated or cooled through the same temperature interval. Table 13.1 lists coefficients of linear expansion for various substances.

Figure 13.3

The Fahrenheit and Celsius temperature scales.

Thermal expansion and contraction change the length of the Verrazano-Narrows Bridge across New York Bay by over a meter between summer and winter.

Table 13.1   Coefficients of linear expansion

Substance	Coefficient	Coefficient
Aluminum	$2.4 \times 10^{-5}/°C$	$1.3 \times 10^{-5}/°F$
Brass	1.8	1.0
Concrete	0.7–1.2	0.4–0.7
Copper	1.7	0.94
Iron	1.2	0.67
Lead	3.0	1.7
Quartz	0.05	0.028
Silver	2.0	1.1
Steel	1.2	0.67

   E X A M P L E   13.2

What is the increase in length of a steel girder that is 10 m long at 5°C when its temperature rises to 30°C?

SOLUTION   The coefficient of linear expansion of steel is $1.2 \times 10^{-5}/°C$. From Eq. (13.2)

$$\Delta L = aL_0 \, \Delta T = (1.2 \times 10^{-5}/°C)(10 \text{ m})(25°C) = 3.0 \times 10^{-3} \text{ m}$$

$$= 3.0 \text{ mm}$$ ◆

E X A M P L E   13.3

How much force would be needed to stretch the girder by the same amount at constant temperature if the girder's cross sectional area were 200 cm²?

SOLUTION   Because the girder increases in length by 3.0 mm, the force is the same as that required to stretch it by 3.0 mm (Fig. 13.4). Equation (9.6) gives the change in length ΔL of a rod that is subjected to a tension or compression force F as

$$\Delta L = \frac{L_0}{Y} \frac{F}{A}$$

where A is the cross-sectional area of the rod and Y is Young's modulus for the material of the rod. The girder has $A = 200 \text{ cm}^2 = 0.020 \text{ m}^2$ and, from Table 9.2, Young's modulus for steel is $2.0 \times 10^{11} \text{ N/m}^2$. Hence

$$F = \frac{YA \, \Delta L}{L_0} = \frac{(2.0 \times 10^{11} \text{ N/m}^2)(0.020 \text{ m}^2)(3.0 \times 10^{-3} \text{ m})}{10 \text{ m}}$$

$$= 1.2 \times 10^6 \text{ N}$$

Figure 13.4

A 10-m steel girder expands 3.0 mm when its temperature is increased by 25°C. At constant temperature, a force of 1.2×10^6 N would be required to produce the same increase in length if the cross-sectional area of the girder were 200 cm² (Example 13.3).

The required force is 1.2 million newtons, which is equivalent to 135 tons! Clearly, thermal expansion can involve quite considerable forces. ◆

Volume Expansion

A formula similar to Eq. (13.2) holds for the changes in volume, ΔV, of a solid or liquid whose temperature changes by an amount ΔT. Here we have

$$\Delta V = b V_0 \, \Delta T \qquad\qquad \textit{Volume expansion} \quad (13.3)$$

where V_0 is the original volume and b is the **coefficient of volume expansion.** Table 13.2 lists coefficients of volume expansion for various substances. In general, the coefficients of linear and volume expansion are related by

$$b = 3a$$

so that we can readily determine the values of b for the materials in Table 13.1.

Table 13.2 Coefficients of volume expansion

Substance	Coefficient	Coefficient
Ethyl alcohol	7.5×10^{-4}/°C	4.2×10^{-4}/°F
Glass (average)	0.2	0.1
Glycerin	5.1	2.8
Ice	0.5	0.3
Mercury	1.8	1.0
Pyrex glass	0.09	0.05
Water	2.1	1.2

Figure 13.5

A hole in a body expands or contracts with a change in temperature precisely as a solid object of the same size, shape, and composition would.

◆   E X A M P L E   13.4

Calculate the volume of water that overflows when a Pyrex beaker filled to the brim with 250 cm³ of water at 20°C is heated to 60°C.

SOLUTION   First we note that a hole in a body expands or contracts by just as much as a solid object whose composition is that of the body and whose volume is that of the hole (Fig. 13.5). The reason is that the material around the hole is not affected by whether the hole is there or not. The boundary of the hole, and therefore the size and shape of the hole, thus behaves as it would if the hole were filled. This means that we can write for the increase in capacity of the beaker

$$\Delta V_b = b_P V_b\, \Delta T = (0.09 \times 10^{-4}/°C)(250\ cm^3)(40°C) = 0.09\ cm^3$$

The increase in the volume of the water is

$$\Delta V_w = b_w V_w\, \Delta T = (2.1 \times 10^{-4}/°C)(250\ cm^3)(40°C) = 2.1\ cm^3$$

and so the volume of water that overflows is

$$\Delta V_w - \Delta V_b = 2.0\ cm^3 \qquad\qquad\qquad ◆$$

Water

The thermal expansion of water is unusual. Above 4°C water expands when heated, just as most other substances do. From 0°C to 4°C, however, the volume of a water sample *decreases* with increasing temperature (Fig. 13.6). Thus water has its maximum density at 4°C. Equally unusual is the expansion of water when it freezes, so that ice floats and exposed water pipes may burst in a severe winter. (The reason for this behavior will be discussed in Chapter 26.)

Because ice floats, a body of water freezes in winter from the top down, not from the bottom up as it would if ice were denser than water and cold water denser than warm water at all temperatures. Ice is a much poorer conductor of heat than water (the difference in heat conductivity is nearly a factor of 4), which means that

Figure 13.6

Water has its greatest density at 4°C. Because water expands when it freezes, ice floats.

the layer of ice that forms initially on the surface of a body of water impedes further freezing. Ice forming at the bottom would not have this effect. Many lakes, rivers, and arms of the sea are therefore able to escape total freezing in winter, and as a result their plant and animal life is able to survive until spring.

13.3 BOYLE'S LAW

At constant temperature, the volume of a gas is inversely proportional to its pressure.

The thermal behavior of a gas is rather different from the behaviors of solids and liquids because, unlike them, a gas does not have a specific volume at a given temperature but expands to fill its container. The only way to change the volume of a gas is to change the capacity of its container. Even though its volume may remain the same, however, another property of a confined gas does vary with its temperature, namely, the pressure it exerts on the container walls. The air pressure in the tires of a car drops in cold weather and increases in warm weather, an illustration of this property.

When the temperature of a sample of gas is held constant, the absolute pressure it exerts on its container is very nearly inversely proportional to the volume of the container. Expanding the container lowers the pressure; shrinking the container raises the pressure. Conversely, increasing the pressure on a gas sample reduces its volume; decreasing the pressure increases its volume (Fig. 13.7). This relationship is

Figure 13.7

Boyle's law states that the volume of a gas sample is inversely proportional to its pressure at constant temperature. Thus $p_1V_1 = p_2V_2 = p_3V_3$, as shown.

called **Boyle's law** after its discoverer, Robert Boyle (1627–1691). Though not exact, Boyle's law is an excellent approximation over a wide range of temperatures and pressures.

Boyle's law can be written as

$$p_1V_1 = p_2V_2 \qquad (T = \text{constant}) \qquad \text{\textit{Boyle's law}} \quad (13.4)$$

where p_1 is the absolute gas pressure when the volume of the gas is V_1, and p_2 is the absolute gas pressure when the volume is V_2.

◆ **E X A M P L E 13.5**

How much air at the sea-level atmospheric pressure of 15 lb/in² can be stored in the 12-ft³ tank of an air compressor which can withstand an absolute pressure of 100 lb/in²?

SOLUTION Letting state 1 represent the air in the compressor tank and state 2 the air at atmospheric pressure, we have from Eq. (13.4)

$$V_2 = \frac{p_1V_1}{p_2} = \frac{(100 \text{ lb/in}^2)(12 \text{ ft}^3)}{15 \text{ lb/in}^2} = 80 \text{ ft}^3$$

A total of 80 ft³ of air at atmospheric pressure can be stored in the 12-ft³ compressor tank at the high pressure. ◆

◆ **E X A M P L E 13.6**

A scuba diver's 12-L tank is filled with air at a gauge pressure of 150 bars. If the diver uses 30 L of air per minute at the same pressure as the water pressure at her depth below the surface, how long can she remain at a depth of 15 m in seawater?

SOLUTION   Since atmospheric pressure is ~1.0 bar, the absolute pressure of the air in the tank is

$$p_1 = p_{atm} + p_{gauge} = 1 \text{ bar} + 150 \text{ bars} = 151 \text{ bars}$$

The absolute pressure at a depth of 15 m is, according to Example 10.3,

$$p_2 = p_{atm} + \rho g h = 1.0 \text{ bar} + 1.5 \text{ bars} = 2.5 \text{ bars}$$

From Eq. (13.4) we have for the volume of air available at a pressure of 2.5 bars

$$V_2 = \frac{p_1 V_1}{p_2} = \frac{(151 \text{ bars})(12 \text{ L})}{2.5 \text{ bars}} = 725 \text{ L}$$

However, 12 L of air remain in the tank, so only 713 L are usable and

$$t = \frac{713 \text{ L}}{30 \text{ L/min}} = 24 \text{ min}$$
◆

13.4 CHARLES'S LAW

At constant pressure, the volume of a gas is proportional to its absolute temperature.

Now let us see what happens to a gas when its temperature is changed. As mentioned earlier, if the volume of a gas is held constant, the pressure it exerts on its container depends on its temperature. According to Boyle's law, then, if we hold the gas pressure constant, its volume should vary with temperature. When this prediction was tested over 150 years ago by Jacques Charles and Joseph Gay-Lussac in France, they found that the change in volume ΔV of a gas sample is in fact related to a change ΔT in its temperature by the same formula, Eq. (13.3), that holds for solids and liquids:

$$\Delta V = b V_0 \, \Delta T$$

The significant thing about gases at constant pressure is that they *all* have very nearly the same coefficient of volume expansion b. By contrast, as Tables 13.1 and 13.2 indicate, the thermal coefficients for solids and liquids may have markedly different values for different substances. At 0°C the coefficient of volume expansion

$-10°C$
$263 K$

$V = 0.963 \text{ m}^3$

$0°C$
$273 K$

$V = 1.000 \text{ m}^3$

$10°C$
$283 K$

$V = 1.037 \text{ m}^3$

Figure 13.8

Charles's law states that the volume of a gas sample is directly proportional to its absolute temperature at constant pressure. Thus $V_1/T_1 = V_2/T_2 = V_3/T_3$ as shown. (The changes in size are greatly exaggerated in the drawing.)

b_0 for all gases is very close to

$$b_0 = \frac{0.0037}{°C} = \frac{1/273}{°C} \qquad \text{(all gases)}$$

If we vary the temperature of a gas sample while holding its pressure constant, its volume changes by 1/273 of its volume at 0°C for each 1°C temperature change. A child's large balloon filled with air whose volume at 0°C is 1.000 m³ has a volume of 1.037 m³ at 10°C and 0.963 m³ at −10°C (Fig. 13.8).

What happens when the balloon is cooled to −273°C? At that temperature the air in the balloon should have lost 273/273 of its volume at 0°C, and therefore should have vanished entirely!

Actually, all gases condense into liquids at temperatures above −273°C, so the question has no physical meaning. But −273°C is still an important temperature. Let us set up a new temperature scale, the **absolute temperature scale**, and designate −273°C as the zero point (Fig. 13.9). Temperatures in the absolute scale are expressed in **kelvins**, denoted K, after Lord Kelvin (1824–1907), a noted British physicist. To convert temperatures from one scale to the other we note that

$$T_K = T_C + 273 \qquad\qquad \textit{Celsius to absolute} \quad (13.5a)$$
$$T_C = T_K - 273 \qquad\qquad \textit{Absolute to Celsius} \quad (13.5b)$$

Thus the equivalent on the absolute scale of room temperature, 20°C, is

$$T_K = T_C + 273 = (20 + 273)K = 293 \text{ K}$$

The reason for setting up the absolute temperature scale is that, provided the pressure is constant, *the volume of a gas sample is directly proportional to its absolute temperature* (Fig. 13.9b). This relationship is called **Charles's law**. Like Boyle's law, Charles's law is not a basic physical principle, but deviations from it are usually quite small.

We can express Charles's law in the form

$$\frac{V_1}{T_1} = \frac{V_2}{T_2} \qquad (p = \text{constant}) \qquad\qquad \textit{Charles's law} \quad (13.6)$$

where V_1 is the volume of a gas sample at the absolute temperature T_1 and V_2 is its volume at the absolute temperature T_2.

If there were a gas that did not liquify before reaching 0 K, then at 0 K its volume would shrink to zero. Because a negative volume has no meaning, it is natural to think of 0 K as **absolute zero**. Indeed 0 K is the lower limit to the temperature anything can have, but on the basis of a stronger argument than one based on imaginary gases. This argument is discussed later in this chapter. To five significant figures the Celsius equivalent of absolute zero is −273.16°C.

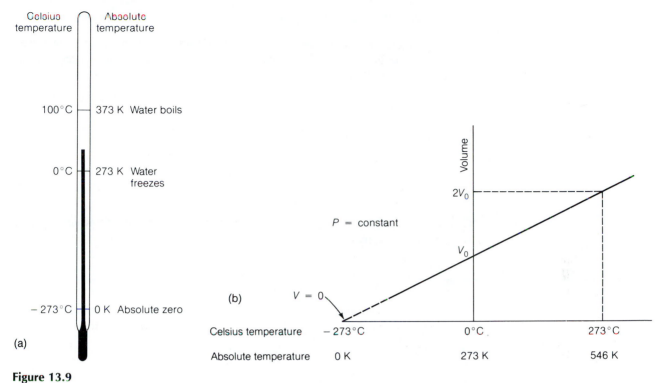

Celsius temperature   Absolute temperature

100°C — 373 K Water boils

0°C — 273 K Water freezes

−273°C — 0 K Absolute zero

(a)

Volume

$2V_0$

P = constant

V_0

(b) $V = 0$

| Celsius temperature | −273°C | 0°C | 273°C |
| Absolute temperature | 0 K | 273 K | 546 K |

Figure 13.9

(a) The absolute temperature scale. (b) The volume of a gas at constant pressure is directly proportional to its absolute temperature.

In the operation of a snow-making machine, a mixture of compressed air and water is blown through a nozzle or set of nozzles. The expansion of the air cools the mixture sufficiently to freeze the water into the ice crystals of snow.

The Rankine Scale

An absolute temperature scale, called the **Rankine scale,** may be constructed based upon the Fahrenheit scale. Temperatures in this scale are designated °R. Absolute zero on the Rankine scale is

$$0°R = \tfrac{9}{5}(-273°C) + 32° = -460°F$$

Hence the conversion formulas for °F to °R and for °R to °F are

$$T_R = T_F + 460° \qquad\qquad \textit{Fahrenheit to Rankine} \quad (13.7a)$$
$$T_F = T_R - 460° \qquad\qquad \textit{Rankine to Fahrenheit} \quad (13.7b)$$

Charles's law holds when temperatures are expressed in the Rankine scale as well as when they are expressed in the Kelvin scale.

   E X A M P L E   13.7

To what temperature must a gas sample originally at 40°F and atmospheric pressure be heated if its volume is to double while its pressure remains the same?

SOLUTION   The Rankine equivalent of 40°F is

$$T_1 = T_F + 460° = 40° + 460° = 500°R$$

Because $V_2 = 2V_1$, we have from Eq. (13.6)

$$T_2 = \frac{V_2 T_1}{V_1} = \frac{(2V_1)T_1}{V_1} = 2T_1 = 2(500°R) = 1000°R$$

The Fahrenheit equivalent of 1000°R is

$$T_2 = T_R - 460° = 1000° - 460° = 540°F \qquad\qquad\qquad ◆$$

Ideal Gas Law

Boyle's law and Charles's law can be combined in a single formula called the **ideal gas law:**

$$\frac{p_1 V_1}{T_1} = \frac{p_2 V_2}{T_2} \qquad\qquad \textit{Ideal gas law} \quad (13.8)$$

When $T_1 = T_2$, the ideal gas law becomes Boyle's law, and when $p_1 = p_2$, it becomes Charles's law.

◆   **E X A M P L E   13.8** _____

(a) A tank with a capacity of 1.0 m³ contains helium gas at 27°C under a pressure of 20 atm. The helium is used to fill a balloon. When the balloon is filled, the gas pressure inside it is 1.0 atm, and its temperature has dropped to −33°C. (The gas has done work in expanding at the expense of its internal energy, and the cooling reflects this loss of internal energy.) What is the volume of the balloon at this time? (b) After a while the helium in the balloon absorbs heat from the atmosphere and returns to its original temperature of 27°C, and it expands further to maintain its pressure at 1.0 atm. What is the final volume of the balloon? (The gas pressure in the balloon is actually slightly greater than 1.0 atm to balance the tendency of the rubber to contract, but this is ignored here for convenience.)

SOLUTION   (a) The equivalents of 27°C and −33°C on the absolute scale are 300 K and 240 K, respectively. Applying the ideal gas law to the initial expansion gives

$$V_2 = \frac{T_2 p_1}{T_1 p_2} V_1 = \left(\frac{240 \text{ K}}{300 \text{ K}}\right)\left(\frac{20 \text{ atm}}{1.0 \text{ atm}}\right)(1.0 \text{ m}^3) = 16 \text{ m}^3$$

Because the tank's capacity is 1.0 m³, the balloon's volume after the initial expansion is 15 m³ (Fig. 13.10).

Figure 13.10

(b) When the helium has reached the outside air temperature of 27°C, which we will call state 3, then $T_1 = T_3$. Hence we need only apply Boyle's law to states 1 and 3 to obtain the eventual volume of the helium:

$$V_3 = \frac{p_1}{p_3} V_1 = \left(\frac{20 \text{ atm}}{1.0 \text{ atm}} \right) (1.0 \text{ m}^3) = 20 \text{ m}^3$$

Again we subtract the 1.0 m³ volume of the tank to find the volume of the balloon itself, which is 19 m³.   ◆

The ideal gas law is obeyed approximately by all gases. The significant thing is not that the agreement with experiment is never perfect, but that *all* gases, no matter what kind, behave almost identically. An **ideal gas** is defined as one that obeys Eq. (13.8) exactly. Although no ideal gases actually exist, they do provide a target for theories of the gaseous state to aim at. It is reasonable to suppose that the ideal gas law is based on the essential nature of gases. Hence the next step is to account for this law and only afterward to seek reasons for its failure to be completely correct.

13.5 ATOMS AND MOLECULES

The ultimate particles of gases.

Elements are the simplest substances of which matter in bulk is composed. Over 100 are known, 92 found in nature and the rest artificially prepared. The ultimate particles of an element are its **atoms.**

Two or more elements may combine chemically to form a **compound**, a substance whose properties are different from those of the elements that compose it. The ultimate particles of a compound in gaseous form are **molecules**, which consist of the atoms of its constituent elements joined together in a definite way. Thus each molecule of water contains two hydrogen atoms and one oxygen atom. While the ultimate particles of elements are atoms, many elemental gases consist of molecules rather than atoms. Oxygen molecules, for instance, contain two oxygen atoms each. The molecules of other gases, such as helium and argon, are single atoms. Figure 13.11 shows schematically the composition of some common molecules.

Atoms and molecules are very small, and even a tiny bit of matter contains huge numbers of them. If each atom in a penny were worth one cent, all the money in the world would not be enough to pay for the penny. Elements in liquid and solid form are usually assemblies of individual atoms. Some compounds in these forms are also assemblies of individual molecules, but more often the situation is more complicated, as we will see in Chapter 26.

Oxygen (O$_2$)   Methane (CH$_4$)

Water (H$_2$O)

Ethyl alcohol
(C$_2$H$_5$OH)

Figure 13.11

Molecular structures of several
common substances.

The masses of atoms and molecules are usually expressed in **atomic mass units** (u), where

$$1 \text{ atomic mass unit} = 1 \text{ u} = 1.66 \times 10^{-27} \text{ kg}$$

A list of the atomic masses of the elements is given in Appendix C. If we know the composition of a compound, we can calculate the corresponding molecular mass.

◆ E X A M P L E   13.9

How many H$_2$O molecules are present in 1.00 g of water? The atomic mass of hydrogen is 1.008 u and that of oxygen is 16.00 u.

SOLUTION   We begin by finding the mass of the H$_2$O molecules in u:

$$
\begin{aligned}
2\text{H} &= 2 \times 1.008 \text{ u} = 2.02 \text{ u} \\
\text{O} &= 1 \times 16.00 \text{ u} = \underline{16.00 \text{ u}} \\
&\phantom{= 1 \times 16.00 \text{ u} = } 18.02 \text{ u}
\end{aligned}
$$

The mass of the H$_2$O molecule in kg is therefore

$$m = (18.02 \text{ u})(1.66 \times 10^{-27} \text{ kg/u}) = 2.99 \times 10^{-26} \text{ kg}$$

and so the number of H$_2$O molecules in 1.00 g $= 1.00 \times 10^{-3}$ kg of water is

$$\text{Molecules of H}_2\text{O} = \frac{\text{mass of H}_2\text{O}}{\text{mass of H}_2\text{O molecule}} = \frac{1.00 \times 10^{-3} \text{ kg}}{2.99 \times 10^{-26} \text{ kg}}$$

$$= 3.34 \times 10^{22} \text{ molecules} \qquad\qquad ◆$$

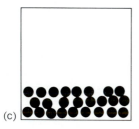

(a)

(b)

(c)

Figure 13.12

(a) The molecules of a gas are in constant, random motion. (b) The constituent particles of a solid are also in motion but oscillate about definite equilibrium positions. (c) The molecules of a liquid keep a more or less constant distance apart but move about freely.

Figure 13.13

(a) An actual gas. (b) Simplified model of the gas.

A considerable amount of experimentation and ingenious reasoning had to be carried out before the reality of atoms and molecules became definitely established. Although the full story of the kinetic-molecular theory of matter, a large part of which involves chemistry, will not be gone into here, it is easy to show that it can account for the ideal gas law. We will also see how the physical properties of solids and liquids, and the deviations of actual gases from an ideal gas, fit into the kinetic-molecular theory.

13.6 KINETIC THEORY OF GASES

Why gases behave as they do.

According to the **kinetic theory of gases**, a gas consists of a great many tiny individual molecules that do not interact with one another except in collisions. The molecules are supposed to be far apart compared with their dimensions and to be in constant motion, incessantly hurtling to and fro, as in Fig. 13.12. They are kept from escaping into space only by the solid walls of a container (or, in the case of the earth's atmosphere, by gravity). A natural consequence of the random motion and large molecular separation is the tendency of a gas to completely fill its container and to be readily compressed or expanded.

The particles of a solid, on the other hand, are close together, and mutual attractive and repulsive forces hold them in place to provide the solid with its characteristic rigidity. In a liquid the intermolecular forces are also sufficient to keep the volume of a sample constant. However, they are not strong enough to prevent nearby molecules from sliding past one another, which allows liquids to flow.

Boyle's law follows directly from the picture of a gas as a group of randomly moving molecules. The pressure the gas exerts comes from the impacts of its molecules. The vast number of molecules in even a tiny gas sample means that their separate blows appear as a continuous force to our senses and measuring instruments. Figure 13.13 shows a simplified model of a gas confined to a box. Although the

(a)

(b)

Pressure falls on top and bottom of
expanded cylinder because molecules spend more time in transit
between collisions

Pressure falls on sides of expanded
cylinder because molecules spread
their impacts over a larger area

Figure 13.14

The origin of Boyle's law according to the kinetic theory of gases.

molecules are actually traveling about in all directions, the effects of their collisions
with the walls of the box are the same as if one-third of them were moving back and
forth between each pair of opposite walls.

When a cylinder containing a gas is doubled in volume, as in Fig. 13.14,
those molecules moving up and down have twice as far to go between impacts. Since
their speed is unchanged, the time between impacts is also doubled, and the pressure
they exert on the top and bottom of the cylinder falls to half its original value. The
expansion of the cylinder also means that the molecules moving horizontally are now
spread over twice their former area, and the pressure on the sides of the cylinder
accordingly falls to half its original value as well. Thus doubling the volume means
halving the pressure, which is Boyle's law. Similar reasoning accounts for a rise in
pressure when the volume is reduced.

Charles's law follows from the kinetic theory of gases provided that

**The average kinetic energy of the random translational motions of the mole-
cules of a gas is proportional to the absolute temperature of the gas.**

This relationship is reasonable: We observe that compressing a gas quickly
(so no heat can enter or leave the container) raises its temperature, and such a
compression must increase the average energy of the molecules because they bounce
off the inward-moving piston more rapidly than they approach it (Fig. 13.15). A
familiar example of the latter effect is a baseball rebounding with greater speed when
struck by a bat. On the other hand, expanding a gas lowers its temperature, and such
an expansion reduces molecular energies because molecules lose speed in bouncing
off an outward-moving piston.

The significance of absolute zero in the kinetic theory of gases is straightforward: It is that temperature at which all molecular translational movement stops

Figure 13.15

The temperature of a gas increases when it is compressed because the average energy of its molecules increases. The temperature of a gas decreases when it is expanded because the average energy of its molecules decreases.

Molecules rebounding from inward-moving piston move faster than before

Molecules rebounding from outward-moving piston move slower than before

Molecules rebounding from fixed wall have unchanged speeds

Figure 13.16

At absolute zero, the kinetic theory of gases predicts that molecular translational motion in a gas will cease. In reality, at absolute zero the molecules would retain a small minimum amount of kinetic energy.

(Fig. 13.16). A more advanced analysis shows that the molecules cannot be totally motionless, but the difference is small and not important here.

The precise relationship between the average molecular kinetic energy KE_{av} and absolute temperature T is found to be

$$KE_{av} = \tfrac{3}{2}kT \qquad \qquad \textit{Molecular energy} \quad (13.9)$$

where k, known as **Boltzmann's constant**, has the value

$$k = 1.38 \times 10^{-23} \text{ J/K}$$

Equation (13.9) holds for the molecules of all gases regardless of the masses of their molecules. It has been verified by direct measurements of molecular velocities.

Thus we have an interpretation of temperature in terms of molecular motion that is much more precise and definite than simply describing temperature as that which is responsible for sensations of hot and cold.

Molecular Speeds

We can use Eq. (13.9) to find the average speed of gas molecules whose mass is m:

$$\overline{\tfrac{1}{2}mv^2} = \tfrac{3}{2}kT$$

$$v_{rms} = \sqrt{\overline{v^2}} = \sqrt{\frac{3kT}{m}} \qquad \textit{Rms molecular speed} \quad (13.10)$$

This speed is denoted v_{rms} because it is the square root of the mean of the squared molecular speeds—the "root-mean-square" speed—and therefore different from the simple arithmetic average speed $\bar{v}$.

◆ **E X A M P L E   13.10** _____

Find the rms speed of oxygen molecules at 0°C.

SOLUTION   Oxygen molecules have two oxygen atoms each. Since the atomic mass of oxygen is 16.00 u, the molecular mass is 32.00 u, and an O_2 molecule has a mass in kg of

$$m = (32.00 \text{ u})(1.66 \times 10^{-27} \text{ kg/u}) = 5.31 \times 10^{-26} \text{ kg}$$

At an absolute temperature of 273 K (which corresponds to 0°C), the rms speed of oxygen molecules is therefore

$$v_{rms} = \sqrt{\frac{3kT}{m}} = \sqrt{\frac{3(1.38 \times 10^{-23} \text{ J/K})(273 \text{ K})}{5.31 \times 10^{-26} \text{ kg}}} = 461 \text{ m/s}$$

This is a little over 1000 mi/h! Evidently molecular speeds are very large compared with those of the objects familiar to us.   ◆

_____

It is important to keep in mind that actual molecular speeds vary considerably on either side of v_{rms}. The graph in Fig. 13.17 shows the distribution of molecular speeds in oxygen at 273 K and in hydrogen at 273 K. The mass of an O_2 molecule is 16 times that of an H_2 molecule. Rms molecular speed decreases with molecular mass, hence at the same temperature molecular speeds in hydrogen are on the average greater than in oxygen. At the same temperature, however, the average molecular *energy* is the same for all gases.

In Fig. 13.18 we see the distributions of molecular speeds in oxygen at 73 K and at 273 K. The average molecular speed indeed increases with temperature. The curves of Figs. 13.17 and 13.18 are not symmetrical, because the lower limit to v is fixed at $v = 0$ whereas there is, in principle, no upper limit. Actually, as the curves show, the likelihood of speeds many times greater than v_{rms} is small.

Figure 13.17

Molecular speeds in oxygen and hydrogen at 273 K (0°C). The smaller masses of H_2 molecules means that they have higher average speeds than O_2 molecules at the same temperature, because the average kinetic energy depends only on temperature.

Figure 13.18

Molecular speeds in oxygen at
73 K (−200°C) and 273 K (0°C).
The higher the temperature, the
greater the average kinetic energy.

13.7 MOLECULAR MOTION IN LIQUIDS

The kinetic theory of evaporation and boiling.

Liquids as well as gases consist of molecules in constant random motion, and the kinetic-molecular theory explains evaporation. At its boiling point the molecules of a liquid have enough energy to break loose from the forces that hold them together, and the liquid becomes a gas. Evaporation, on the other hand, refers to the change of a liquid to a vapor *below* its boiling point. (A **vapor** is a gas whose temperature is below the boiling point of the substance.) Evaporation occurs only at the surface of a liquid, unlike boiling, in which bubbles of gas form throughout the interior of the liquid.

A hint as to what is going on when a liquid evaporates is that it grows cooler as it does so. The faster the evaporation, the more pronounced the cooling effect. Alcohol is effective at chilling the skin because it evaporates rapidly.

Evaporation can be understood on the basis of the distribution of molecular speeds in a liquid. Although not exactly the same as that found in a gas, this distribution resembles those shown in Figs. 13.17 and 13.18 in that a certain fraction of the molecules in any sample have much greater and much smaller speeds than the average. The fastest molecules have enough energy to escape through the liquid surface despite the attractive forces of the other molecules. The molecules left behind redistribute the available energy in collisions among themselves. Because the most energetic molecules have escaped, the average energy now is less than before and the liquid is at a lower temperature (Fig. 13.19).

When molecules from the vapor above a liquid surface happen to strike the surface, they may be trapped there, so that a constant two-way traffic of molecules

The fastest molecules escape

The remaining molecules have lower average energies, hence the liquid temperature is lower

Figure 13.19

After evaporation, the remaining liquid is cooler than before.

Figure 13.21

The variation with temperature of water vapor density in air for various relative humidities. Each curve corresponds to a different relative humidity. The curve for 100% relative humidity represents the maximum vapor density. The table gives the values of saturated vapor densities, which correspond to 100% relative humidity.

to and from the liquid occurs. If the density of the vapor above the liquid is sufficiently great, just as many molecules return as leave it at any time. This situation is described by saying that the region is **saturated** with the vapor (Fig. 13.20). The higher the temperature, the greater the maximum vapor density: At 0°C the density of water vapor at saturation is 5 g/m³, at 20°C it is 17 g/m³, at 100°C it is 598 g/m³, and at 300°C it is all the way up to 45.6 kg/m³ (Fig. 13.21). If for any reason (such as a sudden drop in temperature) the vapor density is more than the saturation value,

Figure 13.20

(a) When the region over a liquid is not saturated with vapor, more molecules leave the liquid than return to it, and the liquid gradually evaporates. (b) When the region is saturated with vapor, as many molecules return to the liquid as leave it, and the liquid level remains the same.

(a)

(b)

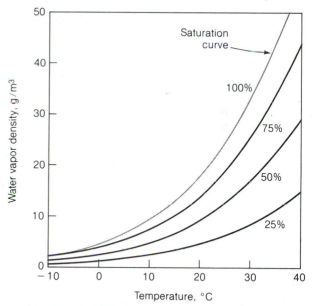

Temperature, °C	Water vapor density, g/m³	Temperature, °C	Water vapor density, g/m³
—10	2.36	20	17.3
—5	3.41	25	23.0
0	4.85	30	30.4
5	6.80	35	39.6
10	9.41	40	51.1
15	12.8		

Clouds form when moist air cools enough for its moisture to condense.

Air expands and cools as it rises.

Figure 13.22

Why mountains often have cloud caps over them.

condensation will be more rapid than evaporation until equilibrium is reestablished. This is why, on a hot day, moisture condenses on the outside of a glass that contains a cold drink. It is also why cloud caps form over mountains (Fig. 13.22).

Relative Humidity

The **relative humidity** of a volume of air describes its degree of saturation with water vapor. Relative humidities of 0, 50%, and 100% mean, respectively, that no water vapor is present; that the air contains half as much moisture as the maximum possible; and that the air is saturated.

On a hot day the evaporation of sweat from the skin is the chief way in which the human body dissipates heat, and a high relative humidity is uncomfortable because it slows down the process. On the other hand, a low relative humidity is not desirable because it leads to the drying of the skin and mucous membranes. The regulation of relative humidity is as important a function of a heating or of an air-conditioning system as the regulation of temperature.

From Fig. 13.21 we can see to what extent heating decreases the relative humidity of an air sample and cooling increases its relative humidity. For instance, between 10°C and 20°C the saturated vapor density (which corresponds to 100% relative humidity) just about doubles. This means that if outside air at 10°C whose relative humidity is, say, 70% is taken inside a house and heated to 20°C, the relative humidity indoors will only be 35% because the actual vapor density stays the same. A way to humidify heated air in winter is clearly desirable.

If the outside air is at 30°C with 70% relative humidity, then cooling it down to 24°C is enough to bring it to saturation, which is 100% relative humidity. Further cooling will cause water to condense out. An air-conditioning system therefore should include a way to remove water vapor from the air being cooled.

The **dew point** of air with a given water vapor content is the temperature at which the air would be saturated. For instance, suppose we have a sample of air at 20°C that contains 9.4 g/m^3 of water vapor. Saturated air at 20°C contains 17.3 g/m^3, so the relative humidity of the sample is

$$\text{Relative humidity} = \frac{\text{actual vapor density}}{\text{saturated vapor density}} = \frac{9.4 \text{ g/m}^3}{17.3 \text{ g/m}^3} = 0.54 = 54\%$$

The dew point of the air sample is 10°C because, from Fig. 13.21, saturated air at 10°C contains 9.4 g/m^3 of water vapor. Dew point is used in meteorology as a guide to the possibility of fog. When air temperature and dew point are far apart, there is no danger of fog, whereas when air temperature is near the dew point and is decreasing, fog is likely.

Important Terms

The **temperature** of a body of matter is a measure of the average kinetic energy of random translational motion of its constituent particles.

A **thermometer** is a device for measuring temperature. The two temperature scales in common use are the **Celsius** (centigrade) scale, in which the freezing point of water is assigned the value 0°C and its boiling point the value 100°C, and the **Fahrenheit** scale, in which these points are assigned the values 32°F and 212°F, respectively.

The **coefficient of linear expansion** is the ratio between the change in length of a solid rod of a particular material and its original length per 1° change in temperature. The **coefficient of volume expansion** is the ratio between the change in volume of a sample of a particular solid or liquid and its original volume per 1° change in temperature.

Boyle's law states that, at constant temperature, the absolute pressure of a sample of a gas is inversely proportional to its volume, so that $pV = $ constant at that temperature.

The **absolute temperature scale** has its zero point at $-273°C$; temperatures in this scale are measured in **kelvins** (K). The **Rankine absolute temperature scale** has its zero point at

$-460°F$; temperatures in this scale are designated °R. **Absolute zero** is 0 K = 0°R = $-273°C$ = $-460°F$.

Charles's law states that, at constant pressure, the volume of a sample of a gas is directly proportional to its absolute temperature, so that $V/T = $ constant at that pressure.

The equation $pV/T = $ constant, a combination of Boyle's and Charles's laws, is called the **ideal gas law** and is obeyed approximately by all gases.

According to the **kinetic theory of gases,** a gas consists of a great many tiny individual molecules that do not interact with one another except when collisions occur. The molecules are far apart compared with their dimensions and are in constant random motion. The ideal gas law may be derived from the kinetic theory of gases. The average kinetic energy of gas molecules is proportional to the absolute temperature of the gas. At absolute zero, gas molecules would have no kinetic energy of translational motion according to elementary kinetic theory.

The **relative humidity** of a volume of air is the ratio between the amount of water vapor it contains and the amount that would be present at saturation.

Important Formulas

Celsius and Fahrenheit scales: $T_F = \frac{9}{5}T_C + 32°$

$$T_C = \frac{5}{9}(T_F - 32°)$$

Thermal expansion: $\Delta L = aL_0 \Delta T$

$$\Delta V = bV_0 \Delta T = 3aV_0 \Delta T$$

Boyle's law: $pV = $ constant $\quad (T = $ constant$)$

Absolute temperature scale: $T_K = T_C + 273$

Rankine scale: $T_R = T_F + 460°$

Charles's law: $\dfrac{V}{T} = $ constant $\quad (p = $ constant$)$

Ideal gas law: $\dfrac{pV}{T} = $ constant $= nR$

Molecular kinetic energy: $\text{KE}_{av} = \frac{3}{2}kT$

MULTIPLE CHOICE

1. Two thermometers, one calibrated in the Celsius scale and the other in the Fahrenheit scale, are used to measure the same temperature. The numerical reading on the Fahrenheit thermometer

    a. is proportional to that on the Celsius thermometer
    b. is greater than that on the Celsius thermometer
    c. is less than that on the Celsius thermometer
    d. may be greater or less than that on the Celsius thermometer

2. The volume of a gas sample is proportional to its

    a. Fahrenheit temperature
    b. Celsius temperature
    c. absolute temperature
    d. pressure

3. Absolute zero may be regarded as that temperature at which

    a. water freezes
    b. all gases become liquids
    c. all substances are solid
    d. molecular motion in a gas would be the minimum possible

4. Which of the following formulas expresses the relationship between the pressure and absolute temperature of a gas sample whose volume is fixed?

    a. $\dfrac{p_1}{T_2} = \dfrac{p_2}{T_1}$ **b.** $\dfrac{p_1}{T_1} = \dfrac{p_2}{T_2}$

    c. $\dfrac{p_1}{p_2} = \dfrac{T_2}{T_1}$ **d.** $\dfrac{p_1}{T_1} = \dfrac{T_2}{p_2}$

5. Which of the following statements is not correct?

    a. Matter is composed of tiny particles called molecules.
    b. These molecules are in constant motion.
    c. All molecules have the same size and mass.
    d. The differences between the solid, liquid, and gaseous states can be attributed to the relative freedom of motion of their respective molecules.

6. The molecules of a compound

    a. are all the same
    b. may be different
    c. always consist of exactly two atoms
    d. never contain more than one atom of a given kind

7. Which of the following is not true of molecular motion in a gas?

    a. There is no order in the motion.
    b. There is no uniformity of speed or direction.
    c. There is a definite average speed at a given temperature.

    d. There is a definite average direction of motion at a given temperature.

8. The kinetic-molecular theory of gases predicts that, at a given temperature,

    a. all of the molecules in a gas have the same average speed
    b. all of the molecules in a gas have the same average energy
    c. light gas molecules have lower average energies than heavy gas molecules
    d. light gas molecules have higher average energies than heavy gas molecules

9. The volume of a gas is held constant while its temperature is raised. The pressure the gas exerts on the walls of its container increases because

    a. the masses of the molecules increase
    b. each molecule loses more kinetic energy when it strikes the wall
    c. the molecules are in contact with the wall for a shorter time
    d. the molecules have higher average speeds and strike the wall more often

10. The temperature of a gas is held constant while its volume is reduced. The pressure the gas exerts on the walls of its container increases because its molecules

    a. strike the container walls more often
    b. strike the container walls with higher speeds
    c. strike the container walls with greater force
    d. have more energy

11. Evaporation cools a liquid because

    a. the slowest molecules tend to escape
    b. the fastest molecules tend to escape
    c. the pressure on the liquid decreases
    d. the pressure on the liquid increases

12. Dry air has a relative humidity of

    a. 0   **b.** 1%
    c. 50%   **d.** 100%

13. The higher the temperature of a volume of air, the

    a. more water vapor it can hold
    b. the less water vapor it can hold
    c. greater its possible relative humidity
    d. lower its possible relative humidity

14. When saturated air is cooled,

    a. its relative humidity goes down
    b. its relative humidity goes up
    c. it becomes able to take up more water vapor

d.   some of its water content condenses out

15. Oxygen boils at $-183°C$. This temperature is
a. $-215°F$ b. $-297°F$
c. $-329°F$ d. $-361°F$

16. A temperature of $100°F$ is almost exactly
a. $38°C$ b. $56°C$
c. $122°C$ d. $212°C$

17. A copper bar is 1.0 m long at $20°C$. At what temperature will it be shorter by 1.0 mm?
a. $-17°C$ b. $-39°C$
c. $-59°C$ d. $-79°C$

18. A platinum bar increases in length by 0.10% when its temperature increases by $111°C$. The coefficient of linear expansion of platinum is
a. $1.1 \times 10^{-6}/°C$ b. $9 \times 10^{-6}/°C$
c. $1.1 \times 10^{-5}/°C$ d. $9 \times 10^{-4}/°C$

19. A certain container holds 1 kg of air at atmospheric pressure. When an additional kg of air is pumped into the container at constant temperature, the new pressure is
a. $\frac{1}{2}$ atm   b. 1 atm
c. 2 atm   d. 4 atm

20. If the absolute pressure on 10 ft^3 of air is increased from 30 lb/in^2 to 120 lb/in^2, the new volume of the air will be
a. 2.5 ft^3 b. 5 ft^3
c. 40 ft^3 d. 900 ft^3

21. An absolute temperature of 100 K is the same as a Celsius temperature of
a. $-173°C$ b. $32°C$
c. $212°C$ d. $373°C$

22. The temperature of an object is raised by $50°C$. This is equivalent to an increase in its absolute temperature of
a. 28 K   b. 50 K

c. 90 K   d. 323 K

23. The boiling point of water on the Rankine scale is
a. $-248°R$ b. $-61°R$
c. $485°R$ d. $672°R$

24. A 1.00-L sample of neon at $0°C$ and 100 kPa pressure is compressed to 0.25 L, and its temperature is raised to $273°C$. The new pressure of the gas is
a. 50 kPa   b. 200 kPa
c. 400 kPa   d. 800 kPa

25. A sample of hydrogen gas is compressed to half its original volume while its temperature is held constant. If the average speed of the hydrogen molecules was originally v, their new average speed is
a. $4v$ b. $2v$
c. v d. $\frac{1}{2}v$

26. The average speed of the molecules in a bottle of gas at the pressure p and absolute temperature T is doubled. The new pressure and temperature are, respectively,
a. $2p$, $2T$ b. $2p$, $4T$
c. $4p$, $2T$ d. $4p$, $4T$

27. At which of the following temperatures would the molecules of a gas have twice the average kinetic energy they have at room temperature, $20°C$?
a. $40°C$ b. $80°C$
c. $313°C$ d. $586°C$

28. The mass of a nitrogen molecule is 14 times greater than that of a hydrogen molecule. The temperature of a sample of hydrogen whose average molecular energy is equal to that in a sample of nitrogen at 300 K is
a. 6.5 K   b. 21 K
c. 300 K   d. 4200 K

QUESTIONS

1. When a mercury-in-glass thermometer is placed in very hot water, the mercury drops a little before rising. Why?

2. An aluminum nut is screwed tightly on a brass bolt. Should the assembly be heated or cooled to make it easier to remove the nut? What if a brass nut were screwed tightly on an aluminum bolt?

3. In the construction of a light bulb, wires are led through the glass at the base by means of airtight seals. If the wires were made of copper, what would happen when the light is turned on and the bulb heats up? What must be true for a wire to be successfully used for this purpose?

4. Actual molecules attract one another slightly. Does this tend to increase or decrease gas pressure from values computed from the ideal gas law? Why?

5. At absolute zero an ideal gas sample would occupy zero volume. Why would an actual gas not occupy zero volume at absolute zero?

6. According to the kinetic theory of gases, molecular motion virtually ceases only at absolute zero. How can this be reconciled with the definite shape and volume of a solid at temperatures well above absolute zero?

7. Is it meaningful to say that an object at a temperature of $200°C$ is twice as hot as one at $100°C$?

8. Give as many methods as you can that will increase the rate of evaporation of a liquid sample and explain why each will have this effect.

9. Molecular speeds are comparable with those of rifle bullets, yet a gas with a strong odor, such as ammonia, takes a few minutes to diffuse through a room. Why?

10. Consider the following gases: CO_2, UF_6, H_2, He, Xe, NH_3. (a) Which has the highest rms molecular speed at a given temperature? (b) Which has the lowest rms molecular speed?

EXERCISES

13.1   Temperature

1. The melting point of lead is 330°C and its boiling point is 1170°C. Express these temperatures on the Fahrenheit scale.

2. The normal temperature of the human body is 98.6°F. What is this temperature on the Celsius scale?

3. At what temperature would Celsius and Fahrenheit thermometers give the same reading?

4. Mercury freezes at −40°C. What is this temperature on the Fahrenheit scale?

5. Dry ice (solid carbon dioxide) vaporizes at −112°F. What is this temperature on the Celsius scale?

13.2   Thermal Expansion

6. A steel bridge is 100 m long at 30°C. What is its change in length when the temperature falls to −10°C?

7. The bimetallic strip of a thermostat consists of strips of copper and steel 10 cm long that are fastened together and bent into a coil. If the strips have the same length at 20°C, what is the difference in their lengths at 200°C?

8. How large a gap should be left between steel rails that are 10 m long when laid at 20°C if they are to just barely touch at 30°C?

9. A rod 2.0 m long expands by 1.0 mm when heated from 8°C to 70°C. What is the coefficient of linear expansion of the material from which the rod is made?

10. The outside diameter of a wheel is 1.000 m. An iron tire for this wheel has an inside diameter of 0.992 m at 20°C. To what temperature must the tire be heated in order for it to fit over the wheel?

11. A steel tape measure is calibrated at 70°F. A reading of 120 ft 0 in is obtained when it is used to determine the width of a building at 0°F. What is the true width of the building at 0°F?

12. A train begins a journey at 100.0 km/h. If the steel wheels

of the locomotive heat up while its engine maintains a fixed number of rpm, what rise in temperature is needed for the train's speed to increase to 100.1 km/h?

13. A Pyrex beaker is filled to the brim with 250 cm^3 of glycerin at 15°C. How much glycerin overflows at 25°C?

14. A 1.000-L Pyrex beaker is filled to the brim with water at 20°C. When the beaker is heated, 15 cm^3 of water overflow. What is the new temperature of the beaker and its contents?

15. A 1 pint 0.00 oz aluminum bottle is filled with ethyl alcohol at 40°F. To what temperature must the bottle be heated in order that 0.25 oz of alcohol overflow?

16. The density of an alloy is 10.0 g/cm^3 at 20°C and 9.8 g/cm^3 at 200°C. What is the alloy's coefficient of linear expansion?

17. The density of lead is 11.0 g/cm^3 at 20°C. Find its density at 200°C.

18. Vodka that is "100 proof" is a mixture of half ethyl alcohol and half water. How much profit per liter will a merchant make if he buys vodka at $10.00 per liter at 0°C and sells it at $10.00 per liter at 25°C?

19. Verify that the force the thermal expansion or contraction of a solid object can produce depends on its cross-sectional area but not on its length.

20. A load of 4000 kg is placed on a vertical steel column 3.0 m long and of cross-sectional area 50 cm^2 when the temperature is 20°C. To what should the temperature be increased if the length of the column is to be the same after the load is applied as it was originally?

21. An aluminum mast whose cross-sectional area is 7.20 in^2 is 60.0 long at 60°F. (a) By how much does it increase in length when its temperature rises to 100°F? (b) How much force would be needed to produce this expansion at constant temperature?

22. A rivet is a headed metal pin often used to fasten metal plates together (Fig. 13.23). If a rivet is heated and then

Figure 13.23

Exercise 22.

inserted and its end hammered over while still hot, it shrinks as it cools and forces the metal plates tightly together. Find the force exerted by a steel rivet 10 mm in diameter that cools from 500°C (just hot enough to glow red) to 20°C.

13.3 Boyle's Law

23. If the pressure on 1.0×10 m^3 of nitrogen is increased from 200 kPa to 800 kPa, what will the new volume of the nitrogen be?

24. How much air at sea-level atmospheric pressure can be stored in the 350-L tank of an air compressor that can withstand an absolute pressure of 750 kPa?

25. An oxygen cylinder used for welding contains 1.3 ft^3 of oxygen at 70°F and a gauge pressure of 2500 lb/in^2. Find the weight of the oxygen given that its density is 0.090 lb/ft^3 at 70°F and 1.0 atm pressure.

26. The containers shown in Fig. 13.24 are filled with air at the same temperature and at the pressures given. What will the pressure in the system be when the valve is opened?

2 L
0.3 MPa

1 L
0.6 MPa

Figure 13.24

Exercise 26.

27. A flat tire contains 5.0 L of air at atmospheric pressure, which means an absolute pressure of very nearly 1.0 bar. How many strokes of a pump whose capacity is 0.5 L are needed to raise the pressure in the repaired tire to a gauge pressure of 2.0 bars? Assume the temperature and volume of the tire do not change in the process.

13.4 Charles's Law

28. A piston maintains constant pressure on a sample of helium in a cylinder. When the temperature of the helium is 27°C, its volume is 1.0 L. When the helium is heated, its volume becomes 1.2 L. What is its new temperature?

29. The tires of a stationary car contain air at a gauge pressure of 24.0 lb/in^2 and a temperature of 40°F. When the car is driven for a while, the temperature of the tires increases to

80°F. If the tire volumes remain unchanged, find their new gauge pressure?

30. An air tank used for scuba diving has a safety valve set to open at an absolute pressure of 280 bars. The normal absolute pressure of the full tank at 20°C is 200 bars. If the tank is heated after being filled to the latter pressure, at what temperature will the safety valve open?

31. A sample of gas occupies 2.0 m^3 at 300K and an absolute pressure of 2.0×10^5 Pa. (a) What is its pressure at the same temperature when it has been compressed to a volume of 1.0 m^3? (b) What is its volume at the same temperature when its pressure has been decreased to 1.5×10^5 Pa? (c) What is its volume at a temperature of 400 K and a pressure of 2.0×10^5 Pa?

32. A sample of gas occupies 100 cm^3 at 0°C and 1.0 bar pressure. What is its volume (a) at 50°C and 1.0 bar pressure; (b) at 0°C and 2.2 bars pressure; (c) at 50°C and 2.2 bars pressure?

33. In a certain diesel engine, air at 20°C and 1.0 bar pressure is drawn into each cylinder and compressed by a piston to $\frac{1}{20}$ of its original volume and a pressure of 55 bars. Diesel oil is then injected and is ignited by the high temperature of the air. What is this temperature?

34. The density of air is 1.29 kg/m^3 at 0°C and 1.00 atm pressure. Find its density at 100°C and 2.00 atm pressure.

13.5 Atoms and Molecules

35. Find the mass of a molecule of glucose, $C_6H_{12}O_6$.

36. Find the mass of a molecule of ethyl alcohol, C_2H_5OH.

37. How many iron atoms are present in 100 g of iron?

38. How many hydrogen atoms are present in 100 g of boric acid, H_2BO_3?

13.6 Kinetic Theory of Gases

39. Compare the average and rms speeds of an assembly of two molecules, one with a speed of 1.0 m/s and the other with a speed of 3.0 m/s.

40. What is the average kinetic energy of the molecules of a gas (a) at 0°C? (b) at 100°C?

41. To what temperature must a gas sample initially at 27°C be raised in order for the average energy of its molecules to triple?

42. Silver is a vapor at 1500 K. What is the rms speed of silver atoms in a vapor at this temperature?

43. The rms speed of a hydrogen molecule at room temperature is about 1.6 km/s. What is the rms speed of an oxygen molecule, whose mass is 16 times greater, at this temperature?

44. (a) Find the rms speed of carbon dioxide (CO_2) molecules at 0°C. (b) At what temperature would this speed be doubled?

45. The rms speed of air molecules is roughly 4×10^2 m/s, and the average distance an air molecule goes between collisions with other molecules is about 10^{-7} m. What is the average number of collisions an air molecule makes per second?

PROBLEMS

47. An aluminum piston 80 mm in diameter moves inside a steel cylinder with a clearance of 0.10 mm at 20°C. At what temperature will the two have exactly the same diameter, so the piston can no longer move?

48. The rails of a railway are sometimes welded together instead of being laid with gaps to allow for thermal contraction and expansion. What percentage of the ultimate strength of steel is the thermal stress associated with a temperature change of 40°C?

49. A steel bar exactly 1 m long at 20°C has its ends fixed in place. Obtain a crude estimate of the deflection of the center of the bar when it is heated to 200°C by assuming that the figure formed by the hot and cold bars is a triangle.

50. A sign is suspended from the middle of a steel cable attached to poles 20.00 m apart. When the temperature is 30°C, the cable sags so that the sign is 2.00 m below a horizontal line between the ends of the cable. Find the amount of sag on a winter day when the temperature is −10°C.

51. A concrete swimming pool 12.0 m × 6.0 m × 2.5 m is filled with water to within 6.0 mm of the top when the temperature is 10°C. The coefficient of linear expansion of the concrete used is 0.90×10^{-5}/°C. What will happen to the water level as the temperature increases? If it rises, at what temperature will the water begin to overflow?

52. A thermometer is being made from a glass tube whose inside diameter is 0.300 mm. The distance between the 0°C and 100°C marks is to be 300 mm. What volume should the mercury bulb at the end of the tube have?

53. A bubble of methane ("marsh gas") doubles in volume when it rises from the bottom of a pond to the surface. If the water temperature is the same throughout the pond, how deep is it?

54. A glass tube closed at its lower end contains a 100-mm column of mercury above 100 mm of trapped air, as shown in Fig. 13.25. When the tube is inverted, the mercury column shifts so that the trapped air occupies 131 mm of the tube. Find the atmospheric pressure at the time these measurements were made.

46. Two vessels of the same size are at the same temperature. One of them holds 1 kg of H_2 gas and the other holds 1 kg of N_2 gas. (a) Which vessel contains more molecules? How many times more? (b) Which vessel is under the greater pressure? How many times greater? (c) In which vessel are the rms molecular speeds greater? How many times greater?

Figure 13.25

Problem 54.

55. A diver blows an air bubble 10 mm in diameter at a depth of 10 m in a freshwater lake where the temperature is 5°C. What is the diameter of the bubble when it reaches the surface of the lake where the temperature is 20°C?

56. Two flasks of the same volume are connected together by a narrow tube. The system contains 5.0 g of nitrogen that is initially at a temperature of 300 K and a pressure of 1.0 bar. Flask 1 is then heated to 600 K, while flask 2 remains at 300 K. (a) Find the mass of nitrogen in each flask. (b) Find the pressure in each flask.

57. A cylinder that contains 10.0 kg of gas has a safety valve that allows gas to escape when its pressure exceeds 10.0 bars. The pressure in the cylinder is 9.5 bars when its temperature is 0°C. What mass of gas escapes when the cylinder's temperature rises to 25°C?

Answers to Multiple Choice

1. d	8. b	15. b	22. b
2. c	9. d	16. a	23. d
3. d	10. a	17. b	24. d
4. b	11. b	18. b	25. c
5. c	12. a	19. c	26. d
6. a	13. a	20. a	27. c
7. d	14. d	21. a	28. c

CHAPTER 14

HEAT

14.1   **Internal Energy and Heat**

14.2   **Specific Heat Capacity**

14.3   **Changes of State**

14.4   **The Triple Point**

14.5   **Heat Conduction**

15.6   **Thermal Resistance**

14.7   **Convection**

14.8   **Radiation**

N ow that we know what temperature is, we can go on to examine heat and how it affects matter. Every body of matter contains **internal energy** as well as any kinetic or potential energy it may have because of its motion or position. The internal energy belongs to the particles of which the body is composed and which are in constant motion. We can think of **heat** as internal energy in transit: A rise in a body's temperature means heat was added to it, and a fall in temperature means heat was lost from it. How much the temperature changes for a given amount of heat depends on the mass and the nature of the body. Changes of state—from solid to liquid, from liquid to gas, from solid to gas, and their reverses—involve changes in internal energy without changes in temperature as the particles of the substance alter their relationships with one another. The last part of this chapter deals with the ways heat can be transferred from place to place. Included is a look at how the greenhouse effect heats the earth's atmosphere and why this is a subject of concern today.

14.1 INTERNAL ENERGY AND HEAT

Temperature difference determines the direction of internal energy flow.

How much internal energy a body has depends on its temperature, its composition, its mass, and its state (solid, liquid, or gas). The temperature of the body alone, however, is what governs whether internal energy will go from it to another body it is in contact with, or vice versa. A large block of ice at 0°C has far more internal energy than a cup of hot water. But when the water is poured on the ice, some of the ice melts and the water becomes cooler, so energy has passed from the water to the ice (Fig. 14.1). This energy is what we are accustomed to calling heat:

Heat is internal energy in transit from one body of matter to another by virtue of a temperature difference between them.

The transfer of heat is not the only way to change a body's internal energy. Work done against friction, for instance, ends up as internal energy.

Because heat is a form of energy, the correct SI unit of heat is the joule. However, two older units, the **kilocalorie** (kcal) and the **British thermal unit** (Btu), are still sometimes used. The kilocalorie is the amount of heat required to raise the temperature of 1 kg of water through 1°C. Similarly 1 kcal of heat must be removed from 1 kg of water to reduce its temperature by 1°C. Because this amount of heat actually varies slightly with temperature, the kilocalorie is formally defined as the amount of heat involved in changing the temperature of 1 kg of water from 14.5°C

Figure 14.1

The internal energy of a body depends on its temperature and mass. However, the direction of internal energy flow depends only on the temperatures of the bodies involved.

Ice

Low temperature but large mass, hence much internal energy

High temperature but small mass, hence little internal energy

Internal energy

Internal energy always flows from hot body to cold body, regardless of internal energy content to 15.5°C; the difference can be ignored for most purposes, however. The relationship between the joule and the kilocalorie is as follows:

$$1\,J = 2.39 \times 10^{-4}\,kcal \qquad 1\,kcal = 4185\,J$$

The **calorie** that dieticians use is the same as the kilocalorie. Thus the energy content of a 150-calorie cupcake is really 150 kcal. The carbohydrate content of foods averages 4.1 kcal/g, the protein content 4.2 kcal/g, and the fat content 9.3 kcal/g; pure ethanol (ethyl alcohol) has an energy content of 7.1 kcal/g. (A heat unit once widely used is also called the calorie. This calorie is equal to the heat needed to raise the temperature of 1 g of water by 1°C, so that 1000 cal = 1 kcal. The dietician's calorie is sometimes written "Calorie" to distinguish it from the ordinary, smaller calorie.)

◆ **E X A M P L E 14.1** _____

A woman eats a cupcake and proposes to work off its 150-kcal energy content by exercising with a 10.0-kg barbell. If each lift of the barbell from the floor to over her head is through 2.00 m and the efficiency of her body is 10.0% under these circumstances, how many times must she lift the barbell?

SOLUTION The work done in each lift of the barbell is

$$W_{out} = mgh = (10.0\,kg)(9.80\,m/s^2)(2.00\,m) = 196\,J$$

At 10.0% efficiency the energy used is

$$W_{in} = \frac{W_{out}}{Eff} = \frac{196\,J}{0.100} = 1960\,J = 1.96\,kJ$$

This is equivalent to

$$W_{in} = (1.96\,kJ)(0.239\,kcal/kJ) = 0.468\,kcal$$

The barbell must therefore be lifted

$$\frac{W_{total}}{W_{in}} = \frac{150\,kcal}{0.468\,kcal} = 320\,times$$

◆

The British thermal unit is the amount of heat needed to raise the temperature of 1 lb of water by 1°F. When 1 Btu of heat is removed from 1 lb of water, its temperature falls by 1°F. The Btu is related to the other units of energy and heat as follows:

$$1\,Btu = 1054\,J = 0.252\,kcal = 778\,ft \cdot lb$$

We note that weight rather than mass is specified in defining the Btu. On the earth's surface this is not an important difference in dealing with heat. All the formulas and procedures that follow can be used with British units by expressing temperature in degrees Fahrenheit and heat in Btu and letting *m* represent weight in pounds.

14.2 SPECIFIC HEAT CAPACITY

Different substances need different amounts of heat for the same temperature change.

If we add a given amount of heat to the same mass of different substances, they respond with different temperature changes. Thus 10 kJ added to 1 kg of water raises its temperature by 2.4°C, but the rise is 3.2°C for 1 kg of helium (its volume held constant), 4.8°C for 1 kg of ice, and no less than 77°C for 1 kg of gold (Fig. 14.2).

The **specific heat capacity** (symbol *c*) of a substance refers to what we might think of as its thermal inertia:

The specific heat capacity of a substance is the amount of heat that must be added or removed from a unit mass of the substance to change its temperature by 1°.

A high specific heat capacity means a relatively small change in temperature for a given change in internal energy content, just as a large inertial mass means a relatively small acceleration when a given force is applied.

In the SI system the unit of *c* is the J/kg · °C, and in this system the specific heat capacity of water is 4185 J/kg · °C. Usually it is more convenient to use the

Figure 14.2

When 10 kJ of heat is added to 1 kg of each of the substances shown, their respective rises in temperature differ considerably.

Table 14.1 Average specific heat capacities of various substances

Substance	Specific heat capacity	
	(kJ/kg · °C)	(kcal/kg · °C) (Btu/lb · °F)
Air (1 atm)	0.70	0.17
Alcohol (ethyl)	2.43	0.58
Aluminum	0.92	0.22
Concrete	2.9	0.7
Copper	0.39	0.093
Glass	0.84	0.20
Gold	0.13	0.030
Granite	0.80	0.19
Human body	3.47	0.83
Ice	2.09	0.50
Iron	0.46	0.11
Lead	0.13	0.030
Mercury	0.14	0.033
Silver	0.23	0.056
Steam	2.01	0.48
Water	4.19	1.00
Wood (pine)	1.76	0.42

kilojoule (kJ) in connection with heat, in which case $c = 4.19$ kJ/kg · °C. Table 14.1 lists specific heats for some common substances. The actual values vary somewhat with temperature, and the ones given in the table represent averages. Water has the highest specific heat of any common substance.

When the kcal is being used as the heat unit, the corresponding unit of c is the kcal/kg · °C. Thus the specific heat capacity of water is $c = 1.00$ kcal/kg · °C since 1 kcal of heat is involved when 1 kg of water changes in temperature by 1°C. In the British system, the unit of c is the Btu/lb · °F. Because of the way the kcal and the Btu are defined, the numerical value of c for a substance is the same whether the kcal/kg · °C or the Btu/lb · °F is the unit; thus the specific heat capacity of water is 1.00 Btu/lb · °F.

With the help of specific heat capacity, we can write a formula for the quantity of heat Q involved when a quantity m of a substance undergoes a change in temperature of ΔT. This formula is simply

$$Q = mc\Delta T \qquad\qquad \textit{Quantity of heat} \quad (14.1)$$

Heat transferred = (mass)(specific heat capacity)(temperature change)

◆ **E X A M P L E 14.2**

How much heat must be removed from 14 lb of aluminum in order to cool it from 80°F to 15°F?

SOLUTION   From Table 14.1 the specific heat capacity of aluminum is 0.22 Btu/lb · °F. Since $\Delta T = -65°F$ here,

$$Q = mc\Delta T = (14 \text{ lb})\left(0.22\frac{\text{Btu}}{\text{lb} \cdot °F}\right)(-65°F) = -200 \text{ Btu}$$

The minus sign means that this quantity of heat is to be removed to achieve the temperature change of $-65°F$.   ◆

Figure 14.3

◆   **E X A M P L E   14.3** _____

Water flows through a 3.5-kW instantaneous-type water heater at a rate of 1.5 L/min. If the initial water temperature is 20°C and all the heat produced is added to the water, find the temperature of the hot water.

SOLUTION   Since 3.5 kW = 3.5 kJ/s, the heat provided to the water in 1 min is

$$Q = (3.5 \text{ kJ/s})(60 \text{ s}) = 210 \text{ kJ}$$

The mass of 1 L of water is 1 kg, so the mass of water that flows through the heater per minute is $m = 1.5$ kg. From Eq. (14.1) we have for the rise in temperature

$$\Delta T = \frac{Q}{mc} = \frac{210 \text{ kJ}}{(1.5 \text{ kg})(4.19 \text{ kJ/kg} \cdot °C)} = 33°C$$

The final water temperature is 20°C + 33°C = 53°C (Fig. 14.3).   ◆

◆   **E X A M P L E   14.4** _____

If 0.20 kg of coffee at 90°C is poured into a 0.30-kg cup at 20°C, and we assume that no heat is transferred to or from the outside, what is the final temperature of the coffee?

SOLUTION   We will take the specific heat of coffee to be that of water and the specific heat of the cup to be that of glass. To solve the problem, we begin by noting that

$$Q_{\text{gained}} = Q_{\text{lost}}$$

Heat gained by cup = heat lost by coffee

If the final temperature of both coffee and cup is T, then the heat gained by the cup is

$$Q_{\text{gained}} = m_{\text{cup}}c_{\text{cup}}(T - 20°C) = (0.30 \text{ kg})\left(0.84\frac{\text{kJ}}{\text{kg} \cdot °C}\right)(T - 20°C)$$

$$= (0.25T - 5.0) \text{ kJ}$$

90°C

20°C

74°C

+   =

Figure 14.4

and the heat lost by the coffee is

$$Q_{lost} = m_{coffee}c_{coffee}(90°C - T)$$

$$= (0.20 \text{ kg})\left(4.19 \frac{\text{kJ}}{\text{kg} \cdot °C}\right)(90°C - T) = (75.4 - 0.84T) \text{ kJ}$$

Now we set the heat gained by the cup equal to the heat lost by the coffee and solve for T:

$$Q_{gained} = Q_{lost}$$
$$0.25T - 5.0 = 75.4 - 0.84T$$
$$T = 74°C$$

The temperature of the coffee drops by 16°C as it warms the cup (Fig. 14.4). Evidently the cup must be preheated to get very hot coffee.   ◆

14.3 CHANGES OF STATE

They involve internal energy changes at constant temperature.

Adding or removing heat from a sample of matter does not always lead to a change in its temperature. Instead the sample may change its state from solid to liquid or from liquid to gas when heat is added, or it may change from gas to liquid or from liquid to solid when heat is taken away. Such changes of state take place at definite temperatures for most substances at any given pressure, but for a few (glass or wax, for instance) there is only a gradual softening or hardening over a range of temperatures. Substances of the latter kind are not true solids, however. Their structures are really those of liquids, and their firmness at room temperature is a kind of exaggerated viscosity.

Figure 14.5 shows what happens as we add heat to 1 kg of ice initially at −50°C. The specific heat capacity of ice is 2.09 kJ/kg · °C, and so 105 kJ is needed to bring the ice to 0°C.

Figure 14.5

A graph of the temperature of 1 kg of water, initially ice at −50°C, as heat is added to it. The heat of fusion of water is 335 kJ, and its heat of vaporization is 2260 kJ.

At 0°C the ice begins to melt. The temperature remains the same until all the ice has melted, which requires a total of 335 kJ.

When all the ice has turned to water, the temperature goes up again as more heat is supplied. Because $c = 4.19$ kJ/kg · °C, there is now a rise of 0.24°C per kJ of heat. This rate of change is less than that of ice: The specific heat capacity of water is greater than that of ice, and so the slope of the graph is less steep than it was for ice.

At 100°C the water begins to turn into steam. The temperature stays the same until a total of 2260 kJ is added, at which time all the water has become steam.

The temperature of the steam now begins to rise. The specific heat of steam is 2.01 kJ/kg · °C, so the temperature increase is 0.50°C per kJ of heat and the slope of the graph is therefore steeper than it was for ice or water.

The **heat of fusion** of a substance is the amount of heat needed to turn 1 kg of the substance at its melting point from the solid to the liquid state. Thus the heat of fusion of water is 335 kJ/kg. The same amount of heat must be removed from 1 kg of the substance in the liquid state at its melting point to turn it into a solid. The symbol for heat of fusion is L_f.

Similarly the **heat of vaporization** of a substance is the amount of heat needed to turn 1 kg of the substance at its boiling point from the liquid to the gaseous state. Thus the heat of vaporization of water is 2260 kJ/kg (at atmospheric pressure). The same amount of heat must be removed from 1 kg of the substance in the gaseous state at its boiling point to turn it into a liquid. The symbol for heat of vaporization is L_v. The heats of fusion and vaporization for a number of substances are listed in Table 14.2, together with their melting and boiling points.

Table 14.2   Heats of fusion and vaporization and melting and boiling points of various substances at atmospheric pressure

Substance	Melting point		Heat of fusion, L_f			Boiling point		Heat of vaporization, L_v		
	°C	°F	kJ/kg	kcal/kg	Btu/lb	°C	°F	kJ/kg	kcal/kg	Btu/lb
Alcohol (ethyl)	−114	−173	105	25	45	78	172	854	204	367
Copper	1083	1981	134	32	58	1187	2169	5069	1211	2181
Lead	330	626	25	5.9	10.6	1170	2138	732	175	315
Mercury	−39	−38	12	2.8	5.0	358	676	297	71	128
Nitrogen	−210	−346	26	6.1	11	−196	−320	201	48	86
Oxygen	−219	−362	14	3.3	5.9	−183	−306	213	51	92
Silver	961	1762	88	21	38	2193	3979	2335	558	1004
Tungsten	3410	6170	184	44	79	5900	10,650	4813	1150	2071
Water	0	32	335	80	144	100	212	2260	540	972

It is clear from Table 14.2 that the heat of vaporization of a substance is much greater than its heat of fusion. This follows from the structures of the various states of matter. To turn a solid into a liquid, each of its particles (atoms or molecules) must gain enough energy to loosen (so to speak) the bonds that hold it to its neighbors. A disorderly, constantly shifting arrangement of particles then replaces the fixed pattern characteristic of a solid, but the particles still stick together sufficiently to give a liquid sample a definite volume. To turn a liquid into a gas, however, requires enough energy to permanently break the bonds between the particles. This is a much harder job. The particles of a gas then can move about freely, and the gas expands indefinitely unless prevented from doing so by the walls of a container or by gravity. At atmospheric pressure 1 L of water becomes 1680 L of steam.

The high heat of vaporization of water is what makes steam so dangerous. If a gram of water (about a third of a teaspoonful) at its boiling point of 100°C falls on a person's skin, the heat given to the skin will be

$$Q = mc\Delta T = (0.0010 \text{ kg})(4.19 \text{ kJ/kg} \cdot °C)(100°C - 37°C) = 0.26 \text{ kJ}$$

because body temperature is 37°C. But if a gram of steam at the same temperature strikes the person's skin, the heat of vaporization given up as the steam condenses into water will be

$$Q = mL_v = (0.0010 \text{ kg})(2260 \text{ kJ/kg}) = 2.26 \text{ kJ}$$

for a total of $(2.26 + 0.26)\text{kJ} = 2.52$ kJ—nearly ten times as much heat. Great care must always be taken near steam, in the kitchen and when opening a car's radiator as well as in industry.

 EXAMPLE 14.5 ⎯⎯⎯⎯⎯⎯⎯⎯⎯⎯⎯⎯⎯⎯⎯⎯⎯⎯⎯⎯

How much steam is produced when 1000 Btu of heat is added to 2.00 lb of water at 120°F?

Figure 14.6

2.00 lb water   1000 Btu   1.16 lb water

SOLUTION   The heat needed to raise the temperature of the water from 120°F to its boiling point of 212°F is, since $\Delta T = 212°F - 120°F = 92°F$,

$$Q_1 = mc\Delta T = (2.00 \text{ lb})(1.00 \text{ Btu/lb} \cdot °F)(92°F) = 184 \text{ Btu}$$

Thus $Q_2 = 1000 \text{ Btu} - Q_1 = 816 \text{ Btu}$ of heat is available to convert water at 212°F to steam at the same temperature. From $Q_2 = m_{steam}L_v$ we have

$$m_{steam} = \frac{Q_2}{L_v} = \frac{816 \text{ Btu}}{972 \text{ Btu/lb}} = 0.84 \text{ lb}$$

Of the 2.00 lb of water we started with, 0.84 lb becomes steam at 212°F and 1.16 lb remains as water at 212°F (Fig. 14.6). (As before, for the sake of convenience we use m in the formulas for heat with British units even though weights are specified rather than masses when using this system of units.)   ◆

◆   **E X A M P L E   14.6** ─────────────────────────────

A 20.0-g lead bullet whose temperature is 50°C and whose speed is 400 m/s strikes a large block of ice at 0°C and stops inside it (Fig. 14.7). Assuming that all the energy lost by the bullet goes into melting ice, how much ice melts?

SOLUTION   As it cools to 0°C, the bullet loses the heat

$$Q_{lost} = mc\Delta T = (0.0200 \text{ kg})(0.13 \text{ kJ/kg} \cdot °C)(50°C) = 0.13 \text{ kJ}$$

Figure 14.7

The kinetic energy lost by the bullet in coming to a stop is

$$KE_{lost} = \tfrac{1}{2}mv^2 = \tfrac{1}{2}(0.0200 \text{ kg})(400 \text{ m/s})^2 = 1600 \text{ J} = 1.60 \text{ kJ}$$

If M is the mass of ice that melts, the heat gained by the ice is

$$Q_{gained} = ML_f = (M)(335 \text{ kJ/kg}) = 335M \text{ kJ/kg}$$

From conservation of energy,

$$\text{Heat lost by bullet} + \text{KE lost by bullet} = \text{heat gained by ice}$$
$$Q_{lost} + KE_{lost} = Q_{gained}$$
$$0.13 \text{ kJ} + 1.60 \text{ kJ} = 335M \text{ kJ/kg}$$
$$M = \frac{(0.13 + 1.60)\text{kJ}}{335 \text{ kJ/kg}} = \frac{1.73 \text{ kJ}}{335 \text{ kJ/kg}}$$
$$= 5.2 \times 10^{-3} \text{ kg} = 5.2 \text{ g} \qquad \blacklozenge$$

◆   **E X A M P L E   14.7** _____

What is the minimum amount of ice at $-10°C$ that must be added to 0.50 kg of water at 20°C in order to bring the temperature of the water down to 0°C?

SOLUTION   We begin as before with the statement of energy conservation,

$$Q_{gained} = Q_{lost}$$
$$\text{Heat gained by ice} = \text{heat lost by water}$$

The heat Q_1 absorbed by the unknown mass of ice in going from $-10°C$ to its melting point of 0°C is

$$Q_1 = m_{ice}c_{ice}\,\Delta T_{ice} = m_{ice}\left(2.09\,\frac{\text{kJ}}{\text{kg}\cdot°C}\right)(10°C) = m_{ice}(21 \text{ kJ/kg})$$

and the heat Q_2 absorbed by the ice in melting at 0°C is

$$Q_2 = m_{ice}L_{fice} = m_{ice}(335 \text{ kJ/kg})$$

Hence the total heat absorbed by the ice is

$$Q_{gained} = Q_1 + Q_2 = 356\, m_{ice} \text{ kJ/kg}$$

The heat lost by the water in cooling from 20°C to 0°C is

$$Q_{lost} = m_{water}c_{water}\,\Delta T_{water} = (0.50 \text{ kg})\left(4.19\,\frac{\text{kJ}}{\text{kg}\cdot°C}\right)(20°C) = 42 \text{ kJ}$$

Figure 14.8

0.50 kg
water

0.12 kg
ice

0.62 kg
water

Equating the heat gained with the heat lost and then solving for m_{ice} yields (Fig. 14.8)

$$Q_{gained} = Q_{lost}$$

$$356\, m_{ice}\frac{kJ}{kg} = 42\,kJ$$

$$m_{ice} = 0.12\,kg$$

This mass is a minimum because some heat will be lost to the outside world.   ◆

Under certain circumstances most substances can change directly from the solid to the vapor state, or vice versa. Both processes are called **sublimation**. For example, "dry ice" (solid carbon dioxide) evaporates directly to gaseous carbon dioxide at temperatures above $-78.5°C$ without passing through the liquid state. With the exception of carbon dioxide and a few other substances, however, sublimation does not occur except at pressures well below that of the atmosphere.

14.4 THE TRIPLE POINT

All three states of matter in equilibrium.

Changes of state are affected by pressure. Figure 14.9 shows how the boiling point of water varies with pressure: The higher the pressure, the higher the boiling point. This is the principle that underlies the pressure cooker. By heating water in a sealed container, the pressure can be raised well above 1 atm, and the temperature at which the water inside boils will be correspondingly higher than 100°C. In this way

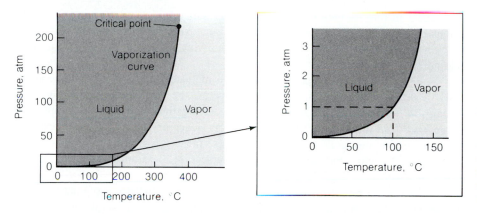

Figure 14.9

The vaporization curve of water, showing how the boiling point of water varies with pressure. Above the temperature of its critical point, a substance cannot exist in the liquid state.

food can be cooked more rapidly than in an open pan. At a pressure of 2 atm, for instance, water boils at 120°C. The converse is also true: Lowering the pressure below 1 atm reduces the boiling point below 100°C. Atmospheric pressure decreases with altitude, so the boiling point of water is lower than 100°C in regions located above sea level. Water boils at 96°C in Denver, for instance.

The upper limit of the **vaporization curve** of water shown in Fig. 14.9, which occurs at a temperature of 374°C and a pressure of 218 atm, is known as the **critical point**. A substance cannot exist in the liquid state at a temperature above that of its critical point, regardless of how great the pressure may be. The gas becomes more and more dense with increasing pressure but does not condense into a liquid with a definite interface between liquid and gas. The average molecular energy at and past the critical temperature is simply too great to allow the molecules to form the bonds with one another characteristic of the liquid state. Helium has the lowest critical temperature, −268°C, and is therefore a gas at all temperatures above that.

The melting points of solids also depend on pressure, although to a smaller extent than the boiling points. The variation of the melting point of ice with pressure is shown in the **fusion curve** of Fig. 14.10. Ice, together with gallium and bismuth,

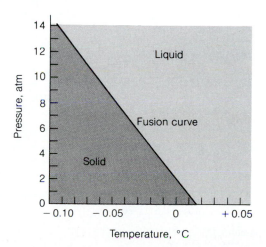

Figure 14.10

The fusion curve of water, showing how the melting point of ice varies with pressure.

Figure 14.11

Because the melting point of ice decreases with pressure, a thin wire with weights at both ends can melt its way through a block of ice.

is unusual in that its melting point *decreases* with increasing pressure. The melting points of all other substances increase with increasing pressure. Hence it is possible to melt ice at temperatures below 0°C by applying sufficient pressure to it without adding heat (Fig. 14.11).

The fusion and vaporization curves of water intersect at a temperature of 0.01°C and a pressure of 4.6 torr, as shown on the combined plot of Fig. 14.12. Along the fusion curve ice and water can exist together, and along the vaporization curve water and water vapor can exist together. Hence under conditions corresponding to those of the intersection of the two curves, the solid, liquid, and vapor states of water can all occur simultaneously. This intersection is accordingly called the **triple point** of water.

At pressures below that of its triple point, no substance can exist as a liquid. The dividing line on a pressure-temperature graph between the solid and vapor states is called the **sublimation curve**. This curve represents the conditions required for a solid to vaporize directly or a vapor to solidify directly. At atmospheric pressure adding heat melts ice, because the triple point of water lies well below 1 atm. However, adding heat causes solid carbon dioxide to sublime, because its triple point lies above 1 atm (Fig. 14.13).

At a temperature below that of its triple point, a substance passes directly from the solid state to the vapor state if the pressure is low enough. This phenomenon is the basis of the **freeze-drying** process widely used for the preservation of foods, blood plasma, and biological samples. Usually the material is cooled below the triple point of water in a gas-tight chamber, and then the chamber is evacuated with a vacuum pump to remove the water vapor given off. Freeze-drying affects the structure of a material of biological origin less than other methods of dehydration do.

Figure 14.12

Triple-point diagram of water. The solid, liquid, and vapor phases of water can exist simultaneously at the temperature and pressure of the triple point.

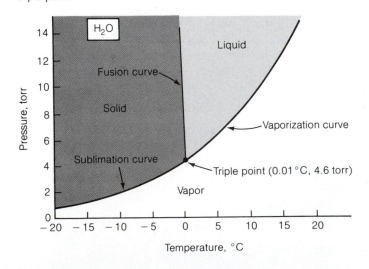

Figure 14.13

Triple-point diagram of carbon dioxide.

14.5 HEAT CONDUCTION

Metals are the best conductors.

Figure 14.14 shows the three mechanisms of heat transfer. When we put one end of a metal rod in a fire, the other end becomes warm as heat is conducted through the metal. **Conduction** is very slow in air, and a stove warms a room chiefly through the actual movement of heated air. This process is called **convection**. Neither conduction nor convection can take place to any real extent in space, which is virtually empty of matter. Instead the heat the earth receives from the sun arrives in the form of **radiation**. These mechanisms all embody a fundamental fact: The natural direction of heat flow is from hot bodies to cold ones.

In most materials conduction is a consequence of the kinetic behavior of matter. Molecules (or atoms, depending on the nature of the rod) at the hot end of a rod vibrate faster and faster as the temperature there increases. When these molecules collide with their less energetic neighbors, some kinetic energy is transferred to them (Fig. 14.15). Through such molecular collisions energy travels down the rod, something we can equally well describe by saying that heat travels down the rod. The average positions of the molecules themselves do not change in conduction.

For heat to be conducted through a body, its ends must be at different temperatures. If the entire body is at the same temperature, all its molecules have the same average energy, and any molecule has as much chance of losing energy in a collision with a nearby molecule as it has of gaining energy. In this case there is no flow of energy from a region of rapidly moving molecules to an adjacent one of slowly moving molecules.

There are wide differences in how well various substances conduct heat. Gases are poor conductors, because their molecules are relatively far apart and so do not collide often. The molecules of liquids and nonmetallic solids are closer together, leading to somewhat higher thermal conductivities. Metals exhibit by far the greatest ability to conduct heat, which is why saucepans are made of metal but have handles of wood or plastic.

(a) Conduction

(b) Convection

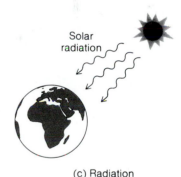

(c) Radiation

Figure 14.14

Mechanisms of heat transfer.

Figure 14.15

Molecules at the hot end of a rod vibrate faster than those at the cold end.

Figure 14.16

The rate Q/t at which heat flows through the slab is equal to $kA \, \Delta T/d$.

The reason for the exceptional thermal conductivity of metals is that a significant number of electrons in metals are able to move about freely instead of being bound permanently to particular atoms. Picking up kinetic energy at the hot end of a metal object, the free electrons can travel past many atoms before giving up their energy in collisions and thereby can speed up the rate of energy transport toward the cold end of the object. Heat conduction by free electrons in a metal compares with heat conduction by molecular interactions like travel by express train compares with travel by local train. In a later chapter we will explore further the important subject of free electrons in metals.

The rate at which heat flows through a slab of some material depends on four quantities (Fig. 14.16):

1. The temperature difference ΔT between the faces of the slab: The greater the temperature difference, the faster the heat transfer.
2. The thickness d of the slab: The thicker the slab, the slower the heat transfer.
3. The area A of the slab: The greater the area, the faster the heat transfer.
4. The **thermal conductivity** k of the material, which is a measure of its ability to conduct heat: The larger the value of k, the faster the heat transfer.

The amount of heat Q that passes through the slab in the period of time t is given by

$$\frac{Q}{t} = \frac{kA\Delta T}{d}$$

Heat conduction   (14.2)

In SI units Q/t is expressed in watts and k in W/m · °C. When Q/t is expressed in kcal/s, the corresponding unit for k is the kcal/m · s · °C.

In the U.S. Customary System, it is normal in problems that involve heat transfer to have Q/t in Btu/h, ΔT in °F, d in in., and A in ft². The British unit for k

Table 14.3 Thermal conductivities

Material	$k, \dfrac{W}{m \cdot °C}$	$k, \dfrac{kcal}{m \cdot s \cdot °C}$	$k, \dfrac{Btu}{ft^2 \cdot h \cdot °F/in}$
Metals			
Aluminum	205	4.9×10^{-2}	1.4×10^3
Brass	109	2.6×10^{-2}	0.76×10^3
Copper	385	9.2×10^{-2}	2.7×10^3
Iron and steel	50	1.2×10^{-2}	0.35×10^3
Silver	406	9.7×10^{-2}	2.8×10^3
Other solids			
Body fat	0.17	0.4×10^{-4}	1.2
Brick	0.6	1.5×10^{-4}	4.4
Concrete	0.8	2×10^{-4}	5.8
Glass	0.8	1.9×10^{-4}	5.5
Ice	1.6	4×10^{-4}	12
Water	0.60	1.4×10^{-4}	4.2
Wood (pine)	0.13	0.3×10^{-4}	0.87
Snow	0.8	2×10^{-4}	5.8
Insulating materials			
Cork	0.04	0.1×10^{-4}	0.3
Glass wool	0.04	0.1×10^{-4}	0.3
Down	0.02	0.06×10^{-4}	0.17
Kapok	0.03	0.08×10^{-4}	0.23
Gases			
Hydrogen	0.13	0.3×10^{-4}	0.9
Air	0.024	0.06×10^{-4}	0.17

is therefore the Btu/(ft^2 · h · °F/in). Table 14.3 is a list of the thermal conductivities of a number of common materials.

◆ **E X A M P L E   14.8** _____

A door of pine wood is to be installed in a brick wall 10 in thick. If the rate of heat conduction is to be no greater than before, what minimum thickness should the door have?

SOLUTION   Since A and ΔT are equal in both cases, the ratio k/d must be the same for Q/t to be the same. Hence

$$\frac{k_{brick}}{d_{brick}} = \frac{k_{pine}}{d_{pine}}$$

$$d_{pine} = \left(\frac{k_{pine}}{k_{brick}}\right)(d_{brick}) = \left(\frac{0.87}{4.4}\right)(10 \text{ in}) = 2.0 \text{ in}$$

◆

◆   E X A M P L E   14.9 _____

An electric water heater has a cylindrical tank 60 cm in diameter and 50 cm long that is insulated with 2.0 cm of glass wool. If the water in the tank is at 80°C and the outside of the insulating jacket is at 25°C, at what rate does the tank lose heat to the air around it? How many kWh (kilowatt-hours) of electric energy must be provided each day to keep the water at 80°C even if no water is drawn from the tank?

SOLUTION   (a) Each end of the tank is a circle 30 cm in radius (Fig. 14.17) whose area is

$$A_{end} = \pi r^2 = \pi (0.30 \text{ m})^2 = 0.283 \text{ m}^2$$

The body of the tank is a cylinder 30 cm in radius and 50 cm long whose area is

$$A_{cylinder} = (\text{circumference})(\text{length}) = (2\pi r)(L) = (2\pi)(0.30 \text{ m})(0.50 \text{ m})$$
$$= 0.942 \text{ m}^2$$

The total area of the tank is

$$A = 2A_{end} + A_{cylinder} = 2(0.283 \text{ m}^2) + 0.942 \text{ m}^2 = 1.51 \text{ m}^2$$

With $k = 0.04$ W/m · °C for glass wool, $\Delta T = 55$°C, and $d = 0.020$ m, the rate at which heat is lost from the tank by conduction is

$$\frac{Q}{t} = \frac{kA\Delta T}{d} = \frac{(0.04 \text{ W/m} \cdot °\text{C})(1.51 \text{ m}^2)(55°\text{C})}{0.020 \text{ m}} = 166 \text{ W}$$

(b) In the course of 24 h the total energy needed to make up the heat loss is

$$Q = \left(\frac{Q}{t}\right)(t) = (0.166 \text{ kW})(24 \text{ h}) = 4.0 \text{ kWh}$$

It is worth noting that the statement of the problem gives the temperature of the outside of the tank. If instead we were only given the room temperature, we would

Figure 14.17

not know by just how much the insulating effect of the air layers outside the tank affects the actual temperature at the tank's surface. This insulating effect can be large enough to reduce the heat flow considerably. Only if there is a very vigorous circulation of air past the tank will the temperature of the outside of the tank equal room temperature.   ◆

Because the rate of heat flow Q/t through an insulating layer depends on $1/d$, in any particular application it usually does not make sense to increase the insulation thickness d beyond a certain value. Consider a freezer that absorbs 8 MJ/day through 5 cm of insulation. Doubling the thickness to 10 cm will decrease the heat flow to 4 MJ/day. A further doubling to 20 cm of insulation will reduce the heat flow to 2 MJ/day. As the thickness increases, the rate of change of heat flow goes down: Going from 5 cm to 10 cm cuts the heat flow by 4 MJ/day, but going from 10 cm to 20 cm brings it down only by an additional 2 MJ/day. Clearly there is a point at which the cost of more insulation averaged over an appropriate period of time will exceed the savings in operating costs. Just what this point is depends, of course, on the relative costs involved, including that of the space occupied by the insulation.

14.6 THERMAL RESISTANCE

What R values mean.

The insulating ability of a layer of some material is often given in terms of its **thermal resistance** R. If the thermal conductivity of the material is k and it is d thick,

$$\text{Thermal resistance} = \frac{\text{thickness}}{\text{thermal conductivity}}$$

$$R = \frac{d}{k} \qquad\qquad \textit{Thermal resistance} \quad (14.3)$$

In terms of R rather than k/d, Eq. (14.2) becomes

$$\frac{Q}{t} = \frac{A\Delta T}{R} \qquad\qquad \textit{Heat conduction} \quad (14.4)$$

The greater the value of R, the greater the resistance to the flow of heat.

Figure 14.18

Two ways to interpret a thermal resistance of $R = 2.3$ ft$^2 \cdot$ °F/(Btu/h).

In customary units, thermal resistance is expressed in ft$^2 \cdot$ °F/(Btu/h). Thus the thermal resistance of 2.0 in of pine wood is

$$R = \frac{d}{k} = \frac{2.0 \text{ in}}{0.87 \text{ Btu/(ft}^2 \cdot \text{h} \cdot \text{°F/in)}} = 2.3 \frac{\text{ft}^2 \cdot \text{°F}}{\text{Btu/h}}$$

We can think of this result in two ways (Fig. 14.18):

1. If the temperature difference across the wood is 2.3°F, then 1 Btu of heat flows through each square foot.
2. If the temperature difference is 1°F, then 1 Btu/h of heat flows through each 2.3 ft^2.

Usually R values are given without units, so the above result would be expressed as "R-2.3."

◆   **E X A M P L E   14.10** —————————————————————

Find the rate at which heat flows through an R-2.3 pine door 7.0 ft high and 2.5 ft wide when the inside temperature is 70°F and the outside temperature is 20°F.

SOLUTION   The temperature difference here is $\Delta T = 50$°F and the area of the door is $A = (7.0 \text{ ft})(2.5 \text{ ft}) = 17.5$ ft^2. From Eq. (14.4) the heat flow through the door from the inside to the outside is

$$\frac{Q}{t} = \frac{A \Delta T}{R} = \frac{(17.5 \text{ ft}^2)(50\text{°F})}{2.3 \text{ ft}^2 \cdot \text{°F/(Btu/h)}} = 380 \text{ Btu/h}$$

                                                            ◆

◆   **E X A M P L E   14.11** —————————————————————

Compare the thermal resistance of a pane of glass $\frac{1}{4}$ in thick with that of a sandwich consisting of two panes of glass $\frac{1}{8}$ in thick separated by $\frac{1}{4}$ in of air (Fig. 14.19).

SOLUTION   (a) The R value of the single pane of glass is

$$R = \frac{d}{k} = \frac{0.25 \text{ in}}{5.5 \text{ Btu/(ft}^2 \cdot \text{h} \cdot \text{°F/in)}} = 0.045 \frac{\text{ft}^2 \cdot \text{°F}}{\text{Btu/h}} = \text{R-0.045}$$

(b) The R value of glass $\frac{1}{8}$ in thick is half that of glass $\frac{1}{4}$ in thick, so here $R_1 = R_3 = 0.023$ ft$^2 \cdot$ °F/(Btu/h). The R value of the air gap is

$$R_2 = \frac{d}{k} = \frac{0.25 \text{ in}}{0.17 \text{ Btu/(ft}^2 \cdot \text{h} \cdot \text{°F/in)}} = 1.47 \frac{\text{ft}^2 \cdot \text{°F}}{\text{Btu/h}} = \text{R-1.47}$$

Thus the thermal resistance of the glass-air-glass sandwich is

$$R = R_1 + R_2 + R_3 = (0.023 + 1.47 + 0.023)\frac{\text{ft}^2 \cdot \text{°F}}{\text{Btu/h}} = 1.52\frac{\text{ft}^2 \cdot \text{°F}}{\text{Btu/h}} = \text{R-1.52}$$

This is nearly 34 times greater than the R value of the same total thickness of glass found in (a). The advantage of double-glazed windows in regions with cold winters is obvious. For even better insulation, triple-glazed windows can be used. Such a window consists of three panes of glass separated by two air spaces. Why not just have a thicker air space between two panes of glass? The reason is that heat flow through a thick layer of air occurs by means of convection as well as conduction, and the insulating effect is thereby reduced. ◆

Figure 14.19

A double-glazed window has a much higher thermal resistance than a single-glazed one.

14.7 CONVECTION

It involves the motion of a heated fluid.

Convection is a much simpler physical process than conduction because it consists of the actual motion of a volume of hot fluid from one place to another. The hot fluid displaces cold fluid in its path, thereby setting up a **convection current**. Convection is the chief mechanism of heat transfer in fluids under most circumstances.

Convection may be either natural or forced. In natural convection, the buoyancy of a heated fluid leads to its motion. When a portion of a fluid (either gas or liquid) is heated, it expands to become of lower density than the surrounding, cooler fluid and hence rises upward (Fig. 14.20). A steam or hot-water heating system uses radiators in each room that heat the rooms with the help of the convection currents they set up.

In forced convection, a blower or pump directs the heated fluid to its destination. In the cooling system of most cars, water is circulated between the engine block and the radiator by a pump. In the radiator, heat is transferred to the atmosphere by conduction through the thin-walled metal tubes of which it is made. A hot-air household heating system uses a fan to blow the air from the furnace through ducts to outlets in each room. In an animal's body, excess metabolic heat is carried by the blood to the skin, through which it passes by conduction. This heat is then dissipated

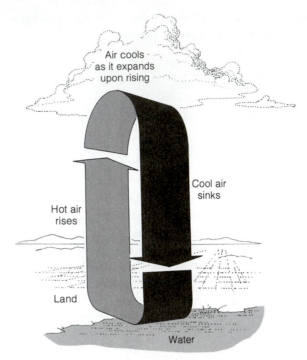

Figure 14.20

The temperature of a land surface rises more rapidly than that of a water surface in sunlight. The resulting convection is responsible for the sea breeze experienced on sunny days near the shore of a body of water.

by the evaporation of moisture from the skin, by convection to the atmosphere, and by radiation.

The rate Q/t at which a hot object transfers heat to a surrounding fluid by convection is approximately proportional to the area A of the object in contact with the fluid and to the temperature difference ΔT between them:

$$\frac{Q}{t} = hA\Delta T \qquad\qquad\qquad\qquad \textit{Convection} \quad (14.5)$$

The convection coefficient h depends on the form and orientation of the object (h is nearly twice as great for the upper surface of a horizontal plate as it is for the lower surface) and on the properties and state of motion of the fluid (in the case of convective loss from the skin to air, h is twice as great when the air is moving at 1 m/s as it is in quiet air).

  E X A M P L E   14.12 _____

A photographic enlarger has as its light source a 25-W lamp in a metal hood that keeps light from escaping except through the enlarger lens. When the darkroom temperature is 20°C, the metal hood is at a temperature of 50°C. If the 25-W lamp is replaced by a 50-W lamp, what is the new temperature of the hood?

SOLUTION Because h and A are each the same in both cases,

$$\frac{(Q/t)_1}{(Q/t)_2} = \frac{\Delta T_1}{\Delta T_2}$$

When $(Q/t)_1 = 25$ W, the temperature difference ΔT_1 between the hood and the surrounding air is $50°C - 20°C = 30°C$. When $(Q/t)_2 = 50$ W,

$$\Delta T_2 = \Delta T_1 \frac{(Q/t)_2}{(Q/t)_1} = (30°C)\left(\frac{50 \text{ W}}{25 \text{ W}}\right) = 60°C$$

Hence the new hood temperature is $20°C + 60°C = 80°C$. ◆

14.8 RADIATION

The hotter an object, the faster it radiates energy in the form of electromagnetic waves.

In the process of radiation, energy is carried by **electromagnetic waves**. These waves travel at the speed of light (3.00×10^8 m/s = 186,000 mi/s) and require no material medium for their passage. Radio and radar waves, light waves, and X- and gamma rays are all electromagnetic waves; they differ only in their wavelength. Figure 14.21 shows the classification of electromagnetic waves according to wavelength. Visible light ranges in wavelength from about 400 nm for violet light

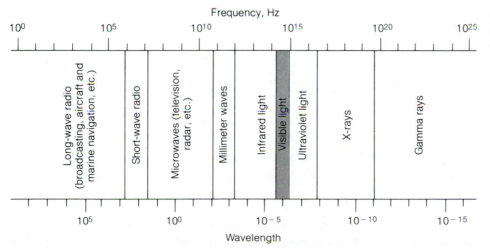

Figure 14.21

The electromagnetic wave spectrum. The boundaries of the various categories are not sharp except in the case of visible light.

through blue, green, yellow, and orange light to about 700 nm for red light (1 nm = 1 nanometer = 10^{-9} m).

Every object radiates electromagnetic waves of all wavelengths, though the intensities of the different wavelengths vary considerably. We are all familiar with the glow of a hot piece of metal, where enough radiation is emitted as visible light for our eyes to respond, but other wavelengths are also given off. An object need not be so hot that it gives off visible light for it to be radiating electromagnetic energy—the radiation from an object at room temperature, for instance, is mainly in the infrared part of the spectrum to which the eye is not sensitive.

The ability of an object to emit radiation is proportional to its ability to absorb radiation: A good absorber is a good emitter, and vice versa. This conclusion follows from the fact that something at the same temperature as its environment must be absorbing and emitting radiation at exactly the same rates. When an object is warmer than its environment, it emits more radiation than it absorbs, and it is this difference we detect. A perfect absorber is called a **blackbody**, and it is accordingly the best possible radiator as well.

The rate R (in W/m^2) at which an object of surface area A and absolute temperature T emits radiation is given by the **Stefan-Boltzmann law** as

$$R = \frac{P}{A} = e\sigma T^4 \qquad\qquad \textit{Stefan-Boltzmann law} \quad (14.6)$$

The value of the constant σ (Greek letter *sigma*) is

$$\sigma = 5.67 \times 10^{-8} \text{ W/m}^2 \cdot \text{K}^4$$

The **emissivity** e depends on the nature of the radiating surface and ranges from 0, for a perfect reflector that does not radiate at all, to 1, for a blackbody. Some typical values of e are 0.07 for polished steel, 0.6 for oxidized copper and brass, and 0.97 for matte black paint.

Figure 14.22 shows how the intensity of the radiation emitted by an object varies with wavelength at various temperatures. The total emitted radiation (which is proportional to the area under each curve) increases with increasing temperature, while the wavelength corresponding to the peak of each curve decreases. An object that glows red is not as hot as one that glows bluish-white, because red light has the longer wavelength. Figure 14.23 shows how the color of a hot object varies with its temperature.

Thermograms of a man riding an exercise bicycle show skin temperatures as shades of gray. The skin is at the highest temperature (lightest gray) over active muscles.

  E X A M P L E 14.13 _____

A **thermograph** is a device that measures the amount of infrared radiation each small portion of a person's skin emits. It presents this information in pictorial form by different shades of gray or different colors in a **thermogram**. The skin over a tumor is warmer than elsewhere (perhaps because of increased blood flow or a higher rate of metabolism), and thus a thermogram can help diagnose such diseases as breast and

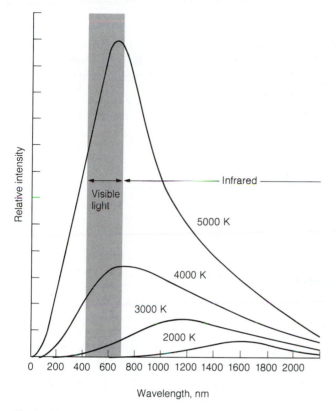

Figure 14.22

The intensity of the electromagnetic radiation emitted by an object at various temperatures, as a function of wavelength. The greater the area under each curve, the greater the rate at which the object radiates energy. Not only does a hot object radiate more power than a cool one, but the predominant wavelength in the radiation is shorter. (1 nm = 10^{-9} m)

Figure 14.23

The color of an object hot enough to glow varies with its temperature approximately as shown here.

thyroid cancer. To verify that a small difference in skin temperature leads to a significant difference in radiation rate, calculate the percentage difference between the radiation from skin at 34°C and at 35°C.

SOLUTION   The emissivity of the skin is the same at both temperatures, whose values on the absolute scale are $T_1 = 34°C + 273 = 307$ K and $T_2 = 35°C + 273 = 308$ K. Because R_1 is proportional to T_1^4 and R_2 is proportional to T_2^4,

$$\frac{R_2 - R_1}{R_1} = \frac{T_2^4 - T_1^4}{T_1^4} = \frac{(308 \text{ K})^4 - (307 \text{ K})^4}{(307 \text{ K})^4} = 0.013$$

which is 1.3%.   ◆

Figure 14.24

A small hole in the wall of a cavity acts as a blackbody since any radiation that falls on it is trapped inside until absorbed. The emissivity of such a hole is e = 1.00.

◆   **E X A M P L E   14.14**

A small hole in the wall of a cavity in an object of any kind behaves like a blackbody because any radiation that falls on it is trapped inside by reflections from the cavity wall until it is absorbed (Fig. 14.24). At what rate does radiation escape from a hole 10 cm² in area in the wall of a furnace whose interior is at a temperature of 700°C?

SOLUTION   The absolute temperature of the interior of the furnace is $T = 700°C + 273 = 973$ K. Since the hole acts as a blackbody, its emissivity is $e = 1.00$. The hole's area is $A = (10 \text{ cm}^2)(10^{-4} \text{ m}^2/\text{cm}^2) = 10^{-3}$ m². From Eq. (14.6) the hole radiates energy at the rate

$$P = RA = e\sigma AT^4 = (1.00)(5.67 \times 10^{-8} \text{W/m}^2 \cdot \text{K}^4)(10^{-3}\text{m}^2)(973 \text{ K})^4$$
$$= 51 \text{ W}$$

◆

The Greenhouse Effect

Owing to its high surface temperature, the sun emits a great deal of visible light. Most of the sunlight that reaches the top of the earth's atmosphere is either reflected back into space or is absorbed by the earth itself (Fig. 14.25). Because the average temperature of the earth's surface is only 15°C, its own radiation is concentrated in the long-wavelength infrared to which the eye is not sensitive. This infrared radiation is readily absorbed in the atmosphere, chiefly by CO_2 (carbon dioxide), whose molecules then transfer energy to the other constituents of the atmosphere. Thus the principal source of the energy that warms the atmosphere and drives its winds is reradiation from the earth, not direct sunlight. This phenomenon is called the **greenhouse effect**. The interior of a greenhouse is warmer than the outside because

Figure 14.25

The greenhouse effect. Much of the short-wavelength visible light from the sun that reaches the earth's surface is reradiated as long-wavelength infrared that is absorbed in the atmosphere, largely by CO_2. Thus the atmosphere is heated chiefly from below by the earth rather than from above by the sun. The total energy the earth and its atmosphere radiate into space on the average equals the total energy they receive from the sun.

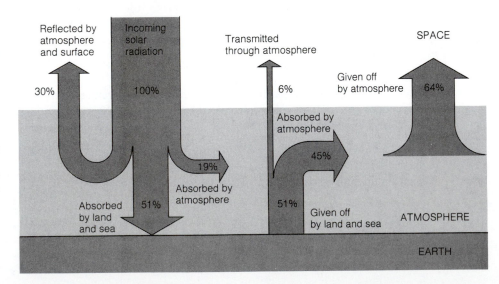

sunlight can enter through its windows but the infrared that the interior gives off cannot get through glass, so the incoming solar energy is trapped.

The CO_2 content of the atmosphere has been increasing for over a hundred years owing to the burning of the fossil fuels coal, oil, and natural gas to heat our homes and to power industry and transportation. At present our chimneys and exhaust pipes pour out 20 billion tons of CO_2 per year. About half is absorbed by vegetation, mainly forests, and by the oceans. The rest stays in the atmosphere, where it enhances the greenhouse effect. The result seems to have been a warming of the atmosphere that may have already changed climates.

If temperatures continue to increase—the six warmest years on record have all occurred since 1980—climate patterns will shift even more, and plants and animals will have a hard time keeping up with their new environments. Many species will disappear and continental interiors will tend to dry out, leaving agriculture unable to keep up with the needs of a human population much larger than today's. Farther ahead, the polar icecaps may melt on a large enough scale to flood coastal areas everywhere.

To be sure, it is possible that natural absorbers of CO_2 may become more effective as the CO_2 concentration increases, so the result may not be worldwide catastrophe. Nobody can say for sure: We are in the midst of a gigantic experiment whose outcome is unknown. To stop the experiment means a drastic cut in our use of fossil fuels, not an easy thing to do, as we will see in Chapter 15.

Important Terms

Heat is internal energy in transit from one body of matter to another by virtue of a temperature difference between them. If a body of matter does not change state during the addition or removal of heat and neither does work nor has work done on it, the change in its internal energy results in a corresponding change in temperature. The SI unit of heat is the joule. Another unit of heat in common use is the **kilocalorie** (kcal), which is the amount of heat required to change the temperature of 1 kg of water by 1°C. The **British thermal unit** (Btu) is the amount of heat required to change the temperature of 1 lb of water by 1°F.

The **specific heat capacity** of a substance is the amount of heat required to change the temperature of 1 kg of the substance by 1°C.

The **heat of fusion** of a substance is the amount of heat that must be supplied to change 1 kg of the substance at its melting point from the solid to the liquid state. The same amount of heat must be removed from 1 kg of the substance in the liquid state at its melting point to change it to a solid.

The **heat of vaporization** of a substance is the amount of heat

that must be supplied to change 1 kg of the substance at its boiling point from the liquid to the gaseous (or vapor) state. The same amount of heat must be removed from the substance at its boiling point to change the substance into a liquid.

Sublimation is the direct conversion of a substance from the solid to the vapor state, or vice versa, without it first becoming a liquid.

The **critical point** is the upper limit of the vaporization curve of a substance; a substance cannot exist in the liquid state at a temperature above that of its critical point.

The **triple point** is the intersection of the vaporization, fusion, and sublimation curves of a substance. All three states of a substance may exist in equilibrium at the temperature and pressure of its triple point.

Heat can be transferred from one place to another by means of **conduction, convection,** or **radiation.** In conduction heat is transported by successive molecular collisions, and in convection by the motion of a volume of hot fluid from one place to another. Heat transfer by radiation takes place by means of

electromagnetic waves, which require no material medium for their passage.

The **thermal conductivity** of a material is a measure of its ability to conduct heat. The **thermal resistance** of a layer of a material is a measure of its insulating ability.

The **greenhouse effect** refers to the heating of the earth's atmosphere not by direct sunlight but by infrared light reradiated by the surface and absorbed mainly by atmospheric carbon dioxide.

Important Formulas

Heat and temperature change: $\quad Q = mc\,\Delta T$

Heat conduction: $\quad Q = \dfrac{kAt\,\Delta T}{d}$

Thermal resistance: $R = \dfrac{d}{k}$

$\qquad\qquad = R_1 + R_2 + R_3 + \cdots$

Convection: $\quad \dfrac{Q}{t} = hA\,\Delta T$

Stefan-Boltzmann law: $\quad R = e\sigma T^4$

MULTIPLE CHOICE

1. The natural direction of the heat flow between two reservoirs depends on
    a. their temperatures
    b. their internal energy contents
    c. their pressures
    d. whether they are in the solid, liquid, or gaseous state

2. Of the following substances, the one that requires the greatest amount of heat per kilogram for a given increase in temperature is
    a. ice   **b.** water
    c. steam   **d.** copper

3. One gram of steam of 100°C causes a more serious burn than 1 g of water at 100°C because the steam
    a. is less dense
    b. strikes the skin with greater force
    c. has a higher specific heat capacity
    d. contains more internal energy

4. The freezing point of water is 0°C. Its melting point is
    a. slightly less than 0°C
    b. 0°C
    c. slightly more than 0°C
    d. 32°C

5. When a vapor condenses into a liquid,
    a. it absorbs heat
    b. it evolves heat
    c. its temperature rises
    d. its temperature drops

6. The heat of vaporization of a substance is
    a. less than its heat of fusion
    b. equal to its heat of fusion
    c. greater than its heat of fusion
    d. any of the above, depending on the nature of the substance

7. A pressure cooker cooks food more rapidly than an ordinary pot with a loose lid because
    a. the pressure forces heat into the food
    b. the higher pressure lowers the boiling point of water
    c. the higher pressure raises the boiling point of water
    d. the higher pressure increases the specific heat capacity of water

8. Sublimation refers to
    a. the vaporization of a solid without first becoming a liquid
    b. the melting of a solid
    c. the vaporization of a liquid
    d. the condensation of a gas into a liquid

9. Under conditions corresponding to its triple point, a substance
    a. is in the solid state
    b. is in the liquid state
    c. is in the gaseous state
    d. may be in any or all of the above states

10. In the process of freeze drying, ice goes directly into water vapor. The temperature at which this can take place is
    a. below the triple point of water
    b. at the triple point of water
    c. above the triple point of water
    d. any of the above, depending on the pressure

11. Heat transfer by conduction occurs
- **a.** only in liquids
- **b.** only in solids
- **c.** only in liquids and solids
- **d.** in gases, liquids, and solids

12. Heat transfer by convection occurs
- **a.** only in gases
- **b.** only in liquids
- **c.** only in gases and liquids
- **d.** in gases, liquids, and solids

13. The rate at which heat flows through a slab of some material does not depend on
- **a.** the temperature difference between the faces of the slab
- **b.** the thickness of the slab
- **c.** the area of the slab
- **d.** the specific heat capacity of the material

14. Metals are good conductors of heat because
- **a.** they contain free electrons
- **b.** their atoms are relatively far apart
- **c.** their atoms collide infrequently
- **d.** they have reflecting surfaces

15. The materials with the highest heat conductivities are the
- **a.** gases
- **b.** liquids
- **c.** woods
- **d.** metals

16. In natural convection a heated portion of a fluid moves because
- **a.** its molecular motions become aligned
- **b.** of molecular collisions within it
- **c.** its density is less than that of the surrounding fluid
- **d.** of currents in the surrounding fluid

17. Four pieces of iron are heated in a furnace to different temperatures. The one at the highest temperature appears
- **a.** white
- **b.** yellow
- **c.** orange
- **d.** red

18. In order to emit electromagnetic radiation, an object must be at a temperature
- **a.** above 0 K
- **b.** above 0°C
- **c.** above that of its surroundings
- **d.** high enough for it to glow

19. The rate at which an object radiates electromagnetic energy does not depend on its
- **a.** surface area
- **b.** mass
- **c.** temperature
- **d.** ability to absorb radiation

20. In an hour the heat produced by a 1-kW heating element equals
- **a.** 3.4 Btu
- **b.** 1054 Btu
- **c.** 3416 Btu
- **d.** 3600 Btu

21. Thirty kilojoules of heat is removed from 1.4 kg of ice initially at $-10°C$. Its new temperature is
- **a.** $-10.2°C$
- **b.** $-20.2°C$
- **c.** $-30°C$
- **d.** $-98°C$

22. Hot water at 100°C is added to 300 g of water initially at 0°C until the mixture is at 40°C. The minimum mass of added hot water is
- **a.** 60 g
- **b.** 75 g
- **c.** 120 g
- **d.** 200 g

23. If 1.0 kg of punch of specific heat capacity 3.4 kJ/kg · °C at a temperature of 5°C is poured into a 1.0-kg glass punch bowl that is at 20°C, the final temperature of the punch is
- **a.** 6°C
- **b.** 8°C
- **c.** 10°C
- **d.** 12°C

24. A 1.0-kg lead bar at 80°C is placed in 2.0 kg of water at 20°C. The final temperature of the lead bar is
- **a.** 22°C
- **b.** 28°C
- **c.** 40°C
- **d.** 50°C

25. Fifty kilojoules of heat is added to a 20-kg block of ice at 0°C. The amount of ice that melts is
- **a.** 0.093 kg
- **b.** 0.15 kg
- **c.** 0.63 kg
- **d.** 2.5 kg

26. Ten lb of ice at 0°F is added to 100 lb of water at 50°F. The temperature of the resulting mixture is
- **a.** 19°F
- **b.** 31°F
- **c.** 32°F
- **d.** 34°F

27. If 10.0 kg of ice at 0°C is added to 2.0 kg of steam at 100°C, the temperature of the resulting mixture is
- **a.** 0°C
- **b.** 23°C
- **c.** 28°C
- **d.** 40°C

28. Water at 50°C can be obtained by mixing together which one or more of the following combinations?
- **a.** 1 kg of ice at 0°C and 1 kg of steam at 100°C
- **b.** 1 kg of ice at 0°C and 1 kg of water at 100°C
- **c.** 1 kg of water at 0°C and 1 kg of steam at 100°C
- **d.** 1 kg of water at 0°C and 1 kg of water at 100°C

29. Fifteen megajoules of heat is removed from 5 kg of steam initially at 200°C. The steam afterward is in the form of
- **a.** a mixture of steam and water
- **b.** water
- **c.** a mixture of water and ice
- **d.** ice

30. The 3.0-mm iron door of an oven is replaced by a brass door 9.0 mm thick. With the oven and room temperatures the same as before, the rate of heat flow through the door relative to its original rate is

    a. 15%   **b.** 73%

    c. 138%   **d.** 300%

31. A brick wall is 5.0 m long, 3.0 m high, and 30 cm thick. The conductivity of brick is 0.60 W/m · °C. When the outside temperature of the wall is 0°C and the inside temperature is 20°C, the rate at which heat flows through the wall is

    a. 54 W   **b.** 60 W

    c. 0.60 kW   **d.** 1.7 kW

32. When the difference in temperature between the sides of a wooden board 80 cm long, 50 cm wide, and 10 mm thick is 40°C, heat flows through the board at the rate of 240 W. The thermal conductivity of the wood is

    a. 0.0015 W/m · °C   **b.** 0.15 W/m · °C

    c. 6.7 W/m · °C   **d.** 6.7 J/m² · °C

33. The thermal conductivity of cork is 0.3 Btu (ft² · h · °F/in). An R-2.5 layer of cork has a thickness of

    a. 0.12 in   **b.** 0.75 in

    c. 2.5 in   **d.** 8.3 in

34. The average temperature of the earth's surface is 15°C. Assuming that the earth radiates as a blackbody, the rate at which it loses energy by radiation is

    a. 0.0029 W/m²   **b.** 251 W/m²

    c. 288 W/m²   **d.** 390 W/m²

35. A brass rod at a temperature of 150°C radiates energy at a rate of 20 W. If its temperature is increased to 300°C, at what rate does it radiate now?

    a. 27 W   **b.** 40 W

    c. 67 W   **d.** 320 W

QUESTIONS

1. Why will the engine of a car whose cooling system is filled with an alcohol antifreeze be more likely to overheat in summer than one whose cooling system is filled with water?

2. How does perspiration give the body a means of cooling itself?

3. In cooling a drink, does it make any difference whether 10 g of water at 0°C is added or 10 g of crushed ice at 0°C? If so, which is more effective and why?

4. When a certain amount of a vapor condenses into a liquid, what happens to its internal energy content and to its temperature?

5. What would the processes of evaporation and sublimation look like when drawn on a graph like that in Fig. 14.5?

6. If you are unhappy about the rate at which eggs cook in a pan of boiling water, would it make sense to turn up the gas flame? To use a pressure cooker?

7. Carbon dioxide is usually shipped in tanks under a pressure of approximately 70 atm. At 20°C, is the carbon dioxide a solid, a liquid, or a gas?

8. The pressure and temperature of the atmosphere at 35,000 m are 4 torr and −23°C, respectively. What is the state of water under those conditions? Of carbon dioxide?

9. Solid wood has a thermal conductivity about twice that of sawdust of the same kind. Can you account for the difference?

10. In the winter, why does the steel blade of a shovel seem colder than its wooden handle?

11. Why is it advantageous to install the heating element of an electric hot water heater near the bottom of the tank?

12. Under what circumstances does an object radiate electromagnetic waves? How is the predominant wavelength in the radiation related to the temperature of the object?

13. By what mechanism or mechanisms does a person seated in front of a fire receive heat from it? A person seated in front of a radiator through which hot water is circulated?

14. A sufficiently hot piece of metal glows brightly, but a transparent piece of quartz at the same temperature hardly glows at all. What is the reason for the difference?

15. A bare copper pipe carries hot water to a faucet. What effect, if any, will polishing the pipe have on the rate of heat flow from the pipe to the room?

16. If a red star and a white star radiate energy at the same rate, can they be the same size? If not, which must be the larger? Why?

17. A Thermos bottle consists of two glass vessels, one inside the other, with the space between them evacuated. The vessels are both coated with thin films of silver. Why is this device so effective in keeping the contents of the bottle at a constant temperature?

EXERCISES

14.2 Specific Heat Capacity

1. How much heat must be removed form 60 lb of ice at 20°F to lower its temperature to −20°F?

2. A 25-kg storage battery has an average specific heat of 0.84 kJ/kg · °C. When fully charged, the battery contains 1.4 MJ of electrical energy. If all of this energy were dissipated within the battery, find the increase in its temperature.

3. Twenty kilojoules of heat are added to a 400-g block of wood at 20°C. The final temperature of the wood is 50°C. Find its specific heat capacity.

4. Find the heat needed to raise the temperature of the air in a room 4.0 m × 5.0 m × 3.0 m from 5°C to 20°C.

5. A person decides to lose weight by eating only cold food. A 100-gram piece of apple pie yields about 1.5 MJ of energy when eaten. If its specific heat capacity is 1.7 kJ/kg · °C, how much greater is its energy content at 50°C than at 20°C? What percentage difference is this?

6. A 50-gal water heater is rated at 55,000 Btu/h. If 20% of the heat escapes, how long does the heater take to raise the temperature of 50 gal of water from 50° to 140°F?

7. In an oil-fired central heating system, 0.60 kg of water is heated in a boiler to 75°C each second. This water is circulated through the radiators in the house and returns to the boiler at 30°C. The heat of combustion of the oil is 45 MJ/kg, of which 2/3 actually goes into heating the water. How much oil does the system use per hour?

8. A space vehicle returning to earth is first slowed by its motors to a speed of 450 m/s and then is slowed further to 100 m/s by air resistance in the atmosphere. Its average specific heat capacity is 0.60 kJ/kg · °C. Find the increase in its temperature on the assumption that half the heat produced is absorbed by the vehicle and that this heat is distributed uniformly throughout its volume.

9. An essential part of a residential solar heating system is some way to store heat for use at night and on cloudy days. In a particular system the water used for storage is initially at 70°C, and it is required to provide an average of 7.5 kW to keep the house at an average of 18°C for three days. (a) What is the minimum amount of water required? (b) How many 200-L drums would be needed to hold the water?

10. Air is blown past the 2.0-kW element of an electric heater at a rate of 0.050 m³/s. What is the increase in temperature of the air that passes through the heater?

11. A 300-g aluminum pot containing 0.50 L of water is placed on the 1.0-kg iron hot plate of an electric stove. The heating element of the hot plate is rated at 1.5 kW. If the hot plate, the pot, and the water are all at 20°C when the heating element is switched on, how long will it take until the water begins to boil? Neglect heat losses to the room.

12. The 30-kg copper tank of a 5.0-kW water heater has a capacity of 100 kg of water. The water temperature in the tank is 15°C when it is switched on and is 50°C an hour later. What percentage of energy input was lost?

13. A bathtub contains 70 kg of water at 26°C. If 10 kg of water at 90°C is poured in, what is the final temperature of the mixture? Neglect the heat absorbed by the bathtub.

14. A 100-g piece of silver is taken from a bath of hot oil and placed in an 80-g glass jar containing 200 g of water at 15°C. The temperature of the water increases by 8°C. What was the temperature of the oil?

15. A 600-g copper container holds 1.5 kg of water at 20°C. A 100-g iron ball at 120°C is dropped into the water. What is the final temperature of the water?

14.3 Changes of State

16. Figure 14.26 shows the heating curve for 1.0 kg of a substance other than water. From the curve find the values of the following thermal properties of the substance: (a) melting point, (b) boiling point, (c) specific heat capacity of the solid, (d) specific heat capacity of the liquid, (e) specific heat capacity of the gas, (f) heat of fusion, and (g) heat of vaporization.

Figure 14.26

Exercise 16.

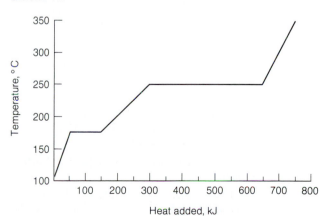

17. A person is sitting in the shade where the air temperature is 37°C, which is the same as his body temperature. Under these circumstances the chief way his body gets rid of the 120 W his metabolic processes liberate is through the evaporation of sweat. How much sweat per hour is required? At 37°C the heat of vaporization of water is 2430 kJ/kg. (In Section 13.7 we saw how it is possible for a liquid to evaporate at a temperature below its boiling point.)

18. Hot water enters a radiator at 70°C and leaves it at 40°C. If instead steam at 100°C is used that leaves the radiator as water at 80°C, how much steam would provide as much heat as 1.0 kg of the original hot water?

19. How much heat is needed to change 4.0 kg of water at 10°C into ice at −10°C?

20. How much heat (in British thermal units) is needed to change 3.0 lb of water at 80°F to steam at 250°F?

21. How much steam is produced when 1.0 MJ of heat is added to 1.0 kg of water at 50°C?

22. One hundred kilojoules of heat is added to 600 g of zinc at 20°C, and 75 g of it melts. If the specific heat capacity of zinc is 0.385 kJ/kg · °C and it melts at 420°C, what is its heat of fusion?

23. One way to produce fresh water from seawater is to boil seawater and condense the steam, leaving the salt behind. (a) Find the minimum amount of heat needed to produce 1.0 L of fresh water in this way from seawater at 18°C. Assume that the specific heat capacity and heat of vaporization of seawater are the same as those of fresh water. (b) At what rate must energy be supplied to produce 100,000 L of fresh water per day by this method?

24. Banana ice cream freezes at −2°C and has a heat of fusion of 290 kJ/kg. Its specific heat capacity is 3.3 kJ/kg·°C when melted and 1.9 kJ/kg · °C when frozen. How much refrigeration capacity, in kilowatts, is needed to make 500 kg of banana ice cream at −12°C in an 8-h working day from a mix initially at 18°C?

25. If all the heat lost by 1.0 kg of water at 0°C when it turns into ice at 0°C could be turned into kinetic energy, what would the speed of the ice be?

26. A 1.0-kg block of ice at 0°C falls into a lake whose water is also at 0°C, and 0.010 kg of ice melts. What was the minimum altitude from which the ice fell?

27. A lead bullet at 100°C strikes a steel plate and melts. What was its minimum speed?

28. A 30-g lead bullet whose temperature is 80°C strikes a large block of ice at 0°C and stops inside it, and 4.5 g of the ice melts. What was the bullet's speed?

14.5 Heat Conduction

29. What thicknesses of (a) concrete, (b) brick, and (c) pine wood have the same insulating ability as 10 cm of glass wool?

30. The walls of a certain refrigerator are equivalent to a 5.6-m² layer of glass wool 75 mm thick. If the interior of the refrigerator is at an average of 4°C and the kitchen temperature is 20°C, what is the rate of heat flow (in watts) through the refrigerator walls?

31. A house window consists of a pane of glass 1.0 m wide, 1.2 m high, and 5.0 mm thick. If the inner and outer faces of the glass are at temperatures of 15°C and 5°C, respectively, at what rate is heat lost by conduction through the window?

32. The neoprene foam material of a scuba diver's wet suit has a thermal conductivity of 0.075 W/m · °C. (a) Find the rate of heat loss of a diver whose skin area is 1.7 m² wearing a suit of such material 6.0 mm thick who is diving in 10°C water. Assume a skin temperature of 32°C. (b) How much heat does the diver lose in a dive lasting 1.0 h? (The suit material when submerged is squeezed by water pressure to a smaller thickness than at the surface. At a depth of 20 m the thickness is about half that at the surface, for instance. The figure quoted above refers to the actual thickness at the diver's depth.)

33. A double boiler has an aluminum upper pan whose bottom is 20 cm in diameter and 0.90 mm thick. If the lower pan contains boiling water and the upper one milk at 5°C, compute the rate of heat transfer to the milk. What happens to this rate as the milk warms up?

34. The hinges of a freezer door 15 cm thick are attached by four brass bolts 4.0 mm in diameter that pass through the entire door and are secured on the inside by nuts. If the interior of the freezer is at −18°C and the room it is in is at 18°C, what is the rate in watts at which heat is lost through the bolts?

35. An oven has a brick wall 5.0 in thick through which heat is found to pass at a rate of 5.0 Btu/ft² per minute. If the outside of the oven is at 200°F, find the temperature of the inside of the oven.

36. Temperatures in the earth's crust in a certain region increase with depth at a rate of 1.0°C per 35 m. If the thermal conductivity of crustal rock in that region is 1.7 W/m · °C, what is the rate of heat flow outward there per square meter?

37. An igloo of compacted snow has a surface area of 24 m² and a wall thickness of 40 cm. The thermal conductivity of the snow is 8.37×10^{-4} kJ/m · s · °C. If the temperature inside the igloo is 20°C and the outside temperature is −10°C, how much blubber (heat of combustion = 39.8 MJ/kg) must be burned per hour to keep the interior temperature constant?

Neglect the melting of the igloo wall and the heat evolved by the igloo's inhabitants.

38. An insulated copper rod 5.0 mm on a side and 1.0 m long has one end in a steam bath and the other in contact with a block of ice at 0°C. How much ice melts per hour?

14.6 Thermal Resistance

39. A sheet of R-0.080 plasterboard is 0.50 in thick. Find the thermal conductivity of the plasterboard.

40. An icebox 4.0 ft × 3.0 ft × 2.0 ft is made of R-0.30 plywood on the inside and outside with R-10.00 foam insulation between the plywood walls. If the box contains ice at 32°F and the outside temperature is 80°F, how much ice melts per hour?

41. A 12-cm brick wall between two rooms in a house has a 10 mm layer of plaster ($k = 0.5$ W/m · °C) on each side. If the temperature difference between the sides of the wall is 3°C, find the rate at which heat is conducted through each square meter of the wall.

42. A double-glazed window 6.0 ft × 6.0 ft consists of two panes of glass 0.25 in thick separated by 0.25 in of air. The temperature of the outside of the window is 40°F and that of the inside is 65°F. (a) Find the heat that passes through the window per hour. (b) The window leaks and the space between the panes has filled up with water. If the outside and inside temperatures are the same as before, find the heat that passes through the window per hour now.

14.7 Convection

43. A roast turkey cools from 65°C to 60°C in 5 min when it is in a room whose temperature is 20°C. How long will it take to cool from 45°C to 40°C in the same room?

44. A glass of beer warms from 7°C to 10°C in 3 min when the air temperature is 35°C. How long will it take to warm from 12°C to 15°C?

PROBLEMS

51. A 1200-kg car whose transmission is in neutral is coasting down a 10° hill with its brakes being used to maintain a constant speed. If the total mass of the car's iron brake drums is 20 kg, what is the increase in their temperature per 100 m of travel? Assume that the only friction acting occurs in the brakes and that all the heat produced is absorbed by the drums, so the resulting figure is a maximum.

52. (a) Show that the increase in length of a metal bar when it absorbs an amount of heat Q does not depend on the original

14.0 Radiation

45. A blackbody is at a temperature of 500°C. What should its temperature be for it to radiate twice as much energy per second?

46. The sun radiates approximately as a blackbody at a temperature of 5800 K, and solar energy arrives at the earth at a rate of about 1.4 kW/m². At what rate would solar energy arrive if the sun were 10% cooler?

47. Sunspots appear dark, although their temperatures are typically 5000 K, because the rest of the sun's surface is even hotter, about 5800 K. Compare the radiation rates of surfaces of the same emissivity whose temperatures are, respectively, 5000 K and 5800 K.

48. A copper sphere 50 mm in diameter whose emissivity is 0.30 is heated in a furnace to 400°C. At what rate does it radiate energy?

49. A naked person whose skin area is 1.5 m² and whose skin temperature is 33°C is sitting in a room at 25°C. Find the rates at which the person loses heat by convection and by radiation. The convection coefficient h for such a person is about 7 W/m² · °C, and at such temperatures the emissivity of human skin is very nearly 1 (the radiation is chiefly in the infrared part of the spectrum). Take into account the heat the person gains by radiation from the room.

50. The color of an object hot enough to glow varies with its temperature approximately as follows:

Temperature, °C	Appearance
500	Just visible
750	Cherry red
1000	Orange
1250	Light yellow

If a certain blackbody radiates at 1.0 kW when its temperature is 500°C, what will it radiate at the other temperatures listed above?

length of the bar. (b) How much heat in kJ must be added to an iron bar 4.0 mm in diameter if its length is to increase by 1.0 mm?

53. A 50-kg block of ice at 0°C is pushed across a wooden floor also at 0°C for a distance of 20 m. A total of 25 g of ice melts as a result of the friction of the block on the floor. What is the minimum coefficient of friction in this case?

54. How much steam at 150°C is needed to melt 200 g of ice at 0°C?

55. How much ice at $-10°C$ is required to cool a mixture of 0.10 kg ethyl alcohol and 0.10 kg water from 20°C to 5°C?

56. Water at 10°C from a river is used to condense spent steam at 120°C from an electric generating plant to water at 50°C. If the cooling water leaves the condenser at 30°C, how many kilograms of river water are needed per kilogram of steam?

57. By mistake, 0.20 kg of water at 0°C is poured into a vessel containing liquid nitrogen at $-196°C$. How much nitrogen vaporizes?

58. If 6.0 kg of ice at $-10°C$ is added to 6.0 kg of water at $+10°C$, what is the temperature of the resulting mixture?

59. If 10.0 kg of steam at 104°C is passed through 10.0 kg of water at 93°C, what are the temperature and the physical state of the mixture afterward?

60. A 5.0-kg iron bar is taken from a forge at a temperature of 1000°C and plunged into a pail containing 10.0 kg of water at 60°C. How much steam is produced?

61. An unknown mass of a certain solid, initially at 60°C, is heated at a uniform rate. After 4.0 min the solid reaches its melting point of 80°C, and it remains at that temperature for 3.0 min until completely melted. After a further 6.0 min of heating, the resulting liquid is at 100°C. If the specific heat capacity of the substance in the solid state is 2.0 kJ/kg · °C, find the heat of fusion of the substance and its specific heat capacity in the liquid state.

62. Three slabs of identical dimensions, one each of aluminum, silver, and brass, are stacked together in that order. The aluminum side of the stack is placed on a block of ice at 0°C, and steam at 100°C is directed at the brass slab, so that the outer surfaces of the stack are at these respective temperatures.

(a) Find the temperature at the aluminum-silver interface. (b) Find the temperature at the silver-brass interface.

63. The layer of ice on a frozen lake is 50 mm thick at 5 P.M. During the night the air temperature is $-2°C$ and the temperature of the water beneath the ice is 0°C. How thick is the ice at 8 A.M. the next day?

64. Water in an outdoor swimming pool 5.0 m × 8.0 m is covered by a 32-mm layer of ice. If the water is at 0°C and the surrounding air is at $-10°C$, how many millimeters of ice form per hour?

65. A woman is taking a sunbath with 0.50 m^2 of her skin exposed to solar radiation, which reaches her at the rate of 1.2 kW/m^2. Her skin temperature is the same as the air temperature. She has light skin whose emissivity is 0.70, so that 0.70 of the radiation reaching her is absorbed. Her metabolic processes result in the liberation of heat at a rate of 80 W. How much sweat must she evaporate per hour in order to get rid of the metabolic heat and absorbed solar energy? The heat of vaporization of sweat is 2430 kJ/kg.

Answers to Multiple Choice

1. a	10. a	19. b	28. d
2. b	11. d	20. c	29. c
3. d	12. c	21. b	30. b
4. b	13. d	22. d	31. c
5. b	14. a	23. b	32. b
6. c	15. d	24. a	33. b
7. c	16. c	25. b	34. d
8. a	17. a	26. d	35. c
9. d	18. a	27. d	

C H A P T E R

15

THERMODYNAMICS

15.1 First Law of Thermodynamics

15.2 Work Done by and on a Gas

15.3 Second Law of Thermodynamics

15.4 Carnot Engine

15.5 Engine Efficiency

15.6 Internal Combustion Engines

15.7 Refrigerators

15.8 Thermodynamics of Refrigeration

15.9 The Energy Problem

Thermodynamics has as its basic concern the transformation of heat into mechanical energy. Thermodynamics thus plays a central role in technology, because almost all the "raw" energy available for our use is in the form of heat. A device or system that converts heat into mechanical energy is called a **heat engine,** and the principles that govern its operation are the same whether it is a car engine whose heat source is the burning of gasoline, a steam turbine whose heat source is a nuclear reactor, the earth's atmosphere whose heat source is the sun, or a human being whose heat source is the food he or she eats.

15.1 FIRST LAW OF THERMODYNAMICS

Heat engines must conserve energy.

Three characteristic processes take place in all heat engines:

1. Heat is absorbed from a source at a high temperature.
2. Mechanical work is done.
3. Heat is given off at a lower temperature.

Different heat engines carry out these processes in different ways, but the general pattern of operation is always the same (Fig. 15.1). Both hot and cold reservoirs are needed to produce a flow of heat from which work can be extracted. Some of the heat input to the engine must pass through it unused or there will be no heat flow and hence no work output.

Two general principles have been found to apply to all heat engines. The **first law of thermodynamics** expresses the conservation of energy: Energy cannot be created or destroyed, but may be converted from one form to another. In general terms this law states that, in any process a system of some kind undergoes,

Net heat input = change in internal energy + net work output

$$Q = \Delta U + W \qquad \textit{First law of thermodynamics} \quad (15.1)$$

Figure 15.1

The work output of a heat engine is the difference between the amount of heat it takes in from a high-temperature reservoir and the heat it exhausts to a low-temperature reservoir. The low-temperature reservoir is needed for a flow of heat to occur, part of which the engine turns into work.

In Eq. (15.1) Q is the net heat added to the system during the process; if the system gives off heat, Q is negative. When the internal energy U of the system increases, ΔU is positive; when U decreases, ΔU is negative. The net work done *by* the system on something else during the process is W, so W is negative if work is done *on* the system.

If the system is an engine that operates in a cycle, energy may be stored and released from storage, but the engine does not undergo a net change in its internal energy. In this case

Net heat input = work output

The net heat input equals the amount of heat the engine takes in from a reservoir at high temperature minus the amount of heat it gives off to a reservoir at low temperature. In a steam engine the high-temperature reservoir is the boiler, and the low-temperature reservoir is the escaping steam. In a gasoline engine the high-temperature reservoir is the exploding mixture of air and gasoline vapor in each cylinder, and the low-temperature reservoir is the exhaust gas. In the earth's atmosphere the ultimate high-temperature reservoir is the sun, and the ultimate low-temperature reservoir is the rest of the universe.

A **refrigerator** is a heat engine operating in reverse. Ordinarily a heat engine absorbs heat from a high-temperature reservoir and exhausts it to a low-temperature reservoir, with mechanical work being done in the process. In a refrigerator, on the other hand, heat is transferred from a low-temperature reservoir (typically a storage chamber) to a high-temperature reservoir (the outside world), and mechanical energy must be supplied in order to do this. In effect, heat must be "pushed" if it is to go from a cold region to a warm one (Fig. 15.2).

High-temperature reservoir

Refrigerator

Energy input

Low-temperature reservoir

Figure 15.2

A refrigerator takes heat from a low-temperature reservoir and transfers it to a high-temperature one. Energy must be supplied to permit this reverse flow of heat to occur.

It is important to keep in mind that a refrigerator does not "produce cold." Cold is a relative deficiency of internal energy and not something in its own right, as heat is. What a refrigerator does is to take internal energy from a specific region and shift it elsewhere.

15.2 WORK DONE BY AND ON A GAS

An expanding gas does work, but work must be done on a gas to compress it.

The work output of most heat engines is produced by an expanding gas: Steam in the case of a steam engine or turbine, a mixture of burning gases in the case of gasoline, diesel, and jet engines. Let us look into the factors that govern the work done by a gas when it expands.

Figure 15.3 shows a gas-filled cylinder with a movable piston. The cross-sectional area of the cylinder is A and the gas pressure is p. From the definition of pressure as $p = F/A$, the force on the piston is

$$F = pA$$

If the distance Δs through which the piston moves is small, the pressure P remains very nearly constant during the expansion, and the work ΔW done by the expanding gas is

$$\text{Work} = (\text{force})(\text{distance})$$
$$\Delta W = F\Delta s = pA\Delta s$$

Since $A\Delta s$ is the change in volume ΔV of the gas,

$$\Delta W = p\Delta V \qquad \qquad \textit{Work done by expanding gas} \quad (15.2)$$

Figure 15.3

The work ΔW done by a gas under the pressure p in expanding by the amount ΔV is $\Delta W = p\Delta V$.

Figure 15.4

The work done by a gas at the constant pressure p when it expands from V_1 to V_2 is $W = p(V_2 - V_1)$, which equals the shaded area under the curve that represents the process.

If the gas is compressed rather than expanded, ΔV is negative and ΔW is negative also. Thus work is done *on* a gas during a compression.

An **isobaric** process is one during which a system remains under constant pressure. Hence an isobaric expansion at the pressure p from an initial volume V_1 to a final volume V_2 involves the work

$$W = p\,(V_2 - V_1) \qquad\qquad\qquad \textit{Isobaric expansion} \quad (15.3)$$

This expansion is plotted on a graph of pressure versus volume in Fig. 15.4. The expansion is the horizontal line at the pressure p from V_1 to V_2. The shaded region under this line has an area of (height)(width) $= (p)(V_2 - V_1)$. But this is the same as the work done in the expansion! We therefore have the very useful result that

---

The work done by an expanding gas equals the area under the corresponding curve on a pressure-volume (p-V) diagram.

What about an expansion in which the pressure changes? In such a case we can imagine the expansion to consist of a series of small expansions. The first expansion involves a volume change of ΔV_1 at the pressure p_1, the second a volume change of ΔV_2 at the pressure p_2, and so on (Fig. 15.5). The total work done is

$$W = p_1\Delta V_1 + p_2\Delta V_2 + p_3\Delta V_3 + \cdots$$

Figure 15.5

Even when the pressure of an expanding gas varies, the work it does equals the area under the curve of the process on a p–V graph. To verify this, the region under the curve can be imagined as divided into thin strips, each of which represents a small expansion at constant pressure.

The actual smooth curve of pressure versus volume has been replaced by a jagged curve, but by making the volume changes ΔV small enough, we can get as close as we want to the actual curve. Since the work done in each little expansion ΔV equals the area under the horizontal line segment at its top, the total work done in the expansion equals the area under the p-V curve that represents it, in accord with the statement given above.

◆   **E X A M P L E   15.1** _____

The heat of vaporization of water at atmospheric pressure is $L_v = 2260$ kJ/kg. (a) How much of this heat represents work done to expand the water into steam against the pressure of the atmosphere? (b) What becomes of the rest of the heat? At $T = 100°C$ and $p = 1.00$ atm, the density of water is 1.00×10^3 kg/m^3 and the density of steam is 0.600 kg/m^3.

SOLUTION   (a) The volumes of 1.00 kg of water and 1.00 kg of steam under the specified conditions are, respectively,

$$V_1 = \frac{m}{\rho_{water}} = \frac{1.00\,\text{kg}}{1.00 \times 10^3\,\text{kg/m}^3} = 1.00 \times 10^{-3}\,\text{m}^3$$

$$V_2 = \frac{m}{\rho_{steam}} = \frac{1.00\,\text{kg}}{0.600\,\text{kg/m}^3} = 1.67\,\text{m}^3$$

Atmospheric pressure is $p = 1.013 \times 10^5$ Pa. Hence the work done during the expansion of 1.00 kg of water to steam is, from Eq. (15.3),

$$W = p(V_2 - V_1) = (1.013 \times 10^5\,\text{Pa})(1.67\,\text{m}^3 - 0.00100\,\text{m}^3)$$
$$= 1.69 \times 10^5\,\text{J}$$
$$= 169\,\text{kJ}$$

This is

$$\frac{W}{L_v} = \frac{169\,\text{kJ}}{2260\,\text{kJ}} = 0.075 = 7.5\%$$

of the heat of vaporization.

(b) The rest of the heat of vaporization becomes internal energy of the steam. From the first law of thermodynamics the internal energy ΔU added to the water as it becomes steam is

$$\Delta U = Q - W = 2260\,\text{kJ} - 169\,\text{kJ} = 2091\,\text{kJ} \qquad ◆$$

Indicator Diagram

The difference between a heat engine that runs continuously and a one-time-only event such as a dynamite blast is that the engine is not permanently changed as it turns heat into work. A sequence of processes brings the engine back to its starting point after each cycle. The p-V graph of a heat engine is therefore a closed curve that the engine traces over and over again in its operation, as in Fig. 15.6 (a).

The cycle shown in Fig. 15.6 (a) has four parts, each representing a different process. In expansions ab and bc the gas in the engine expands, thereby doing the work W_{ab} and W_{bc} on whatever the engine is connected to. This work is equal to the area under the curves ab and bc on the graph (Fig. 15.6b). The compressions cd and da then bring the engine back to its original state. In these compressions the work W_{cd} and W_{da} is done on the engine itself (Fig. 15.6c). Thus the net work done in each cycle is

Net work = work done by engine − work done on engine
$$W = (W_{ab} + W_{bc}) - (W_{cd} + W_{da})$$

As we see in Fig. 15.6(d), W is equal to the area enclosed by the cycle $abcd$. The larger this area, the more work is done. A graph of this kind for a heat engine is called its **indicator diagram.** Indicator diagrams for gasoline and diesel engines are given in Section 15.6.

(a)

(b)

(c)

(d)

Figure 15.6

Indicator diagram of the operating cycle of an imaginary heat engine. The work done in an entire cycle is equal to the area enclosed by the curve $abcd$ of the cycle.

15.3 SECOND LAW OF THERMODYNAMICS

No engine can be perfectly efficient.

Heat is the easiest and cheapest form of energy to obtain, because all we need do to liberate it is to burn a fuel such as wood, coal, or oil. The real problem is to turn heat into mechanical energy so it can power cars, ships, airplanes, electric generators, and machines of all kinds.

To appreciate the problem, we recall that heat consists of the kinetic energies of moving atoms and molecules. In order to change heat into a more usable form, we must extract some of the energy of the random motions of atoms and molecules and convert it into the regular motions of a piston or a wheel. Such conversions cannot take place efficiently, for the same reason that it is easier to shatter a wineglass than to reassemble the fragments: The natural tendency of all physical systems is toward increasing disorder. The **second law of thermodynamics** is an expression of this tendency, whose role in the evolution of the universe is quite as central as are those of the various conservation principles.

According to the first law of thermodynamics, an engine cannot operate without a source of energy, but this law tells us nothing about the character of possible sources of energy. For instance, there is an immense amount of internal energy in the atmosphere, yet we cannot run a car by just taking in air, extracting some of its internal energy, and then exhausting liquid air. Or, to give an even more extreme case, it is energetically possible for a puddle of water to rise by itself into the air, cooling and freezing from water to ice as its internal energy changes into potential energy (Fig. 15.7). After all, a block of ice dropped from high enough melts when it hits the ground, its initial PE first being converted to KE and then into heat. Needless to say, water does not rise upward of its own accord, and we need a way to express this conclusion.

The second law of thermodynamics is the physical principle that supplements the first law in limiting our choice of heat sources for our engines. It can be stated in a number of equivalent ways, a common one being as follows:

It is impossible to construct an engine, operating in a cycle (that is, continuously), that does nothing other than take heat from a source and perform an equivalent amount of work.

According to the second law of thermodynamics, then, no engine can be completely efficient—some of its heat input *must* be ejected. As we will see, the greatest efficiency any heat engine is capable of depends on the temperatures of its

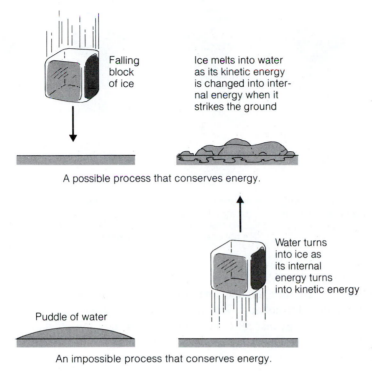

Falling block of ice

Ice melts into water as its kinetic energy is changed into internal energy when it strikes the ground

A possible process that conserves energy.

Water turns into ice as its internal energy turns into kinetic energy

Puddle of water

An impossible process that conserves energy.

Figure 15.7

The second law of thermodynamics provides a way to identify impossible processes that are in accord with all other physical principles.

heat source and of the reservoir to which it exhausts heat. The greater the difference between these temperatures, the more efficient the engine. The second law is a consequence of the fact that

---

The natural direction of heat flow is from a reservoir of internal energy at a high temperature to a reservoir of internal energy at a low temperature, regardless of the total energy content of each reservoir.

This statement, in fact, may be regarded as an alternative expression of the second law (Fig. 15.8).

If we are to utilize the internal energy content of the atmosphere or the oceans, we must first provide a reservoir at a lower temperature than theirs in order to extract heat from them. There is no nearby reservoir in nature suitable for this purpose, for if there were, heat would flow into it until its temperature reached that of its surroundings. To establish a low-temperature reservoir, we must employ a refrigerator (which is a heat engine running in reverse by using up energy to extract heat), and in so doing we will perform more work than we can successfully obtain from the heat of the atmosphere or the oceans.

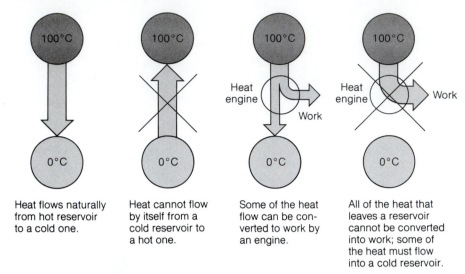

Figure 15.8

The second law of thermodynamics.

Heat flows naturally from hot reservoir to a cold one.

Heat cannot flow by itself from a cold reservoir to a hot one.

Some of the heat flow can be converted to work by an engine.

All of the heat that leaves a reservoir cannot be converted into work; some of the heat must flow into a cold reservoir.

To sum up, the first law of thermodynamics tells us that we cannot get something for nothing. The second law singles out heat from other kinds of energy and recognizes that all conversions of heat into any other form must be inefficient. We cannot break even, either.

15.4 CARNOT ENGINE

Because it is reversible without losing energy, no other engine can be more efficient.

Every heat engine behaves in the same general way: It absorbs heat at a certain temperature, converts some of the heat into work, and exhausts the rest at a lower temperature. The second law of thermodynamics expresses the fact that a heat engine cannot be 100% efficient in turning heat into work. But suppose we have an engine that is not subject to friction or to the loss of stored heat by conduction or radiation. What is the maximum efficiency of such an ideal engine?

Maximum efficiency can be reached only if all the processes that occur in the engine's operation do not involve changes in the engine that need extra work to reverse, because any such changes obviously mean the waste of energy. That is, *every process in the engine must be reversible without losing any energy.*

The flow of heat from a hot reservoir to a cooler one is not reversible in this way, because work would have to be done to "push" the heat back the other way. However, heat flow at constant temperature *is* reversible without losing energy. An **isothermal** process is one in which the temperature of the substance that undergoes a change of some kind stays constant. If the substance is gas in a container, we can

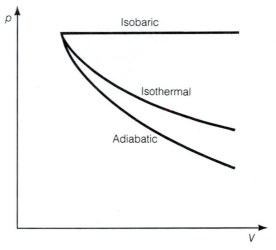

Figure 15.9

Isobaric (p = constant), isothermal (T = constant), and adiabatic (Q = 0) processes.

imagine that the container is a good conductor of heat and is surrounded by a constant-temperature reservoir of heat. No process that occurs in the real world is ever wholly isothermal, just as no actual gas exactly resembles an ideal gas, but many processes are quite close to being isothermal.

In an isothermal process in a system of some kind, all the heat absorbed by the system is turned into work. If the system has work done on it, all the work is given off as heat.

At the other extreme from being in such close contact with a heat reservoir that its temperature never changes, a system might be so completely isolated from its surroundings that heat can neither enter nor leave it. Any process undergone by a system in this situation is called **adiabatic.** An adiabatic process can involve a temperature change and still be reversible without losing energy because no heat transfer is involved that would be subject to the second law of thermodynamics. Most rapid thermodynamic processes are approximately adiabatic because heat transfer takes time and the process may be over before much heat has passed through the walls of the system.

In an adiabatic process any work done by the system comes from the internal energy of the system. Usually this means that the temperature of the system decreases. If the system has work done on it, its internal energy increases by the same amount. Figure 15.9 compares isobaric, isothermal, and adiabatic processes on a p-V diagram.

An imaginary heat engine that uses only isothermal and adiabatic processes was devised in 1824 by the French engineer Sadi Carnot. A **Carnot engine** consists of a cylinder that is filled with an ideal gas and has a movable piston at one end. The four stages in its operating cycle are shown in Fig. 15.10, together with a graph of each stage on a p-V diagram. These stages are as follows:

1. An amount of heat Q_1 is added to the gas, which expands isothermally at its initial temperature T_1. The heat added equals exactly the work done by the gas, which is why its temperature does not change.

Figure 15.10

The Carnot cycle. This imaginary engine, which uses an ideal gas as its working substance, is the most efficient possible operating between reservoirs at the temperatures T_1 and T_2.

2. The heat source is removed, and the expansion is allowed to continue. The second expansion is adiabatic and takes place at the expense of the energy stored in the gas, and so the gas temperature falls from T_1 to T_2. During expansions 1 and 2 the piston exerts a force on whatever it is attached to and thereby performs work.

3. Having done work in pushing the piston outward, the engine must now be returned to its initial state in order for it to be able to do further work. The third stage involves an isothermal compression of the gas at the constant temperature T_2 during which an amount of heat Q_2 is given off. The heat given off exactly equals the work done on the gas by the piston, and so its temperature does not change.

4. The gas is returned to its initial temperature, pressure, and volume by an adiabatic compression in which heat is neither added to it nor removed from it. Work is done on the gas in this compression, which is why its temperature rises.

15.5 ENGINE EFFICIENCY

It depends on the ratio between the temperatures at which the engine absorbs and rejects heat.

In each cycle a Carnot engine performs some net amount of work W, which is the difference between the work it does during the two expansions and the work done on it during the two compressions. It has taken in the heat Q_1 and ejected the heat Q_2 (Fig. 15.11). We note from Fig. 15.10 that the heat Q_2 *must* be ejected in order that the engine return to its initial state, from which it can begin another cycle. According to the first law of thermodynamics,

$$W = Q_1 - Q_2 \qquad (15.4)$$

because there is no internal energy change per cycle. The efficiency of the engine is the ratio between its work output W and its heat input Q_1, so that

$$\text{Eff} = \frac{W}{Q_1} = \frac{Q_1 - Q_2}{Q_1} = 1 - \frac{Q_2}{Q_1} \qquad (15.5)$$

The smaller the ratio of the ejected heat Q_2 to the absorbed heat Q_1, the more efficient the engine.

As it happens, the heat Q transferred to or from a Carnot engine is directly proportional to the absolute temperature T of the reservoir with which it is in contact. That is, for a given Carnot engine,

$$\frac{Q}{T} = \text{constant} \qquad \qquad \textit{Carnot engine} \quad (15.6)$$

(The argument that leads to this conclusion is rather long, although it does not involve any physical principles we have not yet met, and so is omitted here.)

According to Eq. (15.6), the ratio Q_2/Q_1 between the amounts of heat ejected and absorbed per cycle by a Carnot engine is equal to the ratio T_2/T_1 between the temperatures of the respective reservoirs. The efficiency of such an engine is therefore

$$\text{Eff} = 1 - \frac{T_2}{T_1} \qquad \qquad \textit{Carnot efficiency} \quad (15.7)$$

Figure 15.11

The efficiency of a Carnot engine depends on the ratio between Q_2 and Q_1.

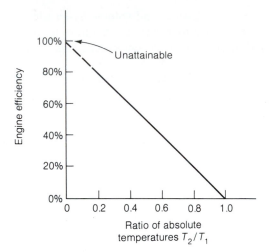

Figure 15.12

The efficiency of a Carnot engine is equal to $1 - T_2/T_1$.

The smaller the ratio between the absolute temperatures T_2 and T_1, the more efficient the engine (Fig. 15.12). No engine can be 100% efficient because no reservoir can have an absolute temperature of 0 K. (Even if such a reservoir could somehow be created, the exhaust of heat to it by the engine would raise its temperature above 0 K immediately.)

As mentioned earlier, the significance of the Carnot engine is that, being reversible, it has the highest efficiency permitted by the laws of thermodynamics. A real engine is never exactly reversible because of friction and heat losses through the engine walls, so its efficiency is less than that of a Carnot engine. We therefore have a third way to express the second law of thermodynamics:

No engine operating between the absolute temperatures T_1 and T_2 can be more efficient than a Carnot engine operating between the same temperatures, whose efficiency is $1 - T_2/T_1$.

◆   **E X A M P L E   15.2** _____

Steam enters a certain steam turbine (Fig. 15.13) at 570°C and emerges into a partial vacuum at 95°C. What is the upper limit to the efficiency of this engine?

SOLUTION   The absolute temperatures equivalent to 570°C and 95°C are, respectively, 843 K and 368 K. The efficiency of a Carnot engine operating between these two absolute temperatures is

$$\text{Eff} = 1 - \frac{T_2}{T_1} = 1 - \frac{368\,\text{K}}{843\,\text{K}} = 0.56 = 56\%$$

Figure 15.13

(a) A primitive steam turbine. (b) In a modern turbine, steam flows past a dozen or more sets of blades on the same shaft to extract as much power as possible. Stationary blades are interleaved between the moving blades to direct the flow of steam in the most advantageous way.

The efficiency of a Carnot engine is the highest possible for an engine operating between a given pair of temperatures. An actual steam turbine operating between 570°C and 95°C would have an efficiency of no more than about 40% because of the inevitable presence of friction and heat losses to the atmosphere. Practical problems rule out steam temperatures higher than about 570°C, which would otherwise improve the efficiency. One of these problems is that, at such temperatures, steam can decompose into hydrogen and oxygen. Steel absorbs hydrogen and becomes brittle as a result, which can reduce the lifetime of the steel components of a turbine to an uneconomic extent. ◆

This steam turbine rotor is 9 m long and turns at 3600 rpm to produce 200 MW of power. Steam enters the turbine at 540°C.

15.6 INTERNAL COMBUSTION ENGINES

Theory into practice.

An internal combustion engine is relatively efficient because it generates the input heat within the engine itself. In a gasoline engine a mixture of air and gasoline vapor is ignited in each cylinder by a spark plug, and the heat given off is turned into mechanical energy by the pressure of the hot gases on a piston. The greater the ratio between the initial and final volumes of the expanding gases, the greater the engine efficiency. In a gasoline engine this ratio is limited to about 8 to 1, because the gasoline-air mixture in the cylinder will otherwise ignite by itself during its compression before the end of the stroke is reached. The result is a theoretical efficiency of about 55% and an actual efficiency of about 30%.

Figure 15.14 shows the operating cycle of a typical four-stroke gasoline engine. Each cylinder in such an engine has one power stroke for every two shaft revolutions.

The operating cycle of a gasoline engine (often called the *Otto cycle* after the inventor of the first practical gasoline engine) is shown on a *p-V* diagram in Fig.

Figure 15.14

The operating cycle of a four-stroke gasoline engine. In the intake stroke, a mixture of gasoline vapor and air from the carburetor is drawn into the cylinder through the intake valve by the suction of the downward-moving piston. In the compression stroke, both valves are closed and the upward-moving piston compresses the fuel-air mixture. At the top of the stroke, the spark plug is fired, which ignites the fuel-air mixture. The burning fuel expands and forces the piston down in the power stroke. At the end of the power stroke, the exhaust valve opens and the upward-moving piston expels the waste gases.

Intake valve   From carburetor   To exhaust   Exhaust valve

Intake stroke   Compression stroke   Power stroke   Exhaust stroke

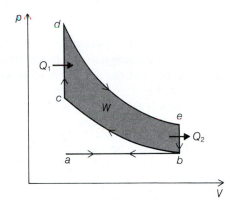

ab: intake of fuel-air mixture
bc: compression stroke
cd: ignition of fuel-air mixture
de: power stroke
e: exhaust valve opens
ba: exhaust stroke

Figure 15.15

Idealized indicator diagram of the Otto cycle of a four-stroke gasoline engine.

15.15. The heat Q_1 enters the engine during the expansion cd when the gasoline-air mixture is ignited by the spark plug. The power stroke de is an adiabatic expansion, and the heat Q_2 is ejected during eb when the exhaust valve is open at the end of the power stroke. (The heat that leaves during the exhaust stroke ba is smaller than Q_2 and is ignored in the ideal cycle of the diagram.) The work W done in each cycle is equal to the area enclosed by $bcdeb$. The greater the compression ratio, the larger the area and the more work done per cycle.

Diesel engines are more efficient than gasoline engines because of their higher compression ratios. Early ignition is avoided by compressing only air and injecting fuel oil into the hot, compressed air at the instant the piston reaches the top of its travel. No spark plug is required. The compression ratio in a diesel engine is typically 20 to 1, for a theoretical efficiency of as much as 70% and an actual efficiency of about 35%.

The operating cycle of a four-stroke diesel engine is shown on the p-V diagram in Fig. 15.16. Air is drawn into the cylinder in the intake stroke ab and is compressed adiabatically, which raises its temperature, in the compression stroke bc. The heat Q_1 enters the engine in the constant-pressure expansion cd during which fuel is injected into the hot air and burned. Then comes an adiabatic expansion to e when the exhaust valve opens and the heat Q_2 is ejected.

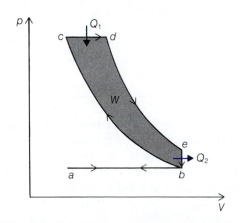

ab: intake of air
bc: compression stroke
cd: fuel injection and burning
ce: power stroke
e: exhaust valve opens
ba: exhaust stroke

Figure 15.16

Idealized indicator diagram of a four-stroke diesel engine.

Power Output

The power output P_c of each cylinder of a gasoline or diesel engine depends on four factors: the average pressure p on the piston during a power stroke; the length L of piston travel; the area A of the piston; and the number N of power strokes per second. Since the force on the piston during a power stroke is $F = pA$, the work done per power stroke is $W = FL = pLA$. Hence

$$\text{Power output per cylinder} = (\text{work done per stroke})\left(\frac{\text{power strokes}}{\text{second}}\right)$$

$$P_c = pLAN \qquad\qquad \textit{Power output} \quad (15.8)$$

In SI units; with p in Pa, L in m, and A in m^2, P is in watts. In British units, with p in lb/in^2, L in ft, and A in $in.^2$, p is in ft · lb/s. If we want P in horsepower, we must divide $pLAN$ by 746 W/hp or by 550 (ft · lb/s)/hp, respectively.

◆ **E X A M P L E 15.3** _____

The Mercedes-Benz OM314 is a four-cylinder, four-stroke diesel engine that develops 80 hp at 2600 rpm. The pistons of this engine are 97 mm in diameter, and they travel 128 mm. Find the average pressure on the pistons during each power stroke.

SOLUTION The area of each piston is

$$A = \frac{\pi d^2}{4} = \frac{\pi (0.097 \text{ m})^2}{4} = 0.0074 \text{ m}^2$$

In a four-stroke engine a power stroke occurs in each cylinder once every two revolutions, so here there are 1300 power strokes per minute, or 1300/60 = 21.7 per second. Because the engine has four cylinders, its total power output is $P = 4P_c = 4pLAN$ and

$$p = \frac{P}{4LAN} = \frac{(80 \text{ hp})(746 \text{ W/hp})}{(4)(0.128 \text{ m})(0.0074 \text{ m}^2)(21.7/s)} = 7.3 \times 10^5 \text{ Pa} = 7.3 \text{ bar}$$

This is slightly over 100 lb/in^2. Of course, the pressure calculated in this way is an average. The actual pressure is greater at the beginning of the stroke just after the fuel is ignited and less at the end of the stroke. ◆

The efficiency of conventional gasoline and diesel engines (nearly as high as that of steam turbines) has to some extent slowed the development of the still more efficient gas turbine. A gas turbine is similar to the steam turbine of Fig. 15.13(b) except that hot gases from the burning fuel pass through its sets of blades instead of steam. A gas turbine is lighter in weight and has fewer moving parts than a piston

Figure 15.17

Turbojet engine. The reaction force of the exhaust gases pushes the aircraft forward.

engine, but the high temperature and high rotational speeds at which it operates present difficulties in manufacture. Gas turbines are nevertheless coming into wider and wider use: "Turboprop" aircraft engines are gas turbines, for instance, and a number of ships are already powered by gas turbines.

The rapidly rotating shaft of a turboprop engine is coupled to a propeller through a reduction gear. In a turbojet engine the propeller is eliminated and the hot gases from the burning fuel are ejected at high speed from the rear of the engine to furnish a reaction force that pushes the aircraft forward (Fig. 15.17). The energy freed by the burning fuel thus goes directly into propulsion with no moving parts except for a turbine that powers the necessary air compressor. Rocket motors are jet engines in which the oxygen or other oxidizing agent needed to burn the fuel comes from an internal reservoir instead of from the atmosphere (see Fig. 5.8). In a solid-fuel rocket, the ultimate in simplicity, both the components required for combustion are combined in a stable mixture whose reaction rate when ignited is relatively slow and steady rather than explosive.

Table 15.1 Typical heats of combustion for common fuels. (Gas volumes at 0°C and atmospheric pressure.)

Substance	Heat of combustion	
Solids	*MJ/kg*	*Btu/lb*
Charcoal	33.9	14,600
Coal	32.6	14,000
Wood	18.8	8,000
Liquids	*MJ/kg*	*Btu/lb*
Diesel oil	44.8	19,400
Domestic fuel oil	45.2	19,500
Ethyl alcohol	32.6	14,000
Gasoline	47.3	20,400
Kerosene	46.0	19,800
Gases	*MJ/m³*	*Btu/ft³*
Acetylene	54.0	1,450
Coal gas	18.0	490
Hydrogen	10.2	275
Natural gas	33–71	900–2000
Propane	86.2	2,330

Heat of Combustion

The **heat of combustion** of a substance is the energy liberated when 1 kg (or 1 m^3) of it is completely burned. Table 15.1 lists heats of combustion for a number of common fuels. If the fuel consumption rate of an engine is known at a certain power output, this table can be used to calculate its efficiency.

◆ **E X A M P L E   15.4** _____

The diesel engine of a fishing boat consumes 90 lb/h of fuel when its power output is 200 hp. Find the efficiency of the engine.

SOLUTION   Since the heat of combustion of diesel oil is 19,400 Btu/lb, the power input of the engine is

$$P_{input} = (90 \text{ lb/h})(19,400 \text{ Btu/lb}) = 1.746 \times 10^6 \text{ Btu/h}$$

To convert this figure into hp we note that 1 Btu = 778 ft · lb, 1 h = 3600 s, and 1 hp = 550 ft · lb/s. Hence

$$P_{input} = \frac{(1.746 \times 10^6 \text{ Btu/h})(778 \text{ ft} \cdot \text{lb/Btu})}{(3600 \text{ s/h})(550 \text{ ft} \cdot \text{lb/s})/\text{hp}} = 686 \text{ hp}$$

The efficiency of the engine is therefore

$$\text{Eff} = \frac{P_{\text{output}}}{P_{\text{input}}} = \frac{200 \text{ hp}}{686 \text{ hp}} = 0.29 = 29\%$$

This is less than half its Carnot efficiency.   ◆

From Table 15.1 it would seem that hydrogen has a very low heat of combustion, well under those of the other gaseous fuels. However, hydrogen is also much less dense than the other gases, and on a mass basis its heat of combustion of 143 MJ/kg is actually by far the highest of any of the fuels listed. The exceptional heat of combustion of hydrogen is the reason for the use of liquid hydrogen in spacecraft engines, where fuel weight is a critical factor. Liquid hydrogen would be an ideal fuel for many other purposes, except for two factors. First, the safe storage and transport of hydrogen in large quantities is not easy. Second, although the hydrogen in the waters of the earth is almost limitless, a great deal of expensive electric energy is needed to extract it, so the overall efficiency of hydrogen as a fuel is today quite low. Research is under way on methods to use sunlight as the energy source for decomposing water into hydrogen and oxygen. If this research is successful and the safety problems can be overcome, hydrogen may well be widely used in the future.

Air Pollution

A kilogram of gasoline needs 15.6 kg of air (about 12 m^3 at atmospheric pressure) to burn up completely. The products are carbon dioxide (CO_2) and water vapor, which are harmless to living things. A "rich" mixture of gasoline and air contains a greater proportion of gasoline than this, and a "lean" mixture contains a smaller proportion. Both mixtures produce polluting exhaust gases.

If the mixture that enters the cylinders of an engine from its carburetor is rich, the amount of air is not enough for complete combustion. As a result the exhaust will contain unburned hydrocarbons and carbon monoxide (CO), which are harmful substances. The hazard of carbon monoxide comes from its tendency to combine with the hemoglobin of the blood in place of oxygen, thereby depriving the body of the oxygen it needs and leading to death if too much CO is inhaled.

If the mixture of gasoline and air is lean, there will be no unburned hydrocarbons or carbon monoxide because combustion will be complete, but in this case the resulting temperature will be high enough to oxidize some of the nitrogen in the air. Nitrogen oxides are poisonous and combine with water to form corrosive nitric acid. Thus vehicle exhausts are one of the causes of the "acid rain" so destructive to agriculture and forests. (The other cause is power plants that burn sulfur-containing coal and oil and whose emissions contain sulfuric acid as a result.) It is very difficult to build an engine in which just the right mixture of gasoline and air is burned at just the right temperature throughout all the variations in speed and power output involved in driving a motor vehicle. For this reason special devices are needed to treat most vehicle exhausts to prevent excessive pollution of the atmosphere.

15.7 REFRIGERATORS

They use outside energy to push heat from a cold reservoir to a hot one.

A refrigerator takes in heat at a low temperature and exhausts it at a higher temperature. According to the second law of thermodynamics, such a process needs an external energy supply to force heat to flow opposite to its normal direction (see Fig. 15.2). In a typical refrigerator two to three times as much heat is taken from the storage chamber as the amount of external energy used.

In nearly all refrigerators the working substance (or **refrigerant**) is a gas that is easily liquefied. Figure 15.18 shows the vaporization curve of Freon 12, one of the most widely used refrigerants. Under conditions of temperature and pressure that correspond to points above the curve, only liquid Freon is present, whereas under conditions that correspond to points under the curve, only Freon vapor is present. Other refrigerants have different vaporization curves. The choice of a refrigerant depends on the kind of refrigerator (or air conditioner) involved and the temperatures between which it is to operate.

Let us examine a refrigeration system that uses Freon 12. As shown in Fig. 15.19, this system consists of a **compressor,** a **condenser,** an **expansion valve,** and an **evaporator.** When the piston of the compressor moves downward, Freon vapor at

Figure 15.18

Vaporization curve of Freon 12, a common refrigerant. Freon 12 is a trade name for dichlorodifluoromethane, CCl_2F_2.

Figure 15.19

A refrigeration system using Freon 12. Gauge pressures are shown. Heat from the refrigerated space enters the system through the evaporator and leaves it through the condenser.

1.4 bars (gauge pressure) and approximately room temperature is sucked into the cylinder. As the piston reaches the bottom of its stroke and begins to move upward, the intake valve closes and the discharge valve opens. The compressed Freon, which is now at a pressure of 7.5 bars and a high temperature, passes into the condenser in which it is cooled until it liquefies. The condenser may be water cooled, as shown in the sketch, or air cooled. It is in this stage that the heat taken from the refrigerated space is dissipated.

The liquid Freon then goes into the expansion valve from which it emerges at a lower pressure (1.4 bars) and temperature ($-8°C$). The amount of Freon supplied by the valve is regulated by a sensing element in the refrigerated space; this may be simply a gas-filled bulb that responds to temperature changes by pressure changes that actuate a bellows in the valve. As the cold liquid Freon flows through the evaporator tubes, it absorbs heat from the region being cooled. From Fig. 15.18 we see that, at an absolute pressure of 2.4 bars, the boiling point of Freon 12 is $-8°C$, so that the heat absorbed by the liquid Freon in the evaporator causes it to vaporize. Farther along in the evaporator, the Freon vapor itself absorbs heat, rising in temperature to perhaps $-2°C$. Finally the Freon vapor leaves the evaporator and enters the compressor to begin another cycle.

Heat Pump

A **heat pump** is a refrigeration system that can take heat from the cold outdoors in winter and deliver it to the interior of a house. In summer the same system can serve as an air conditioner to take heat from the house and exhaust it to the warmer outdoors (Fig. 15.20).

Figure 15.20

A heat pump is a refrigeration system that can be used to heat a house in winter and cool it in summer.

The heat pump is interesting because it transfers more heat than the work done. For example, a heat pump with an energy input of 5 kW might be able to ''pump'' an additional 15 kW of heat to provide a total of 20 kW. An ordinary furnace would have to burn fuel at a rate of 20 kW and so would require four times as much energy input as the heat pump. However, heat pumps have a cost problem to overcome: Energy in the form of fossil fuel (coal, oil, gas) is still quite a bit cheaper than the electrical energy needed to operate a heat pump, and a heat pump is more expensive than a conventional heating plant. Only when a heat pump is needed for air conditioning as well as for space heating is it a practical proposition.

Freon 12 is one of a family of chlorofluorocarbons (CFCs) widely used in insulating foam and as propellants for aerosol sprays as well as in refrigeration. About a million tons of CFCs are manufactured each year, much of which ends up in the upper atmosphere where its chlorine content acts as a catalyst to break up the ozone (O_3) molecules there. Ozone is an excellent absorber of the short-wavelength ultraviolet radiation that arrives from the sun, radiation that is extremely harmful to living things. The dumping of CFCs into the atmosphere is therefore a serious matter and has already led to a measurable ozone loss. In 1987 an international treaty was signed to stabilize and eventually reduce CFC production, but since CFCs remain in the atmosphere for many decades, we can look forward to a continuing rise in skin cancers and crop damage due to solar ultraviolet radiation.

15.8 THERMODYNAMICS OF REFRIGERATION

The greater the temperature difference, the more work is needed to extract heat from the low-temperature reservoir.

As mentioned earlier, a refrigerator is a heat engine operating in reverse. In the most common type of refrigerator, the compressor is the means through which work is done on the refrigerant as it circulates between the evaporator, where it picks up heat, and the condenser, where the heat is given up. If Q_1 is the heat absorbed at the low temperature T_1, and Q_2 is the heat ejected at the high temperature T_2 (Fig. 15.2), then the work done on the refrigerator is

$$W = Q_2 - Q_1$$

The **coefficient of performance** is the ratio between the absorbed heat and the work required:

$$\text{Coefficient of performance} = \frac{\text{heat absorbed}}{\text{work done}}$$

$$CP = \frac{Q_1}{W}$$

$$= \frac{Q_1}{Q_2 - Q_1} \quad \textit{Coefficient of performance} \quad (15.9)$$

The higher the coefficient of performance, the less work is needed to remove a given amount of heat from the cold chamber. Typical coefficient of performance values range from 2 to 3: Two to three times as much heat is removed as the amount of work done.

A quantity called **energy efficiency ratio** (EER) is sometimes used instead of coefficient of performance. The energy efficiency ratio of a refrigerator or air conditioner is the ratio between the heat in BTU extracted per hour and the power in watts the machine consumes:

$$EER = \frac{\text{heat absorbed per hour (Btu/h)}}{\text{power consumption (watts)}} = 3.42\ CP$$

Thus a 3.5-kW air conditioner that removes 25,000 Btu per hour has an energy efficiency ratio of 7.1 Btu/Wh, which is equivalent to a coefficient of performance of 2.1.

An **ideal refrigerator** is just a Carnot engine operating backwards. As in a Carnot engine, the heat transferred to or from a Carnot refrigerator is proportional to the absolute temperature T at which the transfer takes place, and so

$$\frac{Q_2}{Q_1} = \frac{T_2}{T_1}$$

The coefficient of performance of a Carnot refrigerator is accordingly

$$CP = \frac{Q_1}{Q_2 - Q_1} = \frac{T_1}{T_2 - T_1} \qquad\qquad \textit{Ideal refrigerator} \quad (15.10)$$

The more nearly equal the two temperatures are, the less work is needed: The temperature ''hill'' is less steep. Here is where practical difficulties arise, because a low output temperature T_2 is not always easy to arrange. For one thing, the higher T_2 is, the faster is the exchange of heat between refrigerant and cooling water or air in the condenser. A high ideal CP in a refrigerator may therefore mean a low actual coefficient of performance when the power consumed by the water pump or fan in the condenser is taken into account. Equation (15.10) applies to an ideal refrigerator, and actual refrigerators, in which friction and unwanted heat transfer from the environment are present, are less efficient owing to these factors as well.

◆   **E X A M P L E   15.5** _____

A 500-W refrigerator with a CP of 2.0 extracts heat from a storage chamber at $-20°C$ and exhausts it at 40°C. What would the CP of an ideal refrigerator operating between the same two temperatures be? How many joules per hour can the refrigerator extract from the storage chamber?

SOLUTION   (a) The absolute temperatures that correspond to $-20°C$ and 40°C are respectively

$$T_1 = -20°C + 273 = 253 \text{ K} \qquad T_2 = 40°C + 273 = 313 \text{ K}$$

Hence the coefficient of performance of the ideal refrigerator would be

$$CP = \frac{T_1}{T_2 - T_1} = \frac{253 \text{ K}}{313 \text{ K} - 253 \text{ K}} = 4.2$$

The actual refrigerator is a little less than half as efficient as an ideal one would be.
(b) From Eq. (15.9), $Q_1 = (CP)(W)$. Since $P = W/t$,

$$\frac{Q_1}{t} = (CP)\left(\frac{W}{t}\right) = (CP)(P) = (2.0)(500 \text{ W}) = 1000 \text{ W}$$

To express this result in J/h we note that 1000 W = 1000 J/s and 1 h = 3600 s. Hence

$$\frac{Q_1}{t} = \left(1000\frac{J}{s}\right)\left(3600\frac{s}{h}\right) = 3.6 \times 10^6 \text{ J/h} = 3.6 \text{ MJ/h} \qquad \blacklozenge$$

15.9 THE ENERGY PROBLEM

Unlimited demand, limited supply.

The rise of modern civilization has been paralleled by a steady increase in the world's use of energy. This is no coincidence. All that we do needs energy, and the more energy we have at our command, the better we can meet our desires for food, clothing, shelter, warmth, light, transport, communication, and manufactured goods. In the United States, energy is used at a rate of 11 kW per person.

A single source has provided almost all the energy we use today—the sun. Light and heat arrive directly from the sun; food owes its energy content to photosynthesis; water power exists because solar heat evaporates water that later falls as rain on high ground; and wind power comes from convection in the atmosphere due to unequal heating of the earth's surface by the sun. The fossil fuels coal, oil, and natural gas were formed from plants and animals that lived and stored energy derived from sunlight millions of years ago. Only nuclear energy and heat from sources inside the earth cannot be traced to the sun's rays (Fig. 15.21).

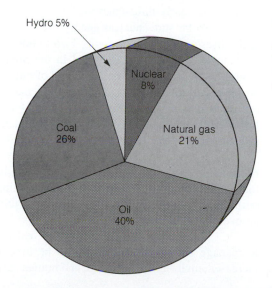

Figure 15.21

A total of nearly 10^{20} J of energy is expected to be produced in the United States in 1995 from the sources shown.

In advanced countries, where the standard of living is already high and populations are stable, energy consumption is unlikely to grow fast. Indeed, it may even decline as energy use becomes more efficient. For more than half the people of the world, however, energy consumption is still low—a rate of less than 1 kW per person. These people seek better lives and their numbers are increasing rapidly, which means much more energy will be needed. Where is it to come from?

The fossil fuels coat, oil, and natural gas, which today furnish by far the greater part of the world's energy, cannot last forever. Oil and natural gas will be the first to be exhausted. At the current rate of consumption, known oil reserves will last only a few decades more. More oil will certainly be found, and better technology will increase the yield from existing wells, but even so in the next century oil will inevitably become scarce. The same is true for natural gas. This is unfortunate because oil and gas burn efficiently and are easy to extract, process, and transport. Half the oil used today goes into fuels that power ships, trains, aircraft, cars, and trucks, and oil and gas are superb feedstocks for synthetic materials of all kinds. Although liquid fuels can be made from coal and coal can serve as the raw material for synthetics, these technologies involve greater expense and greater risk to health and to the environment.

Even though the coal currently consumed each year took about two million years to accumulate, enough remains to last another two or three centuries. But coal is far from a desirable fuel. Not only does its mining leave large tracts of land unfit for further use, but the air pollution due to coal burning has shortened the lives of millions of people through cancer and respiratory diseases. Coal-burning power plans actually expose their surrounding populations to more radioactivity than do normally operating nuclear plants because of the radon gas they liberate.

Nuclear fuel reserves exceed those of fossil fuels. Properly built and operated plants utilizing this abundant fuel supply are in many respects excellent sources of energy. Nuclear energy provides almost a fifth of the electricity generated in the United States, and in a number of other countries the proportion is even higher; in France it is nearly three-quarters.

To be sure, nuclear energy has serious drawbacks. The potential for large-scale disaster is always present. Two major reactor accidents, at Three Mile Island in Pennsylvania and at Chernobyl in the Soviet Union, have already occurred. Although the overall public-health record of nuclear plants is far better than that of coal-burning ones, above-average cancer and leukemia rates have been reported near badly run nuclear installations here and abroad. Further, a reactor produces many tons of waste each year whose radioactivity will remain high for thousands of years, and the safe disposal of nuclear waste is still unsettled.

The sun and stars obtain their energy from nuclear fusion, and in the long run we may do so as well. As described in Chapter 27, a fusion reactor will get its fuel from the sea, will be safe and nonpolluting, and cannot be adapted for military purposes. But nobody can yet say for sure this ultimate energy source will become an everyday reality.

Assuming that fusion power is really on the way, the next question is how to manage until it arrives. Coal can be used more widely, but only at the cost of increased environmental damage and human suffering. And the faster burning of fossil fuels will increase the carbon dioxide in the atmosphere, which, as we saw in Section 14.8, may affect the world's climates. Nuclear energy can help bridge the gap, but memories of Three Mile Island and Chernobyl will have to fade before reactors can be built on a large enough scale.

Renewable Energy Sources

What about the energy of sunlight, of winds and waves, of tides and falling water, of trees and plants? The technology needed to exploit these renewable resources already exists. But even a brief look shows that such energy is unlikely to provide more than a small fraction (though a welcome one) of future needs.

To begin with, a disadvantage of all renewable-energy installations is that they take a lot of space. A city of medium size might use 1000 MW of power. Less than 150 acres is enough for a 1000-MW nuclear plant, whereas 5000 acres might be needed for solar collectors of the same capacity, and over 10,000 acres for windmills. Converting crops into fuel might require 200 square miles of farmland to yield 1000 MW averaged over a year. This is not to say that renewable sources are without value, particularly where local conditions are favorable. Such sources, however, do not seem to be the ultimate answer.

Solar energy. Bright sunlight can deliver over 1 kW of power to each square meter it falls on. At this rate a tennis court receives solar energy equivalent to that in a gallon of gasoline every 10 min or so. Although the supply of sunshine varies with location (Fig. 15.22), time of day, season, and weather, plenty is available. One way to use it involves a series of mirrors that track the sun and concentrate its radiation on a boiler whose steam drives a turbine and thence an electric generator. Better in principle are photovoltaic cells that turn light energy directly into electricity. These have no moving parts and avoid thermodynamic limitations because no heat is involved. Most spacecraft rely on solar cells. Unfortunately today's cells are both inefficient and expensive, although better ones are being developed, and a way to store the energy they produce is needed for a continuous supply of electricity.

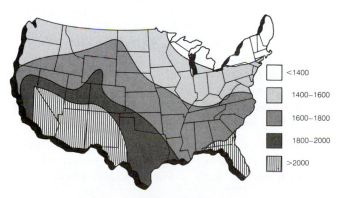

Figure 15.22

Solar energy received per year in the United States expressed in kWh per square meter (1 kWh = 3.6 MJ).

<1400

1400–1600

1600–1800

1800–2000

>2000

Wind.   A large modern windmill with rotor blades 100 m in diameter might produce 3 MW when the wind speed is 12 m/s (about 27 mi/h). A more typical windmill is smaller, with an output of perhaps 100 kW. All are noisy. Windmills are cheaper per watt than solar cells, though still more expensive than conventional power stations, and make sense only where winds are steady and strong. Windmills with a total capacity of over 1500 MW are operating in the United States and elsewhere.

Waves.   As anybody who has stood in the surf or watched waves dash against a rocky shore knows, waves carry energy in abundance. One method of capturing this energy is to have ocean waves run up a sloping funnel-like channel to a reservoir above. Water from the reservoir then powers a turbine generator on its way back to the ocean. Such a system is feasible only where the seabed is so shaped that wave energy is focused on a particular spot on a coast and where the winds that drive the waves are reliable.

Tides.   The twice-daily rise and fall of the tides brings corresponding flows of water into and out of bays and river mouths. The energy involved is enormous when the tidal range is great. Tides in the Rance River estuary in northern France have driven 240-MW generators since 1966, and smaller installations are operating in China, Canada, and the Soviet Union. A tidal power project involving a barrier 16 km long has been proposed for the Severn River in England. The project would produce 7200 MW, a fifth of the country's electricity demand. A few other suitable sites on such a scale exist, but the size of the barriers needed is likely to make them impractical.

Hydroelectricity.   The power of falling water is responsible for 60,000 MW of electricity in the United States alone. Untapped flows exist here and elsewhere in the world, but it is hard to imagine more than a doubling or tripling of the energy available from this ideal source without running into financial and environmental obstacles.

Biomass.   From the time fire came under human control until coal became the dominant fuel a century and a half ago, wood was the chief source of our energy apart from food. Trees grow almost everywhere and ought to be a renewable resource. Over a billion people today in the poorer countries rely on wood for fuel but are burning it up faster than it is being replaced. Wood smoke is even more polluting than coal smoke and is especially high in carcinogens. And wood is too valuable as a material for construction and paper simply to burn it.

In another biomass approach, plant carbohydrates are converted to alcohol for use as vehicle fuel. In Brazil millions of cars run on pure alcohol made from sugar cane, and in the United States a gasoline-alcohol blend is being increasingly used. This technology does not deplete natural resources and causes little pollution. But overall if we include the fuel needed for cultivation and harvesting, it may consume more energy than it provides. In the long run diverting agricultural land from food crops may be unwise: The land needed to run one car on alcohol can provide food for 8 to 16 people.

Geothermal energy.   Temperature increases with depth in the earth's crust, but only in a few places is it hot enough near the surface for useful energy to be extracted. One such place is north of San Francisco, where turbines driven by natural steam produce 665 MW of electric power. Similar plants are operating in New Zealand and Italy, but few other regions seem to have much geothermal potential. Although the natural water supply of a geothermal source may eventually run out, the subsurface heat will remain, and water can be pumped down to produce the needed steam.

Ocean thermal energy. In tropical oceans surface water is typically 27°C, whereas at a depth of a kilometer or so it is 5°C or less. As long ago as 1881 warm surface water was proposed as the heat source for vaporizing a suitable fluid that would drive a turbine and then be condensed by cold water drawn from below. Although ''fuel'' in the form of warm surface water is unlimited and free, the thermodynamic efficiency of such an engine would be only a few percent. Hence a very large plant would be needed relative to its capacity. The economics of ocean thermal energy do not seem very promising and a good many engineering problems must be overcome, yet studies continue because the basic idea is so attractive.

Clearly we can expect no magic solution in the near future to the problem of safe, cheap, and abundant energy. The only sensible course is to practice conservation and try to get the best from the various available technologies while pursuing fusion energy as rapidly as possible. As we have seen, each of these technologies has limitations but may be useful in a given situation. If their full potential is realized and population growth slows down, social disaster (starvation, war) and environmental catastrophe (a planet unfit for life) may well be avoided during the wait for fusion.

Even given fusion energy, however, energy consumption cannot rise indefinitely. The second law of thermodynamics is what stands in the way—the inevitable waste heat must go somewhere, and even now its disposal is difficult in many places. In the United States about 10 percent of the flow of all rivers and streams is already being used as cooling water for power stations. If the resultant heating of inland waters increases still further, the biological impact will be severe. Discharging waste heat into the atmosphere through cooling towers on a large scale will alter the weather and climate of the region, not necessarily for the better. The oceans can absorb much waste heat, but locating power plants on their shores poses the problem of carrying their electricity thousands of miles inland. Only a limit to the world's population will prevent a conflict between the desire for more energy and the need to get rid of waste heat.

Important Terms

A **heat engine** is any device that converts heat into mechanical energy or work.

The **first law of thermodynamics** states that the work output of a heat engine is equal to its net heat input.

The **second law of thermodynamics** states that it is impossible to construct an engine, operating in a repeatable cycle, which does nothing other than take energy from a source and perform an equivalent amount of work.

An **isobaric** process takes place at constant pressure, and an **isothermal** process takes place at constant temperature. No heat enters or leaves a system during an **adiabatic** process.

A **Carnot engine** is an idealized engine that is not subject to such practical difficulties as friction or heat losses by conduction or radiation but that obeys all physical laws. No engine operating between the same two temperatures can be more efficient than a Carnot engine operating between them.

A **refrigerator** is a device that transfers heat from a cold reservoir to a hot one, and it must perform work to do this. In essence, it is a heat engine operating in reverse. The **coef-** **ficient of performance** of a refrigeraor is the amount of heat removed per unit of work done.

Important Formulas

First law of thermodynamics: $Q = \Delta U + W$

Work done by expanding gas: $\Delta W = p\Delta V$

Engine efficiency: $\text{Eff} = \dfrac{W}{Q_1} = 1 - \dfrac{Q_2}{Q_1}$

Carnot efficiency: $\text{Eff} = 1 - \dfrac{T_2}{T_1}$

Refrigeration coefficient of performance:

$$\text{CP} = \frac{Q_1}{W} = \frac{Q_1}{Q_2 - Q_1}$$

$$= \frac{T_1}{T_2 - T_1} \quad \text{(ideal refrigerator)}$$

MULTIPLE CHOICE

1. A heat engine operates by taking in heat at a particular temperature and
   a. converting it all into work
   b. converting some of it into work and exhausting the rest at a lower temperature
   c. converting some of it into work and exhausting the rest at the same temperature
   d. converting some of it into work and exhausting the rest at a higher temperature

2. The natural direction of heat flow is from a high-temperature reservoir to a low-temperature reservoir, regardless of their respective heat contents. This fact is incorporated in the
   a. first law of thermodynamics
   b. second law of thermodynamics
   c. law of conservation of energy
   d. law of conservation of entropy

3. A type of process that does not need outside energy to reverse is one that occurs at constant
   a. temperature   b. pressure
   c. volume   d. speed

4. An adiabatic process in a system is one in which
   a. no heat enters or leaves the system
   b. the system does no work nor is work done on it
   c. the temperature of the system remains constant
   d. the pressure of the system remains constant

5. A process involving a gas that cannot, even in principle, be reversed without work being done is
   a. an isobaric compression
   b. an isothermal compression
   c. an adiabatic compression
   d. an adiabatic expansion

6. In any process the maximum amount of heat that can be converted to mechanical energy
   a. depends only on the intake temperature
   b. depends on the intake and exhaust temperatures
   c. depends on whether kinetic or potential energy is involved
   d. is 100%

7. In any process the maximum amount of mechanical energy that can be converted to heat
   a. depends only on the intake temperature
   b. depends on the intake and exhaust temperatures
   c. depends on whether kinetic or potential energy is involved
   d. is 100%

8. The work output of every heat engine
   a. equals the difference between its heat intake and heat exhaust
   b. equals that of a Carnot engine with the same intake and exhaust temperatures
   c. depends only on its intake temperature
   d. depends only on its exhaust temperature

9. The area enclosed by the *p-V* graph of a complete heat engine cycle equals
   a. the heat intake per cycle
   b. the heat output per cycle
   c. the work done on the engine per cycle
   d. the work done by the engine per cycle

10. A frictionless heat engine can be 100% efficient only if its exhaust temperature is
   a. equal to its input temperature
   b. less than its input temperature
   c. 0°C
   d. 0 K

11. A process not involved in the operating cycle of a Carnot engine is
   a. an isothermal expansion
   b. an isobaric expansion
   c. an adiabatic compression
   d. an adiabatic expansion

12. A Carnot engine turns heat into work
   a. with 100% efficiency
   b. with 0% efficiency
   c. without itself undergoing a permanent change
   d. with the help of expanding steam

13. A Carnot engine that operates between the absolute temperatures T_1 and T_2
   a. is 100% efficient
   b. has the maximum efficiency possible under these circumstances
   c. has an efficiency of T_2/T_1
   d. has the same efficiency as an actual engine that operates between T_1 and T_2

14. Which of the following engines is normally the least efficient?
   a. gasoline engine   b. diesel engine
   c. gas turbine   d. Carnot engine

15. Which of the following engines is the most efficient?
   a. gasoline engine   b. diesel engine
   c. gas turbine   d. Carnot engine

16. The fuel in a diesel engine is ignited by
   a. a spark plug
   b. being compressed until it is hot enough
   c. the hot compressed air into which it is injected
   d. exhaust gases remaining from the previous cycle

17. The physics underlying the operation of a refrigerator most closely resembles the physics underlying
   a. a heat engine
   b. the melting of ice
   c. the freezing of water
   d. the evaporation of water

18. A refrigerator
   a. produces cold
   b. causes heat to vanish
   c. removes heat from a region and transports it elsewhere
   d. changes heat to cold

19. A refrigerator exhausts
   a. less heat than it absorbs from its contents
   b. the same amount of heat it absorbs from its contents
   c. more heat than it absorbs from its contents

   d. any of the above, depending on the circumstances

20. The working substance (or refrigerant) used in most refrigerators is a
   a. gas that is easy to liquefy
   b. gas that is hard to liquefy
   c. liquid that is easy to solidify
   d. liquid that is hard to solidify

21. Heat is absorbed by the refrigerant in a refrigerator when it
   a. melts   b. vaporizes
   c. condenses   d. is compressed

22. A heat pump is basically a
   a. pump for hot air
   b. pump for hot water
   c. heat engine
   d. refrigerator

23. The chief source of energy in the world today is
   a. coal   b. oil
   c. natural gas   d. uranium

24. The source of energy whose reserves are greatest is
   a. coal   b. oil
   c. natural gas   d. uranium

25. A gas is compressed from 5 L to 3 L under an average pressure of 2 bars. During the compression 0.3 kJ of heat is given off. The internal energy of the gas
   a. decreases by 0.1 kJ
   b. is unchanged
   c. increases by 0.1 kJ
   d. increases by 0.4 kJ

26. A Carnot engine absorbs heat at a temperature of 127°C and exhausts heat at a temperature of 77°C. Its efficiency is
   a. 13%   b. 39%
   c. 61%   d. 88%

27. If a Carnot engine exhausts heat at 140°F is to have an efficiency of 33%, it must take in heat at
   a. 200°F   b. 440°F
   c. 660°F   d. 900°F

28. A Carnot engine operates between 800 K and 200 K. If it absorbs 8 kJ of heat in each cycle, the work it does per cycle is
   a. 1 kJ   b. 2 kJ
   c. 2.7 kJ   d. 6 kJ

29. A refrigerator gives off 3 J of heat for every joule of mechanical energy input. Its coefficient of performance is
   a. 1   b. 2
   c. 3   d. 4

QUESTIONS

1. The sun's corona is a very dilute gas at a temperature of about 10^6 K that is believed to extend into interplanetary space at least as far as the earth's orbit. Why can we not use the corona as the high-temperature reservoir of a heat engine in an earth satellite?

2. (a) An attempt is made to cool a kitchen in the summer by switching on an electric fan and closing the kitchen door and windows. What will happen? (b) In another attempt to cool the kitchen, the refrigerator door is left open, again with the kitchen door and windows closed. Now what will happen?

3. A gas sample expands from V_1 to V_2. Does it perform the most work when the expansion takes place at constant pressure, at constant temperature, or adiabatically? In which process does the gas perform the least work? Why?

4. The operation of a steam engine proceeds approximately as follows: (1) Water under pressure is heated to the boiling point; (2) the water turns into steam and expands at constant pressure at its boiling point; (3) the steam enters the cylinder of the engine and expands adiabatically against the piston; (4) the spent steam condenses into water at constant pressure; (5) the water is pumped back into the boiler. Plot the entire cycle on a *p-V* diagram and indicate in what parts of the cycle heat is absorbed, heat is rejected, and work is done on the outside world.

5. Which three fuels provide most of the world's energy today? List their chief advantages and disadvantages.

6. What energy sources, if any, cannot be traced to sunlight falling on the earth?

7. What are some of the disadvantages that have kept renewable energy sources from supplying more than a small fraction of the world's energy?

EXERCISES

15.1   First Law of Thermodynamics

15.2   Work Done by and on a Gas

1. A gas expands by 1.0 L at a constant pressure of 2.0 bars. During the expansion 500 J of heat is added. Find the change in the internal energy of the gas.

2. A gas expands isothermally from 40 L to 60 L while absorbing 5.0 kJ of heat. Find the average pressure on the gas during the expansion.

3. A steam engine built in 1712 had a piston 1.8 ft in diameter that moved back and forth through a distance of 8.0 ft. If the pressure on the piston was a constant 15 lb/in² during each expansion, how much work was done?

Figure 15.23

Exercise 4.

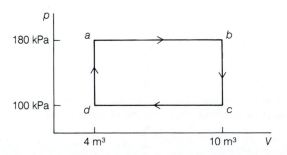

4. The indicator diagram of a certain engine is shown in Fig. 15.23. (a) Find the work done in each part of the cycle—namely, $a{\rightarrow}b$, $b{\rightarrow}c$, $c{\rightarrow}d$, and $d{\rightarrow}a$—and identify it as work done by or on the engine. (b) How much net work is done per cycle by the engine?

5. The engine in Exercise 4 is operated in reverse, with each cycle proceeding as *adcba*. Answer the same questions for the engine in this case.

15.5   Engine Efficiency

6. As mentioned in Section 15.9, a typical temperature for surface water in a tropical ocean is 27°C, whereas at a depth of a kilometer or more it is only about 5°C. It has been proposed to operate heat engines using surface water as the hot reservoir and deep water as the cold reservoir. What would the maximum efficiency of such an engine be? Why might such an engine eventually be a practical proposition even with so low an efficiency?

7. An engine operating between 300°C and 50°C is 15% efficient. What would its efficiency be if it were a Carnot engine?

8. An engine is proposed that is to operate between 200°C and 50°C with an efficiency of 35%. Will the engine perform as predicted? If not, what would its maximum efficiency be?

9. One of the most efficient engines ever developed operates between about 2000 K and 700 K. Its actual efficiency is 40%. What percentage of its maximum possible efficiency is this?

10. A steam engine takes in steam at 200°C. What is the maximum temperature at which spent steam can leave the engine if its efficiency is to be 20%?

11. A Carnot engine absorbs 200 kJ of heat at 500K and exhausts 150 kJ. What is the exhaust temperature?

12. Three designs for a heat engine to operate between 450 K and 300 K are proposed. Design A is claimed to require a heat input of 2.0 kJ for each kJ of work output, design B a heat input of 3.0 kJ, and design C a heat input of 4.0 kJ. Which design would you choose and why?

13. A Carnot engine takes in 1.0 MJ of heat from a reservoir at 327°C and exhausts heat to a reservoir at 127°C. How much work does it do?

14. A Carnot engine whose efficiency is 35% takes in heat at 500°C. What must the intake temperature be if the efficiency is to be 50% with the same exhaust temperature?

15. A Carnot engine absorbs 500 Btu of heat at 500°F and performs 100,000 ft · lb of work. What is the exhaust temperature?

16. A Carnot engine operating between 580°F and 120°F has a power output of 120 hp. How much heat per minute does it absorb? How much does it exhaust?

15.6 Internal Combustion Engines

17. A four-cylinder, four-stroke diesel engine has pistons 108 mm in diameter whose travel is 127 mm. The engine develops 100 kW at 66 rev/s. Find the average pressure on the pistons during the power stroke.

18. The six-cylinder, four-stroke gasoline engine of a car has pistons 3.5 in. in diameter whose travel is 4.0 in. If the average pressure on the pistons during the power stroke is 70 lb/in^2, find the number of horsepower developed by the engine when it operates at 3000 rpm.

19. A turboprop aircraft engine consumes 250 g/h of kerosene for each horsepower it develops. Find the efficiency of the engine.

20. A 4250-hp aircraft engine is used in a power station in England to run a 3.0-MW generator to supply electricity during peak-load periods. The engine consumes 370 g of kerosene per kilowatt-hour of energy produced. What is the overall efficiency of the installation?

21. At 800 rpm a diesel engine develops 60 kW and consumes 18 kg of fuel per hour. At 1800 rpm the same engine develops 120 kW and consumes 32 kg of fuel per hour. Find its efficiency at each speed.

22. A certain power station consumes 2000 metric tons (1 metric ton = 10^3 kg) of coal per day. The overall efficiency of the station is 40%. Find the number of megawatts of electricity the station produces.

23. How many cubic meters of propane must be burned to heat 50 L of water from 5°C to 90°C? Assume that 25% of the heat is wasted.

24. A certain jet aircraft uses 20,000 lb of kerosene in traveling 900 mi at an average speed of 500 mi/h. If the engines of the airplane develop an average of 15,000 hp during the flight, what is their percentage efficiency?

15.8 Thermodynamics of Refrigeration

25. Refrigerators are sometimes rated in **tons,** where 1 ton of refrigeration capacity is that rate of heat removal that can freeze 2000 lb of water at its freezing point to ice per day. (a) Express the refrigeration ton in kW. (b) A $\frac{3}{4}$-ton window air conditioner is found able to maintain a constant temperature of 21°C in an empty room when the outside temperature is 32°C by running half the time. A sitting person liberates about 120 W. How many people can sit reading in the room, each with a 75-W lamp, without exceeding the capacity of the air conditioner?

26. Three designs for a refrigerator to operate between −20°C and 30°C are proposed. Design A is claimed to need 100 J per kilojoule of heat extracted, design B to need 200 J, and design C to need 300 J. Which design would you choose and why?

27. A Carnot refrigerator extracts heat from a freezer at −5°C and exhausts it at 25°C. How much work per joule of heat extracted is required?

28. Find the CP of a refrigerator whose energy supply equals half the amount of heat it extracts from a cold chamber.

29. A refrigerator with a CP of 2.4 extracts heat from a freezer at −10°C and exhausts it at 25°C. How much work is required per J of heat extracted?

30. How much power in kilowatts must be supplied to a central air conditioning system with a CP of 2.3 that removes heat from a house at the rate of 50,000 Btu/h? What is its energy efficiency ratio?

31. A certain refrigerator that absorbs heat at 5°F and exhausts it at 85°F has a CP of 3.0. What would the coefficient of performance of a Carnot refrigerator be that operated between the same two temperatures?

PROBLEMS

32. A coal-fired electric power plant of 400 MW output operates at an efficiency of 39%, which is about three-quarters of its Carnot efficiency. Water from a river flowing at 50 m³/s is used to absorb the waste heat. By how much does the river's temperature rise as a result?

33. The total drop of the Wollomombi Falls in Australia is 482 m. What would be the Carnot efficiency of an engine operating between the top and bottom of the falls if the water temperature at the top were 10°C and all the potential energy of the water at the top were converted to heat at the bottom?

34. A Carnot refrigerator is used to make 1.0 kg of ice at −10°C from 1.0 kg of water at 20°C, which is also the temperature of the kitchen. How many joules of work must be done?

35. In its operation, a refrigerator absorbs heat from a storage chamber at −23°C and exhausts heat at 27°C. The rate at which heat is exhausted is found to be 250 kJ/min, and the coefficient of performance is half that of a Carnot refrigerator. (a) How much heat is absorbed from the storage chamber per minute? (b) How much power does the refrigerator require?

36. Ice cream has a specific heat of 0.78 Btu/lb · °F before it freezes and 0.45 Btu/lb · °F afterward. Its heat of fusion is 126 Btu/lb, and it freezes at 28°F. What refrigeration capacity (in Btu/h) is required to produce 60 lb of ice cream at 8°F from a mix at 60°F per hour?

37. An ice-making plant produces 50 tons of ice per day at 15°F from water at 60°F. Assuming no losses, what must be the refrigeration capacity of the plant in Btu/h?

38. The resistive forces on a car total 1.2 kN when its speed on a level road is 80 km/h. If the car can go 8.5 km per liter of gasoline at that speed, find the efficiency of its engine and drive train.

Answers to Multiple Choice

1. b	**9.** d	**17.** a	**25.** c
2. b	**10.** d	**18.** c	**26.** a
3. a	**11.** b	**19.** c	**27.** b
4. a	**12.** c	**20.** a	**28.** d
5. a	**13.** b	**21.** b	**29.** b
6. b	**14.** a	**22.** d	
7. d	**15.** d	**23.** b	
8. a	**16.** c	**24.** d	

C H A P T E R

16

ELECTRICITY

16.1 Electric Charge

16.2 Coulomb's Law

16.3 Electricity and Matter

16.4 Electrical Conduction

16.5 What Is a Field?

16.6 Electric Field

16.7 Electric Field Lines

We have now looked into force and motion, mass and energy, temperature and heat. The ideas we have found help make sense of much of the world of our experience. Are they enough for us to understand how the entire physical universe works?

For an answer all we need do is run a hard rubber comb through our hair on a dry day. Little sparks occur, and afterward the comb can pick up dust and small bits of paper. The attraction is surely not due to gravity, because the gravitational force between comb and paper is far too small. And gravity should not depend on whether or not the comb is run through our hair. What has been revealed by this experiment is an electrical phenomenon, so called after *elektron*, the Greek word for amber, a substance used in the earliest studies of electricity.

16.1 ELECTRIC CHARGE

Like charges repel; unlike charges attract.

Electricity is familiar to all of us as the name for that which causes our light bulbs to glow, many of our motors to turn, our telephones and radios to communicate sounds, our television screens to communicate pictures. But there is more to electricity than its ability to transmit energy and information. Electrical forces hold electrons to nuclei to form atoms, and they hold atoms together to form molecules, solids, and liquids. All of the chief properties of matter in bulk—with the notable exception of mass—can be traced to the electrical nature of its basic particles.

Let us begin our study of electricity by looking at three basic experiments. The first experiment is shown in Fig. 16.1. By convention, we call whatever it is that a rubber rod has by virtue of having been stroked with a piece of fur **negative electric charge.** Part of the negative charge on the rubber rod in Fig. 16.1 passed to the

Figure 16.1

A rubber rod stroked with fur becomes negatively charged; two negatively charged objects repel each other.

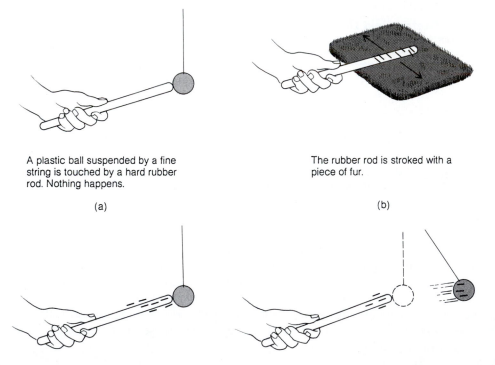

A plastic ball suspended by a fine string is touched by a hard rubber rod. Nothing happens.

(a)

The rubber rod is stroked with a piece of fur.

(b)

The plastic ball is again touched by the rubber rod.

(c)

After the touch, the ball flies away from the rod.

(d)

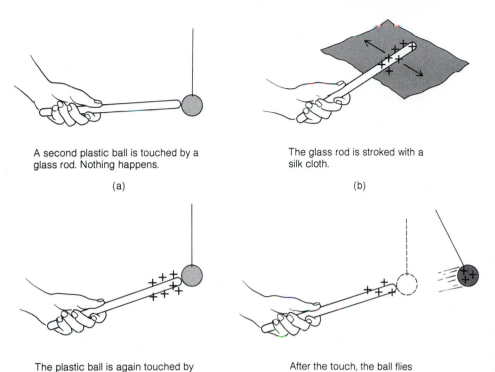

A second plastic ball is touched by a glass rod. Nothing happens.

(a)

The glass rod is stroked with a silk cloth.

(b)

The plastic ball is again touched by the glass rod.

(c)

After the touch, the ball flies away from the rod.

(d)

Figure 16.2

A glass rod stroked with silk becomes positively charged; two positively charged objects repel each other.

plastic ball when it was touched, and the fact that the ball then flew away from the rod suggests that negative electric charges repel each other.

The next experiment, shown in Fig. 16.2, is very similar. By convention, we call whatever it is that a glass rod has by virtue of having been stroked with a silk cloth **positive electric charge.** Part of the positive charge on the glass rod in Fig. 16.2 passed to the plastic ball when it was touched, and the fact that the ball then flew away from the rod suggests that positive electric charges repel each other.

Why is it assumed that the electric charge on the glass rod is different from that on the rubber rod? The reason lies in the result of the third experiment, shown in Fig. 16.3. The attraction of the two plastic balls means (1) that the charges they carry are different, since like charges have already been observed to repel, and (2) that unlike charges attract each other.

Figure 16.3

Objects with unlike charges attract each other.

Figure 16.4

The process of stroking a rubber rod with a piece of fur serves to separate charges so that the rod becomes negatively charged and the fur positively charged.

The preceding results can be summarized very simply:

Like charges repel; unlike charges attract.

Where do the charges come from when one substance is stroked with another? When we charge one plastic ball with a rubber rod and another with the fur the rod was stroked with, we find that the two balls attract. Since the rubber rod is negatively charged, this experiment shows that the fur is positively charged (Fig. 16.4). A similar experiment with a glass rod and a silk cloth shows that the cloth picks up a negative charge during the stroking. Evidently the process of stroking serves to *separate* charges. We conclude that rubber has an affinity of some kind for negative charges and fur an affinity for positive charges, so that, when rubbed together, each tends to acquire a different kind of charge.

A great many experiments with a variety of substances have shown that there are only these two kinds of electric charge, positive and negative. All electrical phenomena involve either or both kinds of charge. An "uncharged" body of matter actually has equal amounts of positive and negative charge, so that appropriate treatment—mere rubbing is sufficient for some substances—can leave an excess of either kind on the body and thereby cause it to show electrical effects.

What is electric charge? All that can be said is that charge, like rest mass, is a fundamental property of certain of the elementary particles of which all matter is composed. Three types of particles are found in atoms, the positively charged **proton,** the negatively charged **electron,** and the neutral (that is, uncharged) **neutron.** The proton and electron have exactly equal amounts of charge, though of opposite sign. Neutrons and protons have nearly the same mass, which is almost 2000 times greater than the electron mass (Table 16.1).

An atom normally contains equal numbers of protons and electrons, so it is electrically neutral unless disrupted in some way. In Figs. 16.1 and 16.2 it is atomic electrons that shift from fur to rubber and from glass to silk to give the observed charge separations. Figure 16.5 shows a simplified model of a helium atom, whose core, or **nucleus** consists of two protons and two neutrons. The two electrons that circle the nucleus some distance away balance the charges of the protons.

Figure 16.5

A simplified model of the helium atom. The size of the atom is much greater than the size of its nucleus. Nearly all the mass of an atom is concentrated in its nucleus.

Helium atom

Helium nucleus

- Electron
- Proton
- Neutron

Table 16.1 Neutrons, protons, and electrons are the constituents of atoms. An atom contains equal numbers of protons and electrons, so it is electrically neutral.

	Neutron	Proton	Electron
Charge	0	$+e = 1.60 \times 10^{-19}$ C	$-e = -1.60 \times 10^{-19}$ C
	Uncharged	Equal in magnitude, opposite in sign	
Mass	1.675×10^{-27} kg	1.673×10^{-27} kg	9.11×10^{-31} kg
	Nearly equal		Electron mass is 1/1836 of proton mass

Conservation of Charge

The **principle of conservation of charge** states:

The net electric charge in an isolated system remains constant.

By ''net charge'' is meant the algebraic sum of the charges present. Net charge can be positive, negative, or zero.

Every known physical process in the universe conserves electric charge. Separating or bringing together charges does not affect their magnitudes, so such rearrangements leave the net charge the same.

16.2 COULOMB'S LAW

What the force between two electric charges depends on.

The law of force between charges was first published by the eighteenth-century French scientist Charles Coulomb. If we use the symbol Q for electric charge, **Coulomb's law** for the magnitude F of the force between two charges Q_A and Q_B that are the distance r apart states that

$$F = k\frac{Q_A Q_B}{r^2} \qquad\qquad \textit{Coulomb's law} \quad (16.1)$$

The force between two charges is proportional to the product of the charges and is inversely proportional to the square of the distance between them. The quantity k is

Figure 16.6

(a) Like charges repel, unlike charges attract. (b) The force between two charges varies inversely as the square of their separation. (c) The force is proportional to the product of the charges.

a constant whose value depends on the units used and on the medium (air, for instance) in which the charges are located. In Eq. (16.1) it is assumed that the charges are either small enough to be considered points or else are spherically symmetric, in which case r is measured from their centers.

Electric force is a vector quantity, and the preceding formula gives only its magnitude. The direction of **F** is always along the line joining Q_A and Q_B, and the force is attractive if Q_A and Q_B have opposite signs and repulsive if they have the same signs (Fig. 16.6).

At the beginning of this chapter we noted a familiar observation: A hard rubber comb that has been charged by being passed through someone's hair on a dry day is able to attract small bits of paper. Since the paper bits were originally uncharged, how could the comb exert a force on them?

The explanation depends on Coulomb's law (Fig. 16.7). When the negatively charged comb is brought near the paper, some of the electrons in the paper that are not tightly bound in place move as far away as they can from the comb. Because electrical forces vary inversely with distance, the attraction between the comb and the closer positive charges is greater than the repulsion between the comb and the farther electrons. The paper therefore moves toward the comb. Only a small amount of charge separation actually occurs, and so, with little force available, only very light objects can be attracted in this way.

Figure 16.7

How a charged object attracts an uncharged one.

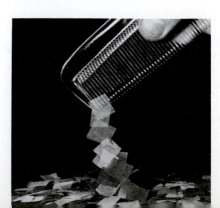

The Coulomb

The unit of electric charge is the **coulomb** (abbreviated C). The formal definition of the coulomb, given in Chapter 19, is in terms of magnetic forces. A more realistic way to think of the coulomb is in terms of the number of individual elementary charges that add up to this amount of charge (Fig. 16.8). Because electric charge always occurs in multiples of $\pm\, 1.60 \times 10^{-19}$ C, this amount of charge has been given a special name, the electron (or electronic) charge, and a special symbol, e:

6.25 × 10¹⁸ protons

$Q = +1$ coulomb

$$e = 1.60 \times 10^{-19}\ \text{C} \qquad\qquad \textit{Electron charge}$$

Thus a charge of $+1.60 \times 10^{-19}$ C is abbreviated $+e$, and one of -3.20×10^{-19} C is abbreviated $-2e$.

In most processes that lead to a net charge on some object, electrons are either added to it or removed from it. Hence we can think of an object whose charge is -1 C as having 6.25×10^{18} electrons more than its normal number, and of an object whose charge is $+1$ C as having 6.25×10^{18} electrons less than its normal number. (By "normal number" is meant a number of electrons equal to the number of protons present, so that the object has no net charge.)

6.25 × 10¹⁸ electrons

$Q = -1$ coulomb

Figure 16.8

Electric charge is not continuous but occurs in multiples of $\pm 1.60 \times 10^{-19}$ C.

When Q_A and Q_B in Coulomb's law are expressed in coulombs and r in meters, the constant k has the value in vacuum of

$$k = 9.0 \times 10^9\ \text{N} \cdot \text{m}^2/\text{C}^2$$

The value of k in air is very slightly greater.

The constant k is often written

$$k = \frac{1}{4\pi\epsilon_0}$$

where ϵ_0, called the permittivity of free space, is equal to

$$\epsilon_0 = 8.85 \times 10^{-12}\ \text{C}^2/\text{N} \cdot \text{m}^2$$

◆ **E X A M P L E   16.1** _____

Find the magnitude of the force between two charges of 1.0 C each that are 1.0 m apart (Fig. 16.9).

+ 1.0 C   + 1.0 C

9.0 × 10⁹ N   9.0 × 10⁹ N

1.0 m

Figure 16.9

SOLUTION   From Coulomb's law

$$F = k\frac{Q_AQ_B}{r^2} = \left(9.0 \times 10^9 \frac{\text{N} \cdot \text{m}^2}{\text{C}^2}\right)\left[\frac{(1.0\,\text{C})(1.0\,\text{C})}{(1.0\,\text{m})^2}\right] = 9.0 \times 10^9\,\text{N}$$

This force is equal to about 2 billion lb. Evidently even the most highly charged objects that can be produced seldom contain more than a tiny fraction of a coulomb of net charge of either sign.   ◆

16.3 ELECTRICITY AND MATTER

Gravitational forces dominate on a large scale, electrical forces on a small scale.

Coulomb's law for the electrical force between charges is very similar in form to Newton's law for the gravitational force between masses. The most striking difference is that electrical forces may be either attractive or repulsive, whereas gravitational forces are always attractive. The latter fact means that matter in the universe tends to come together to form large bodies, such as stars and planets, and these bodies are always found in groups, such as galaxies of stars and families of planets around stars.

There is no comparable tendency for electric charges of a given sign to come together; quite the contrary. Because unlike charges attract strongly, it is hard to separate neutral matter into portions of opposite signs. Furthermore, like charges repel, so it becomes harder and harder to add more charge to an already charged object. Hence the large-scale structure of the universe is largely governed by gravitational forces.

On an atomic scale, though, the relative importance of gravity and electricity is reversed. Elementary particles are so tiny that the gravitational forces between them are insignificant, whereas their electric charges are sufficiently great for electrical forces to govern the structures of atoms, molecules, liquids, and solids.

16.4 ELECTRICITY CONDUCTION

Most substances conduct electricity either very well or very badly.

An electric current is a flow of charge. Nearly all substances fall into two categories: **conductors,** through which charge can flow easily, and **insulators,** through which charge can flow only with great difficulty. Metals, many liquids, and **plasmas** (gases whose particles are charged) are conductors. Nonmetallic solids, certain liquids, and gases whose molecules are electrically neutral are insulators. Several substances, called **semiconductors,** are intermediate in their ability to conduct charge.

In a solid metal each atom gives up one or more electrons to a common ''gas'' of freely moving electrons. These electrons can move easily through the crystal structure of the metal, so if one end of a metal wire is given a positive charge and the other end a negative charge, electrons will flow through the wire from the negative to the positive end. This flow, of course, is an electric current. By supplying new electrons to the negative end of the wire and removing electrons from the positive end as they arrive there—which can be done by connecting the wire to a battery or to a generator—a steady current can be maintained in the wire.

In nonmetallic solids, such as salt, glass, rubber, minerals, wood, and plastics, all the atomic electrons are fixed in particular atoms or groups of atoms and cannot move from place to place. Such solids are therefore insulators. Actually, nonmetallic solids do conduct very small amounts of current, but their abilities to do this are much inferior to those of metals. For instance, when identical bars of copper and sulfur are connected to the same battery, about 10^{23} times more current flows in the copper bar.

As mentioned earlier, there are a few substances called semiconductors through which charge flows more readily than through insulators but with more difficulty than through conductors. Thus about 10^7 times more current flows in a germanium bar connected to a battery than in a sulfur bar of the same size, but this is still about 10^{16} times less current than in a copper bar. The electrical conductivity of solids is discussed in more detail in Chapter 26.

Ions

Electrical conduction in liquids and gases is different from that in metals. The current in a metal consists of a flow of electrons past the stationary atoms in its structure. The current in a fluid other than a liquid metal, however, consists of a flow of entire atoms or molecules that are electrically charged. An atom or molecule that carries a net charge is called an **ion**, and both positive and negative ions participate in conduction in liquids and gases.

An atom or molecule becomes a positive ion when it loses one or more of its electrons. If it gains one or more electrons in addition to its usual number, it becomes a negative ion. The fundamental positive charges in matter are protons, which are very tightly bound in the nucleus of every atom. Atomic electrons, however, are held more loosely, and one or two of them can be detached from an atom with relative ease. Thus the oxygen and nitrogen gases in ordinary air become ionized when a spark occurs, in the presence of a flame, and by the passage of X-rays or even ultraviolet light. These processes so disturb the air molecules that some electrons are dislodged, leaving behind positive ions. The liberated electrons almost at once become attached to other nearby molecules to create negative ions (Fig. 16.10).

The electrical attraction between positive and negative charges in time brings the ions together, and the extra electrons on the negative ions return to the positive

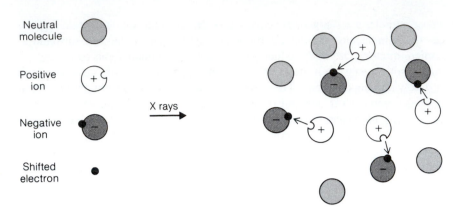

Figure 16.10

Schematic representation of the ionization of air by X-rays. A molecule losing an electron becomes a positive ion; a molecule gaining an electron becomes a negative ion.

ions. The gas molecules are then neutral, as they were originally. This **recombination** is rapid at normal atmospheric pressure and temperature.

In the upper atmosphere, where air molecules are so far apart that the recombination of ions is a slow process, the continual bombardment of X-rays and ultraviolet light from the sun maintains a certain proportion of ions at all times. The layers of ions in the upper atmosphere constitute the **ionosphere,** and they make possible long-range radio communication by their ability to reflect radio waves (see Fig. 22.9). The ionosphere is an example of a plasma. The behavior of a plasma, unlike that of an ordinary gas, is strongly influenced by electrical and magnetic forces because its particles are charged. Most of the universe is in the plasma state.

The earth as a whole is a fairly good conductor. Hence if a charged object is connected to the earth by a piece of metal, the charge is conducted away from the object to the earth. This convenient way to remove excess charge from an object is called **grounding** the object. As a safety measure, the metal shells of electrical appliances are grounded through special wires that give electric charges in the shells paths to the earth. The round post in the familiar three-prong electric plug is the ground connection.

Superconductivity

Electric conductors, even the best, resist to some extent the flow of charge through them at ordinary temperatures. At very low temperatures, however, certain metals, alloys, and chemical compounds permit currents to pass unimpeded through them. This phenomenon is called **superconductivity.**

Superconductivity was discovered by Kamerlingh Onnes in the Netherlands in 1911. Aluminum becomes superconducting at temperatures below 1.2 K and lead at temperatures below 7.2 K, to give two examples. If a current is set up in a closed wire loop at room temperature, it will die out in less than a second even if the wire is made of a good conductor such as copper or silver. If the wire is made of a superconductor and is kept cold enough, on the other hand, the current will continue indefinitely. Currents have persisted unchanged in superconducting loops for several years.

The effect of electrical resistance is to dissipate as heat some of the energy of an electric current in a conductor. An analogy can be made with the presence of friction in a mechanical system, where kinetic energy is dissipated as heat. About 10% of the electric energy generated in the United States is lost as heat in transmission lines. For this reason superconductivity is of great potential importance for the carrying of electric energy. Because magnetic fields are produced by electric currents (Chapter 19), superconductivity is also important in applications where strong magnetic fields are required. Already laboratory electromagnets with superconducting coils are in use, and experimental electric motors with superconducting windings have been built. There is no fundamental reason why superconducting magnets cannot be used to suspend trains above their tracks and thereby both increase their speeds and reduce their energy needs.

Despite much effort, until 1986 no substance was known whose critical temperature (the temperature below which it is superconducting) was higher than 23 K. In that year Alex Müller and Georg Bednorz, working in Switzerland, studied a class of materials that had never before been suspected of superconducting behavior. They discovered a superconducting ceramic whose critical temperature was 35 K, and soon afterward others extended their approach to produce superconductors with critical temperatures over 100 K. Although still extremely cold by everyday standards (100 K = −173°C), such temperatures are above the 77-K boiling point of liquid nitrogen, which is cheap and readily available. The much greater cost of liquid helium, needed to reach the lower critical temperatures of the period up to 1986, prevented widespread commercial application of superconductivity before then.

The new superconductors are not without problems of their own. In particular, their superconductivity tends to fail when they carry very high currents or in the presence of strong magnetic fields. Also, like other ceramics, the new superconductors are relatively weak mechanically. For these reasons the first large-scale industrial use of superconductivity is likely to be in increasing the operating speeds of computers and in other electronic applications.

A material that is superconducting at room temperature would revolutionize the world's technology. By reducing the waste of electric energy, the rate at which the world's resources are being depleted would also be reduced. Such a material, inconceivable until very recently, is now the subject of an active search.

16.5 WHAT IS A FIELD?

A region of space with a continuous distribution of a certain property.

Electrical forces, like gravitational ones, act between objects that may be far apart. A **force field** is a concept that provides a framework for understanding how forces are transmitted from one object to another across empty space.

A successful model does more than just organize our knowledge of a certain phenomenon. Such a model also enables us to predict hitherto unsuspected effects and relationships, which of course must then be verified by experiment. The creative role of a scientific model is beautifully illustrated by the electromagnetic field model proposed by James Clerk Maxwell over a century ago. With the help of this model, Maxwell predicted in 1864 that electromagnetic waves should exist; he calculated that their speed should be $c = 3 \times 10^8$ m/s; and he suggested that light consists of such waves. In 1887, after Maxwell's death, Heinrich Hertz confirmed Maxwell's hypothesis in the laboratory.

A field—in the sense that physicists use the word—is a region of space in which a certain quantity has a definite value at every point. Thus it is appropriate to speak of the temperature field in a room, of the velocity field of the water in a river, and of the gravitational field around the earth. But it is not appropriate to speak of the "chair field" in a room, or of the "rowboat field" in a river, or of the "airplane field" around the earth, because chairs, rowboats, and airplanes are separate objects. Their presence somewhere is not a property of a region of space that varies continuously throughout that region in the way that temperature, water velocity, and gravitational force vary.

To determine the temperature field in a room, we must measure the temperature at a great many points with a thermometer. The results might be displayed by a series of cross-sectional maps of the room with temperature values written in at each point of measurement, as in Fig. 16.11.

Figure 16.11

The temperature field in a room is a scalar field, because the temperature at a point has no direction associated with it. The temperatures here are expressed in degrees Celsius.

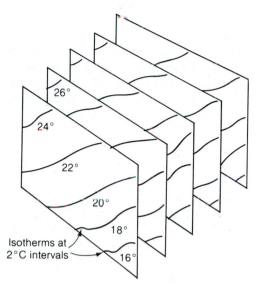

Figure 16.12

An isotherm is a line that joins points whose temperatures are the same. Here the temperature field in the room in Fig. 16.11 is displayed by means of isotherms.

A better way to show the temperature field is to draw a series of lines on each map that connect points having the same temperatures (Fig. 16.12). These lines are called **isotherms** and might be drawn, for example, for temperatures that are 2°C apart. Naturally the actual measurements are not necessarily 16°, 18°, 20°, and so on; we must interpolate between the actual measurements to find the contours of the various isotherms.

Temperature is a scalar quantity because it involves a magnitude only. To say that the temperature somewhere is 20°C describes it completely. The field of a scalar quantity is a **scalar field,** and it can always be pictured by a plot of isolines (lines joining sets of identical values) as in the case of a temperature field. A vector quantity has direction as well as magnitude, so picturing a **vector field** is less easy. In Section 16.7 we will see how to do this for an electric field.

16.6 ELECTRIC FIELD

Its action on a charge leads to a force on the charge.

When an electric charge is present somewhere, the properties of space around it are altered in such a way that another charge brought to this region will experience a force there. The "alteration in space" caused by a charge at rest is called its **electric field,** and any other charge is considered to interact with the field and not directly with the charge that gives rise to it.

All forces, not just electric ones, can be interpreted as arising through the action of a force field of one kind or another. Thus the sun is regarded as having a gravitational field around it, and the forces exerted by this field on the planets are what hold them in their orbits. In the case of the "direct-contact" forces between the solid objects of everyday life, the force fields involved are electric fields.

The electric field **E** at a point in space is defined as the ratio of the force **F** on a charge Q (assumed positive) at that point to the magnitude of Q:

$$\mathbf{E} = \frac{\mathbf{F}}{Q} \qquad\qquad \textit{Electric field} \quad (16.2)$$

The units of **E** are newtons per coulomb. Electric field is a vector quantity that possesses both magnitude and direction.

Once we know what the electric field **E** is at some point, from the definition we see that the force that the field exerts on a charge Q at that point is

$$\mathbf{F} = Q\mathbf{E} \qquad\qquad \textit{Electric force} \quad (16.3)$$

Force = (charge)(electric field)

◆   **E X A M P L E   16.2** _____

The electric field between the electrodes of the gas discharge tube used in a certain neon sign has the magnitude 5.0×10^4 N/C. Find the acceleration of a neon ion of mass 3.3×10^{-26} kg in the tube if it carries a charge of $+e$ (Fig. 16.13).

SOLUTION   Since $e = 1.6 \times 10^{-19}$ C, the force on the ion has the magnitude

$$F = QE = (1.6 \times 10^{-19}\ \text{C})(5.0 \times 10^4\ \text{N/C}) = 8.0 \times 10^{-15}\ \text{N}$$

Hence the acceleration of the ion is

$$a = \frac{F}{m} = \frac{8.0 \times 10^{-15}\ \text{N}}{3.3 \times 10^{-26}\ \text{kg}} = 2.4 \times 10^{11}\ \text{m/s}^2$$

This is 25 billion times greater than the acceleration of gravity. The speed such an ion actually reaches is limited by frequent collisions with the neon atoms in its path.

Figure 16.13

A charge Q in an electric field **E** experiences a force equal to Q**E**.

◆

We can use Coulomb's law to determine the magnitude of the electric field around a single charge Q. First we find the force F that Q exerts on another charge Q' at the distance r away, which is

$$F = k\frac{QQ'}{r^2}$$

Since the electric field that Q' experiences is $E = F/Q'$ we have

$$E = \frac{F}{Q'} = k\frac{Q}{r^2} \qquad \textit{Electric field of a charge} \quad (16.4)$$

This formula tells us that the electric field magnitude at the distance r from a point charge is directly proportional to the magnitude Q of the charge and is inversely proportional to the square of r.

When more than one charge contributes to the electric field at a point P, the net field $\mathbf{E}$ is the *vector sum* of the fields of the individual charges. That is,

$$\mathbf{E} = \mathbf{E}_1 + \mathbf{E}_2 + \mathbf{E}_3 + \cdots = \Sigma\mathbf{E}_n \qquad \textit{Total field} \quad (16.5)$$

◆ **E X A M P L E 16.3** _____

Two charges, $Q_1 = +2.0 \times 10^{-8}$ C and $Q_2 = +3.0 \times 10^{-8}$ C, are 50 mm apart, as in Fig. 16.14. What is the electric field halfway between them?

SOLUTION At a point halfway between the charges, $r_1 = r_2 = 25$ mm. Here the electric fields of the two charges have the respective magnitudes

$$E_1 = k\frac{Q_1}{r_1^2} = \frac{(9.0 \times 10^9 \text{N} \cdot \text{m}^2/\text{C}^2)(2.0 \times 10^{-8}\text{C})}{(2.5 \times 10^{-2}\text{m})^2} = 2.9 \times 10^5 \text{ N/C}$$

$$E_2 = k\frac{Q_2}{r_2^2} = \frac{(9.0 \times 10^9 \text{N} \cdot \text{m}^2/\text{C}^2)(3.0 \times 10^{-8}\text{C})}{(2.5 \times 10^{-2}\text{m})^2} = 4.3 \times 10^5 \text{ N/C}$$

The direction of an electric field at a point is the direction in which a positive charge placed there would move. Hence the direction of $\mathbf{E}_1$ is to the right in Fig. 16.14 and that of $\mathbf{E}_2$ is to the left. Since $\mathbf{E}_1$ and $\mathbf{E}_2$ are in opposite directions, the magnitude of their sum $\mathbf{E}$ is

$$E = E_2 - E_1 = 1.4 \times 10^5 \text{ N/C}$$

The direction of $\mathbf{E}$ is toward Q_1.

Figure 16.14

If the charges had both been negative rather than positive, the field halfway between them would have the same magnitude but would be in the opposite direction. If one charge had been negative and the other positive, $\mathbf{E}_1$ and $\mathbf{E}_2$ would point in the same direction along the line between the charges. The total field $\mathbf{E}$ there would then have the magnitude $E_1 + E_2$ and its direction would be toward the negative charge.   ◆

◆   **E X A M P L E   16.4** _____

Where is the electric field equal to zero in the neighborhood of the charges of the previous example?

SOLUTION   Since the fields $\mathbf{E}_1$ and $\mathbf{E}_2$ are in opposite directions along a line between Q_1 and Q_2, the point where they cancel out each other must lie on this line. The cancellation occurs where the magnitudes E_1 and E_2 are equal:

$$E_1 = E_2 \qquad k\frac{Q_1}{r_1^2} = k\frac{Q_2}{r_2^2} \qquad \frac{Q_1}{r_1^2} = \frac{Q_2}{r_2^2}$$

This last is an equation with two unknown quantities, r_1 and r_2. We need another equation that relates r_1 and r_2 in order to find their values. Such an equation comes from the fact that the distance between Q_1 and Q_2 is $s = 50$ mm, so that

$$r_1 + r_2 = s \qquad r_2 = (s - r_1)$$

Hence

$$\frac{Q_1}{r_1^2} = \frac{Q_2}{r_2^2} = \frac{Q_2}{(s - r_1)^2}$$

To solve for r_1, we rewrite the equation as

$$\frac{(s - r_1)^2}{r_1^2} = \frac{Q_2}{Q_1}$$

Then we take the square root of both sides:

$$\frac{s - r_1'}{r_1} = \sqrt{\frac{Q_2}{Q_1}}$$

Finally we multiply both sides by r_1 and solve for r_1, which gives

$$r_1 = \frac{s}{1 + \sqrt{Q_2/Q_1}}$$

Here $s = 50$ mm and $\sqrt{Q_2/Q_1} = 1.22$, so that

$$r_1 = \frac{s}{1 + \sqrt{Q_2/Q_1}} = \frac{50 \text{ mm}}{1 + 1.22} = \frac{50 \text{ mm}}{2.22} = 23 \text{ mm}$$

The point where $\mathbf{E} = 0$ lies between the charges 23 mm from the charge Q_1. ◆

16.7 ELECTRIC FIELD LINES

Imaginary lines that help us to picture an electric field.

   Electric field lines provide a way to give intuitive form to our thinking about an electric field. These lines are constructed as follows: At several points in space we draw arrows in the direction a positive charge there would experience a force (Fig. 16.15a). Then we connect these arrows to form smooth curves, which are the electric field lines (Fig. 16.15b). Two rules should be kept in mind:

1. Electric field lines leave positive charges and enter negative ones.
2. The spacing of field lines is such that they are close together where the field is strong and far apart where the field is weak.

   If we place a positive charge in an electric field, it will experience a force in the direction of the field line it is on. The magnitude of the force will be proportional

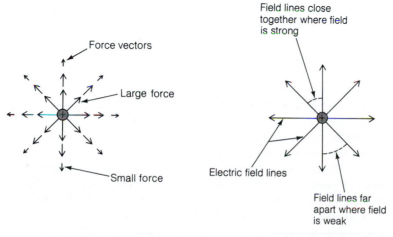

Figure 16.15

How electric field lines around a positive charge are drawn. In (a) the forces whose vectors are shown are those exerted by the charge on a positive test charge.

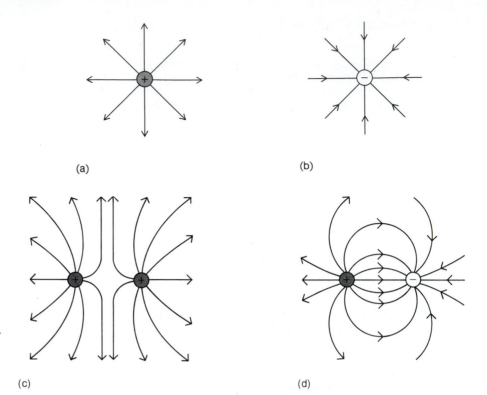

(a)

(b)

(c)

(d)

Figure 16.16

Electric field lines near separate charges and near pairs of charges. At large distances relative to the separation of the charges, the field in (c) approaches that of a single charge.

to the concentration of field lines in its vicinity. Figure 16.16 shows the patterns of field lines around a positive and a negative charge and around pairs of charges.

It is important to keep in mind that field lines do not actually exist as threads in space but are simply a device to help our thinking about force fields. The use of field lines is not limited to electric fields. Gravitational and magnetic fields, for instance, are often described in this way.

The notion of electric field lines is often helpful as a guide to our thinking when electric fields of charge distributions are concerned. For example, let us inquire into the electric field around a spherical distribution of charge whose total amount is Q. Such a distribution is quite easy to arrange; all we need do is to add the charge Q to a metal sphere. Since metals are good conductors of electricity, the mutual repulsion of the individual charges that make up Q causes them to spread out uniformly over the surface of the sphere so that the charges are as far apart as possible. It does not matter whether the sphere is hollow or solid.

Whatever the number of field lines we choose to represent the electric field around a point charge Q, the same number must be associated with a body that carries the same net charge Q. In the case of a charged sphere, the field lines are symmetrically arranged. Therefore the pattern of field lines outside the sphere is identical with that around a point charge of the same magnitude (Fig. 16.17). The same is true for the electric field $\mathbf{E}$ itself, of course.

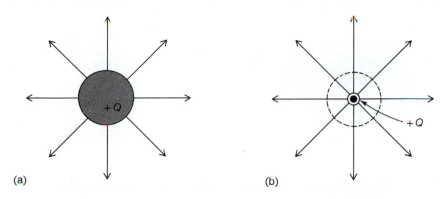

(a)　　　　　　　　　　　　　(b)

Figure 16.17

(a) Electric field lines around a uniformly charged sphere. (b) Electric field lines around a point charge.

Three important properties of a charged metal object are illustrated in Fig. 16.18:

1. Inside the object the electric field is zero everywhere. The electric fields due to the individual charges on the surface all cancel out in the interior, although outside the object they do not. We can see why this must be true even without a formal calculation. Suppose there *were* an electric field **E** in the interior of the object. Then the charges inside the object that are free to move (electrons in the case of a metal) would move under the influence of the field **E.** But no currents are observed in a charged conducting object except for a moment after the charge is placed on it. Since energy is needed to maintain an electric current, a supply of energy would be needed for currents to continue in such an object. The only conclusion is that the interior of an isolated conducting object is always free of electric field.

2. Just outside the object the electric field **E** must be perpendicular to the object's surface. If **E** had a component parallel to the surface, electrons on the surface would be in constant motion. Because they are not, **E** must be perpendicular to the surface.

3. All the excess charge (of either sign) on the object must be on its surface. If any point in the interior had a net charge, an electric field would be present around it. But there are no electric fields inside a metal object, so there can be no excess charge there.

Metal object

Charges reside on surface only

Figure 16.18

There are neither charges nor an electric field inside a charged metal object. Just outside the object the field is perpendicular to its surface.

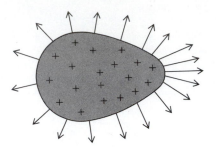

Figure 16.19

The electric field near a charged asymmetric metal object is strongest near the parts of the object that have the smallest radii of curvature.

Electric Field around an Irregular Metal Body

When a nonspherical metal body is charged, the individual charges do not distribute themselves uniformly on its surface. As a general rule, the more highly curved parts have greater concentrations of charge than the gently curved parts. The electric field is accordingly most intense near the highly curved parts (Fig. 16.19). We can see why this must be true from the preceding discussion of the electric field around a charged metal object. There is no field inside the object, and the field just outside it is perpendicular to its surface. In order to bring this about, the charges must be arranged on the object's surface so that the inward-directed parts of the fields of charges on one side of the object exactly balance the inward-directed parts of the fields on the opposite side. Hence the charges at the small end of the egg-shaped object in Fig. 16.19 must be closer together than those at the large end for their respective fields to cancel out in the interior.

At a metal point the electric field may become so great that it causes a separation of charge in nearby air molecules. The resulting electric "discharge" is visible as a luminous glow. If two oppositely charged pointed rods are brought close together, charge flows between them via disrupted air molecules and a spark occurs. If more gently curved charged bodies, such as large spheres, are similarly brought near each other, their separation must be much smaller before a spark can occur.

Important Terms

Electric charge, like mass, is a fundamental property of certain of the elementary particles of which all matter is composed. There are two kinds of electric charge, **positive charge** and **negative charge;** charges of like sign repel, unlike charges attract. The unit of charge is the **coulomb.** All charges, of either sign, occur in multiples of the fundamental **electron charge** of 1.6×10^{-19} coulomb.

The principle of **conservation of charge** states that the net electric charge in an isolated system remains constant.

Coulomb's law states that the force one charge exerts on another is directly proportional to the magnitudes of the charges and inversely proportional to the square of the distance between them.

An **atom** consists of a tiny, positively charged nucleus surrounded at some distance by **electrons.** The **nucleus consists of protons** and **neutrons,** and the number of electrons equals the number of protons, so the atom as a whole is electrically neutral.

An **ion** is an atom or a group of atoms that carries a net electric charge. An atom or a group of atoms becomes a negative ion when it picks up one or more electrons in addition to its normal number, and becomes a positive ion when it loses one or more electrons of its usual number.

A **force field** is a region of space at every point of which an appropriate test object would experience a force.

An **electric field** exists wherever an electrical force acts on a charged particle. The magnitude of an electric field at a point is equal to the force that would act on a charge of $+1$ C placed there. The direction of the field is the direction of the force on the charge. The unit of electric field is the newton/coulomb (N/C).

Important Formulas

Coulomb's law: $F = k \dfrac{Q_A Q_B}{r^2}$

Electric field: $\mathbf{E} = \dfrac{\mathbf{F}}{Q}$

Electric field of a charge: $E = k\dfrac{Q}{r^2}$

MULTIPLE CHOICE

1. A rod with an unknown charge attracts a plastic ball. Which one or more of the following might describe them?
 a. The rod is positive and the ball is negative.
 b. The rod is negative and the ball is positive.
 c. The rod is positive and the ball is uncharged.
 d. The rod is negative and the ball is uncharged.

2. Electric charge
 a. is a continuous quantity that can be subdivided indefinitely
 b. is a continuous quantity but it cannot be subdivided into smaller parcels than $\pm 1.6 \times 10^{-19}$ C
 c. occurs only in separate parcels, each of $\pm 1.6 \times 10^{-19}$ C
 d. occurs only in separate parcels, each of ± 1 C

3. Which one or more of the following statements is not true?
 a. The electron and proton have charges of the same magnitude but of opposite sign.
 b. The electron and proton have the same mass.
 c. The proton and neutron have the same mass.
 d. Atomic nuclei contain only protons and neutrons.

4. The electrons in an atom
 a. are bound to it permanently
 b. are some distance away from the nucleus
 c. have more mass than the nucleus
 d. may be positively or negatively charged

5. An object has a positive electric charge whenever
 a. it has an excess of electrons
 b. it has a deficiency of electrons

 c. the nuclei of its atoms are positively charged
 d. the electrons of its atoms are positively charged

6. In the formula $F = kQ_A Q_B/r^2$, the value of the constant k
 a. is the same under all circumstances
 b. depends on the medium the charges are located in
 c. is different for positive and negative charges
 d. has the numerical value 1.6×10^{-19}

7. Relative to the electrical force between two protons, the gravitational force between them is
 a. weaker
 b. equal in magnitude
 c. stronger
 d. any of the above, depending on how far apart the protons are

8. A solid conductor is one
 a. whose electrons are firmly bound in place
 b. in which some electrons are free to move about
 c. which has no electrons
 d. which has too many electrons

9. Superconductivity occurs in certain substances
 a. only at very low temperatures
 b. only at very high temperatures
 c. at all temperatures
 d. only when they are in the liquid state

10. The electric field at a point in space is equal in magnitude to
 a. the number of electric field lines that pass through the point

    b. the electric charge there
    c. the force a charge of one coulomb would experience there
    d. the force an electron would experience there

11. Electric field lines
    a. leave negative charges and enter positive ones
    b. are close together where E is weak and far apart where E is strong
    c. have the same pattern around a charged metal sphere as they do around a point charge
    d. physically exist

12. Ten million electrons are placed on a solid copper sphere. The electrons become
    a. uniformly distributed on the sphere's surface
    b. uniformly distributed in the sphere's interior
    c. concentrated at the center of the sphere
    d. concentrated at the bottom of the sphere

13. If 10,000 electrons are removed from a neutral plastic ball, its charge is now
    a. $+1.6 \times 10^{-15}$ C   b. $+1.6 \times 10^{-23}$ C
    c. -1.6×10^{-15} C   d. -1.6×10^{-23} C

14. Two charges of $+Q$ are 1 cm apart. If one of the charges is replaced by a charge of $-Q$, the magnitude of the force between them is
    a. zero   b. smaller
    c. the same   d. larger

15. A charge of $+q$ is placed 2 cm from a charge of $-Q$. A second charge of $+q$ is then placed next to the first. The force on the charge of $-Q$
    a. decreases to half its former magnitude
    b. remains the same
    c. increases to twice its former magnitude
    d. increases to four times its former magnitude

16. Two charges repel each other with a force of 10^{-6} N when they are 10 cm apart. When they are brought closer together until they are 2 cm apart, the force between them becomes
    a. 4×10^{-8} N   b. 5×10^{-6} N
    c. 8×10^{-6} N   d. 2.5×10^{-5} N

17. Two charges, one positive and the other negative, are initially 2 cm apart and are then pulled away from each other until they are 6 cm apart. The force between them is now smaller by a factor of
    a. $\sqrt{3}$ b. 3
    c. 9   d. 27

18. The force between two charges of -3.0×10^{-9} C that are 50 mm apart is
    a. 1.8×10^{-16} N   b. 3.6×10^{-15} N
    c. 1.6×10^{-6} N   d. 3.2×10^{-5} N

19. Two equal charges attract each other with a force of 1.0×10^{-5} N. When they are moved 4 mm farther apart, the force between them becomes 2.5×10^{-6} N. The original separation of the charges was
    a. 1 mm   b. 2 mm
    c. 4 mm   d. 8 mm

20. The magnitude of the charges in Multiple Choice 19 is
    a. 6.7×10^{-11} C   b. 1.3×10^{-10} C
    c. 1.1×10^{-9} C   d. 2.1×10^{-9} C

21. The magnitude of the force on an electron in an electric field of 200 N/C is
    a. 8.0×10^{-22} N   b. 3.2×10^{-21} N
    c. 3.2×10^{-17} N   d. 6.4×10^{-15} N

22. The electric field 20 mm from a certain charge has a magnitude of 1.0×10^5 N/C. The magnitude of **E** 10 mm from the charge is
    a. 2.5×10^4 N/C   b. 5.0×10^4 N/C
    c. 2.0×10^5 N/C   d. 4.0×10^5 N/C

QUESTIONS

1. (a) When two objects attract each other electrically, must both of them be charged? (b) When two objects repel each other electrically, must both of them be charged?

2. What reasons might there be for the universal belief among scientists that there are only two kinds of electric charge?

3. How can the principle of charge conservation be reconciled with the fact that a rubber rod can be charged by stroking it with a piece of fur?

4. An insulating rod has a charge of $+Q$ at one end and a charge of $-Q$ at the other. How will the rod tend to behave when it is placed near a fixed positive charge that is initially equidistant from the ends of the rod?

5. How will the rod in Question 4 tend to behave when it is placed in a uniform electric field whose direction is (a) parallel to the rod and (b) perpendicular to the rod?

6. Electricity was once regarded as a weightless fluid, an excess of which was "positive" and a deficiency of which was "negative." What phenomena can this hypothesis still explain? What phenomena can it not explain?

7. What property of the electrons in a metal allows it to conduct electric current readily? What property of the electrons in an insulator prevents it from conducting electric current readily?

8. How does electrical conduction in a metal differ from that in an ionized gas?

9. List several good conductors of electricity and several good insulators. How well do these substances conduct heat? What general relationship between an ability to conduct electricity and an ability to conduct heat can you infer?

10. What aspect of superconductivity has prevented its large-scale application thus far?

11. How could you distinguish between an electric field and a gravitational field?

12. Can electric field lines ever intersect in space? Explain.

13. What can you tell about the force a charged object would experience at a given point in an electric field by looking at a sketch of the field lines?

14. Figure 16.20 shows electric field lines near two charges. (a) What is the sign of each charge? (b) Do the charges have the same magnitude? If not, which has the greater magnitude and what is the ratio between their magnitudes?

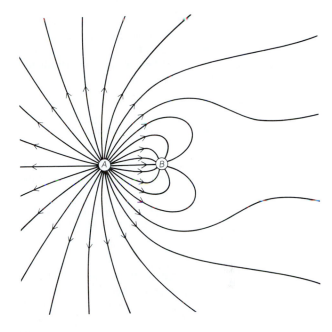

Figure 16.20

Question 14.

EXERCISES

16.2 Coulomb's Law

1. Find the charge of 1.0 μg of protons.

2. Two charges attract each other with a force of 4.0×10^{-6} N when they are 4.0 mm apart. Find the force between them when their separation is increased to 5.0 mm.

3. Two electric charges originally 80 mm apart are brought closer together until the force between them is greater by a factor of 16. How far apart are they now?

4. Two charges of unknown magnitude and sign are observed to repel one another with a force of 0.10 N when they are 50 mm apart. What will the force be when they are (a) 100 mm apart? (b) 500 mm apart? (c) 10 mm apart?

5. The nucleus of a hydrogen atom is a single proton. Find the force between the two protons in a hydrogen molecule, H_2, that are 7.42×10^{-11} m apart. (The two electrons in the molecule spend more time between the protons than outside them, which leads to attractive forces that balance the repulsion of the protons and permit a stable H_2 molecule; see Chapter 26.)

6. A charge of -5.0×10^{-7} C is 10 cm from a charge of $+6.0 \times 10^{-6}$ C. Find the magnitude and direction of the force on each charge.

7. A charge of $+5.00 \times 10^{-9}$ C is attracted by a charge of -3.00×10^{-7} C with a force of 0.135 N. How far apart are they?

8. Two identical metal spheres, one with a charge of $+2.0 \times 10^{-5}$ C and the other with a charge of -1.0×10^{-5} C, are 10 cm apart. (a) What is the force between them? (b) The two spheres are brought into contact, and then separated again by 10 cm. What is the force between them now?

9. How far apart should two electrons be if the force each exerts on the other is to equal the weight of an electron at sea level?

10. The permittivity ϵ of air is only a trifle greater than ϵ_0, the permittivity of free space, but ϵ is considerably greater for some other materials. For instance, the permittivity of water is $\epsilon = 80 \, \epsilon_0$. Find the force, in air and in water, between two charges of $+2.0 \times 10^{-10}$ C that are 1.0 mm apart.

11. A particle carrying a charge of $+6.0 \times 10^{-7}$ C is located halfway between two other charges, one of $+1.0 \times 10^{-6}$ C and the other of -1.0×10^{-6} C, that are 40 cm apart. All three charges lie on the same straight line. What is the magnitude and direction of the force on the $+6.0 \times 10^{-7}$ C charge?

12. A test charge of -5.0×10^{-7} C is placed between two other charges so that it is 50 mm from a charge of -3.0×10^{-7} C and 10 cm from a charge of -6.0×10^{-7} C. The three charges lie along a straight line. What is the magnitude and direction of the force on the test charge?

16.6 Electric Field

13. Find the electric field 40 cm from a charge of $+7.0 \times 10^{-5}$ C.

14. The average distance between the electron and the proton in a hydrogen atom is 5.3×10^{-11} m. How strong is the electric field the electron experiences?

15. Four charges of $+1.0$ μC are at the corners of a square

that measures 1.0 m on each side. Find the electric field at the center of the square.

16. A particle carrying a charge of 1.0×10^{-5} C starts moving from rest in a uniform electric field whose magnitude is 50 N/C. (a) What is the force on the particle? (b) How much kinetic energy will the particle have after it has moved 1.0 m?

17. How strong an electric field is needed to support a proton against gravity at sea level?

18. Two charges of $+4.0$ μC and $+8.0$ μC are 2.0 m apart. What is the electric field halfway between them?

19. Two charges, one of $+1.5$ μC and the other of $+3.0$ μC, are 20 cm apart. Where is the electric field along the line joining them equal to zero?

PROBLEMS

20. Two charges, one of $+2.0 \times 10^{-8}$ C and the other of $+1.0 \times 10^{-8}$ C are 20 cm apart. (a) Where should a charge of -1.0×10^{-7} C be placed in order that there be no resultant force on it? (b) Where should a charge of $+1.0 \times 10^{-7}$ C be placed in order that there be no resultant force on it?

21. Two charges, one of -1.0×10^{-6} C and the other of -3.0×10^{-6} C are 40 cm apart. (a) Where should a charge of -1.0×10^{-7} C be placed on the line between other charges in order that there be no resultant force on it? (b) Where should a charge of $+1.0 \times 10^{-7}$ C be placed in order that there be no resultant force on it?

22. Two 3.0-g balloons are suspended from a nail by strings 50 cm long. Each balloon has a charge of $+Q$, and there is an angle of 40° between the strings. Find Q.

23. According to a simplified model, the hydrogen atom consists of a proton circled by an electron whose orbit has a radius of 5.3×10^{-11} m. How fast must the electron be moving if the required centripetal force is provided by the electric force exerted by the proton?

24. Suppose the force between the earth and the moon were electrical rather than gravitational, with the earth having a positive charge and the moon having a negative one. If the magnitude of each charge were proportional to the respective body's mass, find the Q/m ratio required for the moon to follow its present orbit of 3.84×10^8-m radius with its period of 27.3 days. The earth's mass is 6.0×10^{24} kg, and the moon's mass is 7.3×10^{22} kg.

Answers to Multiple Choice

1. a, b, c, d	**7.** a	**13.** a	**19.** c
2. c	**8.** b	**14.** c	**20.** b
3. b, c	**9.** a	**15.** c	**21.** c
4. b	**10.** c	**16.** d	**22.** d
5. b	**11.** c	**17.** c	
6. c	**12.** a	**18.** d	

17

ELECTRIC ENERGY

17.1 **Electric Potential Energy**

17.2 **Potential Energy of Two Charges**

17.3 **Potential Difference**

17.4 **Cathode-Ray Tube**

17.5 **The Electron Volt**

17.6 **Electric Field Energy**

17.7 **Capacitance**

17.8 **Dielectric Constant**

17.9 **Capacitors in Combination**

In our study of mechanics we found the related concepts of work and energy to be useful in analyzing a wide variety of situations. These concepts are equally useful in the study of electricity—in particular, electric current. Instead of potential energy itself, a related quantity called potential difference turns out to be especially appropriate in describing electrical phenomena. Later in this chapter we will look into how capacitors store electric energy in their electric fields.

17.1 ELECTRIC POTENTIAL ENERGY

A charge located in an electric field has potential energy.

Let us consider the potential energy of a charge in a uniform electric field with the help of a gravitational analogy. At the left in Fig. 17.1(a) is a uniform electric field $\mathbf{E}$ between two parallel, uniformly charged plates A and B. At the right is a region near the earth's surface in which the gravitational field is also uniform. Now we place a particle of charge Q (assumed positive) in the electric field and a particle of mass m in the gravitational field. As shown in Fig. 17.1(b), the charge is acted on by the electrical force

$$\mathbf{F}_{\text{elec}} = Q\mathbf{E}$$

and the mass is acted on by the gravitational force

$$\mathbf{F}_{\text{grav}} = m\mathbf{g}$$

If the charge is on plate A and we want to move it to plate B, we must apply a force of magnitude QE to it because we have to push against a force of this magnitude (and opposite direction) exerted by the electric field. When the charge is at B, we will have done the work

$$\text{Work} = (\text{force})(\text{distance})$$
$$W = QEd$$

on it, where d is the distance the charge has moved (Fig. 17.1(c)). Similarly, to raise the mass from the ground to a height h, we must apply a force of magnitude mg to it, and the work we do is

$$W = mgh$$

At plate B the charge has the potential energy

$$\text{PE} = QEd \qquad\qquad \textit{PE of charge in uniform electric field} \quad (17.1)$$

with respect to A. If we let it go, the PE will become kinetic energy as the electric field $\mathbf{E}$ accelerates the charge, and when the charge is back at A, it will have a KE equal to QEd (Fig. 17.1(d)). In the same way, the mass has PE *with respect to the ground* in its new location. This PE is equal to the work done in raising it through the height h and is

$$\text{PE} = mgh$$

If we let the mass go, it will fall to the ground with a final KE of mgh.

Electric field lines

B +

Electric field **E**

(a)

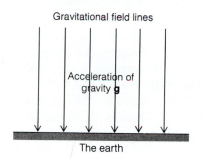

Gravitational field lines

Acceleration of gravity **g**

The earth

Figure 17.1

Analogy between electric and gravitational potential energy.

B +

F = Q**E** Force exerted on Q by electric field

A −

(b)

m

Force exerted on m by gravitational field **F** = m**g**

The earth

B +

Q PE = QEd relative to A
KE = O

Applied force

F = Q**E**

d

A −

(c)

m PE = mgh
relative to ground
KE = O

Applied force

F = m**g**

h

m

The earth

B +

Q

PE = O
KE = QEd

A −

(d)

m

PE = O
KE = mgh

The earth

To summarize: The amount of work QEd must be performed to move a charge Q from A to B, the distance d apart, in a uniform electric field $\mathbf{E}$. At B the charge accordingly has the potential energy QEd. If the charge is released at B, the force $Q\mathbf{E}$ acting on it produces an acceleration such that the charge has the kinetic energy QEd when it is back at A. Thus the work done in moving the charge in the field becomes PE, which in turn becomes KE when it is released.

There is no change in the energy of a charge moved perpendicular to an electric field, just as there is no change in the energy of a mass moved perpendicular to a gravitational field (for instance, along the earth's surface).

17.2 POTENTIAL ENERGY OF TWO CHARGES

It is a property of the system of both charges.

A particle of charge Q_A that is located the distance r from another particle of charge Q_B has electric potential energy there because a force is exerted on it by the electric field of Q_B. When Q_A and Q_B have the same sign, the force is repulsive; when the charges have opposite signs, the force is attractive. In either case, when Q_A is released, it will begin to move and gain KE at the expense of its original PE (Fig. 17.2 shows this effect for two like charges).

We have been assuming that Q_B is fixed in place. If instead Q_A is fixed in place, then we can speak of the PE of Q_B in the electric field of Q_A. This PE is exactly the same as before, since by Newton's third law of motion the force one object exerts on another is equal in magnitude to the force the second object exerts on the first.

In reality, the PE belongs to the *system* of the two particles. If the particles are both released, *both* of them begin to move, and they share the original PE between them. The relative speeds of the two particles will be such as to conserve linear

Figure 17.2

A charge has potential energy when it is in the electric field of another charge. Actually, the potential energy belongs to the system of the two charges. Here Q_B is assumed to be fixed in place, so the PE of the system becomes kinetic energy of Q_A when Q_A is released.

At rest, Q_A has PE only

When released, Q_A gains KE at the expense of its original PE

momentum, so the lighter particle will move faster than the heavier one. (The total amount of energy available depends on the charges of the particles, and the division of the energy depends on their masses.)

If one of the particles has a much greater mass than the other has, its speed will be much smaller than that of the other, and we may then legitimately regard it as being fixed in place. For example, the mass of a proton is 1836 times that of an electron, so in a hydrogen atom, which consists of a proton and an electron, we may think of the proton as being stationary and of the electron as having the PE of the system.

A potential energy of any kind must be specified relative to a reference location. Near the earth, for example, the gravitational PE of an object is usually given with the earth's surface as the reference location for PE = 0. In the case of charges interacting with one another, the reference location is chosen to be infinity, since the electric field of a charge falls to zero an infinite distance away.

Because the electric field $\mathbf{E}$ of the charge Q_B (which we assume to be fixed in place) is not uniform, it is not easy to calculate the PE of Q_A when it is the distance r away. What we must do is start with Q_A at infinity and imagine we move it a short distance toward Q_B. In this short distance we consider $\mathbf{E}$ as having a constant magnitude, so we can find the work needed for this step from the formula $W = QEd$. Then we move Q_A through another short interval closer to Q_B. Again we consider $\mathbf{E}$ as having a constant magnitude, though a greater one than before, and we find the work needed for the second step. By continuing this process until Q_A is r away from Q_B and adding up all the amounts of work needed, we can find the PE of Q_A. With the help of calculus this addition yields the result

$$ PE = k\frac{Q_AQ_B}{r} \qquad \textit{Potential energy of system of two charges} \quad (17.2) $$

If the charges have the same sign, their PE is positive, so a positive PE corresponds to a repulsive force. If the charges have opposite signs, their PE is negative, so a negative PE corresponds to an attractive force. The PE of a charge decreases as it moves away from another charge of the same sign and increases as it moves away from another charge of the opposite sign.

◆   E X A M P L E   17.1 _____

An electron of initial speed 1.0×10^3 m/s is aimed at another electron, whose position is fixed, from a distance of 1.0 mm. How close to the stationary electron will the other one approach before it comes to a stop and reverses its direction?

SOLUTION   Figure 17.3 illustrates the situation. Since $Q_A = Q_B = -e$ here, the potential energies of the moving electron at the initial distance $r_1 (= 1.0 \times 10^{-3}$ m) and the final distance r_2 are, respectively,

$$ PE_1 = \frac{ke^2}{r_1} \qquad PE_2 = \frac{ke^2}{r_2} $$

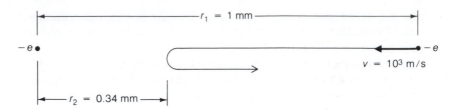

Figure 17.3

Charge separation begins in a thunderstorm when droplets of water and ice crystals collide, and the separation continues on a larger scale mainly through convection. The result is usually a region of negative charge a few kilometers across in the storm center with a region of positive charge in the clouds above it. The density of separated charge is about a coulomb per cubic kilometer. The charge separation produces a potential difference of up to several hundred million volts, and the electrical discharges that then occur appear as lightning flashes.

The potential energies are both positive, corresponding to a repulsive force between the electrons. The difference between the two PE values is equal to the initial kinetic energy $\frac{1}{2}mv^2$ of the moving electron, and therefore

$$\text{KE} = \text{PE}_2 - \text{PE}_1$$

$$\tfrac{1}{2}mv^2 = ke^2\left(\frac{1}{r_2} - \frac{1}{r_1}\right)$$

$$\frac{1}{r_2} = \frac{mv^2}{2ke^2} + \frac{1}{r_1}$$

$$r_2 = \frac{1}{(mv^2/2ke^2) + (1/r_1)} = 3.4 \times 10^{-4}\,\text{m} = 0.34\,\text{mm} \qquad \blacklozenge$$

17.3 POTENTIAL DIFFERENCE

The volt is the unit of potential difference.

The quantity **potential difference** describes the situation of a charge in an electric field in an especially convenient way. The potential difference V_{AB} between two points A and B is defined as the ratio between the work that must be done to take a charge Q from A to B and the value of Q:

$$V_{AB} = \frac{W_{AB}}{Q} \qquad\qquad\qquad \textit{Potential difference} \quad (17.3)$$

Potential difference = work per unit charge

The unit of potential difference is the joule per coulomb. Because this quantity is so often used, its unit has been given a name of its own, the **volt** (V). Thus

$$1 \text{ volt} = 1 \text{ joule/coulomb}$$

The volt is named after the Italian physicist Alessandro Volta (1745–1827), who invented the battery.

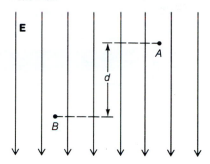

Figure 17.4

The potential difference between two points in a uniform electric field **E** is equal to *Ed*, where *d* is the component parallel to **E** of the distance between the points.

In a uniform electric field, $W_{AB} = QEd$, with the result that the potential difference between *A* and *B* is QEd/Q or

$$V_{AB} = Ed \qquad \textit{Potential difference in a uniform electric field} \quad (17.4)$$

In a uniform field the potential difference between two points is the product of the field magnitude *E* and the separation *d* of the two points in a direction parallel to that of **E** (Fig. 17.4).

From Eq. (17.4) we see that the volt/meter (V/m) is equal to the newton/coulomb (N/C), so that either can be used as the unit of electric field.

A positive potential difference means that the energy of the charge is *greater* at *B* than at *A*. A negative potential difference means that its energy is *less* at *B* than at *A*. If V_{AB} is positive, then a charge at *B* tends to return to *A*. If V_{AB} is negative, the charge tends to move farther away from *A*.

One advantage of specifying the potential difference between two points in an electric field, rather than the magnitude of the field between them, is that an electric field is normally created by imposing a difference of potential between two points. A battery uses chemical reactions to produce a potential difference between two terminals (Fig. 17.5). A ''6-volt'' battery is one that has a potential difference of 6 V between its terminals. A generator is another device for producing a potential difference. Batteries and generators are to electric charges as elevators are to masses: All of them increase the potential energy of what they act on.

When a charge *Q* goes from one terminal of a battery whose potential difference is *V* with respect to the other, Eq. (17.3) tells us that the work

$$W = QV \qquad\qquad \textit{Work done on a charge} \quad (17.5)$$

is done on it regardless of the path taken by the charge and regardless of whether the actual electric field that acted on the charge is strong or weak. Given *V*, we can find *W* at once, no matter what the details of the process are. Hence the notion of potential difference simplifies the analysis of electrical phenomena, just as the notion of potential energy simplifies the analysis of mechanical phenomena.

Figure 17.5

A battery is to a positive electric charge as an elevator is to a mass.

Figure 17.6

◆   E X A M P L E   17.2 _____

Figure 17.6 shows a tube that has a source of electrons at one end and a metal plate at the other. A 100-V battery is connected between the electron source and the plate, so that there is a potential difference of 100 V between them. The negative terminal of the battery is connected to the electron source. What is the speed of the electrons when they arrive at the metal plate? (The tube is evacuated to prevent collisions between the electrons and air molecules.)

SOLUTION   To find the electron's kinetic energy, we note that the work done by the electric field within the tube on an electron is $W = QV = eV$. Since the KE of the electron is equal to the work done on it, its KE after passing through the entire field is eV. Hence

$$\text{KE} = W = eV = \tfrac{1}{2}mv^2$$

$$v = \sqrt{\frac{2eV}{m}} = \sqrt{\frac{2(1.6 \times 10^{-19}\,\text{C})(100\,\text{V})}{9.1 \times 10^{-31}\,\text{kg}}} = 5.9 \times 10^6\,\text{m/s}$$

◆

◆   E X A M P L E   17.3 _____

The electron source and the positive electrode are 400 mm apart in the tube of the preceding example. (a) What is the magnitude E of the electric field (assumed uniform) between them? (b) If this distance is reduced to 200 mm, what is the new value of E and what effect does this change have on the final speed of the electrons?

SOLUTION   (a) The electric field magnitude in a uniform field is given by $E = V/d$ according to Eq. (17.4). Hence

$$E = \frac{100\,\text{V}}{0.400\,\text{m}} = 250\,\text{V/m}$$

(b) If d is reduced to 200 mm, the field magnitude increases to

$$E = \frac{100\,\text{V}}{0.200\,\text{m}} = 500\,\text{V/m}$$

The energy given to an electron by this field when it travels through the entire 200 mm is the same as the energy given to an electron that travels through the 400-mm length of the previous, weaker field, since in both cases QEd is the same. We reach the same conclusion by noting that KE $= QV$ and the potential difference V in this problem is independent of the spacing of the electrodes. The final electron speed is therefore unchanged. ◆

17.4 CATHODE-RAY TUBE

How to display an electrical signal.

Before their nature was understood, beams of electrons were known as "cathode rays." Even today, a tube in which an electron beam can be controlled in such a way as to trace a desired pattern on a fluorescent screen is called a **cathode-ray tube.**

Figure 17.7 shows how such a tube is constructed. Electrons emitted from the heated cathode (negative electrode) are accelerated through a potential difference of 5 kV to 50 kV between the cathode and the anode (positive electrode) and emerge from a hole in the latter. Near the cathode is an electode whose potential difference relative to the cathode controls the number of electrons in the beam and therefore the brightness of the spot on the screen. The entire assembly, which also incorporates another anode to focus the beam, is called an **electron gun.**

Figure 17.7

A cathode-ray tube.

Fluorescent screen

The electron beam then passes through a horizontal electric field set up between a pair of parallel metal plates. The direction of the field determines whether the beam is deflected to the right or to the left, and its magnitude (which is proportional to the voltage applied to the plates) governs the amount of the deflection. The vertical electric field between the next set of plates similarly governs the vertical position of the electron beam when it reaches the screen. The screen is coated with a substance called a **phosphor** that emits light when struck by the electrons. The screen continues to glow for a short time after the electron beam has moved away from a point, which permits a picture to be built up that lasts long enough to be viewed as a whole.

An **oscilloscope** uses a cathode-ray tube to display electrical signals that vary rapidly. The small mass of the electron permits a beam of them to respond at once to a changing voltage applied to a pair of deflecting plates. The horizontal deflection plates are connected to a source of sawtooth pulses (Fig. 17.8(a)), which rise steadily from a certain voltage $-V$ through 0 to $+V$ and then return rapidly to $-V$ to start a new cycle. In the figure the pulses take 0.1 s to go from $-V$ to $+V$, where V is chosen to correspond to a full horizontal deflection in a particular tube. Thus each sawtooth pulse sweeps the electron beam from left to right across the screen in 0.1 s and then returns the beam to the left to start another sweep.

Suppose we connect the vertical deflection plates to a voltage source that alternates in polarity 50 times per second (in other words, whose frequency is 50 Hz). What we will see are five complete cycles on the oscilloscope screen, as in Fig. 17.8(b), because each cycle involves 1/50 s = 0.02 s. A wide range of sweep times is provided on an oscilloscope, and the input signal can be amplified as necessary to give a trace large enough to be inspected and measured.

The picture tube of a television set is a cathode-ray tube, but in this application it is more convenient to use magnetic rather than electric fields to deflect the electron beam; magnetic fields are considered in later chapters. The electron beam is moved horizontally across the screen just as in an oscilloscope, but after each sweep the beam is shifted downward in order to cover a new line on the screen (Fig. 17.9). In the United States, a complete image is divided into 525 lines, but in many other countries more lines are used to give better picture quality. A complete image is built up by two scans of the screen, each covering alternate lines, a process that takes place 30 times per second. During the scans the beam intensity changes to produce

Figure 17.8

(a) A sawtooth voltage with a period of 0.1 s. (b) Oscilloscope trace of a 50-Hz alternating signal applied to the vertical deflection plates when the sawtooth voltage of (a) is applied to the horizontal deflection plates. Each cycle takes 0.02 s, so five appear on the screen.

(a)

(b)

Figure 17.9

The electron beam of a television picture tube covers the screen in a pattern of horizontal lines starting at the upper left. The returns from right to left take about one-tenth as long as the sweeps from left to right that produce the image.

the variations in brightness that make up the image. In color television separate signals are transmitted that correspond to the red, green, and blue contents of the image. The receiver screen consists of red, green, and blue phosphor dots that flash accordingly when the electron beam strikes them.

17.5 THE ELECTRON VOLT

The energy unit of atomic and nuclear physics.

The **electron volt** (abbreviated eV) is the customary energy unit in atomic and nuclear physics. By definition, 1.00 eV is the energy gained by an electron that has been accelerated through a potential difference of 1.00 volt. Hence

$$W = QV$$
$$1.00 \, \text{eV} = (1.60 \times 10^{-19} \, \text{C})(1.00 \, \text{V})$$

and so

$$1.00 \, \text{eV} = 1.60 \times 10^{-19} \, \text{J} \qquad \textit{Electron volt} \quad (17.6)$$

Typical quantities expressed in electron volts are the ionization energy of an atom (the work needed to remove one of its electrons) and the binding energy of a molecule (the work needed to break it apart into separate atoms). Thus the ionization energy of nitrogen is usually given as 14.5 eV, and the binding energy of the hydrogen molecule, which consists of two hydrogen atoms, is usually given as 4.5 eV.

The eV is too small a unit for nuclear physics, where its multiples the MeV (10^6 eV) and the GeV (10^9 eV) are commonly used. The M and G, respectively, signify *mega* (= 10^6) and *giga* (= 10^9) and are used in connection with other units as well, for instance, the megabuck (10^6) and the gigawatt (10^9 watts). A typical quantity expressed in MeV is the energy liberated when the nucleus of a uranium atom splits into two parts. Such **fission** of a uranium nucleus releases an average of 200 MeV; this is the process that powers nuclear reactors and weapons.

◆   **E X A M P L E   17.4** _____

What is the speed of a neutron whose kinetic energy is 50 eV?

SOLUTION   The kinetic energy of the neutron is

$$KE = (50 \text{ eV})(1.6 \times 10^{-19} \text{ J/eV}) = 8.0 \times 10^{-18} \text{ J}$$

Since $KE = \frac{1}{2}mv^2$ and the neutron mass is 1.67×10^{-27} kg, the speed of the neutron is

$$v = \sqrt{\frac{2 \, KE}{m}} = \sqrt{\frac{(2)(8.0 \times 10^{-18} \text{ J})}{1.67 \times 10^{-27} \text{ kg}}} = 9.8 \times 10^4 \text{ m/s} \qquad \blacklozenge$$

17.6 ELECTRIC FIELD ENERGY

A charged capacitor stores energy in its electric field.

Work must be done to create an electric field because work is needed to separate positive and negative charges against the Coulomb forces attracting them together. Where does the work go? The answer is that it is stored as potential energy, which we can think of as residing in the electric field between the charges. A system of conductors that stores energy in the form of an electric field is called a **capacitor**.

An example of a capacitor is a pair of parallel metal plates of area A that are a distance d apart, as in Fig. 17.10. Let us imagine that we create an electric field between them by bringing electrons from one plate to the other until a total charge of Q has been transferred. When we are finished, there will be a potential difference of V between the plates.

At the beginning, the potential difference between the plates is small, and little work is needed to move each electron. As the charge builds up, the potential difference becomes greater, and more work is needed per electron. The average potential difference $\overline{V}$ during the charge transfer is

$$\overline{V} = \frac{V_{\text{final}} + V_{\text{initial}}}{2} = \frac{V + 0}{2} = \frac{1}{2}V$$

Area = A

$\bullet +e$ $\circ -e$

Figure 17.10

As a capacitor is charged, the potential difference between its plates builds up and more work must be done to transfer each successive charge. (In an actual capacitor, the electrons are transferred by means of an electric circuit connecting the plates, not directly.)

Since the total charge transferred is Q, the work W that was done is the product of Q and the average potential difference V, namely,

$$W = Q\overline{V} = \tfrac{1}{2}QV \qquad\qquad \textit{Energy of charged capacitor} \quad (17.7)$$

Because the metal plates are good conductors, the $+$ and $-$ charges on them are spread out evenly, and the field between the plates (except near the edges) is uniform with the magnitude

$$E = 4\pi k\frac{Q}{A} \qquad\qquad \textit{Electric field inside capacitor} \quad (17.8)$$

The derivation of this formula, which follows from the considerations discussed in Chapter 16, is not given here because it involves calculus. We can rewrite Eq. (17.8) in the form

$$Q = \frac{AE}{4\pi k} \qquad\qquad\qquad\qquad (17.9)$$

The potential difference between the plates is equal to

$$V = Ed \qquad\qquad\qquad\qquad (17.10)$$

Hence the electric potential energy of the system of two charged plates is

$$W = \tfrac{1}{2}QV = \frac{Ad}{8\pi k}E^2 \qquad\qquad\qquad (17.11)$$

We note that Ad, the product of the area of each plate and the distance between them, is the volume occupied by the electric field **E** (Fig. 17.11). If we

Figure 17.11

The energy density of the electric field in a region is the total energy in the region divided by its volume.

define the **energy density** w of an electric field as the electric potential energy per unit volume associated with it, $w = W/Ad$, and so

$$w = \frac{E^2}{8\pi k}$$

Electric energy density   (17.12)

The energy density of an electric field is directly proportional to the square of its magnitude E. Even though this formula was derived for a special situation, it is a completely general result.

◆   **E X A M P L E   17.5** _____

Dry air is an insulator provided the electric field in it does not exceed about 3.0×10^6 V/m. What energy density does this correspond to?

SOLUTION   Because $E = 3.0 \times 10^6$ V/m,

$$w = \frac{E^2}{8\pi k} = \frac{(3.0 \times 10^6 \text{ V/m})^2}{8\pi(9.0 \times 10^9 \text{ N} \cdot \text{m}^2/\text{C}^2)} = 40 \text{ J/m}^3$$

At sea level 1 m³ of air has a mass of 1.3 kg, so if this amount of energy were gravitational potential energy, it would correspond to an elevation of about 3 m.   ◆

17.7 CAPACITANCE

The ratio of the charge on either conductor of a capacitor to the potential difference between the conductors.

The potential difference V across a capacitor is always directly proportional to the charge Q on either of its plates (Fig. 17.12). The more the charge, the stronger the electric field between the plates, and the greater the potential difference. The ratio between Q and V is therefore a constant for any capacitor and is known as its **capacitance** (symbol C):

$$C = \frac{Q}{V}$$

Capacitance   (17.13)

$$\text{Capacitance} = \frac{\text{charge on either conductor}}{\text{potential difference between conductors}}$$

Figure 17.12

The ratio between the charge on a particular capacitor and the potential difference across it is a constant called its *capacitance*.

The unit of capacitance is the **farad**, abbreviated F, where

1 farad = 1 coulomb/volt

The farad is so large a unit that, for practical purposes, it is usually replaced by the **microfarad** (μF) or **picofarad** (pf), whose values are

$$1\mu F = 10^{-6} \, F \qquad 1 \, pF = 10^{-12} \, F$$

The capacitance of a pair of separated conductors depends solely on their geometry and on the material between them. In the case of a parallel-plate capacitor in vacuum, from Eqs. (17.9) and (17.10) we have

$$C = \frac{Q}{V} = \frac{AE/4\pi k}{Ed}$$
$$= \frac{1}{4\pi k}\frac{A}{d} \qquad \text{\textit{Parallel-plate capacitor}} \quad (17.14)$$

To a good degree of approximation, the same formula can be used for such a capacitor in air as well.

◆ **E X A M P L E 17.6** _____

The plates of a parallel-plate capacitor are 10 cm square and 1.0 mm apart. Find its capacitance in air and the charge each plate will have when a potential difference of 100 V is applied.

SOLUTION Here $A = 1.0 \times 10^{-2} \, m^2$ and $d = 1.0 \times 10^{-3}$ m. In air the capacitance is

$$C = \left[\frac{1}{(4\pi)(9.0 \times 10^9 \, N \cdot m^2/C^2)}\right]\left(\frac{1.0 \times 10^{-2} \, m^2}{1.0 \times 10^{-3} \, m}\right) = 8.9 \times 10^{-11} \, F$$
$$= 89 \, pF$$

If a potential difference of 100 V is placed across the plates of this capacitor, each plate will have a charge of

$$Q = CV = (8.9 \times 10^{-11} \, F)(100 \, V) = 8.9 \times 10^{-9} \, C \qquad ◆$$

Figure 17.13

Sound waves cause the diaphragm of a capacitor microphone to vibrate, and the resulting changes in its capacitance are transformed into electrical signals.

A parallel-plate capacitor can be used as a microphone if one of its plates is light and flexible enough to respond to the changing air pressure of a sound wave (Fig. 17.13). When the pressure increases, the diaphragm moves closer to the fixed plate, and C increases; when the pressure drops, the diaphragm moves outward, and

C decreases. Because the voltage across the capacitor is fixed, the changes in C vary the charge on it, and the result is an electrical output whose variations match those in the incoming sound wave.

The formula $\frac{1}{2}QV$ for the potential energy of a charged capacitor derived in the previous section can be written in three equivalent ways:

$$W = \tfrac{1}{2}QV \qquad\qquad\qquad \textit{Potential energy of charged capacitor} \quad (17.15)$$

$$W = \tfrac{1}{2}CV^2 \qquad\qquad\qquad\qquad\qquad\qquad\qquad\qquad\qquad (17.16)$$

$$W = \tfrac{1}{2}\frac{Q^2}{C} \qquad\qquad\qquad\qquad\qquad\qquad\qquad\qquad\qquad (17.17)$$

These equations hold for all capacitors, regardless of their construction.

◆   **E X A M P L E   17.7** _____

A heart attack often leads to a condition called **fibrillation** in which the heart's actions lose their synchronization and it is unable to pump blood effectively. This condition can often be corrected by an electric shock to the heart that completely stops it for a moment. The heart may then start again spontaneously in its normal rhythm. An appropriate shock can be provided by the discharge of a 10-μF capacitor that has been charged to a potential difference of 6000 V. (a) How much energy is released in the current pulse? (b) How much charge passes through the patient's body?

SOLUTION

$$\text{(a) } W = \tfrac{1}{2}CV^2 = \tfrac{1}{2}(1.0 \times 10^{-5} \text{ F})(6000 \text{ V})^2 = 180 \text{ J} = 0.18 \text{ kJ}$$

This is the amount of kinetic energy a baseball moving at 50 m/s (113 mi/h) has.

$$\text{(b) } Q = CV = (1.0 \times 10^{-5} \text{ F})(6000 \text{ V}) = 0.060 \text{ C} \qquad\qquad ◆$$

Table 17.1 is a summary of the various formulas for charge Q, potential difference V, capacitance C, and energy W that follow from the definition of capacitance $C = Q/V$ and from the energy formula $W = \frac{1}{2}QV$.

Although capacitors have their most important application in alternating-current circuits, as we shall find in Chapter 21, they have direct-current uses as well, such as the ones described above. Another is to minimize the effects of arcing across switch contacts. Immediately after a switch is opened in a current-carrying circuit, the air gap between the switch contacts is so small that the voltage across them exceeds the insulating ability of the gap and an arc discharge occurs briefly. If the original current in the circuit was large, the contacts can be damaged. The problem is especially severe in direct-current motors and generators, whose brushes and commutators are really switches that may open and close hundreds of times per second.

Table 17.1 Electrical quantities

Unknown quantity	Known quantities					
	C and V	C and Q	Q and V	W and C	W and V	W and Q
$Q =$	CV			$\sqrt{2WC}$	$\dfrac{2W}{V}$	
$V =$		$\dfrac{Q}{C}$		$\sqrt{\dfrac{2W}{C}}$		$\dfrac{2W}{Q}$
$C =$			$\dfrac{Q}{V}$		$\dfrac{2W}{V^2}$	$\dfrac{Q^2}{2W}$
$W =$	$\dfrac{CV^2}{2}$	$\dfrac{Q^2}{2C}$	$\dfrac{QV}{2}$			

A capacitor connected across a switch prevents such arcing (Fig. 17.14). When the switch is closed, the capacitor is uncharged since its plates are at the same potential. When the switch is opened, the uncharged capacitor provides a path for current whose resistance is lower than that of the air gap, so the momentary current after the swtich is opened goes to charge the capacitor rather than to produce an arc in the air gap.

Figure 17.14

A capacitor connected across a switch prevents arcing when the switch is opened by providing an alternate path of lower resistance than the initial small air gap.

17.8 DIELECTRIC CONSTANT

An insulator between its plates reduces the voltage across a charged capacitor.

Let us now see what happens when a slab of an insulating material is placed between the plates of a capacitor.

Although an insulator cannot conduct an appreciable electric current, it can respond to an electric field in another way. The electrons in the molecules of many substances are not uniformly distributed. A molecule of this kind is called a **polar molecule,** and it behaves as though one end were positively charged and the other negatively charged (Fig. 17.15). In the absence of an external electric field, the molecules in an assembly of polar molecules are oriented randomly, as in Fig. 17.16(a). When an electric field is present, however, the molecules become aligned opposite to the field, as in Fig. 17.16(b).

Although nonpolar molecules ordinarily have symmetric charge distributions, an electric field can distort their arrangements of electrons so that an effective sepa-

Figure 17.15

The water molecule is polar because the end where the hydrogen atoms are attached behaves as if positively charged and the opposite end behaves as if negatively charged.

Figure 17.16

An electric field tends to align polar molecules opposite to the field.

(a) No electric field

(b) Electric field present

ration of charge takes place (Fig. 17.17). Again the molecules have their charged ends aligned opposite to the external field. In either case, then, the net electric field between the plates of the capacitor is *less* than it would be with nothing between them.

The **dielectric constant,** symbol K, of a substance is a measure of how effective it is in reducing an electric field set up across a sample of it. For a capacitor with a given charge Q, reducing the electric field means reducing V as well. Since $C = Q/V$, this means an increase in capacitance. If the capacitance of a capacitor is C_0 when there is a vacuum between the plates, its capacitance will be

$$C = KC_0 \qquad\qquad (17.18)$$

when a substance of dielectric constant K is between the plates (Fig. 17.18). Table 17.2 lists dielectric constants for various substances. Water and alcohol molecules are highly polar, and the values of K for water, ice, and ethyl alcohol are accordingly high.

◆   **E X A M P L E   17.8**

A capacitor with air between its plates is connected to a 50-V source and then disconnected. The space between the plates of the charged capacitor is filled with Teflon ($K = 2.1$). What is the potential difference across the capacitor now?

Figure 17.17

An electric field tends to distort the charge distributions in molecules that are ordinarily nonpolar.

(a)   No electric field

(b)   Weak electric field

(c)   Strong electric field

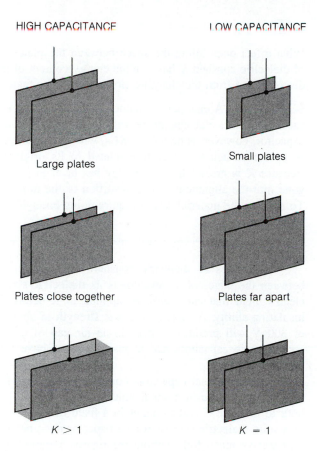

HIGH CAPACITANCE   LOW CAPACITANCE

Large plates   Small plates

Plates close together   Plates far apart

$K > 1$ $K = 1$

Figure 17.18

The capacitance of a parallel-plate capacitor depends on the area and spacing of its plates and on the dielectric constant K of the medium between them.

SOLUTION   The initial charge on the capacitor is $Q = C_1 V_1$. When the Teflon dielectric is inserted, the charge remains the same but the capacitance increases to $C_2 = KC_1$. The new voltage across the capacitor is therefore

$$V_2 = \frac{Q}{C_2} = \frac{C_1 V_1}{KC_1} = \frac{V_1}{K} = \frac{50 \text{ V}}{2.1} = 24 \text{ V}$$

The ratio V_2/V_1 is independent of the original capacitance of the capacitor.   ◆

Table 17.2   Dielectric constants at 20°C except where noted

Substance	K	Substance	K
Air	1.0006	Mica	2.5–7
Air, liquid ($-191°C$)	1.4	Neoprene	6.7
Alcohol, ethyl	26	Sulfur	3.9
Benzene	2.3	Teflon	2.1
Glass	5–8	Water	80
Ice ($-2°C$)	94	Waxed paper	2.2

◆   **E X A M P L E   17.9**

What effect does filling the space between the plates of a capacitor with a material of dielectric constant K have on the energy content of the capacitor? The capacitor is disconnected from the charging circuit before the material is inserted.

SOLUTION   A charged capacitor has an energy of $W = Q^2/2C$. The charge Q on each plate does not change when the material is inserted. The capacitance of the capacitor, however, is now $C = KC_0$, where C_0 was its initial value. The new energy W of the capacitor is therefore related to its original energy W_0 by $W = W_0/K$. Because K is greater than 1, energy has been lost, with the missing energy having gone into the alignment and/or distortion of the molecules of the dielectric material. To remove the material would require doing enough work to make up the difference between W and W_0.   ◆

Besides its dielectric constant, an important property of a material used between the plates of a capacitor is its **dielectric strength,** which is the maximum electric field that can safely be applied to it before it breaks down and loses its insulating ability. Air has a dielectric strength of about 3×10^6 V/m; thus a voltage of 300 V will produce a spark in an air gap of 0.1 mm. Most materials used as separators in capacitors and to insulate wires have dielectric strengths that exceed 10^7 V/m.

Commercial capacitors consist of many interleaved plates to make possible a high capacitance in a small unit (Fig. 17.19). Solid dielectrics are usually used, both to increase C and to maintain a fixed distance between the plates. With sheets of a solid dielectric such as waxed paper or mica between them, the plates can be of inexpensive metal foil, without the rigidity required to prevent accidental contact if only air separated them.

When an alternating electric field is applied to a nonconducting material, the polarizations of its molecules change back and forth. The material is heated as a result. The heat is produced uniformly inside the material, rather than being transferred from the outside. Such **dielectric heating** makes it possible to bring a nonconducting object to a high temperature rapidly with no danger of burning its exterior.

Figure 17.19

Most fixed capacitors are made of interleaved sheets of metal foil separated by layers of dielectric material.

17.9 CAPACITORS IN COMBINATION

Finding the equivalent capacitance of capacitors connected in series and in parallel.

The **equivalent capacitance** of a set of interconnected capacitors is the capacitance of the single capacitor that can replace the set.

Figure 17.20 shows three capacitors in **series,** which means that they are joined end-to-end. Each has charges of the same magnitude Q on its plates, in agreement with the principle of conservation of charge. Hence the potential differences across the capacitors are, respectively,

$$V_1 = \frac{Q}{C_1} \qquad V_2 = \frac{Q}{C_2} \qquad V_3 = \frac{Q}{C_3}$$

If C is the equivalent capacitance of the set and $V = Q/C$ is the potential difference across the set, then

$$V = V_1 + V_2 + V_3 \qquad \frac{Q}{C} = \frac{Q}{C_1} + \frac{Q}{C_2} + \frac{Q}{C_3} \qquad \frac{1}{C} = \frac{1}{C_1} + \frac{1}{C_2} + \frac{1}{C_3}$$

For any number of capacitors in series,

$$\frac{1}{C} = \frac{1}{C_1} + \frac{1}{C_2} + \frac{1}{C_3} + \cdots \qquad\qquad \textit{Capacitors in series} \quad (17.19)$$

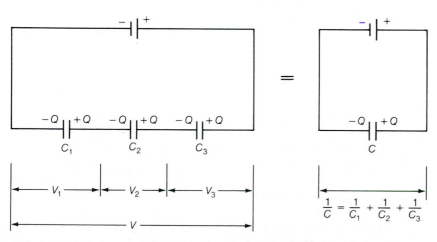

Figure 17.20

Capacitors in series. A charge of the same magnitude Q is present on all the plates of the capacitors. The symbol ⊣⊢ represents a battery and the symbol ⊣⊢ represents a capacitor.

The reciprocal of the equivalent capacitance of a series arrangement of capacitors is equal to the sum of the reciprocals of the individual capacitances. Evidently C is smaller than the capacitance of any of the individual capacitors.

If only two capacitors are in series,

$$\frac{1}{C} = \frac{1}{C_1} + \frac{1}{C_2} = \frac{C_1 + C_2}{C_1 C_2}$$

and so

$$C = \frac{C_1 C_2}{C_1 + C_2}$$

A calculator with a reciprocal $(1/X)$ key makes it easy to solve Eq. (17.19) in an actual problem. If three capacitors are in series, the key sequence would be: $[C_1]$ $[1/X][+][C_2][1/X][+][C_3][1/X][=][1/X]$.

◆   **E X A M P L E   17.10** _____

Two capacitors, one of 10 μF and the other of 20 μF, are connected in series across a 12-V battery, as in Fig. 17.21. Find the equivalent capacitance of the combination, the charge on each capacitor, and the potential difference across it.

Figure 17.21

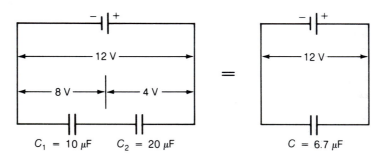

SOLUTION   The equivalent capacitance of the two capacitors is

$$C = \frac{C_1 C_2}{C_1 + C_2} = \frac{(10\ \mu\text{F})(20\ \mu\text{F})}{10\ \mu\text{F} + 20\ \mu\text{F}} = 6.7\ \mu\text{F}$$

and so the charges on them are

$$Q_1 = Q_2 = Q = CV = (6.7 \times 10^{-6}\ \text{F})(12\ \text{V}) = 8.0 \times 10^{-5}\ \text{C}$$

The potential differences across the capacitors are, respectively,

$$V_1 = \frac{Q}{C_1} = \frac{8.0 \times 10^{-5}\,\text{C}}{1.0 \times 10^{-5}\,\text{F}} = 8.0\,\text{V}$$

$$V_2 = \frac{Q}{C_2} = \frac{8.0 \times 10^{-5}\,\text{C}}{2.0 \times 10^{-5}\,\text{F}} = 4.0\,\text{V}$$

The potential difference is greatest across the capacitor of smaller C. The sum of V_1 and V_2 is 12 V, as it should be. ◆

Figure 17.22 shows three capacitors connected in **parallel,** which means that each has one end joined to the corresponding end of all the others. The same potential difference is across all of them, so that the charges on their plates have the respective magnitudes

$$Q_1 = C_1V \qquad Q_2 = C_2V \qquad Q_3 = C_3V$$

The total charge $Q_1 + Q_2 + Q_3$ on either the positive or the negative plates of the capacitors is equal to the charge Q on the corresponding plate of the equivalent capacitor. Hence

$$Q = Q_1 + Q_2 + Q_3$$
$$CV = C_1V + C_2V + C_3V$$
$$C = C_1 + C_2 + C_3$$

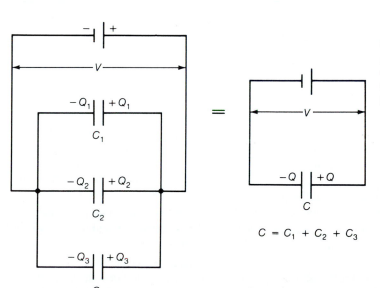

Figure 17.22

Capacitors in parallel. The same potential difference V is across all of them, but each has a charge on its plates whose magnitude is proportional to its capacitance.

This result can be generalized to

$$C = C_1 + C_2 + C_3 + \cdots \qquad\qquad \textit{Capacitors in parallel} \quad (17.20)$$

More energy is stored in a set of capacitors when they are connected in parallel across a given potential difference than when they are connected in series, because the electric fields in them are stronger in the parallel case.

◆   E X A M P L E **17.11** _____

The capacitors of Fig. 17.21 are reconnected in parallel across the same battery. Compare the energies of the capacitors now with what they were when connected in series.

SOLUTION   In the series connection the capacitors had the energies

$$W_1(\text{series}) = \tfrac{1}{2}QV_1 = \tfrac{1}{2}(8.0 \times 10^{-5}\,\text{C})(8.0\,\text{V}) = 3.2 \times 10^{-4}\,\text{J}$$
$$W_2(\text{series}) = \tfrac{1}{2}QV_2 = \tfrac{1}{2}(8.0 \times 10^{-5}\,\text{C})(4.0\,\text{V}) = 1.6 \times 10^{-4}\,\text{J}$$

The capacitor of smaller C had the greater energy. In the parallel connection the capacitors have the same potential difference $V = 12$ V, and their energies are

$$W_1(\text{parallel}) = \tfrac{1}{2}C_1V^2 = \tfrac{1}{2}(1.0 \times 10^{-5}\,\text{F})(12\,\text{V})^2 = 7.2 \times 10^{-4}\,\text{J}$$
$$W_2(\text{parallel}) = \tfrac{1}{2}C_2V^2 = \tfrac{1}{2}(2.0 \times 10^{-5}\,\text{F})(12\,\text{V})^2 = 14 \times 10^{-4}\,\text{J}$$

In this parallel connection the total stored energy is 4.4 times as great as it was in the series connection. Here the capacitor of larger C has the greater energy.   ◆

Important Terms

The electric **potential difference** between two points is the work per unit charge that must be done to take a charge from one of the points to the other. The unit of potential difference is the **volt,** which is equal to 1 J/C.

In a **cathode-ray tube,** an electron beam controlled by electric (or magnetic) fields traces out an image on a fluorescent screen.

The **electron volt** is the energy acquired by an electron that has been accelerated by a potential difference of 1 V. It is equal to 1.6×10^{-19} J.

The **energy density** of an electric field is the electric potential energy per unit volume associated with it.

A **capacitor** is a device that stores electric energy in the form of an electric field. The ratio between the charge on either plate of a capacitor and the potential difference between the plates is called its **capacitance.** The unit of capacitance is the **farad,** which is equal to 1 coulomb per volt.

The **dielectric constant** K of a particular material is a measure of how effective it is in reducing an electric field set up across a sample of it.

A **polar molecule** is one whose charge distribution is not uniform, so that one end is positive and the other negative even though the molecule as a whole is electrically neutral.

Important Formulas

Potential energy of a charge in uniform field: $PE = QEd$

Potential energy of two charges: $PE = k\dfrac{Q_A Q_B}{r}$

Potential difference: $V_{AB} = \dfrac{W_{AB}}{Q}$
$= Ed$ (in uniform electric field)

Energy density of electric field in free space: $w = \dfrac{E^2}{8\pi k}$

Capacitance: $C = \dfrac{Q}{V}$

Parallel-plate capacitor:

$$C = \dfrac{KA}{4\pi kd} \quad (K = \text{dielectric constant})$$

Potential energy of charged capacitor:

$$W = \tfrac{1}{2}QV = \tfrac{1}{2}CV^2 = \dfrac{Q^2}{2C}$$

Capacitors in series: $\dfrac{1}{C} = \dfrac{1}{C_1} + \dfrac{1}{C_2} + \dfrac{1}{C_3} + \cdots$

Capacitors in parallel: $C = C_1 + C_2 + C_3 + \cdots$

MULTIPLE CHOICE

1. Of the following quantities, the one that is vector in character is electric
  a. charge
  b. field
  c. energy
  d. potential difference

2. The magnitude of the electric field in the region between two parallel oppositely charged metal plates is
  a. zero
  b. uniform throughout the region
  c. greatest near the positive plate
  d. greatest near the negative plate

3. From its definition, the unit of electric field E is the N/C. An equivalent unit of E is the
  a. V·m **b.** V·m^2
  c. V/m **d.** V/m^2

4. A system of two charges has a positive potential energy. This signifies that
  a. both charges are positive
  b. both charges are negative
  c. both charges are positive or both are negative
  d. one charge is positive and the other is negative

5. An electron and a proton are accelerated through the same potential difference.
  a. The electron has the greater KE.
  b. The proton has the greater KE.
  c. The electron has the greater speed.
  d. The proton has the greater speed.

6. The electron volt is a unit of
  a. charge **b.** potential difference
  c. energy **d.** momentum

7. The energy content of a charged capacitor resides in its
  a. plates
  b. potential difference
  c. charge
  d. electric field

8. When a slab of insulating material is placed between the plates of a charged capacitor, the electric field there relative to what it was before is
  a. less
  b. the same
  c. more
  d. any of the above, depending on the circumstances

9. The farad is not equivalent to which of the following combination of units?
  a. C^2/J **b.** C/V
  c. C·V^2 **d.** J/V^2

10. The plate area of a parallel-plate capacitor of capacitance C is doubled and the distance between the plates is halved, so that the volume of the dielectric remains the same. The new capacitance of the capacitor is
  a. $C/4$ **b.** C
  c. $2C$ **d.** $4C$

11. A parallel-plate capacitor with air between its plates is charged until a potential difference of V appears across it. Another capacitor, having hard rubber (dielectric constant = 3) between its plates but otherwise identical, is also charged to the same potential difference. If the energy of the first capacitor is W, that of the second is
  a. $\tfrac{1}{3}W$ **b.** W
  c. $3W$ **d.** $9W$

12. An electric field of magnitude 200 V/m can be produced by applying a potential difference of 10 V to a pair of parallel metal plates separated by

- **a.** 20 mm
- **b.** 50 mm
- **c.** 20 m
- **d.** 2000 m

13. A charge of 10^{-10} C between two parallel metal plates 1 cm apart experiences a force of 10^{-5} N. The potential difference between the plates is

- **a.** 10^{-5} V
- **b.** 10 V
- **c.** 10^3 V
- **d.** 10^5 V

14. In charging a storage battery, a total of 2.0×10^5 C is transferred from one set of electrodes to another. The potential difference between the electrodes is 12 V. The energy stored in the battery is

- **a.** 1.7×10^4 J
- **b.** 2.4×10^6 J
- **c.** 2.4×10^7 J
- **d.** 2.9×10^7 J

15. A storage battery is being charged at a rate of 100 W at a potential difference of 13.6 V. The rate at which charge is being transferred between its plates is

- **a.** 0.136 C/s
- **b.** 0.54 C/s
- **c.** 7.35 C/s
- **d.** 1360 C/s

16. An electron between two metal plates 40 mm apart experiences an acceleration of 1.0×10^{12} m/s. The potential difference between the plates is

- **a.** 0.23 V
- **b.** 5.7 V
- **c.** 0.14 kV
- **d.** 2.8×10^{12} V

17. An electron whose speed is 1.0×10^7 m/s has a KE of

- **a.** 4.6×10^{-17} eV
- **b.** 0.28 keV
- **c.** 0.73 keV
- **d.** 5.7 keV

18. An electron whose KE is 150 eV has a speed of

- **a.** 7.3×10^6 m/s
- **b.** 5.1×10^7 m/s
- **c.** 2.3×10^8 m/s
- **d.** 7.3×10^{13} m/s

19. A capacitor acquires a charge of 0.002 C when connected across a 50-V battery. Its capacitance is

- **a.** 1 μF
- **b.** 2 μF
- **c.** 4 μF
- **d.** 40 μF

20. A 50-μF capacitor has a potential difference of 8 V across it. Its charge is

- **a.** 4×10^{-3} C
- **b.** 4×10^{-4} C
- **c.** 6.25×10^{-5} C
- **d.** 6.25×10^{-6} C

21. If a 20-μF capacitor is to have an energy content of 2.5 J, it must be placed across a potential difference of

- **a.** 150 V
- **b.** 350 V
- **c.** 500 V
- **d.** 0.25 MV

22. A parallel-plate capacitor has a capacitance of 50 pF in air and 110 pF when immersed in turpentine. The dielectric constant of turpentine is

- **a.** 0.45
- **b.** 0.55
- **c.** 1.1
- **d.** 2.2

23. Two 50-μF capacitors are connected in series. The equivalent capacitance of the combination is

- **a.** 25 μF
- **b.** 50 μF
- **c.** 100 μF
- **d.** 200 μF

24. The capacitor combination in Multiple Choice 23 is connected across a 100-V battery. The potential difference across each capacitor is

- **a.** 25 V
- **b.** 50 V
- **c.** 100 V
- **d.** 200 V

25. Two 50-μF capacitors are connected in parallel. The equivalent capacitance of the combination is

- **a.** 25 μF
- **b.** 50 μF
- **c.** 100 μF
- **d.** 200 μF

26. The capacitor combination in Multiple Choice 25 is connected across a 100-V battery. The potential difference across each capacitor is

- **a.** 25 V
- **b.** 50 V
- **c.** 100 V
- **d.** 200 V

QUESTIONS

1. The potential energy of a certain system of two charges increases as the charges are moved farther apart. What does this tell us about the signs of the charges?

2. The electric potential V at a point can be defined as the potential energy per coulomb of a charge located there relative to its PE at infinity. What is the potential at a distance r from a charge Q?

3. In the electron gun of a cathode-ray tube, electrons are accelerated by a potential difference between a suitable source and an anode (positive electrode) that has a hole through which

the beam can pass. What effect does doubling the potential difference have on the speed of the electrons that reach the screen?

4. A sheet of mica whose dielectric constant is 5 is placed between the plates of a charged, isolated parallel-plate capacitor. How is the potential difference across the capacitor affected? How is the charge on the capacitor affected?

5. Is there any kind of material that, when inserted between the plates of a capacitor, reduces its capacitance?

EXERCISES

17.2 Potential Energy of Two Charges

1. In an experiment, helium nuclei of charge $+2e$ and kinetic energy 7.7 MeV are directed at a gold foil. The nucleus of a gold atom contains 79 protons, so its charge is $+79e$. Find the closest distance a helium nucleus gets to a gold nucleus when it approaches head-on. (The radius of the gold nucleus is about $\frac{1}{6}$ of this distance.)

2. Two electrons are 1.0×10^{-9} m apart when both are released. What is the speed of the electrons when they are 1.0×10^{-8} m apart?

17.3 Potential Difference

3. Twelve joules of work are needed to transfer 2.0 C of charge from one terminal of a storage battery to another. What is the potential difference between the terminals?

4. An electric field stronger than 3.0×10^6 V/m will cause sparks to occur in air. What is the maximum potential difference that can be applied across two metal plates 1.0 mm apart before sparking begins?

5. A potential difference of 50 V is applied across two parallel metal plates, and an electric field of 1.0×10^4 V/m is produced. How far apart are the plates?

6. The potential difference between two parallel metal plates that are 50 mm apart is 1.0×10^4 V. Find the force on an electron located between the plates.

7. A cloud is at a potential of 8.0×10^6 V relative to the ground. A charge of 40 C is transferred in a lightning stroke between the cloud and the ground. Find the energy dissipated.

8. The electrodes in a neon sign are 1.2 m apart, and the potential difference across them is 8000 V. (a) Find the acceleration of a neon ion of mass 3.3×10^{-26} kg and charge $+e$ in the field. (b) If the ion starts at the positive electrode of the sign and moves unimpeded to its negative electrode, how much energy would it gain? (c) Why is it extremely unlikely that the ion would actually acquire this much energy?

9. Two parallel metal plates are 40 mm apart. If the force on an electron between the plates is to be 1.0×10^{-14} N, what should the potential difference between them be?

10. The storage battery of a car is being charged by an alternator at the rate of 10 C/s. The potential difference across the alternator's terminals is 14 V, and the alternator is 90% efficient. Find the power supplied by the car's engine to charge the battery.

11. In charging a 20-kg storage battery, a total of 2.0×10^5 C is transferred from one set of electrodes to another. The potential difference between the electrodes is 14 V. (a) How much energy is stored in the battery? (b) If this energy were used to raise the battery above the ground, how high would it go? (c) If this energy were used to provide the battery with kinetic energy, what would its speed be?

12. A potential difference of 20 kV is used to accelerate electrons in a television picture tube. (The impact of these electrons on the tube's screen produces the flashes of light that make up the picture we see.) How much kinetic energy do the electrons acquire? What is their speed?

17.5 The Electron Volt

12. What is the kinetic energy in electron volts of an electron whose speed is 1.0×10^6 m/s?

13. What is the speed of an electron whose kinetic energy is 50 eV?

14. What is the speed of a neutron whose kinetic energy is 50 eV?

15. What is the kinetic energy in electron volts of a potassium atom of mass 6.5×10^{-26} kg when its speed is 1.0×10^6 m/s?

16. What potential difference is needed between two metal plates 20 mm apart to accelerate an electron to a speed of 5.0×10^6 m/s? What will the KE of the electron be, expressed in electron volts?

17.6 Electric Field Energy

17. A potential difference of 300 V is applied across a pair of parallel metal plates 10 mm apart. What is the energy density of the electric field between the plates?

18. The electric field near the earth's surface is about 100 V/m. How much electric energy is stored in the lowest kilometer of the atmosphere?

17.7 Capacitance

19. A 25-μF capacitor is connected to a source of potential difference of 1000 V. What is the resulting charge on the capacitor? How much energy does it contain?

20. What is the potential difference between the plates of a 20-μF capacitor whose charge is 0.010 C? How much energy does it contain?

21. What potential difference must be applied across a 10-μF capacitor if it is to have an energy content of 1.0 J?

22. A 0.50-μF capacitor has plates 0.50 mm apart in air. What is the maximum charge the capacitor can have before sparks occur between the plates?

23. The plates of a parallel-plate capacitor are 50 cm^2 in area and 1.0 mm apart. (a) What is its capacitance? (b) When the capacitor is connected to a 45-V battery, what is the charge on either plate? (c) What is the energy of the charged capacitor?

24. A parallel-plate capacitor with plates 0.20 mm apart in air has a charge on each plate of 4.0×10^{-8} C when the potential difference is 250 V. Find its capacitance, the area of each plate, and the stored energy.

25. A 200-μF capacitor requires an energy of 50 J to fire a flashlamp. What voltage is needed to charge the capacitor? How much charge passes through the flashlamp?

26. A variable capacitor set at 200 pF is connected to a 100-V battery. The battery is then disconnected, and the capacitor is adjusted to a capacitance of 10 pF. (a) What is the potential difference across the capacitor now? (b) In the absence of friction, how much work had to be done to change the capacitor to the new setting?

17.8 Dielectric Constant

27. The capacitance of a parallel-plate capacitor is increased from 8.0 μF to 50.0 μF when a sheet of glass is inserted between its plates. What is the dielectric constant of the glass?

28. A capacitor with air between its plates is connected to a battery, and each of its plates receives a charge of 1.0×10^{-4} C. While still connected to the battery, the capacitor is immersed in oil, and a further charge of 1.0×10^{-4} C is added to each plate. What is the dielectric constant of the oil?

29. The space between the plates of the capacitor in Exercise 23 is filled with sulfur. Answer the same questions for this case.

PROBLEMS

38. A potential difference of 1000 V is applied across two parallel metal plates 100 mm apart. An electron leaves the negative plate at the same time as a proton leaves the positive plate. (a) Find the speeds and kinetic energies of the particles when they reach the opposite plates. (b) At what distance from the positive plate do the electron and proton pass each other?

39. A potential difference of 10 V is applied across two parallel metal plates 20 mm apart. An electron is projected at a speed of 1.0×10^7 m/s halfway between the plates and

17.9 Capacitors in Combination

30. Find the equivalent capacitance of a 20-μF capacitor and a 50-μF capacitor that are connected in series.

31. Find the equivalent capacitance of a 20-μF capacitor and a 50-μF capacitor that are connected in parallel.

32. List the capacitances that can be obtained by combining three 10-μF capacitors in all possible ways.

33. Three capacitors whose capacitances are 5.0, 10.0, and 50.0 μF are connected in series across a 12.0-V battery. Find the charge on each capacitor and the potential difference across it.

34. Three capacitors whose capacitances are 2.0, 4.0, and 5.0 μF are connected in series across a 100-V battery. Find the charge on each capacitor and the potential difference across it.

35. The three capacitors in Exercise 33 are connected in parallel across the same battery. Find the charge on each capacitor and the potential difference across it.

36. The three capacitors in Exercise 34 are connected in parallel across the same battery. Find the charge on each capacitor and the potential difference across it.

37. (a) Find the equivalent capacitance of the system shown in Fig. 17.23. (b) The dielectric of the 2.0-μF capacitor breaks down and becomes conducting. Find the equivalent capacitance of the system now. (c) If the dielectric of the 6.0-μF capacitor had broken down instead, what would the equivalent capacitance of the system be?

Figure 17.23

Exercise 37.

2.0 μF   4.0 μF

6.0 μF

parallel to them. How far will the electron travel before striking the positive plate?

40. The electron gun of a television picture tube has an accelerating potential difference of 15 kV and a power rating of 25 W. How many electrons reach the screen per second? At what speed?

41. A parallel-plate capacitor of capacitance C is given the charge Q and then disconnected from the circuit. How much work is required to pull the plates of this capacitor to twice their original separation?

42. A 1.0-μF capacitor and a 2.0-μF capacitor are each charged across a potential difference of 1200 V. The capacitors are disconnected from the source of the potential difference and are then connected with terminals of the same sign together. What is the final charge of each capacitor?

43. The charged capacitors in Problem 42 are instead connected with terminals of opposite sign together. Now what is the final charge of each capacitor?

44. (a) Find the energy stored in an isolated 5.00-μF capacitor when the potential difference across it is 200 V. (b) The capacitor is now connected to an initially uncharged 15.00-μF

capacitor. Find the energy stored in the combination. (c) If there is any difference between the answers to (a) and (b), can you account for it?

Answers to Multiple Choice

1. b	8. a	15. c	22. d
2. b	9. c	16. a	23. a
3. c	10. d	17. b	24. b
4. c	11. c	18. a	25. c
5. c	12. b	19. d	26. c
6. c	13. c	20. b	
7. d	14. b	21. c	

18

ELECTRIC CURRENT

18.1   **Electric Current**

18.2   **Ohm's Law**

18.3   **Resistivity**

18.4   **Electric Power**

18.5   **Resistors in Series**

18.6   **Resistors in Parallel**

18.7   **Electromotive Force**

18.8   **Impedance Matching**

18.9   **Kirchhoff's Rules**

18.10   **Capacitive Time Constant**

An electric current consists of a flow of charge from one place to another. Currents, not static charges, are involved in nearly all practical applications of electricity. In an electric circuit, current is the means by which energy is transferred from a source such as a battery or a generator to a load. The load may be a lamp, a motor, or any other device that absorbs electric energy and converts it into some other form of energy or into work. In this chapter we consider the chief factors that govern direct currents in simple circuits. In later chapters the magnetic effects of currents and how they are used in technology will be examined. The important topic of alternating current is the subject of Chapter 21.

18.1 ELECTRIC CURRENT

A conducting path and a potential difference are both needed for a current to occur.

The magnitude of an electric current, denoted I, is the rate at which charge passes a given point. If the net charge Q goes past in the time interval t, then the average current is

$$I = \frac{Q}{t}$$ *Electric current*   (18.1)

$$\text{Current} = \frac{\text{charge}}{\text{time interval}}$$

The unit of electric current is the **ampere** (A), where

1 ampere = 1 coulomb/second

The current in a 100-W light bulb is a little less than 1 A.

The direction of a current is, by convention, taken as that in which *positive* charges would have to move in order to produce the same effects as the observed current. Thus a current is always assumed to proceed from the positive terminal of a battery or a generator to its negative terminal in an external circuit (Fig. 18.1).

Despite this convention, actual electric currents in metals consist of flows of electrons, which carry negative charges. However, a current that consists of negative particles moving in one direction is electrically the same as a current that consists of positive particles moving the other way. Since there is no overwhelming reason to prefer one way of designating current to the other, we will follow the usual practice of considering current as a flow of positive electric charge.

Figure 18.1

By convention, an electric current is assumed to flow from the positive terminal of a battery or generator to its negative terminal in an external circuit. Actual currents in metals consist of electrons that move in the opposite direction. A complete conducting path such as this one is called a *circuit.*

 E X A M P L E 18.1

About 10^{20} electrons/cm participate in conducting electric current in a certain wire. That is, about 10^{20} electrons in each centimeter of the wire are moving when the wire is carrying a current. What is the average speed of the electrons when there is a current of 1 A in the wire?

SOLUTION   Each electron has a charge of $e = 1.6 \times 10^{-19}$ C. Because there are 10^{20} electrons per centimeter in the wire, the number per meter is 10^{22} and the charge in motion per meter in a length x of the wire is $Q/x = 10^{22}\ e/\text{m} = 1.6 \times 10^3$ C/m. The time needed for the electrons in a length x of the wire to pass a given point at the average speed v is $t = x/v$. The current in the wire is therefore

$$I = \frac{Q}{t} = \frac{Qv}{x} \quad \text{and} \quad v = \frac{I}{(Q/x)} = \frac{1\,\text{A}}{1.6 \times 10^3\,\text{C/m}} = 6 \times 10^{-4}\,\text{m/s}$$

This is less than 1 mm/s, slower than a tired caterpillar. The average speed is so low because the moving electrons, which are accelerated by the electric field in the wire, collide frequently with atoms in their paths and have to start over again each time. ◆

The following two conditions must be met in order for an electric current to exist between two points:

1. There must be a conducting path between the two points along which charge can flow.
2. There must be a difference of potential between the two points. (A superconductor is an exception to this requirement.) Just as the rate of flow of water between the ends of a pipe depends on the difference of pressure between them, so the rate of flow of charge between two points depends on the difference of potential between them (Fig. 18.2). A large potential difference means a large ''push'' given to each charge.

A particular conducting path—for instance, a copper wire, a light bulb, an electric heater—is usually called a conductor, even though this is also the name of

Figure 18.2

The role of potential difference in producing an electric current is analogous to the role of pressure in producing a flow of water.

Greater pressure

Increased flow of water

More current

Higher potential difference

the class of substances through which current flows readily. The **resistance** of a conductor is the ratio between the potential difference V across it and the resulting current I that flows:

$$R = \frac{V}{I}$$ *Resistance*   (18.2)

$$\text{Resistance} = \frac{\text{potential difference}}{\text{current}}$$

The unit of resistance is the **ohm** (Ω), where

1 ohm = 1 volt/ampere

A conductor in which there is a current of 1 A when a potential difference of 1 V exists across it has a resistance of 1 Ω. (The symbol Ω is the Greek capital letter *omega*.)

   E X A M P L E   18.2 _____

A 120-V electric heater draws a current of 15 A. Find its resistance.

SOLUTION   From the definition of resistance

$$R = \frac{V}{I} = \frac{120 \text{ V}}{15 \text{ A}} = 8 \ \Omega$$ ◆

18.2 OHM'S LAW

Current is proportional to voltage in many substances.

The resistance of a conductor depends in general both on its properties—its nature and its dimensions—and on the potential difference applied across it. In some conductors R increases when V increases, in others R decreases when V increases, and in still others R depends on the direction of the current (Fig. 18.3).

In a conductor whose resistance is constant, the current is proportional to the voltage:

$$I = \frac{V}{R} \quad (R = \text{constant})$$ *Ohm's law*   (18.3)

Figure 18.3

The relationship between current and voltage for (a) a metal, (b) a solution of NaCl (table salt), and (c) a semiconductor diode. Only in (a) is I proportional to V, a relationship known as Ohm's law.

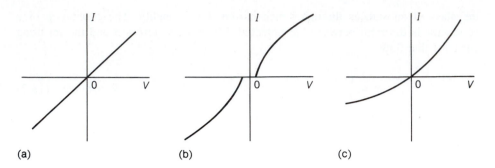

(a)   (b)   (c)

This relationship is called **Ohm's law** because it was first verified experimentally by the German physicist Georg Ohm (1787–1854). Despite its name, Ohm's law is not a fundamental physical principle. Rather, it is a relationship that describes the electrical behavior of many substances, notably most metals, over a wide range of voltages and currents.

Ohm's law is not the same as the definition of resistance, $R = V/I$. Ohm's law holds only for conductors in which the ratio V/I is constant; such conductors are said to be **ohmic.** On the other hand, the resistance of a conductor is given by $R = V/I$ whether or not it is constant as the voltage is changed.

◆   **E X A M P L E   18.3** _____

A light bulb has a resistance of 240 Ω. Find the current in it when it is placed in a 120-V circuit (Fig. 18.4).

Figure 18.4

$V = 120$ V $R = 240 \, \Omega$

$I = 0.50$ A

SOLUTION   From Ohm's law

$$I = \frac{V}{R} = \frac{120 \, \text{V}}{240 \, \Omega} = 0.5 \, \text{A}$$ ◆

◆ **E X A M P L E   18.4** _____

The current in the coil of an 8-Ω loudspeaker is 0.5 A. Find the voltage across its terminals.

SOLUTION We rewrite Ohm's law in the form $V = IR$ and obtain

$$V = IR = (0.5 \text{ A})(8 \text{ Ω}) = 4 \text{ V}$$ ◆

Electrical Hazards

An electric current affects body tissue both by stimulating nerves and muscles and through the heat produced. Tissue is a fairly good conductor because of the ions in solution it contains. Dry skin has more resistance and can protect the rest of the body to some extent in case of accidental exposure to a high voltage, protection that disappears when the skin is wet. A current of as little as 1 mA (0.001 A) can be felt by most people, one of 5 to 10 mA is painful, and one of 15 mA or more causes muscle contractions that may prevent the person involved from letting go of the source of the current. Breathing may then become impossible.

Because a closed conducting path is necessary for a current to occur, touching a single "live" conductor has no effect if the body is isolated. However, if a person at the same time is in contact with a water pipe, is standing on wet soil, or otherwise is grounded, a current will pass through his or her body. The resistance of the human body itself is of the order of magnitude of 1000 Ω, so contact via wet skin with a 120-V line will lead to a current in the neighborhood of $I = 120 \text{ V}/1000 \text{ Ω} = 0.12 \text{ A} - 120$ mA. Such a current is extremely dangerous because it is likely to cause the heart muscles to contract rapidly and irregularly and, if allowed to persist, will cause death.

The water in a bathtub is grounded via the tub's drainpipe, so a person in the tub is at risk if he or she touches any electrical device, even a switch; the moisture on a wet finger may be sufficient to provide a conducting path to the interior of the device. Another situation of great potential danger occurs in hospitals where electrical appliances are often attached to patients to monitor various functions or to control them, as in the case of cardiac pacemakers. Even a minor malfunction of such an appliance, or the failure to ground it properly may have fatal consequences because the conducting path in the body is then short compared with what it is when the contact is made via a finger.

It is sometimes thought that a fuse or a circuit breaker in a circuit eliminates any danger. Such a device "opens" a circuit when the current is greater than a certain value and thus prevents damage to the equipment being used or to the wiring that carries the current. For instance, a 15-A fuse contains a length of special wire that melts when the current is 15 A or more. Since only about 50 mA (1/300 of 15 A) is enough to kill, fuses and circuit breakers do not protect people. A device called a ground-fault circuit interrupter, however, will open a circuit when there is a leakage current to ground of as little as 5 mA, and this does offer useful protection.

18.3 RESISTIVITY

A measure of the ability of a substance to conduct electric current.

The resistance of a conductor that obeys Ohm's law depends on three factors:

1. **The material of which it is composed:** The ability to carry an electric current varies more than almost any other physical property of a material.
2. **Its length L:** The longer the conductor, the greater its resistance.
3. **Its cross-sectional area A:** The thicker the conductor, the less its resistance.

Once again we note the correspondence to water flowing through a pipe. The longer the pipe, the more friction against the pipe wall slows the water, and the wider the pipe, the larger the volume of water that can pass through per second when everything else is the same (Fig. 18.5).

The simple formula

$$R = \rho \frac{L}{A}$$

Resistance of ohmic conductor   (18.4)

Figure 18.5

The way in which the dimensions of a conductor affect the flow of charge in it is analogous to the way in which the dimensions of a pipe affect the flow of water in it.

Short, wide pipe

Large flow of water

Long, narrow pipe

Small flow of water

Short, thick wire

Large current

Long, thin wire

Small current

Table 18.1 Approximate resistivities (at 20°C) and their temperature coefficients

Substance	ρ ($\Omega \cdot$ m)	ρ ($\Omega \cdot$ cmil/ft)	α (/°C)
Conductors			
Aluminum	2.6×10^{-8}	16	0.0039
Constantan (60% Cu, 40% Ni)	49×10^{-8}	295	0.000002
Copper	1.7×10^{-8}	10.4	0.0039
Iron	12×10^{-8}	72	0.0050
Lead	21×10^{-8}	126	0.0043
Manganin (84% Cu, 12% Mn, 4% Ni)	44×10^{-8}	265	0.000000
Mercury	98×10^{-8}	590	0.00088
Nichrome	112×10^{-8}	676	0.0002
Platinum	11×10^{-8}	66	0.0036
Silver	1.6×10^{-8}	9.6	0.0038
Semiconductors			
Carbon	3.5×10^{-5}	2.1×10^4	-0.0005
Germanium	0.5	3×10^8	
Copper oxide (CuO)	1×10^3	6×10^{11}	
Insulators			
Glass	$10^{10} - 10^{14}$	$10^{19} - 10^{23}$	
Quartz	7.5×10^{17}	4.5×10^{26}	
Sulfur	10^{15}	10^{24}	

has been found to hold for the resistance of a conductor that obeys Ohm's law. The quantity ρ (the Greek letter *rho*) is called the **resistivity** of the material from which the conductor is made. Table 18.1 lists the resistivities of various substances at room temperature (20°C). Given the nature of a conductor and its dimensions, the value of R can be calculated at once.

In the SI system, lengths are measured in meters and areas in square meters, and the unit of resistivity is accordingly the ohm-meter ($\Omega \cdot$ m). Metric wire sizes are usually specified by their cross-sectional areas in square millimeters, so a particular wire might be referred to as 2.5 mm², for instance, instead of having its diameter of 1.8 mm quoted.

◆ **E X A M P L E 18.5** _____

A water heater draws 30 A from a 120.0-V power source 10 m away. What is the minimum cross section of the wire that can be used if the voltage applied to the heater is not to be lower than 115.0 V?

SOLUTION The permissible voltage drop is 5.0 V, and the resistance that corresponds to this drop when the current is 30 A is

$$R = \frac{V}{I} = \frac{5.0 \text{ V}}{30 \text{ A}} = 0.167 \ \Omega$$

The total length of wire involved is twice the distance between the source and the heater, so $L = (2)(10 \text{ m}) = 20 \text{ m}$. From Eq. (18.4) we have

$$A = \frac{\rho L}{R} = \frac{(1.7 \times 10^{-8}\, \Omega \cdot \text{m})(20\, \text{m})}{0.167\, \Omega} = 2.0 \times 10^{-6}\, \text{m}^2 = 2.0\, \text{mm}^2 \qquad \blacklozenge$$

In engineering practice in the United States the cross-sectional area of a round conductor is often expressed in terms of a unit called the **circular mil** (cmil). A circular mil is the area of a circle whose diameter is 1 mil, where 1 mil = 0.001 in (Fig. 18.6). The area of a circle in cmil is equal to the square of its diameter in mils:

$$A_{\text{cmil}} = (d_{\text{mil}})^2 \qquad\qquad\qquad\qquad\qquad \textit{Circular mil}$$

Since nearly all wires have circular cross sections, using the cmil as the unit of area simplifies calculations by eliminating the factor $\pi/4$. (We recall that the area in ordinary units of a circle of diameter s is $\pi d^2/4$.) When the length of a wire is in ft and its area in cmils, the unit of resistivity is the $\Omega \cdot$ cmil/ft, as in Table 18.1.

◆   **E X A M P L E   18.6** _____

Find the resistance of 20.0 ft of constantan wire 0.00500 in. in diameter.

SOLUTION   Because 1 mil = 0.001 in, the wire is 5.00 mils in diameter and has a cross-sectional area of

$$A_{\text{cmil}} = (d_{\text{mil}})^2 = (5.00)^2 \text{ cmils} = 25.0 \text{ cmils}$$

Hence the resistance of the wire is

$$R = \rho \frac{L}{A} = \frac{(295\, \Omega \cdot \text{cmil/ft})(20.0\, \text{ft})}{25.0\, \text{cmils}} = 236\, \Omega \qquad \blacklozenge$$

Figure 18.6

The circular mil (cmil) is a unit of area equal to the area of a circle whose diameter is 1 mil, where 1 mil = 0.001 in. The area of a circle whose diameter is d in mils is given by d^2 in cmils.

◆ **E X A M P L E 18.7** _____

The maximum resistance of a copper wire 800 ft long is to be 2.00 Ω. What should the minimum diameter of the wire be?

SOLUTION The minimum area of the wire in cmils is

$$A = \frac{\rho L}{R} = \frac{(10.4\Omega \cdot \text{cmil/ft})(800 \text{ ft})}{2.00 \ \Omega} = 4160 \text{ cmils}$$

Because $A_{\text{cmil}} = (d_{\text{mil}})^2$,

$$d_{\text{mil}} = \sqrt{A_{\text{cmil}}} = \sqrt{4160} \text{ mils} = 64 \text{ mils}$$

This corresponds to No. 14 wire in the American Wire Gage system. ◆

Temperature Variation of Resistivity

The resistivities of most substances vary with temperature. In the case of a metal, this is because its atoms vibrate with greater amplitude as the temperature rises. As a result they interfere to a greater extent with the motion of the free electrons that constitute a current in the metal.

If R is the resistance of a conductor at a particular temperature, then the change ΔR in its resistance when the temperature changes by ΔT is approximately proportional to both R and ΔT, and therefore

$$\Delta R = \alpha R \ \Delta T \tag{18.5}$$

The quantity α is the temperature coefficient of resistivity of the material. In Table 18.1 the temperature coefficient of carbon is labeled negative because its resistivity decreases with increasing temperature.

◆ **E X A M P L E 18.8** _____

A **resistance thermometer** makes use of the temperature variation of resistivity. When a coil of platinum wire whose resistance at 20°C is 11 Ω is placed in a furnace, its resistance triples. Find the temperature of the furnace, assuming that α remains constant.

SOLUTION Since $R = 11 \ \Omega$ and $\Delta R = 33 \ \Omega - 11 \ \Omega = 22 \ \Omega$,

$$\Delta T = \frac{\Delta R}{\alpha R} = \frac{22 \ \Omega}{(0.0036/°\text{C})(11 \ \Omega)} = 556°\text{C}$$

The temperature of the furnace is $T + \Delta T = 20°\text{C} + 556°\text{C} = 576°\text{C}$. ◆

18.4 ELECTRIC POWER

The product of current and voltage.

Electric energy in the form of electric current is converted into heat in an electric stove, into radiant energy in a light bulb, into chemical energy when a storage battery is charged, and into mechanical energy in an electric motor. The widespread use of electric energy is due as much to the ease with which it can be transformed into other kinds of energy as to the ease with which it can be carried through wires.

The work that must be done to take a charge Q through the potential difference V is, by definition,

$$W = QV$$

Since a current I carries the amount of charge $Q = It$ in the time t, the work done is

$$W = IVt \qquad\qquad\qquad \textit{Electric work} \quad (18.6)$$

The energy input to a device of any kind through which the current I flows when the potential difference V is placed across it is equal to the product of the current, the potential difference, and the time interval.

◆   **E X A M P L E   18.9** _____

A battery is rated according to the amount of charge it can circulate, expressed in ampere-hours (A · h). Thus a battery whose capacity is 80 A · h can supply a current of 1 A for 80 h, a current of 2 A for 40 h, a current of 80 A for 1 h, and so forth. The less the current, the longer the time (Fig. 18.7). (a) How many coulombs of charge can a 12-V, 80-A · h battery transfer? (b) How much energy is stored in the battery?

SOLUTION   (A) From Eq. (18.1), $I = Q/t$, so $Q = It$. The charge transferred by a current of 80 A that flows for 1 h = 3600 s is

$$Q = It = (80\ \text{A})(3600\ \text{s}) = 2.9 \times 10^5\ \text{C}$$

(b) From Eq. (18.6), the energy stored in the battery is

$$W = IVt = (80\ \text{A})(12\ \text{V})(3600\ \text{s}) = 3.5 \times 10^6\ \text{J} = 3.5\ \text{MJ} \qquad\qquad ◆$$

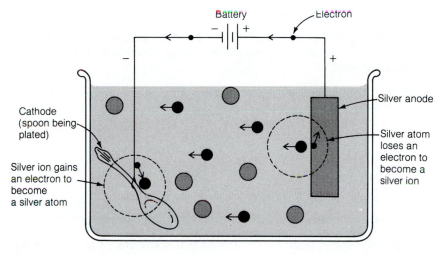

Figure 18.7

Combinations of current and discharge time available from a battery rated at 80 A · h.

Power is the rate at which work is being done: $P = W/t$. When the work is done by an electric current, $W = IVt$, so

$$P = IV \qquad\qquad \textit{Electric power} \quad (18.7)$$

Electric power = (current)(potential difference)

The unit of power is the watt, and when I and V are in amperes and volts, respectively, P will be in watts.

◆ **E X A M P L E   18.10** _____

A solar cell 10 cm in diameter produces a current of 2.15 A at 0.45 V in bright sunlight whose intensity is 0.10 W/cm². Find the efficiency of the cell.

SOLUTION The area of the cell is $A = \pi d^2/4 = 78.5 \text{ cm}^2$, so it receives solar energy at the rate of $P_{\text{input}} = (0.10 \text{ W/cm}^2)(78.5 \text{ cm}^2) = 7.85 \text{ W}$. The cell's output is

$$P_{\text{output}} = IV = (2.15 \text{ A})(0.45 \text{ V}) = 0.97 \text{ W}$$

The efficiency of the cell is therefore

$$\text{Eff} = \frac{P_{\text{output}}}{P_{\text{input}}} = \frac{0.97 \text{ W}}{7.85 \text{ W}} = 0.12 = 12\%$$

◆

◆   E X A M P L E   18.11 _____

How much current is drawn by a $\frac{1}{2}$-hp electric motor operated from a 120-V source of electricity? Assume that 80% of the electric energy absorbed by the motor is turned into mechanical work.

SOLUTION   The power input to the motor is $P_{\text{input}} = IV$ and its power output is $P_{\text{output}} = (0.50 \text{ hp})(746 \text{ W/hp}) = 373 \text{ W}$. Hence

$$P_{\text{input}} = IV = \frac{P_{\text{output}}}{\text{Eff}}$$

$$I = \frac{P_{\text{output}}}{(\text{Eff})(V)} = \frac{373 \text{ W}}{(0.8)(120 \text{ V})} = 3.9 \text{ A}$$

◆

The power consumed by a resistance that obeys Ohm's law ($I = V/R$) may be expressed in the alternative forms

$$P = IV \tag{18.7}$$

$$P = I^2R \qquad\qquad\qquad Electric\ power \tag{18.8}$$

$$P = \frac{V^2}{R} \tag{18.9}$$

Equation (18.7) holds regardless of the nature of the current-carrying device. Depending on which quantities are known in a specific case, any of these expressions for P may be used.

◆   E X A M P L E   18.12 _____

Find the power consumed by a 240-Ω light bulb when the current through it is 0.50 A.

SOLUTION   The formula $P = I^2R$ is easiest to use here. We have

$$P = I^2R = (0.50 \text{ A})^2(240 \text{ }\Omega) = 60 \text{ W}$$

◆

Table 18.2 summarizes the various formulas for potential difference V, current I, resistance R, and power P that follow from Ohm's law, $I = V/R$, and from the power formula, $P = IV$.

Owing to the resistance that all conductors offer, electric power is dissipated whenever a current exists, regardless of whether the current also supplies energy that is converted to some other form. Electrical resistance is much like friction: The power consumed in causing a current to flow is dissipated as heat. If too much current flows in a particular wire, it becomes so hot that it may start a fire or even melt. To prevent this from happening, nearly all electric circuits are protected by fuses or circuit

Table 18.2   Formulas valid for circuits that obey Ohm's law

Unknown quantity	Known quantities					
	V and I	I and R	V and R	P and I	P and V	P and R
$V =$		IR		$\dfrac{P}{I}$		$\sqrt{RP}$
$I =$			$\dfrac{V}{R}$		$\dfrac{P}{V}$	$\sqrt{\dfrac{P}{R}}$
$R =$	$\dfrac{V}{I}$			$\dfrac{P}{I^2}$	$\dfrac{V^2}{P}$	
$P =$	IV	I^2R	$\dfrac{V^2}{R}$			

breakers, as mentioned earlier, which interrupt the current when I exceeds a safe value. For example, a 15-A fuse in a 120-V power line means that the maximum power that can be carried is

$$P = IV = (15 \text{ A})(120 \text{ V}) = 1800 \text{ W} = 1.8 \text{ kW}$$

The maximum safe current in a wire depends both on how its temperature varies with current and on the nature of its insulation. The larger the diameter of a wire, the more current it can safely carry, partly because its resistance is lower and partly because it has more surface area to dissipate heat. Tables have been compiled that give the allowable currents for copper wires of standard sizes. For example, rubber-covered No. 14 wire (diameter 1.63 mm), the smallest size permitted by the National Electrical Code for use in homes, farms, and industry, is limited to 20 A.

18.5 RESISTORS IN SERIES

Their equivalent resistance is the sum of the individual resistances.

In analyzing a circuit, it is convenient to group its various components into individual **resistors** that we imagine as joined together by resistanceless wires. The **equivalent resistance** of a set of interconnected resistors is the value of the single resistor that can replace the entire set without affecting the current in the rest of the circuit.

The symbol for a resistor, ─/\/\/─, represents any circuit component that has electrical resistance and obeys Ohm's law. Although there are an unlimited number of ways in which resistors can be put together, many of them are merely combinations

Resistors
in series

Figure 18.8

The symbol of a resistor is
‑‑‑‑‑‑ .

of the basic **series** and **parallel** arrangements (Fig. 18.8). Resistors in series are joined end-to-end, so the same current is present in all of them. On the other hand, resistors in parallel have their ends connected together, so the total current is split up among them.

The potential difference V across the ends of a series set of resistors is the sum of the potential differences V_1, V_2, V_3, . . . across each one. This statement follows from the principle of conservation of energy: If V is the work done per coulomb in pushing a charge through the set of resistors, then it must equal the sum $V_1 + V_2 + V_3 + \cdots$ of the work done per coulomb in pushing the charge through each resistor in turn. In the case of the three resistors in series in Fig. 18.9,

$$V = V_1 + V_2 + V_3$$

Because the same current I passes through all the resistors, the individual potential drops are

$$V_1 = IR_1 \qquad V_2 = IR_2 \qquad V_3 = IR_3$$

If we let R be the equivalent resistance of the set, the potential difference across it is

$$V = IR$$

We therefore have

$$V = V_1 + V_2 + V_3$$
$$IR = IR_1 + IR_2 + IR_3$$

Dividing through by the current I gives

$$R = R_1 + R_2 + R_3$$

Figure 18.9

Resistors in series. The same current I passes through each resistor.

$$R = R_1 + R_2 + R_3$$

In general, *the equivalent resistance of a set of resistors connected in series is equal to the sum of the individual resistances*:

$$R = R_1 + R_2 + R_3 + \cdots \qquad\qquad \textit{Resistors in series} \quad (18.10)$$

◆   **E X A M P L E   18.13** _____

A 5.0-Ω resistor and a 20.0-Ω resistor are connected in series, and a potential difference of 100 volts is applied across them by means of a generator. Find (a) the equivalent resistance of the circuit, (b) the current that flows in it, (c) the potential difference across each resistor, (d) the power dissipated by each resistor, and (e) the power dissipated by the entire circuit.

SOLUTION   Successive stages in the solution of this problem are shown in Fig. 18.10.

   (a) The equivalent resistance of the two resistors is

$$R = R_1 + R_2 = 5.0 \ \Omega + 20.0 \ \Omega = 25.0 \ \Omega$$

This resistance is more than that of either of the resistors.

   (b) The current in the circuit is, from Ohm's law,

$$I = \frac{V}{R} = \frac{100 \text{ V}}{25.0 \ \Omega} = 4.0 \text{ A}$$

Figure 18.10

(c)

(d)

(c) The potential differences across R_1 and R_2 are

$$V_1 = IR_1 = (4.0 \text{ A})(5.0 \text{ }\Omega) = 20 \text{ V}$$

$$V_2 = IR_2 = (4.0 \text{ A})(20 \text{ }\Omega) = 80 \text{ V}$$

We note that the sum of V_1 and V_2 is 100 V, the same as the applied potential difference, as it must be.

(d) The power dissipated by R_1 is

$$P_1 = IV_1 = (4.0 \text{ A})(20 \text{ V}) = 80 \text{ W}$$

We can also find P_1 from the alternative formulas $P = I^2R$ and $P = V^2/R$. In each case we get the same answer:

$$P_1 = I^2R_1 = (4.0 \text{ A})^2(5.0 \text{ }\Omega) = 80 \text{ W}$$

$$P_1 = \frac{V^2}{R_1} = \frac{(20 \text{ V})^2}{5.0 \text{ }\Omega} = 80 \text{ W}$$

The power dissipated by R_2 is

$$P_2 = IV_2 = (4.0 \text{ A})(80 \text{ V}) = 320 \text{ W} = 0.32 \text{ kW}$$

(e) The power dissipated by the entire circuit is

$$P = IV = (4 \text{ A})(100 \text{ V}) = 400 \text{ W} = 0.40 \text{ kW}$$

and is equal to $P_1 + P_2$. The dissipated power appears as heat.   ◆

◆   **E X A M P L E   18.14** _____

A 20-Ω load whose maximum power rating is 5.0 W is to be connected to a 24-V battery. What is the minimum resistance of the series resistor (Fig. 18.11) that will prevent the power rating from being exceeded?

SOLUTION   The maximum allowable current in the resistor R_1 may be found from the formula $P = I^2R_1$ to be

$$I = \sqrt{\frac{P}{R_1}} = \sqrt{\frac{5.0 \text{ W}}{20 \text{ }\Omega}} = 0.50 \text{ A}$$

For a current of 0.50 A, the equivalent resistance R of the circuit should be

$$R = \frac{V}{I} = \frac{24 \text{ V}}{0.50 \text{ A}} = 48 \text{ }\Omega$$

Figure 18.11

Because $R = R_1 + R_2$, the minimum resistance R_2 of the required series resistor is

$$R_2 = R - R_1 = 48\ \Omega - 20\ \Omega = 28\ \Omega \qquad \blacklozenge$$

18.6 RESISTORS IN PARALLEL

Their equivalent resistance is less than any of the individual resistances.

Let us now consider three resistors, R_1, R_2, and R_3, connected in parallel, as in Fig. 18.12. The total current I through the set is the sum of the currents through the separate resistors, so that

$$I = I_1 + I_2 + I_3$$

The potential difference V is the same across all the resistors. By applying Ohm's law to each of the resistors, we find that

$$I_1 = \frac{V}{R_1} \qquad I_2 = \frac{V}{R_2} \qquad I_3 = \frac{V}{R_3}$$

The smaller the resistance, the greater the proportion of the total current that flows through it.

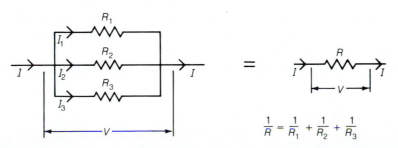

Figure 18.12

Resistors in parallel. The same potential difference V exists across each resistor.

The total current flowing through the set of three resistors is given in terms of their equivalent resistance R by

$$I = \frac{V}{R}$$

Hence we have

$$I = I_1 + I_2 + I_3$$

$$\frac{V}{R} = \frac{V}{R_1} + \frac{V}{R_2} + \frac{V}{R_3}$$

The final step is to divide through by V, which gives

$$\frac{1}{R} = \frac{1}{R_1} + \frac{1}{R_2} + \frac{1}{R_3}$$

In general, *the reciprocal of the equivalent resistance of a set of resistors connected in parallel is equal to the sum of the reciprocals of the individual resistances*:

$$\frac{1}{R} = \frac{1}{R_1} + \frac{1}{R_2} + \frac{1}{R_3} + \cdots \qquad\qquad \textit{Resistors in parallel} \quad (18.11)$$

The formulas for the equivalent resistances of series and parallel arrangements of resistors are in accord with the basic formula

$$R = \rho\frac{L}{A}$$

for the resistance of a conductor. When several resistors are connected in series, the effect is the same as increasing the length L. When they are connected in parallel, the effect is the same as increasing the cross-sectional area A. Thus a series set of resistors lets through less current than any of the individual resistors, and a parallel set of resistors lets through more current than any of the individual resistors, in each case assuming the same potential difference.

As a check after making a calculation, it is worth recalling that in a series circuit, the equivalent resistance is always *greater* than any of the individual resistances. In a parallel circuit, on the other hand, the equivalent resistance is always *smaller* than any of the individual resistances.

In the case of two resistors in parallel, the equivalent resistance is

$$\frac{1}{R} = \frac{1}{R_1} + \frac{1}{R_2} = \frac{R_1 + R_2}{R_1 R_2}$$

Taking the reciprocal of both sides of this equation gives the convenient result

$$R = \frac{R_1 R_2}{R_1 + R_2} \qquad \textit{Two resistors in parallel} \quad (18.12)$$

When using a calculator with a reciprocal [1/X] key, it is easier to use Eq. (18.11) directly, as we saw in Section 17.9 in the case of the similar formula Eq. (17.19).

◆ **E X A M P L E   18.15** ─────────────────────────────

A 5.0-Ω resistor and a 20.0-Ω resistor are connected in parallel, and a potential difference of 100 volts is applied across them by means of a generator. Find (a) the equivalent resistance of the circuit, (b) the current that flows in each resistor and in the circuit as a whole, (c) the power dissipated by each resistor, and (d) the power dissipated by the entire circuit.

SOLUTION   Successive stages in the solution of this problem are shown in Fig. 18.13.

(a) The equivalent resistance of the resistors is

$$R = \frac{R_1 R_2}{R_1 + R_2} = \frac{(5.0\ \Omega)(20.0\ \Omega)}{5.0\ \Omega + 20.0\ \Omega} = 4.0\ \Omega$$

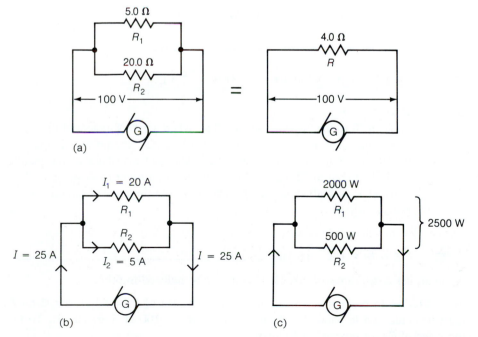

Figure 18.13

Alternatively, we can find R from

$$\frac{1}{R} = \frac{1}{5\,\Omega} + \frac{1}{20\,\Omega}$$

with the following key sequence on a calculator: [5] [1/X] [+] [20] [1/X] [=] [1/X]. The generalization to three or more resistors is obvious. For three resistors, for instance, the sequence would be: $[R_1]$ [1/X] [+] R_2 [1/X] [+] $[R_3]$ [1/X] [=] [1/X].

(b) The currents that flow in R_1 and R_2 are

$$I_1 = \frac{V}{R_1} = \frac{100\text{ V}}{5.0\,\Omega} = 20\text{ A}$$

$$I_2 = \frac{V}{R_2} = \frac{100\text{ V}}{20.0\,\Omega} = 5\text{ A}$$

The total current in the circuit is

$$I = \frac{V}{R} = \frac{100\text{ V}}{4.0\,\Omega} = 25\text{ A}$$

and is equal to $I_1 + I_2$.

(c) The powers dissipated by R_1 and R_2 are

$$P_1 = I_1 V = (20\text{ A})(100\text{ V}) = 2000\text{ W} = 2.0\text{ kW}$$
$$P_2 = I_2 V = (5\text{ A})(100\text{ V}) = 500\text{ W} = 0.5\text{ kW}$$

(d) The power dissipated by the entire circuit is

$$P = IV = (25\text{ A})(100\text{ V}) = 2500\text{ W} = 2.5\text{ kW}$$

and is equal to $P_1 + P_2$.

We recall from the previous section that, when a 100-V potential difference is applied across the same two resistors connected in series, the power dissipated is only 400 W. The difference is due to the lower equivalent resistance of the parallel combination (4 Ω instead of 25 Ω), which permits more current to flow. Since $P = IV$ and V is the same in both cases, the greater current in the parallel circuit leads to a greater power dissipation.   ◆

   E X A M P L E   18.16 _____

A circuit has a resistance of 100 Ω. How can it be reduced to 60 Ω?

SOLUTION   In order to have an equivalent resistance of $R = 60\ \Omega$, a resistor R_2 must be connected in parallel with the circuit of $R_1 = 100\ \Omega$ (Fig. 18.14). To find the value of R_2 we proceed as follows:

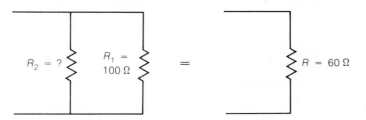

Figure 18.14

$$\frac{1}{R} = \frac{1}{R_1} + \frac{1}{R_2}$$

$$\frac{1}{R_2} = \frac{1}{R} - \frac{1}{R_1} = \frac{R_1 - R}{R_1 R}$$

$$R_2 = \frac{R_1 R}{R_1 - R} = \frac{(100\ \Omega)(60\ \Omega)}{100\ \Omega - 60\ \Omega} = 150\ \Omega$$

With a calculator we would work from

$$\frac{1}{R_2} = \frac{1}{60\ \Omega} - \frac{1}{100\ \Omega}$$

and use this key sequence:

$$[60][1/X][-][100][1/X][=][1/X] \qquad\qquad \blacklozenge$$

Series-Parallel Combinations

Complicated combinations of resistors may often be broken down into series and parallel groupings that can be analyzed in the usual way. Two examples will make clear the procedure.

◆ **E X A M P L E 18.17** _____

Find the equivalent resistance of the set of resistors shown in Fig. 18.15.

Figure 18.15

$R_1 = 3.0\ \Omega$
$R_2 = 12.0\ \Omega$
$R_3 = 8.0\ \Omega$
$R_4 = 6.0\ \Omega$

Figure 18.16

Successive steps in determining the equivalent resistance R of the resistor network shown in Fig. 18.15.

SOLUTION   Figure 18.16 shows how the original circuit is broken up into its series and parallel groupings. The first step is to find the equivalent resistance R' of the parallel resistors R_2 and R_3:

$$R' = \frac{R_2 R_3}{R_2 + R_3} = \frac{(12.0\ \Omega)(8.0\ \Omega)}{12.0\ \Omega + 8.0\ \Omega} = 4.8\ \Omega$$

This pair of resistors is in series with R_1. The equivalent resistance R'' of the upper branch of the circuit is therefore

$$R'' = R' + R_1 = 4.8\ \Omega + 3.0\ \Omega = 7.8\ \Omega$$

The upper and lower branches of the circuit are in parallel, which means that the equivalent resistance R of the entire circuit is

$$R = \frac{R'' R_4}{R'' + R_4} = \frac{(7.8\ \Omega)(6.0\ \Omega)}{7.8\ \Omega + 6.0\ \Omega} = 3.4\ \Omega \qquad \blacklozenge$$

◆   **E X A M P L E   18.18** _____

A potential difference of 12.0 V is applied across the set of resistors shown in Fig. 18.15. Find the current in each resistor.

SOLUTION   The full potential difference is applied across R_4, and so, by Ohm's law, the current I_4 in it is

$$I_4 = \frac{V}{R_4} = \frac{12.0\ \text{V}}{6.0\ \Omega} = 2.0\ \text{A}$$

The current I_1 in R_1 also flows through the entire upper branch of the circuit. Since the equivalent resistance of this branch is R'',

$$I_1 = \frac{V}{R''} = \frac{12.0 \text{ V}}{7.8 \text{ }\Omega} = 1.54 \text{ A}$$

The potential difference V' across the parallel resistors R_2 and R_3 is equal to the current I_1 through the equivalent resistance R' multiplied by R':

$$V' = I_1R' = (1.54 \text{ A})(4.8 \text{ }\Omega) = 7.4 \text{ V}$$

Another way to find V' is to subtract the potential difference

$$V_1 = I_1R_1 = (1.54 \text{ A})(3.0 \text{ }\Omega) = 4.6 \text{ V}$$

from the total of 12.0 volts to obtain

$$V' = V - V_1 = 12.0 \text{ V} - 4.6 \text{ V} = 7.4 \text{ V}$$

Hence the currents I_2 and I_3 through resistors R_2 and R_3 are

$$I_2 = \frac{V'}{R_2} = \frac{7.4 \text{ V}}{12.0 \text{ }\Omega} = 0.62 \text{ A} \qquad I_3 = \frac{V'}{R_3} = \frac{7.4 \text{ V}}{8.0 \text{ }\Omega} = 0.92 \text{ A} \qquad \blacklozenge$$

18.7 ELECTROMOTIVE FORCE

The no-load voltage of a source.

The potential difference across a battery, a generator, or other source of electric energy when it is not connected to any external circuit is called its **electromotive force.** Electromotive force is usually referred to simply as emf, and its symbol is $\mathscr{E}$.

As charges pass through a source of electric energy, work is done on them. The emf of the source is the work done per coulomb on the charges. The emf of a car's storage battery is 12 V, which means that 12 J of work are done on each coulomb of charge that passes through the battery. In the case of a battery, chemical energy is converted into electric energy by means of the work done on the charges in transit through it; in a generator, mechanical energy is converted into electric energy; in a thermocouple, heat energy is converted into electric energy; and so on. The emf of an electric source bears a relationship to its power output similar to that of applied force to mechanical power in a machine, which is the reason for its name.

Figure 18.17

The terminal voltage V of a battery in a circuit is always less than its emf $\mathscr{E}$ because of the potential drop Ir within the battery itself.

When a source of electric energy is part of a complete circuit, a current I flows. The potential difference across the terminals of the source is then less than its emf owing to its **internal resistance.** Every electric source has a certain amount of internal resistance r, which means that a potential drop Ir occurs *within* the source. Hence the actual terminal voltage V across a source of emf $\mathscr{E}$ and internal resistance r is

$$V = \mathscr{E} - Ir \qquad\qquad Terminal\ voltage \quad (18.13)$$

Terminal voltage = emf − potential drop within source

If the source is disconnected, no current flows, and $V = \mathscr{E}$. The existence of a current lowers the value of V by an amount proportional to I.

Figure 18.17 shows a battery of emf $\mathscr{E}$ connected to a circuit whose equivalent resistance is R. The total resistance in the entire circuit is R plus the internal resistance r of the battery, so that the current I that flows is

$$I = \frac{\mathscr{E}}{R + r} \qquad\qquad (18.14)$$

The actual potential difference across the battery is given by Eq. (18.13).

The internal resistance of a battery governs the maximum current it can supply. One of the major advantages of lead-acid batteries is their low internal resistances, which permit large currents to be drawn for short periods to operate the starting motors of gasoline and diesel engines. The lower the temperature of the electrolyte in a battery, the more slowly its ions move and the higher the internal resistance. As a result the current available for the starting motor in freezing weather may be less than half that available on a warm day.

◆ **E X A M P L E 18.19** _____

A "D" cell of emf 1.5 V and internal resistance 0.3 Ω is connected to a flashlight bulb whose resistance is 3.0 Ω. Find the current in the circuit and the terminal voltage of the cell.

SOLUTION   The current in the circuit is

$$I = \frac{\mathscr{E}}{R + r} = \frac{1.5 \text{ V}}{3.0 \ \Omega + 0.3 \ \Omega} = 0.45 \text{ A}$$

The terminal voltage of the cell is

$$V = \mathscr{E} - Ir = 1.5 \text{ V} - (0.45 \text{ A})(0.3 \ \Omega) = 1.4 \text{ V} \qquad \blacklozenge$$

◆   **E X A M P L E   18.20** _____

A storage battery whose emf is 12 V and whose internal resistance is 0.20 Ω is to be charged at a rate of 20 A. What applied voltage is required?

SOLUTION   The applied voltage V must exceed the battery's emf $\mathscr{E}$ by the amount Ir to provide the charging current I. Hence

$$V = \mathscr{E} + Ir = 12 \text{ V} + (20 \text{ A})(0.20 \ \Omega) = 16 \text{ V} \qquad \blacklozenge$$

Cells in Combination

When the emf of a single cell is too small for an application, two or more cells can be connected in series (Fig. 18.18). The emf of the set is the sum of the emf's of the individual cells, and the internal resistance of the set is the sum of the individual internal resistances:

$$\mathscr{E}_{\text{series}} = \mathscr{E}_1 + \mathscr{E}_2 + \mathscr{E}_3 + \cdots$$
$$r_{\text{series}} = r_1 + r_2 + r_3 + \cdots$$

A familiar example of such an arrangement is the use of six lead-acid cells in series to make the 12-V battery of a car.

When the emf of a cell or a battery is sufficient but its capacity is too small, two or more cells with the same emf can be connected in parallel. (If the cells have different emf's, currents will circulate among them that waste energy.) The emf and internal resistance of the set are now as follows:

$$\mathscr{E}_{\text{parallel}} = \mathscr{E}_1 = \mathscr{E}_2 = \mathscr{E}_3 = \cdots$$
$$\frac{1}{r_{\text{parallel}}} = \frac{1}{r_1} + \frac{1}{r_2} + \frac{1}{r_3} + \cdots$$

If all the cells have the same internal resistance r, r_{parallel} is just r divided by the number of cells. The reduced internal resistance of a parallel set of cells means that more current can be drawn by a given load. When a car's battery is too weak to start the car on a cold day, the remedy is another battery put in parallel with it by using jumper cables, which increases the available current. The ampere-hour capacity of a series set of cells is the same as that of each cell; the capacity of a parallel set is the sum of the capacities of the individual cells.

Figure 18.18

Cells in series and in parallel. In series, terminals of opposite polarity are connected in line. In parallel, terminals of the same polarity are connected together.

Series

Parallel

Figure 18.19

What value of the load resistance R will produce the maximum rate of energy transfer in this circuit?

18.8 IMPEDANCE MATCHING

Energy transfer from one system to another is most efficient when both have the same impedance.

When a source of electric energy is connected to a load, as in Fig. 18.19, the rate at which energy is transferred to the load is a maximum when both source and load have the same resistance. If R is the load resistance and I is the current, then according to Eq. (18.14), $I = \mathcal{E}/(R + r)$, and the power in the load is

$$P = I^2 R = \frac{\mathcal{E}^2 R}{(R + r)^2} \qquad\qquad \textit{Power transfer} \quad (18.15)$$

The maximum value of P occurs when $R = r$:

$$P_{\max} = \frac{\mathcal{E}^2 R}{(R + R)^2} = \frac{\mathcal{E}^2 R}{(2R)^2} = \frac{\mathcal{E}^2}{4R} \qquad (R = r) \qquad\qquad (18.20)$$

Figure 18.20 shows how $P/P_{\max}$ varies with the resistance ratio R/r.

   E X A M P L E   18.21

Verify that the power transfer from a 100-V source of internal resistance $r = 10\ \Omega$ to a load resistance R is a maximum when $R = 10\ \Omega$.

Table 18.3   Current *I* and load power *P* for various values of *R* for the circuit of Fig. 18.19 when $\mathcal{E}$ = 100 V, r = 10 Ω

$R(\Omega)$	I(A)	P(W)		$R(\Omega)$	I(A)	P(W)
2	8.3	139		12	4.5	248
4	7.1	204		14	4.2	243
6	6.3	234		16	3.8	237
8	5.6	247		18	3.6	230
10	6.0	250		20	3.3	222

SOLUTION   Table 18.3 shows the current I and power P calculated from Eqs. (18.14) and (18.15) for values of R from 2 Ω to 20 Ω. The maximum power of 250 W indeed occurs for $R = r = 10\ \Omega$. Although the current is greater when R is less than r, the smaller R means that P is less than $P_{\max}$. When R is greater than r, the current is then sufficiently low that P is less than $P_{\max}$ in this case as well.   ◆

Figure 18.20

The power transferred to an external load of resistance R is a maximum when R equals the internal resistance r of the source.

The conclusion reached in the preceding discussion is an example of **impedance matching**. When energy is being transferred from one system to another (here from the battery to the external resistance), the efficiency is greatest when both systems have the same **impedance,** which is a general term for resistance to the flow of energy in whatever form it may take in a particular case.

We met impedance matching earlier, the first time in Section 5.7. There we saw that when a moving object strikes a stationary one, the maximum energy transfer takes place when both have the same mass. In this situation the inertia of each object, as measured by its mass, represents its impedance. The curve of relative energy transfer versus mass ratio in Fig. 5.17 is identical in form to that in Fig. 18.20.

Another example of impedance matching occurs when a pulse or wave moves down a stretched string, as discussed in Section 12.1. When two strings having the same mass per unit length are joined together, a pulse in one passes to the other with no reflection. However, if the second string has either a greater or a smaller mass per unit length, reflection occurs at the junction and not all the energy of the pulse is transmitted (see Figs. 12.5 and 12.6). Though best known for its application to electric circuits, impedance matching is an important concept in many other branches of physics and engineering as well.

18.9 KIRCHHOFF'S RULES

They make it possible to analyze complex circuits.

It is often difficult or impossible to figure out the currents that flow in the various branches of a complex network just by computing equivalent resistances. Two rules formulated by Gustav Kirchhoff (1824–1887) make it possible to find the current

Figure 18.21

A *junction* is a point where three or more wires come together; a *loop* is any closed conducting path in the network. Here the internal resistances of the batteries are included in R_1 and R_3. The directions of the currents I_1, I_2, and I_3 are chosen arbitrarily; the value of I_3 turns out to be negative, which means that its actual direction is opposite to the one shown.

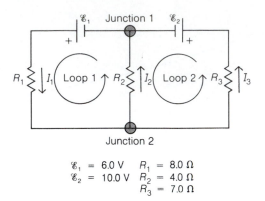

$$\mathscr{E}_1 = 6.0\ \text{V} \qquad R_1 = 8.0\ \Omega$$
$$\mathscr{E}_2 = 10.0\ \text{V} \qquad R_2 = 4.0\ \Omega$$
$$R_3 = 7.0\ \Omega$$

in each part of a direct-current circuit, no matter how complicated, if we are given the emf's of the sources of potential difference and the resistances of the various circuit elements. These rules apply to **junctions,** which are points where three or more wires come together, and to **loops,** which are closed conducting paths that are part of the circuit (Fig. 18.21).

Kirchhoff's first rule follows from the conservation of electric charge. Charge is never found to collect at any point in a circuit, nor can it be created there, and so there cannot be any net current into or out of a junction. Hence the first rule:

---

1. **The sum of the currents flowing into a junction is equal to the sum of the currents flowing out of the junction.**

The second rule follows from the conservation of energy. The sum of the emf's in a loop equals the amount of work done per coulomb by the sources of emf *on* a charge that moves once around the loop. The work done per coulomb *by* the charge as it moves around the loop equals the sum of the *IR* potential drops along the way. To conserve energy, the work done on the charge must be the same as the work done by the charge. Thus we have Kirchhoff's second rule:

---

2. **The sum of the emf's around a loop is equal to the sum of the *IR* potential drops around the loop.**

A definite procedure must be followed when Kirchhoff's rules are applied to a network. Let us call a section of a loop between two junctions a **branch.** The first step is to give the current in each branch a symbol of its own (I_1, I_2, I_3, and so on) and to assume a direction for each current. These symbols and directions should be marked on the circuit diagram, as in Fig. 18.21. The current is the same in all the resistors and sources of emf in a given branch, but of course the currents are different in the different branches.

It does not matter which direction is chosen for each current. If we have guessed correctly about the direction of the current in a branch, the solution of the problem will give a positive value for the current. If we have guessed wrong, a negative value will show that the actual current in the branch is in the opposite direction.

When the currents have been assigned to each branch, we then apply Kirchhoff's first rule to the junctions in the network. This will give us as many equations as the number of junctions. However, we will always find that one of these equations is a combination of the others and so contains no new information. Thus the number of junction equations we can use is always one less than the number of junctions.

The third step is to apply Kirchhoff's second rule to the loops in the network. In going around a loop, we must follow a consistent path, either clockwise or counterclockwise. These paths should be marked on the diagram, as in Fig. 18.21. An emf is reckoned positive if we meet the negative terminal of its source first. If instead we meet the positive terminal first, the emf is reckoned negative. An *IR* drop is considered positive if the assumed current in the resistor is in the same direction as the path we are following. The drop is considered negative if the current is opposite in direction to that of our path. The internal resistance of an emf source is treated as if it were a separate resistor in series with the source.

As in the case of the junction equations, it is important to avoid loop equations that merely duplicate information in the other loop equations. As a guide, the number of useful loop equations will equal the number of separate areas enclosed by branches of the network. (We can think of these areas as adjacent plots of land.)

Thus we have three steps to follow when using Kirchhoff's rules:

1. Assume directions for the branch currents.
2. Apply the first rule to the currents at the junctions.
3. Apply the second rule to the emf's and *IR* drops in the loops.

◆ **E X A M P L E   18.22** ————————————————————

Find the current in each of the resistors in the network shown in Fig. 18.21.

SOLUTION   The assumed directions of the unknown currents I_1, I_2, and I_3 are shown in the figure. Applying Kirchhoff's first rule to junction 1, we obtain

$$I_1 = I_2 + I_3 \qquad\qquad\qquad Junction\ 1$$

This rule applied to junction 2 yields the same result.

In applying Kirchhoff's second rule to the two loops, we will follow counterclockwise routes. For loop 1,

$$\mathscr{E}_1 = I_1R_1 + I_2R_2 \qquad\qquad\qquad Loop\ 1$$

and for loop 2,

$$-\mathscr{E}_2 = -I_2R_2 + I_3R_3 \qquad\qquad\qquad Loop\ 2$$

In loop 2 we consider $\mathscr{E}_2$ as negative because we meet its positive terminal first, and I_2 as negative because its direction is opposite to our counterclockwise path.

There is also a third loop, namely, the outside one in Fig. 18.21, which similarly must obey Kirchoff's second rule. Proceeding counterclockwise yields

$$-\mathscr{E}_2 + \mathscr{E}_1 = I_1 R_1 + I_3 R_3 \qquad \textit{Outside loop}$$

This last equation is just the sum of the two preceding loop equations. (With two enclosed areas in the network, we expect only two independent loop equations.) Hence we may use the junction equation and any two of the loop equations to solve for the unknown currents. Nothing will be gained by using all three loop equations.

There are three unknown currents, and we have three separate equations relating them. Hence the problem can be solved by routine algebra. One way to do this is to start by substituting $I_2 + I_3$ for I_1 in the first loop equation to obtain

$$\mathscr{E}_1 = I_2 R_1 + I_3 R_1 + I_2 R_2$$

We now solve both this equation and the second loop equation for I_3:

$$I_3 = \frac{\mathscr{E}_1 - I_2 R_1 - I_2 R_2}{R_1} \qquad I_3 = \frac{-\mathscr{E}_2 + I_2 R_2}{R_3}$$

Setting the two expressions for I_3 equal enables us to solve for I_2 as follows:

$$\frac{\mathscr{E}_1 - I_2 R_1 - I_2 R_2}{R_1} = \frac{-\mathscr{E}_2 + I_2 R_2}{R_3}$$

$$\mathscr{E}_1 R_3 - I_2 R_1 R_3 - I_2 R_2 R_3 = -\mathscr{E}_2 R_1 + I_2 R_1 R_2$$

$$I_2(R_1 R_3 + R_2 R_3 + R_1 R_2) = \mathscr{E}_2 R_1 + \mathscr{E}_1 R_3$$

$$I_2 = \frac{\mathscr{E}_2 R_1 + \mathscr{E}_1 R_3}{R_1 R_3 + R_2 R_3 + R_1 R_2}$$

Finally we insert the values of $\mathscr{E}_1$, $\mathscr{E}_2$, R_1, R_2, and R_3 given in Fig. 18.21 to obtain

$$I_2 = 1.05 \text{ A}$$

With the value of I_2 known, we can find the value of I_3 from the second loop equation:

$$I_3 = \frac{-\mathscr{E}_2 + I_2 R_2}{R_3} = -0.83 \text{ A}$$

The minus sign means that the direction of I_3 is opposite to that shown in Fig. 18.21. The junction equation finally gives us the value of I_1:

$$I_1 = I_2 + I_3 = 0.22 \text{ A}$$

The same results would have been found had we chosen other directions for the currents or taken other routes around the loops in applying Kirchhoff's second rule. The important thing is not what choice of directions is made but to follow that choice consistently in working out the problem.

The outside loop equation, which was not used in the calculation, gives us a handy way to check the results. We have

$$-\mathcal{E}_2 + \mathcal{E}_1 = I_1R_1 + I_3R_3 \qquad \text{\textit{Outside loop}}$$
$$-10\,\text{V} + 6\,\text{V} = (0.22\,\text{A})(8\,\Omega) + (-0.83\,\text{A})(7\,\Omega)$$
$$-4\,\text{V} = -4\,\text{V}$$

In this calculation symbols for the various quantities were used until the very end to make clear the procedure. In working out problems of this kind, however, it is sometimes more convenient to substitute numerical values from the start. ◆

18.10 CAPACITIVE TIME CONSTANT

To charge or discharge a capacitor takes time.

A circuit that contains capacitance as well as resistance does not respond instantly to changes in the applied emf. For instance, when a capacitor is connected to a battery, as in Fig. 18.22, it does not at once become fully charged.

At first the only limit to the current to the capacitor is the resistance R in the circuit. Thus the initial current is $I = \mathcal{E}R$, where $\mathcal{E}$ is the emf of the battery. As the capacitor becomes charged, however, a potential difference appears across it, whose polarity is such as to tend to oppose the further flow of current. When the charge on the capacitor has built up to some value Q, this opposing potential difference is $V = Q/C$. Hence the net potential difference is $\mathcal{E} - Q/C$, and the current is

$$I = \frac{\mathcal{E} - Q/C}{R} \qquad (18.17)$$

$$\text{Current} = \frac{\text{applied emf} - \text{voltage across capacitor}}{\text{resistance}}$$

Figure 18.22

A circuit that contains a capacitor, a resistor, and a source of emf in series. When the switch is closed, the charge on the capacitor increases gradually to its ultimate value of $C\mathcal{E}$.

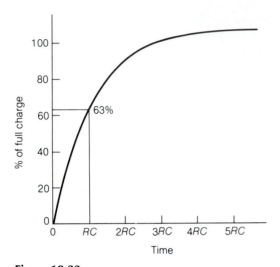

Figure 18.23

The growth of charge in the capacitor of Fig. 18.22 after the switch is closed.

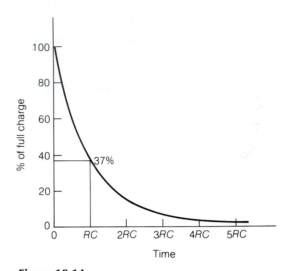

Figure 18.14

The decay of charge in a circuit containing capacitance and resistance when the battery is short-circuited.

As Q increases, then, its *rate* of increase drops. This gives a steadily decreasing slope in the curve of Fig. 18.23, which is a graph showing how Q varies with time when a capacitor is being charged; the capacitor is connected to the battery at $t = 0$.

A mathematical analysis of Eq. (18.17) shows that after a time interval of RC (the product of the resistance R in the circuit and the capacitance C of the capacitor), the charge on the capacitor reaches 63% of its ultimate value of $Q_0 = C\mathscr{E}$. The time RC is therefore a convenient measure of how rapidly the capacitor becomes charged and is accordingly called the **time constant** of the circuit:

$$T_C = RC \qquad\qquad \text{\textit{Capacitive time constant}} \quad (18.18)$$

In principle the capacitor acquires its ultimate charge Q_0 only after an infinite time has elapsed. As we can see from Fig. 18.23, however, this value is very nearly reached after only a few time constants. At $t = 3RC$, the charge is 95% of Q_0, and at $t = 4RC$ it is 98% of Q_0.

If a capacitor with an initial charge is discharged through a resistance, its charge decreases with time as shown in Fig. 18.24. After the time RC the charge on the capacitor is reduced to 37% of its original value, a drop of 63%.

◆ **E X A M P L E   18.23** _____

A 5.0-μF capacitor is charged by being connected to a 1.5-V dry cell. The total resistance of the circuit is 2.0 Ω. (a) What is the final charge on the capacitor and how long does it take to reach 63% of this charge? (b) The battery is then disconnected from the capacitor. If the resistance of the dielectric between the capacitor plates is $1.0 \times 10^{10}\ \Omega$, after what period of time will the charge on the capacitor drop to 37% of its initial value?

SOLUTION   (a) The ultimate charge on the capacitor is

$$Q_0 = CV = (5.0 \times 10^{-6}\ \text{F})(1.5\ \text{V}) = 7.5 \times 10^{-6}\ \text{C}$$

The time constant of the circuit is

$$T_C = RC = (2.0\ \Omega)(5.0 \times 10^{-6}\ \text{F}) = 1.0 \times 10^{-5}\ \text{s}$$

and so this is the time needed for the capacitor to acquire 63% of its ultimate charge.

(b) When the battery is disconnected from the capacitor, charge gradually leaks through the dielectric of the latter. The time constant for discharge is

$$T_C = RC = (1.0 \times 10^{10}\ \Omega)(5.0 \times 10^{-6}\ \text{F}) = 5.0 \times 10^{4}\ \text{s} = 14\ \text{h}$$

and in this period the charge on the capacitor will drop to 37% of its original value.   ◆

Important Terms

A flow of electric charge from one place to another is called an **electric current.** The unit of electric current is the **ampere,** which is equal to a flow of 1 coulomb/second.

Ohm's law states that the current in a metallic conductor is proportional to the potential difference between its ends. The **resistance** of a conductor is the ratio between the potential difference across its ends and the current that flows. The unit of resistance is the **ohm,** which is equal to 1 volt/ampere. The resistance of a conductor is proportional to its length and to the **resistivity** ρ of the material of which it is made, and inversely proportional to its cross-sectional area.

The **equivalent resistance** of a set of interconnected resistors is the value of the single resistor that can be substituted for the entire set without affecting the current that flows in the rest of any circuit of which it is a part. Resistors in **series** are connected consecutively so that the same current flows through all of them, whereas resistors in **parallel** have their terminals connected together so that the total current is split up among them.

The **electromotive force** (emf) of a battery, a generator, or other source of electric energy is the potential difference across its terminals when no current flows. When a current is flowing, the terminal voltage is less than the emf owing to the potential drop in the **internal resistance** of the source.

Kirchhoff's rules for network analysis are: (1) The sum of the currents flowing into a junction of three or more wires is equal to the sum of the currents flowing out of the junction; (2) the sum of the emf's around a closed conducting loop is equal to the sum of the *IR* potential drops around the loop.

The **capacitive time constant** of a circuit that contains capacitance and resistance is a measure of the time needed for the charge on the capacitor to respond to changes in the applied emf.

Important Formulas

Electric current: $I = \dfrac{Q}{t}$

Resistance: $R = \dfrac{V}{I}$

Ohm's law: $I = \dfrac{V}{R}$ (holds for most metals)

Resistance of ohmic conductor: $R = \rho\dfrac{L}{A}$

Power: $P = IV$

$\qquad = I^2R = \dfrac{V^2}{R}$ (ohmic conductor)

Resistors in series: $R = R_1 + R_2 + R_3 + \cdots$

Resistors in parallel: $\dfrac{1}{R} = \dfrac{1}{R_1} + \dfrac{1}{R_2} + \dfrac{1}{R_3} + \cdots$

$\qquad\qquad R = \dfrac{R_1R_2}{R_1 + R_2}$ (two resistors)

Emf and terminal voltage: $V = \mathscr{E} - IR$

Capacitive time constant: $T_C = RC$

MULTIPLE CHOICE

1. The resistance of a conductor does not depend on its
   a. mass   **b.** length
   c. cross-sectional area   **d.** resistivity

2. A certain wire has a resistance R. The resistance of another wire, identical with the first except for having twice its diameter, is
   a. $\frac{1}{4}R$ **b.** $\frac{1}{2}R$
   c. $2R$ **d.** $4R$

3. Which of the following combinations of length and cross-sectional area will give a certain volume of copper the least resistance?
   a. L and A
   b. $2L$ and $\frac{1}{2}A$
   c. $\frac{1}{2}L$ and $2A$
   d. Does not matter, because the volume of copper remains the same

4. If the diameter d of a circle is expressed in mils, its area in units of circular mils is equal to
   a. d **b.** d^2
   c. πd **d.** πd^2

5. The temperature of a copper wire is raised. Its resistance
   a. decreases
   b. remains the same
   c. increases
   d. any of the above, depending on the temperatures involved

6. Of the following combinations of units, the one that is not equal to the watt is the
   a. J/s   **b.** AV
   c. $A^2\Omega$ **d.** Ω^2/V

7. A resistor R_1 dissipates the power P when connected to a certain generator. If a resistor R_2 is inserted in series with R_1, the power dissipated by R_1
   a. decreases
   b. increases
   c. remains the same
   d. may do any of the above, depending on the values of R_1 and R_2

8. If the resistor R_2 in Multiple Choice 7 is placed in parallel with R_1, the power dissipated by R_1
   a. decreases
   b. increases
   c. remains the same
   d. may do any of the above, depending on the values of R_1 and R_2

9. A battery of emf $\mathscr{E}$ and internal resistance r is connected to an external circuit of equivalent resistance R. If $R = r$,
   a. the current in the circuit will be a minimum
   b. the current in the circuit will be a maximum
   c. the power dissipated in the circuit will be a minimum
   d. the power dissipated in the circuit will be a maximum

10. A battery is connected to an external circuit. The potential drop within the battery is proportional to
   a. the emf of the battery
   b. the equivalent resistance of the circuit
   c. the current in the circuit
   d. the power dissipated in the circuit

11. Which of the following is neither a basic physical law nor derivable from one?
   a. Coulomb's law   **b.** Ohm's law

c. Kirchhoff's first law
d. Kirchhoff's second law

12. If the wrong direction is assumed for a current I in a network being analyzed by Kirchhoff's rules, the value of the current that is obtained will be
a. I
b. $-I$
c. incorrect
d. 0

13. A wire has a diameter of 0.050 in. Its cross-sectional area in circular mils is
a. 0.020 cmil
b. 0.0025 cmil
c. 50 cmil
d. 2500 cmil

14. A 200-m length of copper wire has a resistance of 2.0 Ω. Its cross-sectional area is
a. 0.0017 mm^2
b. 1.7 mm^2
c. 3.4 mm^2
d. 5.3 mm^2

15. An electric heater draws a current of 20 A when connected to a 120-V power source. Its resistance is
a. 0.17 Ω
b. 6 Ω
c. 8 Ω
d. 2400 Ω

16. The 8.0-Ω coil of a loudspeaker carries a current of 0.80 A. The potential difference across its terminals is
a. 0.1 V
b. 5.1 V
c. 6.4 V
d. 10 V

17. The power rating of an electric motor that draws a current of 3.00 A when operated at 120 V is
a. 40 W
b. 0.36 kW
c. 0.54 kW
d. 1.08 kW

18. When a 100-W, 240-V light bulb is operated at 200 V, the current that flows in it is
a. 0.35 A
b. 0.42 A
c. 0.50 A
d. 0.58 A

19. A 115-V, 1-kW electric oven is by mistake connected to a 230-V power line that has a 15-A fuse. The oven will
a. give off less than 1 kW of heat
b. give off 1 kW of heat
c. give off more than 1 kW of heat
d. blow the fuse

20. A 230-V, 1-kW electric oven is by mistake connected to a 115-V power line that also has a 15-A fuse. The oven will
a. give off less than 1 kW of heat
b. give off 1 kW of heat
c. give off more than 1 kW of heat
d. blow the fuse

21. The energy content of a fully charged 12-V storage battery of capacity 50 A-h is
a. 50 J
b. 0.60 kJ
c. 36 kJ
d. 2.2 MJ

22. A 20-V potential difference is applied across a series combination of a 10-Ω resistor and a 30-Ω resistor. The current in the 10-Ω resistor is
a. 0.5 A
b. 0.67 A
c. 1 A
d. 2 A

23. The potential difference across the 10-Ω resistor in Multiple Choice 22 is
a. 5 V
b. 10 V
c. 15 V
d. 20 V

24. The equivalent resistance of a 10-Ω resistor and a 30-Ω resistor connected in parallel is
a. 0.13 Ω
b. 7.5 Ω
c. 20 Ω
d. 40 Ω

25. A 20-V potential difference is applied across the resistors in Multiple Choice 24. The current in the 10-Ω resistor is
a. 0.5 A
b. 1 A
c. 2 A
d. 2.67 A

26. The equivalent resistance of a network of three 2-Ω resistors cannot be
a. 0.67 Ω
b. 1.5 Ω
c. 3 Ω
d. 6 Ω

27. Two identical resistors in parallel have an equivalent resistance of 2 Ω. If the resistors were in series, their equivalent resistance would be
a. 2 Ω
b. 4 Ω
c. 8 Ω
d. 16 Ω

28. A 12-V potential difference is applied across a series combination of four 6-Ω resistors. The current in each resistor is
a. 0.5 A
b. 2 A
c. 8 A
d. 18 A

29. A 12-V potential difference is applied across a parallel combination of four 6-Ω resistors. The current in each resistor is
a. 0.5 A
b. 2 A
c. 8 A
d. 18 A

30. A resistor of unknown resistance is in parallel with a 12-Ω resistor. A battery of emf 24 V and negligible internal resistance is connected across the combination. The battery provides a current of 3 A. The unknown resistance is
a. 8 Ω
b. 12 Ω
c. 24 Ω
d. 36 Ω

31. A 10-Ω resistor and a 20-Ω resistor are connected in parallel to a battery. If heat is produced in the 10-Ω resistor at the rate P, the rate at which heat is produced in the 20-Ω resistor is
a. $P/4$
b. $P/2$
c. P
d. $2P$

32. A 10-Ω resistor and a 20-Ω resistor are connected in series to a battery. If heat is produced in the 10-Ω resistor at the rate P, the rate at which heat is produced in the 20-Ω resistor is

    a. $P/4$ **b.** $P/2$

    c. P **d.** $2P$

33. A 50-V battery is connected across a 10-Ω resistor and a current of 4.5 A flows. The internal resistance of the battery is

    a. 0   **b.** 0.5 Ω

    c. 1.1 Ω **d.** 5 Ω

34. A 4.0-Ω resistor is connected to a battery of emf 20 V and internal resistance 1.0 Ω. The potential difference across the resistor is

    a. 4 V   **b.** 16 V

    c. 19 V   **d.** 20 V

35. Two batteries of emf 6 V and internal resistance 1 Ω are connected in parallel. The load that will absorb the greatest power from the combination has a resistance of

    a. 0.5 Ω **b.** 1 Ω

    c. 2 Ω **d.** 6 Ω

QUESTIONS

1. Why are two wires used to carry electric current instead of a single one?

2. Which of the I-V graphs shown in Fig. 18.25 correspond to a substance that obeys Ohm's law?

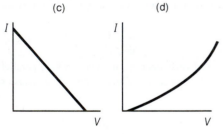

(a) (b) (c) (d)

Figure 18.25

Question 2.

3. When a metal object is heated, both its dimensions and its resistivity increase. Is the increase in resistivity likely to be a consequence of the increase in length?

4. It is sometimes said that an electrical appliance "uses up" electricity. What does such an appliance actually use in its operation?

5. Alice and Fred are discussing whether an electric heater with a large or with a small resistance will yield the greater heat output. Alice favors a small value of R because $P = V_2/R$; Fred favors a large value of R because $P = I^2R$. What is your conclusion?

Figure 18.26

Question 6.

Figure 18.27

Question 8.

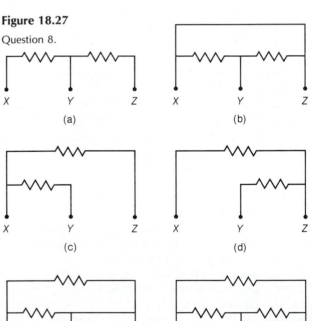

(a) (b) (c) (d) (e) (f)

6. The light bulbs in Fig. 18.26 are identical. Which bulb gives off the most light? The least light?

7. Why is it undesirable to connect cells of different emf in parallel?

8. A black box with three terminals marked X, Y, and Z contains two or three resistors of unknown resistance that may be connected in any of the six ways shown in Fig. 18.27. A battery is connected across the terminals X and Z, and the potential differences V_{xy} and V_{yz} between terminals x and y and between y and z, respectively, are measured. It is found that $V_{xy} = 2$ V and $V_{yz} = 4$ V. (a) Which, if any, of the circuits shown could give this result? (b) If more than one circuit is possible, what further test could tell which of them is correct?

EXERCISES

18.1 Electric Current

1. Sensitive instruments can detect the passage of as few as 60 electrons/s. To what current does this correspond?

2. Cosmic rays are atomic nuclei, mainly protons, that move at high speeds through the Milky Way galaxy, of which the sun is a member. An average of about 1500 of them arrive per square meter per second at the top of the atmosphere; because the earth's magnetic field affects their paths, their intensity varies somewhat around the earth. Assuming that all the cosmic ray particles are protons, find the total cosmic-ray current reaching the earth.

18.2 Ohm's Law

3. Find the resistance of a 120-V electric toaster that draws a current of 8.0 A.

4. A 240-V water heater has a resistance of 24 Ω. What must be the minimum rating of the fuse in the circuit to which the heater is connected?

5. What potential difference must be applied across a 1500-Ω resistor in order that the resulting current be 50 mA?

6. In resistance welding, a pair of electrodes presses two sheets of metal together. Heat from the passage of current melts the metal sheets where they are in contact between the electrodes to form a weld. What potential difference is needed to produce a current of 10 kA when the resistance of the metal between the electrodes is 0.20 mΩ?

18.3 Resistivity

7. How does the resistance per meter of 5.0-mm^2 copper wire compare with that of 2.0-mm^2 copper wire?

8. The resistance of No. 8 copper wire is 0.21 Ω/100 m. What is the resistance per 100 m of No. 8 aluminum wire?

9. American Wire Gage (AWG) 000 wire has a cross-sectional area of 85 mm^2. Find the resistance of 10 km of copper wire of this size at 20°C.

10. A silicon wafer 1 cm square and 2 mm thick has a resistance of 1.2 kΩ. What is the resistivity of silicon?

11. A silver wire 2.0 m long is to have a resistance of 0.50 Ω. What should its diameter be?

12. A 40-Ω resistor is to be wound from platinum wire 0.10 mm in diameter. How much wire is needed?

13. A metal rod 1.0 m long and 10 mm in diameter is drawn out into a wire 1.0 mm in diameter. (a) What is the length of the wire? (b) Compare the resistance of the rod with the resistance of the wire.

14. A metal bar of length a has a square cross section of length b on each edge. (a) Compare the resistance between its square ends with the resistance between opposite long faces. (b) What is the resistance ratio when $a = 20$ cm and $b = 10$ mm?

15. AWG 18 wire has a diameter of 0.0403 in. What is its cross-sectional area in circular mills?

16. Find the diameter of a wire whose cross-sectional area is 6400 cmil.

17. How long should a copper wire 20 mils in diameter be for it to have a resistance of 5.0 Ω?

18. The maximum resistance of an aluminum wire 2200 ft long is to be 1.8 Ω. Find the minimum diameter of the wire in mils.

19. The maximum allowable voltage drop in a cable that carries a current of 20 A from a power source to a shed 70 m away is 3.0 V. What is the minimum cross-sectional area of the copper wires in the cable in square millimeters?

20. An electric motor draws 10 A from a 240-V source 120 ft away. What is the minimum diameter of the copper wire in mils if the voltage drop is not to exceed 2%?

21. Aluminum wires are sometimes used instead of copper wires to transmit electric power. What is the ratio between the masses of an aluminum and a copper wire of the same length whose cross-sectional areas are such that they have the same resistance?

22. A copper wire 1.0 mm in diameter carries a current of 12 A. Find the potential difference between two points in the wire that are 100 m apart.

23. The resistance of a copper wire is 100 Ω at 20°C. Find its resistance at 0°C and at 80°C.

24. Motors, generators, transformers, and other electrical devices should not be operated above certain temperatures, whose values depend on their construction. The temperature in the interior of such a device can be found by measuring the resistance of one of its windings before it is run and after it has been run for some time. If the copper field windings of an electric motor have resistances of 100 Ω at 20°C and of 115 Ω when the motor is operating, what is their temperature under the latter conditions?

25. An iron wire has a resistance of 2.00 Ω at 0°C and a resistance of 2.46 Ω at 45°C. Find the temperature coefficient of resistivity of the wire.

18.4 Electric Power

26. Currents of 3.0 A flow through two wires, one that has a potential difference of 60 V across its ends and another that has a potential difference of 120 V across its ends. Compare (a) the rates at which charge passes through the wires and (b) the rates at which energy is dissipated in the wires.

27. An electric drill rated at 400 W is connected to a 240-V power line. How much current does it draw?

28. Find the maximum current in a 50-Ω, 10-W resistor if its power rating is not to be exceeded.

29. An electric motor whose power output is 0.50 hp draws 4.0 A at 120 V. What is its efficiency?

30. The alternator on a car's engine is 83% efficient. Find the mechanical power the alternator takes when it is delivering 20 A at 14 V.

31. A 240-V electric motor whose efficiency is 80% draws 10 A when it is used to operate a milling machine. Find the power output of the motor.

32. How much current is drawn by a 0.50-hp electric motor operated from a 120-V source of electricity? Assume that 80% of the electric energy absorbed by the motor is turned into mechanical work.

33. A 240-V clothes dryer draws a current of 15 A. How much energy, in kilowatt-hours and in joules, does it use in 45 min of operation?

34. A fully charged 12-V storage battery has an energy content of 3.5 MJ. What is its capacity in ampere-hours?

35. When a certain 1.5-V battery is used to power a 3.0-W flashlight bulb, it is exhausted after an hour's use. (a) How much charge has passed through the bulb in this period of time? (b) If the battery costs $0.50, what is the cost of a

kilowatt-hour of electric energy obtained in this way? How does this compare with the cost of the electric energy supplied to your home?

36. A trolley car of mass 1.0×10^4 kg takes 10 s to reach a speed of 8.0 m/s starting from rest. It operates from a 5.0-kV power line and is 50% efficient. Find the average current it draws during the acceleration.

18.5 Resistors in Series

37. A 200-Ω resistor and a 500-Ω resistor are in series as part of a larger circuit. If the voltage across the 200-Ω resistor is 2.0 V, what is the voltage across the 500-Ω resistor?

38. It is desired to limit the current in an 80-Ω resistor to 0.50 A when it is connected to a 50-V power source. What is the value of the series resistor that is needed?

39. A stingy American moves to Europe, where 240 V is the normal household voltage, and proposes to use his old 120-V light bulbs by connecting them two at a time in series. How good is this idea when both bulbs of each pair have the same power rating? When the bulbs have different ratings—for instance, 10 W and 100 W at 120 V?

40. A set of Christmas tree lights consists of 12 bulbs connected in series to a 120-V power source. Each bulb has a resistance of 5.0 Ω. (a) What is the current in the circuit? (b) How much power is dissipated in the circuit?

41. A 5.0-Ω light bulb and a 10.0-Ω light bulb are connected in series with a 12.0-V battery. (a) What is the current in each bulb? (b) What is the voltage across each bulb? (c) What is the power dissipated by each bulb and the total power dissipated by the circuit?

42. A 20-Ω resistor and a 60-Ω resistor are in series as part of a larger circuit. The two resistors dissipate a total of 20 W. (a) How much power does each resistor dissipate? (b) What is the potential difference across each resistor?

43. A 5.0-Ω resistor and a 10.0-Ω resistor are connected in series. If the power dissipated in the 5.0-Ω resistor is 125 W, what is the potential difference across the combination?

18.6 Resistors in Parallel

44. How can the resistance of a 20.0-Ω circuit be reduced to 5.0 Ω?

45. List the resistances that can be obtained by combining three 100-Ω resistors in all possible ways.

46. Three identical resistors connected in parallel have an equivalent resistance of $1\frac{2}{3}$ Ω. What would their equivalent resistance be if connected in series?

47. A 20-Ω resistor, a 40-Ω resistor, and a 50-Ω resistor are connected in parallel across a 60-V power source. Find the equivalent resistance of the set and the current in each resistor.

48. In a certain bus forty 15-W, 30-V light bulbs are connected in parallel to a 30-V power source. (a) What is the current in each bulb? (b) What is the current provided by the source? (c) How much power is provided by the source?

49. The light bulbs in Exercise 41 are connected in parallel across the same battery. Answer the same questions for this arrangement.

50. What is the potential difference between the points *A* and *B* in the circuit shown in Fig. 18.28?

Figure 18.28

Exercise 50.

51. (a) Find the equivalent resistance of the circuit in Fig. 18.29. (b) What is the current in the 8.0-Ω resistor when a potential difference of 12.0 V is applied to the circuit?

Figure 18.29

Exercise 51.

52. Each of the resistors in the circuit in Fig. 18.30 can safely dissipate 10 W. What is the maximum power the entire circuit can dissipate?

Figure 18.30

Exercise 52.

53. A 60-V potential difference is applied to the circuit in Fig. 18.31. Find the current in the 10-Ω resistor. (*Hint:* Redraw

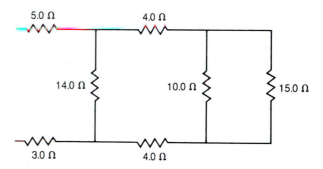

Figure 18.31

Exercise 53.

the circuit to bring out the series and parallel combinations of resistors more clearly.)

54. (a) Find the equivalent resistance of the circuit in Fig. 18.32. (b) What is the current in the 12-Ω resistor when a potential difference of 100 V is applied to the circuit?

Figure 18.32

Exercise 54.

18.7 Electromotive Force

55. A 12-V battery of internal resistance 1.5 Ω is connected to an 8.0-Ω resistor. Find the total power produced by the battery and the percentage of this power dissipated as heat within it.

56. A battery of emf 24 V is connected to an 11-Ω resistor. The potential difference across the battery terminals is 22 V. What is the internal resistance of the battery?

57. A battery having an emf of 24 V is connected to a 10-Ω load, and a current of 2.2 A flows. Find the internal resistance of the battery and its terminal voltage.

58. A generator has an emf of 240 V and an internal resistance of 0.30 Ω. When the generator is supplying a current of 20 A, find (a) its terminal voltage, (b) the power supplied to the load, and (c) the power dissipated in the generator itself.

59. The brightness of a light bulb depends on the power dissipated by its filament. As a dry cell ages, its internal resistance increases while its emf remains approximately unchanged at 1.5 V. A fresh No. 6 dry cell might have an internal resistance of 0.05 Ω and an old one an internal resistance of 0.20 Ω. Find the ratio between the powers dissipated in a 0.25-Ω bulb when it is connected to a fresh and to an old dry cell.

60. Four batteries, each of emf 6.0 V and internal resistance 0.30 Ω, are connected in series with a load of 2.0 Ω. Find the current in the load.

61. If the batteries in Exercise 60 are connected in parallel with the same load, what is the current in the load?

62. Twelve cells, each of emf 2.1 V and internal resistance 0.20 Ω, are connected as shown in Fig. 18.33. Find the emf and internal resistance of the combination.

Figure 18.33

Exercise 62.

63. A 12-V storage battery with an internal resistance of 0.012 Ω delivers 80 A when used to crank a gasoline engine.

If the battery mass is 20 kg and it has an average specific heat of 0.84 kJ/kg · °C, what is its rise in temperature during 1 min of cranking the engine?

64. A 5.0-Ω resistor and a 10.0-Ω resistor are connected in parallel. This combination is connected in series with another pair of parallel resistors whose resistances are both 8.0 Ω. (a) What is the equivalent resistance of the network? (b) The network is connected to a 24-V battery whose internal resistance is 1.5 Ω. Find the current in each of the resistors.

18.10   Capacitive Time Constant

65. A reusable flash bulb requires an energy of 100 J for its discharge. A 450-V battery is used to charge a capacitor for this purpose. The resistance of the charging circuit is 15 Ω. (a) What is the required capacitance? (b) What is the time constant of the circuit?

66. A 100-μF electrolytic capacitor has a leakage current of 5 μA when the potential difference across its terminals is 12 V. If the capacitor is connected to a 12-V battery and then removed, how long will it take for the charge to fall to 37% of its original value?

67. A 5-μF capacitor is connected across a 1000-V battery with wires whose resistance is a total of 5000 Ω. (a) What is the time constant of this circuit? (b) What is the initial current that flows when the battery is connected? (c) How long would it take to charge the capacitor if this current remained constant?

PROBLEMS

68. Large electromagnets are sometimes wound with copper tubing through which water is circulated to carry away the heat produced. If such an electromagnet has a coil whose resistance is 0.80 Ω and through which 30 L/min of water flows, find the rise in temperature of the water when a current of 300 A is present in the coil. Assume that all the heat produced is absorbed by the water.

69. The insulation in an electric cord is damaged, and a short circuit occurs 2.0 m from a 120-V outlet. If all the power developed in the copper wires goes into heating them, how long will it take for the wires to melt? (Neglect the change in resistivity with temperature.)

70. (a) Find the potential difference between points a and b of the circuit in Fig. 18.34. (b) What resistor should be connected to the 16-Ω resistor to make $V_{ab} = 0$? How should the connection be made?

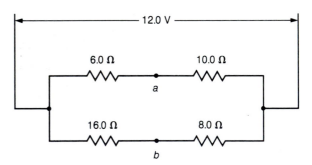

Figure 18.34

Problem 70.

71. Find the values of V_1 and V_2 and the current through the middle resistor of the circuit in Fig. 18.35. The internal resistances of the batteries are included in the values of the resistors.

Figure 18.35

Problem 71.

Figure 18.37

Problem 76.

72. A capacitor is charged by connecting it through a resistance to a battery. Verify that half the work done by the battery is dissipated as heat.

73. A certain 16-cell 32-V storage battery has an emf of 33.5 V when charged to 75% of its 250 A-h capacity. The internal resistance of the battery is 0.1 Ω. It is desired to charge the battery to its full capacity, when its emf will be 34.3 V. (a) What potential difference must be applied to the battery if it is to be charged at the initial rate of 40 A? (b) If this potential difference is held constant, what will be the rate of charge at the end of the process? (c) The emf of the battery arises from the conversion of chemical to electric energy; the higher potential difference is required for charging in order to pass a current through the battery and thereby produce chemical changes that store energy. Find the proportion of the power supplied during the charging process that is stored as chemical energy and the proportion that is dissipated as heat. (Assume an average battery emf during charging of 33.9 V.)

74. Three identical 1.5-V dry cells with internal resistances of 0.15 Ω are connected in parallel with an external 0.5-Ω resistor. How much current flows through the resistor?

75. Find the current in the 20-Ω resistor of the circuit in Fig. 18.36.

Figure 18.36

Problem 75.

76. Find the current in the 10-Ω resistor of the circuit in Fig. 18.37.

77. Find the values of R_1 and R_2 of the circuit in Fig. 18.38.

$V_1 = 50$ V	$I_1 = 5.0$ A	$R_1 = ?$
$V_3 = 20$ V	$I_3 = 2.0$ A	$R_2 = ?$
		$R_3 = 10.0 \Omega$

Figure 18.38

Problem 77.

78. Find the currents in the three resistors of the circuit in Fig. 18.39. The internal resistances of the batteries are included in the values of the resistors.

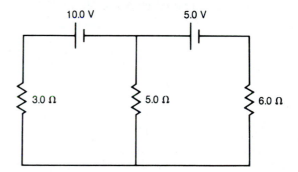

Figure 18.39

Problem 78.

79. Find the currents in each of the resistors of the circuit in Fig. 18.40. The internal resistances of the batteries are included in the values of the resistors.

Figure 18.40

Problem 79.

80. (a) Find the current in the 5.0-Ω resistor of the circuit in Fig. 18.41. (b) Find the potential difference between points a and b.

Figure 18.41

Problem 80.

81. Find the potential difference between points a and b and between points a and c of the circuit in Fig. 18.42.

Figure 18.42

Problem 81.

82. If points a and c of the circuit in Fig. 18.42 are connected, find the potential difference between points a and b.

83. A 2-μF capacitor is connected in series with a 6-μF capacitor and a 100-Ω resistor. (a) What is the time constant of the combination? (b) The combination is connected to a 24-V source for a time equal to the time constant and then is disconnected. What is the voltage across each capacitor?

Answers to Multiple Choice

1. a	**10.** c	**19.** d	**28.** a
2. a	**11.** b	**20.** a	**29.** b
3. c	**12.** b	**21.** d	**30.** c
4. b	**13.** d	**22.** a	**31.** b
5. c	**14.** b	**23.** a	**32.** d
6. d	**15.** b	**24.** b	**33.** c
7. a	**16.** c	**25.** c	**34.** b
8. c	**17.** b	**26.** b	**35.** a
9. d	**18.** a	**27.** c	

19

MAGNETISM

19.1   The Nature of Magnetism

19.2   Magnetic Field

19.3   Magnetic Field of a Current

19.4   Magnetic Properties of Matter

19.5   Force on a Moving Charge

19.6   Force on a Current

19.7   Force Between Two Currents

19.8   Torque on a Current Loop

19.9   Magnetic Poles

W e are all familiar with magnets and have marveled at their ability to attract nails and other iron or steel objects. Not all of us, however, appreciate how closely related electricity and magnetism are.

Charges at rest relative to an observer appear to exert only electrical forces on one another. When the charges are moving relative to the observer, though, the forces between them are different from before. These differences are by tradition said to arise from "magnetic" forces. Thus magnetic forces represent modifications to electrical forces, modifications that come about because of the motion of charges. Because electrical and magnetic effects are distinct from each other, we can legitimately consider electrical and magnetic forces separately and think in terms of separate electric and magnetic fields.

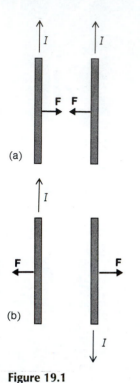

(a)

(b)

Figure 19.1

(a) Parallel electric currents in the same direction attract each other. (b) When the currents are in opposite directions, they repel each other.

19.1 THE NATURE OF MAGNETISM

Magnetic forces arise from the interactions of moving charges.

The forces that moving charges exert on one another are different from those the same charges exert when at rest. For instance, if we place a current-carrying wire parallel to another current-carrying wire, with the currents in the same direction, we find that the wires attract each other (Fig. 19.1). If the currents are in opposite directions, the wires repel each other.

Electrical forces cannot be responsible for these observations, because there is no net charge on a wire when it carries a current. The forces that come into being when electric currents interact are called **magnetic forces.** All magnetic effects can ultimately be traced to currents or, more exactly, to moving charges.

The gravitational force between two masses and the electrical force between two charges at rest are both **fundamental forces** in the sense that they cannot be accounted for in terms of anything else. On the other hand, the force a bat exerts on a ball is not fundamental, because it can be traced to the electrical forces between the atomic electrons of the bat and those of the ball.

What about magnetic forces? It is an important fact that whatever it is in nature that shows itself as an electrical force between stationary charges *must* also show itself as a magnetic force between moving charges. One effect is not possible without the other. Thus there is only a single fundamental interaction between charges, the **electromagnetic interaction,** which has two aspects, electric and magnetic.

Every electric charge has an electromagnetic field around it. The electric part of this field is always there, but the magnetic part appears only when relative motion is present. In the case of a current-carrying wire, there is only a magnetic field because the wire itself is electrically neutral. The electric fields of the electrons are canceled out by the opposite electric fields of the positive ions in the wire, but the ions are stationary and have no magnetic fields to cancel the magnetic fields of the moving electrons. If we simply move a wire that has no current flowing in it, the electric and magnetic fields of the electrons are canceled out by the electric and magnetic fields of the positive ions.

19.2 MAGNETIC FIELD

It is defined in terms of the force it exerts on a moving charge; its unit is the tesla.

We recall from Chapter 16 that the electric field **E** somewhere is defined in terms of the force **F** the field exerts on a stationary positive charge Q located there.

Because F is proportional to Q, it made sense to specify the magnitude of **E** by the ratio

$$E = \frac{F}{Q} \qquad \textit{Electric field magnitude}$$

The direction of **E** is that of **F**. Once we know what **E** is, we can easily find the magnitude and direction of the electric force on *any* charge that happens to be there.

The symbol for **magnetic field** is **B**. Because magnetic forces act only on moving charges, **B** is defined in terms of the magnetic force **F** exerted on a positive charge Q whose velocity is **v**. Experiment and theory both show that the magnetic force on a moving charge is proportional to two factors. One is the product Qv. The larger the charge and the faster it moves, the greater the magnetic force.

The other factor concerns direction. When a charge is in a magnetic field, there is always a certain line along which it can move with no magnetic force acting on it. The direction of **B** is taken to lie along this line. When **v** is at the angle θ with respect to **B,** the magnetic force on the charge is found to be proportional to $\sin \theta$. Thus F is a maximum at $\theta = 90°$—the maximum force occurs for motion perpendicular to **B.** Because F depends on $Qv \sin \theta$, the magnitude B of the magnetic field responsible for **F** is defined, by analogy with the definition of electric field $E = F/Q$, as

$$B = \frac{F}{Qv \sin \theta} \qquad \textit{Magnetic field magnitude} \quad (19.1)$$

We note that $v \sin \theta$ is the component of **v** perpendicular to **B.** When **v** is itself perpendicular to **B,** $\sin \theta = \sin 90° = 1$ and

$$B = \frac{F}{Qv} \qquad (\mathbf{v} \perp \mathbf{B}) \qquad\qquad\qquad (19.2)$$

Although **B** at a point lies along the line through the point where $F = 0$, two opposite directions are possible along that line. Hence more has to be said about the direction of **B.** Let us consider a charge moving perpendicularly to that line. The direction of **B** is, by convention, given by a right-hand rule (Fig. 19.2):

Open your right hand so that the fingers are together and the thumb sticks out. When your thumb is in the direction of v and your palm faces in the direction of F, your fingers are in the direction of B.

(An easy way to remember this rule is to associate the outstretched thumb with hitchhiking, and so with velocity; the palm with pushing on something, and so with force; and the parallel fingers with magnetic field lines.)

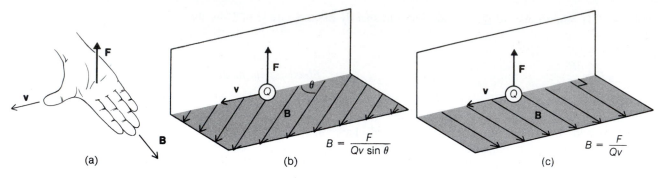

Figure 19.2

(a) Right-hand rule for the direction of the magnetic field **B** that exerts the force **F** on a positively charged particle of velocity **v.** If the particle has a negative charge, the force is in the opposite direction. (b) The magnitude of **B** is defined in terms of F, Q, and v. (c) When $\mathbf{v} \perp \mathbf{B}$, $\sin \theta = \sin 90° = 1$.

This completes an operational definition of **B,** since we now can find **B** in any region of space by performing suitable experiments. Such an experiment might use a cathode-ray tube in which the electron beam is deflected by the magnetic field as well as by an electric field, as described in Example 19.1 below.

The unit of magnetic field is, from its definition, the newton/ampere-meter (N/A · m), because the units of Qv are coulomb-meter/second = ampere-meter. The name **tesla,** abbreviated T, has been given to this unit:

$$1 \text{ tesla} = 1 \text{ T} = 1 \text{ N/A} \cdot \text{m}$$

Figure 19.3

Some representative values of magnetic field. The earth's magnetic field is discussed in Section 19.9.

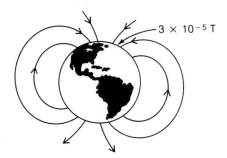

The magnitude of the earth's magnetic field at sea level is about 3×10^{-5} T.

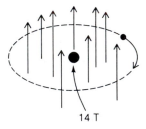

The magnetic field produced at the nucleus of a hydrogen atom by the electron circling around it is about 14 T.

The magnetic field near a strong permanent magnet is about 0.1 T.

The most powerful magnetic fields achieved in the laboratory have magnitudes in the neighborhood of 100 T.

Thus a force of 1 N will be exerted on a charge of 1 C when it is moving at 1 m/s perpendicular to a magnetic field whose magnitude is 1 T. Another unit of **B** in common use is the **gauss,** where

$$1 \text{ gauss} = 10^{-4} \text{ T} \quad \text{or} \quad 1 \text{ T} = 10^4 \text{ gauss}$$

Nikola Tesla (1856–1943), a pioneering electrical engineer, was born in what is now Yugoslavia and moved to the United States in 1884; Karl Friedrich Gauss (1777–1855) was a German mathematician and physicist.

Figure 19.3 shows some representative values of magnetic field that may help to give a feeling for the magnitude of the tesla.

◆   **E X A M P L E   19.1** _____

Figure 19.4 shows a cathode-ray tube (see Section 17.4) whose beam consists of electrons that have been accelerated through a potential difference of 1000 V. A magnetic field **B** applied to the region between the deflection plates bends the beam out of the paper, but this effect is canceled out when the electric field E between the plates is 10^4 V/m. Find the direction and magnitude of **B**.

SOLUTION   Since electrons have negative charges, the direction of **B** must be opposite to that given by the right-hand rule, so **B** is downward. To find B, we start with the speed v of the electrons. Since the electron kinetic energy $\frac{1}{2}mv^2$ is equal to the energy eV they gain in the electron gun,

$$\frac{1}{2}mv^2 = eV$$

$$v = \sqrt{\frac{2eV}{m}}$$

(We must be careful not to confuse eV, the product of the electron charge e and the potential difference V, with eV, the abbreviation for the electron volt, which is a unit of energy.)

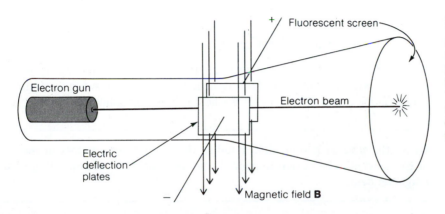

Figure 19.4

Experimental arrangement to determine **B.** The electric field between the plates is adjusted until the electron beam is undeflected. The same procedure can be used with a known magnetic field to determine the charge-to-mass ratio e/m of the electron.

When the electrical force eE on the electron balances the magnetic force evB (since $\theta = 90°$ here),

$$eE = evB$$

$$B = \frac{E}{v}$$

Substituting for v gives

$$B = \frac{E}{v} = E \sqrt{\frac{m}{2eV}} = (10^4 \, \text{V/m}) \sqrt{\frac{9.1 \times 10^{-31} \, \text{kg}}{2(1.6 \times 10^{-19} \, \text{C})(10^3 \, \text{V})}}$$

$$= 5.3 \times 10^{-4} \, \text{T}$$

The same experimental arrangement can be used with a known B to find the ratio e/m between the charge and mass of the electron. This was first done in 1897 by the English physicist J. J. Thomson, whose finding that e/m is always the same provided the first definite evidence that "cathode rays" are actually streams of particles. Later work by Robert A. Millikan in the United States showed that electrons all have the same charge $-e = -1.60 \times 10^{-19}$ C, from which the electron mass could be established.   ◆

19.3 MAGNETIC FIELD OF A CURRENT

The stronger the current, the stronger the magnetic field.

Unfortunately it is usually rather difficult to calculate the magnetic field of a given current configuration. The results in the most important cases are given here.

Long, Straight Current

Figure 19.5 shows the form of the magnetic field around a long, straight wire that carries the current I. The magnetic field lines are concentric circles with the current at the center. The sense of the field is given by another right-hand rule:

Grasp the wire with your right hand so that your thumb points in the direction of the current. The curled fingers of that hand point in the direction of the magnetic field.

I $B = \dfrac{\mu I}{2\pi s}$

Figure 19.5

The field lines of the magnetic field around a long, straight current consist of concentric circles. The sense of the field is given by the right-hand rule.

The magnetic field a distance s from the wire has the magnitude

$$B = \frac{\mu I}{2\pi s} \qquad \qquad \textit{Long, straight current} \quad (19.3)$$

The greater the current and the closer one is to it, the stronger the magnetic field.

The constant μ is called the **permeability** of the medium in which the magnetic field exists. In free space

$$\mu_0 = 4\pi \times 10^{-7}\,\text{T} \cdot \text{m/A} = 1.257 \times 10^{-6}\,\text{T} \cdot \text{m/A}$$

so that

$$\frac{\mu_0}{4\pi} = 1.000 \times 10^{-7}\,\text{T} \cdot \text{m/A}$$

The value of μ in air is very close to μ_0; we will assume they are the same here. The unit of permeability is sometimes expressed in other ways—for instance, as N/A^2.

◆ **E X A M P L E   19.2** _____

Find the magnetic field in air 10 mm from a wire that carries a current of 1.0 A.

SOLUTION Since 10 mm $= 1.0 \times 10^{-2}$ m, we have (Fig. 19.6)

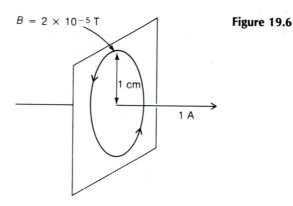

$B = 2 \times 10^{-5}\,\text{T}$

1 cm

1 A

Figure 19.6

$$B = \frac{\mu_0 I}{2\pi s} = \frac{(4\pi \times 10^{-7}\,\text{T} \cdot \text{m/A})(1\,\text{A})}{(2\pi)(1.0 \times 10^{-2}\,\text{m})} = 2.0 \times 10^{-5}\,\text{T}$$

This is only a little smaller than the magnitude of the earth's magnetic field. For this reason great care is taken aboard ships to keep current-carrying wires away from magnetic compasses.   ◆

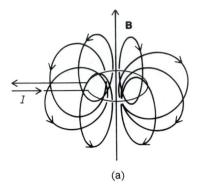

(a)   (b)   (c)

Figure 19.7

(a) The magnetic field around a circular current loop (b) At the center of the loop, **B** is perpendicular to the plane of the loop and its direction is given by the right-hand rule shown. (c) If there are N loops, the magnetic fields of the individual loops add up to give a field N times as strong as each one produces by itself.

Current Loop

The magnetic field around a circular current loop has the form shown in Fig. 19.7. At the center of the loop **B** is perpendicular to the plane of the loop with the magnitude

$$B = \frac{\mu I}{2r}$$ *Center of current loop*   (19.4)

where I is the current in the loop and r is its radius. The direction of **B** is given by still another right-hand rule:

Grasp the loop so that the curled fingers of your right hand point in the direction of the current; the thumb of that hand then points in the direction of B.

In the case of a flat coil of more than one loop, as in Fig. 19.7(c), the magnetic fields of each individual loop add up to give a proportionately stronger field. If there are N turns, then

$$B = \frac{\mu N I}{2r}$$ *Center of flat coil*   (19.5)

Solenoid

A **solenoid** is a coil of wire in the form of a helix (Fig. 19.8). If the turns are close together and the solenoid is long relative to its diameter, then the magnetic field inside it is uniform and parallel to its axis except near the ends. The direction of the field inside a solenoid is given by the same right-hand rule that gives the direction of **B** inside a current loop.

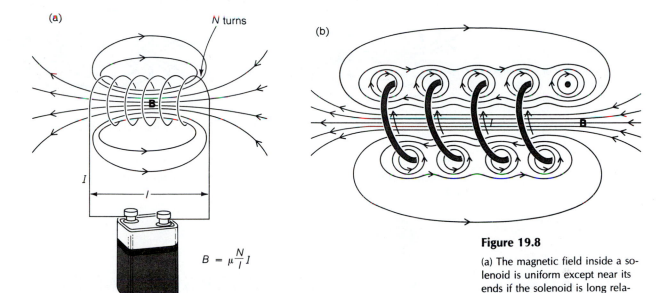

(a) *N* turns

(b)

$$B = \mu \frac{N}{l} I$$

Figure 19.8

(a) The magnetic field inside a solenoid is uniform except near its ends if the solenoid is long relative to its diameter and if its turns are close together. (b) An expanded view of a solenoid showing how the magnetic fields of the individual turns add together to give a uniform field inside it.

The magnetic field in the interior of a solenoid of the length l that has N turns of wire and carries the current I has the magnitude

$$B = \mu \frac{N}{l} I \qquad\qquad \textit{Interior of solenoid} \quad (19.6)$$

The diameter of the solenoid does not matter, provided it is small compared with the length l.

◆ **E X A M P L E 19.3** _____

A solenoid 20 cm long and 40 mm in diameter with an air core is wound with a total of 200 turns of wire. The solenoid's axis is parallel to the earth's magnetic field at a place where the latter is 3.0×10^{-5} T in magnitude. What should the current in the solenoid be for its field to exactly cancel the earth's field inside the solenoid?

SOLUTION In air $\mu = \mu_0$, and so, since $l = 0.20$ m here, the required current is

$$I = \frac{Bl}{\mu_0 N} = \frac{(3.0 \times 10^{-5}\,\text{T})(0.20\,\text{m})}{(4\pi \times 10^{-7}\,\text{T} \cdot \text{m/A})(2.0 \times 10^2)} = 0.024\,\text{A} = 24\,\text{mA}$$

The solenoid diameter here has no significance except as a check that the solenoid is long relative to its diameter. ◆

Figure 19.9

The magnetic fields of a bar magnet and of a solenoid are the same.

The magnetic field of a bar magnet is identical with that of a solenoid, which is not surprising since all permanent magnets owe their character to an alignment of atomic current loops no different in principle from the alignment of the current loops in a solenoid (Fig. 19.9). The behavior of permanent magnets is discussed in Section 19.9.

19.4 MAGNETIC PROPERTIES OF MATTER

Diamagnetism, paramagnetism, and ferromagnetism.

The strength of the magnetic field of a current-carrying solenoid is changed when a rod of almost any material is used as a core. Some materials increase B (for instance, oxygen and aluminum), others decrease B (for instance, mercury and bismuth), but in almost all cases the difference is very small. A few substances, however, give a dramatic increase in B—the new field may be hundreds or thousands of times stronger. Such substances are called **ferromagnetic** (Fig. 19.10). Iron is the most familiar example, but nickel, cobalt, and certain alloys are also ferromagnetic, as are ceramics called **ferrites.**

Figure 19.10

(a) Magnetic field of a solenoid with no core. (b) Using a ferromagnetic core increases the magnetic field considerably.

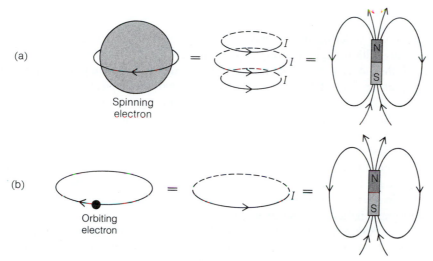

Figure 19.11

Sources of atomic magnetism.

The magnetic properties of matter can be traced almost entirely to atomic electrons; the contribution of the nucleus is very minor. The magnetic behavior of an atomic electron has two origins:

1. An electron in certain respects resembles a spinning charged sphere, which we may imagine as a series of ultrasmall current loops, as in Fig. 19.11(a). Hence every electron has the magnetic field of a tiny bar magnet.
2. When it is part of an atom, an electron may be thought of as circling the nucleus much like a planet circling the sun. As in the case of electron spin, this is a crude model of the actual situation, but it is adequate for many purposes. An orbiting electron is a current loop, and so the result is again the magnetic field of a tiny bar magnet (Fig. 19.11b).

In most substances whose atoms or molecules have an even number of electrons, the various magnetic fields of the electrons cancel each other out in pairs when there is no external magnetic field. (If we shake an even number of small bar magnets together in a box, they will also end up paired off with opposite poles together.) When an external field is present, however, the orbital motions of the electrons are affected. By Lenz's law, described in Section 20.3, the resulting changes in the orbital magnetic fields are such as to oppose the external field, an effect called **diamagnetism.** Diamagnetism thus reduces the strength of a magnetic field.

In other substances one or more electrons per atom or molecule have spin magnetic fields that are not canceled out. In an external magnetic field, the fields of these electrons tend to line up to enhance the external field. This effect is called **paramagnetism.** The increase in B due to paramagnetism (when it is present) is always greater than the decrease due to diamagnetism (which is always present), so the net result is an increase in B. The increase is usually small, however, because the constant thermal agitation of the atoms or molecules prevents complete alignment of the spin magnetic fields with the external magnetic field.

Electromagnet used to load scrap iron and steel.

Ferromagnetism

In a ferromagnetic material the unpaired electrons in each atom interact strongly with those in adjacent atoms. This causes the unpaired spin magnetic fields in all the atoms to be locked together. Atoms in a ferromagnetic material are accordingly grouped together in assemblies called **domains,** typically about 5×10^{-5} m across and just visible in a microscope. In an unmagnetized sample the directions of magnetization of the domains are random, although in each domain the unpaired electron spins are parallel. When such a sample is placed in an external magnetic field, either the spins within the domains turn to line up with the field or, in pure and homogeneous materials, the domain walls change so that those domains already lined up with the field grow at the expense of the others (Fig. 19.12). When all the unpaired spins in a ferromagnetic sample are lined up, no further increase in B is possible, and the sample is said to be **saturated.**

Above a certain temperature (770°C in the case of iron), the atoms in a domain have enough kinetic energy to overcome the interatomic forces that align their spins. The spins and their magnetic fields then become randomly oriented, and the ferromagnetic material loses its special magnetic properties. Thus heating a ''permanent magnet'' can demagnetize it.

Iron, unlike steel (which is an alloy, or mixture, of iron with carbon and other elements), loses nearly all of its magnetization when an external magnetic field is removed. Hence if an iron rod is placed inside a solenoid, we have a very strong magnet that can be turned on and off just by switching the current in the solenoid on and off. Such an **electromagnet** is much stronger than the solenoid itself and can be

Figure 19.12

(a) Magnetization of a ferromagnetic material by domain alignment. (b) Magnetization by domain growth.

Figure 19.13

A relay is a switch operated by an electromagnet. A small current in the solenoid is enough to pull the armature down against the spring and close the contact points. When the current in the solenoid is cut off, the spring pushes the armature up to break the secondary circuit. The relay of a heating system is actuated by a thermostat whose contacts close when the room temperature drops below a preset value. Because little current is needed to operate the relay, the thermostat can be quite small, and light wiring is sufficient between it and the relay. Large contact points can be provided on the relay to carry the current needed by the oil burner and the blowers or pumps that circulate the hot air or hot water it produces.

stronger than a permanent magnet as well. Also, unlike a permanent magnet, its field can be controlled at will by adjusting the current in the solenoid.

Electromagnets are among the most widely used electrical devices. They range in size from the tiny one in a telephone receiver that causes a steel plate to vibrate and thus produce the sounds we hear to the giant electromagnets used to pick up cars in scrap yards. A **relay** is a switch operated by an electromagnet (Fig. 19.13).

Hysteresis

Let us see what happens to its magnetic field when the current in an iron-core solenoid is increased.

Figure 19.14 is a plot of B (the total field, including the contribution of the iron core) against B_0 (the field due to the solenoid itself, without the core). Because B is so much greater than B_0, the respective scales on the graph are different. At first, B_0 is too weak to align the domains in the iron to any great extent, so B increases slowly with increasing B_0. When B_0 is made stronger, it is more effective in aligning the domains, and B rises rapidly until it is over 5500 times greater than B_0. Eventually, when the domains are virtually all aligned with B_0, B levels off to a nearly constant saturated value. Clearly μ, the permeability of the iron, is not a fixed quantity but varies with the magnetizing field B_0.

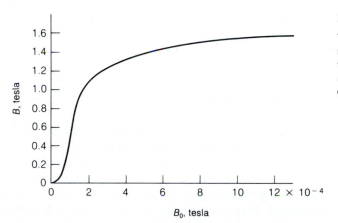

Figure 19.14

The magnetization curve of annealed iron. B_0 is the magnetic field without the iron present, and B is the total field including the effect of the iron.

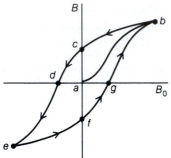

Figure 19.15

Hysteresis loop.

A ferromagnetic material tends to keep some of its magnetization even when the magnetic field that originally aligned its domains is removed. Hence μ does not even have a fixed value at a given B_0 but may take on different values depending on its past history. This phenomenon is called **hysteresis.**

Suppose we vary the current in a ferromagnetic core solenoid from zero to a maximum in one direction, down through zero to a maximum in the other direction, back through zero to the first maximum, and so on. Figure 19.15 is a plot of B versus B_0 for this cycle. When we first turn on the current, the B-B_0 curve is the same as in Fig. 19.14. At the point b we reduce the current, so that B_0 drops, but now B does *not* retrace its original path. From b to c the values of B are higher than they were from a to b at corresponding values of B_0 owing to the magnetic "memory" of the core. At c there is no current in the coil and $B_0 = 0$, yet B still has the magnitude B_c. The ferromagnetic sample is now permanently magnetized.

Reversing the direction of B_0 does not at first reverse B but merely reduces it, until finally, at d, B_0 is sufficiently negative to bring B to $B = 0$. A further increase in $-B_0$ takes the B-B_0 curve to e, where B and B_0 are equal in magnitude and opposite in sign to what they were at b. When B_0 is again brought to $B_0 = 0$, B is at f, where $B_f = -B_c$. Increasing B_0 in the positive sense returns B_0 to point b, but along a curve on which B is always less than it was from a to b. Further cycles simply retrace the curve $bcdefgb$.

The hysteresis curve of the iron used in electromagnet cores is narrow, so that ac is short and the field B_c that remains when B_0 is reduced to 0 is very small.

19.5 FORCE ON A MOVING CHARGE

The magnetic force is a sideways push.

A magnetic field can exert a force on an electric current, whether it is a current in a wire, a moving charged particle, or an atomic current such as those in a piece of iron. Whenever we switch on an electric motor, we see this property in action.

According to Eq. (19.1) the force on a particle of charge Q and velocity $\mathbf{v}$ in the magnetic field $\mathbf{B}$ has the magnitude

$$F = QvB \sin \theta \qquad\qquad\qquad \textit{Force on moving charge} \quad (19.7)$$

where θ is the angle beween $\mathbf{v}$ and $\mathbf{B}$ (Fig. 19.2b). The direction of the force $\mathbf{F}$ is given by the right-hand rule shown in Fig. 19.2(a).

The work done by a force on a body depends on the component of the force in the direction the body moves. Because the force on a charged particle in a magnetic field is perpendicular to its direction of motion, the force does no work on it. Hence the particle keeps the same speed v and energy it had when it entered the field, even

Figure 19.16

The path of a charged particle moving perpendicularly to a uniform magnetic field is a circle.

though it is deflected. On the other hand, the speed and energy of a charged particle in an *electric* field are always affected by the interaction between the field and the particle, unless **v** is perpendicular to **E.**

A particle of charge Q and velocity **v** that is moving in a uniform magnetic field so that **v** is perpendicular to **B** experiences a force of magnitude

$$F = QvB \qquad (\mathbf{v} \perp \mathbf{B}) \tag{19.8}$$

since $\sin 90° = 1$. This force is perpendicular to both **v** and **B,** so the particle travels in the circular path shown in Fig. 19.16.

To find the radius R of the circular path of the charged particle, we note that the magnetic force QvB provides the particle with the centripetal force mv^2/R that keeps it moving in a circle. Setting equal the magnetic and centripetal forces gives

$$F_{\text{magnetic}} = F_{\text{centripetal}}$$

$$QvB = \frac{mv^2}{R}$$

The radius of the particle's circular path is therefore

$$R = \frac{mv}{QB} \qquad\qquad \textit{Orbit radius} \tag{19.9}$$

The greater the momentum mv, the larger the circle, and the stronger the field, the smaller the circle.

A charged particle that moves parallel to a magnetic field experiences no force and is not deflected. The same particle moving perpendicularly to the field follows a circular path. What about the case when the particle moves at any other angle with respect to **B**? If we call $\mathbf{v}_\parallel$ the component of the particle's velocity **v** that is parallel to **B** and $\mathbf{v}_\perp$ the component of **v** perpendicular to **B,** then the motion of the particle is the resultant of a forward motion at the velocity $\mathbf{v}_\parallel$ and a circular motion perpendicular to this whose radius is $mv_\perp/QB$. As shown in Fig. 19.17, the path is a helix.

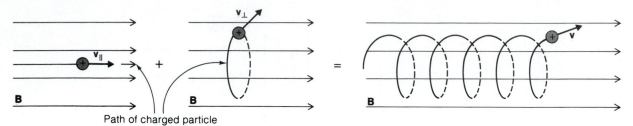

Figure 19.17

A charged particle that has velocity components both parallel and perpendicular to a magnetic field follows a helical path in the field.

◆   **E X A M P L E   19.4**

A long, straight wire carries a current of 100 A. (a) What is the force on an electron moving parallel to the wire, in the opposite direction to the current, at a speed of 1.0×10^7 m/s when it is 10 cm from the wire? (b) What is the force on the electron under the same circumstances when it is moving perpendicularly toward the wire?

SOLUTION   (a) The magnetic field 0.10 m from the wire is, from Eq. (19.3),

$$B = \frac{\mu_0 I}{2\pi s} = \frac{(4\pi \times 10^{-7}\,\text{T}\cdot\text{m/A})(100\,\text{A})}{(2\pi)(0.10\,\text{m})} = 2.0 \times 10^{-4}\,\text{T}$$

Since $\mathbf{v} \perp \mathbf{B}$ here, the force on the electron has the magnitude

$$\begin{aligned} F = evB &= (1.6 \times 10^{-19}\,\text{C})(1.0 \times 10^7\,\text{m/s})(2.0 \times 10^{-4}\,\text{T}) \\ &= 3.2 \times 10^{-16}\,\text{N} \end{aligned}$$

Because the electron has a negative charge, the direction of the force is opposite to that given by the right-hand rule of Fig. 19.2 and so is toward the wire (Fig. 19.18).

Figure 19.18

(b) Here $\mathbf{v} \perp \mathbf{B}$ also, so the force has the same magnitude. The direction of the force, however, is now the same as that of the current.   ◆

◆   **E X A M P L E   19.5**

A **mass spectrometer** is an instrument that measures atomic and molecular masses. The first step in the operation of the simple mass spectrometer shown in Fig. 19.19

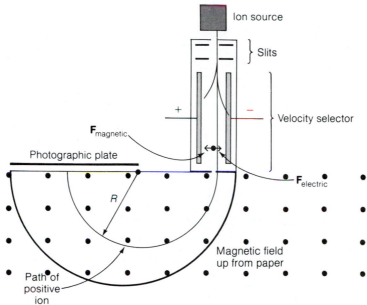

Ion source

Slits

Velocity selector

$\mathbf{F}_{\text{magnetic}}$

Photographic plate

$+$ $-$

$\mathbf{F}_{\text{electric}}$

R

Magnetic field
up from paper

Path of
positive
ion

Figure 19.19

A simple mass spectrometer.
Modern instruments use electrical
ion detectors and more complicated fields.

is to produce positive ions of the substance under study, which is often done by
bombardment with electrons. The ions have charges of $+e$ and are then accelerated
by an electric field. (Ions with other charges are sometimes present but are easy to
take into account.) A pair of slits eliminates ions not moving in the desired direction.
Then the beam passes through a **velocity selector** that consists of uniform electric
and magnetic fields perpendicular to each other and to the ion beam. Only ions with
a specific speed v that depends on the values of E and B are not deflected. Once past
the velocity selector, the ions follow curved paths in the magnetic field whose radii
depend on their masses. (a) The velocity selector of a mass spectrometer has crossed
fields of $E = 40.0$ kV/m and $B = 0.0800$ T. What is the speed of the ions that pass
through it? (b) Ions of a certain lithium isotope have radii of curvature in the same
magnetic field of 390 mm after leaving the velocity selector. What is their mass?

SOLUTION   (a) The electric field $\mathbf{E}$ exerts the force eE on the ions to the right, and
the magnetic field $\mathbf{B}$ exerts the force evB on them to the left. The condition for no
deflection is then $F_{\text{electric}} = F_{\text{magnetic}}$, $eE = evB$, and so

$$v = \frac{E}{B} = \frac{4.00 \times 10^4 \text{ V/m}}{8.00 \times 10^{-2} \text{ T}} = 5.00 \times 10^5 \text{ m/s}$$

(b) From Eq. (19.9) the mass of the ions is

$$m = \frac{QBR}{v} = \frac{(1.60 \times 10^{-19} \text{ C})(8.00 \times 10^{-2} \text{ T})(0.390 \text{ m})}{5.00 \times 10^5 \text{ m/s}}$$
$$= 9.98 \times 10^{-27} \text{ kg} \qquad \blacklozenge$$

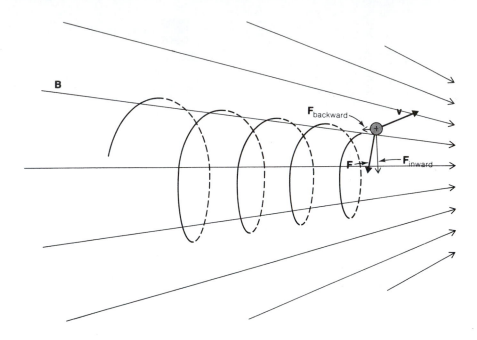

Figure 19.20

The principle of the magnetic mirror.

Magnetic Mirrors

Something very interesting happens when a charged particle moving in a magnetic field approaches a region where the field becomes stronger. The magnetic field lines that describe such a field converge, because their spacing is proportional to the magnitude of the field they describe. The force the particle experiences now has a backward component as well as the inward component that leads to its helical path, as shown in Fig. 19.20. The backward force may be strong enough and extend over a long enough distance to reverse the particle's direction of motion. A converging magnetic field can thus act as a **magnetic mirror.**

Magnetic mirrors are found both in the laboratory and in nature. In the laboratory a pair of them can be used as a ''magnetic bottle,'' as in Fig. 19.21, to contain a hot plasma in research on thermonuclear fusion. If a solid container were

Figure 19.21

A ''magnetic bottle.''

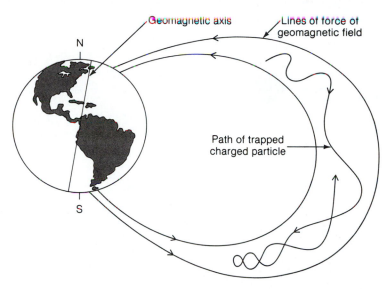

Geomagnetic axis

Lines of force of geomagnetic field

N

Path of trapped charged particle

S

Figure 19.22

Protons and electrons are trapped by the earth's magnetic field.

used, contact with its walls would contaminate the plasma and also cool it so that the ions would not have enough energy to interact. Magnetic bottles of this kind are somewhat leaky because ions moving along the axis of a magnetic mirror experience no backward force and hence can escape.

The earth's magnetic field traps electrons and protons from space in the **magnetosphere,** a giant doughnut-shaped magnetic bottle that surrounds the earth and extends from about 1000 km above the equator out to perhaps 65,000 km. The magnetosphere contains large numbers of particles with relatively high energies (100 MeV, for instance). Figure 19.22 shows a typical particle path in the magnetosphere.

19.6 FORCE ON A CURRENT

A magnetic field has the same effect on a current element as on a moving charge.

Because an electric current is a flow of charge, we would expect a current-carrying wire to be affected by a magnetic field in the same way a moving charged particle is. According to Eq. (19.1) the force on a charge Q whose velocity is $\mathbf{v}$ when it is in the magnetic field $\mathbf{B}$ has the magnitude

$$F = QvB \sin \theta \tag{19.1}$$

where θ is the angle between $\mathbf{v}$ and $\mathbf{B}$. What we have to do is replace the Qv of this formula with the quantity appropriate for a current.

(a)

(b)

Figure 19.23

The force on a charge Q moving with the speed *v* in a magnetic field is the same as that on a wire ΔL long carrying the current *I*, where *I* ΔL = Qv.

Figure 19.23 (a) shows a particle of charge Q and speed v. In the time t the particle travels the distance

$$\Delta L = vt$$

and while it does so, it is equivalent to a current of

$$I = \frac{Q}{t}$$

Hence

$$v = \frac{\Delta L}{t} \quad \text{and} \quad Q = It$$

so that

$$Qv = I\, \Delta L$$

Figure 19.24

The right-hand rule for the direction of the force on a current-carrying wire in a magnetic field.

We conclude that the force on an element ΔL long of current I when it is in a magnetic field **B** has the magnitude

$$F = I\, \Delta LB \sin \theta \qquad \textit{Force on current element} \quad (19.10)$$

where θ is the angle between the direction of I and that of **B** (Fig. 19.23b).

There are two ways to figure out the direction of the force on a current element in a magnetic field. Both give the same result, of course, and deciding which one to use in a particular case is largely a matter of personal preference. The first is the same as the right-hand rule used for a moving charge in Section 19.2, except that now the thumb points in the direction of the current. This version of the rule is illustrated in Fig. 19.24.

The other way to find the direction of **F** is based on the pattern of magnetic field lines around a current in a magnetic field. Figure 19.25(a) shows the field lines of a uniform field **B**, and (b) shows the field lines around a wire carrying a current I. Since the current is into the paper, the field lines are concentric circles in the clockwise sense. When field (a) is added vectorially to field (b), the resulting pattern of field lines is like that shown in (c): The lines are closer together in the region above the wire where the field of the current is in the same direction as **B** and farther

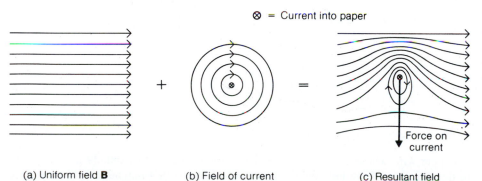

Figure 19.25

The direction of the force on a current element in a magnetic field is from the region of strong field to the region of weak field.

(a) Uniform field **B**   (b) Field of current   (c) Resultant field apart under the wire where the field of the current is opposite to **B.** *The direction of the force on the current element is from the region of strong field to the region of weak field,* as though the magnetic field lines were rubber bands that try to straighten out when distorted by the presence of the current.

The pictorial method for establishing the direction of **F** is easy to use and appeals to the intuition. However, we must keep in mind that field lines do not physically exist but are only a way to visualize the magnitude and direction of a force field. It is the field itself that exists in space as a continuous property of the region it occupies, not a series of strings. Despite the artificial nature of field lines, they can still be very helpful in representing various aspects of the interaction between magnetic fields and electric currents, and we can use them for this purpose whenever appropriate.

◆ E X A M P L E **19.6** ———————————————————————

A wire carrying a current of 100 A due west is suspended between two towers 50 m apart. The lines of force of the earth's magnetic field enter the ground there in a northerly direction at a 45° angle; the magnitude of the field at that location is 5.0×10^{-5} T. Find the force on the wire exerted by the earth's field.

SOLUTION   The wire is perpendicular to **B,** and so the magnitude of the force is

$$F = I \, \Delta LB \sin \theta = (100 \text{ A})(50 \text{ m})(5.0 \times 10^{-5} \text{ T})(\sin 90°) = 0.25 \text{ N}$$

This force is about an ounce. By either of the methods described above, the force acts downward at a 45° angle with the ground toward the south (Fig. 19.26).

Figure 19.26

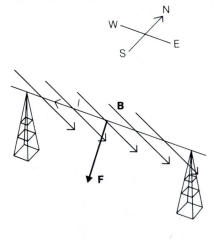

◆ E X A M P L E **19.7** ———————————————————————

A copper wire whose linear density is 10 g/m is stretched horizontally perpendicular to the direction of the horizontal component of the earth's magnetic field at a place where the magnitude of that component is 2.0×10^{-5} T. What must the current in the wire be for its weight to be supported by the magnetic force on it? What do you think would happen to such a wire if this current were passed through it?

SOLUTION   Here $\theta = 90°$ and $\sin \theta = 1$, so Eq. (19.10) becomes $F = I \, \Delta LB$. The magnetic force per meter on the wire is $F/\Delta L$ and must be equal and opposite to the wire's weight per meter $(m/L)g$. Hence $F/\Delta L = IB = (m/L)g$ and

$$I = \frac{(m/L)g}{B} = \frac{(1.0 \times 10^{-2}\,\text{kg/m})(9.8\,\text{m/s}^2)}{2.0 \times 10^{-5}\,\text{T}} = 4.9 \times 10^3\,\text{A}$$

A current of nearly 5000 A is very high for such a small-diameter wire, which leads to the question of whether the wire can carry it without melting. It is left for the reader's amusement to show that the wire would indeed melt a fraction of a second after the current is switched on.   ◆

Magnetic forces are being used to support (''levitate'') and propel high-speed experimental trains. By eliminating friction (except for air resistance), a ''maglev'' train has already exceeded 500 km/h, over 300 mi/h. Two systems are currently being developed. In Germany attractive magnets lift the undercarriage of the train to about a centimeter from a guiding T-shaped track. Very precise control is needed to keep the train floating correctly. In Japan repulsive magnets are used, a more stable method because the closer the train gets to its guiding track, the more the upward push. Stronger magnetic fields are required, however, which means more expensive superconducting electromagnets to produce them. In both systems an alternating current passed through electromagnets under the track creates an attractive force in front of the train's own magnets and a repulsive force behind them. The higher the frequency of the current, the higher the train's speed.

The German *Transrapid* train, using magnetic forces for both support and propulsion, has reached a speed of 413 km/h.

19.7   FORCE BETWEEN TWO CURRENTS

Parallel currents attract, antiparallel currents repel.

Every current has a magnetic field around it, and as a result nearby currents exert magnetic forces on one another.

Figure 19.27 shows two parallel wires a distance s apart that carry the currents I_1 and I_2 respectively. The magnetic fields a distance s from each of the wires are, from Eq. (19.3),

$$B_1 = \frac{\mu I_1}{2\pi s} \qquad B_2 = \frac{\mu I_2}{2\pi s}$$

The fields are perpendicular to the wires, which means that $\theta = 90°$ and $\sin \theta = 1$. Therefore the force F_{12} on a length L of current 1 exerted by the magnetic field of current 2 is

$$F_{12} = I_1 \, \Delta L B_2 \sin \theta = \frac{\mu I_1 I_2}{2\pi s} L \qquad (19.11)$$

The force F_{21} on current 2 exerted by the magnetic field of current 1 has the same magnitude. Both forces can be expressed in the form

$$\frac{F}{L} = \frac{\mu I_1 I_2}{2\pi s} \qquad \textit{Force between parallel currents} \quad (19.12)$$

where F/L is the *force per unit length* each wire exerts on the other by virtue of its magnetic field. From the pattern of lines of force around each wire, it is clear that the forces are always opposite in direction and hence obey Newton's third law of motion, as they must. The forces are attractive when the currents are in the same direction and repulsive when they are in opposite directions.

◆ **EXAMPLE 19.8**

The cables that connect the starting motor of a car with its battery are 10 mm apart for a distance of 40 cm. Find the forces between the cables when the current in them is 300 A.

SOLUTION The currents in the cables are opposite in direction, and hence the forces are repulsive. Their magnitudes, taking $\mu = \mu_0$, are

$$F = \frac{\mu_0 I^2}{2\pi s} L = \frac{(4\pi \times 10^{-7}\,\text{T} \cdot \text{m/A})(300\,\text{A})^2(0.40\,\text{m})}{(2\pi)(1.0 \times 10^{-2}\,\text{m})} = 0.72\,\text{N}$$

This is 0.16 lb.   ◆

Equation (19.12) is used to define the ampere: An ampere is that current in each of two parallel wires 1 m apart in free space that produces a force on each wire of exactly 2×10^{-7} N per meter of length. (Thus $\mu_0 = 4\pi \times 10^{-7}$ T · m/A.) In turn, the coulomb is defined in terms of the ampere as that amount of charge transferred per second by a current of 1 A. The ampere is chosen as the primary electrical unit instead of the coulomb because it can be defined in terms of a more direct experiment than would be possible with the coulomb.

19.8 TORQUE ON A CURRENT LOOP

How an electric motor works.

No net force acts on a loop of current in a uniform magnetic field. Instead, a torque occurs that tends to turn the loop to bring its plane perpendicular to **B**. This effect underlies the operation of all electric motors, from the tiniest one in a clock to the many-thousand-horsepower giant in a locomotive.

Let us look at the forces on each side of a rectangular current-carrying wire loop whose plane is parallel to a uniform magnetic field **B**, as in Fig. 19.28(a). The sides *A* and *C* of the loop are parallel to **B**, so there is no magnetic force on them. Sides *B* and *D* are perpendicular to **B**, however, and each therefore experiences a force. To find the directions of the forces on *B* and *D*, we can use the right-hand rule: With the fingers of the right hand in line with **B** and the outstretched thumb in line with **I**, the palm faces the same way as **F**. What we find is that $\mathbf{F}_B$ is opposite in direction to $\mathbf{F}_D$. The same conclusion follows from the pattern of lines of force.

Figure 19.28

(a) A current-carrying wire loop whose plane is parallel to a magnetic field experiences a torque. (b) If the plane of the loop is perpendicular to the magnetic field, there is no torque on the loop. In both cases there is no net force on the loop.

⊙ Current out of paper
⊗ Current into paper

(a)

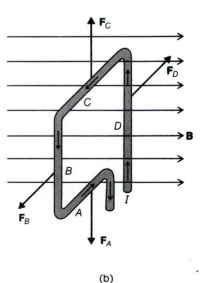

(b)

The forces $\mathbf{F}_B$ and $\mathbf{F}_D$ are the same in magnitude, so there is no net force on the current loop. Because $\mathbf{F}_B$ and $\mathbf{F}_D$ do not act along the same line, they exert a torque on the loop that tends to turn it. Although the origin of this torque is easiest to understand in the case of a rectangular loop, the torque occurs for a loop of any shape in a magnetic field when the plane of the loop is not perpendicular to the magnetic field.

If the plane of the loop is perpendicular to the magnetic field instead of parallel to it, there is neither a net force nor a net torque on it. This is easy to verify by applying the right-hand rule in Fig. 19.28(b). Evidently $\mathbf{F}_A$ and $\mathbf{F}_C$ cancel out each other out, and $\mathbf{F}_B$ and $\mathbf{F}_D$ also cancel out each other. There is no torque now because $\mathbf{F}_A$ and $\mathbf{F}_C$ have the same line of action, and $\mathbf{F}_B$ and $\mathbf{F}_D$ have the same line of action.

These results can be summarized by saying:

When the plane of a current loop in a magnetic field B is not perpendicular to B, the loop tends to turn so that its plane becomes perpendicular to B.

A detailed calculation shows that the torque τ on a wire loop in a magnetic field is given by

$$\tau = IAB \cos \theta \qquad \qquad \textit{Torque on current loop} \quad (19.13)$$

where

I = current in loop
A = area enclosed by loop
B = magnetic field
θ = angle between plane of loop and direction of $\mathbf{B}$

Although a rectangular loop is shown in Fig. 19.28, the loop's shape does not matter, only its area A.

If a coil of N loops is in the magnetic field instead of a single loop, the torque on it is

$$\tau = INAB \cos \theta \qquad \qquad \textit{Torque on coil} \quad (19.14)$$

The value of $\cos \theta$ is 1 when $\theta = 0$, and the torque has its maximum value of $INAB$ when the plane of the coil is parallel to $\mathbf{B}$, as in Fig. 19.28(a). When $\theta = 90°$, the plane of the coil is perpendicular to $\mathbf{B}$, as in Fig. 19.28(b), and $\cos \theta = \cos 90° = 0$, $\tau = 0$. The quantity INA is sometimes called the **magnetic moment** of the coil.

Galvanometers and Motors

A **galvanometer** is a device based on the effect just described that is used to measure electric current. As in Fig. 19.29, a U-shaped permanent magnet is used to provide a magnetic field, and between its poles is a small coil wound on an iron

Figure 19.29

The construction of a common type of galvanometer. Such galvanometers can respond to currents as small as 0.1 microampere (10^{-7} A). Even more sensitivity can be attained if the moving coil is suspended by a thin wire to which a small mirror is attached: Bearing friction is avoided in this way, and the mirror deflects a light beam so that the "pointer" may be a meter or more long instead of a few centimeters. Laboratory galvanometers like this can be used to measure currents of 10^{-10} A.

Figure 19.30

A simple dc electric motor. The commutator automatically reverses the current in the rotating loop twice per rotation so that the torque will stay in the same direction.

core to enhance the torque developed when the unknown current is passed through it. The coil assembly is held in place by two bearings, and a pair of hairsprings keeps the pointer at 0 when there is no current in the coil.

When a current flows, there is a torque on the coil because of the interaction between the current and the magnetic field, and the coil rotates as far as it can against the opposing torque of the springs. The more the current, the stronger the torque and the farther the coil turns. The restoring torque of the hairsprings is proportional to the angle through which they are twisted, and as a result the deflection of the pointer is directly proportional to the current I in the coil.

How does an electric motor achieve continuous motion? The torque a magnetic field exerts on a current loop disappears when the loop turns so that its plane is perpendicular to the field direction. If the loop swings past this position, the torque on it will be in the opposite sense and will return the loop to the perpendicular orientation. In order for a motor to rotate continuously, then, the current in the loop must be reversed each time it turns through 180°. The way this is done is shown in Fig. 19.30. The current is led to the loop by means of graphite rods called **brushes** that press against a split ring called a **commutator**. As the loop rotates, the current is reversed twice per turn as the commutator segments make contact alternately with the brushes. The torque is always in the same direction, except at the moments of switching when it is zero because the loop is perpendicular to the field. However, the angular momentum of the loop carries it past this point so it can go on turning.

Windings for
electromagnet

Armature

Lines of
force of
magnetic field

Wires embedded
in armature

Figure 19.31

Actual dc electric motors employ
various means to increase the
available torque.

While actual **direct-current** (dc) electric motors, such as the starter motor
of a car, are the same in principle as the simple device in Fig. 19.30, they use a
number of methods to increase the available torque. Electromagnets rather than
permanent magnets provide the field, and there are six or more different coils with
many turns each on a slotted iron core called an **armature,** instead of a single loop
(Fig. 19.31). A commutator with a pair of segments for each coil is provided so that
only those coils approximately parallel to the magnetic field receive current at any
time, which means that maximum torque is developed continuously.

Electric energy for industrial and domestic purposes is usually transmitted
by **alternating current** (ac), whose direction periodically reverses itself. In the United
States the frequency of ordinary ac is 60 Hz, which means that the current changes
direction 120 times per second. This frequency is 50 Hz in much of the rest of the
world. In ac electric motors, commutators and brushes are not needed because the
current itself does the required reversing, which makes such motors easier to build
and more reliable than dc motors.

Stationary windings of a large
electric motor.

In order to start the armature of an ac electric motor turning, and to ensure that it begins to turn the desired way, an auxiliary stationary winding is used. This winding, together with the operating winding, creates a magnetic field whose direction rotates about the motor's axis. This rotating field pulls the armature around when the motor is switched on. As the armature approaches its normal running speed, the alternating magnetic fields of the operating windings are sufficient to keep it going, and (depending on the motor design) the starting winding may then be cut out of the circuit.

19.9 MAGNETIC POLES

They come in pairs.

The magnetic field of a bar magnet is the same as that of a solenoid, as we saw in Fig. 19.9. Hence we can use the ideas of this chapter to understand how such magnets behave.

Because the external magnetic field of a bar magnet seems to originate in its ends, these are by custom called its **poles.** At one time it was believed that the poles were "magnetic charges" analogous to electric charges, and that the field of the magnet was due to these poles. This belief was reinforced by the repulsion of like poles and the attraction of unlike ones, effects that have their true explanation in the forces between parallel and antiparallel currents (Fig. 19.32).

An important difference between magnetic poles and electric charges is that magnetic poles always occur in pairs of equal strength and opposite polarity. If a magnet is sawed in half, the poles are not separated, but instead two new magnets

Figure 19.32

Interactions between magnets can be traced to interactions between current loops.

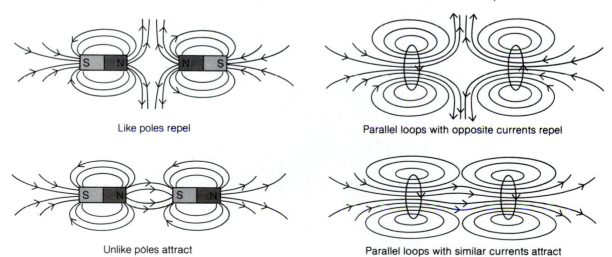

Like poles repel

Parallel loops with opposite currents repel

Unlike poles attract

Parallel loops with similar currents attract

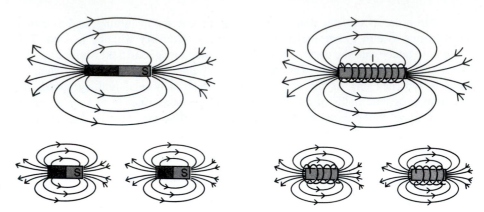

Figure 19.33

Cutting a magnet in half produces two new magnets.

are created, as in Fig. 19.33. Magnetic poles are therefore not really like electric charges, and all effects they are supposed to cause can be explained in terms of current-carrying solenoids.

Measurements of the earth's magnetic field show that it is very much like the field that would be produced by a powerful current loop whose center is a few hundred kilometers from the earth's center and whose plane is tilted by 11° from the plane of the earth's equator (Fig. 19.34). On the basis of geological evidence, the earth is thought to have a core of molten iron 3470 km (2160 mi) in radius, a little over half the earth's radius, and there is no doubt today that electric currents in this core are responsible for the observed geomagnetic field. Exactly how these currents came into being and how they are maintained are still uncertain, however.

As we saw, a current loop tends to turn in a magnetic field until its axis is parallel to the field. A bar magnet does the same. Because of the earth's magnetic field, a magnet suspended by a string therefore turns so as to line up in an approximately north-south direction. A compass consists of a pivoted magnetized iron needle together with a card that permits directions relative to magnetic north to be determined

Figure 19.34

The earth's magnetic field originates in currents in its core of molten iron. The magnetic axis is tilted by 11° from the axis of rotation.

North geomagnetic pole

North-seeking pole

South-seeking pole

Figure 19.35

A magnetic compass.

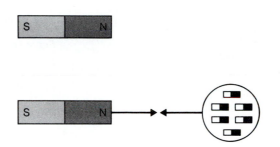

Figure 19.36

How a permanent magnet attracts a ferromagnetic object.

Ferromagnetic substance

(Fig. 19.35). The end of a freely swinging magnet that points toward the north is called its north-seeking pole, usually shortened to just **north pole.** The other end is its south-seeking pole, or **south pole.** *Magnetic field lines leave the north pole of a magnet and enter its south pole.* (The north geomagnetic pole is thus a south pole, and the south geomagnetic pole is a north pole. This has confused people for several hundred years.)

The mechanism by which a magnet or a solenoid attracts an iron object (or an object of any other ferromagnetic material such as cobalt or nickel) is very similar to the way in which an electric charge attracts an uncharged object. First the presence of the magnet causes the atomic magnets in the iron object to line up with its field by one of the mechanisms shown in Fig. 19.12, and then the attraction of opposite poles leads to a force on the object that draws it toward the magnet, as in Fig. 19.36.

Important Terms

Charged particles in motion relative to an observer exert forces on one another that are different from the electrical forces they exert when at rest. These differences are by custom said to arise from **magnetic forces.** In reality, magnetic forces represent modifications of electrical forces due to the motion of the charges involved.

A **magnetic field** exists wherever a magnetic force would act on a moving charged particle. The direction of a magnetic field **B** at a point is such that a charged particle would experience no force if it moved in that direction at the point. The magnitude of **B** is numerically equal to the force that would act on a charge of 1 C moving at 1 m/s perpendicularly to **B.** The unit of magnetic field is the **tesla** (T), equal to 1 N/A · m. When different substances are inserted in a current-carrying wire coil, the magnetic field in its vicinity changes. Those

substances that lead to a slight increase in B are called **paramagnetic,** those that lead to a great increase in B are called **ferromagnetic,** and those that lead to a slight decrease in B are called **diamagnetic.** A **permanent magnet** is an object composed of ferromagnetic material whose atomic current loops have been aligned.

The **permeability** of a medium is a measure of its magnetic properties. Paramagnetic and ferromagnetic substances have higher permeabilities than free space, and diamagnetic ones that lower permeabilities. The permeability of a ferromagnetic material in a given magnetizing field B_0 depends on its past history as well as on B_0, a phenomenon called **hysteresis.**

The ends of a permanent magnet are called its **poles.** Magnetic lines of force leave the **north pole** of a magnet and enter its **south pole.**

Important Formulas

Magnetic field: $B = \dfrac{F}{Qv \sin \theta}$

Field around long, straight current: $B = \dfrac{\mu I}{2\pi s}$

Field at center of flat coil: $B = \dfrac{\mu N I}{2r}$

Field inside a solenoid: $B = \mu \dfrac{N}{I} I$

Force on moving charge: $F = QvB \sin \theta$

Force on current element: $F = I\,\Delta LB \sin \theta$

Force between parallel currents: $\dfrac{F}{L} = \dfrac{\mu I_1 I_2}{2\pi s}$

MULTIPLE CHOICE

1. All magnetic fields originate in
  a. iron atoms
  b. permanent magnets
  c. magnetic domains
  d. moving electric charges

2. An observer moves past a stationary electron. His instruments measure
  a. an electric field only
  b. a magnetic field only
  c. both electric and magnetic fields
  d. any of the above, depending on his speed

3. Magnetic fields do not interact with
  a. stationary electric charges
  b. moving electric charges
  c. stationary permanent magnets
  d. moving permanent magnets

4. Magnetic field lines provide a convenient way to visualize a magnetic field. Which of the following statements is not true?
  a. The path followed by an iron particle moving in a magnetic field corresponds to a field line.
  b. The path followed by an electric charge moving in a magnetic field corresponds to a field line.
  c. A compass needle in a magnetic field lines up parallel to the field lines around it.
  d. Magnetic field lines do not actually exist.

5. A drawing of the field lines of a magnetic field provides information on
  a. the direction of the field only
  b. the magnitude of the field only
  c. both the direction and the magnitude of the field
  d. the source of the field

6. In a drawing of magnetic field lines, the stronger the field is,
  a. the closer together the field lines are
  b. the farther apart the field lines are
  c. the more nearly parallel the field lines are
  d. the more divergent the field lines are

7. The magnetic field near a strong permanent magnet might be
  a. 10^{-9} T   **b.** 10^{-5} T
  c. 0.1 T   **d.** 100 T

8. The earth's magnetic field at sea level typically is
  a. 3×10^{-9} T   **b.** 3×10^{-5} T
  c. 3×10^{5} T   **d.** 3×10^{9} T

9. A current is flowing east along a power line. If we neglect the earth's field, the direction of the magnetic field below it is
  a. north   **b.** east
  c. south   **d.** west

10. The magnetic field lines around a long, straight current are in the form of
  a. straight lines parallel to the current
  b. straight lines that radiate perpendicularly from the current, like the spokes of a wheel
  c. concentric circles centered on the current
  d. concentric helixes whose axis is the current

11. Inside a solenoid the magnetic field
  a. is zero
  b. is uniform
  c. increases with distance from the axis
  d. decreases with distance from the axis

12. The magnitude of the magnetic field inside a solenoid of N turns does not depend on
  a. the nature of the medium inside the solenoid
  b. the solenoid's length
  c. the solenoid's diameter
  d. the current in the solenoid

13. When a ferromagnetic substance is inserted in a current-carrying solenoid, the magnetic field is
  a. slightly decreased   **b.** greatly decreased
  c. slightly increased   **d.** greatly increased

14. Which of the following statements about the magnetic field of a solenoid with an iron core is (are) not always true?
   a. Increasing I increases **B**.
   b. Decreasing I decreases **B**.
   c. **B** = 0 when I = 0.
   d. Changing the direction of I changes the direction of **B**.

15. When a magnetized iron bar is strongly heated, its magnetic field
   a. becomes weaker
   b. becomes stronger
   c. reverses its direction
   d. is unchanged

16. An electron enters a magnetic field parallel to **B**. The electron's
   a. motion is unaffected
   b. direction is changed
   c. speed is changed
   d. energy is changed

17. An electron enters a magnetic field perpendicularly to **B**. The electron's
   a. motion is unaffected
   b. direction is changed
   c. speed is changed
   d. energy is changed

18. An ion moves in a circular orbit of radius R in a magnetic field. If the particle's speed is doubled, the orbit radius will become
   a. $R/2$
   b. R
   c. $2R$
   d. $4R$

19. An ion that moves through crossed electric and magnetic fields (that is, so that **E** $\perp$ **B**) perpendicularly to both **E** and **B** is not deflected when its speed is equal to
   a. EB
   b. E/B
   c. B/E
   d. B/E^2

20. Two parallel wires carry currents in the same direction.
   a. The wires attract each other.
   b. The wires repel each other.
   c. Torques act on the wires that tend to turn them so that they are perpendicular.
   d. Neither forces nor torques act on the wires.

21. A current-carrying wire is in a uniform magnetic field with the direction of the current the same as that of the field.
   a. There is a force on the wire that tends to move it parallel to the field.
   b. There is a force on the wire that tends to move it perpendicularly to the field.
   c. There is a torque on the wire that tends to rotate it until it is perpendicular to the field.
   d. There is neither a force nor a torque on the wire.

22. A current-carrying loop in a magnetic field always tends to rotate until the plane on the loop is
   a. parallel to the field
   b. perpendicular to the field
   c. either parallel or perpendicular to the field, depending on the direction of the current
   d. at a 45° angle with the field

23. The nature of the force responsible for the operation of an electric motor is
   a. electrical
   b. magnetic
   c. a combination of electrical and magnetic
   d. either electrical or magnetic, depending on the design of the motor

24. The magnetic field of a bar magnet most closely resembles the magnetic field of
   a. a straight current-carrying wire
   b. a stream of electrons moving parallel to one another
   c. a current-carrying wire loop
   d. a horseshoe magnet

25. A permanent magnet does not exert a force on
   a. an unmagnetized iron bar
   b. a magnetized iron bar
   c. a stationary electric charge
   d. a moving electric charge

26. The magnetic field shown in Fig. 19.37 is produced by
   a. a north pole and a south pole
   b. two north poles
   c. two south poles
   d. a south pole and an unmagnetized iron bar

Figure 19.37

Multiple Choice 26.

27. At different places on the earth's surface, the earth's magnetic field
   a. is the same in direction and magnitude
   b. may be different in direction but not in magnitude
   c. may be different in magnitude but not in direction
   d. may be different in both magnitude and direction

28. The magnetic field 2 cm from a long, straight wire is 10^{-6} T. The current in the wire is
   a. 0.01 A
   b. 0.1 A
   c. 1 A
   d. 10 A

29. The magnetic field inside a 100-turn solenoid 20 mm long that carries a 10-A current is

    a. 0.00063 T   **b.** 0.0013 T

    c. 0.063 T   **d.** 0.13 T

30. In a certain electric motor, wires that carry a current of 6 A are perpendicular to a magnetic field of 0.5 T. The force per centimeter on these wires is

    a. 0.03 N   **b.** 0.2 N

QUESTIONS

1. An experimenter is able to measure electric, magnetic, and gravitational fields. Which will be detected when (a) a proton moves past her, and (b) when she moves past a proton?

2. When you face the screen of a TV tube, in what direction is the magnetic field of the electron beam?

3. Figure 19.38 shows a current-carrying wire and a compass. In what direction does the compass needle point?

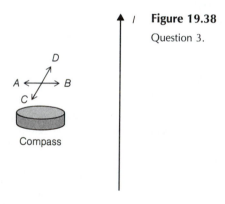

Figure 19.38

Question 3.

4. A solenoid has two windings, one on top of the other, each with the same length, number of turns, and resistance. The windings are such that their magnetic fields are in the same direction. Compare the magnetic field inside the solenoid when the windings are connected (a) in series and (b) in parallel to the same battery of negligible internal resistance.

5. An alternating current whose variation with time follows a sine curve is sent through an iron-core solenoid. Does the resulting magnetic field also vary sinusoidally with time? Explain.

6. What aspect of the magnetic field of a ''magnetic mirror'' enables it to reflect approaching charged particles?

7. An electron is moving vertically upward when it enters a magnetic field directed to the east. In what direction is the force on the electron?

8. What path is followed by an electron sent into a current-

    c. 3 N   **d.** 0.3 kN

31. Two parallel wires in free space are 10 cm apart and carry currents of 10 A each in the same direction. The force each wire exerts on the other per meter of length is

    a. 2×10^{-7} N, attractive

    b. 2×10^{-7} N, repulsive

    c. 2×10^{-4} N, attractive

    d. 2×10^{-4} N, repulsive

carrying solenoid along its axis? What is its path when the electron enters the solenoid at an angle relative to its axis?

9. A current-carrying wire is in a magnetic field. (a) What angle should the wire make with **B** for the force on it to be zero? (b) What should the angle be for the force to be a maximum?

10. A length of copper wire *AB* rests across a pair of parallel copper wires that are connected to a battery through a switch, as in Fig. 19.39. The arrangement is placed between the poles of a magnet, and the switch is closed. Does the wire move to the left or to the right, or rise upward, or press downward?

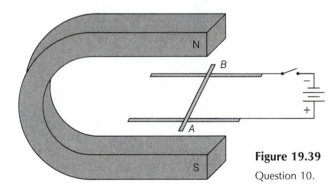

Figure 19.39

Question 10.

11. A stream of protons is moving parallel to a stream of electrons. Is the force between the two streams necessarily attractive? Explain.

12. A beam of protons, initially moving slowly, is accelerated to higher and higher speeds. What happens to the diameter of the beam during this process?

13. The electron beam in a cathode-ray tube is aimed parallel to a nearby wire carrying a current in the same direction. Is the electron beam deflected? If so, in what direction?

14. What, if anything, happens to the length of a helical spring when a current is passed through it?

15. A current is passed through a loop of highly flexible wire. What shape does the loop assume? Why?

16. What is the direction of the magnetic field that causes the wire loop in Fig. 19.40 to experience a clockwise torque about the z-axis?

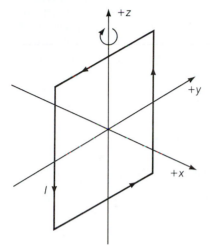

Figure 19.40

Question 16.

17. A current-carrying wire loop is in a uniform magnetic field. Under what circumstances, if any, will there be no torque on the loop? No net force? Neither torque nor net force?

18. A bar magnet is placed in a uniform magnetic field so that it is parallel to the field with its north pole in the field direction. Sketch the field lines of the resulting magnetic field **B** around the magnet. Indicate where **B** = 0.

19. The bar magnet in Question 18 is turned so that its north pole is now opposite to the direction of the uniform field. Sketch the field lines of the resulting magnetic field **B** around the magnet for this situation. Indicate where **B** = 0.

20. Why is a piece of iron attracted by *either* pole of a magnet?

21. Would you expect a compass to be more accurate near the equator or in the polar regions? Why?

22. What is believed to be the origin of the earth's magnetic field?

EXERCISES

19.3 Magnetic Field of a Current

1. At what distance from a long, straight wire carrying a current of 12 A does the magnetic field equal that of the earth, approximately 3.0×10^{-5} T?

2. A power line 10 m above the ground carries a current of 5000 A. Find the magnetic field magnitude of the current on the ground directly under the cable.

3. Two parallel wires 100 mm apart carry currents in the same direction of 8.0 A. Find the magnetic field magnitude midway between the wires.

4. Two parallel wires 100 mm apart carry currents in opposite directions of 8.0 A. Find the magnetic field magnitude midway between the wires.

5. Two parallel wires 200 mm apart carry currents in the same direction of 5.0 A. Find the magnetic field, magnitude between the wires 50 mm from one of them and 150 mm from the other.

6. Two parallel wires 200 mm apart carry currents in opposite directions of 5.0 A. Find the magnetic field magnitude between the wires 50 mm from one of them and 150 mm from the other.

7. What should the current be in a wire loop 10 mm in diameter if the magnetic field magnitude at the center of the loop is to be 0.0010 T?

8. A 10-turn circular coil of radius 20 mm carries a current of 0.50 A. Find the magnetic field magnitude at its center.

9. A long solenoid is wound with 30 turns/cm. What is the magnetic field magnitude inside the solenoid when the current is 3.8 A?

10. A solenoid 10 cm long is meant to have a magnetic field magnitude of 0.0020 T inside it when the current is 3.0 A. How many turns are needed?

11. The magnetic field inside a long current-carrying spring has the magnitude B. What is the field there when the spring is pulled out to twice its normal length?

12. A toroidal coil is a solenoid bent around into a doughnut shape with its axis in a circle of radius R. Find the magnetic field magnitude inside a toroidal coil of $R = 100$ mm that has 800 turns when the current is 0.50 A.

13. A tube 200 mm long is wound with 400 turns of wire in one layer and with 200 turns in a second layer in the opposite sense to the first. A current of 0.10 A is passed through the coils. What is the magnetic field magnitude in the interior of the tube?

14. A solenoid is wound with 18 turns/cm and carries a current of 1.2 A. Another layer of turns is wound over the solenoid with 10 turns/cm, and a current of 5.0 A is passed through the new coil, opposite to the direction of the current in the solenoid. What is the magnetic field magnitude in the interior of the solenoid?

15. A current of 0.50 A is passed through a solenoid wound with 20 turns/cm. A single loop of wire 10 mm in radius is bent around the middle of the solenoid. What should the current in this loop be in order for the magnetic field at its center to cancel out the field of the solenoid there?

19.5 Force on a Moving Charge

16. An electron in a television picture tube travels at 3×10^7 m/s and is acted on both by gravity and by the earth's magnetic field. Which exerts the greater force on the electron?

17. What is the radius of the path of an electron whose speed is 1.0×10^7 m/s in a magnetic field of magnitude 0.020 T when the electron's path is perpendicular to the field?

18. Find the minimum radius of curvature in the earth's magnetic field at sea level, assuming $B = 3.0 \times 10^{-5}$ T, of (a) a proton whose speed is 2.0×10^7 m/s and (b) an electron of the same speed.

19. A charge of $+1.0 \times 10^{-6}$ C is moving at 500 m/s along a path parallel to a long, straight wire and 100 mm from it. The wire carries a current of 2.0 A in the same direction as that of the charge. What is the magnitude and direction of the force on the charge?

20. A charge of $+2.0 \times 10^{-6}$ C is moving at 1.0×10^3 m/s at a distance of 120 mm away from a straight wire carrying a current of 4.0 A. Find the magnitude and direction of the force on the charge when it is moving parallel to the wire (a) in the same direction as the current and (b) in the opposite direction to the current.

21. An electron is moving at 6.0×10^7 m/s at a distance of 50 mm from a long, straight wire carrying a current of 40 A. Find the magnitude and direction of the force on the electron when it is moving parallel to the wire (a) in the same direction as the current and (b) in the opposite direction to the current.

22. The charge in Exercise 20 is moving in a direction perpendicular to the same wire. Find the magnitude and direction of the force on the charge when it is moving (a) toward the wire and (b) away from the wire.

23. The electron in Exercise 21 is moving in a direction perpendicular to the same wire. Find the magnitude and direction of the force on the electron when it is moving (a) toward the wire and (b) away from the wire.

24. An electron moving through an electric field of 500 V/m and a magnetic field of 0.10 T experiences no force. The two

PROBLEMS

33. Two long, parallel wires a distance s apart each carry the current I in opposite directions. Verify that the magnetic field

fields and the electron's direction of motion are all mutually perpendicular. What is the speed of the electron?

25. A velocity selector uses a magnet to produce a 0.050-T magnetic field and a pair of parallel metal plates 10 mm apart to produce a perpendicular electric field. What potential difference should be applied to the plates to permit singly charged ions of speed 5.0×10^6 m/s to pass through the selector?

26. A mass spectrometer has a velocity selector that consists of a magnetic field of 0.0400 T perpendicular to an electric field of 5.00×10^4 V/m. The same magnetic field is then used to deflect the ions. Find the radius of curvature of singly charged lithium ions of mass 1.16×10^{-26} kg in this spectrometer.

19.6 Force on a Current

27. A wire 1.0 m long is perpendicular to a magnetic field of 0.10 T. (a) What is the force on the wire when it carries a current of 10 A? (b) What is the force if the wire is parallel to the magnetic field?

28. A horizontal wire 600 mm long whose mass is 4.0 g is to be supported magnetically against the force of gravity. The current in the wire is 12.0 A and goes from south to north. Find the magnitude and direction of the magnetic field of least magnitude that will support the wire.

29. A horizontal north-south wire 5.0 m long is in a 0.020-T magnetic field whose direction is northeast. (a) What is the magnitude and direction of the force on the wire when a 4.0-A current flows north in it? (b) When the same current flows south in it?

30. A vertical wire 2.0 m long is in a 1.0×10^{-2} T magnetic field whose direction is northeast. (a) What is the magnitude and direction of the force on the wire when a 5.0-A current flows upward in it? (b) When the same current flows downward in it?

19.7 Force Between Two Currents

31. The parallel wires in a lamp cord are 2.0 mm apart. What is the force per meter between them when the cord is used to supply power to a 120-V, 200-W light bulb?

32. A certain electric transmission line consists of two wires 4.0 m apart that carry currents of 1.0×10^4 A. If the towers supporting the wires are 200 m apart, how much force does each current exert on the other between the towers?

at a point equidistant from the wires and x away from the plane in which they lie is given by $2\mu_0 Is/\pi(4x^2 + s^2)$.

34. The table below lists corresponding values of B_0 and B for a type of carbon steel, where B_0 is the magnetic field of a current-carrying coil with no core and B is the magnetic field of the same coil with a carbon steel core. (a) Plot a graph of B versus B_0 for this material. (b) Plot a graph of the relative permeability μ/μ_0 of this material versus B_0. Approximately what is the maximum value of μ/μ_0, and at approximately what value of B_0 is this maximum reached?

B_0	B
0.0×10^{-5} T	0.0 T
4.2	0.2
6.3	0.4
7.7	0.6
9.0	0.8
12	1.0
20	1.2
36	1.4
75	1.6

35. According to the Bohr model, a hydrogen atom consists of an electron that circles a proton at a speed of 2.2×10^6 m/s in an orbit of radius 5.3×10^{-11} m. (a) Find the magnetic field the proton experiences as a result of the motion of the electron around it. (b) Find the magnetic field that would be needed by an electron with this speed to move in an orbit of this radius.

36. Verify that the time required for a charged particle in a magnetic field to make a complete revolution is independent of its speed and the radius of its orbit.

37. In a typical loudspeaker a permanent magnet creates a radial magnetic field in which a wire coil attached to the apex of a paper cone can move perpendicularly to the field (Fig. 19.41). An alternating current in the coil causes the cone to oscillate and thereby produce sound waves. If the 40-turn coil

Figure 19.41

A loudspeaker (Problem 37).

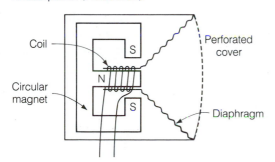

of a certain loudspeaker is 8.0 mm in radius and is located in a 0.40-T magnetic field, what is the force on the coil when the current in it is 0.050 A?

38. Two parallel wires 100 cm long each of mass 20 g are suspended by strings 10 cm long from an overhead rod (Fig. 19.42). When a current I is passed through the wires in opposite directions, the wires swing out so that the strings are 5° apart. Find I.

10 cm

5°

Figure 19.42

Problem 38.

39. (a) The current loop in Fig. 19.28(a) has the width x and the length y, as in Fig. 19.43. Verify that the magnetic torque on the loop about the axis shown in $\tau = IAB$, where $A = xy$ is the area enclosed by the loop. (b) If θ is the angle between the plane of the loop and **B**, verify that $\tau = IAB \cos \theta$, which is Eq. (19.13).

Axis

x

y

B

Figure 19.43

Problem 39.

Answers to Multiple Choice

1.	d	9.	a	17.	b	25.	c
2.	c	10.	c	18.	c	26.	c
3.	a	11.	b	19.	b	27.	d
4.	b	12.	c	20.	a	28.	b
5.	c	13.	d	21.	d	29.	c
6.	a	14.	a, b, c, d	22.	b	30.	a
7.	c	15.	a	23.	b	31.	c
8.	b	16.	a	24.	c		

C H A P T E R

20

ELECTROMAGNETIC INDUCTION

20.1 **Electromagnetic Induction**

20.2 **Moving Wire in a Magnetic Field**

20.3 **Faraday's Law**

20.4 **The Generator**

20.5 **The Transformer**

20.6 **Inductance**

20.7 **Inductive Time Constant**

20.8 **Magnetic Energy**

20.9 **Electrical Oscillations**

We have seen that one way in which electricity and magnetism are related is the presence of a magnetic field around every electric current. Are there any other ways? In particular, since a current produces a magnetic field, can a magnetic field in its turn somehow produce a current? Many of the early workers in electricity and magnetism tackled this problem without success until, in 1831, Michael Faraday in England and Joseph Henry in the United States independently discovered electromagnetic induction. Three familiar applications of electromagnetic induction are the electric generator, which is the source of most electric energy; the transformer, which enables the emf of an alternating current to be changed easily; and the playback head of a tape deck, which turns the pattern of magnetization on a tape into electical signals.

20.1 ELECTROMAGNETIC INDUCTION

An emf is produced when there is relative motion between a conductor and a magnetic field.

A galvanometer connected to a stationary wire in an unchanging magnetic field shows no sign of a current. What Faraday and Henry found is that *moving* the wire produces a current (Fig. 20.1). It does not matter whether the wire or the source of the magnetic field is being moved, provided that a component of the motion is perpendicular to the field. The origin of the current lies in the *relative motion* between a conductor and a magnetic field. This effect is known as **electromagnetic induction.**

Electromagnetic induction occurs for the following reason:

An electromotive force (emf) is produced in a conductor whenever there is relative motion between the conductor and a magnetic field, provided the motion has a component across the field direction.

It is this emf that leads to the current that flows whenever there is relative motion across the field lines between a conductor and a magnetic field. In fact, no actual motion of either a wire or a source of magnetic field is needed, because a magnetic field that changes in strength has moving field lines associated with it.

Moving a wire *parallel* to a magnetic field does not give rise to a current. Electromagnetic induction occurs only when there is a component of the wire's velocity perpendicular to the field.

Electromagnetic induction should not come as a surprise to us because it is in accord with the force a magnetic field exerts on a moving charge. As we saw in the previous chapter, a wire carrying a current is pushed sideways in a magnetic field because of the forces on the moving electrons. In Faraday's and Henry's experiments, electrons are also moved through a magnetic field, but now by shifting the entire wire. As before, the electrons are pushed to the side, which here moves them along the wire. The motion of these electrons along the wire is the electric current we find.

The right-hand rule for the force on a moving positive charge in a magnetic field can also be used to give the direction of the induced current in a wire moving across a magnetic field. Hold your right hand so that the fingers point in the direction of **B** and the outstretched thumb is in the direction of **v**. The palm then faces in the same direction as the force on the positive charges in the wire, and hence in the direction of the conventional current (Fig. 20.2).

In the arrangement of Fig. 20.1, a current will also flow if the magnet is moved vertically past the wire, instead of the wire being moved past the magnet.

Figure 20.1

(a) There is no current in a stationary wire in a magnetic field. (b) When the wire is moved across the field, a current is produced. Reversing the direction of motion also reverses the direction of the current.

(a)

(b)

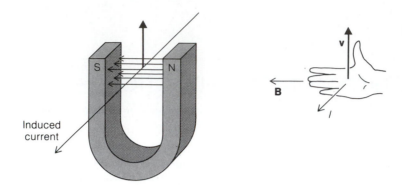

Figure 20.2

Right-hand rule for the direction of an induced current.

This effect is the basis of the playing head of a tape recorder (Fig. 20.3). The tape used is a plastic film with a ferromagnetic coating, and it is magnetized by being passed close to an electromagnet whose current varies with the signal to be recorded. For instance, the recording head would be connected via an amplifier to a microphone if music or other sound were to be registered on the tape. The playing head contains a small coil, and the current induced in it by the changing magnetic fields of the moving tape is amplified and fed into a loudspeaker to reconstruct the original sound.

Figure 20.3

A tape recorder. (a) Recording. (b) Playback. The polarity and degree of magnetization of the ferromagnetic coating on the tape correspond to the pressure fluctuations of the original sound wave.

20.2 MOVING WIRE IN A MAGNETIC FIELD

The emf depends on the wire's length and velocity and on the field strength.

When a wire moves through a magnetic field **B** with a velocity **v** that is not parallel to **B,** an induced emf $\mathscr{E}_i$ comes into being between the ends of the wire. It is not hard to figure out what $\mathscr{E}_i$ is in terms of **B, v,** and the length L of the wire.

Let us assume that the wire is perpendicular to **B** and that **v** has the component $v_\perp$ perpendicular to **B,** as in Fig. 20.4. From Eq. (19.8) the force on a positive charge Q in the moving wire has the magnitude

$$F = Qv_\perp B$$

The direction of this force is along the wire, as shown.

Now we consider a charge Q that moves from one end of the wire to the other under the influence of the magnetic force $Qv_\perp B$. The work done on the charge by whatever moves the wire is

$$\text{Work} = (\text{force})(\text{distance})$$
$$W = FL = Qv_\perp BL$$

since the wire is L long. By definition the potential difference between two points is the work done in moving a unit charge between these points. The potential difference between the ends of the wire is therefore W/Q or

$$\mathscr{E}_i = BLv_\perp \qquad\qquad \textit{Motional emf} \quad (20.1)$$

This potential difference is the emf induced in the wire by its motion through the magnetic field (Fig. 20.5).

If the moving wire is part of a complete circuit, how much current will flow? The answer is given by Ohm's law, since an induced emf is like any other emf in its

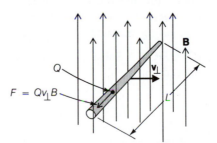

Figure 20.4

The force on a charge Q within a conductor when the conductor is moved through a magnetic field is $Qv_\perp B$, where $v_\perp$ is the component of the conductor's velocity perpendicular to **B**.

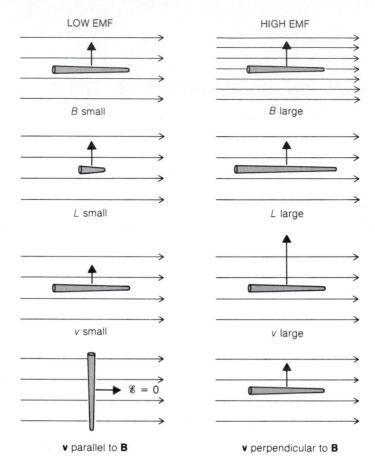

Figure 20.5

The emf induced in a conductor moving through a magnetic field depends on the magnetic field strength, on the length and speed of the conductor, and on the direction of **v** relative to **B.** The conductors shown here are oriented perpendicular to **B;** that is, perpendicular to the page.

ability to cause a current. If the total resistance in the circuit is R, the resulting current will be $I = \mathcal{E}/R$ as long as the emf exists.

◆   **E X A M P L E   20.1**

Figure 20.6

Find the potential difference between the wing tips of a jet airplane induced by its motion through the earth's magnetic field. The total wing span of the airplane is 40 m, and its speed is 300 m/s in a region where the vertical component of the earth's field has the magnitude 3.0×10^{-5} T (Fig. 20.6).

SOLUTION   Substituting the given values of $v_{\perp}$, L, and B in Eq. (20.1) for the induced emf yields

$$\mathcal{E}_i = BLv_{\perp} = (3.0 \times 10^{-5}\,\text{T})(40\,\text{m})\left(300\,\frac{\text{m}}{\text{s}}\right) = 0.36\,\text{V}$$

◆

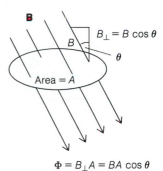

$$\Phi = B_\perp A = BA \cos \theta$$

Figure 20.7

The magnetic flux Φ through a flat wire loop of area A is proportional to A and to the component $B_\perp$ of the magnetic field perpendicular to the plane of the loop. The emf induced in the loop is equal to the rate at which the flux Φ it encloses is changing.

20.3 FARADAY'S LAW

The induced emf in a loop equals the rate of change of the magnetic flux through it.

A convenient way to look at electromagnetic induction makes use of the idea of **magnetic flux.** As in Fig. 20.7, we consider a flat wire loop of area A in a magnetic field **B**, where θ is the angle between a perpendicular to the plane of the loop and the direction of **B.** The loop can have any shape. The total magnetic flux Φ (Greek capital letter *phi*) through the loop is defined as

$$\Phi = B_\perp A = BA \cos \theta \qquad \text{\emph{Magnetic flux}} \quad (20.2)$$

where $B_\perp = B \cos \theta$ is the component of **B** perpendicular to the plane of the loop. The unit of flux is the **weber** (Wb), where $1 \text{ Wb} = 1 \text{ T} \cdot \text{m}^2$.

Faraday found that the electromotive force $\mathcal{E}_i$ in such a wire loop is *equal to the rate of change of the flux through it.* That is,

$$\mathcal{E}_i = -\frac{\Delta \Phi}{\Delta t} \qquad \text{\emph{Induced emf}} \quad (20.3)$$

Here $\Delta \Phi$ is the change in the flux Φ that takes place during a period of time Δt. Equation (20.3) is **Faraday's law of electromagnetic induction.** The reason for the minus sign is Lenz's law, which will be described shortly.

The induced emf in a wire loop is equal to the potential difference that would be found between the ends of an identical *open* wire loop. The notion of emf is useful here because it is meaningless to speak of the potential difference around a closed circuit. The emf in a closed circuit is the potential difference that would be found between the ends of the circuit if it were cut anywhere.

If a coil of N turns replaces the single loop of Fig. 20.7, the induced emf's in the turns are added together, and the total emf in the entire coil is

$$\mathcal{E}_i = -N\frac{\Delta \Phi}{\Delta t} \qquad (20.4)$$

Lenz's Law

The minus sign in Eqs. (20.3) and (20.4) follows from the law of conservation of energy. If the sign of $\mathcal{E}$ were the same as that of $\Delta \Phi / \Delta t$, the induced electric current would be in such a direction that its own magnetic field would *add* to that of the external field **B.** This additional changing field would then increase the rate of change of the flux Φ, and more and more current would flow even if the original

Figure 20.8

The current induced in the wire
loop is such that its own magnetic
field opposes the field change due
to the motion of the permanent
magnet, in accordance with
Lenz's law.

Induced current

$\Delta\Phi/\Delta t$ were to stop. But energy is associated with every current, and no current can increase without energy being supplied. The only possibility, then, is that $\mathcal{E}$ be opposite in sign to $\Delta\Phi/\Delta t$, which means:

The direction of an induced current is always such that its own magnetic field opposes the change in flux responsible for producing it.

This observation is known as **Lenz's law.** An example of Lenz's law is illustrated in Fig. 20.8.

◆   **E X A M P L E   20.2**

A 12-turn coil 100 mm in diameter has its axis parallel to a magnetic field of 0.50 T that is produced by a nearby electromagnet. The current in the electromagnet is cut off, and as the field collapses, an average emf of 8.0 V is induced in the coil. What is the length of time required for the field to disappear?

SOLUTION   The change in the flux through the coil is

$$\Delta\Phi = BA = B\pi r^2 = (-0.50 \text{ T})(\pi)(0.050 \text{ m})^2 = -0.0039 \text{ Wb}$$

Hence the flux drops to zero in the time

$$\Delta t = -\frac{N\Delta\Phi}{\mathcal{E}_i} = -\frac{(12)(-0.0039 \text{ Wb})}{8.0 \text{ V}} = 0.0059 \text{ s}$$ ◆

A Derivation of Faraday's Law

It is not hard to derive Faraday's law of electromagnetic induction from the formula for the emf induced in a wire moved across a magnetic field. Figure 20.9 shows a moving wire that slides across the legs of a U-shaped metal frame, so that the frame completes the loop. At the moment (say $t = 0$) when the moving wire is the distance x from the closed end of the frame, the flux enclosed by the loop is

$$\Phi = BA = BLx$$

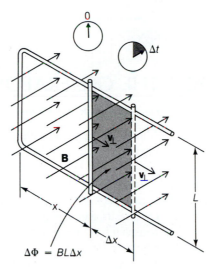

Figure 20.9

The flux enclosed by the wire frame and movable wire increases as the wire moves to the right. An emf is induced in the wire by its motion through the magnetic field, which is equal to the rate of change of the enclosed flux, in agreement with Faraday's law of electromagnetic induction.

At the later time $t = \Delta t$, the moving wire is $x + \Delta x$ from the closed end of the frame, so the flux now enclosed by the loop is

$$\Phi + \Delta\Phi = BLx + BL\,\Delta x$$

Hence the increase in the enclosed flux during the time Δt is

$$\Delta\Phi = BL\,\Delta x$$

The speed of the moving wire is $v_\perp$, and so

$$\Delta x = v_\perp\,\Delta t$$

and

$$\Delta\Phi = BLv_\perp\,\Delta t$$
$$\frac{\Delta\Phi}{\Delta t} = BLv_\perp$$

But $BLv_\perp$ is the emf induced in the moving wire. Hence we have, inserting a minus sign to remind ourselves of Lenz's law,

$$\mathcal{E}_i = -\frac{\Delta\Phi}{\Delta t}$$

which is Faraday's law.

20.4 THE GENERATOR

Electromagnetic induction underlies its operation.

Almost all the world's electric energy is produced by electromagnetic induction, something to keep in mind whenever we switch on a lamp. In a **generator** a coil of wire is turned in a magnetic field so that the flux through the coil changes constantly. The resulting emf in the coil causes a current to flow in an external circuit, and cables can carry this current for long distances from its origin.

Despite its name, a generator does not *create* electric energy. What it does is *convert* mechanical energy into electric energy, just as a battery converts chemical energy into electric energy.

The construction of an idealized simple generator is shown in Fig. 20.10. A wire loop is rotated in a magnetic field, and the ends of the loop are connected to two **slip rings** on the shaft. Brushes pressing against the slip rings join the loop to an external circuit. Only sides *B* and *D* of the loop contribute to the induced emf.

Because the sides of the loop reverse their direction of motion through the magnetic field twice per rotation, the induced current is also reversed twice per rotation and for this reason is called an **alternating current.** The emf of this generator varies with time as in Fig. 20.11. The shape of the curve is that of a cosine (or sine) curve. In a complete cycle, which corresponds to one turn of the loop, the emf increases to a maximum, falls to zero, continues to a negative maximum, and then returns to zero. If the loop rotates *f* times per second, the generator output will go through *f* cycles per second. An ac generator is often called an **alternator.**

Figure 20.10

A simple ac generator.

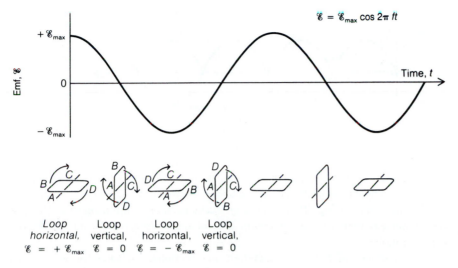

$$\mathscr{E} = \mathscr{E}_{max} \cos 2\pi ft$$

Loop horizontal, $\mathscr{E} = +\mathscr{E}_{max}$

Loop vertical, $\mathscr{E} = 0$

Loop horizontal, $\mathscr{E} = -\mathscr{E}_{max}$

Loop vertical, $\mathscr{E} = 0$

Figure 20.11

The variation with time of the emf of a simple ac generator.

In nearly all actual generators, electromagnets produce the magnetic field. (If a permanent magnet is used, the generator is called a **magneto**.) In an alternator the armature and field windings are often exchanged so that, in effect, the field rotates inside a stationary armature. This makes no difference to the physics involved because there is still relative movement of a conductor and a magnetic field. The advantage of a stationary armature is that a high-voltage output can be taken directly from it without any moving electrical contacts while the field current is supplied at a modest voltage through such contacts.

The ac output of a simple generator can be changed to dc by substituting a commutator for the slip rings that connect the rotating coil with the external circuit (Fig. 20.12). The emf of such a dc generator varies with time as in Fig. 20.13. The

Figure 20.12

A simple dc generator. The current in the brushes is always in the same direction.

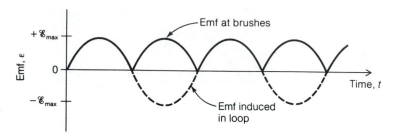

Figure 20.13

The variation with time of the emf of a simple dc generator.

emf is always in the same direction, but it rises to a maximum and drops to zero twice per complete rotation. The resemblance between the dc generator in Fig. 20.12 and the dc motor in Fig. 19.30 is not an accident: One is the inverse of the other. If the output terminals of a dc generator are connected to a battery, the generator will run as a motor.

The emf of Fig. 20.13, though unidirectional, pulsates too much for most applications. A steadier emf can be obtained by using a pair of coils on the armature set perpendicular to each other. When one of the coils has its maximum emf, the other has an emf of zero. Figure 20.14 shows the emf's of each coil and also what happens when the outputs of the two are connected in series: The resulting emf has only a moderate ripple. Commercial generators have many coils in their armatures to produce very nearly constant emf's.

Another approach is to use a **rectifier** in the output circuit of an alternator (Fig. 20.15). A rectifier is a device that permits current to pass through it in only one direction; the operation of semiconductor rectifiers is described in Chapter 26. Because alternators are simpler to construct than dc generators and are more reliable, they are often used with semiconductor rectifiers to produce the direct current needed in cars and other vehicles.

Figure 20.14

The output of a dc generator whose armature has two perpendicular coils shows only a slight ripple.

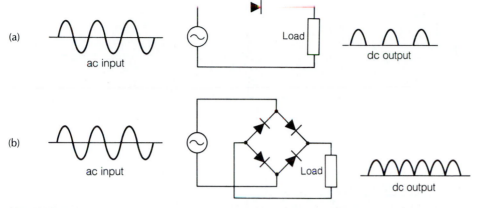

Figure 20.15

The symbol of a rectifier is ——▶|——. (a) Half-wave rectifier circuit. (b) Full-wave rectifier circuit. Current flows through one pair of rectifiers when it is in one direction and through the other pair when it is in the opposite direction to give the pulsating output shown. Ripple-free direct current can be obtained by using a combination of capacitors and inductors (see Section 21.10) to absorb energy at the top of each half cycle and then release it at the bottom.

Back Emf

As we just saw, a coil rotated in a magnetic field is a source of emf, which implies that an emf is generated in the rotating armature of an electric motor. This induced emf occurs because wires are being moved through a magnetic field. The presence of a current of external origin in the armature at the same time, which is what causes the armature to turn in the first place, does not matter. Every electric motor is also a generator.

By Lenz's law all induced emf's are such as to oppose the changes that bring them into being. This means that the induced emf in a motor armature is opposite in direction to the external voltage applied to the armature. The effect of this **back emf** (or **counter emf**) is to reduce the current in the armature.

Back emf is equal to $-\Delta\Phi/\Delta t$ and so is proportional to the angular speed of the armature. It is therefore zero when the motor is turned on and increases as the speed increases. As a result the current is a maximum when the motor starts, and it drops afterward. The existence of back emf in a motor helps to keep its speed constant. At the normal speed,

$$V = \mathscr{E}_b + IR \qquad (20.5)$$

Applied voltage = back emf + potential drop in armature

If the speed falls, $\mathscr{E}_b$ decreases and I goes up, and the greater current means more torque on the armature to restore the speed to its original value. If the speed rises above normal, $\mathscr{E}_b$ increases and I goes down, with the result that the torque is reduced and the motor loses speed until it assumes its normal value.

Figure 20.16

(a) Shunt-wound dc motor.
(b) Series-wound dc motor.

◆   E X A M P L E   20.3 _____

A 120-V dc electric motor has its armature and field windings connected in parallel. (Such a motor is said to be "shunt wound." When armature and field are in series, the motor is said to be "series wound." See Fig. 20.16.) The armature resistance is 2.0 Ω and the field resistance is 200 Ω. When the motor is at its operating speed, the total current is 3.0 A. Find (a) the back emf of the motor at its operating speed, (b) the efficiency of the motor, and (c) the current in the motor at the moment its switch is turned on.

SOLUTION   (a) The current in the field winding is

$$I_f = \frac{V}{R_f} = \frac{120 \text{ V}}{200 \text{ } \Omega} = 0.6 \text{ A}$$

Since the total current is 3.0 A, the armature current is

$$I_a = I - I_f = 3.0 \text{ A} - 0.6 \text{ A} = 2.4 \text{ A}$$

The magnitude of the back emf is therefore

$$\mathcal{E}_b = V - I_a R_a = 120 \text{ V} - (2.4 \text{ A})(2.0 \text{ } \Omega) = 115 \text{ V}$$

The direction of the back emf is, of course, opposite to that of the impressed potential difference V.

(b) The power lost as heat in the field and armature windings is

$$P_{\text{loss}} = I_f^2 R_f + I_a^2 R_a = (0.6 \text{ A})^2(200 \text{ } \Omega) + (2.4 \text{ A})^2(2.0 \text{ } \Omega) = 83.5 \text{ W}$$

The electrical power input of the motor is equal to IV, the total current multiplied by the applied potential difference:

$$P_{input} = IV = (3.0 \text{ A})(120 \text{ V}) = 360 \text{ W}$$

Hence the motor's efficiency is

$$\text{Eff} = \frac{P_{input} - P_{loss}}{P_{input}} = \frac{360 \text{ W} - 83.5 \text{ W}}{360 \text{ W}} = 0.77 = 77\%$$

(c) At the moment the motor is started, there is no back emf, because the armature is still stationary. The total resistance of the armature and field windings is

$$R = \frac{R_a R_f}{R_a + R_f} = \frac{(2.0 \text{ }\Omega)(200 \text{ }\Omega)}{2.0 \text{ }\Omega + 200 \text{ }\Omega} = 1.98 \text{ }\Omega$$

The current is therefore

$$I = \frac{V}{R} = \frac{120 \text{ V}}{1.98 \text{ }\Omega} = 61 \text{ A}$$

The starting current is more than 20 times greater than the operating current! ◆

20.5 THE TRANSFORMER

Stepping voltage up and down.

Earlier we noted that a current can be induced in a wire or other conductor by a changing magnetic field as well as by relative motion between a wire and a constant magnetic field. The essential condition for an induced emf is that magnetic field lines cut across the wire (or vice versa). It does not matter exactly how this comes about.

A change in the current in a wire loop is accompanied by a change in its magnetic field. If another wire loop is nearby, an emf will therefore be induced in it. If an alternating current flows in the first loop, the magnetic field around it will vary periodically and an alternating emf will be induced in the second loop. A **transformer** is a device based on this effect that is used to produce an alternating emf in a secondary circuit larger or smaller than the alternating emf in the primary circuit. A transformer thus permits electric energy to be transferred from one ac circuit to another without a direct connection between them.

With the help of Fig. 20.17, we can see how a changing current in a wire loop induces a changing current in another wire loop not connected to it.

(a) The switch connecting loop 1 with the battery is open, and no current is in either loop.

Figure 20.17

When the current in a wire loop is turned on or off, the magnetic field around it changes. The changing field can induce a current in another nearby loop, even though both loops are stationary.

(b) The switch has just been closed, and the expanding field lines resulting from the increasing current in loop 1 cut across loop 2, inducing a current in it. The current in loop 2 is in the *opposite* direction to that in loop 1 because of Lenz's law: The direction of an induced current is always such that its own magnetic field opposes the change in flux that is inducing it.

(c) Now there is a constant current in loop 1, and since the flux through it does not change, no current is induced in loop 2.

(d) The switch has just been opened, and the contracting field lines resulting from the decreasing current in loop 1 cut across loop 2, inducing a current in it. The current in loop 2 is now in the *same* direction as that in loop 1. This is because it is the decreasing current in loop 1 that leads to the current in loop 2, and Lenz's law requires that an induced current oppose the change in flux that brings it about.

(e) Finally the current in loop 1 disappears, and again no current is in either loop.

Figure 20.18

A simple transformer.

Instead of being switched on and off, the current input to a transformer is alternating current that reverses its direction regularly. The construction of a simple transformer is shown in Fig. 20.18. Coils are used instead of single loops, and they are wound together on an iron core so that the alternating magnetic flux set up by an alternating current in one coil induces an alternating emf in the other coil. The **primary winding** of a transformer is the coil that is fed with an alternating current, and the **secondary winding** is the coil to which power is transferred via the changing magnetic flux. Either winding may be the primary, and the other is then the secondary. A transformer may have more than one secondary winding so that it can provide several different secondary emf's.

Step-up and Step-down Transformers

An ideal transformer has no "leakage" of flux outside the iron core and no losses within it, so that $\Delta\Phi/\Delta t$ is the same for each of the turns of both windings. The emf *per turn* is therefore the same in both windings, and the total emf in each winding is proportional to the number of turns it contains. Hence

$$\frac{\mathscr{E}_1}{\mathscr{E}_2} = \frac{N_1}{N_2}$$

(20.6)

$$\frac{\text{Primary emf}}{\text{Secondary emf}} = \frac{\text{total primary turns}}{\text{total secondary turns}}$$

When the secondary winding has more turns than the primary winding, $\mathscr{E}_2$ is greater than $\mathscr{E}_1$. The result is a **step-up transformer.** When the reverse is the case and $\mathscr{E}_2$ is less than $\mathscr{E}_1$, the result is a **step-down transformer.**

In an ideal transformer the power output is equal to the power input. If the primary current is I_1 and the second current is I_2, then

$$\mathscr{E}_1 I_1 = \mathscr{E}_2 I_2$$

Power input = power output

and

$$\frac{I_1}{I_2} = \frac{\mathscr{E}_2}{\mathscr{E}_1} = \frac{N_2}{N_1} \qquad\qquad\qquad \textit{Transformer} \quad (20.7)$$

Increasing the voltage with a step-up transformer means a decrease in the current, whereas dropping the voltage leads to a greater current.

◆ **E X A M P L E   20.4** ────────────────────────────

A transformer connected to a 120-V ac power line has 200 turns in its primary winding and 50 turns in its secondary winding. The secondary is connected to a 100-Ω light bulb. How much current is drawn from the 120-V power line?

SOLUTION   The voltage across the secondary of the transformer is

$$V_2 = \frac{N_2}{N_1}V_1 = \left(\frac{50 \text{ turns}}{200 \text{ turns}}\right)(120 \text{ V}) = 30 \text{ V}$$

and so the current in the secondary circuit is

$$I_2 = \frac{V_2}{R} = \frac{30 \text{ V}}{100 \text{ } \Omega} = 0.30 \text{ A}$$

The current in the primary circuit is

$$I_1 = \frac{N_2}{N_1}I_2 = \left(\frac{50 \text{ turns}}{200 \text{ turns}}\right)(0.30 \text{ A}) = 0.075 \text{ A}$$

◆

──

Alternating current owes its wide use largely to the ability of transformers to change the voltage at which it is transmitted. Electricity generated at perhaps 11,000 V is stepped up by transformers at the power station to as much as 750,000 V (or even more) for transmission, and then is stepped down by transformers at substations near the point of use to less than a thousand volts. Local transformers then step this down further to 240 or 120 V. High voltages are desirable because, since $P = IV$, the higher the voltage, the smaller the current for a given amount of power. Power is dissipated as heat at the rate I^2R, so the smaller the current, the less the power lost due to resistance in the transmission line.

◆ **E X A M P L E   20.5** ────────────────────────────

Find the power lost as heat when a 10.0-Ω cable is used to transmit 1000 W of electricity at 240 V and at 240,000 V.

SOLUTION   The current in the cable in each case is

$$I_a = \frac{P}{V_a} = \frac{1000\ \text{W}}{240\ \text{V}} = 4.17\ \text{A}$$

$$I_b = \frac{P}{V_b} = \frac{1000\ \text{W}}{240{,}000\ \text{V}} = 4.17 \times 10^{-3}\ \text{A}$$

The respective rates of heat production per kilowatt are therefore

$$I_a^2 R = (4.17\ \text{A})^2 (10.0\ \Omega) = 174\ \text{W}$$

$$I_b^2 R = (4.17 \times 10^{-3}\ \text{A})^2 (10.0\ \Omega) = 1.74 \times 10^{-4}\ \text{W}$$

a difference of a factor of 10^6. Transmission at 240 V therefore means a million times more power lost as heat than does transmission at 240,000 V. ◆

Transformers at Morro Bay, California, step up electric power generated at 18,000 V to 230,000 V for efficient transmission.

Eddy Currents

When a piece of metal is placed in a coil that carries an alternating current, the changing magnetic flux causes **eddy currents** to flow in the metal. The energy of these currents is dissipated as heat, and the metal becomes hot as a result. This effect is called **induction heating.** Induction furnaces heat metals rapidly and are often used in foundries to melt casting metal before it is poured into molds.

Eddy currents in a transformer core are wasteful partly because of the power lost as heat and partly because the flux they establish interferes with the proper operation of the transformer. To minimize eddy currents, actual transformer cores are usually laminated, with thin sheets arranged parallel to the magnetic field, as in Fig. 20.18. The laminations are insulated from each other by natural oxide layers on their surfaces or by varnish coatings. Only very weak eddy currents are induced in a core of this kind. In small transformers an alternative technique is to use a core of compressed powdered iron, so that each grain of iron is fairly well insulated from the others.

20.6 INDUCTANCE

A self-induced emf occurs whenever the current in a wire loop changes.

Let us look at what happens when a wire loop is connected to a battery. As the current starts to flow, a magnetic field begins to build up. This changing magnetic field induces an emf in the *same* wire loop whose current led to the field in the first place. By Lenz's law the **self-induced emf** is such as to oppose the change in flux responsible for it. Thus an increasing current produces a self-induced emf opposite to the current, which slows down its rise. If instead we switch off an existing current in the loop, the self-induced emf will be in the same direction as the current, which slows down its fall.

The magnitude of the self-induced emf $\mathscr{E}_i$ depends on the rate of change $\Delta\Phi/\Delta t$ of the flux through the loop. For a loop whose geometry is fixed, Φ is proportional to the current I in the loop, and $\Delta\Phi/\Delta t$ is proportional to $\Delta I/\Delta t$, the rate at which the current is changing. Thus we can write

$$\mathscr{E}_i = -L\frac{\Delta I}{\Delta t} \qquad\qquad \textit{Self-induced emf} \quad (20.8)$$

where L, the constant of proportionality, is called the **inductance** of the loop. The minus sign, as in Eq. (20.3), follows from Lenz's law. A high value of L means a large $\mathscr{E}_i$ for a given rate of change $\Delta I/\Delta t$, and a low value of L means a small $\mathscr{E}_i$.

The unit of inductance is the **henry** (H), where

$$1\,\text{H} = 1\,\frac{\text{J}}{\text{A}^2} = 1\,\frac{\text{V}\cdot\text{s}}{\text{A}}$$

The henry, like the farad, is usually too large a unit for convenience. Inductances are accordingly often expressed in **millihenries** (mH) or **microhenries** (μH), where

$$1\,\text{mH} = 10^{-3}\,\text{H} \qquad 1\,\mu\text{H} = 10^{-6}\,\text{H}$$

◆   **E X A M P L E   20.6** _____

An average self-induced emf of -0.75 V is produced in a 25-mH coil when the current in it falls to 0 in 0.010 s. What was the original current in the coil?

SOLUTION   From Eq. (20.8)

$$\Delta I = \frac{-\mathscr{E}_i\,\Delta t}{L} = \frac{-(-0.75\,\text{V})(0.010\,\text{s})}{2.5\times10^{-2}\,\text{H}} = 0.30\,\text{A}$$

Since $\Delta I = I_0 - 0$, the original current was $I_0 = 0.30$ A.   ◆

_____

A circuit element in which a self-induced emf accompanies a changing current is called an **inductor.** A wire loop is an example of an inductor. The inductance of an inductor depends on its geometry and on the presence of a core with magnetic properties. A coil with many turns has more inductance than one with few turns because its magnetic field is stronger for a given current (Fig. 20.19). The longer the coil and the larger its cross-sectional area, the more its inductance, because it has a larger volume of magnetic field.

HIGH INDUCTANCE

LOW INDUCTANCE

Closely spaced turns

Widely spaced turns

Long coil

Short coil

Large cross section

Small cross section

Iron core

Air core

Figure 20.19

The inductance of a coil depends on how closely wound it is, on its length, its cross-sectional area, and on the nature of its core.

A core in a coil changes its inductance by changing the flux through the coil (see Section 19.4). Thus a diamagnetic core decreases the inductance slightly from its value in free space or air, a paramagnetic core increases it slightly, and a ferromagnetic core increases it considerably—by a factor of as much as 10^4 in some cases—but by an amount that is not constant.

◆ **E X A M P L E 20.7** _____

Find the inductance of an N-turn solenoid of length l and cross-sectional area A.

SOLUTION   From Eq. (20.4) the emf induced in a coil of N turns through which the magnetic flux changes at the rate $\Delta\Phi/\Delta t$ is

$$\mathcal{E}_i = -N\frac{\Delta\Phi}{\Delta t}$$

Since the magnetic field inside a solenoid is perpendicular to its turns, $\Phi = BA$ and

$$\mathcal{E}_i = -NA\frac{\Delta B}{\Delta t}$$

In the case of a solenoid, according to Eq. (19.6), $B = \mu(N/l)I$, so that

$$\frac{\Delta B}{\Delta t} = \mu \frac{N}{l} \frac{\Delta I}{\Delta t}$$

and

$$\mathscr{E}_i = -\mu N^2 \frac{A}{l} \frac{\Delta I}{\Delta t}$$

Comparing this result with Eq. (20.8), which defines L, gives

$$L = \mu N^2 \frac{A}{l} \qquad\qquad\qquad \textit{Inductance of a solenoid} \quad (20.9)$$

for the inductance of a solenoid. ◆

◆ E X A M P L E 20.8

Find the inductance in air of a 1000-turn solenoid 10 cm long that has a cross-sectional area of 20 cm^2.

SOLUTION Since 20 cm^2 = 2.0×10^{-3} m^2 and $\mu = \mu_0$ in air (very nearly),

$$L = \mu_0 N^2 \frac{A}{l} = \left(1.26 \times 10^{-6} \frac{\text{T} \cdot \text{m}}{\text{A}}\right)(10^3)^2 \left(\frac{2.0 \times 10^{-3}\,\text{m}^2}{1.0 \times 10^{-1}\,\text{m}}\right)$$

$$= 2.5 \times 10^{-2}\,\text{H} = 25\,\text{mH}$$ ◆

Figure 20.20

A circuit with an inductor, a resistor, and a source of emf in series. When the switch is closed, the current increases gradually to its ultimate value of $\mathscr{E}/R$ because of the opposing self-induced emf in the inductor.

20.7 INDUCTIVE TIME CONSTANT

The current in an inductor needs time to build up or to disappear.

When a circuit containing inductance is connected to a battery, the current in the circuit does not rise at once to its final value $I = \mathscr{E}/R$, where $\mathscr{E}$ is the emf of the battery and R is the total resistance in the circuit. As the switch in Fig. 20.20 is closed, the current I starts to grow. The induced emf $-L(\Delta I/\Delta t)$ then comes into being in the opposite direction to the battery emf $\mathscr{E}$. The net emf acting to establish current in the circuit is therefore $\mathscr{E} - L(\Delta I/\Delta t)$, and the current reaches its final value of I_0 in a gradual manner.

When the current in the circuit of Fig. 20.20 is I and is changing at the rate $\Delta I / \Delta t$,

$$\mathcal{E} - L\frac{\Delta I}{\Delta t} = IR \tag{20.10}$$

Applied emf $-$ induced emf $=$ net voltage

Hence the rate at which the current is increasing is

$$\frac{\Delta I}{\Delta t} = \frac{\mathcal{E} - IR}{L}$$

The larger the inductance L, the more slowly the current increases. At the moment the switch is closed, $I = 0$ and $\Delta I / \Delta t$ has its maximum value of

$$\left(\frac{\Delta I}{\Delta t}\right)_{\text{max}} = \frac{\mathcal{E}}{L}$$

Eventually the current reaches its final value of I_0 and $\Delta I / \Delta t = 0$. From then on

$$\frac{\Delta I}{\Delta t} = 0 = \frac{\mathcal{E} - I_0 R}{L}$$

$$I_0 = \frac{\mathcal{E}}{R}$$

Thus the effect of having inductance in the circuit is to delay the establishment of the final current.

The graph in Fig. 20.21 shows how I varies with time when a current is being established in a circuit containing inductance. A mathematical analysis shows that, after a time interval of L/R, the current reaches 63% of I_0. The time $T_L = L/R$ is therefore a convenient measure of how rapidly a current rises in a circuit containing

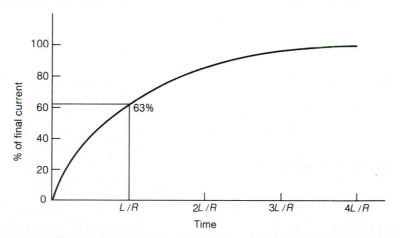

Figure 20.21

The growth of current in a circuit containing inductance and resistance.

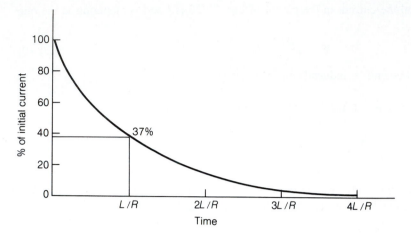

Figure 20.22

The decay of current in a circuit containing inductance and resistance when the battery is replaced by a wire.

inductance. T_L is called the **inductive time constant** of the circuit:

$$T_L = \frac{L}{R}$$

Inductive time constant (20.11)

When the battery in Fig. 20.20 is replaced by a wire, the current I drops slowly in the manner shown in Fig. 20.22, since the induced emf now tends to maintain the existing current. The current falls to 37% of its original value in the time L/R after the battery is removed.

◆　**E X A M P L E　20.9** _____

A 2.0-H inductor whose resistance is 100 Ω is connected to a 24-V battery of negligible internal resistance. Find the initial current in the circuit and the initial rate at which the current is increasing, the time required for the current to reach 63% of its ultimate value, and the magnitude of the final current.

SOLUTION　The initial current is 0, and the initial rate of current increase is

$$\frac{\Delta I}{\Delta t} = \frac{\mathscr{E}}{L} = \frac{24\text{ V}}{2.0\text{ H}} = 12\text{ A/s}$$

The time needed for the current to rise to 63% of the final value is

$$T_L = \frac{L}{R} = \frac{2.0\text{ H}}{100\ \Omega} = 0.02\text{ s}$$

and the final current will be

$$I_0 = \frac{\mathscr{E}}{R} = \frac{24\text{ V}}{100\ \Omega} = 0.24\text{ A}$$

　◆

20.8 MAGNETIC ENERGY

Energy is stored in the magnetic field of an inductor.

Energy is associated with magnetic as well as with electric fields. To appreciate this, we return to what happens when an inductor is connected to a battery.

By Lenz's law the self-induced emf $\mathscr{E}_i$ in the inductor opposes the rising current. The battery therefore has to push electrons through the inductor against $\mathscr{E}_i$, so that the battery does work against $\mathscr{E}_i$. Where does the work go? Since the self-induced emf has nothing to do with the resistance of the inductor, the work must go into the magnetic field of the current that is starting to flow. When the current in the inductor reaches its final value, there is no more self-induced emf. The work done to establish the inductor's magnetic field has become magnetic energy.

Let us see how much magnetic energy is contained in an inductor when a steady current I flows through it. This energy is equal to the work W done against $\mathscr{E}_i$ to establish the current starting from $I = 0$.

If we suppose the current rises at a uniform rate from 0 to I in a time interval Δt, then the average current during Δt is $\bar{I} = \frac{1}{2}I$. In this time the total charge Q that passes through the inductor is

$$Q = \bar{I}\Delta t = \tfrac{1}{2}I\,\Delta t$$

The work done to build up the current (and thus to create the final magnetic field) is

$$W = Q\mathscr{E} = \tfrac{1}{2}I\mathscr{E}\,\Delta t$$

where $\mathscr{E}$ is the applied emf. Since $\mathscr{E} = -\mathscr{E}_i$, from Eq. (20.8) we have

$$\mathscr{E} = -\mathscr{E}_i = L\frac{\Delta I}{\Delta t}$$

Here the change in current ΔI in the time Δt is the final current I, so $\Delta I = I$. The work done is therefore

$$W = \tfrac{1}{2}I\left(L\frac{I}{\Delta t}\right)\Delta t$$

and so

$$W = \tfrac{1}{2}LI^2 \qquad\qquad \textit{Energy of inductor} \quad (20.12)$$

A circuit of inductance L has the magnetic energy of $\frac{1}{2}LI^2$ stored in it when a current I is present in it. When the potential difference responsible for the current is removed, the energy $\frac{1}{2}LI^2$ is what powers the self-induced emf that slows down the drop in current. The gradual rise of current in a circuit containing inductance may be thought of as the result of absorbing $\frac{1}{2}LI^2$ of magnetic energy, and its gradual drop may be thought of as the result of the transfer of the inductor's energy to the circuit.

◆   E X A M P L E   20.10 _____

How much current is needed in a 50-mH coil for it to have 1.0 J of magnetic energy?

SOLUTION   From Eq. (20.12) we have

$$I = \sqrt{\frac{2W}{L}} = \sqrt{\frac{(2)\,(1.0\text{ J})}{50 \times 10^{-3}\,\text{H}}} = 6.3\text{ A} \qquad\qquad ◆$$

Magnetic Energy Density

In Section 17.6 the energy density w of an electric field of magnitude E was shown to be

$$w = \frac{E^2}{8\pi k}$$ *Electric energy density*

The corresponding formula for the energy density of a magnetic field turns out to be

$$w = \frac{B^2}{2\mu}$$ *Magnetic energy density*   (20.13)

The energy density of a magnetic field is directly proportional to the square of its magnitude B. This formula holds for all magnetic fields, not just for those inside solenoids.

◆   E X A M P L E   20.11 _____

The earth's magnetic field in a certain region has the magnitude 6.0×10^{-5} T. Find the magnetic energy per cubic kilometer in this region.

SOLUTION   The magnetic energy density corresponding to $B = 6.0 \times 10^{-5}$ T is, since here $\mu = \mu_0$,

$$w = \frac{B^2}{2\mu_0} = \frac{(6.0 \times 10^{-5}\,\text{T})^2}{2(1.26 \times 10^{-6}\,\text{T}\cdot\text{m/A})} = 1.4 \times 10^{-3}\text{ J/m}^3$$

A cubic kilometer is equal to $(10^3\text{ m})^3 = 10^9\text{ m}^3$, so the total magnetic energy in a cubic kilometer is

$$W = wV = (1.4 \times 10^{-3}\text{ J/m}^3)(10^9\text{ m}^3) = 1.4 \times 10^6\text{ J}$$

This amount of energy is enough to raise the 365,000-ton Empire State Building by 0.42 mm.

♦

20.9 ELECTRICAL OSCILLATIONS

An LC circuit is the electrical analog of a harmonic oscillator.

Let us connect a charged capacitor to an inductor, as in Fig. 20.23. The following sequence of events occurs:

(a) Each of the capacitor's plates starts with the charge Q and its electric energy is $\frac{1}{2}Q^2/C$, whereas the inductor has no energy since $I = 0$.

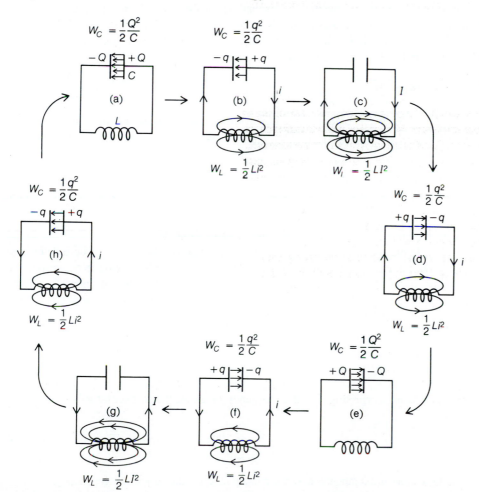

Figure 20.23

The various stages in the oscillation of a circuit containing the inductance L and the capacitance C. The maximum charge on the capacitor is Q, and the maximum current in the circuit is I.

(b) The capacitor is partly discharged, and the current in the inductor leads to a magnetic field whose energy is $\frac{1}{2}Li^2$.

(c) No charge is left on the capacitor's plates, and all the energy of the circuit is magnetic energy $\frac{1}{2}LI^2$.

(d) Now charge begins to build up on the capacitor's plates with polarities the reverse of the original ones, and the electric energy of the capacitor grows at the expense of the magnetic energy of the inductor as the current drops.

(e) At the instant the capacitor is fully charged again, the current is zero and the energy of the circuit is once more entirely electric.

(f) Next the capacitor begins to discharge, and eventually the entire circuit returns to its original state at (a). The cycle will continue to repeat itself indefinitely if no resistance is present.

This sequence is wholly analogous to the behavior of a harmonic oscillator. As we learned in Chapter 11, in a harmonic oscillator energy is constantly interchanged between KE and PE. Here the same sort of interchange occurs, with the electric energy of a charged capacitor corresponding to PE and the magnetic energy of an inductor through which a current flows corresponding to KE.

A harmonic oscillator whose mass is m and whose spring constant is k has the potential and kinetic energies

$$\text{PE} = \tfrac{1}{2}kx^2 \qquad \text{KE} = \tfrac{1}{2}mv^2 \qquad\qquad \textit{Harmonic oscillator}$$

where x is the displacement of the object from its equilibrium position and the speed v is the rate of change of x. The electric and magnetic energies in an electrical oscillator are

$$W_e = \tfrac{1}{2}\frac{1}{C}Q^2 \qquad W_m = \tfrac{1}{2}LI^2 \qquad\qquad \textit{Electrical oscillator}$$

where the current I is the rate of change of the charge Q.

In the above formulas we note that $1/C$ corresponds to k, since both determine the amount of energy present at the moment when nothing is moving in the respective oscillators. Similarly L corresponds to m. Both are measures of inertia—a large inductance tends to slow down changes in current, and a large m tends to slow down changes in speed. The frequency of a harmonic oscillator is

$$f = \frac{1}{2\pi}\sqrt{\frac{k}{m}}$$

and so it is not surprising that the frequency of an electrical oscillator is

$$f = \frac{1}{2\pi\sqrt{LC}} \qquad\qquad \textit{Oscillator frequency} \quad (20.14)$$

The larger the inductance L and the capacitance C, the lower the frequency.

The parallel between electrical and mechanical oscillations is a very close one. Just as every mechanical system has certain specific natural frequencies of vibration that depend on its properties, so every electric circuit has a natural frequency of oscillation that depends on L and C. Just as the energy of every actual mechanical vibration is eventually dissipated as heat owing to the inevitable presence of friction, so the energy of every electrical oscillation is eventually dissipated as heat owing to the inevitable presence of resistance in the wires of the circuit.

Electrical resonance is much like mechanical resonance. If a source of emf that alternates at the natural frequency of a circuit is connected to the circuit, energy is fed into the oscillations that can maintain or increase their amplitude. At any other frequency the energy absorbed is small.

Until recently all radio communication was based on the ability of an LC circuit to oscillate only at the frequency given by Eq. (20.14). Such a circuit was used to ''tune'' a transmitter so that only a single frequency of radio waves was produced, and another similar circuit ''tuned'' a receiver to the same frequency. Today a different method is often used that is based on the multiplication of a standard frequency by special circuits.

◆   **E X A M P L E   20.12** _____

A 5.0-μF capacitor with a charge of 1.0×10^{-4} C is connected to a 20-mH coil. (a) What is the frequency of the oscillations that result? (b) What is the maximum current in the circuit?

SOLUTION   (a) From Eq. (20.14) the charge in the circuit will oscillate back and forth through the inductor at the frequency

$$f = \frac{1}{2\pi\sqrt{LC}}$$

$$= \frac{1}{2\pi\sqrt{(20 \times 10^{-3}\,\text{H})(5.0 \times 10^{-6}\,\text{F})}} = 503\text{ Hz}$$

(b) The maximum current occurs when all the charge has left the plates of the capacitor, which corresponds to (c) or (g) in Fig. 20.23. At such times the magnetic energy $\frac{1}{2}LI^2$ of the inductor equals the electric energy $\frac{1}{2}Q^2/C$ of the fully charged capacitor, so that

$$\tfrac{1}{2}LI^2 = \tfrac{1}{2}\frac{Q^2}{C}$$

$$I = \frac{Q}{\sqrt{LC}}$$

$$= \frac{1.0 \times 10^{-4}\,\text{C}}{\sqrt{(20 \times 10^{-3}\,\text{H})(5.0 \times 10^{-6}\,\text{F})}} = 0.32\text{ A}$$

◆

Important Terms

Electromagnetic induction refers to the production of an electric field in a conductor whenever magnetic field lines move across it.

Lenz's law states that the direction of an induced current must be such that its own magnetic field opposes the changes in flux that are inducing it.

A **generator** is a device that converts mechanical energy into electric energy.

The direction of an **alternating current** reverses itself periodically.

A **back** (or **counter**) **emf** is induced in the rotating coils of an electric motor and is opposite in direction to the external voltage applied to them.

An alternating current flowing in the primary coil of a **transformer** induces another alternating current in the secondary coil. The ratio of the emf's is proportional to the ratio of turns in the coils.

The **inductance** L of a circuit is the ratio between the magnitude of the self-induced emf $\mathscr{E}$ due to a changing current in the circuit and the rate of change $\Delta I/\Delta t$ of the current. The unit of inductance is the **henry.**

The inductive time constant of a circuit that contains inductance and resistance is a measure of the time needed for a current to be established or to disappear in the circuit.

The **energy density** of a magnetic field is the magnetic energy per unit volume associated with it.

Important Formulas

Motional emf: $\mathscr{E}_i = BLv_\perp$

Magnetic flux: $\Phi = B_\perp A = BA \cos \theta$

Induced emf: $\mathscr{E}_i = -N\dfrac{\Delta\Phi}{\Delta t}$

Transformer: $\dfrac{I_1}{I_2} = \dfrac{\mathscr{E}_2}{\mathscr{E}_1} = \dfrac{N_2}{N_1}$

Self-induced emf: $\mathscr{E}_i = -L\dfrac{\Delta I}{\Delta t}$

Inductive time constant: $T_L = \dfrac{L}{R}$

Magnetic energy of inductor: $W = \frac{1}{2}LI^2$

Magnetic energy density: $w = \dfrac{B^2}{2\mu}$

Oscillator frequency: $f = \dfrac{1}{2\pi\sqrt{LC}}$

MULTIPLE CHOICE

1. The emf produced in a wire by its motion across a magnetic field does not depend on
- **a.** the length of the wire
- **b.** the diameter of the wire
- **c.** the orientation of the wire
- **d.** the flux density of the field

2. A bar magnet is passed through a coil of wire. The induced current is greatest when
- **a.** the magnet moves slowly, so that it is inside the coil for a long time
- **b.** the magnet moves fast, so that it is inside the coil for a short time
- **c.** the north pole of the magnet enters the coil first
- **d.** the south pole of the magnet enters the coil first

3. The magnetic flux through a wire loop in a magnetic field **B** does not depend on
- **a.** the area of the loop
- **b.** the shape of the loop
- **c.** the angle between the plane of the loop and the direction of **B**
- **d.** the magnitude B of the field

4. The unit of magnetic flux is the weber, where 1 Wb =
- **a.** 1 T·m²
- **b.** 1 T/m²
- **c.** 1 A·m²
- **d.** 1 A/m²

5. A wire loop is moved parallel to a uniform magnetic field. The induced emf in the loop
- **a.** depends on the area of the loop
- **b.** depends on the shape of the loop
- **c.** depends on the magnitude of the field
- **d.** is 0

6. According to Lenz's law an induced current
- **a.** occurs whenever a moving conductor has a component of its velocity parallel to a magnetic field

b.   occurs whenever a moving conductor has a component of its velocity perpendicular to a magnetic field

c.   gives rise to a magnetic field of its own that reinforces the flux change that induces it

d.   gives rise to a magnetic field of its own that opposes the flux change that induces it

7. Lenz's law is a consequence of the law of conservation of

a.   charge
b.   momentum
c.   field lines
d.   energy

8. When a wire loop is rotated in a magnetic field, the direction of the induced emf changes once in each

a. $\frac{1}{4}$ revolution
b. $\frac{1}{2}$ revolution
c.   1 revolution
d.   2 revolutions

9. The back emf in an electric motor is a maximum when the motor

a.   is switched on
b.   is increasing in speed
c.   has reached its maximum speed
d.   is decreasing in speed

10. The alternating current in the secondary coil of a transformer is induced by

a.   a varying electric field
b.   a varying magnetic field
c.   the iron core of the transformer
d.   motion of the primary coil

11. The ratio between primary and secondary currents in a transformer does not depend on the

a.   ratio of turns in the two windings
b.   resistance of the windings
c.   nature of the core
d.   primary voltage

12. Eddy currents occur only in

a.   ferromagnetic materials
b.   insulators
c.   conductors
d.   coils

13. The unit of inductance is the henry, where 1 H =

a. $1 \text{ J} \cdot \text{A}^2$
b. 1 J/A^2
c. $1 \text{ V} \cdot \text{A}$
d. 1 V/A

14. A large increase in the inductance of a coil can be achieved by using a core that is

a.   diamagnetic
b.   paramagnetic
c.   ferromagnetic
d.   polar

15. The time constant of an RL circuit is the time needed for the current to reach which percentage of its final value?

a.   50%
b.   63%
c.   90%
d.   100%

16. Magnetic fields invariably contain

a.   a ferromagnetic material

b.   inductance
c.   electric current
d.   energy

17. A wire coil carries the current I. The magnetic energy of the coil does not depend on

a.   the value of I
b.   the number of turns in the coil
c.   whether or not the coil has an iron core
d.   the resistance of the coil

18. The horizontal aluminum main boom of a sailboat sailing downwind at 5.0 m/s is 7.0 m long. The boom is at an angle of 75° relative to the boat's direction of motion. If the vertical component of the earth's magnetic field there is 4.0×10^{-5} T, the potential difference between the ends of the boom is

a.   0.03 mV
b.   0.36 mV
c.   1.35 mV
d.   1.40 mV

19. A square wire loop 5 cm on a side is perpendicular to a magnetic field of 0.08 T. If the field drops to 0 in 0.2 s, the average emf induced in the loop during that time is

a.   0.04 mV
b.   0.5 mV
c.   1 mV
d.   8 V

20. A 400-turn coil of resistance 6.0 Ω encloses an area of 30 cm². How rapidly should a magnetic field parallel to the coil axis change in order to induce a current of 0.30 A in the coil?

a.   0.0045 T/s
b.   0.25 T/s
c.   0.67 T/s
d.   1.5 T/s

21. The primary winding of a transformer has 200 turns and its secondary winding has 50 turns. If the current in the secondary winding is 40 A, the current in the primary is

a.   10 A
b.   80 A
c.   160 A
d.   8.0 kA

22. A transformer has 100 turns in its primary winding and 300 turns in its secondary. If the power input to the transformer is 60 W, the power output is

a.   20 W
b.   60 W
c.   0.18 kW
d.   0.54 kW

23. A 60-W, 24-V light bulb is to be powered from a 120-V ac line. Which of the following combinations of primary and secondary transformer windings is needed?

a.   20 turns primary, 100 turns secondary
b.   50 turns primary, 100 turns secondary
c.   100 turns primary, 20 turns secondary
d.   100 turns primary, 50 turns secondary

24. The self-induced emf in a 0.10-H coil when the current in it is changing at the rate of 200 A/s is

a.   10 V
b.   20 V
c.   0.10 kV
d.   2.0 kV

25. The current in a circuit falls to 0 from 16 A in 0.010 s. The average emf induced in the circuit during the drop is 64 V. The inductance of the circuit is

   a. 0.032 H   **b.** 0.040 H

   c. 0.25 H   **d.** 4.0 H

26. A 0.20-H inductor of resistance 8.0 Ω is connected to a 12-V battery of 1.0 Ω internal resistance. The initial rate at which the current in the inductance increases is

   a. 2.4 A/s   **b.** 7.7 A/s

   c. 30 A/s   **d.** 60 A/s

27. The time needed for the current in the inductor of Multiple Choice 26 to reach 63% of its final value is

   a. 0.022 s   **b.** 0.025 s

   c. 1.6 s   **d.** 1.8 s

28. A 2.0-mH coil carries a current of 10 A. The energy stored in its magnetic field is

   a. 0.050 J   **b.** 0.10 J

   c. 1.0 J   **d.** 0.10 kJ

29. The energy contained in a cubic meter of space in which the magnetic field is 1.0 T is

   a. 6.3×10^{-7} J   **b.** 4.0×10^{-5} J

   c. 4.0×10^{5} J   **d.** 7.9×10^{5} J

30. The inductance needed with a capacitor of 200 pF in order to have an *LC* circuit that oscillates at 8.0 kHz is

   a. 2.0 mH   **b.** 4.0 mH

   c. 12 mH   **d.** 2.0 H

QUESTIONS

1. A car is traveling from New York to Florida. Which of its wheels are positively charged and which are negatively charged?

2. One end of a bar magnet is thrust into a coil, and the induced current is clockwise as seen from the side of the coil into which it was thrust. (a) Is the end of the magnet its north or south pole? (b) What will be the direction of the induced current when the magnet is withdrawn?

3. A bar magnet held vertically with its north pole downward is dropped through a wire loop whose plane is horizontal. Is the current induced in the loop just before the magnet enters it clockwise or counterclockwise as seen by an observer above? What is the direction of the current when the magnet passes through the center of the loop? Just after it leaves the loop?

4. A loop of copper wire is rotated in a magnetic field about an axis along a diameter. (a) Why does the loop resist this rotation? (b) If the loop were made of aluminum wire, would the resistance to rotation be different?

5. Explain how the loudspeaker in Fig. 19.41 can also be used as a microphone to convert sound waves into alternating current.

6. The following are basic electromagnetic phenomena: (a) An electric current produces a magnetic field; (b) a current-carrying wire experiences a force in a magnetic field (unless aligned with the field); and (c) an emf is induced in a wire when there is relative motion between it and a magnetic field. Which of these phenomena is involved in the operation of the following devices: an electric motor, a generator, a transformer, a relay, a tape-recording head, a tape-playback head?

7. Why is it easy to turn the shaft of a generator when it is not connected to an outside circuit, but much harder when such a connection is made?

8. Why does an electric motor require more current when it is turned on than when it is running continuously?

9. A coil is rotated 100 times per second in a magnetic field. What time interval separates successive instants when the induced emf is zero?

10. For which of the following can alternating current be used without first being rectified into direct current: an electromagnet, a light bulb, an electric heater, the charging of a storage battery?

11. How is the back emf in the armature of a dc motor at a given speed of rotation related to the emf developed in the same armature when the motor is used as a generator at the same speed of rotation and with the same current in its field coils?

12. In a series-wound generator the magnetic field is produced by an electromagnet whose windings are connected in series with the armature windings. What happens to the emf in such a generator when the current drawn by the external circuit increases? When the external circuit current decreases? Why?

13. In a shunt-wound generator the electromagnet windings are connected in parallel with the armature windings. What happens to the emf in such a generator when the current drawn by the external circuit increases? When the current decreases? Why?

14. What would happen if the primary winding of a transformer were connected to a battery?

15. What acts on the secondary winding of a transformer to cause an alternating potential difference to occur across its ends even though the primary and secondary windings are not connected?

16. Why are transformer cores made from thin sheets of steel?

Would there be any advantage in making permanent magnets in this way?

17. What is the purpose of the air gaps in the recording and playback heads of a tape recorder?

18. The greater its capacitance, the less energy is stored in a capacitor when it is given a certain charge. On the other hand, the greater its inductance, the more energy is stored in an inductor when a certain current is present in it. Explain the difference.

19. What is the direction of the self-induced emf in a coil when the current in it increases? When the current decreases? What is the reason in each case?

20. What becomes of the work done against the back emf in an inductive circuit when a current is being established in it?

EXERCISES

20.2 Moving Wire in a Magnetic Field

1. A car is traveling at 90 km/h in a region where the horizontal component of the earth's magnetic field has the magnitude 2.0×10^{-5} T. Find the emf induced in the car's vertical radio antenna, which is 1.2 m long.

2. A train is traveling at 130 km/h in a region where the vertical component of the earth's magnetic field is 3.0×10^{-5} T. The railway is standard gauge with its rails 1.435 m apart. Find the potential difference between the wheels on each axle of the train's cars.

3. A car is traveling at 30 m/s on a road where the vertical component of the earth's magnetic field is 3.0×10^{-5} T. What is the potential difference between the ends of its axles, which are 2.0 m long?

4. A potential difference of 1.8 V is found between the ends of a 2.0-m wire moving in a direction perpendicular to a magnetic field at a speed of 12 m/s. What is the magnitude of the field?

20.3 Faraday's Law

5. A rectangular wire loop 50 mm $\times$ 100 mm in size is perpendicular to a magnetic field of 1.0×10^{-3} T. (a) What is the flux through the loop? (b) If the magnetic field drops to zero in 3.0 s, what is the potential difference induced between the ends of the loop during that period?

6. The wire loop in Exercise 5 is turned so that a perpendicular to its plane makes an angle of 40° with **B**. Answer the same questions for this situation.

7. A wire loop is 50 mm in diameter and is oriented with its plane perpendicular to a magnetic field. How rapidly should the field change if a potential difference of 1.0 V is to appear across the ends of the loop?

8. A flat 100-turn coil that encloses an area of 50 cm² is turned so that its plane goes from being parallel to a magnetic field of 1.0×10^{-3} T to being perpendicular to the field in 0.040 s. Find the average emf induced in the coil.

9. The magnetic field inside a solenoid 15 mm in radius is increasing at a rate of 0.060 T/s. Find the emf induced in a 400-turn coil wound around the solenoid.

10. A circular wire loop 10 cm in radius with a resistance of 0.40 Ω is in a magnetic field of 0.60 T whose direction is perpendicular to the plane of the loop. The loop is grasped at opposite ends of a diameter and pulled out into a rectangle 30 cm long and 1.4 cm wide in a time of 0.20 s. Find the average current in the wire while it is being deformed.

11. A square wire loop 10 cm on a side is oriented with its plane perpendicular to a magnetic field. The resistance of the loop is 5.0 Ω. How rapidly should the magnetic field change if a current of 2.0 A is to flow in the loop?

20.4 The Generator

12. A 1.50-kW dc generator has a full-load potential difference of 35.0 V and a no-load potential difference of 38.0 V. How much power is dissipated as heat in the armature at full load?

13. The armature of a dc generator has a resistance of 0.20 Ω. The terminal potential difference of the generator is 120 V with no load and 115 V at full load. How much power is delivered at full load?

14. A shunt-wound 120-V, 3.0-kW generator has an armature resistance of 0.10 Ω and a field resistance of 120 Ω. Find its efficiency.

20.5 The Transformer

15. A transformer has 100 turns in its primary winding and 500 turns in its secondary winding. If the primary voltage and current are, respectively, 120 V and 3.0 A, what are the secondary voltage and current? (Assume 100% efficiency in Exercises 15–18.)

16. An electric welding machine employs a current of 400 A. The device uses a transformer whose primary winding has 400 turns and that draws 4.0 A from a 200-V power line. (a) How many turns are there in the secondary winding of the transformer? (b) What is the potential difference across the secondary?

17. A transformer rated at a maximum power of 10 kW is used to couple a 5000-V transmission line to a 240-V circuit. (a) What is the ratio of turns in the windings of the transformer? (b) What is the maximum current in the 240-V circuit?

18. A transformer connected to a 120-V power line has 100 turns in its primary winding and 40 turns in its secondary winding. The primary current is 0.24 A when a light bulb is connected to the secondary winding. What is the resistance of the bulb?

20.6 Inductance

19. The current in a circuit drops from 5.0 A to 1.0 A in 0.10 s. If an average emf of 2.0 V is induced in the circuit while this is happening, what is the inductance of the circuit?

20. What is the self-induced emf in a 0.40-H coil when the current in it is changing at a rate of 500 A/s?

21. Find the inductance of an air-core coil 40 cm long and 40 mm in diameter that has 1000 turns of wire.

22. An air-core solenoid 20 cm long and 20 mm in diameter has an inductance of 0.178 mH. How many turns of wire does it contain?

23. An inductor consists of an iron ring 50 mm in diameter and 1.0 cm^2 in cross-sectional area that is wound with 1000 turns of wire. (Such an inductor is essentially a solenoid bent into a circle.) The permeability of the iron is constant at 400 times that of free space at the magnetic fields at which the inductor will be used. What is the inductance of the inductor?

20.7 Inductive Time Constant

24. Show that the unit of L/R is the second.

25. What is the inductance of a coil whose resistance is 14 Ω and whose time constant is 0.10 s?

26. A 50-mH coil with a resistance of 20 Ω is connected to a 90-V battery of negligible internal resistance. What is the time constant of the circuit?

20.8 Magnetic Energy

27. A magnetic field of 1.0×10^2 T is very strong. (a) How much energy is contained in 1.0 liter of such a field? (b) What electric field would have the same energy density?

28. A 20-mH coil carries a current of 0.20 A. (a) How much

energy is stored in its magnetic field? (b) What should the current be in order that it contain 1.0 J of energy?

29. A solenoid 200 mm long and 24.0 mm in diameter is wound with 1200 turns of wire whose total resistance is 40.0 Ω. The solenoid is connected to a 12.0-V battery whose internal resistance is 2.00 Ω. Find the inductance of the solenoid, the final current in it, and its energy content when the final current flows.

30. The solenoid in Exercise 29 is immersed in liquid oxygen whose permeability is 1.0049 times greater than that of free space. Find the inductance of the solenoid, the final current in it, and its energy content when the final current flows.

31. A 2.0-H coil carries a current of 0.50 A. (a) How much energy is stored in it? (b) In how much time should the current drop to 0 if an emf of 100 V is to be induced in it?

20.9 Electrical Oscillations

32. Show that the unit of $1/2\pi\sqrt{LC}$ is the hertz.

33. In the antenna circuit of a certain radio receiver, the inductance is fixed at 4.0 mH but the capacitance can be varied. When the capacitance is 10 pF, what is the frequency of the radio waves the receiver responds to?

34. What inductance is needed in a circuit in which $C = 60$ μF if its natural frequency is to be 30 Hz?

35. What capacitance is needed in a circuit in which $L = 2$ H if its natural frequency is to be 200 Hz?

36. The frequencies used in the commerical radio broadcasting range from 550 to 1600 kHz. What range of capacitance should a variable capacitor have if it is connected to a coil of inductance 1.0 mH in a circuit designed to respond to frequencies in this band?

37. A 1.0-μF capacitor is charged by being connected to a 10-V battery. The battery is then removed and the capacitor connected to a 10-mH coil. (a) Find the frequency of the resulting oscillations. (b) Find the maximum value of the charge on the capacitor. (c) Find the maximum current that flows through the inductor.

38. A capacitor with the charge Q is connected to an inductor, and electrical oscillations with the frequency f occur as a result. How is the maximum current I that flows in the circuit related to Q and f?

PROBLEMS

39. An air-core solenoid 20 cm long and 20 mm in radius is wound with 1000 turns and has a secondary coil of 300 turns wound over it. (a) If the current in the solenoid is changing at

a rate of 0.50 A/s, what is the emf in the secondary coil? (b) An iron core with a constant permeability of $\mu = 400\mu_0$ is placed in the solenoid. What is the induced emf now?

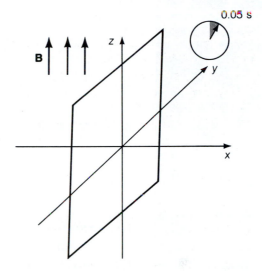

Figure 20.24

Problem 40.

40. A square wire loop 100 mm on a side rotates at a constant angular velocity in a uniform magnetic field **B** so that it turns through 90° in 0.050 s, as shown in Fig. 20.24. Initially the loop lies in the xy-plane and afterward in the yz-plane. The field **B** has a magnitude of 0.40 T and points in the $+z$-direction. (a) Find the average emf $\mathscr{E}_{av}$ induced in the loop during the turn. (b) In what part of the turn is $\mathscr{E}$ a minimum? A maximum? What are the values of $\mathscr{E}_{min}$ and $\mathscr{E}_{max}$?

41. A shunt-wound 120-V dc electric motor has an armature resistance of 1.0 Ω and a field resistance of 200 Ω. The motor has a power input of 420 W at 2000 rpm. (a) Find the power output and efficiency of the motor. (b) What is the starting current? (c) What series resistance is required if the starting current is to be limited to 25 A?

42. Find the power output, the power input, and the efficiency of the motor in Problem 41 when it is operating at 1800 rev/min.

43. A series-wound 24-V dc electric motor has an armature resistance of 0.10 Ω and a field resistance of 0.40 Ω. The motor draws 20 A at its normal operating speed. Find the power output, the power input, and the efficiency of the motor.

44. Show that the total inductance L of three inductors in parallel is given by $1/L = 1/L_1 + 1/L_2 + 1/L_3$. (*Hint:* The self-induced emf is the same for all the inductors, because they are connected in parallel.)

45. Show that the total inductance L of three inductors in series is given by $L = L_1 + L_2 + L_3$. (*Hint:* The total self-induced emf of the inductors is equal to the sum of the self-induced emf's of the individual inductors.)

46. A coil 20 cm long and 30 mm in diameter is tightly wound with one layer of copper wire 1.0 mm in diameter. Find its time constant.

47. A potential difference of 50 V is suddenly applied across a 12.0-mH, 8.0-Ω coil. Find (a) the initial current and the initial rate of change of current, (b) the current when the rate of change of current is 2000 A/s, and (c) the final current and final rate of change of current.

48. A potential difference of 100 V is suddenly applied across a 0.50-H, 20.0-Ω inductor. Find (a) the initial rate of increase of current, (b) the rate of increase of current when the current is 3.0 A, (c) the final current, and (d) the energy content of the inductor when the final current flows in it.

49. A 2.0-μF capacitor is charged by being connected to a 24-V battery. The battery is then removed and the capacitor connected to a 50-mH coil. After $\frac{1}{8}$ cycle the initial energy will be equally divided between the electric field in the capacitor and the magnetic field in the coil, as in Fig. 20.23(b). (a) Find the time interval required for this to occur. (b) Find the charge on the capacitor and the current in the inductor at this moment.

Answers to Multiple Choice

1. b	9. c	17. d	25. b
2. b	10. b	18. c	26. d
3. b	11. d	19. c	27. a
4. a	12. c	20. d	28. b
5. d	13. b	21. a	29. c
6. d	14. c	22. b	30. d
7. d	15. b	23. c	
8. b	16. d	24. b	

C H A P T E R

ALTERNATING CURRENT

21.1   **Effective Current and Voltage**

21.2   **Phasors**

21.3   **Phase Relationships**

21.4   **Inductive Reactance**

21.5   **Capacitive Reactance**

21.6   **Impedance**

21.7   **Resonance**

21.8   **Power in AC Circuits**

21.9   **Impedance Matching**

21.10   **Parallel AC Circuits**

Nearly all the world's electric energy is carried by alternating current. Mainly this is due to the economy of high-voltage transmission, which minimizes I^2R heat losses. With step-up transformers at the production end and step-down transformers at the consumption end, the transmission voltage is limited only by insulation and atmospheric discharge problems. Alternating current is also preferred in industry because ac electric motors are, as a class, cheaper, more durable, and less in need of maintenance than dc motors. (Dc motors, however, have certain characteristics, such as better speed regulation, that make them more suitable for specialized duties.) Alternating currents are also involved in all aspects of modern communication: The electrical equivalent of a sound wave of a certain frequency is an alternating current of that frequency, and radio waves are produced by antennas fed with high-frequency ac. Alternating current behaves in a circuit in a very different way from direct current, and some insight into ac circuit behavior is needed to understand much of modern technology.

21.1 EFFECTIVE CURRENT AND VOLTAGE

Each is less than the corresponding maximum value.

A direct current is described in terms of its direction and magnitude. In a simple circuit there might be a current of 6 A from the positive terminal of a battery through a resistance network to its negative terminal. An alternating current has neither a constant direction nor a constant magnitude. How do we describe it?

Since alternating current flows back and forth in a circuit, it has no "direction" in the same sense as a direct current has. However, the oscillations have a certain frequency in each case, and the value of this frequency—that is, how many times per second the current goes through a complete cycle—is part of the description of the current.

An alternating current varies with time as shown in Fig. 21.1, and it would seem natural to specify its maximum value, I_{max}, as well as its frequency. The trouble with doing this is that I_{max} is not a measure of the ability of a current to do work or produce heat. A 6-A direct current is not equivalent to an alternating current in which I_{max} = 6 A. A better procedure is to define an **effective current** I_{eff} such that a direct current of this magnitude produces heat in a resistor at the same rate as the alternating current.

An alternating current varies with time according to the formula

$$I = I_{max} \sin 2\pi ft \qquad\qquad (21.1)$$

where f is the frequency of the current. Here the current is assumed to be $I = 0$ and is increasing when $t = 0$. Figure 21.1 is a graph of this formula. The rate at which heat is dissipated in a resistance R by an alternating current is, at any time t, given by

$$I^2R = I^2_{max}R \sin^2 2\pi ft$$

The average value of I^2R over a complete cycle is

$$[I^2R]_{av} = I^2_{eff}R = I^2_{max}R[\sin^2 2\pi ft]_{av} \qquad\qquad (21.2)$$

What we must find is the average value of $\sin^2 2\pi ft$ over a complete cycle. (The average value of I over a cycle is 0, since I is positive for half the cycle and negative for the other half. However, I^2 is always positive, and its average is a positive quantity. See Fig. 21.2.)

We begin with the trigonometric identity

$$\sin^2 \theta = \tfrac{1}{2}(1 - \cos 2\theta)$$

The average value of $\sin^2 \theta$ over a complete cycle is therefore

$$[\sin^2 \theta]_{av} = \tfrac{1}{2}[1 - \cos 2\theta]_{av} = \tfrac{1}{2} - \tfrac{1}{2}[\cos 2\theta]_{av} = \tfrac{1}{2}$$

Figure 21.1

The variation of an alternating current with time. The frequency of the current is the number of cycles that occur per second.

Figure 21.2

(a) The average value of I in an ac circuit is zero. (b) The average value of I^2 in an ac circuit is $\frac{1}{2}I^2_{max}$.

(a)

(b)

because, over a complete cycle, the average value of cos 2θ is 0 by the same reasoning as in the case of sin θ. As a result

$$I^2_{\text{eff}}R = I^2_{\text{max}}R[\sin^2 2\pi ft]_{\text{av}} = \tfrac{1}{2}I^2_{\text{max}}R$$

and

$$I_{\text{eff}} = \frac{I_{\text{max}}}{\sqrt{2}} = 0.707I_{\text{max}} \qquad\qquad \textit{Effective current} \quad (21.3)$$

The effective magnitude of an alternating current is 70.7% of its maximum value.

In a similar way the effective voltage in an ac circuit turns out to be

$$V_{\text{eff}} = \frac{V_{\text{max}}}{\sqrt{2}} = 0.707V_{\text{max}} \qquad\qquad \textit{Effective voltage} \quad (21.4)$$

It is customary to express currents and voltages in ac circuits in terms of their effective values. Thus the potential difference across a "120-volt, 60-Hz" power line actually varies from

$$+ V_{\text{max}} = + \frac{V_{\text{eff}}}{0.707} = + \frac{120 \text{ volts}}{0.707} = + 170 \text{ volts}$$

through 0 to -170 volts and back to $+170$ volts a total of 60 times per second (Fig. 21.3).

In what follows, when current, potential difference, and emf values are given for an ac circuit without other qualification, they will refer to the effective magnitudes of these quantities.

By analogy with circular motion, angular frequency ω (in radians per second) is often used instead of frequency f (in hertz) in discussing alternating currents, where

$$\omega = 2\pi f$$

In this notation, which will not be used here, the instantaneous current in an ac circuit is written

$$I = I_{\text{max}} \sin \omega t$$

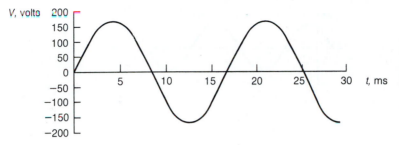

Figure 21.3

The voltage in a "120-V, 60-Hz" power line actually varies from −170 V to +170 V. Each complete cycle takes 16.7 ms.

What are called here "effective" values of current and voltage are elsewhere sometimes called "root-mean-square" or "rms" values from their definitions as the square roots of the average values of I^2 and V^2.

21.2 PHASORS

Rotating vectors used to represent alternating currents or voltages.

A convenient way to represent ac currents and voltages is in terms of **phasors.** This scheme is based on the fact that the component in any direction of a uniformly rotating vector varies sinusoidally—that is, as $\sin \theta$ varies with θ—with time. In the case of an alternating current, the length of the phasor $\mathbf{I}_{max}$ corresponds to I_{max}, and we imagine it to rotate f times per second in a counterclockwise sense (Fig. 21.4). Since the angle θ is equal to $2\pi ft$, the vertical component of the phasor at any time

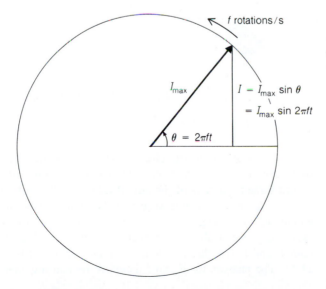

Figure 21.4

Phasor representation of I_{max} and I in an alternating current of frequency f. The vertical component of the phasor I_{max} at any time t is equal to the instantaneous value I of the current at that time.

Figure 21.5

(a) In a certain ac circuit, the oscillations in V occur earlier than those in I by $\frac{1}{6}$ cycle. (b) In a phasor diagram, the phasor $\mathbf{V}_{max}$ leads the phasor $\mathbf{I}_{max}$ by $360°/6 = 60°$. The orientations of the phasors correspond to the time T in the graphs of part (a).

t corresponds to the instantaneous current I of Eq. (21.1):

$$I = I_{max} \sin \theta = I_{max} \sin 2\pi ft$$

In a similar way a phasor $\mathbf{V}_{max}$ can be used to represent an ac voltage V.

Phasors are helpful because the current and voltage in an ac circuit or circuit element always have the same frequency but may differ in phase, so that the peaks in each quantity do not occur at the same times. Suppose, for instance, that the voltage in a certain circuit leads the current by $\frac{1}{6}$ cycle, as in Fig. 21.5(a). (That is, the voltage peaks occur $\frac{1}{6}$ cycle earlier than the current peaks do.) In a phasor diagram, such as Fig. 21.5(b), this situation is shown by having the phasor $\mathbf{V}_{max}$ at an angle of 60° counterclockwise from the phasor $\mathbf{I}_{max}$, since $360°/6 = 60°$. We must imagine the two phasors rotating together, always with the same angle of 60° between them, and generating the changing values of V and I shown in Fig. 21.5(a) as times goes on.

21.3 PHASE RELATIONSHIPS

In a resistor, V is in phase with I; in an inductor, V leads I; in a capacitor, V lags behind I.

All actual electric circuits exhibit resistance, capacitance, and inductance to some degree. When direct current flows in a circuit, only its resistance is significant, but all three properties of the circuit affect the flow of alternating current.

A pure-resistance ac circuit is an idealized circuit whose capacitance and inductance are negligible. The instantaneous values of the voltage and current are in the same phase at all times in such a circuit. Both V and I are zero at the same time, both V and I pass through their maximum values at the same time, and so on (Fig. 21.6). The phasors for V and I therefore remain together.

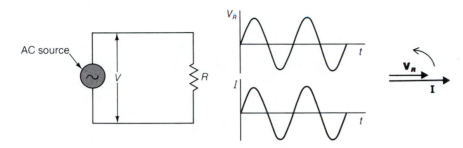

Figure 21.6

Current and voltage are in phase in a pure resistance.

A pure-inductance ac circuit is an idealized circuit whose resistance and capacitance are negligible. There is no IR potential drop across the inductance, and the potential difference across it is therefore proportional to $\Delta I/\Delta t$, the rate of change of the current (see Section 20.6). In this situation V and I cannot be in phase with each other. The current I changes most rapidly when $I = 0$, so $V = \pm V_{max}$ when $I = 0$. On the other hand, $\Delta I/\Delta t = 0$ at I_{max}, so $V = 0$ when $I = \pm I_{max}$. As shown in Fig. 21.7,

---

The voltage across a pure inductor leads the current in the inductor by $\frac{1}{4}$ cycle.

That is, the variations in voltage occur $\frac{1}{4}$ cycle *earlier than* the corresponding variations in current.

In a phasor diagram a difference in phase of $\frac{1}{4}$ cycle means that the angle between the phasors $\mathbf{V}$ and $\mathbf{I}$ is $360°/4 = 90°$. Since $\mathbf{V}$ leads $\mathbf{I}$ in a pure inductor, $\mathbf{V}$ is $90°$ counterclockwise from $\mathbf{I}$, as in Fig. 21.7.

The potential difference across a capacitor depends on the amount of charge stored on its plates. The charge is a maximum when $I = 0$, which is when the current is about to reverse direction and carry away the stored charge. The potential difference across a capacitor is

$$V = \frac{Q}{C}$$

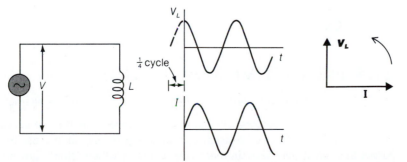

Figure 21.7

The voltage across a pure inductor leads the current in the inductor by $\frac{1}{4}$ cycle.

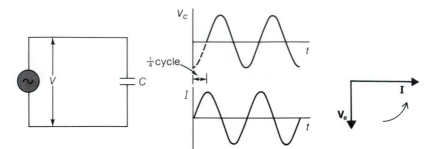

Figure 21.8

The voltage across a pure capacitor lags behind the current into and out of the capacitor by $\frac{1}{4}$ cycle.

and so $V = \pm V_{\max}$ when $I = 0$. The stored charge is 0 when $I = \pm I_{\max}$, because at these times the former stored charge is all gone and charge of the opposite sign is about to build up. Hence $V = 0$ when $I = \pm I_{\max}$. As shown in Fig. 21.8,

---

The voltage across a pure capacitor lags behind the current into and out of the capacitor by $\frac{1}{4}$ cycle.

That is, the variations in the voltage occur $\frac{1}{4}$ cycle *later than* the corresponding variation in the current. In a phasor diagram **V** is 90° clockwise from **I,** as in Fig. 21.8.

In a pure-capacitance circuit the phase relationship between current and voltage is different from what it is in pure-resistance and pure-inductance circuits. Although alternating current does not flow *through* a capacitor, it does flow *into one plate and out of the other* because changes in the amount of charge stored on one of the plates are mirrored by changes in the charge stored on the other plate. A flow of $+Q$ into one plate means that $+Q$ flows out of the other plate to leave the latter with a net charge of $-Q$. If direct current were involved, eventually enough charge would accumulate to prevent the arrival of any more, and the current would stop. In the case of an alternating current, however, the current always stops and reverses itself periodically anyway, so the presence of a series capacitor does not prevent ac from flowing in the circuit.

21.4 INDUCTIVE REACTANCE

A measure of the opposition of an inductor to the flow of alternating current.

Resistors, inductors, and capacitors all impede the flow of alternating current in a circuit. The effect of resistance is to dissipate part of the electric energy that passes through it into heat. In those conductors in which Ohm's law holds for direct

Low L High L

High I_L Low I_L

Figure 21.9
At a given frequency, the lower
the inductance L, the lower the
inductive reactance X_L and the
higher the current I_L.

current, it holds for alternating current as well, and

$$I = \frac{V_R}{R} \tag{21.5}$$

Here V_R is the effective potential difference across the resistance R, and I is the effective current through it.

   The opposition of an inductor to the flow of alternating current comes from the self-induced back emf produced in it by the changing current. The back emf represents a potential drop across the inductor, and the current in the circuit is correspondingly reduced.

   The **inductive reactance** X_L of an inductor is a measure of its effect on an alternating current passing through it. The effective current I in an inductor is related to the effective potential difference V_L across it and the inductive reactance X_L by

$$I = \frac{V_L}{X_L} \tag{21.6}$$

The unit of inductive reactance is the ohm. Equation (21.6) is analogous to Ohm's law, but there is a basic distinction between reactance and resistance in that there is no power loss in an inductor, whereas power is dissipated as heat in a resistor.

   The inductive reactance of an inductor is given by the formula

$$X_L = 2\pi f L \qquad\qquad \textit{Inductive reactance} \quad (21.7)$$

where f is the frequency of the current in hertz (cycles/second) and L is the inductance in henries. The direct dependence of X_L on f and L is reasonable. The self-induced back emf, which is what opposes the current, is proportional to both $\Delta I/\Delta t$ and L, and so the more rapidly the current changes and the larger the value of L, the greater the back emf (Fig. 21.9).

◆   E X A M P L E   **21.1** _____

What is the current in a coil of negligible resistance and inductance 0.40 H when it is connected to a 120-V, 60-Hz power line?

SOLUTION   The reactance of the coil is

$$X_L = 2\pi fL = (2\pi)(60 \text{ Hz})(0.4 \text{ H}) = 151 \ \Omega$$

and the current in it is

$$I = \frac{V_L}{X_L} = \frac{120 \text{ V}}{151 \ \Omega} = 0.80 \text{ A}$$

Both the 120-V and 0.80-A figures represent effective values.   ◆

21.5 CAPACITIVE REACTANCE

A measure of the opposition of a capacitor to the flow of alternating current.

The extent to which a capacitor opposes the flow of alternating current depends on its **capacitive reactance** X_C. If V_C is the effective potential difference across a capacitor whose reactance is X_C, the effective current into and out of the capacitor is

$$I = \frac{V_c}{X_c} \tag{21.8}$$

The capacitive reactance of a capacitor is given by the formula

$$X_C = \frac{1}{2\pi fC} \qquad\qquad\qquad\qquad \textit{Capacitive reactance} \quad (21.9)$$

If f is in hertz and the capacitance C in farads, the unit of X_C is the ohm.

◆   **E X A M P L E   21.2** _____

A capacitor whose reactance is 80 Ω at 50 Hz is used in a 60-Hz circuit. What is its reactance in the latter circuit?

SOLUTION   The capacitance of the capacitor, from Eq. (21.9), is

$$C = \frac{1}{2\pi fX_C} = \frac{1}{(2\pi)(50 \text{ Hz})(80 \ \Omega)} = 4 \times 10^{-5} \text{ F}$$

Figure 21.10

At a given frequency, the higher the capacitance C, the lower the capacitive reactance X_C and the higher the current I_C.

which is 40 μF. Its reactance at 60 Hz is

$$X_C = \frac{1}{2\pi f C} = \frac{1}{(2\pi)(60 \text{ Hz})(4 \times 10^{-5} \text{ F})} = 66 \ \Omega$$

♦

A capacitor impedes the flow of alternating current through the reverse potential difference that appears across it as charge builds up on its plates. Thus there is a potential drop across a capacitor in an ac circuit that affects the current just as the potential drop in a resistor does.

The inverse dependence of X_C on f and C can be understood from the following argument. If the charge on the plates of a capacitor changes by ΔQ in the time Δt, then the instantaneous current is $I_{\text{inst}} = \Delta Q/\Delta t$. If V_C is the change in the potential difference across the capacitor that leads to ΔQ, then $\Delta Q = C(\Delta V_C)$, and $I_{\text{inst}} = C(\Delta V_C/\Delta t)$. The higher the frequency, the greater the rate of change $\Delta V_C/\Delta t$ of the potential difference and the greater the current. The larger C is, also, the greater the current (Fig. 21.10). Since a high current means a low reactance, increasing f and C decreases X_C.

Capacitive and inductive reactances vary differently with frequency, as Fig. 21.11 shows: X_C decreases with increasing f, whereas X_L increases with increasing f.

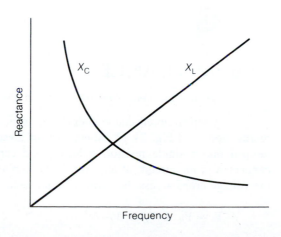

Figure 21.11

The reactance X_L of an inductor increases with frequency, whereas the reactance X_C of a capacitor decreases with frequency. In the dc limit of $f = 0$, $X_L = 0$ and $X_C = \infty$.

(a)

(b)

Figure 21.12

A capacitor can be used to favor the passage of either ac or dc between two circuits, depending on how it is connected.

In the limit of $f = 0$, which means direct current, $X_L = 0$ and $X_C = \infty$. When the current is steady, there is no self-induced back emf in an inductor, and no inductive reactance to impede current. On the other hand, a capacitor completely obstructs direct current because the charge that builds up on its plates remains there instead of surging back and forth as it does when an ac potential is applied.

Because capacitors pass ac but not dc, they can be used to favor the transmission of signals of one kind or the other between two circuits. In Fig. 21.12(a) the capacitor permits an ac signal to go from circuit 1 to circuit 2 while stopping any direct current. In Fig. 21.12(b) a direct current can pass between the circuits, but the capacitor provides a path to ground for alternating current, and most of the ac will take this path if the reactance is low.

21.6 IMPEDANCE

The ac equivalent of resistance in a dc circuit.

A series circuit that contains resistance, inductance, and capacitance can be represented as in Fig. 21.13, where each of these circuit properties is considered as lumped into a single resistor, inductor, and capacitor. If an ac source of emf is connected to the circuit, at any instant the applied voltage V is equal to the sum of the voltage drops across the various circuit elements:

$$V = V_R + V_L + V_C \qquad \textit{Instantaneous voltage} \quad (21.10)$$

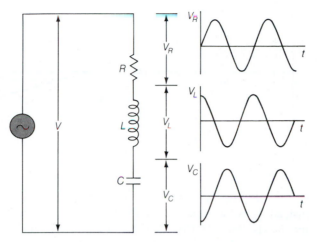

Figure 21.13

The instantaneous value of the voltage V applied to a series RLC circuit is equal to the sum of the instantaneous values of V_R, V_L, and V_C. This relationship does not hold for the effective values of the various voltages because of the phase differences among them.

However, V_R, V_L, and V_C are *out of phase with one another*. The voltage across the resistor, V_R, is always in phase with the current I, but V_L is $\frac{1}{4}$ cycle ahead of I, and V_C is $\frac{1}{4}$ cycle behind I. This situation is shown in Fig. 21.13.

While Eq. (21.10) is always correct when V, V_R, V_L, and V_C refer to the instantaneous values of the various voltages, it is *not* correct when effective (or maximum) values are involved. We must use a vectorial approach to take the phase differences into account.

Figure 21.14 is a phasor diagram that shows the phase differences between $\mathbf{V}_R$, $\mathbf{V}_L$, and $\mathbf{V}_C$. We see that $\mathbf{V}_L$ is 90° ahead of $\mathbf{V}_R$ (hence 90° counterclockwise from $\mathbf{V}_R$, since phasors rotate counterclockwise) and $\mathbf{V}_C$ is 90° behind $\mathbf{V}_R$. The vector sum $\mathbf{V}$ of $\mathbf{V}_R$, $\mathbf{V}_L$, and $\mathbf{V}_C$ represents the effective voltage across the terminals of the circuit. The magnitude of $\mathbf{V}$ is

$$V = \sqrt{V_R^2 + (V_L - V_C)^2} \qquad \textit{Effective voltage} \quad (21.11)$$

The angle ϕ between $\mathbf{V}$ and $\mathbf{V}_R$ is called the **phase angle** because it is a measure of how much the voltage in the circuit leads or lags behind the current. The direction of the effective current phasor $\mathbf{I}$ is always the same as the direction of $\mathbf{V}_R$.

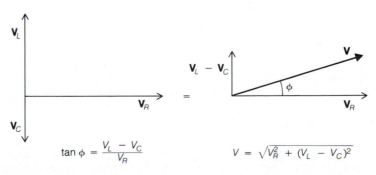

Figure 21.14

Phasor diagram of the effective voltages across the resistor, inductor, and capacitor of Fig. 21.13. The magnitude of the vector sum of the voltage phasors equals the effective voltage across the entire circuit. The angle ϕ is the phase angle. The direction of the effective current phasor $\mathbf{I}$ is always the same as that of $\mathbf{V}_R$.

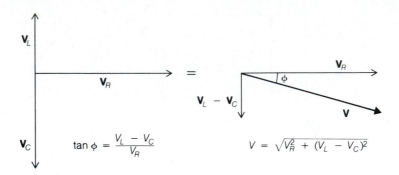

Figure 21.15

The procedures for finding **V** and ϕ are the same whether V_L is greater than V_C or the reverse.

$$\tan \phi = \frac{V_L - V_C}{V_R}$$

$$V = \sqrt{V_R^2 + (V_L - V_C)^2}$$

In Fig. 21.14 V_L is greater than V_C, and $(V_L - V_C)$ is a positive quantity. If instead V_L is the smaller quantity, $(V_L - V_C)$ is negative and the vector **V** is below the x-axis (Fig. 21.15). However, Eq. (21.11) still applies, since $(V_L - V_C)^2 = (V_C - V_L)^2$. The phase angle in either case is specified by

$$\tan \phi = \frac{V_L - V_C}{V_R} \qquad\qquad \textit{Phase angle} \quad (21.12)$$

If $V_C > V_L$, the result will be a negative value for ϕ, which signifies that **V** lags behind **V**$_R$ (and behind the current **I**).

♦ **E X A M P L E   21.3** _____

A resistor, a capacitor, and an inductor are connected in series across an ac power source. The effective voltages across the circuit components are $V_R = 5$ V, $V_C = 10$ V, and $V_L = 12$ V. Find the effective voltage of the source and the phase angle in the circuit.

SOLUTION (a) From Eq. (21.11)

$$V = \sqrt{V_R^2 + (V_L - V_C)^2} = \sqrt{(5 \text{ V})^2 + (12 \text{ V} - 10 \text{ V})^2} = 5.4 \text{ V}$$

We note that V_C and V_L are both greater in magnitude than the source voltage in this case.

(b) Since

$$\tan \phi = \frac{V_L - V_C}{V_R} = \frac{12 \text{ V} - 10 \text{ V}}{5 \text{ V}} = 0.4$$

we have for the phase angle $\phi = 22°$. ♦

Because

$$V_R = IR \qquad V_L = IX_L \qquad \text{and} \qquad V_C = IX_C$$

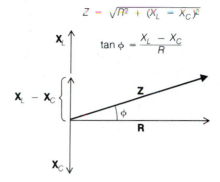

Figure 21.16

The phasor impedance diagram that corresponds to the phasor voltage diagram shown in Fig. 21.14.

we can rewrite Eq. (21.11) in the form

$$V = I\sqrt{R^2 + (X_L - X_C)^2}$$

The quantity

$$Z = \sqrt{R^2 + (X_L - X_C)^2} \qquad \textit{Impedance} \quad (21.13)$$

is known as the **impedance** of a series circuit containing resistance, inductance, and capacitance. The unit of impedance is evidently the ohm. Impedance in an ac circuit plays the same role that resistance does in a dc circuit. In an ac circuit

$$I = \frac{V}{Z} \qquad\qquad \textit{Current in ac circuit} \quad (21.14)$$

is the effective current that flows when the effective voltage V is applied. It is important to keep in mind that Z not only depends on the circuit parameters R, L, and C but also varies with the frequency f.

The current I is the same in all parts of the circuit at all times, and

$$Z = \frac{V}{I} \qquad R = \frac{V_R}{I} \qquad X_L = \frac{V_L}{I} \qquad X_C = \frac{V_C}{I}$$

Hence the phasor voltage diagram of Fig. 21.14 can be replaced by the phasor impedance diagram of Fig. 21.16. The phase angle ϕ is the same in both cases, of course, and can be calculated from the relation

$$\tan \phi = \frac{X_L - X_C}{R} \qquad\qquad \textit{Phase angle} \quad (21.15)$$

◆   **E X A M P L E   21.4** _____

Analyze a series circuit that consists of a 10.0-mH inductor, a 10.0-μF capacitor, and a 30.0-Ω resistor connected to a source of 100-V, 400-Hz alternating current.

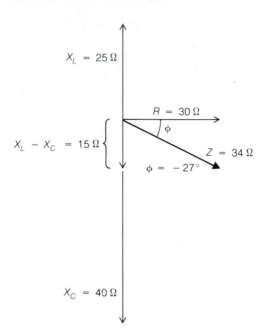

Figure 21.17

A negative phase angle occurs when X_C is greater than X_L. This signifies that the voltage in the circuit lags behind the current (Example 21.4).

SOLUTION   The reactances of the inductor and capacitor at 400 Hz are

$$X_L = 2\pi fL = (2\pi)(400\ \text{Hz})(0.010\ \text{H}) = 25.1\ \Omega$$

$$X_C = \frac{1}{2\pi fC} = \frac{1}{(2\pi)(400\ \text{Hz})(1.00 \times 10^{-5}\ \text{F})} = 39.8\ \Omega$$

The phasor impedance diagram for this circuit is shown in Fig. 21.17. The impedance Z is

$$Z = \sqrt{R^2 + (X_L - X_C)^2} = \sqrt{(30.0\ \Omega)^2 + (25.1\ \Omega - 39.8\ \Omega)^2} = 33.4\ \Omega$$

The phase angle ϕ is found as follows:

$$\tan \phi = \frac{X_L - X_C}{R} = -\frac{14.7\ \Omega}{30.0\ \Omega} = -0.49 \qquad \phi = -26°$$

A negative phase angle signifies that the voltage lags behind the current. Here the lag is 26°, which is $\frac{26}{360}$ or 0.072 of a complete cycle.

The current in the circuit is

$$I = \frac{V}{Z} = \frac{100\ \text{V}}{33.4\ \Omega} = 2.99\ \text{A}$$

The effective potential difference across each of the circuit elements is

$$V_L = IX_L = (2.99 \text{ A})(25.1 \ \Omega) = 75 \text{ V}$$
$$V_C = IX_C = (2.99 \text{ A})(39.8 \ \Omega) = 119 \text{ V}$$
$$V_R = IR = (2.99 \text{ A})(30.0 \ \Omega) = 90 \text{ V}$$

We see again that the effective voltage across an inductor or a capacitor in an ac circuit can exceed the effective voltage applied to the entire circuit.

The arithmetic sum of the above voltages is 284 V, but this sum means nothing because the voltages are not in phase with one another. The vector sum of the voltages, which takes into account the phase differences among them, is

$$V = \sqrt{V_R^2 + (V_L - V_C)^2} = \sqrt{(90 \text{ V})^2 + (75 \text{ V} - 119 \text{ V})^2} = 100 \text{ V}$$

as it should be. ♦

21.7 RESONANCE

Impedance is a minimum when $X_L = X_C$.

When an ac voltage is applied to a series circuit, the current depends on the frequency. The greatest current flows when the impedance Z is a minimum. Since

$$Z = \sqrt{R^2 + (X_L - X_C)^2}$$

the condition for minimum impedance in a given circuit is that the frequency be such that the inductive and capacitive reactances are equal. When this is true,

$$X_L = X_C \quad \text{and} \quad 2\pi f_0 L = \frac{1}{2\pi f_0 C}$$

Thus

$$f_0 = \frac{1}{2\pi\sqrt{LC}} \qquad\qquad \textit{Resonance frequency} \quad (21.16)$$

When the applied voltage has this frequency, the current is a maximum and is limited only by the resistance R. The frequency f_0 is called the **resonance frequency** of the circuit. **Resonance** occurs in a series circuit when the impressed voltage oscillates with the resonance frequency.

We note that Eq. (21.16) is the same as Eq. (20.14) for the "natural" frequency at which an LC circuit will oscillate if the capacitor is initially charged.

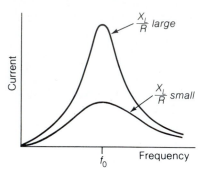

Figure 21.18

The variation with frequency of the current in a series ac circuit. The maximum current occurs at the resonance frequency f_0.

When a circuit is in resonance with an applied voltage, the energy alternately stored in and discharged from the capacitor is the same as the energy stored in and discharged from the inductor. At other frequencies the energy contents of the capacitor and the inductor are different. The difference interferes with the back-and-forth flow of power and gives rise to an impedance that exceeds the resistance R. The current that flows in an *RLC* circuit at resonance is

$$I = \frac{V}{R}$$

just as in a dc circuit.

Figure 21.18 shows how the current in a series circuit varies with frequency. The less the resistance R relative to the inductive reactance X_L, the sharper the peak in the curve. The antenna circuit of one kind of radio receiver is tuned to respond to a particular frequency of radio waves by adjusting a variable capacitor or inductor until the resonance frequency of the circuit is equal to the signal frequency. A circuit in which X_L/R is large has a narrow response curve and can separate signals from two stations very close together in frequency.

At resonance, $X_L = X_C$, and there is no phase difference between current and voltage. Another way to describe resonance, then, is to say that current and voltage are in phase at all times, just as in a pure-resistance circuit.

◆　　**E X A M P L E　21.5** _____

The antenna circuit of a certain radio receiver consists of a 10-mH coil and a variable capacitor; the resistance in the circuit is 50 Ω. An 880-kHz (kilohertz) radio wave produces a potential difference of 1.0×10^{-4} V across the circuit. Find the capacitance required for resonance and the current at resonance.

SOLUTION　　The capacitance required for resonance at 880 kHz is

$$C = \frac{1}{(2\pi f)^2 L} = \frac{1}{(2\pi \times 8.8 \times 10^5 \,\text{Hz})^2 (1.0 \times 10^{-2} \,\text{H})}$$
$$= 3.3 \times 10^{-12} \,\text{F} = 3.3 \,\text{pF}$$

At resonance, inductive and capacitive reactances cancel out each other, and the current that flows is

$$I = \frac{V}{R} = \frac{1.0 \times 10^{-4}\,\text{V}}{50\,\Omega} = 2.0 \times 10^{-6}\,\text{A} = 2.0\,\mu\text{A} \qquad \blacklozenge$$

21.8 POWER IN AC CIRCUITS

Because inductors and capacitors do not absorb power, the apparent power IV may exceed the actual power.

No power is consumed in a pure inductor or capacitor in an ac circuit. These elements act merely as temporary reservoirs of energy and return whatever energy they absorb in one quarter of a cycle to the circuit in the next quarter of that cycle. No power is therefore needed to maintain an ac current in the inductive and capacitive parts of a circuit. The resistance in the circuit, however, dissipates power as heat at the rate $P = IV_R$, where, as usual, I and V_R are effective values.

From Fig. 21.14 we see that

$$V_R = V \cos \phi \tag{21.17}$$

since $V \cos \phi$ is the component of $\mathbf{V}$ that is in phase with the current phasor $\mathbf{I}$. Hence the effective power delivered to an ac circuit is

$$P = IV \cos \phi \qquad \qquad \textit{Power in ac circuit} \quad (21.18)$$

The quantity $\cos \phi$ is called the **power factor** of the circuit. The power factor is equal to 1 only at resonance, when current and voltage are in phase; then $\phi = 0$ and $\cos \phi = 1$. Under other circumstances the voltage is not in phase with the current, and the actual power is less than IV.

There is an interesting analogy between Eq. (21.18) and the mechanical power $P = Fv \cos \theta$ developed by a force $\mathbf{F}$ that acts on an object moving at the velocity $\mathbf{v}$ when the angle between $\mathbf{F}$ and $\mathbf{v}$ is θ. When $\mathbf{F}$ is parallel to $\mathbf{v}$, $\cos \theta = \cos 0 = 1$ and the power is a maximum. When $\mathbf{F}$ is perpendicular to $\mathbf{v}$, $\cos \theta = \cos 90° = 0$ and $P = 0$. In the electrical case, we can think of V as corresponding to F and of I as corresponding to v.

From Fig. 21.16 we see that

$$\cos \phi = \frac{R}{Z} = \frac{R}{\sqrt{R^2 + (X_L - X_C)^2}} \qquad \qquad \textit{Power factor} \quad (21.19)$$

The power factor of a circuit is the ratio between its resistance and its impedance. Often power factors are expressed as percentages rather than as decimals or fractions. Thus a phase angle of 60° means a power factor of

$$\cos 60° = 0.50 = 50\%$$

Apparent Power

Instruments called **wattmeters** respond directly to the effective product of V and I. An ac wattmeter connected in a circuit gives a lower value for the power than the product of the effective values of V and I obtained from a separate voltmeter and ammeter in the same circuit (except at resonance). This is because the separate meters are not affected by the phase difference between current and voltage. Alternating-current generators, transformers, and power lines are usually rated in **volt-amperes,** the product of effective voltage and current without regard to actual power. This is because higher values of V and/or I must be supplied to a circuit whose power factor is less than 1 than is reflected in its power consumption. We can think of the power factor as having the unit of watts per volt-ampere.

Alternating-current devices of various kinds—for example, ac electric motors and fluorescent lamps—may have net inductive or capacitive reactances and so have power factors of less than 100%. A power factor of 70%, for instance, means that 1 kVA (kilovolt-ampere) of apparent power must be supplied for every 700 W of power actually consumed. This is an uneconomical situation, partly because of the additional generator capacity required, and partly because of the additional heat losses in the transmission lines as the unused power circulates between the generator and the device. The remedy is to introduce capacitors or inductors into the power line to increase the power factor to an acceptable figure.

◆ **E X A M P L E   21.6** _____

An inductive load connected to a 24-V, 400-Hz power source draws a current of 2.0 A and dissipates 40 W. (a) Find the power factor of the load and the phase angle ϕ. (b) What series capacitance is needed to make the power factor 100%? (c) What would the current be? (d) How much power would the load then dissipate?

SOLUTION   (a) Since $P = IV \cos \phi$, the power factor is

$$\cos \phi = \frac{P}{IV} = \frac{40 \text{ W}}{(2.0 \text{ A})(24 \text{ V})} = 0.83 = 83\%$$

The phase angle is therefore $\phi = 34°$.

(b) The power factor will be 100% when $X_C = X_L$. In the original circuit $X_C = 0$, and so, from Eq. (21.15), $X_L = R \tan \phi$. The resistance in the circuit is

$$R = \frac{P}{I^2} = \frac{40 \text{ W}}{(2.0 \text{ A})^2} = 10 \text{ }\Omega$$

Hence the inductive reactance is

$$X_L = R \tan \phi = (10\ \Omega)(\tan 34°) = 6.75\ \Omega$$

Since the required capacitive reactance X_C must equal X_L when $f = 400$ Hz,

$$C = \frac{1}{2\pi f X_C} = \frac{1}{(2\pi)(400\ \text{Hz})(6.75\ \Omega)} = 5.9 \times 10^{-5}\ \text{F} = 59\ \mu\text{F}$$

(c) When $X_C = X_L$, $Z = R$ and

$$I = \frac{V}{Z} = \frac{V}{R} = \frac{24\ \text{V}}{10\ \Omega} = 2.4\ \text{A}$$

(d) $P = I^2 R = (2.4\ \text{A})^2 (10\ \Omega) = 57.6$ W. ◆

21.9 IMPEDANCE MATCHING

Power transfer is a maximum between circuits with the same impedance.

In Section 18.8 we saw that the maximum power transfer between two dc circuits occurs when their resistances are equal. The same applies to ac circuits with the impedance of the circuits rather than their resistances as the quantities to be matched. An additional consideration with ac circuits is that an impedance mismatch may result in a distorted signal by altering the properties of the circuits—for instance, by changing their resonance frequencies.

Ac circuits have the advantage that a transformer can be used to correct impedance mismatches. A transformer is often used for this purpose in an audio system where the amplifier circuit might have an impedance of several thousand ohms whereas the voice coil of the loudspeaker has an impedance of only a few ohms. Connecting the voice coil directly to the amplifier would be grossly inefficient.

Let us calculate the ratio of turns N_1/N_2 between the primary and secondary windings of a transformer needed to couple a circuit of impedance Z_1 to a circuit of impedance Z_2. If the voltages and currents in the circuits are, respectively, V_1, I_1 and V_2, I_2, then from Eq. (20.7)

$$\frac{V_1}{V_2} = \frac{N_1}{N_2} \qquad \text{and} \qquad \frac{I_2}{I_1} = \frac{N_1}{N_2}$$

Since $Z_1 = V_1/I_1$ and $Z_2 = V_2/I_2$,

$$\frac{Z_1}{Z_2} = \frac{V_1 I_2}{V_2 I_1} = \left(\frac{N_1}{N_2}\right)^2$$

and the ratio of the turns is

$$\frac{N_1}{N_2} = \sqrt{\frac{Z_1}{Z_2}} \qquad\qquad\qquad \textit{Impedance matching} \quad (21.20)$$

Thus if a loudspeaker whose voice coil has an impedance of 10 Ω is to be used with an amplifier whose load impedance is 9000 Ω, a transformer should be used whose ratio of turns is

$$\frac{N_1}{N_2} = \sqrt{\frac{Z_1}{Z_2}} = \sqrt{\frac{9000\ \Omega}{10\ \Omega}} = 30$$

21.10 PARALLEL AC CIRCUITS

The branch currents in a parallel circuit are not in phase.

When a resistor, an inductor, and a capacitor are connected in parallel, as in Fig. 21.19, the potential difference is the same across each circuit element:

$$V = V_R = V_L = V_C$$

The total instantaneous current is the sum of the instantaneous currents in each branch, as in the case of a dc parallel circuit. However, this is not true of the total effective current, because the branch currents are not in phase. The current I_R is always in phase with the voltage V, but I_C leads V by 90° and I_L lags behind V by 90°.

A phasor diagram of the situation is shown in Fig. 21.20. The magnitudes of the currents in the branches are

$$I_R = \frac{V}{R} \qquad I_C = \frac{V}{X_C} \qquad I_L = \frac{V}{X_L}$$

and their vector sum is

$$I = \sqrt{I_R^2 + (I_C - I_L)^2} \qquad\qquad \textit{Effective current} \quad (21.21)$$

Figure 21.19

A parallel *RLC* circuit. The potential differences across the circuit elements are the same.

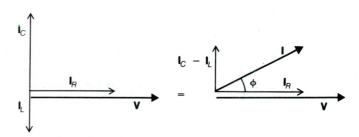

Figure 21.20

A phasor diagram of the effective currents in the resistor, inductor, and capacitor of Fig. 21.19. The magnitude *I* of the vector sum of the current vectors is equal to the effective current through the entire circuit.

The phase angle ϕ specified by

$$\tan \phi = \frac{I_C - I_L}{I_R} \qquad\qquad \textit{Phase angle} \quad (21.22)$$

is the angle between the current and the voltage. A positive phase angle means that the current leads the voltage, a negative phase angle means that the current lags behind the voltage. The power dissipated in a parallel alternating-current circuit is given by the same formula as in the case of a series circuit, namely, $P = IV \cos \phi$.

◆ **E X A M P L E   21.7** _____

A 15.0-Ω resistor, a 5.00-μF capacitor, and a 2.00-mH inductor are connected in parallel to a 50.0-V, 2.50-kHz ac source. Find (a) the current in each component, (b) the total current in the circuit, (c) the impedance of the circuit, (d) the phase angle in the circuit, and (e) the power dissipated by the circuit.

SOLUTION (a) The reactances of the capacitor and inductor are respectively

$$X_C = \frac{1}{2\pi f C} = \frac{1}{2\pi (2.50 \times 10^3 \, \text{Hz})(5.00 \times 10^{-6} \, \text{F})} = 12.7 \, \Omega$$

$$X_L = 2\pi f L = 2\pi (2.5 \times 10^3 \, \text{Hz})(2.00 \times 10^{-3} \, \text{H}) = 31.4 \, \Omega$$

The currents in the three circuit elements are therefore

$$I_R = \frac{V}{R} = \frac{50.0 \, \text{V}}{15.0 \, \Omega} = 3.33 \, \text{A}$$

$$I_C = \frac{V}{X_C} = \frac{50.0 \, \text{V}}{12.7 \, \Omega} = 3.94 \, \text{A}$$

$$I_L = \frac{V}{X_L} = \frac{50.0 \, \text{V}}{31.4 \, \Omega} = 1.59 \, \text{A}$$

(b) From Eq. (21.21) we have for the total current

$$I = \sqrt{I_R^2 + (I_C - I_L)^2} = \sqrt{(3.33\,\text{A})^2 + (3.94\,\text{A} - 1.59\,\text{A})^2} = 4.08\,\text{A}$$

(c) The impedance of the circuit is

$$Z = \frac{V}{I} = \frac{50.0\,\text{V}}{4.08\,\text{A}} = 12.3\,\Omega$$

(d) The phase angle is found from Eq. (21.22):

$$\tan\phi = \frac{I_C - I_L}{I_R} = \frac{3.94\,\text{A} - 1.59\,\text{A}}{3.33\,\text{A}} = 0.706 \qquad \phi = 35°$$

Because ϕ is positive, the current leads the voltage by 35°.

(e) Power is dissipated here only in the resistor. Hence

$$P = I_R^2\,R = (3.33\,\text{A})^2\,(15.0\,\Omega) = 166\,\text{W} \qquad\qquad \blacklozenge$$

Filters

In a series *RLC* circuit, the impedance is a minimum at resonance and increases at higher and lower frequencies than f_0. At resonance, $X_L = X_C$, $Z = R$, and $I = V/R$. In a parallel *RLC* circuit, resonance again corresponds to $X_L = X_C$, $Z = R$, and $I = V/R$, but now the impedance is a *maximum* at f_0 since at higher and lower frequencies some current can pass through the inductor and capacitor as well as through the resistor. Thus a series circuit can be used as a selector to favor a particular frequency, and a parallel circuit with the same *L* and *C* can be used as a selector to discriminate against the same frequency. Such circuits form the basis of **band-pass** and **band-reject** filters.

Figure 21.21 shows two kinds of band-pass filters. At frequencies near f_0 the impedance of the series *LC* circuit of Fig. 21.21(a) is low, so the load current is high. (From Fig. 21.18 we can see why a band of frequencies rather than a single one is favored.) At frequencies much above or below f_0 the impedance is high, so the load current is low. In Fig. 21.21(b) the impedance of the parallel *LC* circuit is high for

Figure 21.21

Band-pass filters. The impedance Z of a series *LC* circuit is a minimum near f_0, whereas Z is a maximum near f_0 for a parallel *LC* circuit.

(a)

(b)

(a)

(b)

Figure 21.22

Band-reject filters.

frequencies near f_0, and most of the current flows through the load. At other frequencies current is diverted through the LC circuit and thus reduces the load current. The band-reject filters of Fig. 21.22 behave in just the opposite ways to minimize the load current for frequencies near f_0.

Band-pass filters make it possible to send a number of different signals at the same time through a single pair of wires. What is done is to have each signal modulate a "carrier" alternating current of constant frequency, so that the resulting variations in the amplitude of the carrier correspond to the signal. This procedure is illustrated in Fig. 22.6 for radio transmission. Modulated carriers of different frequencies can be transmitted at the same time from one place to another along the same pair of wires. At the receiving end band-pass filters separate the carriers and special circuits then demodulate them to reconstruct the original signals. Telephone systems have used this technique for over 60 years. Up to 32 simultaneous telephone conversations can be carried on an ordinary pair of wires, and a coaxial cable, in which one of the conductors is in the form of a tube insulated from a central wire, can carry thousands of them.

Other arrangements of inductors and capacitors can act as **high-pass** filters to discriminate against low-frequency currents, and as **low-pass** filters to favor them. In the high-pass filter in Fig. 21.23(a), since $X_C = 1/(2\pi fC)$, the lower the frequency, the greater the reactance of the capacitor. In addition, since $X_L = 2\pi fL$, the lower the frequency, the less the reactance of the inductor and the more the current diverted through it instead of passing through the load. These effects are used in the opposite ways in the low-pass filter in Fig. 21.23(b), which is widely used to smooth out ripples in the output of alternating-current to direct-current rectifiers (see Fig. 20.15). More elaborate circuits than those of Figs. 21.21 to 21.23 can be designed that provide sharper cutoffs between the passed and rejected frequency bands.

(a)

(b)

Figure 21.23

(a) High-pass filter. At low frequencies, X_C is large and X_L is small. (b) Low-pass filter. At high frequencies, X_C is small and X_L is large.

Important Terms

The **effective value** of an alternating current is such that a direct current of this magnitude produces heat in a resistor at the same rate as the alternating current.

A **phasor** is a rotating vector whose projection can represent either current or voltage in an ac circuit.

The **phase relationships** between the instantaneous voltage and instantaneous current in ac circuit components are as follows: The voltage across a pure resistor is in phase with the current; the voltage across a pure inductor leads the current by $\frac{1}{4}$ cycle; the voltage across a pure capacitor lags behind the current by $\frac{1}{4}$ cycle.

The **inductive reactance** X_L of an inductor is a measure of its effect on an alternating current. The **capacitive reactance** X_C of a capacitor is a measure of its effect on an alternating

current. Both X_L and X_C vary with the frequency of the current. The **impedance** Z of an ac circuit is analogous to the resistance of a dc circuit. The **resonance frequency** of an ac circuit is that frequency for which the impedance is a minimum.

The **power factor** of an ac circuit is the ratio between the power consumed in the circuit and the product of the effective current and voltage there; this ratio is equal to that between the resistance and the impedance of the circuit, and is less than 1, except at resonance. The unit of the apparent power $V_{eff}I_{eff}$ is the **volt-ampere,** as distinct from the watt, which is the unit of consumed power.

A **filter** is a circuit that discriminates against the passage of currents in a certain frequency range.

Important Formulas

Effective current: $I_{eff} = \dfrac{I_{max}}{\sqrt{2}} = 0.707 I_{max}$

Effective voltage: $V_{eff} = \dfrac{V_{max}}{\sqrt{2}} = 0.707 V_{max}$

Inductive reactance: $X_L = 2\pi f L$

Capacitive reactance: $X_C = \dfrac{1}{2\pi f C}$

Impedance: $Z = \sqrt{R^2 + (X_L - X_C)^2}$

Current in ac circuit: $I = \dfrac{V}{Z}$

Phase angle: $\tan \phi = \dfrac{X_L - X_C}{R}$ $\cos \phi = \dfrac{R}{Z}$

Resonance frequency: $f_0 = \dfrac{1}{2\pi\sqrt{LC}}$

Power in ac circuit: $P = IV \cos \phi$

Impedance-matching transformer: $\dfrac{N_1}{N_2} = \sqrt{\dfrac{Z_1}{Z_2}}$

MULTIPLE CHOICE

1. In an ac circuit the voltage
   a. leads the current
   b. lags behind the current
   c. is in phase with the current
   d. is any of the above, depending on the circumstances

2. The voltage cannot be exactly in phase with the current in a circuit that contains
   a. only resistance
   b. only inductance
   c. inductance and capacitance
   d. inductance, capacitance, and resistance

3. The voltage lags behind the current by $\frac{1}{4}$ cycle in

   a. a pure capacitor
   b. a pure inductor
   c. a pure resistor
   d. a circuit with capacitance and inductance

4. The unit of inductive reactance is the
   a. henry   b. tesla
   c. weber   d. ohm

5. When voltage and current are in phase in an ac circuit, the
   a. impedance is 0   b. reactance is 0
   c. resistance is 0   d. phase angle is 90°

6. The impedance of a circuit does not depend on
  a. I
  b. f
  c. R
  d. C

7. A resistor, a capacitor, and an inductor are connected in series to a source of ac power. If the inductance is decreased, the impedance of the circuit
  a. decreases
  b. increases
  c. decreases or increases
  d. decreases, increases, or remains the same

8. The power dissipated as heat in an ac circuit depends on its
  a. resistance
  b. inductive reactance
  c. capacitive reactance
  d. impedance

9. A coil of inductance L has an inductive reactance of X_L in an ac circuit in which the effective current is I. The coil is made from a superconducting material and has no resistance. The rate at which power is dissipated in the coil is
  a. 0
  b. IX_L
  c. I^2X_L
  d. IX_L^2

10. The power factor of a circuit is equal to
  a. RZ
  b. R/Z
  c. X_L/Z
  d. X_C/Z

11. At resonance, it is not true that
  a. $R = Z$
  b. $X_L = 1/X_C$
  c. $P = IV$
  d. $I = V/R$

12. The impedance of a parallel RLC circuit at resonance is
  a. less than R
  b. equal to R
  c. more than R
  d. any of the above, depending on the circumstances

13. Impedance is a maximum at resonance in
  a. a series RLC circuit
  b. a parallel RLC circuit
  c. all RLC circuits
  d. no RLC circuits

14. The power factor of a circuit in which $X_L = X_C$
  a. is 0
  b. is 1
  c. depends on the ratio X_L/X_C
  d. depends on the value of R

15. A voltmeter across an ac circuit reads 50 V, and an ammeter in series with the circuit reads 5 A. The power consumption of the circuit
  a. is less than or equal to 250 W

  b. is exactly equal to 250 W
  c. is equal to or more than 250 W
  d. may be less than, equal to, or more than 250 W

16. The power factor of a certain circuit in which the voltage lags behind the current is 80%. To increase the power factor to 100%, it is necessary to add to the circuit additional
  a. resistance
  b. capacitance
  c. inductance
  d. impedance

17. A simple low-pass filter can be made by using
  a. an inductor in series with the load
  b. a capacitor in series with the load
  c. a resistor in series with the load
  d. an inductor in parallel with the load

18. The inductive reactance of a 1.0-mH coil in a 5.0-kHz circuit is
  a. 3.1 Ω
  b. 6.3 Ω
  c. 10 Ω
  d. 31 Ω

19. A current of 0.50 A flows in an electromagnet of negligible resistance when it is connected to a source of 120-V, 60-Hz alternating current. The inductance of the electromagnet is
  a. 0.011 mH
  b. 0.16 H
  c. 0.64 H
  d. 0.24 kH

20. The capacitive reactance of a 5.0-μF capacitor in a 20-kHz circuit is
  a. 0.63 Ω
  b. 1.6 Ω
  c. 5.0 Ω
  d. 16 Ω

21. A 2.0-μF capacitor is connected to a 50-V, 400-Hz power source. The current that flows is
  a. 0.20 mA
  b. 0.25 A
  c. 2.5 A
  d. 3.5 A

22. A resistor, a capacitor, and an inductor are connected in series to an ac source of frequency f. The effective voltages across the circuit components are $V_R = 10$ V, $V_C = 20$ V, and $V_L = 14$ V. The effective voltage of the source is
  a. 8 V
  b. 12 V
  c. 16 V
  d. 35 V

23. The phase angle in the circuit in Multiple Choice 22 is
  a. $-31°$
  b. $+31°$
  c. $-53°$
  d. $+53°$

24. The resonance frequency of the circuit in Multiple Choice 22 is
  a. less than f
  b. equal to f
  c. greater than f
  d. any of the above, depending on the values of the components

25. In a series ac circuit $R = 10.0\ \Omega$, $X_L = 8.0\ \Omega$, and $X_C = 6.0\ \Omega$ when the frequency is f. The impedance at this frequency is

a. $10.2\ \Omega$　　　　　b. $12\ \Omega$
c. $24\ \Omega$　　　　　d. $104\ \Omega$

26. The phase angle in the circuit in Multiple Choice 25 is
a. $0.2°$　　　　　b. $2°$
c. $11°$　　　　　d. $45°$

27. The resonance frequency of the circuit in Multiple Choice 25 is

a. less than f
b. equal to f
c. greater than f
d. any of the above, depending on the applied voltage

28. The resonance frequency of a circuit that contains a 50-mH inductor and a 0.20 μF capacitor is
a. 16 kHz　　　　　b. 10 kHz
c. 63 kHz　　　　　d. 16 MHz

29. A coil connected to a 120-V ac source draws a current of 0.50A and dissipates 50 W. If a capacitor were connected in series to bring the power factor to 1, the circuit would dissipate
a. 50 W　　　　　b. 60 W
c. 72 W　　　　　d. 100 W

30. A 2.00-μF capacitor, a 5.00-mH inductor, and a 30.0-Ω resistor are connected in series across a 100-V, 1.00-kHz power source. The current in the circuit is
a. 1.76A　　　　　b. 2.65A
c. 3.33A　　　　　d. 5.50A

31. The power dissipated in the circuit in Multiple Choice 30 is
a. 93 W　　　　　b. 176 W
c. 202 W　　　　　d. 211 W

32. The minimum rating in volt-amperes needed by the power source of the circuit in Multiple Choice 30 is
a. 93 V·A　　　　　b. 176 V·A
c. 265 V·A　　　　　d. 333 V·A

33. What should the capacitance of the circuit in Multiple Choice 30 be for it to have a resonance frequency of 1.00 kHz?
a. 5.07 μF　　　　　b. 31.8 μF
c. 50.7 μF　　　　　d. 71 μF

34. With that capacitance, the power dissipated in the circuit in Multiple Choice 30 would be
a. 144 W　　　　　b. 176 W
c. 230 W　　　　　d. 333 W

35. A 40.0-μF capacitor, a 40.0-mH inductor, and a 25.0-Ω resistor are connected in parallel to a 24.0-V, 100-Hz power source. The total current in the circuit is
a. 0.61 A　　　　　b. 1.02 A
c. 1.31 A　　　　　d. 2.52 A

36. In the circuit of Multiple Choice 35
a. I leads V by 20°
b. I leads V by 68°
c. I lags V by 20°
d. I lags V by 68°

QUESTIONS

1. What happens to the reactance of a coil when an iron core is inserted in it?

2. What happens to the reactance of a capacitor with air between its plates when the space between its plates is filled with a substance of dielectric constant $K > 1$?

3. A coil has the inductive reactance A when an ac voltage of frequency F is applied to it. What is the reactance when the frequency is changed to $F/2$? to $2F$?

4. A capacitor has the capacitive reactance B when an ac voltage of frequency F is applied to it. What is the reactance when the frequency is changed to $F/2$? To $2F$?

5. What properties of an ac circuit are described by its resistance, capacitive reactance, inductive reactance, and impedance? What are the similarities and differences among these quantities?

6. You have dc and ac sources of the same emf and dc and ac ammeters. How would you use them to decide whether a circuit element in a black box with two terminals is a resistor, a capacitor, an inductor, or a rectifier?

7. The frequency of the alternating potential difference applied to a series RLC circuit is halved. What happens to the resistance, the inductive reactance, and the capacitive reactance of the circuit? What further information is needed to establish what happens to the impedance of the circuit?

8. An alternating potential difference whose frequency is higher than the resonance frequency of a series RLC circuit is applied to it. Does the voltage in the circuit lead or lag behind the current?

9. What is the significance of the power factor of a circuit? Is it independent of frequency? Under what circumstances (if any) can it be zero? Under what circumstances (if any) can it be 100%?

10. What kind of filter is shown in Fig. 21.24?

Figure 21.24

Question 10.

11. What kind of filter is shown in Fig. 21.25?

Figure 21.25

Question 11.

12. Loudspeakers are limited in their frequency responses, with large speakers ("woofers") providing more accurate sound reproduction at low frequencies and small ones ("tweeters") at high frequencies. A crossover circuit that divides the output of an audio amplifier into a low-frequency and a high-frequency component is needed when a pair of such speakers is used. Which output of the circuit shown in Fig. 21.26 should be connected to the woofer and which to the tweeter?

Figure 21.26

Question 12.

EXERCISES

21.1 Effective Current and Voltage

1. The graph in Fig. 21.27 shows how the current in a certain ac circuit varies with time. What is the frequency of the current?

Figure 21.27

Exercise 1.

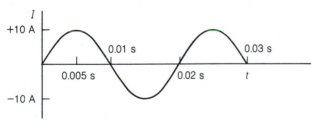

2. What is the effective current in the circuit shown in Fig. 21.27?

3. An ammeter in series with an ac circuit reads 10 A and a voltmeter across the circuit reads 60 V. (a) What is the maximum current in the circuit? (b) What is the maximum potential difference across the circuit? (c) Does the maximum current necessarily occur at the same moments as the maximum voltage?

4. The dielectric used in a certain capacitor breaks down at

a voltage of 300 V. Find the highest effective sinusoidal ac voltage that can be applied to it.

5. The potential difference across a source of alternating current varies sinusoidally with time with $V_{max} = 100$ V. Find the value of the instantaneous potential difference (a) $\frac{1}{8}$ cycle, (b) $\frac{1}{4}$ cycle, (c) $\frac{3}{8}$ cycle, and (d) $\frac{1}{2}$ cycle after $V = 0$.

6. A certain 250-Hz ac power source has a maximum potential difference of 24 V. Find the value of the instantaneous potential difference (a) 0.0005 s, (b) 0.001 s, (c) 0.002 s, (d) 0.0025 s, (e) 0.004 s, and (f) 0.005 s after $V = 0$.

21.4 Inductive Reactance

7. What is the reactance of a 5.0-mH inductor at 10 Hz? At 10 kHz?

8. The reactance of an inductor is 80 Ω at 500 Hz. Find its inductance.

9. Find the current that flows when a 3.0-mH inductor is connected to a 15-V, 5.0-kHz power source.

10. A current of 0.80 A flows through a 50-mH inductor of negligible resistance that is connected to a 120-V power source. What is the frequency of the source?

11. An inductor of negligible resistance whose reactance is 120 Ω at 200 Hz is connected to a 240-V, 60-Hz power line. What is the current in the inductor?

21.5 Capacitive Reactance

12. What is the reactance of an 80-pF capacitor at 10 kHz? At 10 MHz?

13. The reactance of a capacitor is 50 Ω at 200 Hz. What is its capacitance?

14. Find the current that flows when a 10-μF capacitor is connected to a 15-V, 5.0-kHz power source.

15. A capacitor of unknown capacitance is found to have a reactance of 120 Ω at 200 Hz. The capacitor is connected to a 240-V, 60-Hz power line. (a) What is the current in the circuit? (b) Why is it not the same as that in Exercise 11?

21.6 Impedance

16. A pure capacitor, a pure inductor, and a pure resistor are connected in series across an ac power source. A voltmeter placed in turn across each circuit element reads 15 V, 20 V, and 20 V, respectively. What is the potential difference of the source?

17. A 5.0-μF capacitor is connected in series with a 300-Ω resistor, and a 120-V, 50-Hz potential difference is applied. Find the current in the circuit and the power dissipated.

18. A 30-mH, 60-Ω inductor is connected to a 20-V, 400-Hz power source. Find the current in the inductor and the power dissipated in it.

19. The current in a resistor is 2.0 A when it is connected across a 240-V, 50-Hz line. How much inductance should be connected in series with the resistor to reduce the current to 1.0 A?

20. The current in a resistor is 2.0 A when it is connected across a 240-V, 50-Hz line. How much capacitance should be connected in series with the resistor to reduce the current to 1.0 A?

21. A capacitor of unknown capacitance is connected in series with an 80-Ω resistor. The combination is found to draw 0.50 A when connected to a 120-V, 60-Hz power source. Find (a) the capacitance of the capacitor, (b) the power dissipated in the capacitor, and (c) the power dissipated in the resistor.

22. A coil of unknown inductance and resistance is observed to draw 50 mA when a dc potential difference of 5.0 V is applied. When the coil is connected to a source of 400-Hz ac, however, a potential difference of 8.0 V is required to yield the same current. (a) Find the inductance and resistance of the coil. (b) Find the power dissipated in the coil in each situation.

23. A coil of unknown inductance and resistance is observed to draw 8.0 A when a 40-V dc potential difference is applied, and 5.0 A when a 40-V, 60-Hz ac potential difference is applied. (a) Find the inductance and resistance of the coil. (b) Find the power dissipated in the coil in each situation.

24. A series circuit has a resistance of 40 Ω, an inductive reactance of 30 Ω, and a capacitive reactance of 50 Ω when connected to a certain ac source. Find (a) the impedance of the circuit, (b) the phase angle, and (c) the potential difference required for a current of 1.5 A to flow.

25. A pure capacitor, a pure inductor, and a pure resistor are connected in series across a 60-V ac power source. The potential difference across the capacitor is 60 V, and that across the inductor is also 60 V. What is the potential difference across the resistor?

26. The voltage leads the current in a certain ac circuit by 30°. The effective current in the circuit is 5.0 A. (a) Is the capacitive reactance greater or less than the inductive reactance? (b) What is the instantaneous value of I when the instantaneous value of V is 0?

27. A series circuit has a resistance of 20 Ω, an inductive reactance of 20 Ω, and a capacitive reactance of 20 Ω when connected to an ac source. Find (a) the impedance of the circuit, (b) the phase angle, and (c) the potential difference required for a current of 20 A to flow.

28. A 10-μF capacitor, a 0.10-H inductor, and a 60-Ω resistor are connected in series across a 120-V, 60-Hz power line. Find (a) the current in the circuit, (b) the power dissipated in it, and (c) the potential difference across each of the circuit elements.

29. The circuit in Exercise 28 is connected across a 120-V, 30-Hz power line. Answer the same questions for this case.

21.7 Resonance

30. In the antenna circuit of a radio receiver that is tuned to a certain station, $R = 5.0$ Ω, $L = 5.0$ mH, and $C = 5.0$ pF. (a) Find the frequency of the station. (b) If the voltage applied to the circuit is 0.50 mV at this frequency, what is the resulting current? (c) What should the capacitance be in order to receive an 800-kHz radio signal?

31. In a series ac circuit, $R = 20$ Ω, $X_L = 10$ Ω, and $X_C = 25$ Ω when the frequency is 400 Hz. Find the resonance frequency of the circuit.

32. In a series ac circuit $R = 10$ Ω, $X_L = 20$ Ω, and $X_C = 30$ Ω when the applied frequency is 1.0 kHz. Find the resonance frequency of the circuit.

33. A resistor, a capacitor, and an inductor are in series with an ac power source of frequency f. The effective voltages across the components are $V_R = 96$ V, $V_C = 64$ V, and $V_L =$

136 V. Find the effective voltage of the source and the phase angle in the circuit. Is the resonance frequency of the circuit less than f, equal to f, or greater than f?

21.8 Power in AC Circuits

34. A 10-kW electric motor has an inductive power factor of 70%. What minimum rating in kVA must the power line have? A capacitor is connected in series with the motor to increase the power factor to 100%. How does this affect the required rating of the power line?

35. An inductive load connected to a 120-V, 60-Hz power source draws a current of 5.0 A and dissipates 450 W. (a) Find the capacitance needed in series with the load to increase the power factor to 100%. (b) Find the power dissipated when this capacitor is connected.

36. A 10-μF capacitor, a 30-mH inductor, and a 15-Ω resistor are connected in series with a 10-V, 250-Hz power source. Find (a) the impedance of the circuit, (b) the current in it, (c) the power factor, (d) the power dissipated, and (e) the minimum apparent-power rating of the source. (f) What current will flow if the circuit is connected to a 10-V power line whose frequency is equal to its resonance frequency?

37. A 60-μF capacitor, a 0.30-H inductor, and a 50-Ω resistor are connected in series with a 120-V, 60-Hz power source. Find (a) the impedance of the circuit, (b) the current in it, (c) the power factor, (d) the power dissipated, and (e) the minimum apparent-power rating of the source. (f) What current will flow if the circuit is connected to a 120-V power line whose frequency is equal to its resonance frequency?

PROBLEMS

45. A circuit that contains inductance and resistance has an impedance of 50 Ω at 100 Hz and an impedance of 100 Ω at 500 Hz. What are the values of the inductance and the resistance of the circuit?

46. A circuit that contains capacitance and resistance has an impedance of 30 Ω at 80 Hz and an impedance of 14 Ω at 240 Hz. What are the values of the capacitance and resistance of the circuit?

47. An inductor dissipates 75 W of power when it draws 1.0 A from a 120-V, 60-Hz power line. (a) What is its power factor? (b) What capacitance should be connected in series with it to increase its power factor to 100%? (c) How much current would the circuit then draw? (d) How much power would it then dissipate? (e) What should the minimum volt-ampere rating of the power line be then?

48. A circuit consisting of a capacitor in series with a resistor

21.9 Impedance Matching

38. A microphone whose impedance is 20 Ω is to be used with an amplifier whose input impedance is 50,000 Ω. What should the ratio of turns be in the required transformer?

21.10 Parallel AC Circuits

39. In a parallel ac circuit $R = 40$ Ω, $X_L = 20$ Ω, and $X_C = 30$ Ω when the circuit is connected to a 60-V, 1.0-kHz power source. What is the total current in the circuit?

40. The reactances of an inductor and a capacitor connected in parallel to an ac power source are $X_L = 20$ Ω and $X_C = 30$ Ω at the frequency of the source. The resistance in the circuit is negligible. Find the impedance of the circuit and the phase angle.

41. A 10-Ω resistor and an 2.0-mH inductor are connected in parallel across a 10-V, 1.0-kHz power source. Find (a) the current in each component, (b) the total current, (c) the impedance of the circuit, (d) the phase angle, and (e) the power dissipated by the circuit.

42. A 10-Ω resistor and an 8.0-μF capacitor are connected in parallel across a 10-V, 1.0-kHz power source. Answer the same questions for this circuit.

43. A 25-Ω resistor, an 40-μF capacitor, and a 40-mH inductor are connected in parallel across a 24-V, 200-Hz power source. Answer the same questions for this circuit.

44. A 10-Ω resistor, an 8.0-μF capacitor, and a 2.0-mH inductor are connected in parallel across a 10-V, 1.0-kHz power source. Answer the same questions for this circuit.

draws 3.6 A from a 50-V, 100-Hz power line. The circuit dissipates 120 W. (a) What is the precise phase relationship between voltage and current in the circuit? (b) What series inductance should be inserted in the circuit if the current and voltage are to be in phase? (c) How much current would the circuit then draw? (d) How much power would it dissipate?

49. Figure 21.28 shows a simple high-pass filter. (a) What inductance is needed to discriminate against frequencies under

Figure 21.28

Problem 49.

20 kHz when the load is 50 Ω? (b) What proportion of the total current passes through the load when the frequency of the applied signal is 2.0 kHz? (c) When it is 200 kHz?

50. Figure 21.29 shows a simple low-pass filter. (a) What capacitance is needed to discriminate against frequencies above 20 kHz when the load is 50 Ω? (b) What proportion of the

total current passes through the load when the frequency of the applied signal is 2.0 kHz? (c) When it is 200 kHz?

Answers to Multiple Choice

1. d	10. b	19. c	28. a
2. b	11. b	20. b	29. c
3. a	12. b	21. b	30. a
4. d	13. b	22. b	31. a
5. b	14. b	23. a	32. b
6. a	15. a	24. c	33. a
7. c	16. c	25. a	34. d
8. a	17. a	26. c	35. b
9. a	18. d	27. a	36. c

Figure 21.29

Problem 50.

CHAPTER 22

LIGHT

22.1   Electromagnetic Waves

22.2   Types of Em Waves

22.3   Huygens' Principle

22.4   Reflection

22.5   Refraction

22.6   Index of Refraction

22.7   Apparent Depth

22.8   Total Internal Reflection

22.9   Luminous Flux

22.10 Illumination

The discovery that light consists of electromagnetic waves is among the most noteworthy achievements of science. The existence of electromagnetic waves was proposed in 1864 by the British physicist James Clerk Maxwell, who went on to suggest that light waves were of this kind. Experimental confirmation came some years later. Although the wave behavior of light can be analyzed directly from Maxwell's theory, such calculations require advanced mathematics. A simpler, though less comprehensive, approach to optics had been devised in 1678 by the Dutch physicist Christian Huygens. In this chapter the reflection and refraction of light at plane surfaces are considered with the help of Huygens' method.

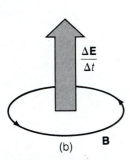

Figure 22.1

(a) According to Faraday's law, a changing magnetic field gives rise to an electric field. (b) According to Maxwell's hypothesis, a changing electric field gives rise to a magnetic field.

22.1 ELECTROMAGNETIC WAVES

Coupled electric and magnetic fields that move with the speed of light.

From the earlier work of Faraday, Maxwell knew that a changing magnetic field causes charges to move in a wire loop. An electric field can also set charges in motion, which means that a changing magnetic field is equivalent in its effects to an electric field (Fig. 22.1(a)). Maxwell proposed the converse (Fig. 22.1(b)):

A changing electric field is equivalent in its effects to a magnetic field.

The electric fields produced by electromagnetic induction are easy to observe because metals offer little resistance to the flow of electrons. Even a weak electric field can lead to a measurable current in a metal. Weak magnetic fields are much harder to detect, however. Maxwell's hypothesis was based on an indirect argument rather than on experimental findings, unlike Faraday's hypothesis of electromagnetic induction.

To appreciate Maxwell's train of thought, imagine a capacitor being charged by a battery. Current flows in the connecting wires, and electric charges of opposite sign build up on the capacitor plates. As the charges grow, the electric field E increases with them. The greater the current, the faster the field changes. The strength I of the current is thus mirrored by the rate of change $\Delta E/\Delta t$ of the electric field. What Maxwell did was to extend this correspondence to include magnetic effects. Just as a current I has a magnetic field associated with it, said Maxwell, so does a changing electric field $\Delta E/\Delta t$. The result was a particularly symmetric set of four equations describing electricity and magnetism that remain valid to this day.

If Maxwell was right, electromagnetic (or simply em) waves should occur in which fluctuating electric and magnetic fields are coupled together both by electromagnetic induction and by the converse mechanism he had proposed. The energy of the waves would be shared by their electric and magnetic fields. Maxwell was able to show that the speed c in free space of em waves is given by

$$c = \frac{1}{\sqrt{\epsilon_0 \mu_0}} \qquad\qquad \textit{Speed of light} \quad (22.1)$$

$$= \frac{1}{\sqrt{(8.85 \times 10^{-12}\,\mathrm{C^2/N \cdot m^2})(1.26 \times 10^{-6}\,\mathrm{T \cdot m/A})}}$$

$$= 3.00 \times 10^8\,\mathrm{m/s}$$

where ϵ_0 is the permittivity of free space and μ_0 is its permeability. This is the same speed that had been measured for light waves! The agreement was too great to be accidental, and Maxwell concluded that light consists of em waves. To five significant figures the value of c is 2.9979×10^8 m/s.

During Maxwell's lifetime the idea of em waves remained without direct experimental support. Finally, in 1888 the German physicist Heinrich Hertz showed that em waves indeed do exist and have just the properties Maxwell had predicted. Hertz generated the waves by applying an alternating emf to an air gap between two metal balls. The width of the gap was small enough for a spark to occur each time the emf reached a peak. A wire loop with a small gap was the detector; em waves set up oscillations in the loop that gave rise to sparks in the gap. Hertz determined the wavelength and speed of the waves he generated, showed that they had both electric and magnetic components, and found that they exhibited the same optical behavior as light waves.

To see how an em wave can come into being, imagine connecting a pair of metal rods to an oscillator that produces an alternating emf, as in Fig. 22.2. Suppose that at first the upper rod has a single positive charge that can move, and that the lower rod has a single negative charge that can move. (Only electrons move in actual metals, to be sure, and a great many of them do so in even a small metal rod. The model used here leads to the same result that a more realistic model would yield, but is easier to visualize.)

(a) When the oscillator is switched on, the positive charge in the upper rod begins to move upward and the negative charge in the lower rod begins to move downward. The electric field lines around the charges are indicated by the

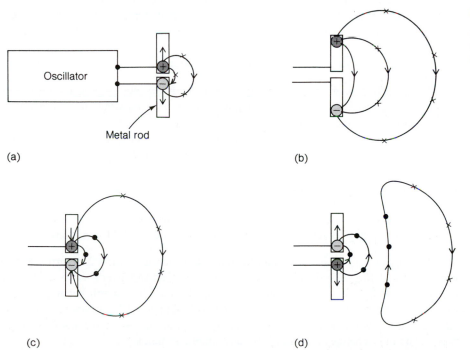

(a)

(b)

(c)

(d)

Figure 22.2

A pair of metal rods connected to an electrical oscillator emit electromagnetic waves.

colored lines, and the magnetic field lines due to the motion of the charges (which are coaxial circles perpendicular to the paper) are indicated by crosses when their direction is into the paper and by dots when their direction is out of the paper. (The dots represent arrowheads and the crosses represent the tail feathers of arrows.)

(b) The charges have reached the limit of their motion and have stopped, so that they no longer produce a magnetic field. The outer magnetic field lines do not disappear but continue to travel outward.

(c) The emf of the oscillator now begins to decrease, and the charges move toward each other. The result is a magnetic field in the opposite direction to the earlier field. The electric field is in the same direction as before.

(d) The emf has passed through 0 and begun to increase in the opposite sense. As a result there is now a negative charge in the upper rod and a positive one in the lower rod that begin to move apart. The electric field is therefore opposite in direction to the earlier field, but the magnetic field is in the same direction because magnetically a positive charge moving downward is equivalent to a negative charge moving upward.

Owing to this sequence of changes in the fields, the outermost electric and magnetic field lines, respectively, form into closed loops. These loops, which lie in perpendicular planes, are divorced from the oscillating charges that gave rise to them and continue moving outward, making up an em wave. As the charges continue oscillating back and forth, further associated loops of electric and magnetic field lines are emitted that form an expanding pattern of loops.

Figure 22.3(a) shows the form of the electric and magnetic fields that spread outward from a pair of oscillating charges. The actual fields are three-dimensional, so that the magnetic field lines form loops in planes perpendicular to the line joining the charges.

Four properties of em waves are worth noting:

1. The variations occur simultaneously in both fields (except close to the oscillating charges), so that the electric and magnetic fields have maxima and minima at the same times and in the same places.

2. The directions of the electric and magnetic fields are perpendicular to each other and to the direction in which the waves are moving. Light waves are therefore transverse.

3. The magnitudes of the electric and magnetic fields are proportional to each other (in fact, $E = cB$).

4. The speed of the waves depends only on the electric and magnetic properties of the medium they travel in, not on the amplitudes of the field variations.

Figure 22.3(b) is an attempt to illustrate properties 1 and 2 in terms of field lines of **E** and **B** a long distance from a source of electromagnetic waves. Closer to the source the field lines are curved, as in Fig. 22.3(a). It is worth keeping in mind that, unlike the other types of waves considered in Chapter 12—waves in a stretched string, water waves, sound waves—nothing material moves in the path of an em wave. The only changes are in electric and magnetic fields.

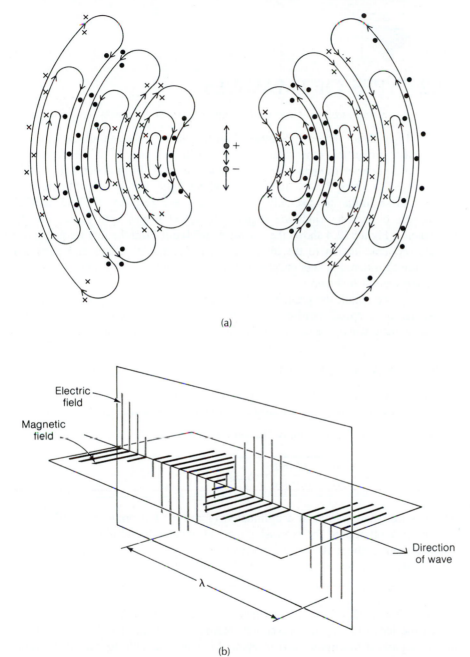

(a)

(b)

Figure 22.3

(a) The configuration of electric and magnetic fields that spread outward from a pair of oscillating charges. (b) The electric and magnetic fields of an electromagnetic wave far from its source vary as shown here. The field directions are perpendicular to each other and to the direction of propagation.

We have been looking at a special kind of em wave source. Actually, *all accelerated charges radiate em waves,* regardless of how the acceleration occurs. For instance, a charge that moves in a curved path is accelerated and so gives off em waves, losing energy as it does so.

22.2 TYPES OF EM WAVES

X rays and radio waves as well as light consist of em waves.

Light is not the only example of an em wave. Although all em waves share certain basic properties, other features of their behavior depend on their frequencies. Light waves themselves span a short frequency interval, from about 4.3×10^{14} Hz for red light to about 7.5×10^{14} Hz for violet light. Em waves with frequencies between these limits are the only ones that the eye responds to, and specialized instruments of various kinds are required to detect waves with higher and lower frequencies. Figure 22.4 shows the em wave **spectrum** from the low frequencies used in radio communication to the high frequencies found in X-rays and gamma rays (which are considered in later chapters). The wavelengths corresponding to the various frequencies are also shown.

As we recall, the product of the frequency f of a wave and its wavelength λ is just the wave speed, here c. Given the wavelength or frequency of a particular electromagnetic wave, we can immediately find the other quantity.

 E X A M P L E   22.1 _____

Find the wavelengths of yellow light whose frequency is 5.00×10^{14} Hz and of radio waves whose frequency is 1.00 MHz. (1 MHz = 1 megahertz = 10^6 Hz.)

SOLUTION   The wavelength of the yellow light is

$$\lambda = \frac{c}{f} = \frac{3.00 \times 10^8 \, \text{m/s}}{5.00 \times 10^{14} \, \text{Hz}} = 6.00 \times 10^{-7} \, \text{m} = 600 \, \text{nm}$$

which is less than 1/1000 of a millimeter. (1 nm = 1 nanometer = 10^{-9} m. The nanometer is often used for expressing the wavelengths found in light.) The width of a human hair is about 70 of these wavelengths. By the same procedure we find that a 1.00-MHz radio wave has a wavelength of 300 m.   ◆

_____

Shortly after Hertz's experiments with em waves, Guglielmo Marconi conceived the idea of using the waves for communication. In Hertz's work only a few meters separated transmitter and receiver. Marconi was able to extend the range to many kilometers, and in 1899 sent a radio message from France to England over the English Channel. Two years later he sent signals from England across the Atlantic Ocean to Newfoundland with the help of kites to raise his antennas. Morse code was used initially; radio communication using speech was first achieved in 1906.

The two chief ways to incorporate information in an em wave are **amplitude modulation** and **frequency modulation.** (The term *modulation* means alteration or

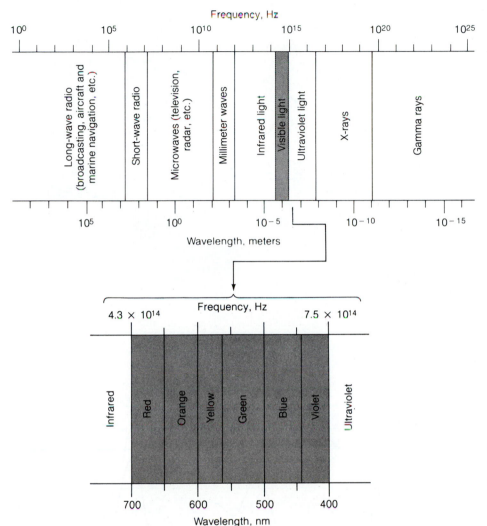

Figure 22.4

The electromagnetic wave spectrum. The boundaries of the various categories are not sharp.
(1 nm = 1 nanometer = 10^{-9} m.)

change.) In amplitude modulaton (AM) a "carrier wave" of constant frequency is varied in amplitude, with the variations constituting the signal. The simplest example is a flashlight switched on and off to give a series of dots and dashes. Another example is the radiotelegraph, in which coded sequences of dots and dashes are used to represent letters of the alphabet, numbers, or other specific information (Fig. 22.5).

A carrier wave can also be modulated so that the amplitude variations correspond to sound waves, as in Fig. 22.6. A microphone converts sound waves to

Figure 22.5

In radiotelegraphy, a constant-frequency electromagnetic wave is switched on and off in a coded sequence to transmit information. The letter v, which is · · · — in the Morse code, is shown.

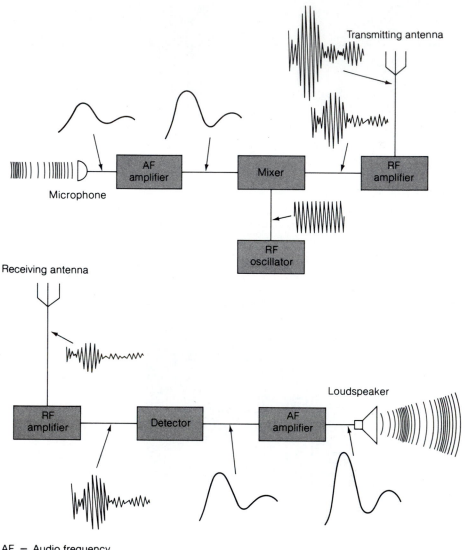

Figure 22.6

In amplitude modulation, the amplitude of a constant-frequency carrier wave is varied in accordance with an audio signal. The transmitting and receiving systems shown are highly simplified.

AF = Audio frequency
RF = Radio frequency

an equivalent electric signal, which is then amplified and combined with the carrier. The final wave is broadcast from an antenna, and at the receiving station the carrier is removed to leave an audio signal. The latter is then amplified and used to generate sound waves in a loudspeaker.

One problem with amplitude modulation is that such sources of random em waves as electric storms and electric machinery can interfere with the broadcast waves to cause "static." In frequency modulation (FM) the frequency, not the amplitude, of the carrier is varied in accordance with the audio signal (Fig. 22.7). The information content of a frequency-modulated em wave is virtually immune to disturbance.

Figure 22.7

The information content of a frequency-modulated wave resides in its frequency variations rather than in its amplitude variations.

In television the desired scene is focused on the light-sensitive screen of a special tube, which is then scanned in a zig-zag fashion by an electron beam. The signal extracted from the screen varies with the image brightness at each successive point that is scanned. This signal is then used to modulate a carrier wave for broadcasting. At the receiver, suitable circuits demodulate the wave. In the picture tube the image is reproduced with the help of an electron beam that moves in step with the electron beam in the camera tube (Fig. 17.9).

Radio waves in the short-wave band of Fig. 22.4 have the useful property of being reflected from the earth's ionosphere (Section 16.5), a region high in the atmosphere where ions are relatively abundant. Without the ionosphere, radio communication would be limited to short distances because em waves travel in straight lines and would be shielded from remote receivers by the curvature of the earth. However, radio waves of appropriate frequencies can bounce one or more times between the ionosphere and the earth's surface. Long-range transmission is therefore possible, even to the opposite side of the earth (Fig. 22.8).

LF	30–300	kHz
MF	0.3–3.0	MHz
HF	3.0–30	MHz
VHF	30–300	MHz
UHF	0.3–30	GHz

Figure 22.8

(a) The ionosphere is a region in the upper atmosphere whose ionized layers make possible long-range radio communication by their ability to reflect radio waves. (b) Only radio waves in the medium- and high-frequency (MF and HF) bands are reflected by the ionosphere. Low-frequency (LF) waves are absorbed there. Very-high- and ultrahigh-frequency (VHF and UHF) waves pass through the ionosphere and so are used to communicate with satellites and spacecraft.

The rotating scanner of a radar set sends out pulses of microwaves in a narrow beam and then detects their reflections from objects in their paths. These reflections are then displayed on a screen. The center of the display corresponds to the position of the scanner.

The ion content of the ionosphere and the heights of the individual ion layers vary with time of day and are influenced by streams of fast protons and electrons emitted by the sun during periods of sunspot activity. Hence no absolute rules can be given for which frequencies are best for which distances. Generally speaking, though, the higher the frequency, the greater the range. Thus a frequency of 10 MHz would ordinarily give good results between 300 and 2400 km, whereas 20 MHz would be better for 2400 to 11,000 km.

The higher-frequency waves used for FM and television broadcasting are not reflected by the ionosphere, so their reception is limited to the line of sight between transmitting and receiving antennas. When the frequencies are up around 10^{10} Hz, corresponding to wavelengths in the centimeter range, the waves can readily be focused into narrow beams. Such beams are reflected by objects such as ships and airplanes, which is the basis of **radar** (from *ra*dio *d*etection *a*nd *r*anging). A rotating antenna is used to send out a pulsed beam, and the distance of a particular target is established by the time needed for the echo to return to the antenna. A picture tube is used to display the echoes, with the center of the image representing the position of the antenna.

22.3 HUYGENS' PRINCIPLE

Secondary wavelets spread out from every point on a wavefront.

Most optical phenomena can be understood on the basis of the wave nature of light alone, with no reference to its em character. Three centuries ago Huygens devised an approach to the behavior of waves of all kinds that is particularly easy to use. His method concerns **wavefronts.** A wavefront is an imaginary surface that joins points where all the waves are in the same phase of oscillation. As in Fig. 22.9,

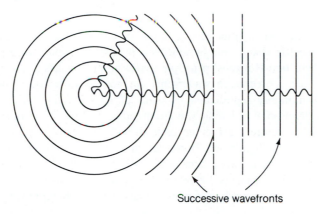

Successive wavefronts

Figure 22.9

The spherical wavefronts of a point source become plane wavefronts at a long distance from the source.

waves from a point source spread out in a series of spherical wavefronts. (In the case of water waves, the wavefronts that result when a stone is dropped in a lake are circular.) At a long distance from a point source, the curvature of the wavefronts is so small that they can be considered as a series of planes.

Huygens' principle states:

Every point on a wavefront can be considered as a source of secondary wavelets that spread out in all directions with the wave speed of the medium. The wavefront at any time is the envelope of these wavelets.

Figure 22.10 shows how Huygens' principle is applied to the motion of plane and spherical wavefronts in a uniform medium. It does not matter just where on the initial wavefront we imagine the sources of the secondary wavelets to be.

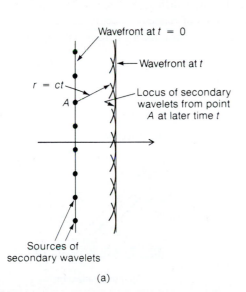

(a)

(b)

Figure 22.10

Huygens' principle applied to the propagation of (a) plane and (b) spherical wavefronts.

Despite its name, Huygens' principle is not in the same category as such fundamental principles as those of conservation of mass, energy, momentum, and electric charge, but is rather a convenient approach to the study of wave motion. We will find Huygens' principle useful in understanding a variety of optical phenomena.

Although wavefronts are needed to analyze wave motion, the results are commonly represented in terms of **rays.** A light ray is an imaginary line in the direction in which the wavefronts advance, and so it is perpendicular to the wavefronts. The picture most of us have of a light ray is a narrow pencil of light, which is perfectly legitimate. However, an approach based only on rays does not reveal such characteristic wave behavior as diffraction (the bending of waves around an obstacle into the ''shadow'' region). The motion of wavefronts is what counts, although we can both properly and conveniently use rays to summarize our conclusions.

22.4 REFLECTION

Mirror, mirror on the wall.

Every object reflects some of the light that falls on it, and this reflected light enables us to see it. In most cases the surface of the object has irregularities that spread out an initially parallel beam of light in all directions to produce **diffuse reflection** (Fig. 22.11). A surface so smooth that any irregularities in it are small relative to the wavelength of the light falling on it behaves differently. When a parallel beam of light is directed at such a surface, it is **specularly reflected** in only one direction (Fig. 22.12). Reflection from a brick wall is diffuse, whereas reflection from a mirror is specular. Often a mixture of both kinds of reflection occurs, as in the case of a surface coated with varnish or glossy enamel. Our concern here is with specular reflection only.

Figure 22.11

Diffuse reflection.

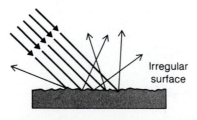

Diffuse reflection

Figure 22.12

Specular reflection

Specular reflection

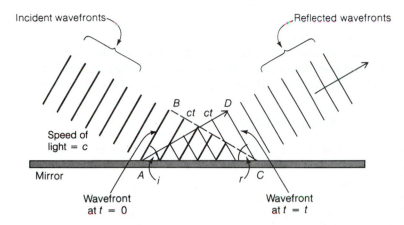

Figure 22.13

The behavior of successive wave-fronts during reflection from a plane mirror. The angle of incidence i and the angle of reflection r are equal.

When we see an object, what enters our eyes are light waves reflected from its surface. What we perceive as the object's color therefore depends on two things: the kind of light falling on it, and the nature of its surface. If white light is used to illuminate an object that absorbs all colors other than red, the object will appear red. If green light is used instead, the object will look black because it absorbs green light. A white object reflects light of all wavelengths equally well, and its apparent color is the same as the color of the light reaching it. A black object, on the other hand, absorbs light of all wavelengths, and it looks black no matter what color light reaches it.

Figure 22.13 shows the specular reflection of a series of plane wavefronts, which corresponds to a parallel beam of light. At $t = 0$ the wavefront AB just touches the mirror, and secondary wavelets start to spread out from A. After a time t the end B of the wavefront reaches the mirror at C. The secondary wavelets from A are now at the point D. The wavefront that was AB at $t = 0$ is therefore CD at the later time t.

The angle i between an approaching wavefront and the reflecting surface is called the **angle of incidence,** and angle r between a receding wavefront and the reflecting surface is called the **angle of reflection.** Here

$$\sin i = \frac{BC}{AC} \quad \text{and} \quad \sin r = \frac{AD}{AC}$$

The wavelet that comes from A takes the time t to reach D. Hence $AD = ct$, where c is the velocity of light. The point B of the wavefront AB also takes the time t to reach C, and so $BC = ct$ as well. Therefore $AD = BC$ and $\sin i = \sin r$, from which we conclude that the angles i and r are equal. Thus we have the basic law of reflection:

The angle of reflection of a plane wavefront with a plane mirror is equal to the angle of incidence.

Figure 22.14

Ray picture of reflection. The angles of incidence and reflection are measured with respect to the normal to the reflecting surface, which is a line perpendicular to it. The incident ray, the normal, and the reflected ray all lie in the same plane.

In terms of rays, the angles of incidence and reflection are measured with respect to the **normal** to the reflecting surface at the point where the light strikes it (Fig. 22.14). The normal is a line drawn perpendicular to the surface at that point. In this representation, too, the angles of incidence and reflection are equal. The incident ray, the reflected ray, and the normal all lie in the same plane.

Mirror Image

When we look at the image of an object in a plane mirror, the image seems to be the same size as the object and as far behind the mirror as the object is in front of it. Why?

In Fig. 22.15 three light rays from a point object at A are reflected by a mirror. To the eye these rays apparently come from the point A' behind the mirror. The rays that seem to come from the image do not actually pass through it, and for this reason the image is said to be **virtual.** (A **real image** is formed by light rays that pass through it.) From the figure it is clear that $AM = A'M$, so that the object and the image are the same distance on either side of the mirror.

In Fig. 22.16 an actual object is reflected by a plane mirror. Again simple geometry shows that the object and its virtual image are the same size. Every point in the image is directly behind the corresponding point on the object, and so the orientation of the image is the same as that of the object. The image is therefore

Figure 22.15

Rays from a point object at A appear to come from A' after reflection from a plane mirror. The image is as far behind the mirror as the object is in front of it.

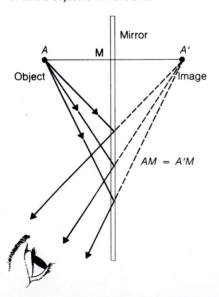

Figure 22.16

Image formation by a plane mirror. The image is erect and is the same size and shape as the object.

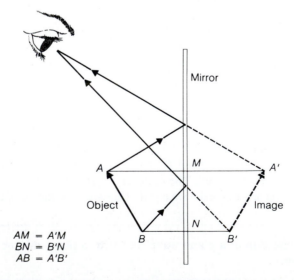

erect. However, left and right are interchanged in a mirror image, as in Fig. 22.17, because front and back have been reversed by the reflection. Thus a printed page appears backward in a mirror, and what seems to be one's left hand in a mirror is actually one's right hand.

Figure 22.17

Left and right are interchanged in reflection.

22.5 REFRACTION

The bending of light when it goes from one medium to another.

Experience tells us that a light ray that passes at a slanting angle from one medium to another, say from air to water, is deflected at the surface between the two media. This phenomenon is called **refraction.** Refraction gives rise to such familiar effects as the apparent distortion of objects partly submerged in water.

Several important aspects of refraction are illustrated in Fig. 22.18. When a light ray goes from air into water along the normal to the surface between them, it simply continues along the same path. However, when it enters the water at any other angle, it is bent toward the normal. The paths are reversible; so a light ray emerging from the water is bent *away* from the normal as it enters the air.

Refraction occurs because light travels at different speeds in the two media. Let us see what happens to the plane wavefront AB in Fig. 22.19 when it passes obliquely from a medium in which its speed is v_1 to a medium in which its speed is v_2, where v_2 is less than v_1. At $t = 0$ the wavefront AB just touches the interface between the two media. After a time t, the end B of the wavefront reaches the interface at C, and the secondary wavelets from A are now at D. Since v_2 is less than v_1, the secondary wavelets generated at the interface go a shorter distance in the same time interval than do wavelets in the first medium. The distance AD is therefore shorter than BC, and the refracted wavefronts move in a different direction from that of the incident wavefronts.

Something similar to refraction occurs when a tracked vehicle such as a bulldozer makes a turn. To turn a bulldozer to the right, which corresponds to the situation in Fig. 22.19, the right-hand track is slowed down, and the greater speed of the left-hand track then swings the bulldozer to the right.

Figure 22.18

The bending of a light beam when it passes obliquely from one medium to another is called *refraction*.

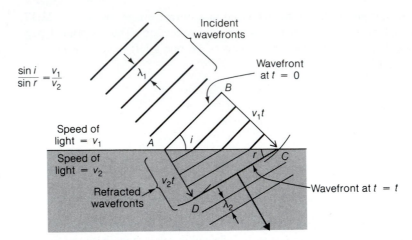

$$\frac{\sin i}{\sin r} = \frac{v_1}{v_2}$$

Figure 22.19

The behavior of successive wavefronts during refraction. The speed of light v_1 in the first medium is greater than that in the second medium.

The angle i between an approaching wavefront and the interface between two media is called the **angle of incidence** of the wavefront. The angle r between a receding wavefront and the interface it has passed through is called the **angle of refraction**. From Fig. 22.19 we see that

$$\sin i = \frac{BC}{AC} \qquad \text{and} \qquad \sin r = \frac{AD}{AC}$$

Because $BC = v_1 t$ and $AD = v_2 t$,

$$\frac{\sin i}{\sin r} = \frac{BC}{AD} = \frac{v_1 t}{v_2 t}$$

and so we have

$$\frac{\sin i}{\sin r} = \frac{v_1}{v_2} \qquad\qquad \textit{Snell's law} \quad (22.2)$$

This useful result is known as **Snell's law** after its discoverer, the seventeenth-century Dutch astronomer Willebrord Snell. It states:

The ratio of the sines of the angles of incidence and refraction is equal to the ratio of the speeds of light in the two media.

Refraction, like reflection, can be described in terms of the ray model of light. As in Fig. 22.20, the angles i and r are taken with respect to the normal to the interface at the point where the rays meet it. The incident ray, the normal, and the refracted ray all lie in the same plane.

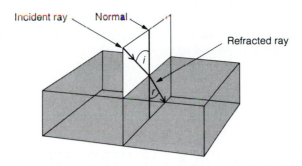

Figure 22.20

Ray picture of refraction. The incident ray, the normal, and the refracted ray all lie in the same plane.

Why does light travel more slowly in a material medium than in a vacuum? The medium consists of atoms that contain electrons. When a light wave arrives at an atom, its electrons absorb energy from the wave as they begin to oscillate at the same frequency. These oscillations then reradiate light waves with this frequency. Because of the inertia of the electrons and because they are bound to atoms, their oscillations lag slightly behind those of the incoming wave. The result is a wave speed in the medium that is less than its speed in free space.

Refraction is not confined to light waves. For example, water waves that approach a sloping beach are refracted. Regardless of the direction of the waves in open water, their direction becomes more and more perpendicular to the shoreline as they come nearer (Fig. 22.21). The reason is that the speed of a water wave decreases

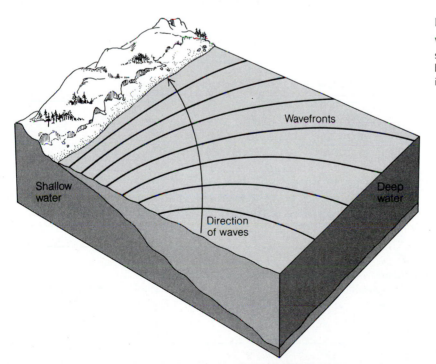

Figure 22.21

Water waves that approach the shore at an angle are refracted because they move more slowly in shallow water.

Figure 22.22

The greater the index of refraction *n* of a medium, the greater the deflection of a light beam on entering or leaving it.

in shallow water because of friction with the bottom, so a wavefront moving at an angle shoreward is progressively affected by the change in speed and swings around until it is parallel to the shore.

22.6 INDEX OF REFRACTION

The greater its value in a medium, the more slowly light travels there.

The ratio between the speed of light *c* in free space and its speed *v* in a particular medium is called the **index of refraction** of the medium. The greater the index of refraction, the greater the extent to which a light beam is deflected on entering or leaving the medium (Fig. 22.22). The symbol for index of refraction is *n*, so that

$$n = \frac{c}{v}$$ *Index of refraction*   (22.3)

Table 22.1 is a list of the values of *n* for a number of substances.

It is easy to rewrite Snell's law in terms of the indexes of refraction n_1 and n_2 of two successive media. In these media light has the respective speeds

$$v_1 = \frac{c}{n_1} \qquad \text{and} \qquad v_2 = \frac{c}{n_2}$$

and so Snell's law of Eq. (22.2) becomes

$$\frac{\sin i}{\sin r} = \frac{v_1}{v_2} = \frac{c/n_1}{c/n_2} = \frac{n_2}{n_1}$$

Table 22.1   Indexes of refraction. The values of *n* vary slightly with the frequency of the light.

Substance	*n*	Substance	*n*
Air	1.0003	Glass, flint	1.63
Benzene	1.50	Ice	1.31
Carbon disulfide	1.63	Lucite™ and Plexiglas™	1.51
Diamond	2.42	Quartz	1.46
Ethyl alcohol	1.36	Water	1.33
Glass, crown	1.52	Zircon	1.92

This is usually written in the form

$$n_1 \sin i = n_2 \sin r \qquad\qquad Snell's\ law \quad (22.4)$$

 E X A M P L E 22.2 _____

A beam of parallel light enters a block of ice at an angle of incidence of 30° (Fig. 22.23). What is the angle of refraction in the ice?

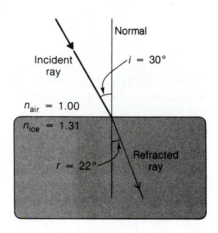

Figure 22.23

SOLUTION The indexes of refraction of air and ice are, respectively, 1.00 and 1.31. From Eq. (22.4)

$$\sin r = \frac{n_{air}}{n_{ice}} \sin i = \frac{1.00}{1.31} \sin 30° = 0.382$$

$$r = 22°$$

◆

 E X A M P L E 22.3 _____

A light ray is incident at an angle of 45° on one side of a glass plate of index of refraction 1.6. Find the angle at which the ray emerges from the other side of the plate.

SOLUTION The geometry of this problem is shown in Fig. 22.24. The angle of refraction r_1 at the first side of the plate is found as follows:

$$\sin r_1 = \frac{n_{air}}{n_{glass}} \sin i_1 = \frac{1.0}{1.6} \sin 45° = 0.442$$

$$r_1 = 26°$$

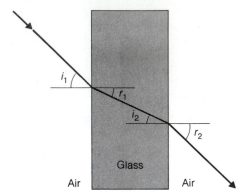

Figure 22.24

Light entering a glass plate with parallel sides emerges parallel to its original direction but is displaced to one side.

Because the sides of the plate are parallel, the angle of incidence i_2 at the second side is equal to r_1. If we let r_2 be the angle of refraction at which the ray emerges from the plate,

$$\sin r_2 = \frac{n_{\text{glass}}}{n_{\text{air}}} \sin i_2 = \frac{1.6}{1.0} \sin 26° = 0.701$$

$$r_2 = 45°$$

The ray leaves the glass plate parallel to its original direction but displaced to one side. This is a general result for parallel-sided plates that holds for all angles of incidence and all indexes of refraction; of course, when $i_1 = 0$, the ray is not displaced.   ◆

Dispersion

The index of refraction of a medium depends to some extent on the frequency of the light involved, with the highest frequencies having the highest values of n. In ordinary glass the index of refraction for violet light is about 1% greater than that for red light, for example.

Because a different index of refraction means a different deflection when a light beam enters or leaves a medium, a beam containing more than one frequency is split into a corresponding number of different beams when it is refracted. This effect, called **dispersion,** is illustrated in Fig. 22.25, which shows the result of directing a narrow pencil of white light at one face of a glass prism. The initial beam separates into beams of various colors, from which we conclude that white light is actually a mixture of light of these different colors. The band of colors that emerges from the prism is known as a **spectrum.**

Dispersion is especially conspicuous in diamond, which is the reason for the vivid play of color when white light shines on a cut diamond. The sparkle of a cut diamond is due partly to its high refractive index and partly to the way it is cut (Fig. 22.26).

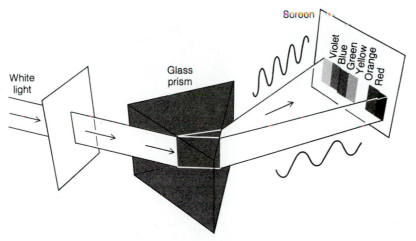

Figure 22.25

Dispersion of white light by a prism. The various colors blend into one another smoothly.

Dispersion in water droplets is responsible for rainbows, which are seen when the sun is behind an observer who is facing falling rain. Figure 22.27 shows what happens when a ray of sunlight enters a raindrop. The ray is first refracted, then reflected at the far surface, and finally refracted again when it emerges. Dispersion occurs at each refraction, with the result that the angles between the incoming sunlight and the violet and red ends of the spectrum are, respectively, 40° and 42°. Thus the light that reaches the observer comes from a ring in the sky between 40° and 42° from the direction of the sunlight. In this ring red light is on the outside because, being deviated most, it originates in droplets farther away. Between the red outside of the rainbow and the violet inside appear all the other colors of the spectrum. A person in an airplane can see the entire ring, but from the ground only the upper part appears to give the familiar arc-shaped spectrum.

Light arriving at raindrops at a larger angle than that shown in Fig. 22.27 can undergo two reflections inside each drop before emerging, which leads to an outer secondary rainbow whose colors are in the reverse order. Because some light is refracted out of a water drop at each reflection, the double reflection makes the secondary rainbow much fainter than the primary one.

Figure 22.26

(a) A diamond cut in the "brilliant" style has 33 facets in its upper part and 25 in its lower part. The proportions of the facets are critical in giving the maximum of sparkle. (b) Dispersion gives a cut diamond its fire.

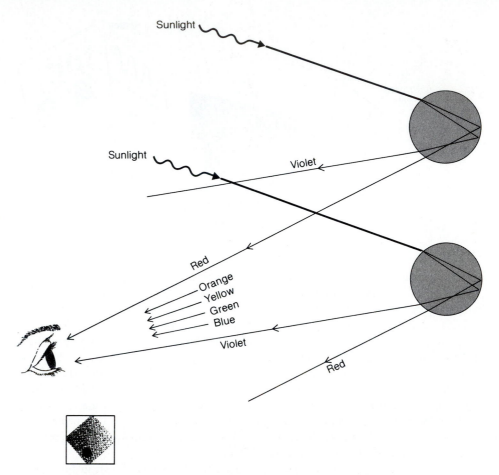

Figure 22.27

Rainbows are created by the dispersion of sunlight by raindrops. Red light arrives at the eye of the observer from the upper drop shown here; violet light arrives from the lower drop. Other raindrops yield the other colors and produce a continuous arc in the sky.

22.7 APPARENT DEPTH

Submerged objects seem closer to the surface than they actually are.

Dip an oar in a lake, and it will seem bent upward where it enters the water because its submerged portion appears to be closer to the surface than it actually is. Figure 22.28 shows how this effect comes about. Light leaving the tip of the oar at *B* is bent away from the normal upon entering the air. To an observer above, who instinctively interprets what he or she sees in terms of the straightline propagation of light, the tip of the oar is a *C*, and the submerged part of the oar seems to be *AC* and not *AB*.

It is not hard to relate the apparent and actual depths of a submerged object. Figure 22.29 shows a fish at the point *F*, an actual depth *h* below the surface of a body of water. To an observer in the air the fish is at *F'*, only *h'* below the surface. If we restrict ourselves to rays that are nearly vertical,

$$\frac{\sin i}{\sin r} \approx \frac{\tan i}{\tan r}$$

Small-angle approximation   (22.5)

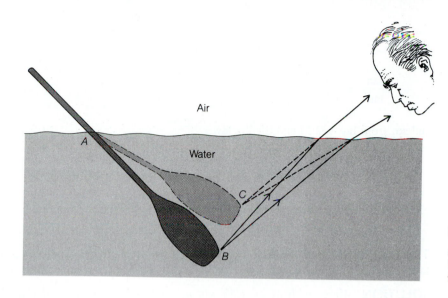

Figure 22.28

An oar appears bent when partly immersed in water because of refraction.

because $\sin \theta \approx \tan \theta$ when θ is small. From Fig. 22.29 we see that

$$\tan i = \frac{x}{h} \qquad \text{and} \qquad \tan r = \frac{x}{h'}$$

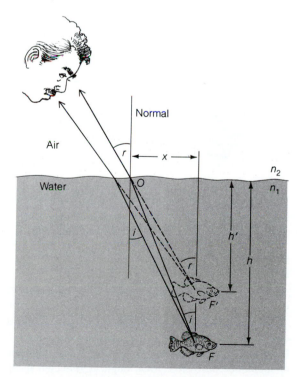

Figure 22.29

Submerged bodies seem closer to the surface than they actually are.

where x is the horizontal distance between the position of the fish and the point Q where the ray under consideration leaves the water. Hence

$$\frac{\tan i}{\tan r} = \frac{x/h}{x/h'} = \frac{h'}{h}$$

and so, from Eq. (22.5) and Snell's law,

$$\frac{h'}{h} = \frac{n_2}{n_1}$$ *Apparent depth*   (22.6)

◆   **E X A M P L E   22.4** _____

To a sailor standing on the deck of a yacht, the water depth appears to be 3.0 m. If this estimate is correct in terms of what she sees, what is the true depth?

SOLUTION   Here, with $n_1 = 1.33$ and $n_2 = 1.00$ (the indexes of refraction of water and air, respectively),

$$\frac{h'}{h} = \frac{1.00}{1.33} = 0.752$$

The apparent depth h' is only about three-quarters of the true depth h. Hence

$$h = \frac{h'}{0.752} = \frac{3.0\,\text{m}}{0.752} = 4.0\,\text{m}$$ ◆

22.8 TOTAL INTERNAL REFLECTION

Past the critical angle, light is reflected rather than refracted.

An interesting effect called **total internal reflection** can occur when light goes from one medium to another whose index of refraction is lower—for instance, from glass or water to air.

In a situation of this kind, the angle of refraction is greater than the angle of incidence, and a light ray is bent *away* from the normal (ray 2 in Fig. 22.30). As the angle of incidence is increased, a certain **critical angle** i_c is reached for which the angle of refraction is 90°. The ''refracted'' ray now travels along the interface between the two media and cannot escape (ray 3). A ray approaching the boundary at an angle of incidence past the critical angle is reflected back into the medium it comes from,

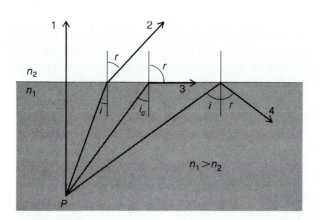

Figure 22.30

Total internal reflection occurs when the angle of refraction of a light ray going from one medium to another of lower index of refraction equals or exceeds 90°.

Figure 22.31

An underwater observer sees a circle of light at the surface. All the light reaching him or her is concentrated in a cone 98° wide.

with the angle of reflection being equal to the angle of incidence as in any other instance of reflection (ray 4).

To find the value of the critical angle, we set $i = i_c$ and $r = 90°$ in Snell's law. Because $\sin 90° = 1$, we obtain

$$n_1 \sin i_c = n_2 \sin 90°$$

$$\sin i_c = \frac{n_2}{n_1} \qquad \text{Critical angle} \quad (22.7)$$

For a ray going from water to air,

$$\sin i_c = \frac{n_{\text{air}}}{n_{\text{water}}} = \frac{1.00}{1.33} = 0.752$$

$$i_c = 49°$$

What does a person (or a fish) who is underwater see when looking upward? By reversing all the rays of light in Fig. 22.30 (the paths taken by light rays are always reversible), it is clear that light from everywhere above the water's surface reaches the eyes through a circle on the surface. The rays are all concentrated in a cone whose angular width is twice i_c, which is 98°. Outside this cone is darkness (Fig. 22.31).

The sharpness and brightness of a light beam are better preserved by total internal reflection than by reflection from an ordinary mirror. Optical instruments accordingly use prisms instead of mirrors whenever light is to be changed in direction. Figure 22.32 shows three types of totally reflecting prisms in common use.

Figure 22.32

Three applications of total internal reflection.

90° deflection

180° deflection

Image inversion

Fiber Optics

Total internal reflection makes it possible to ''pipe'' light from one place to another with a rod of glass or transparent plastic. As in Fig. 22.33, successive internal reflections occur at the surface of the rod, and nearly all the light entering at one end emerges at the other. If a cluster of narrow glass fibers is used instead of a single thick rod, an image can be transferred from one end to the other with each fiber carrying a part of the image. Because a tube of glass fibers is flexible, it can be used for such purposes as examining a person's stomach by passing the tube in through the mouth. Some of the fibers are used to provide light for illumination, and the others carry the reflected light back outside for viewing.

Glass fibers are coming into increasing use in telephone systems, where they have a number of advantages over electric wires for carrying information. In a fiber-optic system, the electric signals that would normally be sent directly through copper wires are first converted into a series of pulses according to a suitable code. The pulses are then transmitted as flashes of light down a thin glass fiber (5–50 μm in diameter), and at the other end the flashes are reconverted into normal electric signals.

Modern electronic methods permit a maximum of 32 telephone conversations to be carried at the same time by a pair of wires, but the theoretical maximum for a glass fiber is in the millions. As a practical matter, several thousand conversations can today be carried on the same fiber. For instance, a six-fiber transatlantic cable over 5500 km long that links the United States and Europe can carry 40,000 conversations at the same time. Optical fibers have the further advantage that signals in them

Figure 22.33

Light can be ''piped'' from one place to another by means of successive internal reflections in a glass rod.

Glass rod

Cluster of
glass fibers

fade less rapidly than electric ones do in copper wires, and so fewer amplifiers are needed over long distances. Also, electrical interference from nearby circuits is impossible.

22.9 LUMINOUS FLUX

How much visible light a source emits.

Only part of the radiation emitted by most light sources is visible light, perhaps 10% in the case of an ordinary light bulb. Also, the eye does not respond with equal sensitivity to light of different colors; the sensitivity is a maximum for green light and it decreases toward both ends of the visible spectrum (see Section 23.6). Simply stating the power output of a light source in watts therefore does not give much information about the illumination it actually provides. A separate system of units has been devised for this purpose that is more closely related to the visual repose of the human eye.

The brightness of a light source is referred to as its **luminous intensity** I, for which the **candela** (cd) is the SI unit. The candela is defined in terms of the light emitted by a small pool of platinum at its melting point. A candle has a luminous intensity of about 1 cd, and in fact the former standard of this quantity was an actual candle of specified composition and dimensions.

The **luminous flux** F emitted by a light source describes the total amount of visible light it gives off. A small bright source may give off less light than a large dim source, just as a small hot object may give off less heat than a large warm one. The unit of luminous flux is the **lumen** (lm), which equals the luminous flux that falls on each square meter of a sphere 1 m in radius at whose center is a 1-cd light source that radiates equally well in all directions (Fig. 22.34). Such a source is said to be **isotropic.** A lumen of flux is equivalent to about 0.0015 W of light of wavelength 555 mm, which corresponds to the yellow-green light the eye is most sensitive to.

Because the area of a sphere of radius r is $4\pi r^2$, a sphere whose radius is exactly 1 m has a total area of 4π m². The total luminous flux radiated by a 1-cd source is thus 4π lm. The luminous flux F radiated by an isotropic source of intensity I is accordingly given by the formula

$$F = 4\pi I \qquad\qquad \textit{Isotropic source} \quad (22.8)$$

Luminous flux $= 4\pi$ (luminous intensity)

Thus the luminous flux emitted by a 100-cd lamp is

$$F = 4\pi I = 4\pi(100 \text{ cd}) = 1.26 \times 10^3 \text{ lm}$$

Although the dimensions of the candela and the lumen are the same, as Eq. (22.8) indicates, these units are given different names because they refer to different quantities.

Solid Angle

The above definition of luminous flux can be generalized to nonisotropic light sources, which are sources that do not radiate equally in all directions. To do this we need the concept of the **solid angle,** which is the three-dimensional counterpart of an ordinary angle in two dimensions. As in Fig. 22.34, the solid angle Ω (Greek capital letter *omega*) subtended by an area A on the surface of a sphere of radius r is defined as

$$\Omega = \frac{A}{r^2} \qquad\qquad\qquad \textit{Solid angle} \quad (22.9)$$

$$\text{Solid angle} = \frac{\text{area on surface of sphere}}{(\text{radius of sphere})^2}$$

The unit of solid angle is the **steradian** (sr). Since the total surface area of a sphere is $4\pi r^2$, the total solid angle subtended by a sphere is $4\pi r^2/r^2$ sr $= 4\pi$ sr. Like the degree and the radian, the steradian is a ratio that has no dimensions and so disappears in calculations.

In terms of solid angle, the definition of luminous flux becomes

$$F = I\Omega \qquad\qquad\qquad \textit{Luminous flux} \quad (22.10)$$

$$\text{Luminous flux} = (\text{luminous intensity})(\text{solid angle})$$

Thus a lumen equals the luminous flux that a 1-cd source radiates per steradian of solid angle, and the candela is equivalent to the lumen per steradian.

Figure 22.34

How the luminous flux emitted by an isotropic light source and the illumination it produces are related to the intensity of the source.

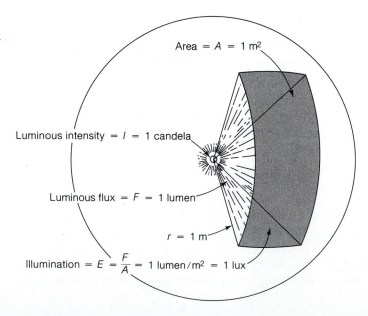

Area $= A = 1$ m²

Luminous intensity $= I = 1$ candela

Luminous flux $= F = 1$ lumen

$r = 1$ m

Illumination $= E = \dfrac{F}{A} = 1$ lumen/m² $= 1$ lux

◆   **E X A M P L E   22.5** _____

A spotlight concentrates the light from a 200-cd isotropic lamp into a circle 1.20 m in radius at a distance of 20.0 m. Find the luminous intensity of the source looking into the beam. (This is the luminous intensity an isotropic source would need to provide the same luminous flux on the illuminated circle.)

SOLUTION   The total luminous flux emitted by the lamp is

$$F = 4\pi I = (4\pi)\ (200\ cd) = 2513\ lm$$

Since the radius $R = 1.20$ m of the illuminated circle is small compared with the 20.0-m radius of a sphere centered on the spotlight, we can ignore the difference between the circle's plane area and its area as measured on the surface of the sphere. Hence

$$A = \pi R^2 = \pi\ (1.20\ m)^2 = 4.52\ m^2$$

and the solid angle of the spotlight beam is

$$\Omega = \frac{A}{r^2} = \frac{4.52\ m^2}{(10.0\ m)^2} = 0.0452\ sr$$

The luminous intensity of the light source looking into the beam is, from Eq. (22.10),

$$I' = \frac{F}{\Omega} = \frac{2513\ lm}{0.0452\ sr} = 5.56 \times 10^4\ cd$$

The optical system of the spotlight has therefore produced an increase in source brightness of $I'/I = 278$.   ◆

Light Sources

The luminous flux emitted by a light source per watt of power input is its **luminous efficiency.** Ordinary light sources fall into three categories:

1. An *incandescent lamp* has a filament of tungsten (a high melting point metal) that is heated to a high temperature by the passage of electric current. Most of the power supplied to an incandescent lamp is wasted as heat. Such a lamp increases in efficiency with power, because the higher the power, the higher the filament temperature and the greater the proportion of visible light in the total radiation. A 10-W lamp, for instance, has a luminous efficiency of about 8 lm/W, whereas it is 16 lm/W for a 100-W lamp.
2. A *vapor lamp* contains a gas or vapor that is caused to radiate by the passage of an electric current. The color of the light depends on the nature of the gas or

vapor, as discussed in Chapter 25. Vapor lamps are more efficient than incandescent lamps; a typical 100-W mercury vapor lamp emits 3600 lm for an efficiency of 36 lm/W.

3. A *fluorescent lamp* consists of a glass tube filled with a mixture of mercury vapor and an inert gas such as argon. When a current is passed through the mixture, ultraviolet radiation is produced. The inside of the tube is coated with a material called a *phosphor* that emits visible light when it is excited by the ultraviolet radiation. Fluorescent lamps produce little heat and so are very efficient; a 100-W fluorescent lamp might give off 5500 lm for an efficiency of 55 lm/W, figures that are 3.4 times greater than those for a 100-W tungsten-filament lamp.

22.10   ILLUMINATION

Better to light a candle than to curse the darkness.

The more the luminous flux that reaches a given surface, such as a book, the more brightly it is lit. To put this idea on a definite basis, the **illumination** E of a surface is defined as the luminous flux per unit area that reaches it:

$$E = \frac{F}{A}$$

$$\text{Illumination} = \frac{\text{luminous flux}}{\text{area}}$$

Illumination   (22.11)

Table 22.2 summarizes the various quantities introduced in this section and the

Table 22.2   For an isotropic source, $F = 4\pi I$.

Quantity	Symbol	Meaning	Formula	Unit
Luminous intensity	I	Brightness of light source		candela (cd)
Solid angle	Ω	Three-dimensional equivalent of an angle in a plane	$\Omega = \dfrac{A}{r^2}$	steradian (sr)
Luminous flux	F	Quantity of visible light	$F = I\Omega$	lumen (lm)
Luminous efficiency		Luminous flux emitted relative to power input to light source	F/P	lumen/watt (lm/W)
Illumination	E	Luminous flux per unit area	$E = F/A$	lux $(= \text{lm/m}^2)$ footcandle $(= \text{lm/ft}^2)$

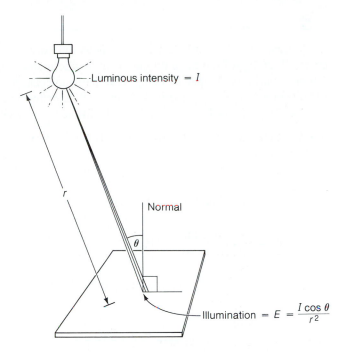

Luminous intensity $= I$

Normal

θ

Illumination $= E = \dfrac{I \cos \theta}{r^2}$

Figure 22.35

The illumination of a surface is equal to the luminous flux per unit area that reaches it.

previous one. In the SI system the unit of illumination is the lumen per square meter, or **lux** (lx). In the British system it is the lumen per square foot, or **footcandle,** where 1 footcandle = 10.8 lux.

About 500 lx is a reasonable figure for general interior lighting, with at least 800 lx recommended for reading and 1100 lx for electrical and machine shop work. The full moon produces an illumination of about 0.50 lx, a clear daytime sky about 11,000 lx, and direct sunlight about 100,000 lx.

The illumination on a surface a distance r away from an isotropic light source of intensity I is given by

$$E = \frac{I \cos \theta}{r^2} \qquad \text{\textit{Illumination due to isotropic source}} \quad (22.12)$$

In this formula θ is the angle between the direction of the light and a normal to the surface where the light strikes it (Fig. 22.35). The $\cos \theta$ factor arises because E depends on the component $E \cos \theta$ of the flux F that is perpendicular to the surface. Thus the illumination is a maximum of I/r^2 for light striking a surface at an angle of $\theta = 0$ because $\cos 0 = 1$, and it drops to $E = 0$ for light parallel to the surface when $\theta = 90°$ and $\cos 90° = 0$.

The illumination from an isotropic source varies as $1/r^2$, just as in the case of sound waves. Doubling the distance from such a source reduces the illumination to $1/(2)^2 = \frac{1}{4}$ of its original value; cutting the distance in half increases the illumination to $1/(\frac{1}{2})^2 = (2)^2 = 4$ times its original value (Fig. 22.36).

Figure 22.36

The illumination from an isotropic light source varies as the inverse square $(1/r^2)$ of the distance from the source because the area covered by a given flux varies as r^2.

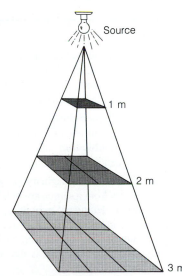

Source

1 m

2 m

3 m

◆   E X A M P L E   22.6 _____

(a) A certain 60-W light bulb has a luminous intensity of 70 cd. A reflector concentrates all the light from the bulb into a circle whose area is 0.80 m² when the bulb is 1.0 m from a book that is being held perpendicular to the light rays. Find the illumination of the book. (b) If the lamp had no reflector and radiated isotropically, how far from the book would it have to be located to produce the same illumination?

SOLUTION   (a) The bulb emits a total luminous flux of

$$F = 4\pi I = (4\pi)(70 \text{ cd}) = 880 \text{ lm}$$

The illumination of the book is therefore

$$E = \frac{F}{A} = \frac{880 \text{ lm}}{0.80 \text{ m}^2} = 1100 \text{ lx}$$

(b) Here $\theta = 0$, $\cos \theta = 1$, and $E = I/r^2$. Hence

$$r = \sqrt{\frac{I}{E}} = \sqrt{\frac{70 \text{ cd}}{1100 \text{ lx}}} = 0.25 \text{ m}$$

The reflection of light from the walls of the room is neglected here, but obviously it would contribute to the illumination of the book.   ◆

Important Terms

Maxwell's hypothesis is that a changing electric field produces a magnetic field.

Electromagnetic waves consist of coupled electric and magnetic field oscillations. Radio waves, microwaves, light waves, X-rays, and gamma rays are all electromagnetic waves differing only in their frequency.

In **amplitude modulation,** information is contained in variations in the amplitude of a constant-frequency carrier wave. In **frequency modulation,** information is contained in variations in the frequency of the carrier wave.

A **wavefront** is an imaginary surface that joins points where all the waves from a source are in the same phase of oscillation. According to **Huygens' principle,** every point on a wavefront can be considered as a point source of secondary wavelets that spread out in all directions with the wave speed of the medium. The wavefront at any time is the envelope of these wavelets.

In **diffuse reflection** an incident beam of parallel light is spread out in many directions, whereas in **specular reflection** the angle of reflection is equal to the angle of incidence. A **mirror** is a specular reflecting surface of regular form that can produce an image of an object placed before it.

A **real image** of an object is formed by light rays that pass through the image; the image would therefore appear on a properly placed screen. A **virtual image** can only be seen by the eye because the light rays that seem to come from the image actually do not pass through it.

The bending of a light beam when passing from one medium to another is called **refraction.** The quantity that governs the degree to which a light beam will be deflected in entering a medium is its **index of refraction,** defined as the ratio between the speed of light in free space and its speed in the medium. **Dispersion** refers to the splitting up of a beam of light containing different frequencies by passage through a substance whose index of refraction varies with frequency.

Snell's law states that the ratio between the sine of the angle of incidence of a light ray on an interface between two media and the sine of the angle of refraction is equal to the ratio of the speeds of light in the two media.

In **total internal reflection,** light arriving at a medium of lower index of refraction at an angle greater than the **critical angle** of incidence is reflected back into the medium it came from.

A **solid angle** is the counterpart in three dimensions of an angle in two dimensions.

The **luminous intensity** of a light source refers to its brightness with respect to the visual response of the eye, and the **luminous flux** it emits is a measure of its light output. The **illumination** of a surface is equal to the flux reaching it per unit area.

Important Formulas

Index of refraction: $\quad n = \dfrac{c}{v}$

Snell's law: $\quad n_1 \sin i = n_2 \sin r$

Apparent depth: $\quad \dfrac{h'}{h} = \dfrac{n_2}{n_1}$

Critical angle: $\quad \sin i_c = \dfrac{n_2}{n_1}$

Solid angle: $\quad \Omega = \dfrac{A}{r^2}$

Luminous flux: $\quad F = I\Omega$
$$F = 4\pi I \quad \text{(isotropic source)}$$

Illumination: $\quad E = \dfrac{F}{A}$
$$E = \dfrac{I \cos \theta}{r^2} \quad \text{(isotropic source)}$$

MULTIPLE CHOICE

1. A changing electric field gives rise to
   a. a magnetic field
   b. electromagnetic waves
   c. sound waves
   d. nothing in particular

2. In an electromagnetic wave the electric field is
   a. parallel to both the magnetic field and the wave direction
   b. perpendicular to both the magnetic field and the wave direction
   c. parallel to the magnetic field and perpendicular to the wave direction
   d. perpendicular to the magnetic field and parallel to the wave direction

3. In a vacuum the speed of an electromagnetic wave
   a. depends on its frequency
   b. depends on its wavelength
   c. depends on its electric and magnetic fields
   d. is a universal constant

4. The highest frequencies are found in
   a. X-rays
   b. ultraviolet light
   c. radio waves
   d. radar waves

5. Electromagnetic waves transport
   a. wavelength
   b. frequency
   c. charge
   d. energy

6. The ionosphere is a region of ionized gas in the upper atmosphere. The ionosphere is responsible for
   a. the blue color of the sky
   b. rainbows
   c. long-distance radio communication
   d. the ability of satellites to orbit the earth

7. Light of which of the following colors has the shortest wavelength?
   a. red
   b. blue
   c. yellow
   d. green

8. The Spanish flag is yellow and red. When viewed with yellow light it appears
   a. all yellow
   b. yellow and red
   c. yellow and black
   d. white and red

9. The quality in sound that corresponds to color in light is
   a. amplitude
   b. resonance
   c. waveform
   d. pitch

10. Most of the effects an em wave produces when it interacts with matter are due to its
    a. electric field
    b. magnetic field
    c. speed
    d. spectrum

11. All real images
  a. are erect
  b. are inverted
  c. can appear on a screen
  d. cannot appear on a screen

12. When you look at yourself in a plane mirror, what you see is a
  a. real image behind the mirror
  b. real image in front of the mirror
  c. virtual image behind the mirror
  d. virtual image in front of the mirror

13. When a beam of light enters one medium from another, a quantity that never changes is its
  a. direction   **b.** speed
  c. frequency   **d.** wavelength

14. The index of refraction of a material medium
  a. is less than 1
  b. is equal to 1
  c. is greater than 1
  d. may be any of the above

15. Relative to the angle of incidence, the angle of refraction
  a. is smaller
  b. is the same
  c. is larger
  d. may be any of the above

16. A light ray enters one medium from another along the normal. The angle of refraction
  a. is 0
  b. is 90°
  c. equals the critical angle
  d. depends on the indexes of refraction of the two media

17. Light goes from medium *A* to medium *B* at an angle of incidence of 40°. The angle of refraction is 30°. The speed of light in *B*
  a. is less than that in *A*
  b. is the same as that in *A*
  c. is greater than that in *A*
  d. may be any of the above, depending on the specific medium

Figure 22.37

Multiple Choice 18.

18. Figure 22.37 shows a ray of light going from water to air. The refracted ray is
  a. *A*   **b.** *B*
  c. *C*   **d.** *D*

19. Which of the diagrams in Fig. 22.38 could represent the path of a light ray through a glass block in air?
  a. *A*   **b.** *B*
  c. *C*   **d.** *D*

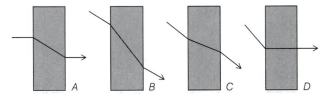

Figure 22.38

Multiple Choice 19.

20. Dispersion is the term used to describe
  a. the splitting of white light into its component colors in refraction
  b. the propagation of light in straight lines
  c. the bending of a beam of light when it goes from one medium to another
  d. the bending of a beam of light when it strikes a mirror

21. The depth of an object submerged in a transparent liquid
  a. always seems less than its actual depth
  b. always seems more than its actual depth
  c. may seem less or more than its actual depth, depending on the index of refraction of the liquid
  d. may seem less or more than its actual depth, depending on the angle of view

22. Total internal reflection can occur when light passes from one medium to another
  a. that has a lower index of refraction
  b. that has a higher index of refraction
  c. that has the same index of refraction
  d. at less than the critical angle

23. When a light ray approaches a glass-air interface from the glass side at the critical angle, the angle of refraction is
  a. 0
  b. 45°
  c. 90°
  d. equal to the angle of incidence

24. Which of the diagrams in Fig. 22.39 could represent the path of a light ray through a glass prism in air? The critical angle in glass is about 42°.

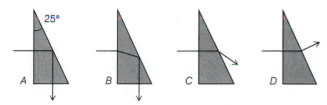

Figure 22.39

Multiple Choice 24.

  a. *A*   **b.** *B*
  c. *C*   **d.** *D*

25. The brightness of a light source is called its luminous intensity, whose unit is the
  a. candela   **b.** lumen
  c. lux   **d.** footcandle

26. Luminous efficiency is least for a
  a. low-wattage light bulb
  b. high-wattage light bulb
  c. mercury vapor lamp
  d. fluorescent tube

27. The minimum illumination recommended for reading is
  a. 800 cd   **b.** 800 lm
  c. 800 lx   **d.** 800 W

28. The index of refraction of benzene is 1.5. The speed of light in benzene is
  a. 1.5×10^8 m/s   **b.** 2×10^8 m/s
  c. 3×10^8 m/s   **d.** 4.5×10^8 m/s

29. Light enters a glass plate whose index of refraction is 1.6 at an angle of incidence of 30°. The angle of refraction is
  a. 18°   **b.** 19°
  c. 48°   **d.** 53°

30. Light leaves a slab of transparent material whose index

of refraction is 2.0 at an angle of refraction of 0°. The angle of incidence is
  a. 0°   **b.** 30°
  c. 45°   **d.** 90°

31. Light enters a glass plate at an angle of incidence of 40° and is refracted at an angle of refraction of 25°. The index of refraction of the glass is
  a. 0.625   **b.** 0.66
  c. 1.52   **d.** 1.6

32. An underwater swimmer shines a flashlight beam upward at an angle of incidence of 40°. The angle of refraction is 60°. The index of refraction of water is
  a. 0.67   **b.** 0.74
  c. 1.3   **d.** 1.5

33. A fish frozen in the clear ice ($n = 1.3$) of a frozen lake appears to be 40 cm below the surface. Its actual depth is
  a. 31 cm   **b.** 46 cm
  c. 52 cm   **d.** 80 cm

34. The critical angle of incidence for light going from crown glass ($n = 1.5$) to ice ($n = 1.3$) is
  a. 12°   **b.** 42°
  c. 50°   **d.** 60°

35. The solid angle subtended by a hemisphere about its center
  a. is $\pi/2$ sr
  b. is π sr
  c. is 2π sr
  d. depends on the radius of the hemisphere

36. The luminous flux emitted by a 60-cd isotropic light source is concentrated on an area of 0.5 m². The illumination of the area is
  a. 9.6 lx   **b.** 120 lx
  c. 377 lx   **d.** 1508 lx

QUESTIONS

1. Why was electromagnetic induction discovered much earlier than its converse, the production of a magnetic field by a varying electric field?

2. Under what circumstances does a charge radiate electromagnetic waves?

3. Why are light waves able to travel through a vacuum, whereas sound waves cannot?

4. Light is said to be a transverse wave phenomenon. What is it that varies at right angles to the direction in which a light wave travels?

5. What is the relationship between the directions of the

electric and magnetic fields of an em wave?

6. Are the magnitudes of the electric and magnetic fields of an em wave related? If so, in what way?

7. A radio transmitter has a vertical antenna. Does it matter whether the receiving antenna is vertical or horizontal? Explain.

8. What is a wavefront? What is the relationship between a light ray and the wavefronts whose motion it is used to describe?

9. Huygens' principle is useful for studying the behavior of light only in certain situations, whereas the em theory of light has a much broader scope. Huygens' principle, however, has one major advantage over the em theory of light. What do you think it is?

10. What is the significance of the ionosphere for radio communication?

11. Of the following colors, which corresponds to the lowest frequency? To the shortest wavelength? Red, yellow, blue, green.

12. The Swedish flag has a yellow cross on a blue background. How does it appear in red light? In blue light?

13. The Danish flag has a red cross on a white background. How does it appear in red light? In blue light?

14. Under what circumstances do the incident ray, the reflected ray, and the normal all lie in the same plane?

15. What is the difference between a real image and a virtual image?

EXERCISES

22.2   Types of Em Waves

1. The marine radiotelephone station on Tahiti transmits at the frequency 4390.2 kHz. What wavelength does this correspond to?

2. A radar sends out 0.050-μs pulses of microwaves whose wavelength is 2.5 cm. What is the frequency of these microwaves? How many waves does each pulse contain?

3. A nanosecond is 10^{-9} s. (a) What is the frequency of an electromagnetic wave whose period is 1.0 ns? (b) What is its wavelength? (c) To what class of electromagnetic waves does it belong?

22.4   Reflection

4. What is the height of the smallest mirror in which a woman 160 cm tall can see herself at full length? Does it matter how far from the mirror she stands?

5. A plane mirror is mounted on the back of a truck that is traveling at 30 km/h. How fast does the image of a man standing in the road behind the truck seem to be moving away from him?

22.5   Refraction

22.6   Index of Refraction

6. Using the indexes of refraction given in Table 22.1 and taking the speed of light in free space as 3.0×10^8 m/s, find the speed of light in (a) air, (b) diamond, (c) crown glass, and (d) water.

16. Do electromagnetic waves have the same speed in all transparent media? If not, how do their speeds in a transparent medium compare with their speed in free space?

17. Explain why a cut diamond held in white light shows flashes of color. What would happen if it were held in red light?

18. Why is a beam of white light not dispersed into its component colors when it passes perpendicularly through a pane of glass?

19. A glass of water ($n = 1.33$) and a glass of ethyl alcohol ($n = 1.36$) seem to be filled to the same depth when viewed from above. Is the liquid level actually the same in both glasses? If not, which liquid is deeper?

20. In Fig. 22.25 the incident light is refracted at the right-hand face of the prism. Why is it not internally reflected, as in the cases shown in Fig. 22.32?

7. A flashlight is frozen into a block of ice. If its beam strikes the surface of the ice at an angle of incidence of 37°, what is the angle of refraction?

8. What is the angle of refraction of a beam of light that enters the surface of a lake at an angle of incidence of 50°?

9. A beam of light strikes a pane of glass at an angle of incidence of 60°. If the angle of refraction is 35°, what is the index of refraction of the glass?

10. A beam of light enters a liquid of unknown composition at an angle of incidence of 30° and is deflected by 5° from its original path. Find the index of refraction of the liquid.

11. The index of refraction of ordinary crown glass for red light is 1.51 and for violet light, 1.53. A beam of white light falls on a cube of such glass at an incident angle of 40°. What is the difference between the angles of refraction of the red and the violet light?

12. A ray of light passes through a plane boundary separating two media whose indexes of refraction are $n_1 = 1.5$ and $n_2 = 1.3$. (a) If the ray goes from medium 1 to medium 2 at an angle of incidence of 45°, what is the angle of refraction? (b) If the ray goes from medium 2 to medium 1 at the same angle of incidence, what is the angle of refraction?

13. A double-glazed window consists of two panes of glass separated by an air space for better insulation. (a) Assuming that the glass has an index of refraction of 1.5, find the angle at which a ray of sunlight enters a room if it arrives at the window at an angle of incidence of 70°. (b) The window leaks, and the space between the panes becomes filled with water.

What is the angle at which the ray enters the room now?

14. A layer of benzene is floating on water. Find the angle of refraction in the water of a light ray whose angle of incidence from air to the benzene is 45°.

15. A glass prism has an apex angle of 60° and an index of refraction of 1.45. If a light ray enters the prism parallel to its base, at what angle of refraction does it leave the prism?

16. The prism in Exercise 15 is placed in water. Find the angle of refraction in this case.

22.7 Apparent Depth

17. The olive in a martini cocktail ($n = 1.35$) appears to be 40 mm below the surface. What is the actual depth of the olive?

18. A glass paperweight in the form of a 60-mm cube is placed on a letter. How far underneath the top of the cube does the letter appear if the index of refraction of the glass is 1.5?

19. A barrel of ethyl alcohol is 80 cm high. How high does it appear to somebody looking into it from above?

20. A glass block has the dimensions 10 cm × 10 cm × 16 cm. When it rests on one of its 10-cm × 10-cm faces and is examined from above, the block appears to be a 10-cm cube. What is the index of refraction of the glass?

22.8 Total Internal Reflection

21. Find the critical angle for total internal reflection in a diamond when the diamond is (a) in air, and (b) immersed in water.

22. The critical angle for total internal reflection in Lucite is 41°. Find its index of refraction.

23. Snell's law also applies to the refraction of sound waves. (a) Can sound waves be totally reflected when they reach an air-water interface from the air side? From the water side? (b) Find the critical angle for the total reflection of sound waves that arrive at an air-seawater interface. The speed of sound is 343 m/s in air and 1531 m/s in seawater.

PROBLEMS

34. A light ray enters one end of a glass fiber at an angle of incidence of θ, as in Fig. 22.40. (a) If the index of refraction of the glass is n, what is the maximum value of θ that will permit the ray to be totally reflected from the wall of the fiber? (b) What is the value of θ_{max} for $n = 1.35$?

35. A ray of light strikes a glass plate at an angle of incidence of 55°. Some of the light is reflected and the rest is refracted.

22.9 Luminous Flux

22.10 Illumination

24. Find the solid angle subtended at the center of a sphere of radius 8.0 m by an area of 200 cm² on its surface.

25. About 29% of the earth's surface is above sea level. If all the land areas were concentrated in a single region, what solid angle would it subtend at the center of the earth?

26. What area on the surface of a sphere of radius 4.0 m is cut by a solid angle of 0.20 sr?

27. A projector produces an illumination of 10,000 lx on a wall perpendicular to its beam that is 6.0 m away. What is the luminous intensity looking into the beam?

28. A reflector concentrates the light emitted by a 60-W lamp wose luminous efficiency is 14 lm/W into a beam of solid angle 0.10 sr. What is the luminous intensity of the source looking into its beam?

29. A reflector concentrates the light emitted by an isotropic 20-W lamp on an area perpendicular to the beam of 5.0 ft² at a distance of 10 ft. If the luminous intensity looking into the beam is 2500 cd, find the luminous efficiency of the lamp.

30. A certain 100-W light bulb has a luminous intensity of 120 cd. (a) What is its luminous efficiency? (b) If the bulb radiates isotropically and is suspended 1.2 m above a table, what is the illumination directly below it?

31. A spotlight concentrates all the light from a 200-cd bulb in a beam 1.0 m in diameter. Find the illumination it produces on a surface perpendicular to the beam.

32. On a certain night the full moon produced an illumination of 0.60 lx. Find the luminous intensity of the moon that night. The distance between the earth and the moon is 3.84×10^8 m.

33. A skating rink 30 m by 60 m is to have an average illumination at night of 300 lx. How many 1.0-kW, 2500-cd lamps are needed if they are equipped with reflectors that direct $\frac{2}{3}$ of their luminous flux on the rink?

Figure 22.40

Problem 34.

If the reflected and refracted rays are perpendicular to each other, what is the index of refraction of the glass?

36. A light bulb is on the bottom of a swimming pool 1.8 m below the water surface. A person on a diving board above the pool sees a circle of light. What is its diameter?

37. Prisms are used in optical instruments such as binoculars instead of mirrors because total internal reflection better preserves the sharpness and brightness of a light beam. Find the minimum index of refraction of the glass to be used in a prism that is meant to change the direction of a light beam by 90°.

38. A 250-cd unshaded light bulb is 1.4 m above a table. Find the illumination of the table directly under the bulb and at a point 1.0 m to one side.

39. An unshaded light bulb 0.80 m above a workbench produces an illumination of 1200 lx directly below it. What is the illumination at a point on the bench 1.00 m to one side?

Answers to Multiple Choice

1. a	**10.** a	**19.** c	**28.** b
2. b	**11.** c	**20.** a	**29.** a
3. d	**12.** c	**21.** a	**30.** a
4. a	**13.** c	**22.** a	**31.** c
5. d	**14.** c	**23.** c	**32.** c
6. c	**15.** d	**24.** c	**33.** c
7. b	**16.** a	**25.** a	**34.** d
8. c	**17.** a	**26.** a	**35.** c
9. d	**18.** d	**27.** c	**36.** d

23

LENSES AND MIRRORS

23.1   Lenses

23.2   Image Formation

23.3   The Lens Equation

23.4   Magnification

23.5   The Camera

23.6   The Eye

23.7   Defects of Vision

23.8   The Microscope

23.9   The Telescope

23.10   Lens Aberrations

23.11   Spherical Mirrors

A lens is a piece of glass or other transparent material so shaped that it can produce an image by refracting light that comes from an object. The image may be real or virtual; erect or inverted; and larger, smaller, or the same size as the object. Lenses have many uses: in eyeglasses to improve vision, in cameras to record scenes, in projectors to show images on a screen, in microscopes to enable us to see small things, in telescopes to enable us to see distant things, and so forth. Mirrors that have properly curved reflecting surfaces also can form images of objects. As we will find, the optics of lenses and those of mirrors have many similarities.

23.1 LENSES

A converging lens has a positive focal length, a diverging lens has a negative focal length.

A simple way to appreciate how a lens produces its effects is to begin with a pair of prisms, as in Fig. 23.1(a). When the prisms have their bases together, parallel light that approaches is deviated so that the various rays intersect. They do not intersect at a single focal point, however, because each prism merely changes the direction of the rays without affecting the shape of the wavefronts. If properly curved rather than flat surfaces are employed, as in Fig. 23.1(b), the result is called a **converging lens,** and it acts to bring an incoming parallel beam of light to a single focal point F. Because the rays pass through F, the focal point is real.

A **diverging lens** may similarly be thought of as a development of a pair of prisms with their apexes together (Fig. 23.2). The prisms spread incoming parallel light outward, but the diverging rays cannot be projected back to a single point. The corresponding lens has curved surfaces so shaped that the diverging rays appear to originate from a single point, the virtual focal point F.

Figure 23.3 shows how lenses affect wavefronts. The speed of light in glass (or other transparent material) is less than in air, and light takes more time to pass through a certain thickness of glass than through the same thickness of air. Light that enters the center of a converging lens is thus retarded more than light that enters toward the edge. The result is that the wavefronts converge after passing through the lens. Exactly the opposite effect occurs with a diverging lens.

A lens with spherical surfaces, or one plane and one spherical surface, gives an undistorted image of something in front of it, provided the lens is relatively thin. Such lenses are also the easiest to manufacture. **Aspherical** (nonspherical) lens surfaces are used in special applications to avoid certain kinds of distortion that occur with spherical lenses. The various forms of simple lenses are shown in Fig. 23.4. Converging lenses are always thickest in the center, whereas diverging lenses are always thinnest in the center. A **meniscus** lens has one concave and one convex surface.

Figure 23.1

A converging lens may be thought of as a development of two prisms placed base to base. Such a lens brings parallel rays to the real focal point *F*, which is real because the refracted rays pass through it.

(a)

(b)

(a)

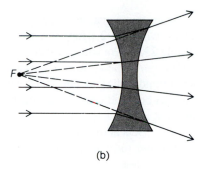

(b)

Figure 23.2

A diverging lens may be thought of as a development of two prisms placed apex to apex. Such a lens spreads out parallel rays as though they came from the virtual focal point F, which is virtual because the refracted rays do not pass through it.

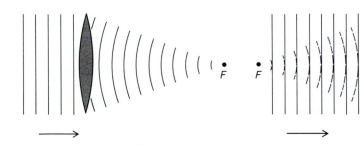

Figure 23.3

How converging and diverging lenses affect plane wavefronts.

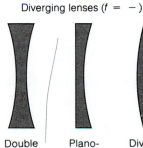

Converging lenses ($f = +$)   Diverging lenses ($f = -$)

Double convex (a)   Plano-convex (b)   Converging meniscus (c)   Double concave (d)   Plano-concave (e)   Diverging meniscus (f)

Figure 23.4

Some simple lenses. The focal length of a converging lens is reckoned positive, that of a diverging lens is reckoned negative.

Focal Length

The distance from a lens to its focal point is called its **focal length** f. The focal length of a particular lens depends both on the index of refraction n of its material relative to that of the medium it is in and on the radii of curvature R_1 and R_2 of its surfaces.

A **thin lens** is one whose thickness is small compared with R_1 and R_2. The **lensmaker's equation** holds for such a lens:

$$\frac{1}{f} = (n - 1)\left(\frac{1}{R_1} + \frac{1}{R_2}\right) \qquad \textit{Lensmaker's equation} \quad (23.1)$$

In using this equation, it does not matter which surface of the lens is considered as 1 and which as 2. However, the sign given to each radius of curvature *is* important (Table 23.1). We will consider a radius as $+$ if the surface is convex (curved outward), and as $-$ if the surface is concave (curved inward). A plane surface has, in effect, an infinite radius of curvature, and $1/R$ for a plane surface is therefore 0. In Fig. 23.4(a) both radii are $+$; in (b) the first radius is infinite and the second is $+$; in (c) the first radius is $-$ and the second is $+$; in (d) both radii are $-$; and so on.

Depending on its shape, a lens may have a positive or a negative focal length. A positive focal length signifies a converging lens, and a negative one signifies a diverging lens.

◆ E X A M P L E   23.1

A meniscus lens has a convex surface whose radius of curvature is 25 cm and a concave surface whose radius of curvature is 15 cm. The index of refraction is 1.52. Find the focal length of the lens and whether it is converging or diverging.

SOLUTION   Here $R_1 = +25$ cm and $R_2 = -15$ cm, in accord with the sign convention. From the lensmaker's equation, Eq. (23.1),

$$\frac{1}{f} = (n - 1)\left(\frac{1}{R_1} + \frac{1}{R_2}\right) = (1.52 - 1.00)\left(\frac{1}{25\,\text{cm}} - \frac{1}{15\,\text{cm}}\right)$$
$$= (0.52)(0.040 - 0.067)\,\text{cm}^{-1} = -0.014\,\text{cm}^{-1}$$

and so

$$f = -71 \text{ cm}$$

The negative focal length indicates a diverging lens.   ◆

◆ E X A M P L E   23.2

The lens in Example 23.1 is placed in water, whose index of refraction is 1.33. Find its focal length there.

Table 23.1 **Sign conventions for the lensmaker's equation. A plane surface is considered to have an infinite radius, so for it $1/R = 0$.**

Quantity	Positive	Negative
Radius	Convex	Concave
Focal length	Converging lens	Diverging lens

SOLUTION The index of refraction of the glass relative to water is

$$n' = \frac{\text{index of refraction of glass}}{\text{index of refraction of water}} = \frac{1.52}{1.33} = 1.14$$

From the lensmaker's equation, since R_1 and R_2 are the same in both air and water, the ratio between the focal length f' of the lens in water and its focal length f in air is

$$\frac{f'}{f} = \frac{n - 1}{n' - 1} = \frac{1.52 - 1}{1.14 - 1} = 3.7$$

Hence

$$f' = 3.7f = (3.7)(-71 \text{ cm}) = -263 \text{ cm}$$

The focal length of *any* lens made of this glass is 3.7 times longer in water than in air. ◆

23.2 IMAGE FORMATION

How to find an image by tracing rays through a lens.

A scale drawing gives us a simple way to find the position, size, and nature of an image formed by a lens. What we do is consider two different light rays that come from a certain point on an object and then trace their paths until they (or their extensions) come together again after being refracted by the lens.

It is worth noting that a lens has *two* focal points, one on each side the distance f from its center. The focal point on the side of the lens from which the light comes is the **near focal point,** and the one on the other side of the lens is the **far focal point.**

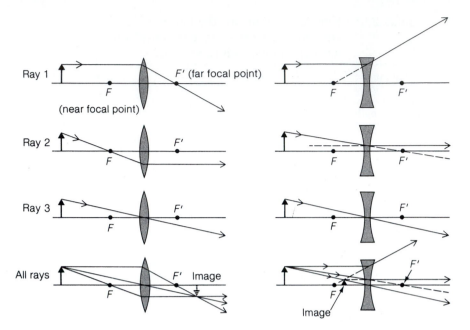

Figure 23.5

The position and size of an image produced by a thin lens can be determined by tracing any two of the rays shown. In tracing the rays, any deviations produced by the lens are assumed to occur at its central plane.

The three easiest rays to trace, as shown in Fig. 23.5, are:

1. A ray that leaves the object parallel to the lens axis. When this ray is refracted by the lens, it passes through the far focal point of a converging lens or seems to come from the near focal point of a diverging lens.
2. A ray that leaves the object and passes through the near focal point of a converging lens or is aimed at the far focal point of a diverging lens. When this ray is refracted, it continues parallel to the axis.
3. A ray that leaves the object and goes through the center of the lens. If the lens is thin relative to the radii of curvature of its surfaces, this ray is not deviated.

Ordinarily only two of these rays are enough to establish the image of an object. Figure 23.6 shows how the properties of the image produced by a converging lens depend on the position of the object. As usual, solid lines represent the actual paths taken by light rays, and dashed lines represent virtual paths.

When the object is closer to the lens than F, the near focal point, as in (a), the image is erect, enlarged, and virtual. The image seems to be behind the lens because the refracted rays diverge as though coming from a point behind it. (A virtual image can be seen by the eye, but it cannot appear on a screen because no light rays actually pass through such an image.)

No image is formed of an object at the focal point, because the refracted rays are all parallel and hence never meet, as in (b). We can think of the image in this case as being at infinity.

An object farther from the lens than F always has an inverted real image that may be smaller than, the same size as, or larger than the object, depending on

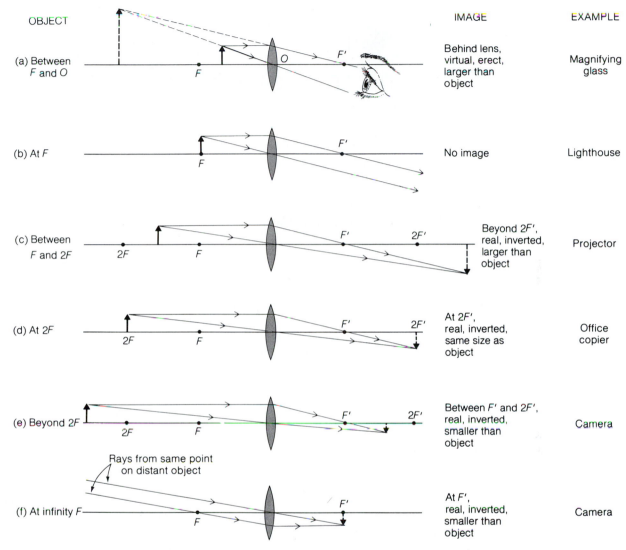

Figure 23.6

Image formation by a converging lens.

whether the object distance is between *f* and 2*f*, as in (c); equal to 2*f*, as in (d); or greater than 2*f*, as in (e) and (f).

In contrast to the diversity of image sizes, natures, and locations produced by a converging lens, the image of a real object formed by a diverging lens is always virtual, erect, and smaller than the object (Fig. 23.7).

Figure 23.7

The image of a real object formed by a diverging lens is always virtual, erect, and smaller than the object.

23.3 THE LENS EQUATION

How object and image distances are related to the focal length of a lens.

A simple formula relates the positions of the image and the object of a thin lens to the lens's focal length f. In terms of the symbols

p = distance of object from lens

q = distance of image from lens

f = focal length of lens

this formula is

$$\frac{1}{p} + \frac{1}{q} = \frac{1}{f}$$ *Lens equation*   (23.2)

When any two of the three quantities are known, the third can be found. The sign conventions to be observed when using the lens equation are as follows:

1. The focal length f is considered $+$ for a converging lens, $-$ for a diverging lens.
2. Object distance p and image distance q are considered $+$ for real objects and images, $-$ for virtual objects and images. (When the virtual image produced by a lens is used as the object of another lens, the image is said to be the virtual object of the second lens.)

It is often convenient to solve the lens equation for p or q before using it in a calculation. We find that

$$\frac{1}{p} = \frac{1}{f} - \frac{1}{q} = \frac{q-f}{qf}$$

$$p = \frac{qf}{q-f}$$ (23.3)

and, in a similar way, that

$$q = \frac{pf}{p-f}$$ (23.4)

$$f = \frac{pq}{p+q}$$ (23.5)

If a calculator with a reciprocal [1/X] key is employed, Eq. (23.2) can be used directly to solve a numerical problem. The sequence of key strokes to find f, for instance, would be $(p)[1/X][+](q)[1/X][=][1/X]$.

◆   **E X A M P L E   23.3** _____

A planoconvex lens of focal length 5.0 cm is used in a reading lamp to focus light from a bulb on a book (Fig. 23.8). If the lens is 60.0 cm from the book, how far should it be from the bulb's filament?

Figure 23.8

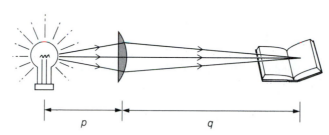

SOLUTION   Here the focal length of the lens is $+5.0$ cm and the image distance is $+60.0$ cm. From the thin-lens equation the object distance is

$$p = \frac{qf}{q-f} = \frac{(60.0\ \text{cm})(5.0\ \text{cm})}{60.0\ \text{cm} - 5.0\ \text{cm}} = 5.5\ \text{cm}$$

◆

◆   **E X A M P L E   23.4** _____

Movie directors often use diverging lenses to see how scenes would appear on the screen without having to move the camera into place first. If a double-concave lens of focal length -50 cm is used for this purpose, where does the image of a scene 10 m away seem to be located?

SOLUTION   Since $f = -0.50$ m and $p = 10$ m,

$$q = \frac{pf}{p-f} = \frac{(10\ \text{m})(-0.50\ \text{m})}{10\ \text{m} - (-0.50\ \text{m})} = -0.48\ \text{m}$$

The image is located 48 cm behind the lens. The negative image distance signifies a virtual image, which is always on the same side of a lens as its object. This situation was pictured in Fig. 23.7.   ◆

Deriving the Lens Equation

The lens equation can be derived geometrically with the help of Fig. 23.9. We observe that the triangles ABO and $A'B'O$ are similar, which means that corre-

Figure 23.9

A ray diagram for deriving the lens equation.

$$\frac{1}{p} + \frac{1}{q} = \frac{1}{f}$$

$\triangle ABO$ and $\triangle A'B'O$ are similar
$\triangle OB''F$ and $\triangle A'B'F$ are similar

sponding sides of these triangles are proportional. Hence

$$\frac{A'B'}{AB} = \frac{A'O}{AO} = \frac{q}{p} \tag{23.6}$$

The triangles $OB''F$ and $A'B'F$ are also similar, and

$$\frac{A'B'}{OB''} = \frac{A'F}{OF} = \frac{q-f}{f} = \frac{q}{f} - 1 \tag{23.7}$$

Since the light ray BB'' is parallel to the axis,

$$OB'' = AB$$

Therefore

$$\frac{A'B'}{OB''} = \frac{A'B'}{AB}$$

which means the right-hand sides of Eqs. (23.6) and (23.7) are equal:

$$\frac{q}{f} - 1 = \frac{q}{p}$$

Dividing each term of this equation by q gives

$$\frac{1}{f} - \frac{1}{q} = \frac{1}{p}$$

and, rearranging terms, we obtain the lens equation:

$$\frac{1}{p} + \frac{1}{q} = \frac{1}{f} \tag{23.2}$$

The preceding derivation used a converging lens with an object distance greater than its focal length. Equation (23.2) is perfectly general, however, and holds for any object distance and for diverging lenses as well.

23.4 MAGNIFICATION

The greater the ratio of object and image distances, the greater the magnification.

The **linear magnification** m of an optical system of any kind is the ratio between the sizes of the image and of the object. A magnification of exactly 1 means that the image and the object are the same size; a magnification of more than 1 means that the image is larger than the object; and a magnification of less than 1 means that the image is smaller than the object. By "size" is meant any transverse linear dimension—for instance, height or width. Thus we can write

$$m = \frac{h'}{h}$$

$$\text{Linear magnification} = \frac{\text{image height}}{\text{object height}}$$

where h = height of object and h' = height of image. A positive m corresponds to an erect image, a negative m to an inverted image.

Figure 23.10 shows a converging lens that forms the image $A'B'$ of the object AB. The object height is $AB = h$, and that of the image is $A'B' = -h'$ because the image is inverted. The triangles ABO and $A'B'O$ are similar, and their corresponding sides are proportional. Hence

$$\frac{A'B'}{AB} = \frac{A'O}{AO}$$

Because $A'B' = -h'$, $AB = h$, $A'O = q$, and $AO = p$, we find that

$$\frac{h'}{h} = -\frac{q}{p}$$

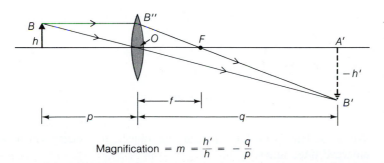

Figure 23.10

The magnification of a lens is equal to minus the ratio of image and object distances. A positive magnification means an erect image; a negative magnification means an inverted image.

Table 23.2   Sign conventions for lenses

Quantity	Positive	Negative
Focal length f	Converging lens	Diverging lens
Object distance p	Real object	Virtual object
Image distance q	Real image	Virtual image
Magnification m	Erect image	Inverted image

and the magnification produced by the lens is

$$m = \frac{h'}{h} = -\frac{q}{p} \qquad \qquad \textit{Linear magnification of lens} \quad (23.8)$$

$$\frac{\text{Linear}}{\text{magnification}} = \frac{\text{image height}}{\text{object height}} = -\frac{\text{object distance}}{\text{image distance}}$$

This formula is a general one that holds for diverging as well as converging lenses and for any object distance.

A useful feature of Eq. (23.8) is that it automatically tells us whether an image is erect (m positive) or inverted (m negative). Table 23.2 summarizes the various sign conventions we have been using.

◆   **E X A M P L E   23.5** _____

A ''magnifying glass'' is a converging lens held less than its focal length from an object being examined (see Fig. 23.6a). How far should a double-convex lens whose focal length is 15 cm be held from an object to produce an erect image three times larger?

SOLUTION   The required magnification is $+3$ since an erect image is required. From Eq. (23.8)

$$m = -\frac{q}{p} \quad \text{or} \quad q = -mp = -3p$$

Using this value of q in the thin-lens formula, Eq. (23.2), gives for $f = +15$ cm

$$\frac{1}{f} = \frac{1}{q} + \frac{1}{p} = -\frac{1}{3p} + \frac{1}{p} = \frac{2}{3p}$$

$$p = \frac{2}{3}f = \frac{2}{3}(15\text{ cm}) = 10\text{ cm}$$

When the lens is held 10 cm from the object, the image will be virtual, erect, and enlarged three times.   ◆

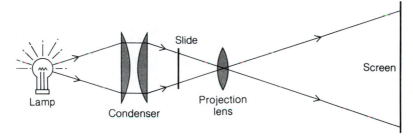

Figure 23.11
An optical projection system. The condenser causes the slide to be evenly illuminated, thereby making possible an image of uniform brightness on the screen.

EXAMPLE 23.6

A slide projector is to be used with its lens 6.0 m from a screen. If a projected image 1.5 m square of a slide 50 mm square is desired, what should the focal length of the lens be?

SOLUTION Projectors use converging lenses to produce enlarged, inverted, real images of transparencies that are illuminated from behind (Fig. 23.11). Here the image distance is given as $q = 6.0$ m, and the required magnification is

$$m = \frac{h'}{h} = \frac{-1.5 \text{ m}}{0.050 \text{ m}} = -30$$

The image height is considered negative as it is inverted. From Eq. (23.8)

$$p = -\frac{q}{m} = -\frac{6.0 \text{ m}}{-30} = 0.20 \text{ m}$$

From Eq. (23.5) we find that the lens should have a focal length of

$$f = \frac{pq}{p + q} = \frac{(0.20 \text{ m})(6.0 \text{ m})}{0.20 \text{ m} + 6.0 \text{ m}} = 0.19 \text{ m} = 19 \text{ cm} \qquad \blacklozenge$$

23.5 THE CAMERA

The larger the f-number, the smaller the aperture and the less the light reaching the film.

A camera uses a converging lens to form an image on a light-sensitive photographic film (Fig. 23.12). The distance between the lens and the film is adjustable to permit bringing an object at any distance to a sharp focus on the film. The film is

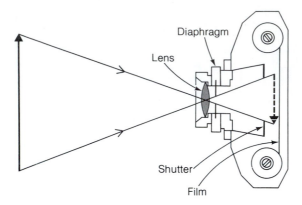

Figure 23.12

A camera. The light-sensitive film is exposed by opening the shutter for a fraction of a second. The adjustable diaphragm permits the intensity of the light entering the camera to be varied to suit the shutter speed and film used.

exposed by opening the shutter for a fraction of a second. The shorter the exposure time, the better we can ''freeze'' a moving object. The longer the exposure time, the more light reaches the film and hence the better we can capture a dimly lit scene.

For a given shutter speed, the amount of light reaching the film from a scene depends on the area of the lens aperture, which is regulated by an iris diaphragm (Fig. 23.13). Apertures are specified in terms of the focal length f of the lens. Thus $f/8$ means that the diameter d of the opening is $\frac{1}{8}$ of f. The ''8'' in this case is referred to as the ''f-number'' of the aperture, so that

$$f\text{-number} = \frac{\text{focal length}}{\text{aperture diameter}} = \frac{f}{d}$$

The larger the f-number, the smaller the aperture and the less the light that reaches the film.

The amount of light per unit area arriving at the film depends on the ratio between the area $\pi d^2/4$ of the aperture and q^2, the square of the image distance q. (If we double q, the light that enters the camera is spread over four times the original area. See Fig. 22.36.) Because q is usually quite close to f in a camera, an aperture of a given f-number gives nearly the same light per unit area on the film for a given scene whatever the focal length of the lens. This is the reason for the f-number scheme. The diaphragm of a camera lens is calibrated using a sequence of f-number

Figure 23.13

The larger the f-number of a camera diaphragm, the smaller the aperture. The sequence of values is such that changing the aperture by one f-number changes its area and the amount of light reaching the film by a factor of 2. An aperture of $f/4$ admits four times as much light as one of $f/8$.

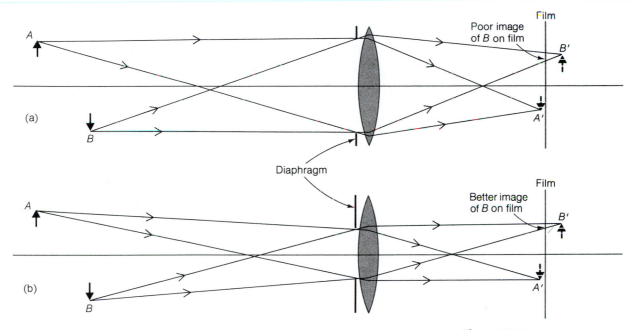

Figure 23.14

Reducing the aperture of a lens increases the depth of field. Here arrowhead *A* is in focus on the film, whereas arrowhead *B* is in focus behind the film because it is closer to the lens. (a) With a large aperture, the image of *B* on the film is very indistinct. (b) With a small aperture, the image of *B* on the film is sharper. When a 50-mm lens is set to a distance of 5 m, objects from 4.5 to 5.6 m are in reasonably sharp focus at *f*/1.8, but at *f*/16 the range is 2.6 to 138 m.

values (for instance, 2.8, 4, 5.6, 8, 11, 16) such that their squares are approximately in the ratios 1:2:4:8 and so on. This means that changing the setting by one *f*-number changes the illumination of the film by the factor 2. The same amount of light reaches the film of a camera during exposures of $\frac{1}{50}$ s at *f*/8, $\frac{1}{100}$ s at *f*/5.6, and $\frac{1}{200}$ s at *f*/4.

Besides helping to govern how much light reaches the film, the size of a camera's lens aperture affects the **depth of field,** which is the range of object distances at which reasonably sharp images are produced. As Fig. 23.14 shows, reducing the aperture (increasing the *f*-number) leads to an increased depth of field. Thus more of a scene will appear sharp if a slow shutter speed and small aperture are used (provided motion in the scene is not a problem, of course).

The film size and the desired angle of view are what determine the appropriate focal length of a camera lens. A ''normal'' lens usually gives an angle of view of about 45°. This means a focal length of 50 mm for a camera using 35-mm film on which the image size is 24 × 36 mm. A lens of shorter focal length is a **wide-angle lens** because it captures more of a given scene, though at the expense of reducing the sizes of details in the scene. Typical wide-angle lenses for a 35-mm camera have focal lengths of 35 and 28 mm, which give angles of view, respectively, of 63° and 75°. A **telephoto lens** has a long focal length to give larger images of distant objects, but this means a reduced angle of view. A 135-mm telephoto lens for a 35-mm camera has an angle of view of 18°, and it is only 8° for a 300-mm telephoto lens.

◆ **E X A M P L E   23.7**

(a) What range of motion should the 50-mm lens of a 35-mm camera have if the camera is to be able to photograph objects as close as 50 cm from the lens? (b) How wide would the image on the film be of a flower 60 mm across when photographed from a distance of 50 cm? (c) What is the aperture diameter when the diaphragm is set at $f/2.8$? At $f/16$?

SOLUTION   (a) Here the image distance q for an object distance of p = 50 cm = 500 mm is

$$q = \frac{pf}{p - f} = \frac{(500 \text{ mm})(50 \text{ mm})}{500 \text{ mm} - 50 \text{ mm}} = 56 \text{ mm}$$

An object at infinity is brought to a sharp focus on the film when the image distance is equal to the focal length (see Fig. 23.6f), which in this case is 50 mm. Hence a range of adjustment of 6 mm will permit the camera to photograph objects from 50 cm away to infinity.

(b) From Eq. (23.8) the linear magnification here is

$$m = -\frac{q}{p} = -\frac{56 \text{ mm}}{500 \text{ mm}} = -0.11$$

Hence the width of the flower on the film would be

$$h' = mh = (-0.11)(60 \text{ mm}) = -6.6 \text{ mm}$$

The minus sign means that the image is inverted.

(c) At $f/2.8$ the aperture diameter of the lens is

$$d = \frac{f}{2.8} = \frac{50 \text{ mm}}{2.8} = 18 \text{ mm}$$

and at $f/16$ it is

$$d = \frac{50 \text{ mm}}{16} = 3 \text{ mm} \qquad\qquad \blacklozenge$$

23.6 THE EYE

A remarkable optical instrument.

The structure of the human eye is shown in Fig. 23.15. Its diameter is typically a little less than 3 cm, and its more or less spherical shape is maintained by internal pressure. Incoming light first passes through the **cornea,** the transparent outer

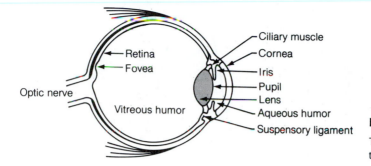

Figure 23.15

The human eye slightly larger than life-size.

membrane, and then enters a liquid called the **aqueous humor.** The **lens** of the eye is a jellylike assembly of tiny transparent fibers that slide over one another when the shape of the lens is altered by the **ciliary muscle** to which it is attached by ligaments. In front of the lens is the colored **iris** whose aperture is the **pupil.** Behind the lens is a cavity filled with another liquid, the **vitreous humor,** and the lining of this cavity is the **retina.**

The retina has millions of tiny structures called **rods** and **cones** that are sensitive to light. The cones are responsible for vision in bright conditions, the rods for vision in dim conditions. Because the sensation of color is produced by the cones, colors are difficult to distinguish in a faint light. The rods and cones respond to the image formed by the lens on the retina and transmit the information to the brain through the **optic nerve.**

No photoreceptors are near the optic nerve, so that region is known as the ''blind spot.'' The blind spot covers a field of view about 8° high and 6° wide. We are not normally aware of the blind spot for two reasons: The blind spots of the two eyes obscure different fields of vision, and the eyes are never completely at rest but are in constant scanning motion. Vision is most acute for the image focused on the **fovea,** a small region near the center of the retina. The rest of the field of view appears less distinct. Since the fovea contains only cones, it is easier to see something in a dim light by looking a bit to one side instead of directly at it.

In bright light the pupil contracts to reduce the amount of light reaching the retina, and it widens in faint light. A fully opened pupil lets in about 16 times as much light as a fully contracted one, since the respective pupil diameters are about 8 and 2 mm for a ratio of 4:1. The retina itself can also cope with a considerable range of brightnesses.

The sensitivity of the eye varies with wavelength, so light of the same intensity but of different wavelengths will give different impressions of brightness. Maximum sensitivity in bright light occurs in the middle of the visible spectrum at about $\lambda = 555$ nm, which corresponds to green (Fig. 23.16). Maximum sensitivity in dim light, when the rods rather than the cones are chiefly involved, occurs at about 510 nm. This wavelength is closer to the blue end of the spectrum and accounts for the reduced ability of the eye to respond to a weak red light.

Refraction at the cornea is responsible for most of the focusing power of the eye. The lens is less effective because its index of refraction is not very different

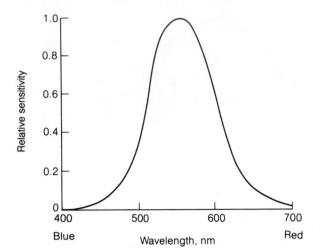

Figure 23.16

How the sensitivity of a normal human eye to bright light varies with wavelength. In dim light the wavelength of maximum sensitivity shifts from about 555 nm (green light) to about 510 nm.

from those of the aqueous and vitreous humors on both sides of it. The shape of the lens is controlled by the ciliary muscle. When this relaxes, the lens brings objects at infinity to a sharp focus on the retina. To permit a closer object to be viewed, the ciliary muscle forces the lens into a more convex shape whose focal length is appropriate for the object distance involved. The image distance in this situation, which is the lens-retina distance, is fixed, and focusing is thus accomplished by changing the focal length of the lens. This process is called **accommodation.** In the case of a camera, on the other hand, the focal length of the lens is fixed, and focusing is done by changing the image distance.

The limited range of accommodation of the lens is not enough to make up for the loss of refractive power that occurs when the eye is immersed in water. This loss is considerable because the index of refraction of the cornea (1.38) is close to that of water (1.33). Hence underwater objects cannot be brought to a sharp focus unless a face mask or goggles are used to keep water away from the cornea. Fish are able to see clearly when submerged because the lenses of their eyes are spherical and have high indexes of refraction. Focusing in most fish eyes is carried out by shifting the lens closer to or farther from the retina, in the same way a camera lens is focused.

23.7 DEFECTS OF VISION

How corrective lenses work.

Two common defects of vision are **myopia** (nearsightedness) and **hyperopia** (farsightedness). In myopia, the eyeball is too long, and light from something at infinity comes to a focus in front of the retina, as in Fig. 23.17(b). Accommodation permits nearby objects to be seen clearly, but not more distant ones. A diverging lens

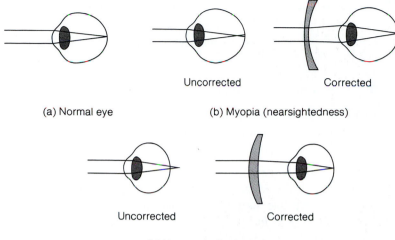

(a) Normal eye

(b) Myopia (nearsightedness)

Uncorrected   Corrected

Uncorrected   Corrected

(c) Hyperopia (farsightedness)

Figure 23.17

Myopia and hyperopia are common defects of vision that can be remedied with corrective lenses.

of the proper focal length can correct this condition. In hyperopia, the eyeball is too short, and light from something at infinity does not come to a focus within the eyeball at all. Its power of accommodation permits a hyperopic eye to focus on distant objects, but the range of accommodation is not enough for nearby objects to be seen clearly. The correction for hyperopia is a converging lens, as in Fig. 23.17(c).

Opticians often use the *power D* of a lens, expressed in **diopters,** in place of its focal length. If the focal length f is given in meters, then

$$D \text{ (diopters)} = \frac{1}{f} \qquad\qquad \textit{Dioptric power} \quad (23.9)$$

Thus a converging lens of $f = +50$ cm $= +0.50$ m could also be described as having a power of $+2$ diopters.

◆ E X A M P L E 23.8

A certain nearsighted eye cannot see objects distinctly when they are more than 25 cm away. Find the power in diopters of a correcting lens that will enable this eye to see distant objects clearly.

SOLUTION The purpose of the lens is to form an image 25 cm in front of the eye of an object that is infinitely far away. Because the image is to be on the same side of the lens as the object, the image must be virtual, so $q = -25$ cm. With an object distance of $p = \infty$,

$$\frac{1}{f} = \frac{1}{p} + \frac{1}{q} = \frac{1}{\infty} = \frac{1}{25 \text{ cm}} = 0 - \frac{1}{25 \text{ cm}}$$
$$f = -25 \text{ cm} = -0.25 \text{ m}$$

The minus sign means that the lens is diverging. Its power in diopters is

$$D = \frac{1}{f} = \frac{-1}{0.25\ \text{m}} = -4\ \text{diopters}$$

◆

◆   **E X A M P L E   23.9** _____

A certain farsighted eye cannot see objects distinctly when they are closer than 1.00 m away. Find the power in diopters of a correcting lens that will enable this eye to read a letter 25 cm away.

SOLUTION   The purpose of the lens is to form an image 1.00 m in front of the eye of an object that is 25 cm away. Since the image is to be on the same side of the lens as the object, the image must be virtual, and so $q = -1$ m. With an object distance of $p = 25$ cm $= 0.25$ m,

$$f = \frac{pq}{p+q} = \frac{(0.25\ \text{m})(-1.00\ \text{m})}{0.25\ \text{m} + (-1.00\ \text{m})} = +0.33\ \text{m}$$

The plus sign means that the lens is converging. Its power in diopters is

$$D = \frac{1}{f} = \frac{1}{0.33\ \text{m}} = +3\ \text{diopters}$$

◆

The range of accommodation decreases with age as the lens hardens, a condition known as **presbyopia.** Distance vision is not affected, but the "near point" of the eye (the closest position at which objects can be seen distinctly) gets increasingly far away. From perhaps 7 cm at age 10, the near point recedes until it is often 2 m or more at age 60. The correction for presbyopia is a converging lens. If the range of accommodation is severely limited, more than one set of corrective lenses may be required. These are often combined in "bifocal" and "trifocal" lenses.

The closer an object is to the eye, the larger the object seems and more of its detail can be made out. Since the eye cannot focus on objects closer than the near point, this sets a limit to the magnification of which the eye is capable. The distance of most distinct vision is usually taken as 25 cm (about 10 in.), which is a comfortable object distance for most people. For purposes of calculation, an optical instrument such as a microscope or a telescope is therefore assumed to form a virtual image 25 cm behind the lens (or lens system) nearest the eye.

Astigmatism is a defect of vision caused by the cornea (or sometimes the lens) having different curvatures in different planes. When light rays that lie in one plane are in focus on the retina of an astigmatic eye, those in other planes will be in focus either in front or in back of the retina. As a result only one of the bars of a cross can be in focus at the same time (Fig. 23.18). Astigmatism is a source of eyestrain because the mechanism of accommodation continually varies the focus of the lens in an effort to produce a completely sharp image. The remedy is a corrective lens that has a cylindrical curvature, as in Fig. 23.19.

Figure 23.18

How a cross is seen by a normal eye and by an astigmatic eye.

Normal eye

Astigmatic eye

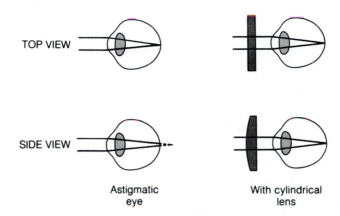

TOP VIEW

SIDE VIEW

Astigmatic
eye

With cylindrical
lens

Figure 23.19

How a cylindrical lens can improve the image formed by an astigmatic eye.

23.8 THE MICROSCOPE

It produces enlarged images of nearby objects.

In many applications a system of two or more lenses is better than a single lens. An ordinary magnifying glass, for instance, cannot produce images enlarged beyond 3× or so without severe distortion. However, two converging lenses arranged as a **microscope** can give good images enlarged hundreds of times.

The optical system of a microscope is shown in Fig. 23.20. The **objective** is a lens of short focal length that forms an enlarged real image of the object. This image is further enlarged by the **eyepiece,** which acts as a simple magnifier to form a virtual final image. The image produced by the objective is thus the object of the eyepiece, and the final image has been magnified twice.

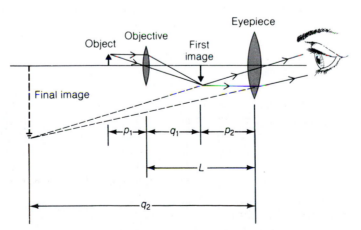

Figure 23.20

In a microscope the image formed by the objective is further magnified by the eyepiece. In this diagram the rays used to locate the final image are not the same as those used to locate the first image.

The total magnification produced by a two-lens system such as a microscope is the product of the magnification m_1 of the first lens (the objective) and the magnification m_2 of the second lens (the eyepiece):

$$m = m_1 m_2 \qquad\qquad \textit{Total magnification} \quad (23.10)$$

Typical objectives yield magnifications of $10\times$ to $100\times$, and standard eyepiece magnifications are $5\times$, $10\times$, and $15\times$. Total magnifications of up to $1500\times$ are therefore possible with a good laboratory microscope. As discussed in the next chapter, however, the wave nature of light limits the useful magnification of a microscope to a maximum of perhaps $500\times$. Higher magnifications give larger images but do not show finer details. Actual microscopes use compound lenses that consist of two to six elements in place of the single lenses shown in Fig. 23.20, but their optical behavior is the same.

23.9 THE TELESCOPE

Its angular magnification equals the ratio of focal lengths of objective and eyepiece.

A **telescope** is a lens system used to examine distant objects. As in a microscope, two lenses are involved, with an eyepiece to enlarge the image produced by the objective. A telescope objective, however, has a long focal length, whereas that of a microscope is very short.

Figure 23.21 shows a simple telescope. The image produced by the objective is real, inverted, and smaller than the object. The eyepiece then acts as a simple magnifier to form an enlarged virtual image whose object is the initial image. In practice the object distance is usually very long relative to the focal length of the objective, and the final image is smaller than the object itself. But the image seen by

Figure 23.21

Ray diagram of a simple telescope. The rays used to locate the final image are not the same as those used to locate the first image.

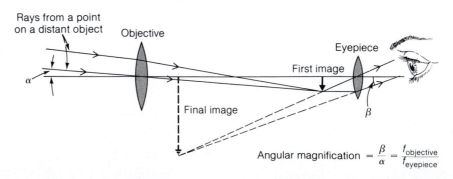

the eye is larger than it would be without the telescope, so the effect is the same as if the object were closer to the eye than it actually is.

Telescopes are described in terms of the **angular magnification** they produce. This quantity is the ratio between the angle β subtended at the eye by the image and the angle α subtended at the eye by the object seen directly. For distant objects ($p \gg f$) the angular magnification of a telescope is simply the ratio between the focal lengths of its objective and eyepiece:

$$m_{ang} = \frac{f_{objective}}{f_{eyepiece}} = \frac{\beta}{\alpha} \qquad \text{Angular magnification} \quad (23.11)$$

◆ **E X A M P L E 23.10** _____

A birdwatcher uses a telescope with an objective of focal length 60 cm and an eyepiece of focal length 1.5 cm to examine a hummingbird 5.0 cm long that is 20 m away. If the image distance of the eyepiece is 25 cm (the distance of most distinct vision), what is the apparent length of the hummingbird?

SOLUTION   The angular magnification of the telescope is

$$m_{ang} = \frac{f_{objective}}{f_{eyepiece}} = \frac{60 \text{ cm}}{1.5 \text{ cm}} = 40$$

The angle α subtended by the bird from the location of the telescope is

$$\alpha = \frac{\text{object length}}{\text{object distance}} = \frac{0.05 \text{ m}}{20 \text{ m}} = 0.0025 \text{ radian}$$

If L is the length of the bird's image as seen through the telescope, the angle β this image subtends is

$$\beta = \frac{\text{image length}}{\text{image distance}} = \frac{L}{25 \text{ cm}}$$

Since the angular magnification of the telescope is 40, from Eq. (23.11) we have

$$\beta = m_{ang}\alpha$$

$$\frac{L}{25 \text{ cm}} = (40)(0.0025 \text{ radian})$$

$$L = 2.5 \text{ cm}$$

The hummingbird seems to be 2.5 cm long, half its actual length, and to be located 25 cm from the viewer's eye.   ◆

When a telescope is used as a camera, it needs no eyepiece—its objective lens (or mirror; see Section 23.11) simply acts as a giant telephoto lens.

◆   **E X A M P L E**   **23.11** _____

The diameter of the planet Mars is 6.8×10^6 m. What focal length must a telescope objective have in order to produce a photographic image of Mars 1.0 mm in diameter at a time when Mars is 8.0×10^{10} m from the earth?

SOLUTION   Here $h = 6.8 \times 10^6$ m and $h' = 1.0$ mm $= 1.0 \times 10^{-3}$ m, so the required magnification is

$$m = \frac{h'}{h} = -\frac{1.0 \times 10^{-3}\,\text{m}}{6.8 \times 10^6\,\text{m}} = -1.5 \times 10^{-10}$$

A minus sign is used because the image will be inverted. The object distance p is the Mars-earth distance of 8×10^{10} m. Since the focal length f is going to be much smaller than p, $p - f \approx p$ and the image distance q is

$$q = \frac{pf}{p - f} \approx f$$

The formula $m = -q/p$ therefore becomes $m = -f/p$ here, and

$$f = -mp = -(-1.5 \times 10^{-10})(8.0 \times 10^{10}\,\text{m}) = 12\,\text{m}$$

The telescope objective should have a focal length of 12 m. Of course, further enlargement of the image can be made from the negative.   ◆

A telescope gives an erect image if a third lens is used between the objective and the eyepiece (Fig. 23.22). The disadvantage is that the telescope tube is longer. A better scheme uses a pair of prisms that both invert the image so that it is erect and shorten the instrument length (Fig. 23.23). This is the arrangement used in binoculars.

Figure 23.22

In a terrestrial telescope an intermediate lens is used to produce an erect image.

Figure 23.23

A pair of prisms is used to erect the image in each half of a prism binocular.

23.10 LENS ABERRATIONS

Simple lenses do not give perfect images.

The image formed by a simple lens is never an exact replica of its object. Of the variety of imperfections to which such an image is subject, the most familiar is the presence of fringes of color around whatever is being looked at. This **chromatic aberration** occurs because the index of refraction of glass varies with wavelength. As a result the focal length of a lens is slightly different from light of different colors (Fig. 23.24a). The remedy is to combine a converging and a diverging lens made of different glass so that the dispersion produced by one is canceled by the other while leaving a net converging or diverging power. Such an **achromatic lens** is illustrated in Fig. 23.24(b).

The lens equation was derived on the basis of light rays that made only small angles with the axis. When a simple lens is used to form images of objects some distance from the axis, a variety of aberrations arise even if the light used is monochromatic. One of them is **spherical aberration,** in which rays passing near the lens rim come to a focus closer to the lens than rays near the axis (Fig. 23.25). Another

Figure 23.24

A compound lens made of different types of glass can correct for chromatic aberration.

Figure 23.25

Spherical aberration.

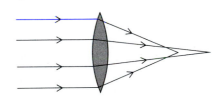

is distortion of the image because the magnification of a simple lens varies with the distance of an object from the axis. By using several lenses of different types of glass and different curvatures to replace a single lens, the various aberrations can be minimized. Compound lenses that consist of two or more elements are always used in high-performance optical systems such as those in microscopes, prism binoculars, and quality cameras.

23.11 SPHERICAL MIRRORS

Concave mirrors are converging, convex mirrors are diverging.

Mirrors with spherical reflecting surfaces form images in much the same way that lenses do. Figure 23.26(a) shows how a concave mirror converges a parallel beam of light to a real focal point, and Fig. 23.26(b) shows how a convex mirror diverges such a beam so that the reflected rays seem to originate in a virtual focal point behind the mirror. If R is the radius of the reflecting surface in each case, the focal lengths of these mirrors are

$$f = \frac{R}{2} \qquad\qquad \textit{Concave mirror} \quad (23.12)$$

$$f = -\frac{R}{2} \qquad\qquad \textit{Convex mirror} \quad (23.13)$$

Figure 23.26

(a) A concave spherical mirror converges a parallel beam of incident light. (b) A convex spherical mirror diverges a parallel beam of incident light.

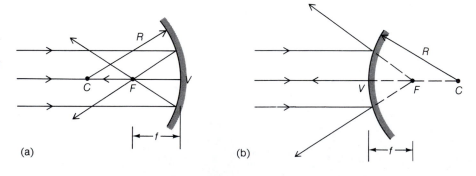

(a)   (b)

R = radius of curvature
V = vertex
C = center of curvature
F = focal point
f = focal length

A compound lens that consists of several individual lenses, as in this camera, minimizes aberrations to produce a better image than a single lens can.

The position, size, and nature of the image produced by a spherical mirror of an object in front of it can be found with the help of a scale drawing. As in the case of a lens, what we do is consider two different light rays that come from a certain point on the object and trace their paths until they (or their backward extensions) come together again after reflection. The three easiest rays to trace, as shown in Fig. 23.27, are:

1. A ray that leaves the object parallel to the mirror axis. When this ray is reflected, it passes through the focal point of a concave mirror or seems to come from the focal point of a convex mirror.
2. A ray that leaves the object and passes through the focal point of a concave mirror, or is aimed at the focal point of a convex mirror. When this ray is reflected, it proceeds parallel to the axis.
3. A ray that leaves the object along a radius of the mirror. When this ray is reflected, it returns along its original path.

Only two of these rays are needed to locate the image of a reflected object. Figure 23.28 shows how the properties of the image produced by a concave mirror vary with the position of the object. Solid lines represent the actual paths taken by light rays, and dashed lines represent virtual paths.

Real image

Virtual image

Figure 23.27

The position and size of an image produced by a spherical mirror can be determined by tracing any two of the rays shown.

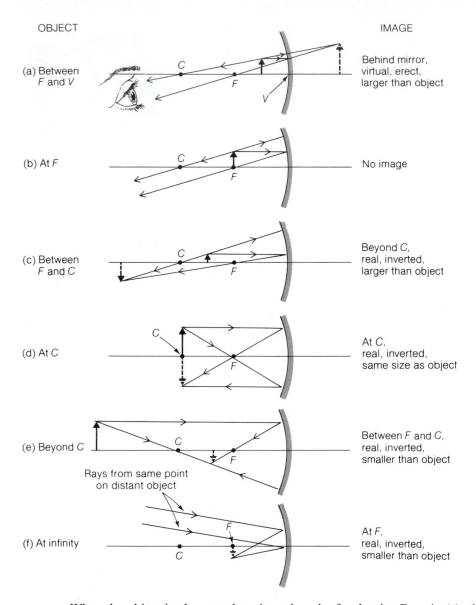

Figure 23.28

Image formation by a concave mirror.

When the object is closer to the mirror than the focal point F, as in (a), the image is erect, enlarged, and virtual. The image seems to be behind the mirror because the reflected light rays from the mirror diverge as though coming from a point behind the mirror.

When the object is at the focal point, as in (b), no image is formed because the reflected rays are all parallel and so do not meet. We can say that the image in this case is at infinity.

An object between the focal point F and the center of curvature C, as in (c), has an inverted, enlarged image that is real. The image would appear on a screen placed at its position.

Figure 23.29

The image of a real object formed by a convex mirror is always virtual, erect, and smaller than the object.

When the object is at C, as in (d), its image is at the same place and is the same size but is inverted. An object past C, as in (e) and (f), has a real image that is smaller and is inverted.

The image formed by a convex mirror of a real object is always erect, smaller than the object, and virtual (Fig. 23.29). The field of view of a convex mirror is wider than that of a plane mirror, which accounts for a number of its applications, such as at blind corners in roads and in rear-view mirrors in cars.

The Mirror Equation

The object and image distances p and q of a spherical mirror are related to its focal length by the same formula that holds for a thin lens:

$$\frac{1}{p} + \frac{1}{q} = \frac{1}{f}$$
  Mirror equation   (23.14)

Equations (23.3), (23.4), and (23.5) hold here as well as for lenses. The linear magnification of a mirror also follows the same formula as for a lens:

$$m = \frac{h'}{h} = -\frac{q}{p}$$
  Magnification of a mirror   (23.15)

The sign conventions for a mirror are given in Table 23.3.

Table 23.3   Sign conventions for spherical mirrors

Quantity	Positive	Negative
Focal length f	Concave mirror	Convex mirror
Object distance p	Real object	Virtual object
Image distance q	Real image	Virtual image
Magnification m	Erect image	Inverted image

◆   **E X A M P L E   23.12** _____

A candle 5.0 cm high is placed 40 cm from a concave mirror whose radius of curvature is 60 cm. Find the position, size, and nature of the image.

SOLUTION   The focal length of the mirror is

$$f = \frac{R}{2} = \frac{60 \text{ cm}}{2} = 30 \text{ cm}$$

The situation therefore corresponds to that shown in Fig. 23.28(c). The image distance q is

$$q = \frac{pf}{p - f} = \frac{(40 \text{ cm})(30 \text{ cm})}{40 \text{ cm} - 30 \text{ cm}} = 120 \text{ cm}$$

The image distance is positive, so the image is a real one. From Eq. (23.15) the size of the image is

$$h' = -h\frac{q}{p} = (-5.0 \text{ cm})\left(\frac{120 \text{ cm}}{40 \text{ cm}}\right) = -15 \text{ cm}$$

The image of the candle is three times as large as the candle itself and is inverted.   ◆

◆   **E X A M P L E   23.13** _____

A concave shaving mirror has a focal length of 40 cm. How far away from it should one's face be for the reflected image to be erect and twice its actual size?

SOLUTION   Here the magnification is $+2$ since the image is erect. Hence

$$m = -\frac{q}{p} = 2 \qquad q = -2p$$

The negative image distance signifies that the image is virtual; see Fig. 23.28(a). Now we substitute $f = 40$ cm and $q = -2p$ into the mirror equation and solve for the object distance p:

$$\frac{1}{p} + \frac{1}{q} = \frac{1}{f}$$

$$\frac{1}{p} - \frac{1}{2p} = \frac{1}{40 \text{ cm}}$$

$$\frac{1}{2p} = \frac{1}{40 \text{ cm}}$$

$$p = 20 \text{ cm}$$

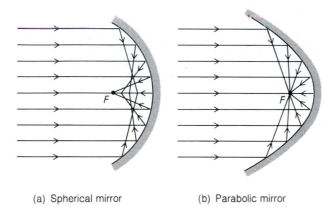

(a) Spherical mirror   (b) Parabolic mirror

Figure 23.30

(a) Spherical aberration. The more highly curved the mirror, the less sharp the image. (b) A mirror with a parabolic cross section avoids spherical aberration.

When one's face is 20 cm from the mirror, the virtual image one sees is magnified twice and is erect. ♦

Figure 23.31

One type of reflecting telescope.

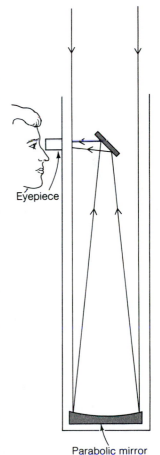

Eyepiece

Parabolic mirror

Mirrors do not suffer from chromatic aberration but do exhibit spherical aberration. In the case of a concave mirror, rays reflected from the outer part of the mirror cross the axis closer to the vertex than rays reflected from the central part (Fig. 23.30). When a high-quality image is required, as in the case of an astronomical telescope, a concave mirror whose cross-sectional shape is a parabola is the answer. Despite spherical aberration, however, spherical mirrors are common because they are the easiest to manufacture.

Reflecting Telescopes

The higher the magnification of a telescope, the greater in diameter the objective must be in order to gather in enough light for the image to be visible. This sets a limit to the size of a refracting telescope, since a large glass lens tends to distort under its own weight. The largest refracting telescope in the world is the 1.02-m-diameter instrument at Yerkes Observatory in Wisconsin, whose objective has a focal length of nearly 20 m.

Modern astronomical telescopes use concave parabolic mirrors as their objectives. Such a mirror produces a real image of a distant object and can be supported from behind. Further advantages are that a parabolic mirror is not subject to spherical or chromatic aberrations and that it needs only one surface to be ground to a precise shape. A small secondary mirror reflects the image outside the telescope tube for viewing or, more often, for photographing (Fig. 23.31).

The largest single-mirror telescope, located in the Soviet Union, has a mirror 6 m in diameter, but technical problems have prevented it from reaching its full potential. Next in size is a 5-m reflector at Mount Palomar in California that has been enlarging the horizons of astronomy since 1948. The latest generation of astronomical telescopes does not rely on single large mirrors, which practical difficulties limit to at most 8 m. Instead, a number of individual mirrors are linked to produce a single image. In one approach the Keck telescope in Hawaii uses hexagonal segments to give a collecting surface 10 m across. In another approach separate circular mirrors are used. A proposed new American telescope will have four 7.5-m mirrors mounted in one structure for the equivalent of a 15-m mirror. A third scheme, which is being developed in Europe, uses four individual 8-m telescopes whose images are added together electronically.

Important Terms

A **lens** is a transparent object that can produce an image of an object placed before it. A **converging lens** brings parallel light to a single **real focal point,** whereas a **diverging lens** deviates parallel light outward as though it originated at a single **virtual focal point.** The distance from a lens to its focal point is its **focal length.**

The **magnification** of an optical system is the ratio between the size of the image and that of the object.

A **microscope** is a lens system used to produce enlarged images of nearby objects. A **telescope** is a lens system used to produce larger images of distant objects than would be seen by the unaided eye, although the image itself may be smaller than the actual object.

A **concave mirror** curves inward toward its center and converges parallel light to a single real focal point.

A **convex mirror** curves outward toward its center and diverges parallel light as though the reflected light came from a single virtual focal point behind the mirror. The distance from a mirror to its focal point is the focal length of the mirror.

Important Formulas

Lensmaker's equation:

$$\frac{1}{f} = (n - 1)\left(\frac{1}{R_1} + \frac{1}{R_2}\right)$$ (R is + for convex, − for concave surface)

Focal length of concave mirror: $f = \dfrac{R}{2}$

Focal length of convex mirror: $f = -\dfrac{R}{2}$

Lens and mirror equation: $\dfrac{1}{p} + \dfrac{1}{q} = \dfrac{1}{f}$

Alternative forms of lens and mirror equation:

$$p = \frac{qf}{q - f} \qquad q = \frac{pf}{p - f} \qquad f = \frac{pq}{p + q}$$

Linear magnification: $m = \dfrac{h'}{h} = -\dfrac{q}{p}$

$$m = m_1 m_2$$

Angular magnification of telescope: $m_{\text{ang}} = \dfrac{f_{\text{objective}}}{f_{\text{eyepiece}}}$

MULTIPLE CHOICE

1. A negative focal length corresponds to which one or more of the following?

   a. Double-convex lens   **b.** Planoconcave lens

   c. Convex mirror   **d.** Concave mirror

2. An object farther from a converging lens than its focal point always has an image that is

   a. inverted   **b.** virtual

   c. the same in size   **d.** smaller in size

3. An object closer to a converging lens than its focal point always has an image that is

   a. inverted   **b.** virtual

   c. the same in size   **d.** smaller in size

4. When the distance of a real object from a converging lens equals the focal length of the lens,

   a. the image is virtual, erect, and larger than the object

   b. the image is real, inverted, and larger than the object

c. the image is real, inverted, and the same size as the object
d. no image is formed

5. Relative to its object, a real image formed by a lens is always
a. erect
b. inverted
c. smaller
d. larger

6. Relative to its object, a real image formed by a spherical mirror is always
a. erect
b. inverted
c. smaller
d. larger

7. The image of a real object produced by a diverging lens is never
a. real
b. virtual
c. erect
d. smaller than the object

8. A negative magnification corresponds to an image that is
a. erect
b. inverted
c. smaller than the object
d. larger than the object

9. A negative image distance signifies an image that is
a. real
b. virtual
c. erect
d. inverted

10. The image a camera forms on the film is
a. always real
b. always virtual
c. sometimes real and sometimes virtual
d. neither real nor virtual

11. Enlarging the lens aperture of a camera
a. increases the f-number
b. increases the depth of field
c. enlarges the image
d. permits a faster shutter speed

12. Which of the following combinations of shutter speed and lens opening will admit the most light to the film of a camera?
a. 1/125 s at $f/8$
b. 1/125 s at $f/16$
c. 1/250 s at $f/4$
d. 1/250 s at $f/5.6$

13. The lens of the eye forms an image on the retina that is
a. real and erect
b. real and inverted
c. virtual and erect
d. virtual and inverted

14. When a converging lens of focal length F is used as a magnifying glass, the object distance must be
a. less than F
b. equal to F
c. between F and $2F$
d. more than $2F$

15. Four lenses with the listed focal lengths are being considered for use as a microscope objective. The one that will produce the greatest magnification with a given eyepiece has

the focal length
a. −5 mm
b. +5 mm
c. −5 cm
d. +5 cm

16. Four lenses with the listed focal lengths are being considered for use as a telescope objective. The one that will produce the greatest magnification with a given eyepiece has the focal length
a. −1 m
b. +1 m
c. −2 m
d. +2 m

17. A concave mirror produces an erect image when the object distance is
a. less than f
b. equal to f
c. between f and $2f$
d. greater than $2f$

18. The image formed by a concave mirror is larger than the object
a. when p is less than $2f$
b. when p is more than $2f$
c. for no values of p
d. for all values of p

19. The image formed by a convex mirror is larger than the object
a. when p is less than $2f$
b. when p is more than $2f$
c. for no values of p
d. for all values of p

20. An object is located 10 cm from a converging lens of focal length 12 cm. The image distance is
a. +5.5 cm
b. −5.5 cm
c. +60 cm
d. −60 cm

21. An object is located 12 cm from a converging lens of focal length 10 cm. The image distance is
a. +5.5 cm
b. −5.5 cm
c. +60 cm
d. −60 cm

22. The image of an object 10 cm from a lens is located 10 cm behind the object. The focal length of the lens is
a. +6.7 cm
b. −6.7 cm
c. +20 cm
d. −20 cm

23. A pencil 10 cm long is placed 70 cm in front of a lens of focal length +50 cm. The image is
a. 4 cm long and erect
b. 4 cm long and inverted
c. 25 cm long and erect
d. 25 cm long and inverted

24. A pencil 10 cm long is placed 100 cm in front of a lens of focal length +50 cm. The image is
a. 5 cm long and erect
b. 5 cm long and inverted

c.   10 cm long and erect
d.   10 cm long and inverted

25. A pencil 10 cm long is placed 175 cm in front of a lens of focal length +50 cm. The image is
  a.   4 cm long and erect
  b.   4 cm long and inverted
  c.   25 cm long and erect
  d.   25 cm long and inverted

26. A magnifying glass is to be used at the fixed object distance of 10.0 mm. If it is to produce an erect image magnified 5.00 times, its focal length should be
  a.   +2.0 mm   **b.**   +8.0 mm
  c.   +12.5 mm   **d.**   +50.0 mm

27. A lens placed 9.00 cm from a postage stamp produces a virtual image of it 3.00 cm from the lens. The lens has a focal length of
  a.   −2.25 cm   **b.**   +2.25 cm
  c.   −4.50 cm   **d.**   +4.50 cm

28. An object 40 cm from a converging lens has an image 40 cm away from the lens on the other side. The focal length of the lens is
  a.   20 cm   **b.**   40 cm
  c.   60 cm   **d.**   80 cm

29. A lens held 20.0 cm from an object forms a virtual image of it 10.0 cm from the lens. The focal length of the lens is
  a.   −6.7 cm   **b.**   +6.7 cm
  c.   −20 cm   **d.**   +20 cm

30. A projector is intended to produce an image 140 cm wide of a slide 35.0 mm wide. If the projector is to be located 5.00 m from the screen, the focal length of the projector's lens should be
  a.   +122 mm   **b.**   +125 mm
  c.   −128 mm   **d.**   +5.13 m

31. A convex mirror is ground with a radius of curvature of 12 cm. Its focal length is
  a.   6 cm   **b.**   24 cm
  c.   −6 cm   **d.**   −24 cm

32. A pencil 10 cm long is placed 30 cm in front of a mirror of focal length +50 cm. The image is
  a.   2.5 cm long and erect
  b.   25 cm long and erect
  c.   0.25 m long and erect
  d.   25 cm long and inverted

33. A pencil 10 cm long is placed 100 cm in front of a mirror of focal length +50 cm. The image is
  a.   3 cm long and erect
  b.   10 cm long and erect
  c.   3 cm long and inverted
  d.   10 cm long and inverted

34. A pencil 10 cm long is placed 30 cm in front of a mirror of focal length −50 cm. The image is
  a.   25 cm long and erect
  b.   6.3 cm long and erect
  c.   25 cm long and inverted
  d.   6.3 cm long and inverted

QUESTIONS

1. A fortune-teller's crystal ball is 15 cm in diameter. (a) Would you expect the lensmaker's equation to hold for the focal length of this ball? (b) Would you expect the ball to form undistorted images?

2. Does a spherical bubble of air in a volume of water act to converge or to diverge light passing through it?

3. A double-convex lens made of crown glass is placed in a tank of benzene. Will it act as a converging or as a diverging lens there?

4. Under what circumstances, if any, is a light ray that passes through a converging lens not deflected? Under what circumstances, if any, is a light ray that passes through a diverging lens not deflected?

5. Under what circumstances, if any, will a diverging lens form an inverted image of a real object? Under what circumstances, if any, will a converging lens form an erect image of a real object?

6. Is the mercury column in a thermometer wider or narrower than it appears?

7. Is there any way in which a diverging lens, used by itself, can form a real image of a real object? Is there any way in which a converging lens, used by itself, can form a virtual image of a real object?

8. When it emerges from the lens, which path will the incident ray shown in Fig. 23.32 follow?

Figure 23.32

Question 8.

9. If the screen is moved farther away from a projector, how should the projector's lens be moved to keep the image in focus?

10. A gnu moves toward a camera. Should the camera lens be moved toward or away from the film to keep the gnu in focus?

11. What can you say about the properties of an image when the magnification is (a) less than -1? (b) between -1 and 0? (c) between 0 and $+1$? (d) greater than $+1$?

12. What are the characteristics of the image formed on the retina by the lens of the eye?

13. Is there any way in which a convex mirror, used by itself, can form a real image of a real object? Is there any way in which a concave mirror, used by itself, can form a virtual image of a real object?

EXERCISES

23.1 Lenses

1. A double-concave lens has surfaces whose radii of curvature are both 400 mm. The lens is made from flint glass whose index of refraction is 1.55. Find its focal length.

2. A meniscus lens has a concave surface of radius 30 cm and a convex surface of radius 25 cm. The index of refraction of the glass is 1.50. Find the focal length of the lens.

3. A converging meniscus lens has surfaces whose radii of curvature are 20 cm and 30 cm. The lens is made from crown glass whose index of refraction is 1.50. Find its focal length.

4. A diverging meniscus lens has surfaces whose radii of curvature are 20 cm and 30 cm. The lens is made from crown glass whose index of refraction is 1.50. Find its focal length.

5. The index of refraction of crown glass is 1.523 for blue light and 1.517 for red light. How far apart are the focal points for blue and red light of a planoconvex lens of crown glass whose radius of curvature is 200 mm?

6. A double-convex lens whose focal length is 35 cm has surfaces whose radii of curvature are 25 cm and 50 cm, respectively. Find the index of refraction of the glass.

7. Both surfaces of a double-concave lens whose focal length is -18 cm have radii of 20 cm. Find the index of refraction of the glass.

8. A planoconcave lens of focal length 90 mm is to be ground from quartz of index of refraction 1.55. Find the required radius of curvature.

9. A converging lens made from glass of $n = 1.60$ has a focal length in air of 150 mm. (a) Is it converging or diverging when immersed in water? (b) What is its focal length in water?

10. A diverging lens made from glass of $n = 1.55$ has a focal length of -80 mm in air. (a) Is it converging or diverging when immersed in carbon disulfide ($n = 1.63$)? (b) What is its focal length in carbon disulfide?

23.3 The Lens Equation

11. A lens is to be used to focus sunlight on a piece of paper to ignite it. What kind of lens should be used? If the focal length of the lens is 80 mm, how far should it be held from the paper?

12. In order to photograph herself for a passport picture, a woman stands beside her camera 60 cm from a mirror. If the focal length of the camera lens is 40 mm, how far from the film should it be?

13. A lens whose focal length is -60 cm is used to look at a duck 4.00 m away. Where is the image located? Is it real or virtual?

14. A lens held 20 cm from a sardine produces a virtual, erect image of it that appears to originate 10 cm in front of the sardine. What is the focal length of the lens? Is it converging or diverging?

15. A lens held 20 cm from a sardine produces a real, inverted image of it 30 cm on the other side. What is the focal length of the lens? Is it converging or diverging?

23.4 Magnification

16. A candle is placed with its flame 10 cm from a lens whose focal length is $+15$ cm. Find the location of the image of the flame. Is the image larger or smaller than the actual flame? What is the character of the image—that is, is it erect or inverted, real or virtual?

17. The flame of the candle in Exercise 16 is 15 cm from the lens. Answer the same questions for this case.

18. The flame of the candle in Exercise 16 is 25 cm from the lens. Answer the same questions for this case.

19. The flame of the candle in Exercise 16 is 30 cm from the lens. Answer the same questions for this case.

20. The flame of the candle in Exercise 16 is 50 cm from the lens. Answer the same questions for this case.

21. A movie director holds a diverging lens 5.00 m from an actress and sees her one-tenth of her normal size. What is the focal length of the lens?

22. A camera whose lens has a focal length of 90 mm is used to photograph a wall 4.00 m away. (a) How far in front of the film should the lens be placed? (b) If the image on the film covers a square 40 mm on a side, how much of the wall does it include?

23. An aerial camera whose lens has a focal length of 1.0 m is used to photograph a military base from an altitude of 7.0 km. How long is the image on the film of a tank 9.0 m long?

24. The moon's diameter is 3476 km and its average distance from the earth is 3.8×10^5 km. (a) What is the diameter of its image when a 35-mm camera with a 300-mm telephoto lens is used to photograph it? (b) What would the image diameter be if a camera using 9×12-cm film were used with the same lens?

25. A magnifying glass of 10-cm focal length is held 80 mm from a stamp. What is the actual length of a feature of the stamp that appears to be 10 mm long?

26. A magnifying glass of 50-mm focal length is held 35 mm from a spider egg 0.50 mm long. What is the apparent size of the egg?

23.5   The Camera

27. What is the diameter of an $f/8$ telephoto lens whose focal length is 300 mm?

28. The lens of a 35-mm camera is marked $f/1.8$ and is 15 mm in diameter. What is its focal length? Is it a wide-angle, normal, or telephoto lens?

29. An exposure meter indicates that a camera should be set for a shutter speed of $\frac{1}{125}$ s and an aperture of $f/8$ to photograph a certain scene. If the shutter speed is changed to $\frac{1}{500}$ s, what should the aperture be changed to?

30. The settings of a camera are changed from $\frac{1}{125}$ s at $f/16$ to $\frac{1}{250}$ s at $f/5.6$. What is the difference in the amount of light reaching the film?

23.6   The Eye

23.7   Defects of Vision

31. The image distance in a certain normal eye is 20 mm. What range of focal lengths does the eye have if its near point is 20 cm?

32. A myopic eye cannot bring to a focus objects farther than 15 cm away. What type of lens is needed to permit this eye to see clearly objects at infinity, and what focal length should such a lens have?

33. A presbyopic eye has a near point 100 cm away. What type of lens is needed to permit the eye to see clearly objects 25 cm away, and what focal length should such a lens have?

34. A hyperopic eye has a near point of 60 cm. What is its near point when a correcting lens of $+3.33$ diopters is used?

35. A myopic person whose eyes have far points of 60 cm is lent a pair of glasses whose power is -1.5 diopters. How far can she see clearly with these glasses?

23.9   The Telescope

36. A telescope with an objective of focal length 90 cm and an eyepiece of focal length 30 mm is used to look at the moon, whose angular diameter as seen from the earth is about 0.5°. If the moon appears to be 25 cm in front of the eyepiece, what is its apparent diameter?

37. A telescope has an objective whose focal length is 60 cm. (a) Find the eyepiece focal length needed for an angular magnification of 20. (b) The telescope is used to watch a young zebra 500 m away. If the zebra's image is 20 mm long and is located 25 cm in front of the eyepiece, find the zebra's actual length.

23.11   Spherical Mirrors

38. A convex mirror has a radius of curvature of 40 cm. What is its focal length? Where is its focal point?

39. A butterfly is 20 cm in front of a concave mirror whose focal length is 40 cm. Find the location of the image. Is the image larger or smaller than the butterfly? What is the character of the image—that is, is it erect or inverted, real or virtual?

40. A dime is 40 cm in front of the mirror in Exercise 39. Answer the same questions for the image of the dime.

41. A peanut is 50 cm in front of the mirror in Exercise 39. Answer the same questions for the image of the peanut.

42. A caterpillar is 80 cm in front of the mirror in Exercise 39. Answer the same questions for the image of the caterpillar.

43. A button is 100 cm in front of the mirror in Exercise 39. Answer the same questions for the image of the button.

44. An object 6 cm high is 30 cm in front of a convex mirror whose focal length is 50 cm. What are the height and the character of the image?

45. A dentist's concave mirror has a diameter of 10 mm and a focal length of 25 mm. What magnification does it produce when held 18 mm from a tooth?

46. A coin 15.0 mm in diameter is placed 15.0 cm from a spherical mirror. The coin's image is 5.0 mm in diameter and is erect. Is the mirror concave or convex? What is its radius of curvature?

47. The coin in Exercise 46 is placed 15.0 cm from another spherical mirror. The coin's image is now 30.0 mm in diameter and is again erect. Is the new mirror concave or convex? What is its radius of curvature?

48. A man stands 6.0 m from a concave mirror, and an inverted image of himself of the same height is formed on a screen beside him. What is the radius of curvature of the mirror?

49. A worm crawls toward a polished metal ball 60 cm in diameter lying on a lawn. How far from the surface of the ball is the worm when its image appears to be 10 cm behind the surface?

50. A mirror in an amusement park produces an erect image four times enlarged of anyone standing 3.0 m away. (a) Is the mirror concave or convex? (b) What is its radius of curvature?

51. What should the radius of curvature of a convex mirror be if it is to produce an image one-fifth the size of an object 150 cm away?

52. The moon is 3476 km in diameter. What radius of curvature should a concave mirror have if it is to produce a lunar image 10 mm in diameter when the moon is 3.84×10^5 km away?

PROBLEMS

53. A 150-mm camera lens, originally focused on a scene 10 m away, is moved 5 mm farther from the film. Scenes at what distance are now in focus?

54. Verify that the effective focal length f of two thin lenses of focal lengths f_1 and f_2 that are in contact is given by

$$\frac{1}{f} = \frac{1}{f_1} + \frac{1}{f_2}$$

(To do this, let the image produced by the first lens be the object of the second.)

55. A converging lens 30.0 cm from a candle produces an inverted image of the candle on a screen 30.0 cm from the other side of the lens. A diverging lens is then placed midway between the candle and the other lens. When the candle is moved back 10.0 cm from its original position, its image reappears on the screen. Find the focal lengths of the two lenses.

56. A frog is 80 cm in front of a lens whose focal length is +50 cm. Another lens, of focal length +20 cm, is 100 cm behind the first lens. Where is the image of the frog located? Is it real or virtual?

57. A photographic enlarger is being designed to produce prints 40 cm × 50 cm from negatives 10 cm × 12.5 cm. (a) If the maximum distance from negative to print paper is to be 60 cm, what should the focal length of the lens be? (b) How far would such a negative be from the print paper using the same lens if the enlargement size were 20 cm × 25 cm?

58. The actual size of each frame of 16-mm motion picture film is 7.5 mm × 10.5 mm. A projector whose lens has a focal length of 25 mm is to be used with a screen 1.5 m wide. How far from the projector should the screen be located?

59. An object should be about 25 cm from a normal eye for maximum distinctness of vision. (a) Find a formula for the magnification of a converging lens in terms of its focal length when the lens is used as a magnifying glass with an image distance of −25 cm; see Fig. 23.6(a). (b) Find the magnification of a lens whose focal length is 50 mm.

60. A myopic eye with near and far points of 12 and 20 cm, respectively, is given a corrective lens that permits distant vision. What is the new near point?

61. A hyperopic person whose eyeglass lenses are converging with a power of +2.5 diopters can see objects distinctly as close as 25 cm away. What is the distance of the near point without the eyeglasses?

62. A nearsighted person whose eyeglass lenses are diverging with a power of −2.5 diopters can see distant objects distinctly. What is the distance of the far point without the eyeglasses?

63. A telescope has an objective of 750-mm focal length and an eyepiece of 25-mm focal length. The telescope is focused on a distant albatross. (a) What is the angular magnification? (b) How far apart are the lenses? Assume that the final image is 25 cm in front of the eyepiece.

64. A telescope with an objective of focal length 1.0 m and an eyepiece of focal length 5.0 cm is used to examine a penguin 40 cm high. Find the apparent height of the penguin when it is (a) 50 m, and (b) 5.0 m away from the telescope. Assume that the image is 25 cm in front of the eyepiece.

65. A pawn 50 mm high is placed in front of a concave mirror whose radius of curvature is 100 cm. What are the two object distances that will lead to images 200 mm high? What is the character of the image in each case?

66. A virtual image 60 mm long is formed of a paper clip 20 mm long placed 100 mm in front of a concave mirror of unknown curvature. Where else can the paper clip be placed for an image 60 mm long to be formed? What is the character of the image in the latter case?

Answers to Multiple Choice

1. b, c	**10.** a	**19.** c	**28.** a
2. a	**11.** d	**20.** d	**29.** c
3. b	**12.** c	**21.** c	**30.** a
4. d	**13.** b	**22.** c	**31.** c
5. b	**14.** a	**23.** d	**32.** b
6. b	**15.** b	**24.** d	**33.** d
7. a	**16.** d	**25.** b	**34.** b
8. b	**17.** a	**26.** c	
9. b	**18.** a	**27.** c	

C H A P T E R 24

PHYSICAL OPTICS

24.1 Interference of Light

24.2 Double Slit

24.3 Diffraction Grating

24.4 X-Ray Diffraction

24.5 Thin Films

24.6 Diffraction

24.7 Polarization

24.8 Scattering

In the previous chapter the wave nature of light played only a small role, because lenses and mirrors are easier to analyze with the help of rays. But other aspects of optics directly involve the wave nature of light. The study of such phenomena is called **physical optics,** whereas the study of effects that can be treated by using rays is called **geometrical optics.** Since the laws of reflection and refraction follow from Huygens' principle, geometrical optics is really an approximation of physical optics. In this chapter we will find that a purely wave approach is necessary to account for the interference, diffraction, polarization, and scattering of light.

24.1 INTERFERENCE OF LIGHT

What happens when light waves from different sources meet.

When light waves from one source are mixed with those from another source, the two groups of waves are said to **interfere.** We recall from Chapter 12 the principle of superposition, which governs interference: When two or more waves of the same nature go past a point at the same time, the instantaneous amplitude at that point is the sum of the instantaneous amplitudes of the individual waves. Constructive interference refers to the reinforcement of waves in phase (in step) with one another, and destructive interference refers to the partial or complete cancellation of waves out of phase with one another (Fig. 24.1).

Anyone with a pan of water can see how interference between water waves can lead to a water surface disturbed in a variety of characteristic patterns. Two people who hum fairly pure tones slightly different in frequency will hear beats as the result of interference in the sound waves. But if we shine light from two flashlights at the same place on a screen, there is no evidence of interference: The region of overlap is merely uniformly brighter.

It is hard to observe interference in light for two reasons. First, the wavelengths in light are very short, about 1% of the width of a human hair. Second, every natural source of light emits light waves only as short groups of random phase, so that any interference that occurs is usually averaged out during even the briefest period of observation by the eye or photographic film unless special procedures are used. Interference in light is nevertheless just as real as interference in water or sound waves. One example of it is familiar to everybody—the bright colors of a thin film of oil spread out on a water surface.

Coherence

Two sources of waves are said to be **coherent** if there is a fixed phase relationship between the waves they emit during the time the waves are being observed. It does not matter whether the waves are exactly in step when they leave the sources, or exactly out of step, or anything in between. The important thing is that

Figure 24.1

(a) Constructive interference. (b) Destructive interference.

(a)   (b)

Figure 24.2

(a) Radio waves from an antenna are coherent. (b) Light waves from a gas discharge tube are incoherent.

the phase relationship stays the same. If the sources shift back and forth in relative phase while the observation is made, the phase differences average out, and there will be no interference pattern. Such sources are **incoherent.**

The question of coherence is especially significant for light waves because an excited atom radiates for 10^{-8} s or less, depending on its environment. Therefore a monochromatic light source such as a gas discharge tube (a neon sign is an example) does not emit a continuous series ("train") of waves as a radio antenna does but instead emits separate short wave groups whose phases are random. The light from such a tube actually comes from a great many individual, uncoordinated sources, namely the gas atoms, which are in effect being switched on and off rapidly and irregularly (Fig. 24.2).

Suppose we have two point sources of monochromatic light—for instance, a discharge tube with a cover that has two pinholes close together. Different atoms are behind each pinhole, so they are independent sources. Therefore we have at most 10^{-8} s to observe the interference of waves from the two sources. If our detecting instruments are fast enough, as some modern electronic devices are, interference can be demonstrated and the sources can be considered coherent. If we are limited to the eye and to photographic film, which average light signals over times far greater than 10^{-8} s, no interference can be observed in the light from the two sources. The sources must then be considered incoherent. Like beauty, coherence lies in the eye of the beholder.

Does the brief lifetime of an excited atom mean that interference patterns can never be actually seen but can only be recorded by instruments? Not at all. There are three ways to construct separate sources of light coherent for long enough periods of time to produce visible interference patterns:

1. Illuminate two (or more) slits with light from one slit behind them. Then the light waves from the secondary slits are automatically coordinated.
2. Obtain coherent virtual sources from a single source by reflection or refraction. This is how interference is produced by thin films of oil.
3. Coordinate the radiating atoms in each separate source so that they always have the same phase even though different atoms are radiating at successive instants. This is done in the **laser.**

The first two of these methods will be discussed in this chapter and the third in Chapter 25.

24.2 DOUBLE SLIT

It produces an interference pattern of light and dark lines.

The interference of light waves was demonstrated in 1801 by Thomas Young, who used an arrangement like that shown in Fig. 24.3. A source of monochromatic light (that is, light consisting of only a single wavelength) is placed behind a narrow slit *S* in a barrier screen. Another barrier with two similar slits *A* and *B* is placed on the other side. Light from *S* passes through both *A* and *B* and then to the viewing screen.

If light were not a wave phenomenon, we would expect to find the viewing screen completely dark, since no light ray can reach it from the source along a straight path. What actually happens is that each slit acts as a source of secondary wavelets—

Figure 24.3

(a) Young's double-slit experiment. In accord with Huygens' principle, each slit acts as a source of secondary wavelets. (b) The appearance of the screen in Young's experiment.

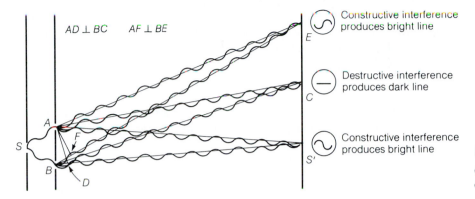

AD ⊥ BC   AF ⊥ BE

Constructive interference
produces bright line

Destructive interference
produces dark line

Constructive interference
produces bright line

Figure 24.4

Origin of the double-slit interference pattern.

we recall Huygens' principle from Section 22.3—so that the entire screen is illuminated. Even the point S', separated from S by the barrier between the slits A and B,
turns out to be bright rather than dark.

Owing to interference the screen is not evenly lit but shows a pattern of
alternate bright and dark lines. Light waves from slits A and B are exactly in phase,
since A and B are the same distance from S (Fig. 24.4). The centerline S' of the
screen is equally distant from A and B, so light waves from these slits interfere
constructively there to produce a bright line.

What happens at the position C on the screen located to one side of S'? The
distance BC is longer than the distance AC by the amount BD, which is equal to
exactly half a wavelength of the light being used. That is,

$$BD = \tfrac{1}{2}\lambda \qquad\qquad\qquad\qquad\qquad \textit{Dark line}$$

When a crest from A reaches C, this difference in path length means that a trough
from B arrives there at the same time, since $\tfrac{1}{2}\lambda$ separates a crest and a trough in the
same wave. The two cancel out each other, the light intensity at C is zero, and the
result is a dark line.

Double-slit interference pattern in
water waves.

If we go past *C* on the screen, we will come to a point *E* such that the distance *BE* is greater by exactly one wavelength than the distance *AE*. That is, the difference *BF* between *BE* and *AE* is

$$BF = \lambda \qquad\qquad\qquad\qquad\textit{Bright line}$$

When a crest from *A* reaches *E*, a crest from *B* also arrives there. Waves arriving at *E* from both slits are always in the same part of their cycles, and they constructively interfere to produce a bright line at *E*.

In this way we find that the alternate bright and dark lines on the screen correspond, respectively, to locations where constructive and destructive interference occurs. Waves reaching the screen from *A* and *B* along paths that are equal or that differ by a whole number of wavelengths (λ, 2λ, 3λ, and so on) reinforce. Waves whose paths differ by an odd number of half wavelengths ($\frac{1}{2}\lambda$, $\frac{3}{2}\lambda$, $\frac{5}{2}\lambda$, and so on) cancel. At intermediate locations on the screen the interference is only partial, so that the light intensity on the screen varies gradually between the bright and dark lines.

24.3 DIFFRACTION GRATING

It produces sharper and brighter interference patterns than a double slit.

There are two difficulties in using a double slit for measuring wavelengths. First, the "bright" lines on the screen are actually quite faint, and an intense light source is therefore needed. Second, the lines are relatively broad, and it is hard to locate their centers accurately. A **diffraction grating** that consists, in essence, of a large number of parallel slits overcomes both of these difficulties.

Gratings are made by ruling grooves on a glass or metal plate with a diamond; the clear bands between the grooves are the "slits" (Fig. 24.5). Replica gratings, made by allowing a transparent liquid plastic to harden in contact with an original grating, are ordinarily used in practice. Replica gratings are often given a thin coating of silver or aluminum and produce their characteristic diffraction patterns by the interference of reflected rather than transmitted light. A phonograph record or compact disk held at a glancing angle acts as a reflection grating by virtue of its closely spaced

Figure 24.5

(a) A diffraction grating. Each space between the ruled grooves acts as a slit. (b) The greater the number of slits, the sharper the line a single wavelength produces.

Few slits

Many slits

(a)   (b)

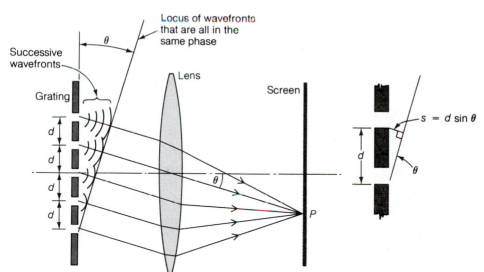

Figure 24.6

The plane diffraction grating.

grooves. Opals are natural diffraction gratings whose constituent particles (tiny spheres of silicon dioxide) are spaced about 250 nm apart in regular arrays.

Gratings are ruled with from 200 to 1000 lines/mm, and a lens is used to focus the light from the slits between them on a screen. The effect of phase differences among the rays from the various slits is enhanced by their great number, and the intensity of light on the screen falls rapidly on either side of the center of each bright line. The bright lines are therefore sharp, and, because there are so many slits, they are also bright in a literal sense. Very accurate wavelength determinations can be made with the help of a grating, and wavelengths that are close together can be resolved. The analysis of a light beam in terms of the particular wavelengths it contains is today almost invariably carried out by using a grating.

Figure 24.6 shows a diffraction grating that forms a bright line on a screen at a deviation angle of θ from the original beam of light. The condition for a bright line is that $s = n\lambda$, where $n = 1, 2, 3$, and so on. Since $s = d \sin \theta$, where d is the spacing of the slits, bright lines occur at those angles for which

$$\sin \theta = n \frac{\lambda}{d} \qquad n = 0, 1, 2, 3, \ldots \qquad \textit{Bright lines} \quad (24.1)$$

When polychromatic light is incident on a grating, a series of spectra is formed on each side of the original beam corresponding to $n = 1$, $n = 2$, and so on. The **first-order spectrum** contains bright lines for which $n = 1$, the **second-order spectrum** contains bright lines for which $n = 2$, and so on (Fig. 24.7). In some gratings the higher-order spectra overlap, so that, for example, the blue end of the third-order spectrum may be deviated by less than the red end of the second-order spectrum. According to Eq. (24.1) the angle θ increases with wavelength. Hence blue light is deviated least and red light most, which is the reverse of what occurs when a prism is used to form a spectrum.

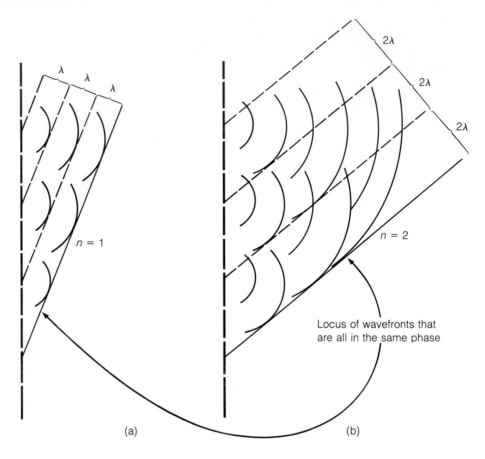

Figure 24.7

(a) In a first-order spectrum, the paths of diffracted rays from successive slits differ by λ in length. (b) In a second-order spectrum, the path differences are 2λ.

Locus of wavefronts that are all in the same phase

(a)   (b)

◆ E X A M P L E   24.1 _____

Visible light includes wavelengths from approximately 400 nm (violet light) to 700 nm (red light). Find the angular width of the first-order spectrum produced by a grating ruled with 8000 lines/cm.

SOLUTION   The slit spacing d that corresponds to 8000 lines/cm is

$$d = \frac{10^{-2}\,\text{m/cm}}{8.00 \times 10^{3}\,\text{lines/cm}} = 1.25 \times 10^{-6}\,\text{m}$$

Because $n = 1$ for a first-order spectrum, the angular deviations of violet and red light, respectively, are given by Eq. (24.1) as

$$\sin \theta_v = \frac{\lambda_v}{d} = \frac{4.00 \times 10^{-7}\,\text{m}}{1.25 \times 10^{-6}\,\text{m}} = 0.320 \qquad \theta_v = 19°$$

$$\sin \theta_r = \frac{\lambda_r}{d} = \frac{7.00 \times 10^{-7}\,\text{m}}{1.25 \times 10^{-6}\,\text{m}} = 0.560 \qquad \theta_r = 34°$$

Red — Violet — Violet — Red

40°   30°   20°   10°   0   10°   20°   30°   40°

Figure 24.8

First-order spectra of white light produced by a grating ruled with 800 lines/mm.

The total width of the spectrum is therefore $34° - 19° = 15°$ (Fig. 24.8).   ◆

24.4 X-RAY DIFFRACTION

How to determine crystal structures.

The atoms of a solid are arranged in regular patterns, and so its surface ought to act as a reflection grating. Then why does every mirror not produce a spectrum?

Light waves are reflected when atomic electrons at the surface of a reflecting material are caused to oscillate by the incoming em waves. The reradiated waves have their central maximum ($n = 0$) at an angle of reflection equal to the angle of incidence. Because atoms in a solid are only about 0.1 nm apart, however, λ/d equals several thousand for visible light. Thus spectra do not occur, because $\sin \theta$ cannot be more than 1. But wavelengths comparable to atomic spacings in solids are found in X-rays, and indeed X-ray diffraction is the means whereby the structures of many solids have been determined. The double-helix form of the DNA molecule was discovered with the help of X-ray interference patterns.

Figure 24.9 shows a beam of monochromatic X-rays directed at a crystal. Most of them go right through without interacting, but some are scattered by atoms in their paths. The conditions for a scattered ray to emerge are the same as those for the reflection of light: The incident and scattered rays must lie in a plane perpendicular to the layers of atoms, and the angle of incidence must equal the angle of scattering.

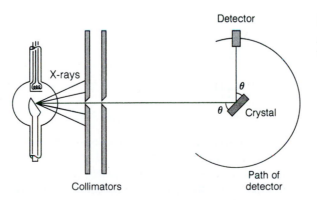

Detector

X-rays

θ

θ Crystal

Collimators

Path of detector

Figure 24.9

A simple X-ray diffraction apparatus.

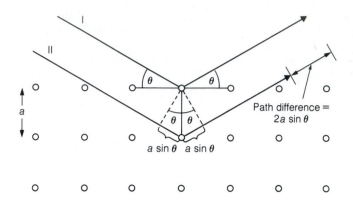

Figure 24.10

X-ray scattering from a crystal with evenly spaced atoms.

In addition, unless all of the X-rays scattered in a given direction are in phase, destructive interference will result, and they will cancel out one another.

We can derive the condition for constructive interference by considering the scattering of a beam of X-rays from two successive layers of atoms a distance a apart, as in Fig. 24.10. Ray II must travel the distance $a \sin \theta$ farther than ray I to reach its scattering atom, and then a further $a \sin \theta$ to leave the crystal. Thus ray II goes $2a \sin \theta$ farther than ray I. If this path difference is a whole number of X-ray wavelengths, the two rays will reinforce each other. The condition for constructive interference is therefore

Figure 24.11

The atoms in a crystal fall into several series of parallel layers.

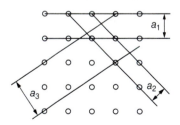

$$2a \sin \theta = n \lambda \qquad n = 1, 2, 3, \ldots$$

Path difference = whole number of wavelengths

(24.2)

By directing a beam of X-rays of known wavelength at a crystal and varying the angle of incidence to find the angles at which scattering occurs, the spacing a between adjacent layers of atoms can be established. Of course, the atoms fall into more than one series of parallel layers, as in Fig. 24.11, and a different spacing will be found for each of them. Most crystals have more complicated three-dimensional structures than the simple one shown in Fig. 24.10, and developments of the procedure just described have been able to establish what the structures are.

24.5 THIN FILMS

Where the colors come from.

We have all seen the marvelous rainbow colors that appear in soap bubbles and thin oil films. Some of us may also have observed the patterns of light and dark bands that occur when two glass plates are almost (but not quite) in perfect contact. Both effects owe their origins to a combination of reflection and interference.

Figure 24.12

Reflection occurs at both surfaces of a soap film.

Let us consider a beam of monochromatic light that strikes a thin film of soapy water. Figure 24.12 shows a ray picture of what happens. We notice that some reflection takes place at both the air-soap and soap-air interfaces. This is a general result: Waves are always partially reflected when they go from one medium to another in which their speed is different. (See the discussion in Section 12.1 on the reflection of pulses in a string under similar circumstances.)

A light ray consists of a succession of wavefronts. Figure 24.13 is the same diagram with the wavefronts drawn in. In (a) the two reflected wave trains are out of phase, and they interfere destructively to cancel each other partially or completely. Most or all of the light reaching this part of the soap bubble therefore passes right through.

Another part of the soap film may have a different thickness. When the film is a little thinner than in (a), the waves in the two reflected trains are exactly in phase, and they interfere constructively to reinforce one another, as in Fig. 24.13(b). Light reaching this part of the soap bubble is strongly reflected. Shining monochromatic light on a soap bubble therefore gives a pattern of light and dark that results from the varying thickness of the bubble.

When the white light shines on a soap bubble, light waves of each wavelength present pass through the soap film without reflection at those places where the film is exactly the right thickness for the two reflected rays to interfere destructively. Light waves of the other wavelengths are reflected to at least some extent and give rise to the vivid colors seen. The varying thickness of the bubble means that the color of the light reflected from the bubble changes from place to place. Exactly the same effect

Figure 24.13

(a) Destructive and (b) constructive interference in a thin film.

(a)   (b)

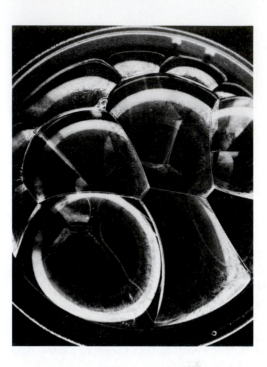

Interference of light in soap bubbles gives rise to patterns of light and dark.

is responsible for the coloration of thin oil films. Generally speaking, soap or oil films whose thickness is comparable to the wavelengths in visible light give rise to the most striking color effects.

Coated Lenses

About 4% of the light striking a glass-air interface is reflected. This is not a lot, but there may be many glass-air interfaces in an optical instrument, and the total amount of light lost through reflection may be considerable. There are ten such interfaces in each of the optical systems in a pair of binoculars, for instance, so only about two-thirds of the incoming light actually gets through to the observer's eyes. Even in a camera, where there are fewer glass-air interfaces, reflections are a nuisance because they may lead to secondary images that blur the picture.

To reduce reflections at a glass-air interface, the glass can be coated with a very thin layer of a transparent substance (usually magnesium fluoride) whose index of refraction is intermediate between those of glass and of air. If the layer is exactly $\frac{1}{4}\lambda$ thick, light reflected at its bottom will have traveled $\frac{1}{2}\lambda$ farther when it rejoins light reflected at the top of the layer, and the two will cancel out exactly, as in Fig. 24.13(a). (Light waves reflected at both the air-fluoride and fluoride-glass surfaces are shifted in phase by $\frac{1}{2}\lambda$, so these shifts have no effect on the cancellation.)

But the cancellation just described is exact only for a particular wavelength λ, whereas white light contains a range of wavelengths. What is therefore done is to choose a wavelength in the middle of the visible spectrum, which corresponds to green light, so that at least partial cancellation occurs over a wide range. The red and violet ends of the spectrum are accordingly least affected, and the light reflected from a coated lens is a mixture of these colors, a purplish hue. The average reflectivity of a glass surface coated in this way is only about 1%. Multiple coatings are sometimes used to reduce reflection even further; a triple coating brings the average reflectivity below 0.5%. Despite their lack of perfection at suppressing reflections, coated lenses transmit appreciably more light than uncoated ones, and they are universally used in fine optical instruments.

24.6 DIFFRACTION

Why shadows are never completely dark.

Waves bend around the edge of an obstacle in their path, a behavior called **diffraction.** We all have heard sound that originated around the corner of a building from where we were standing, for example. These sound waves cannot have traveled in a straight line to our ears. Figure 24.14 shows water waves being diffracted at a gap in a barrier. The waves on the far side of the gap spread out into the geometrical "shadow" of the gap's edges. The diffracted waves spread out as though they originated at the gap, in accord with Huygen's principle.

Diffraction in light is harder to observe because the wavelengths in visible light are so short, less than 10^{-6} m, and the extent of the bending into the shadow zone is correspondingly small. (In contrast, a typical audible sound wave might have a wavelength of 1 m, and a typical wave in a pan of water might have a wavelength of 10 cm; it is easy to notice diffraction effects with such waves.) In fact, because he could not discover any diffraction with his relatively crude apparatus, Newton felt sure that light could not consist of waves.

Figure 24.15 shows what the shadow of a razor blade looks like with the use of monochromatic light from a pinhole. Such patterns of light and dark fringes are the result of interference between secondary wavelets from different parts of the same wavefront, not from different sources as in Young's double-slit experiment. The

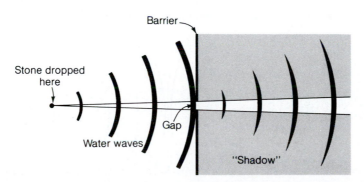

Figure 24.14

Diffraction in water waves. The waves on the far side of the gap spread out as though they had originated at the gap.

Figure 24.15

The shadow of a razor blade.

wavefronts in a beam of unobstructed light produce secondary wavelets that interfere in such a way as to produce new wavefronts exactly like the old ones (see Fig. 22.15). By obstructing part of the wavefronts, points in the shadow region are not reached by secondary wavelets from the entire initial wavefronts but only from part of them, and the result is an interference pattern.

Resolving Power

Diffraction sets a limit to the useful magnification of an optical system such as that of a telescope or microscope. Diffraction occurs whenever wavefronts of light are obstructed, and the light that enters a lens (or mirror) is affected by the limited opening that admits only part of each incident wavefront. No matter how perfect a lens is, the image of a point source of light it produces is always a tiny disk of light with a bright and dark fringes around it (Fig. 24.16). Only if the lens has an infinite diameter can a point source give rise to a point image. The smaller the lens, the larger the image of a point source. The angular width of the radius of this disk of light is about

$$\theta_0 = 1.22 \frac{\lambda}{D} \qquad \text{\textit{Diffraction disk}} \quad (24.3)$$

in radians, where λ is the wavelength of the light used and D is the lens diameter.

Two objects separated by less than θ_0 cannot be **resolved,** that is, distinguished apart, no matter how high the magnification employed, because their images will overlap (Fig. 24.17). Hence there is no advantage in using a higher magnification than will just reveal features that subtend the angle θ_0 at the position of the lens. Although Eq. (24.3) was derived for the image of a point source of light an infinite distance from a lens, it is a reasonable approximation of the resolving power of a telescope or microscope when D is taken as the diameter of the objective lens.

Figure 24.16

The image a lens produces of a point source of light is always a tiny disk with bright and dark fringes around it. The smaller the lens, the larger the image.

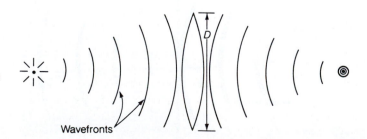

Wavefronts

If two objects d apart are the distance L from an observer, the angle between them, in radians, is

$$\theta = \frac{d}{L}$$

Hence Eq. (24.3) can be rewritten

$$d_0 = 1.22\frac{\lambda L}{D} \qquad\qquad \textit{Resolving power} \quad (24.4)$$

In this formula

Small lens

Very small lens

Figure 24.17

A large lens or mirror is better able to resolve nearby objects than a small one.

d_0 = minimum separation of objects that can be resolved

λ = wavelength of the light used

L = object distance

D = diameter of object lens or mirror

  E X A M P L E   24.2

The pupils of a person's eyes under ordinary conditions of illumination are about 3 mm in diameter, and the distance of most distinct vision is 25 cm for most people. What is the resolving power of the eye at this distance under the assumption that it is limited only by diffraction?

SOLUTION Using λ = 550 nm, which is in the middle of the visible spectrum, Eq. (24.4) gives

$$d_0 = 1.22\frac{\lambda L}{D} = \frac{(1.22)(5.5 \times 10^{-7}\,\text{m})(0.25\,\text{m})}{0.003\,\text{m}} = 6 \times 10^{-5}\,\text{m} = 0.06\,\text{mm}$$

The photoreceptors in the retina are not quite close enough together to permit this degree of resolution, and 0.1 mm is a more realistic figure under good conditions. In terms of angular resolving power, $\theta_0 \approx 5 \times 10^{-4}$ radians for the human eye. ◆

The dependence of resolving power on λ/D poses a severe problem for radiotelescopes, which are antennas designed to receive radio waves from astronomical objects such as stars, galaxies, and gas clouds. Because radio waves from space might have wavelengths as much as a million times greater than those in visible light, an antenna would have to be tens or hundreds of kilometers across to be able to separate sources as close together in the sky as optical telescopes can. In fact, modern electronics permits such resolution to be achieved by combining the signals received by a series of widely spaced antennas. Radio astronomers are currently setting up such an antenna array extending from the Virgin Islands across the Continental United

States to Hawaii, a span of about 8000 km—nearly the diameter of the earth. This array should provide better resolution than is possible with even the largest optical telescope.

24.7 POLARIZATION

Polarized transverse waves lie in a single plane of polarization.

A **polarized** beam of transverse waves is one whose vibrations occur in only a single direction perpendicular to the direction in which the beam travels. The entire wave motion is confined to a plane called the **plane of polarization** (Fig. 24.18).

When many different directions of polarization are present in a beam of transverse waves, vibrations occur equally often in all directions perpendicular to the direction of motion, and the beam is then said to be **unpolarized.** Since the vibrations that make up longitudinal waves can take place in only one direction—namely, that in which the waves travel—longitudinal waves cannot be polarized.

Light waves are transverse, and it is possible to produce and detect polarized light. To clarify the ideas involved, let us first consider the behavior of transverse waves in a stretched string. If the string passes through a hole in a barrier, as in Fig.

Figure 24.18

An unpolarized and a polarized beam of transverse waves.

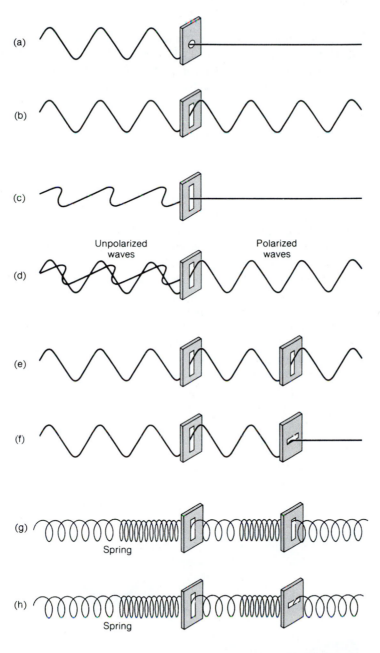

Figure 24.19

Mechanical analogies of fundamental polarization phenomena.

24.19(a), waves traveling down the string are stopped because the string cannot vibrate there. When the hole is replaced by a vertical slot, waves whose vibrations are vertical can get through the barrier, but waves with vibrations in other directions cannot; see Fig. 24.19(b) and (c). In a situation in which several waves vibrating in different directions move down the string, the slot stops all but vertical vibrations, as in (d). An initially unpolarized series of waves has become polarized.

Such an approach can be used to determine whether a particular wave can be polarized or not. In the case of a stretched string, what we do is put another barrier a short distance from the first, as in Fig. 24.19(e). If the slot in the new barrier is also vertical, those waves that can get through the first one can also get through the second. If the slot in the new barrier is horizontal, however, it will stop all waves that reach it from the first barrier, as in (f).

Should longitudinal waves (say in a spring) go through the barrier, it is possible that their amplitudes might decrease in passing through the slots, but the relative alignments of the slots would not matter; see Fig. 24.19(g) and (h). On the other hand, the alignment of the slots is the critical factor in the case of transverse waves.

This chain of reasoning made it possible for the polarization of light waves to be demonstrated in the last century. A number of substances—for instance, quartz, calcite, and tourmaline—have different indexes of refraction for light with different planes of polarization relative to their crystal structures. Prisms can be made from these substances that transmit light in only a single plane of polarization. When a beam of unpolarized light is incident upon such a prism, only those of its waves whose planes of polarization are parallel to a particular plane in the prism emerge from the other side. The remainder of the waves are absorbed or deflected.

Polaroid is an artificially made polarizing material in wide use that transmits light with only a single plane of polarization. To exhibit the transverse nature of light waves, we first place two Polaroid disks in line so that their axes of polarization are parallel (Fig. 24.20(a)) and note that all light passing through one disk also passes through the other. Then we rotate the second disk. As we do so, less and less light gets through. Finally, when the axis of polarization of the second disk is perpendicular to that of the first, as in Fig. 24.20(b), all the light that passes through the first disk is stopped by the second.

Figure 24.20

Experiment showing the transverse nature of light waves.

Overlapping Polaroid sheets whose directions of polarization ·are perpendicular.

Just what is it whose vibrations are aligned in a beam of polarized light? As we know, light waves consist of oscillating electric and magnetic fields perpendicular to each other. Because it is the electric fields of light waves whose interactions with matter produce nearly all common optical effects, the plane of polarization of a light wave is considered to be the plane in which lie both the direction of its electric field and the direction of the wave [the vertical plane in Fig. 22.3(b)]. Even though nothing material moves during the passage of a light wave, it is possible to establish its transverse nature and identify its plane of polarization.

Polarization by Reflection

Light can be polarized when it is reflected from the interface between two media whose indexes of refraction are different. Figure 24.21 shows a beam of unpolarized light arriving at such an interface at the angle of incidence p. Two planes of polarization in the incoming beam are indicated, one in the plane of the paper and one perpendicular to it. (We can think of the beam as having its various polarizations resolved into these components.) Part of the beam is reflected at the angle p, and the rest is refracted at the angle r. If the angle between the reflected and refracted beams is a right angle, none of the waves whose planes of polarization are parallel to the paper is reflected. Thus the reflected wave is completely polarized and the refracted wave is partly polarized.

To find the angle of incidence for which the reflected beam is polarized, we use the fact that $p + r = 90°$ together with Snell's law (Eq. 22.4):

$$n_1 \sin p = n_2 \sin r = n_2 \sin (90° - p) = n_2 \cos p$$

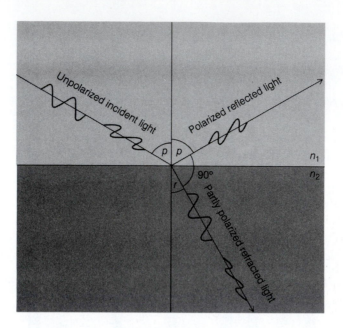

Figure 24.21

Polarization by reflection.

Hence

$$\frac{\sin p}{\cos p} = \frac{n_2}{n_1}$$

$$\tan p = \frac{n_2}{n_1} \qquad\qquad\qquad\qquad \textit{Polarizing angle} \quad (24.15)$$

The angle p is called **Brewster's angle.**

◆   **E X A M P L E   24.3**

At what angle of incidence is sunlight reflected from the surface of a lake fully polarized?

SOLUTION   Here $n_1 = n_{air} = 1.00$ and $n_2 = n_{water} = 1.33$, so

$$\tan p = \frac{n_2}{n_1} = 1.33 \qquad p = 33°$$

The polarization of sunlight reflected by a lake can be verified by looking at the image of the sun through Polaroid sunglasses while rotating them.   ◆

24.8 SCATTERING

Why the sky is blue.

As mentioned earlier, the oscillating electric field in a light wave causes electrons in the atoms and molecules it encounters to oscillate at the same frequency. The electrons then reradiate light waves with this frequency. As a result, some of the incoming light is **scattered** to the side of its original path.

In general, the intensity of light of wavelength λ scattered by something small compared with λ is proportional to λ^{-4}. The shorter the wavelength, the greater the proportion of the incoming light that is scattered. This is the reason the sky is blue. When we look at the sky, what we see is light from the sun that has been scattered by molecules in the upper atmosphere. Blue light, which consists of the shortest wavelengths, is scattered about ten times more readily than red light, so the scattered light is chiefly blue in color (Fig. 24.22(a)).

At sunrise or sunset, when sunlight must make a very long passage through the atmosphere to reach an observer, much of its short-wavelength content is scattered out along the way, and the sun accordingly appears red in color (Fig. 24.22(b)). The water droplets and ice crystals in clouds are larger than λ, and the scattering they

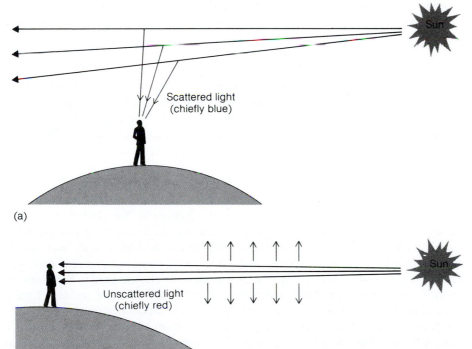

Figure 24.22

(a) Blue light is scattered the most by the earth's atmosphere, red light the least. Hence the sky appears blue. (b) Why the sun appears red at sunrise and sunset.

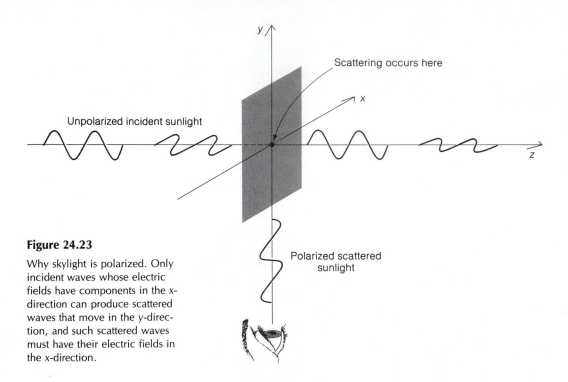

Figure 24.23

Why skylight is polarized. Only incident waves whose electric fields have components in the x-direction can produce scattered waves that move in the y-direction, and such scattered waves must have their electric fields in the x-direction.

produce is independent of λ. Hence clouds do not appear colored. Above the atmosphere the sky appears black, and astronauts can see the moon, stars, and planets in the daytime.

Skylight not only is blue but is partly polarized as well. To check this, all one has to do is hold up a piece of Polaroid against the sky and rotate it. Sunglasses are often made with Polaroid lenses so oriented as to discriminate against the polarized skylight, which reduces glare while affecting other light to a lesser extent. The human eye itself responds equally to all states of polarization. The eyes of bees, however, can detect the polarization of scattered sunlight from a patch of blue sky. This enables bees to find the direction of the sun for navigation when the sun itself is obscured by clouds.

Figure 24.23 shows why scattered light is polarized. A beam of unpolarized light heading in the $+z$-direction strikes some air molecules and is scattered. In this process electrons in the molecules are set in vibration by the electric fields of the light waves, and the vibrating electrons then reradiate. Because the electric field of an electromagnetic wave is perpendicular to its direction of motion, the electric fields in the initial beam lie in the xy-plane only. A scattered wave that proceeds downward, which is the $-y$-direction in Fig. 24.23, can have its electric field in the x-direction only, so it is polarized. Only two planes of polarization for the incident light are shown for clarity. In the case of intermediate planes of polarization, the components of $\mathbf{E}$ in the x-direction can also lead to scattered waves traveling downward that are similarly polarized. Because skylight arrives at our eyes from a variety of directions, the polarization is not complete, but enough occurs to be easily demonstrated.

Important Terms

Two sources of waves are **coherent** if there is a fixed phase relationship between the waves they emit during the time the waves are being observed. Interference can be observed only in waves from coherent sources.

The ability of waves to bend around the edges of obstacles in their paths is called **diffraction**. A **diffraction grating** is a series of parallel slits that produces a spectrum through the interference of light that is diffracted by them.

The **resolving power** of an optical system refers to its ability to produce separate images of nearby objects; resolving power is limited by diffraction, and the larger the objective lens of an optical system, the greater its resolving power.

A **polarized** beam of transverse waves is one whose vibrations occur in only a single direction perpendicular to the direction in which the beam travels, so that the entire wave motion is confined to a plane called the **plane of polarization**. An **unpolarized** beam of transverse waves is one whose vibrations occur equally often in all directions perpendicular to the direction of motion.

Important Formulas

Diffraction grating: $\sin \theta = n\dfrac{\lambda}{d}$ (bright lines)

X-ray diffraction: $2a \sin \theta = n \lambda$

Resolving power: $\theta_0 = 1.22\dfrac{\lambda}{D}$

$$d_0 = 1.22\dfrac{\lambda L}{D}$$

Polarizing angle: $\tan p = \dfrac{n_2}{n_1}$

MULTIPLE CHOICE

1. In order to produce an interference pattern, the waves used must
   a.   all have the same wavelength
   b.   be in a narrow beam
   c.   be coherent
   d.   be electromagnetic

2. Coherent electromagnetic waves are not emitted by
   a.   two antennas connected to the same radio transmitter
   b.   two pinholes in an opaque shield over a sodium-vapor lamp
   c.   a pinhole in an opaque shield over a sodium-vapor lamp and its reflection in a mirror
   d.   two lasers

3. In a double-slit experiment the maximum intensity of the first bright line on either side of the central one occurs on the screen at locations where the arriving waves differ in path length by
   a. $\lambda/4$ b. $\lambda/2$
   c. λ d. 2λ

4. When a spectrum is formed by a double slit or a grating, the light deviated most

   a.   is red
   b.   is green
   c.   is blue
   d.   depends on the slit spacing

5. A diffraction grating is superior to a double slit for wavelength measurements for which one or more of the following reasons?
   a.   The line of a single wavelength is sharper.
   b.   The line of a single wavelength is brighter.
   c.   Light of a larger range of wavelengths can be analyzed.
   d.   The angular width of the spectrum is much wider.

6. The greater the number of lines that are ruled on a grating of given width,
   a.   the shorter the wavelengths that can be diffracted
   b.   the longer the wavelengths that can be diffracted
   c.   the narrower the spectrum that is produced
   d.   the broader the spectrum that is produced

7. X-rays are used to determine crystal structures because
   a.   their wavelengths are much shorter than atomic spacings in crystals

b. their wavelengths are comparable to atomic spacings in crystals

c. their wavelengths are much longer than atomic spacings in crystals

d. they are easily polarized

8. Thin films of oil and soapy water owe their brilliant colors to a combination of reflection and

a. refraction   b. interference

c. diffraction   d. polarization

9. A factor not involved in the production of the colored patterns exhibited when white light is incident on a thin oil film is

a. the presence of different wavelengths in white light

b. the varying thickness of the oil film

c. interference between incident and reflected light

d. reflection at both the upper and lower surfaces of the oil film

10. Coating a glass surface with a thin layer of a substance whose index of refraction is between those of air and glass completely prevents the reflection of light

a. of one wavelength only

b. of two wavelengths only

c. of purple color only

d. of all wavelengths

11. One way to improve the resolving power of a lens is to increase the

a. wavelength of the light used

b. brightness of the light used

c. diameter of the lens

d. object distance

12. The wavelength of light plays no role in

a. interference   b. diffraction

c. resolving power   d. polarization

13. Longitudinal waves do not exhibit

a. refraction   b. reflection

c. diffraction   d. polarization

14. An unpolarized beam of transverse waves is one whose vibrations

a. are confined to a single plane

b. occur in all directions

c. occur in all directions perpendicular to their direction of motion

d. have not passed through a Polaroid disk

15. It is impossible to polarize

a. white light   b. radio waves

c. X-rays   d. sound waves

16. The sky is blue because

a. air molecules are blue

b. the lens of the eye is blue

c. the scattering of light is more efficient the shorter its wavelength

d. the scattering of light is more efficient the longer its wavelength

17. Light scattered perpendicularly to an incident beam is partly

a. diffracted   b. polarized

c. monochromatic   d. coherent

18. Seen from above the atmosphere, the sky is

a. blue   b. red

c. white   d. black

19. Polarized light cannot be produced by

a. reflection   b. scattering

c. diffraction   d. selective absorption

20. The wavelength of light that is deviated in the first order by 15° by a 5000-line/cm grating is

a. 52 nm   b. 259 nm

c. 518 nm   d. 773 nm

21. A pair of binoculars designated "7 × 50" has a magnification of 7 and an objective lens diameter of 50 mm. The smallest detail that in principle can be perceived with such an instrument when viewing an object 1 km away in light of wavelength 500 nm has a linear dimension of approximately

a. 1 mm   b. 1 cm

c. 10 cm   d. 1 m

QUESTIONS

1. Can light from incoherent sources interfere? If so, then why is a distinction made between coherent and incoherent sources?

2. What becomes of the energy of the light waves whose destructive interference leads to dark lines in an interference pattern?

3. What advantages has a diffraction grating over a double slit for determining wavelengths of light?

4. What governs the angular width of the first-order spectrum of white light produced by a grating?

5. What is the difference between the first-order and the second-order spectra produced by a grating? Which is wider?

6. Does a prism produce spectra of different orders?

7. Why is visible light not useful for studying the structure of a crystal, whereas X-rays are?

8. What do diffraction and interference have in common? How do they differ? In Young's double-slit experiment, which effects are due to diffraction and which to interference?

9. Radio waves diffract pronouncedly around buildings, whereas light waves, which are also electromagnetic waves, do not. Why?

10. Explain the peculiar appearance of a distant light source when seen through a piece of finely woven cloth.

11. Give two advantages a large-diameter telescope objective has over a small-diameter one.

12. A camera can be made by using a pinhole instead of a lens. What happens to the sharpness of the picture if the hole is too large? If it is too small?

13. Since light consists of transverse waves, why is not every light beam polarized?

14. What is the relationship between the plane of polarization of a transverse wave and its direction of propagation?

15. Which of the following can occur in (a) transverse waves and (b) longitudinal waves: refraction, dispersion, interference, diffraction, polarization?

EXERCISES

24.3   Diffraction Grating

1. Light of wavelength 750 nm is directed on a grating ruled with 4000 lines/cm. What is the angular deviation of this light in (a) the first order and (b) the third order?

2. A 5500-line/cm diffraction grating produces an image deviated by 27° in the second order. Find the wavelength of the light.

3. How many diffracted images are formed on either side of the central image when radiation of wavelength 600 nm falls on a 4000-line/cm grating?

4. Light containing wavelengths of 500 and 550 nm is directed at a 2000-line/cm grating. How far apart are the lines formed by these wavelengths on a screen 4.0 m away in the second order?

5. White light that contains wavelengths from 400 to 700 nm is directed at a 3000-line/cm grating. How wide is the first-order spectrum on a screen 2.0 m away?

24.4   X-Ray Diffraction

6. A beam of X-rays with wavelengths down to 20 pm is directed at a crystal whose atomic planes are 0.20 nm apart at an angle of 7°. X-rays of what wavelengths will be diffracted?

7. The distance between atomic planes in calcite is 0.30 nm. What is the smallest angle between these planes and an incident beam of 30-pm X-rays at which diffracted X-rays can be detected?

24.6   Diffraction

8. A radar has a resolving power of 30 m at a range of 1.0 km. What is the minimum width of its antenna if its operating frequency is 9500 MHz?

9. The Hale telescope at Mount Palomar in California has a concave mirror 5.0 m in diameter. How many meters apart must two features of the moon's surface be in order to be resolved by this telescope? Take the distance from the earth to the moon as 3.8×10^5 km and use 500 nm for the wavelength of the light.

10. An astronaut circles the earth in a satellite at an altitude of 150 km. If the diameter of his pupils is 2.0 mm and the average wavelength of the light reaching his eyes is 550 nm, is it conceivable that he can distinguish sports stadiums on the earth's surface? Private houses? Cars?

11. The Jodrell Bank radiotelescope has a parabolic reflecting "dish" 76 m in diameter. (a) What is the angular diameter in degrees of a point source of radio waves of wavelength 21 cm as seen by this telescope? (b) How does this figure compare with the angular width of the moon, whose diameter is 3476 km and whose average distance from the earth is 3.8×10^5 km?

12. According to a famous battle command, "Don't fire until you see the whites of their eyes." (a) If the diameter of the white of an eye is 20 mm, the diameter of the pupil in bright sunlight is 2.0 mm, and the light has a wavelength of 500 nm, what is this distance on the basis of the theoretical resolution formula? (b) The actual resolving power of the eye, which is limited by the structure of the retina, is about 5.0×10^{-4} rad. Find the maximum distance at which the white of an eye can be distinguished on the basis of this figure.

13. At night the pupils of a person's eyes are 8.0 mm in diameter. (a) How many km away from a car facing a woman will she be able to distinguish its headlights from each other? (b) If her pupils were 4.0 mm in diameter (say at twilight), how far away from the car could she distinguish its headlights from each other? Assume the headlights are 1.5 m apart, that the average wavelength of their light is 600 nm, and that her eyes are capable of attaining their theoretical resolving power.

PROBLEMS

14. A radio transmitter operating at 15 MHz has a number of vertical antennas 50 m apart on an east-west line. How many intensity maxima in the horizontal plane are there? In what directions?

15. The antennas of Problem 14 are moved so that they are 15 m apart on the same east-west line. How many intensity maxima in the horizontal plane are there now? In what directions?

16. White light that contains wavelengths from 400 to 700 nm is directed at a 5000-line/cm grating. Do the first- and second-order spectra overlap? The second- and third-order spectra? If there is an overlap, will the use of a grating with a different number of lines/cm change the situation?

17. The index of refraction of magnesium fluoride is 1.38. How thick should an antireflection coating of this material be for maximum cancellation at 550 nm, which is the middle of the visible spectrum?

18. A certain laser produces a beam of monochromatic light whose wavelength is 550 nm and whose initial diameter is 1.0 mm. (a) In what distance will diffraction have caused the beam to double its diameter? (b) If the initial diameter of the beam were 10 mm, would the doubling distance be different? If so, what would it be?

19. The smaller the aperture of a camera lens, the greater the depth of field. However, a small aperture means reduced resolution. The criterion for an enlarged print to show sharp detail from a small negative is that the image of a point object on the negative be not more than 0.010 mm across. Find the maximum f-number of a camera lens for this criterion to be met for a distant object in 550-nm light.

Answers to Multiple Choice

1. a	**7.** b	**13.** d	**19.** c
2. b	**8.** b	**14.** c	**20.** c
3. c	**9.** c	**15.** d	**21.** b
4. a	**10.** a	**16.** c	
5. a, b	**11.** c	**17.** b	
6. d	**12.** d	**18.** d	

25

ATOMIC PHYSICS

25.1   Quantum Theory of Light

25.2   X-Rays

25.3   Matter Waves

25.4   Mass and Energy

25.5   The Hydrogen Atom

25.6   Energy Levels

25.7   Atomic Spectra

25.8   Complex Atoms

25.9   Atomic Excitation

25.10   The Laser

Particles and waves are separate concepts in everyday life. A stone thrown into a lake and the ripples that spread out from where it lands resemble each other only in their ability to carry energy from one place to another. But in the world of the atom the situation is different. Electromagnetic waves—which, as we know, show such typical wave behavior as diffraction and interference—turn out to have important properties in common with particles. And electrons—whose particle nature is evident whenever we look at a television picture tube—turn out to have important properties in common with waves. On a very small scale of size, a wave-particle duality replaces the distinction between waves and particles so obvious on a large scale. This duality, as we will see, is the key to understanding the world of the atom.

25.1 QUANTUM THEORY OF LIGHT

Photons are "particles" of light.

When light (particularly ultraviolet light) falls on a metal surface, electrons are found to be given off (Fig. 25.1). This phenomenon is known as the **photoelectric effect.** The photoelectric cell that measures light intensity in a camera, the solar cell that produces electric current when sunlight falls on it, and the television camera tube that converts the image of a scene into an electric signal are all based on the photoelectric effect.

Since light is electromagnetic in nature and carries energy, there seems nothing unlikely about the photoelectric effect; the energy absorbed by the metal may somehow concentrate on individual electrons and reappear as kinetic energy. The situation should be like water waves dislodging pebbles from a beach. But three experimental findings show that no such simple explanation is possible:

1. The electrons (called **photoelectrons**) are emitted at once, even when the light is faint. However, because the energy in an em wave is supposed to be spread across the wavefronts, a period of time should elapse before an individual electron accumulates enough energy to leave the metal. Several months might be needed for a really weak light beam.

2. A bright light yields more photoelectrons than a dim light, but their average KE remains the same. The em theory of light, on the contrary, predicts that the more intense the light, the greater the KE of the electrons.

3. The higher the frequency of the light, the more KE the electrons have. Blue light yields faster electrons than red light (Fig. 25.2). At frequencies below a critical one f_0 (which is characteristic of the metal used), no electrons are given off. Above f_0 the photoelectrons range in energy from 0 to a maximum value that increases with increasing frequency (Fig. 25.3).

Figure 25.1

(a) A method of detecting the photoelectric effect. The photoelectrons ejected from the irradiated metal plate are attracted to the positive collection electrode at the other end of the tube, and the current that results is measured with an ammeter. (b) A method of detecting the maximum energy of the photoelectrons. Note polarity opposite to that in (a). As the stopping electrode is made more negative, the slower photoelectrons are repelled before they can reach it. Finally a voltage will be reached at which no photoelectrons whatever are received at the stopping electrode, as indicated by the current dropping to zero, and this voltage corresponds to the maximum photoelectron energy.

(a)   (b)

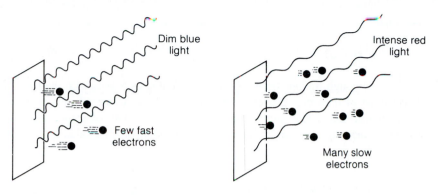

Figure 25.2

The higher the frequency of the light, the more KE the photoelectrons have. The brighter the light, the more photoelectrons are emitted.

Until the discovery of the photoelectric effect a century ago, the em theory of light had been completely successful in explaining the behavior of light. But no amount of ingenuity could bring experiment and theory together here. The result was the creation of the entirely new **quantum theory of light** by Albert Einstein in 1905, the same year that saw the birth of his equally revolutionary theory of relativity. All of modern physics has its roots in these two theories. (The term quantum comes from a Latin word meaning ''how much.'')

Photons

The photoelectric effect is not the only optical phenomenon that is at odds with the idea that light consists of electromagnetic waves, or, indeed, with the idea that light consists of waves of any kind. All of these other phenomena, however, can be understood if light is regarded as consisting instead of little bursts of energy. Each burst is called a **photon** (or **quantum**) and its energy content E is related to the light frequency f by the equation

$$E = hf \qquad \textit{Photon energy} \quad (25.1)$$

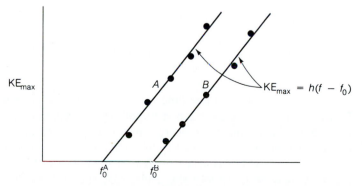

Figure 25.3

The variation of maximum photoelectron energy with the frequency of the incident light for two target metals. No photoelectrons are emitted for frequencies below f_0^A in the case of metal A and below f_0^B in the case of metal B.

The quantity h is a universal constant, called **Planck's constant,** whose value is

$$h = 6.63 \times 10^{-34} \text{ J} \cdot \text{s} \qquad \qquad \textit{Planck's constant}$$

Although the photon has no mass and always moves with the speed of light, it has most of the other properties of particles—it is localized in a small region of space, it has energy and momentum, and it interacts with other particles in more or less the same way a billiard ball interacts with other billiard balls.

◆   E X A M P L E   25.1

The eye can detect as little as 1×10^{-18} J of electromagnetic energy. How many photons of orange light whose wavelength is 600 nm does this energy represent?

SOLUTION   The frequency of this light is

$$f = \frac{c}{\lambda} = \frac{3 \times 10^8 \text{ m/s}}{6 \times 10^{-7} \text{ m}} = 5 \times 10^{14} \text{ Hz}$$

Each photon has the energy hf. Hence

$$\text{Number of photons} = \frac{\text{detectable energy}}{\text{energy/photon}}$$

$$= \frac{1 \times 10^{-18} \text{ J}}{(6.63 \times 10^{-34} \text{ J} \cdot \text{s})(5 \times 10^{14} \text{ Hz})} = 3 \text{ photons} \qquad ◆$$

The experimental data on the photoelectric effect that were plotted in Fig. 25.3 can be understood on the basis of the quantum theory of light. If light consists of photons that each have the energy hf, then

$$hf = \text{KE}_{\text{max}} + hf_0 \qquad \qquad \textit{Photoelectric effect} \quad (25.2)$$

What this equation states is that

Quantum energy = maximum electron energy + energy needed to
eject an electron

The reason for a threshold energy hf_0 (often called **work function**) is clear: It is the minimum energy required to pull an electron from the metal surface, as in Fig. 25.4. (Such a minimum energy must exist or electrons would leave metals all the time.) There are several reasons why not all photoelectrons have the same energy even though a single frequency of light is used. For instance, not all the quantum energy hf may be given to a single electron, and an electron may lose some of its initial energy in collisions with other electrons inside the metal before it gets out.

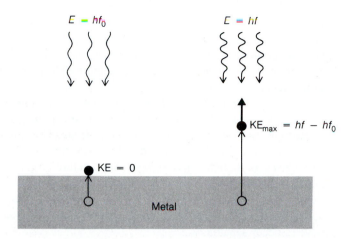

Figure 25.4

If the energy hf_0 is required to remove an electron from a metal surface, the maximum electron kinetic energy will be $hf - hf_0$ when light of frequency f is directed at the surface.

◆ **E X A M P L E   25.2** _____

The photoelectric work function for copper is 4.5 eV. Find the maximum energy of the photoelectrons when ultraviolet light of 1.5×10^{15} Hz falls on a copper surface.

SOLUTION   The quantum energy of the incident photons in eV is

$$hf = \frac{(6.63 \times 10^{-34} \text{ J} \cdot \text{s})(1.5 \times 10^{15} \text{ Hz})}{1.6 \times 10^{-19} \text{ J/eV}} = 6.2 \text{ eV}$$

Hence the maximum photoelectron energy is

$$\text{KE}_{\text{max}} = hf - hf_0 = (6.2 - 4.5)\text{eV} = 1.7 \text{ eV} \qquad ◆$$

What Is Light?

The notion that light travels as a series of little packets of energy is directly opposed to the wave theory of light (Fig. 25.5). Both views have strong experimental support, as we have seen. According to the wave theory, light waves move away

Wave theory
of light

Quantum theory
of light

Figure 25.5

The wave theory of light is needed to explain diffraction and interference, which the quantum theory cannot explain. The quantum theory of light is needed to explain the photoelectric effect, which the wave theory cannot explain.

from a source with the energy they carry spread out continuously through the wave pattern. According to the quantum theory, light is a series of individual photons, each small enough to be absorbed by a single electron. Yet, despite the particle picture of light it presents, the quantum theory needs the frequency f of the light to describe the photon energy.

Which theory are we to believe? A great many scientific ideas have had to be revised or discarded when they were found to disagree with new data. Here, for the first time, two different theories are both needed to explain a single phenomenon. In thinking about this, it is important for us to note that, in a particular situation, light behaves as though it has *either* a wave nature *or* a particle nature. While light sometimes assumes one form and sometimes the other, there is no physical process in which both appear simultaneously. The same light beam can diffract around an obstacle and then impinge on a metal surface to eject photoelectrons, but these two processes occur separately.

Evidently light has a dual character. The wave theory and the quantum theory complement each other. Either theory by itself is only part of the story and can explain only certain effects. The "true nature" of light includes both wave and particle characters, even though there is nothing in everyday life like that to help us visualize it.

25.2 X-RAYS

They consist of high-energy photons.

If photons of light can give up their energy to electrons, can the kinetic energy of moving electrons be converted into photons? The answer is that such a transformation not only is possible but had in fact been discovered (though not understood) prior to the work of Planck and Einstein.

In 1895 the German physicist Wilhelm Roentgen found that a mysterious, highly penetrating radiation is emitted when high-speed electrons impinge on matter. The X-rays (so called because their nature was then unknown) caused phosphorescent substances to glow, exposed photographic plates, traveled in straight lines, and were not affected by electric or magnetic fields. The more energetic the electrons, the more penetrating the X-rays. The greater the number of electrons, the greater the intensity of the resulting X-ray beam.

After more than ten years of study it was finally established that X-rays exhibit both interference and polarization effects, leading to the conclusion that they are em waves. From the interference experiments their wavelengths were found to be very short, shorter than those in ultraviolet light. Electromagnetic radiation in the approximate wavelength interval from 0.01 to 10 nm is today classed as X-radiation.

Figure 25.6 is a diagram of an idealized X-ray tube. The dc source A sends a current through the filament, heating it until it emits electrons. These electrons are

Figure 25.6

An idealized X-ray tube. *A* and *B* are sources of direct current.

then accelerated toward a metal target by the potential difference *V* provided by the dc source *B*. The tube is evacuated to permit the electrons to reach the target freely. The impact of the electrons causes the evolution of X-rays from the target.

What is the physical process involved in the production of X-rays? It is known that charged particles emit em waves when accelerated, and so we may reasonably identify X-rays as the radiation that accompanies the slowing down of fast electrons when they strike matter. The great majority of the electrons, to be sure, lose their kinetic energy too gradually for X-rays to be given off, and merely heat the target. (Hence the targets in X-ray tubes are made of metals with high melting points, and a means for cooling the target is often provided.) A few electrons, however, lose much or all of their energy in single collisions with target atoms, and this is the energy that appears as X-rays. In other words, we may regard X-ray production as an inverse photoelectric effect.

In a CAT (computerized axial tomography) scanner, a series of X-ray exposures of a patient taken from different directions are combined by a computer to give cross-sectional images of the part of the body being examined. The computer, in effect, slices up the tissues on the basis of the X-ray exposures, and any desired slice can be displayed.

Since the threshold energy hf_0 needed to remove an electron from a metal is only a few electron volts, whereas the accelerating potential V in an X-ray tube usually exceeds 10,000 volts, we can neglect the work function hf_0 here. The highest frequency f_{max} found in an X-ray beam should therefore correspond to a quantum energy of hf_{max}, where hf_{max} equals the kinetic energy KE $= Ve$ of an electron that has been accelerated through a potential difference of V. We conclude that

$$hf_{max} = Ve \qquad\qquad \textit{X-ray energy} \quad (25.3)$$

in the operation of an X-ray tube.

◆ **E X A M P L E   25.3** _____

Find the highest frequency present in the radiation from an X-ray machine whose operating potential is 50,000 volts.

SOLUTION   From Eq. (25.3) we find that

$$f_{max} = \frac{Ve}{h} = \frac{(5.0 \times 10^4 \,\text{V})(1.6 \times 10^{-19}\,\text{C})}{6.6 \times 10^{-34}\,\text{J} \cdot \text{s}} = 1.2 \times 10^{19}\,\text{Hz}$$

The corresponding wavelength is $\lambda = c/f = 2.5 \times 10^{-11}$ m $= 0.025$ nm.   ◆

25.3 MATTER WAVES

A moving body behaves in certain ways as though it has a wave nature.

As we have seen, em waves under certain circumstances behave like particles. Is it possible that what we normally think of as particles have wave properties, too? This speculation was first made by Louis de Broglie in 1924. Soon afterward de Broglie's idea was taken up and developed by a number of other physicists (notably Heisenberg. Schrödinger, Born, Pauli, and Dirac) into the elaborate, mathematically difficult—but very beautiful—theory called **quantum mechanics.**

The coming of quantum mechanics did more than provide an accurate and complete description of atomic phenomena. It also altered the way in which physicists approach nature, so that they now think in terms of probabilities instead of in terms of certainties. The atomic world is closer in many respects to a roulette wheel than to a clock—but it is a roulette wheel that obeys certain rules, otherwise known as laws of physics.

Reasoning by analogy with the properties of photons, de Broglie suggested that any object whose momentum is mv behaves like a wave whose wavelength is

$$\lambda = \frac{h}{mv} \qquad\qquad \textit{De Broglie wavelength} \quad (25.4)$$

The more momentum something has, the shorter its wavelength.

How can de Broglie's idea be verified? Perhaps the most striking example of wave behavior is a diffraction pattern, which depends upon the ability of waves both to bend around obstacles and to interfere constructively and destructively with one another. Several years after de Broglie's work, a number of experiments confirmed that a stream of electrons is diffracted by the regular arrays of atoms in a crystal. The diffraction patterns that were observed were in complete accord with the electron wavelengths predicted by Eq. (25.4).

In certain aspects of its behavior, a moving object resembles a wave, and in other aspects it resembles a particle. Which type of behavior is most conspicuous depends on how the object's de Broglie wavelength compares with its dimensions and with the dimensions of whatever it interacts with. Two examples will help us appreciate this statement.

◆   **E X A M P L E   25.4** _____

Find the wavelength of an electron whose speed is 1.0×10^7 m/s.

SOLUTION   From Eq. (25.4)

$$\lambda = \frac{h}{mv} = \frac{6.63 \times 10^{-34}\,\text{J}\cdot\text{s}}{(9.1 \times 10^{-31}\,\text{kg})(1.0 \times 10^7\,\text{m/s})} = 7.3 \times 10^{-11}\,\text{m}$$

This wavelength is comparable to the spacing of the atoms in a crystal. It is not surprising, then, that beams of electrons with speeds of around 10^7 m/s show diffraction effects when directed at crystals, just as X-rays of such wavelengths do.   ◆

◆   **E X A M P L E   25.5** _____

Find the de Broglie wavelength of a 1500-kg car whose speed is 30 m/s.

SOLUTION   The car's wavelength is

$$\lambda = \frac{h}{mv} = \frac{6.63 \times 10^{-34}\,\text{J}\cdot\text{s}}{(1.5 \times 10^3\,\text{kg})(30\,\text{m/s})} = 1.5 \times 10^{-38}\,\text{m}$$

The wavelength is so small compared with the car's dimensions that no wave behavior is to be expected.   ◆

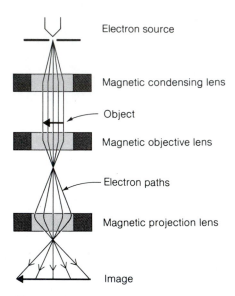

Electron source

Magnetic condensing lens

Object

Magnetic objective lens

Electron paths

Magnetic projection lens

Image

Figure 25.7

An electron microscope uses electrons instead of light waves to produce an enlarged image. Because electron de Broglie wavelengths are shorter than optical wavelengths, an electron microscope has a higher resolving power and hence can be used to produce higher magnifications. The magnetic ''lenses'' are current-carrying coils whose magnetic fields focus an electron beam, just as the glass lenses in an ordinary microscope focus a light beam.

An electron microscope and an electron micrograph of red blood cells.

As we know, the resolving power of any optical system is proportional to the wavelength of the light used. This limits ordinary microscopes to useful magnifications of less than $500\times$. Moving electrons, however, may have wavelengths much shorter than those of light waves. X-rays also have short wavelengths, but it is difficult to focus them. On the other hand, electrons, by virtue of their charge, are easily controlled by electric and magnetic fields.

In an **electron microscope** a beam of electrons passes through a very thin specimen in a vacuum chamber, to prevent scattering of the beam and hence blurring of the image. The electron beam is then brought to a focus on a fluorescent screen or a photographic plate by a system of magnetic fields that act as lenses (Fig. 25.7). For technical reasons the full theoretical resolution of electron waves cannot be realized in practice. For example, a 100-keV electron has a wavelength of 3.7×10^{-12} m, which is 0.0037 nm, but the actual resolution available ordinarily would not exceed perhaps 0.1 nm. However, this is still a great improvement on the 200-nm or more resolution of an optical microscope, and magnifications of $100{,}000\times$ or more are commonly achieved with electron microscopes.

Waves of What?

In water waves, the quantity that varies periodically is the height of the water surface. In sound waves, it is pressure in the medium the waves travel through. In light waves, electric and magnetic fields vary. What is it that varies in the case of matter waves?

The quantity whose variations constitute the matter waves of a moving object is known as its **wave function.** The symbol for wave function is ψ, the Greek letter *psi*.

(a)

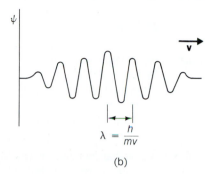

$$\lambda = \frac{h}{mv}$$

(b)

Figure 25.8

(a) Particle description of moving object. (b) Wave description of moving object.

The value of ψ^2 for a particular object at a certain place and time is proportional to the probability of finding the object at that place at that time.

For this reason the quantity ψ^2 is called the **probability density** of the object. A large value of ψ^2 means that it is likely to be found at the specified place and time; a small value of ψ^2 means that it is unlikely to be found at that place and time. Thus matter waves may be regarded as waves of probability.

A group or packet of matter waves is associated with every moving object. The packet travels with the same speed as the object does. The waves in the packet have the average wavelength $\lambda = h/mv$ given by de Broglie's formula (Fig. 25.8). Even though we cannot visualize what is meant by ψ and so cannot form a mental image of matter waves, the agreement between theory and experiment signifies that the notion of matter waves is a meaningful way to describe moving objects.

25.4 MASS AND ENERGY

Matter is a form of energy.

In 1905 a young physicist of 26 named Albert Einstein introduced both the quantum theory of light and the theory of relativity. The latter theory began with an analysis of how measurements of time and space are affected by relative motion between an observer and what he or she is examining. From relativity came some remarkable predictions, all of which have been confirmed by experiment.

The most far-reaching conclusion of special relativity is that mass and energy are closely related—so closely that matter can be converted into energy and energy into matter. The **rest energy** of a body is the energy equivalent of its mass. If a body has the mass m_0 when it is at rest, its rest energy is

$$E_0 = m_0 c^2 \qquad\qquad \textit{Rest energy} \quad (25.5)$$

where c is the speed of light.

Why are we not aware of rest energy as we are aware of kinetic and potential energies? After all, a 1.5-kg object—such as this book—contains the rest energy

$$E_0 = m_0 c^2 = (1.5 \text{ kg})(3.0 \times 10^8 \text{ m/s})^2 = 1.4 \times 10^{17} \text{ J}$$

which is more than enough energy to send a payload of a million tons to the moon. How can so much energy be bottled up without revealing itself in some manner?

In fact, all of us *are* familiar with processes in which rest energy is liberated, only we do not usually think of them in these terms. In every chemical reaction in which energy is given off—for instance, a fire—matter is converted into energy in the form of heat, which is molecular kinetic energy. But the amount of matter that vanishes in such reactions is so small that it escapes our notice. When 1.0 kg of dynamite explodes, 6.0×10^{-11} kg of matter is transformed into energy. The lost mass is so tiny a fraction of the total mass that it is impossible to detect directly (hence the ''law'' of conservation of mass in chemistry). However, the result is the evolution of

$$m_0 c^2 = (6.0 \times 10^{-11} \text{ kg})(3.0 \times 10^8 \text{ m/s})^2 = 5.4 \times 10^6 \text{ J}$$

of energy, which is very hard not to detect.

The radiation of the sun and stars is powered by the conversion of matter into energy in reactions that involve atomic nuclei, as discussed in Chapter 27. The immense destructive force of nuclear weapons has the same source.

Relativity of Mass

Another important finding of the theory of relativity is that the mass of an object is not constant but depends on the object's speed with respect to the observer. The greater the speed, the greater the mass. The variation of mass with speed is not noticeable for ordinary moving things, but in the subatomic world the variation is conspicuous because such tiny particles as electrons, protons, and neutrons may move very fast.

The variation of mass with speed is given by the formula

$$m = \frac{m_0}{\sqrt{1 - v^2/c^2}} \qquad\qquad \textit{Mass increase} \quad (25.6)$$

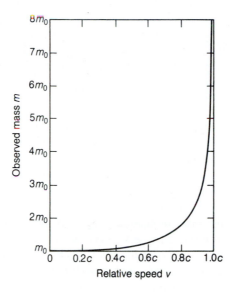

Figure 25.9

The relativity of mass. The closer the relative speed v is to the speed of light c, the greater the observed mass m is compared with the rest mass m_0.

whose symbols have these meanings:

m_0 = mass measured when object is at rest (''rest mass'')

m = mass measured when object is in relative motion

v = speed of relative motion

c = speed of light

Since the denominator of Eq. (25.6) is always less than 1, an object will always appear more massive when in relative motion than when at rest. This mass increase acts both ways. To a measuring device on the rocket ship in flight, its twin ship on the ground also appears to have a mass m greater than its own mass m_0.

Relativistic mass increases are significant only at speeds approaching that of light (Fig. 25.9). The rest mass of the *Apollo II* spacecraft was about 63,070 kg. On its way to the moon, the spacecraft's speed was about 10,840 m/s relative to the earth. Its mass in flight, as measured by an observer on the earth therefore increased to

$$m = \frac{m_0}{\sqrt{1 - \dfrac{v^2}{c^2}}} = \frac{63{,}070 \text{ kg}}{\sqrt{1 - \dfrac{(1.084 \times 10^4 \text{ m/s})^2}{(2.998 \times 10^8 \text{ m/s})^2}}} = 63{,}070.000041 \text{ kg}$$

This is not much of a change.

Smaller objects can be given much higher speeds. It is not very hard to accelerate electrons (rest mass 9.109×10^{-31} kg) to speeds of, say, $0.9999c$. The

mass of such an electron is

$$m = \frac{m_0}{\sqrt{1 - \dfrac{v^2}{c^2}}} = \frac{9.109 \times 10^{-31}\,\text{kg}}{\sqrt{1 - \dfrac{(0.9999c)^2}{c^2}}} = 644 \times 10^{-31}\,\text{kg}$$

The electron's mass is 71 times its rest mass! The mass increases predicted by the relativistic mass formula, even such remarkable ones as this, agree with experimental findings.

Equation (25.6) has something to say about the greatest speed an object can have. The closer v approaches c, the closer v^2/c^2 approaches 1, and the closer $\sqrt{1 - (v^2/c^2)}$ approaches zero. As the denominator of Eq. (25.6) becomes smaller, the mass m becomes larger. If the relative speed v actually were equal to the speed of light, the object's mass would be infinite. The idea of an infinite mass is, of course, nonsense in many ways. Such a mass would have needed an infinite force to reach c, for instance, and it would exert an infinite gravitational force on all other bodies in the universe. Hence we interpret Eq. (25.6) to mean that no material body can ever reach the speed of light.

The mass given by Eq. (25.6) must be used to calculate the de Broglie wavelength of a moving particle when its speed is great enough for an appreciable mass increase. For a speed of $0.1c$ the mass increase is only 0.5%, and so for speeds less than this the mass increase can often be ignored.

25.5 THE HYDROGEN ATOM

Standing waves in the atom.

Every atom consists of a tiny, positively-charged nucleus surrounded by electrons. The electrons cannot be at rest, because the electric attraction of the nucleus would pull them in at once. If the electrons are moving, however, stable orbits like those of the planets around the sun would seem to be possible.

A hydrogen atom, the simplest of all, has a single electron outside a nucleus that consists of only a proton. Experiments indicate that 13.6 eV of work is needed to break apart a hydrogen atom into a proton and an electron that go their separate ways (Fig. 25.10). With the help of calculations based on what we already know about mechanics and electricity, this figure leads to an orbital radius of $r = 5.3 \times$

Figure 25.10

To break a hydrogen atom apart requires 13.6 eV.

Figure 25.11

The hydrogen atom consists of an electron circling a proton. The electrical force exerted by the proton on the electron provides the centripetal force required to hold it in a circular path. The proton is 1836 times as heavy as the electron, so its motion under the influence of the electrical force that the electron exerts is small.

10^{-11} m and an orbital speed of $v = 2.2 \times 10^6$ m/s for the electron in the hydrogen atom. The proton mass is 1836 times the electron mass, so we can consider the proton as stationary with the electron revolving around it, the required centripetal force being provided by the electrical force exerted by the proton (Fig. 25.11).

What such an analysis overlooks is that, according to electromagnetic theory, all accelerated electric charges radiate electromagnetic waves—and an electron moving in a circular path is certainly accelerated. Thus an atomic electron circling its nucleus should radiate away energy, spiraling inward until it is swallowed up by the nucleus (Fig. 25.12). Clearly the ordinary laws of physics cannot explain the stability of the hydrogen atom, whose electron must be whirling around the nucleus to keep from being pulled into it yet must be losing energy all the time. However, because other phenomena similarly impossible to understand on the basis of ordinary laws of physics can be explained in terms of quantum concepts, let us see whether this might also be true for the atom.

Let us begin by looking into the wave properties of the electron in the hydrogen atom. The de Broglie wavelength λ of an object of mass m and speed v is $\lambda = h/mv$ where h is Planck's constant. The speed of the electron in a hydrogen atom is $v = 2.2 \times 10^6$ m/s, and so the wavelength of its matter waves is

$$\lambda = \frac{h}{mv} = \frac{6.63 \times 10^{-34} \text{ J} \cdot \text{s}}{(9.1 \times 10^{-31} \text{ kg})(2.2 \times 10^6 \text{ m/s})} = 3.3 \times 10^{-10} \text{ m}$$

This is an exciting result, because the electron's orbit has a circumference of exactly

$$2\pi r = (2\pi)(5.3 \times 10^{-11} \text{ m}) = 3.3 \times 10^{-10} \text{ m}$$

Mechanical model

Electromagnetic model

Figure 25.12

The mechanical and electromagnetic models of the hydrogen atom are in conflict. Only quantum theory can account for the stability of atoms.

(a)

(b)

Figure 25.13

(a) The electron orbit in a ground-state hydrogen atom is exactly one de Broglie wavelength in circumference. (b) The orbit corresponds to a complete electron wave joined on itself.

We therefore conclude that *the orbit of the electron in a hydrogen atom corresponds to one complete electron wave joined on itself* (Fig. 25.13).

The fact that the electron orbit in a hydrogen atom is one electron wavelength in circumference is just the clue we need to construct a theory of the atom. If we look at the vibrations of a wire loop, as in Fig. 25.14, we see that their wavelengths always fit a whole number of times into the loop's circumference, each wave joining smoothly with the next. With no dissipative effects, such vibrations would go on forever.

Why are these the only vibrations possible in a wire loop? A fractional number of wavelengths cannot be fitted into the loop and still allow each wave to join smoothly with the next (Fig. 25.15). The result would be destructive interference as the waves travel around the loop, and the vibrations would die out rapidly.

By considering the behavior of electron waves in the hydrogen atom as analogous to standing waves in a wire loop, then, we may postulate the following:

Figure 25.14

Three possible modes of vibration of a wire loop.

Circumference =
2 wavelengths

Circumference =
4 wavelengths

Circumference =
9 wavelengths

An electron can circle an atomic nucleus only if its orbit is a whole number of electron wavelengths in circumference.

Figure 25.15

Unless a whole number of wavelengths fits into the wire loop, destructive interference causes the vibrations to die out rapidly.

This postulate is the decisive one in our understanding of the atom. We note that it combines both the particle and the wave characters of the electron into a single statement. Although we can never observe these antithetical characters at the same time in an experiment, they are inseparable in nature.

It is easy to express in a formula the condition that a whole number of electron wavelengths fit into the electron's "orbit." The circumference of a circular orbit of radius r is $2\pi r$, and so the condition for orbit stability is

$$n\lambda = 2\pi r_n \qquad n = 1, 2, 3, \ldots \qquad \textit{Condition for orbit stability} \quad (25.7)$$

where r_n designates the radius of the orbit that contains n wavelengths. The quantity n is called the **quantum number** of the orbit.

A straightforward calculation shows that the stable electron orbits are those whose radii are given by the formula

$$r_n = n^2 r_1 \qquad n = 1, 2, 3, \ldots \qquad \textit{Orbital radii in hydrogen atom} \quad (25.8)$$

where $r_1 = 5.3 \times 10^{-11}$ m is the radius of the innermost orbit (Fig. 25.16).

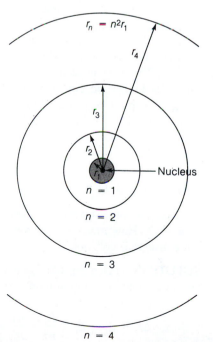

Figure 25.16

Electron orbits in the hydrogen atom according to the wave model. The orbit radii are proportional to n^2.

25.6 ENERGY LEVELS

Each orbit has a characteristic energy.

The total energy of a hydrogen atom is not the same in the various permitted orbits. The energy E_n of a hydrogen atom whose electron is in the nth orbit is given by

$$E_n = \frac{E_1}{n^2} \qquad n = 1, 2, 3, \ldots \qquad \textit{Energy levels of hydrogen atom} \quad (25.9)$$

where $E_1 = -13.6 \text{ eV} = -2.18 \times 10^{-18}$ J is the energy corresponding to the innermost orbit. The energies specified by this formula are called the **energy levels** of the hydrogen atom and are the only ones possible. The situation is rather like that of a person on a ladder, who can stand only on its steps, not in between them.

The energy levels of the hydrogen atom are all less than zero, which means that the electron does not have enough energy to escape from the atom. The lowest energy level E_1, corresponding to the quantum number $n = 1$, is called the **ground state** of the atom. The higher levels E_2, E_3, E_4, and so on are called **excited states** (Fig. 25.17).

As the quantum number n increases, the energy E_n approaches zero. In the limit of $n = \infty$, $E_\infty = 0$ and the electron is no longer bound to the proton to form an atom. An energy greater than zero corresponds to an unbound electron. Because such an electron has no closed orbit that must satisfy quantum conditions, it may have any positive energy whatever.

The work needed to remove an electron from an atom in its ground state is called its **ionization energy.** The ionization energy is therefore equal to $-E_1$, the energy that must be provided to raise an electron from its ground state to an energy of $E = 0$, when it is free. In the case of hydrogen, the ionization energy is 13.6 eV, because the ground-state energy of the hydrogen atom is -13.6 eV.

◆   **E X A M P L E   25.6** _____

An electron collides with a hydrogen atom in its ground state and excites it to a state of $n = 3$. How much energy was given to the hydrogen atom in this inelastic (KE not conserved) collision?

SOLUTION   From Eq. (25.9) the energy change of a hydrogen atom that goes from an initial state of quantum number n_i to a final state of quantum number n_i is

$$\Delta E = E_f - E_i = \frac{E_1}{n_f^2} - \frac{E_1}{n_i^2} = E_1 \left(\frac{1}{n_f^2} - \frac{1}{n_i^2} \right)$$

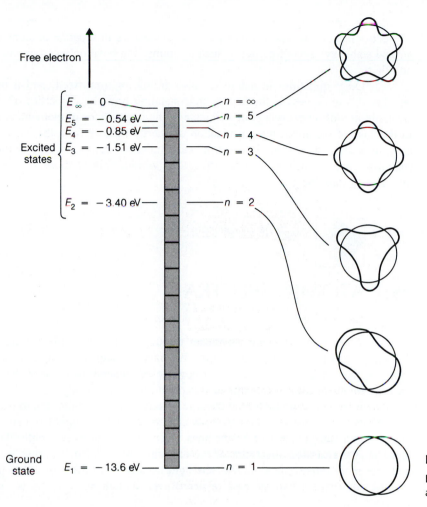

Free electron

Excited states

$E_\infty = 0$ ————————— $n = \infty$

$E_5 = -0.54$ eV ————— $n = 5$

$E_4 = -0.85$ eV ————— $n = 4$

$E_3 = -1.51$ eV ————— $n = 3$

$E_2 = -3.40$ eV ————— $n = 2$

Ground state $E_1 = -13.6$ eV ————— $n = 1$

Figure 25.17

Energy levels of the hydrogen atom.

Here $n_i = 1$ and $n_f = 3$, and $E_1 = -13.6$ eV, so

$$\Delta E = E_1\left(\frac{1}{3^2} - \frac{1}{1^2}\right) = -13.6\left(\frac{1}{9} - 1\right)\text{eV} = 12.1 \text{ eV}$$

◆

Quantization in the Atomic World

The presence of energy levels in an atom—which is true for all atoms, not just the hydrogen atom—is another example of the fundamental graininess of physical quantities on a microscopic scale. In the everyday world, matter, electric charge, energy, and so on seem continuous and capable of being cut up, so to speak, into parcels of any size we like. In the world of the atom, however, matter consists of elementary particles of fixed rest masses that join together to form atoms of fixed rest masses; electric charge aways comes in multiples of $+e$ and $-e$; energy in the form of electromagnetic waves of frequency f always comes in separate photons of energy hf; and stable systems of particles, such as atoms, can have only certain energies and no others.

Other quantities in nature are also grainy, or **quantized,** and it has turned out that this graininess is the key to understanding how the properties of matter we are familiar with in everyday life originate in the interactions of elementary particles. In the case of the atom, the quantization of energy follows from the wave nature of moving bodies: In an atom the electron wave functions can occur only in the form of standing waves, much as a violin string can vibrate only at those frequencies that give rise to standing waves.

25.7 ATOMIC SPECTRA

Each element has a characteristic line spectrum.

When an electric current passes through a glass tube filled with neon gas, the gas gives off a bright orange light. We have all seen neon signs based on this effect, perhaps without being aware of how closely related the color of the light is to the way in which the electrons in each neon atom are arranged.

Every element has a characteristic spectrum that contains certain wavelengths only. Figure 25.18 shows how a spectrometer spreads out the light from an "excited" gas or vapor into a series of bright lines, each of a single wavelength. These lines make up the **emission spectrum** of the element (Fig. 25.19). Because some of the lines are more intense than the rest, the light usually gives the impression of being a specific color (orange in the case of neon) even though other colors are present as

Figure 25.18

The spectrum of a substance may be obtained by "exciting" a sample of its vapor in a tube by means of an electric discharge and then directing the light it gives off through a prism or grating.

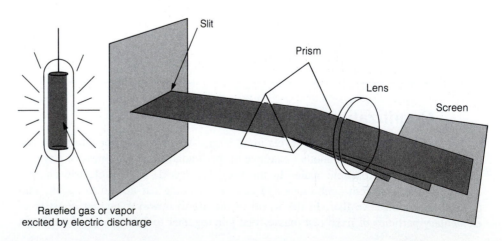

Slit

Prism

Lens

Screen

Rarefied gas or vapor
excited by electric discharge

Figure 25.19

The most prominent lines in the emission spectra of hydrogen, helium, and mercury in the visible region.

Streams of protons and electrons from the sun excite atoms in the upper atmosphere to produce the eerie glow of an aurora. The green colors originate in oxygen, the reds in both oxygen and nitrogen. The earth's magnetic field deflects the paths of the solar ions so that most auroras occur in belts around each pole.

well. Spectral lines are found in the infrared and ultraviolet regions as well as in the visible region. An emission spectrum is different from a **continuous spectrum**—that is, the rainbow band produced when light from a hot object passes through a spectrometer—because a continuous spectrum contains all wavelengths and not just a few.

Spectra of another kind, **absorption spectra**, occur when light from a hot source passes through a cool gas before entering the spectrometer. The light source alone would give a continuous spectrum, but the gas absorbs certain wavelengths out of the light that goes through it. Hence the continuous spectrum appears to be crossed by dark lines, each line corresponding to one of the wavelengths absorbed by the gas.

If the bright-line spectrum of an element is compared with its absorption spectrum, the dark lines in the latter have the same wavelengths as a number of the bright lines in the emission spectrum (Fig. 25.20). Thus a cool gas absorbs some of the wavelengths of the light that it emits when excited. The spectrum of sunlight contains dark lines because the luminous part of the sun, which radiates much like a blackbody heated to 5800 K, has around it an envelope of cooler gas. Because every element has a unique spectrum, the spectrometer is a valuable tool in chemistry and astronomy. Helium was discovered in the solar atmosphere through its spectrum before being identified on the earth (*helios* is Greek for "sun").

Figure 25.20

(a) Absorption spectrum of sodium vapor. (b) Emission spectrum of sodium vapor. Each dark line in the absorption spectrum corresponds to a bright line in the emission spectrum.

Figure 25.21

The line spectrum of hydrogen with the various series of spectral lines indicated. The wavelength scale is not linear in order to cover the entire spectrum.

Origin of Spectral Lines

A century ago the wavelengths in the spectrum of an element were found to fall into sets called **spectral series.** The spectral series of hydrogen are shown in Fig. 25.21, and a simple formula turns out to relate the wavelengths in each series.

The presence of separate energy levels in the hydrogen atom suggests a connection with line spectra. Let us suppose that when an electron in an excited state drops to a lower state, the difference in energy between the states is emitted as a single photon of light. Because electrons cannot exist in an atom except in certain specific energy levels, a jump from one level to the other, with the energy difference being given off all at once in a photon rather than gradually, fits in well with this model.

If the quantum number of the initial (higher energy) state is n_i and the quantum number of the final (lower energy) state is n_f, what we assert is that

$$E_i - E_f = hf \qquad \textit{Origin of spectral lines} \quad (25.10)$$

Initial energy − final energy = quantum energy

where f is the frequency of the emitted photon and h is Planck's constant. This formula agrees with the experimental data on the spectrum of hydrogen.

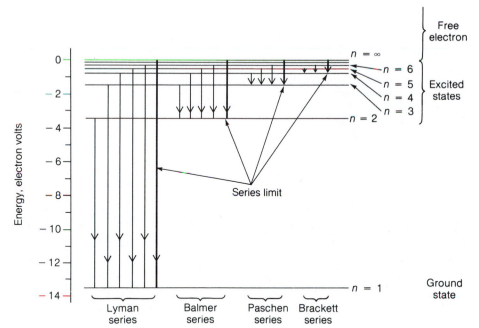

Figure 25.22

Energy-level diagram of the hydrogen atom with some of the transitions that give rise to spectral lines indicated.

Figure 25.22 is an energy-level diagram of the hydrogen atom showing the possible transitions from initial quantum states to final ones. Each transition—or "jump"—involves a characteristic amount of energy and hence a photon of a certain characteristic frequency. The larger the energy difference between initial and final energy levels, as indicated by the lengths of the arrows, the higher the frequency of the emitted photon. The origins of the various series of spectral lines are marked on the diagram.

25.8 COMPLEX ATOMS

A different set of four quantum numbers for each atomic electron.

The above theory of the atom agrees nicely with experiment, but it has some serious limitations. For example, it cannot be applied successfully to the atoms of elements other than hydrogen, that is, to atoms with more than one electron. These limitations are absent from the more general **quantum theory of the atom.** Not only has this theory provided a framework for understanding the structures of all atoms, no matter how complex, but it has also given key insights that explain how and why atoms join together to form molecules, liquids, and solids.

In the simple wave model, an atomic electron is supposed to move in a circular orbit, and all that changes as it revolves around its nucleus is its position on

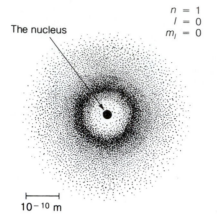

$n = 1$
$l = 0$
$m_l = 0$

The nucleus

10^{-10} m

Figure 25.23

Cross section of the probability
cloud for the ground state of the
hydrogen atom.

this circle. A single quantum number is all that is needed to describe the physical
state of such an electron.

In the quantum theory of the atom, an electron is not limited to a fixed orbit
but moves about in three dimensions. We can think of the electron as circulating in
a probability cloud specified by a wave function ψ. Where the cloud is most dense
(ψ^2 has the highest value), the electron is most likely to be found. Where the cloud
is least dense (ψ^2 has the lowest value), the electron is least likely to be found. Figure
25.23 shows a cross section of the probability cloud for the ground state of the
hydrogen atom, which is the state in which the electron has the lowest energy and is
the one usually occupied.

Three quantum numbers, not just a single one, are needed to specify the size
and shape of the probability cloud of an atomic electron. One of them, designated n
as in the earlier theory, is the chief factor that governs the electron's energy and its
average distance from the nucleus. The other two, designated l and m_l, together with
n determine the details of the probability cloud.

A fourth quantum number, m_s, also turns out to be needed to completely
specify the state of an atomic electron. An electron behaves in certain respects as
though it is a spinning charged sphere, so that it can be thought of as a tiny bar
magnet. The quantum number m_s describes how the electron aligns itself in a magnetic
field, either parallel (''spin up,'' corresponding to $m_s = +\frac{1}{2}$) or antiparallel (''spin
down,'' corresponding to $m_s = -\frac{1}{2}$) (Fig. 25.24).

Figure 25.24

The spin magnetic quantum number m_s of an atomic electron has
two possible values, $+\frac{1}{2}$ and $-\frac{1}{2}$.
Electrons behave in certain re-
spects like spinning charged
spheres, but experiments suggest
they are more likely to be point
particles with no extension in
space.

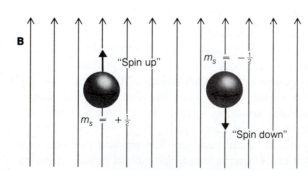

B

"Spin up" $m_s = -\frac{1}{2}$

$m_s = +\frac{1}{2}$

"Spin down"

The Exclusion Principle

In a hydrogen atom the electron is normally in its ground state of $n = 1$. What about more complex atoms? Are all 92 electrons of a uranium atom in the same $n = 1$ state, jammed into a single probability cloud? Many lines of evidence make this idea unlikely.

An example of such evidence is the great difference in chemical behavior shown by certain elements whose atomic structures differ by just one electron. Thus the elements that have the atomic numbers 9, 10, and 11 are, respectively, the corrosive gas fluorine, the inert gas neon, and the metal sodium. Because the electron structure of an atom controls how it interacts with other atoms, it makes no sense that the chemical properties of the elements should change so sharply with a small change in atomic number if all the electrons in an atom were in the same probability cloud.

The **exclusion principle** shows us the way to understand how the electrons are arranged in an atom with more than one electron. It states:

No two electrons in an atom can exist in the same quantum state.

That is, each electron in a complex atom must have a different set of the quantum numbers n, l, m_l, and m_s. Thus each probability cloud in an atom can be occupied by two electrons with their spins in opposite directions. The exclusion principle can be generalized to refer to the electrons in *any* small region of space, regardless of whether or not they constitute an atom, as we will find in Chapter 26.

The electrons in an atom that share the same quantum number n are said to occupy the same **shell.** These electrons average about the same distance from the nucleus and usually have comparable, though not identical, energies. The number of electrons in a shell of given n is limited to $2n^2$. Thus there can be at most 2 electrons in the $n = 1$ shell, 8 in the $n = 2$ shell, 18 in the $n = 3$ shell, and so on. To minimize the atom's energy, the electrons in an atom in its ground state occupy the innermost shells to the maximum extent possible. For example, the chlorine atom, which contains a total of 17 electrons, has 2 electrons in its $n = 1$ shell, 8 in its $n = 2$ shell, and 7 in its $n = 3$ shell. Figure 25.25 is a highly schematic representation of the electron structures of the 18 lightest elements.

Figure 25.25

Schematic representation of the electron structures of the 18 lightest elements. The electrons in the filled inner shells of the second and third rows of atoms are not shown. Metals are at the left, nonmetals at the right; hydrogen is in a category by itself.

Figure 25.26

In (a) both atoms are in their ground states. During the collision some kinetic energy is transformed into excitation energy, and in (b), the target atom is in an excited state. In (c) the target atom has returned to its ground state by emitting a photon.

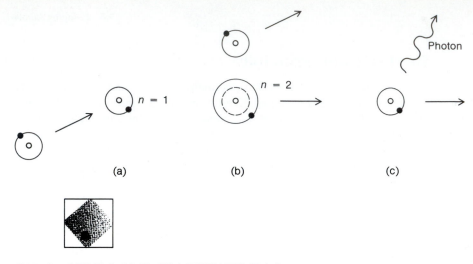

(a)   (b)   (c)

25.9 ATOMIC EXCITATION

The origins of emission and absorption spectra.

There are two principal ways in which an atom may be excited to an energy level above that of its ground state and thereby become able to radiate. One of these ways is by a collision with another atom during which part of their kinetic energy is transformed into electron energy within either or both of the participating atoms. An atom excited in this way will then lose its excitation energy by emitting one or more photons in the course of returning to its ground state (Fig. 25.26). In an electric discharge in a gas, an electric field accelerates electrons and ions until their kinetic energies are great enough to excite atoms with which they happen to collide. A neon sign is a familiar example of this effect.

Another excitation mechanism is the absorption by an atom of a photon of light whose energy is just the right amount to raise it to a higher energy level. A photon of wavelength 121.7 nm is emitted when a hydrogen atom in the $n = 2$ state drops to the $n = 1$ state. The absorption of a photon of wavelength 121.7 nm by a hydrogen atom initially in the $n = 1$ state will therefore bring it up the $n = 2$ state. This process explains the origin of absorption spectra (Fig. 25.27).

Figure 25.27

The origins of emission and absorption spectra.

Origin of emission spectra

Origin of absorption spectra

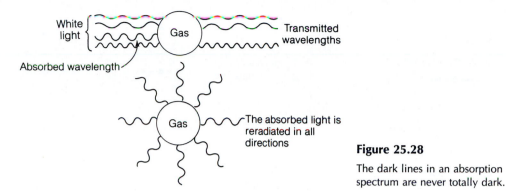

White
light
Gas
Transmitted
wavelengths

Absorbed wavelength

Gas
The absorbed light is
reradiated in all
directions

Figure 25.28

The dark lines in an absorption
spectrum are never totally dark.

When white light (which contains all wavelengths) is passed through hydrogen gas, photons of those wavelengths that correspond to transitions between hydrogen energy levels are absorbed. The resulting excited hydrogen atoms reradiate their excitation energy almost at once, but these photons come off in random directions, not all in the same direction as in the original beam of white light (Fig. 25.28). The dark lines in an absorption spectrum are therefore never totally dark, but only appear so by contrast with the bright background of transmitted light. We would expect the lines in the absorption spectrum of a substance to be the same as lines in its emission spectrum, which is what is found. As Fig. 25.20 shows, only a few of the emission lines appear in absorption. The reason is that reradiation is so rapid that essentially all the atoms in an absorbing gas or vapor are initially in their lowest energy states and therefore take up and reemit only photons that represent transitions to and from that state.

25.10 THE LASER

How to produce light waves all in step.

The **laser** is a device that produces a light beam with some remarkable properties:

1. The beam is extremely intense, more intense by far than the light from any other source. To achieve an energy density equal to that in some laser beams, a hot object would have to be at a temperature of 10^{30} K.
2. There is very little divergence, so that a laser beam from the earth that was reflected by a mirror left on the moon during the *Apollo 11* expedition remained narrow enough to be detected upon its return to the earth—a total journey of more than 750,000 km.
3. The light is essentially monochromatic.

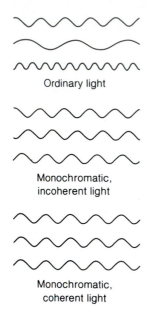

Ordinary light

Monochromatic,
incoherent light

Monochromatic,
coherent light

Figure 25.29

A laser produces a beam of light whose waves all have the same frequency and are in phase with one another. Such a beam is said to be monochromatic and coherent. The beam is also narrow and spreads out very little even over a long distance.

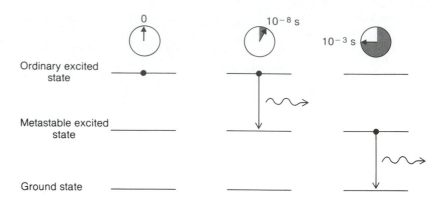

Figure 25.30

An atom can exist in a metastable energy level for a longer time before radiating than it can in an ordinary energy level.

4. The light is coherent, with the waves all exactly in phase with one another (Fig. 25.29). It is possible to obtain interference patterns not only by placing two slits in a laser beam but also by using beams from two separate lasers.

The term *laser* stands for *l*ight *a*mplification by *s*timulated *e*mission of *r*adiation. Let us look into how a laser works.

Left to itself, an atom always stays in its ground state, the quantum state in which it has the least energy. When an atom is raised to an excited state, it usually falls to its ground state almost at once, sometimes first dropping to an intermediate excited state. A photon is normally emitted in each transition. The lifetime of most excited states is only about 10^{-8} s, but certain states are **metastable** (temporarily stable). An atom may remain in a metastable state for 10^{-3} s or more before radiating (Fig. 25.30), provided it does not undergo a collision in the meantime in which its excitation energy would be lost. The operation of lasers depends on the existence of metastable states in atoms.

Three kinds of transition involving electromagnetic radiation are possible between two energy levels in an atom whose energy difference is $E_2 - E_1 = hf$ (Fig. 25.31):

1. **Induced absorption** occurs when the atom is in the lower level and is raised to the upper level by absorbing a photon of energy hf.

Figure 25.31

The three types of transition between energy levels in an atom.

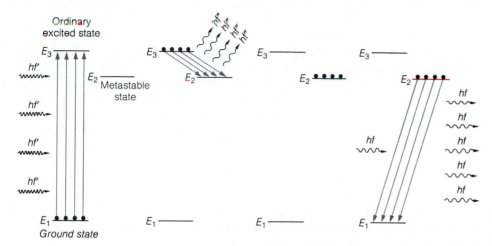

Figure 25.32

How a laser operates.

2. **Spontaneous emission** occurs when the atom is in the upper level and falls by itself to the lower level by emitting a photon of energy hf.
3. **Induced emission** occurs when the atom is in the upper level and radiation of frequency f causes a transition from the upper level to the lower level. In induced emission the emitted light waves are exactly in phase with the incoming ones, so that the result is an enhanced beam of coherent light.

Now suppose we have a group of atoms of some kind that have metastable states of excitation energy hf. If we can raise most of the atoms to the metastable level and then shine light of frequency f on the group, more induced emission than absorption will occur and the original light will be amplified (Fig. 25.32). This is the concept that underlies the operation of the laser. The term **population inversion** is given to an assembly of atoms most of which are in excited states, because normally the ground states are the more highly populated.

Population inversions can be produced in a number of ways. One of them, called optical pumping, uses an external light source some of whose photons have the right frequency to raise the atoms to excited states that decay into the desired metastable ones. This is the method used in the ruby laser, in which the chromium ions Cr^{3+} in a ruby crystal are excited by light from a xenon-filled flash lamp (Fig. 25.33). (A ruby is a crystal of Al_2O_3 a small number of whose Al^{3+} ions are replaced

Figure 25.33

The ruby laser produces pulses of light.

by Cr^{3+} ions, which are responsible for the reddish color. Such ions are Cr atoms that have lost three electrons each.)

The Cr^{3+} ions are raised in the pumping process to a level E_3 from which they decay to the metastable level E_2 by losing energy to other atoms in the crystal. This metastable level has a lifetime of about 0.003 s. Because there are comparatively few Cr^{3+} ions in the ruby rod and a xenon flash lamp emits a great deal of light, most of the Cr^{3+} ions can be pumped to the E_2 level in this way. A few Cr^{3+} ions in the E_2 level then spontaneously fall to the E_1 level, and some of the resulting photons bounce back and forth between the reflecting ends of the ruby rod. The presence of light of the right frequency now stimulates the other Cr^{3+} ions in the E_2 level to radiate, and the result is an avalanche that produces a large pulse of red light that emerges from the partly silvered end of the rod. The length of the rod is made to be exactly a whole number of half-wavelengths long, so the radiation trapped in it forms an optical standing wave. Since the induced emissions are stimulated by the standing wave, their waves are all in step with it.

The helium-neon gas laser achieves a population inversion in a different way. A mixture of helium and neon is placed in a glass tube that has parallel mirrors at both ends. An electric discharge is then produced in the gas by electrodes connected to a source of high-frequency alternating current. Collisions with electrons from the discharge then excite He and Ne atoms to metastable states respectively 20.61 and 20.66 eV above their ground states (Fig. 25.34). Some of the excited He atoms give their energy to ground-state Ne atoms in collisions, with the 0.05 eV of additional energy coming from the kinetic energy of the atoms. The purpose of the He atoms is thus to help produce a population inversion in the Ne atoms.

An excited Ne atom emits a photon of wavelength $\lambda = 632.8$ nm in the transition to another excited state, which is the laser transition. Then another photon is spontaneously emitted, which produces only incoherent radiation because it is not

Figure 25.34

Energy levels involved in the helium-neon laser, which operates continuously.

A laser produces an intense beam of monochromatic, coherent light from the cooperative radiation of excited atoms or molecules. Shown is part of a laser system that can deliver thousands of joules of energy to a target the size of a grain of sand in a pulse lasting less than 10^{-9} s for experiments in controlled fusion energy.

induced. The rest of the excitation energy is lost in collisions with the tube walls. Because the electron impacts that produce the population inversion occur all the time, an He-Ne laser operates continuously, unlike a ruby laser.

Many other kinds of lasers of have been built and still more are under development. Some are very small and produce the coded flashes of light that carry telephone signals along glass fiber transmission lines. More powerful lasers are used in surgery where a laser beam has the advantage of sealing small blood vessels it cuts through. Lasers find numerous applications in industry, from cutting fabric for clothing and making holes in nipples for babies' bottles to welding pipelines and heat-treating the surfaces of engne crankshafts and cylinder walls to harden them.

Important Terms

The **photoelectric effect** is the emission of electrons from a metal surface when light shines on it.

The **quantum theory of light** states that light travels in tiny bursts of energy called **quanta** or **photons.** The quantum theory of light is required to account for the photoelectric effect.

X-rays are high-frequency electromagnetic waves emitted when fast electrons impinge on matter.

A moving body behaves as though it had a wave character. The waves representing such a body are **matter waves,** also called **de Broglie waves.** The wave variable in a matter wave is its **wave function,** whose square is the **probability density** of the body. The value of the probability density of a particular body at a certain place and time is proportional to the probability of finding the body at that place and time. Matter waves may thus be regarded as waves of probability.

Rest energy is the energy an object has by virtue of its mass alone.

The **relativity of mass** refers to the increase in the measured mass of an object when it is moving relative to an observer. The object's **rest mass** is its mass measured when it is at rest relative to the observer.

An **emission spectrum** consists of the various wavelengths of light emitted by an excited substance. An **absorption spectrum** consists of the various wavelengths of light absorbed by a substance when white light is passed through it.

According to the **wave theory of the atom,** an electron can circle an atomic nucleus indefinitely without radiating energy if the circumference of its orbit is a whole number of electron wavelengths. The number of wavelengths that fit into a particular permitted orbit is called the **quantum number** of that orbit. The electron energies corresponding to the various quantum numbers constitute the **energy levels** of the atom, of which the lowest is the **ground state** and the rest are **excited states.**

When an electron in an excited state drops to a lower state, the difference in energy between the states is emitted as a single photon of light; when a photon of the same wavelength is absorbed, the electron goes from the lower to the higher state. These two processes account for the properties of **emission** and **absorption spectra,** respectively.

In the **quantum theory of the atom,** three quantum numbers are needed to specify the probability cloud that an electron occupies in an atom. The greater the probability density in a certain part of the cloud, the more likely the electron is to be found there. The electron's energy and average distance form the nucleus are chiefly determined by the quantum number n. The electrons in an atom that have the same value of n are said to occupy the same **shell.**

Every electron has a certain amount of angular momentum called its **spin.** The quantum number m_1 that governs electron spin has two possible values, $+\frac{1}{2}$ and $-\frac{1}{2}$. Owing to its spin, every electron acts like a tiny bar magnet.

According to the **exclusion principle,** each electron in an atom must have a different set of quantum numbers. Because of this principle, atoms with different numbers of electrons have different properties.

A **laser** is a device for producing a narrow, monochromatic, coherent beam of light. The term stands for *light amplification by stimulated emission of radiation.* Laser operation depends upon the existence of **metastable states,** which are excited atomic states that can persist for unusually long periods of time.

Important Formulas

Quantum energy: $E = hf$

Photoelectric effect: $hf = KE_{max} + hf_0$

X-ray energy: $hf_{max} = Ve$

de Broglie wavelength: $\lambda = \dfrac{h}{mv}$

Rest energy: $E_0 = m_0c^2$

Mass increase: $m = \dfrac{m_0}{\sqrt{1 - v^2/c^2}}$

Condition for orbit stability: $n\lambda = 2\pi r_n$

Energy levels of hydrogen atom:

$$E_n = \frac{E_1}{n^2} \quad n = 1, 2, 3, \ldots$$

Origin of spectral lines: $E_i - E_f = hf$

MULTIPLE CHOICE

1. A surface emits photoelectrons only when the light shone on it exceeds a certain
- **a.** frequency
- **b.** wavelength
- **c.** intensity
- **d.** speed

2. When light is directed on a metal surface, the kinetic energies of the photoelectrons
- **a.** vary with the intensity of the light
- **b.** vary with the frequency of the light
- **c.** vary with the speed of the light
- **d.** are random

3. An increase in the brightness of the light directed at a metal surface causes an increase in the photoelectrons'
- **a.** wavelength
- **b.** speed
- **c.** energy
- **d.** number

4. The photoelectric effect can be understood on the basis of
  a. the electromagnetic theory of light
  b. the special theory of relativity
  c. the principle of superposition
  d. none of the above

5. In a vacuum all photons have the same
  a. frequency   b. wavelength
  c. energy   d. speed

6. The rest mass of a photon
  a. is zero
  b. is the same as that of an electron
  c. depends on its frequency
  d. depends on its energy

7. An increase in the voltage applied to an X-ray tube causes an increase in the X-rays'
  a. wavelength   b. speed
  c. energy   d. number

8. According to the theories of modern physics,
  a. all particles exhibit wave behavior
  b. only moving particles exhibit wave behavior
  c. only charged particles exhibit wave behavior
  d. only uncharged particles exhibit wave behavior

9. The description of a moving body in terms of matter waves is legitimate because
  a. it is based on common sense
  b. matter waves have actually been seen
  c. the analogy with electromagnetic waves is plausible
  d. theory and experiment agree

10. The wave packet that corresponds to a moving particle
  a. has the same size as the particle
  b. has the same speed as the particle
  c. has the speed of light
  d. consists of X-rays

11. A moving body is described by the wave function ψ at a certain time and place; ψ^2 is proportional to the body's
  a. electric field
  b. speed
  c. energy
  d. probability of being found

12. In the formula $E = m_0 c^2$, the symbol c represents
  a. the speed of the body
  b. the speed of sound
  c. the speed of light
  d. the rest energy of 1 kg of matter

13. Classical physics is unable to account for
  a. the presence of electrons in atoms

  b. why atomic electrons are in motion
  c. why atoms can radiate electromagnetic waves
  d. why atoms can be stable

14. In the wave model of the hydrogen atom, the electron revolves around the nucleus in order to
  a. emit spectral lines
  b. produce X-rays
  c. form energy levels that depend on its speed
  d. keep from falling into the nucleus

15. In the wave model, an electron can revolve around the nucleus of a hydrogen atom indefinitely if its orbit is
  a. a perfect circle
  b. sufficiently far from the nucleus to avoid capture
  c. less than one de Broglie wavelength in circumference
  d. exactly one de Broglie wavelength in circumference

16. The electron of a ground-state hydrogen atom
  a. has left the atom
  b. is at rest
  c. is in its orbit of lowest energy
  d. is in its orbit of highest energy

17. The energy difference between adjacent energy levels
  a. is the same for all quantum numbers
  b. is smaller for small quantum numbers
  c. is larger for small quantum numbers
  d. shows no regularity

18. An atom emits a photon when one of its electrons
  a. collides with another of its electrons
  b. is removed from the atom
  c. shifts to a quantum state of lower energy
  d. shifts to a quantum state of higher energy

19. The bright-line spectrum produced by the excited atoms of an element contains wavelengths that
  a. are the same for all elements
  b. are characteristic of the particular element
  c. are evenly distributed throughout the entire visible spectrum
  d. are different from the wavelengths in its dark-line spectrum

20. A neon sign does not produce
  a. a line spectrum
  b. an emission spectrum
  c. an absorption spectrum
  d. photons

21. In an emission spectral series,
  a. the lines are evenly spaced
  b. the excited electrons all started from the same energy level

    c. the excited electrons all return to the same energy level

    d. the excited electrons leave the atoms

22. The sun's spectrum consists of a light background crossed by dark lines. This suggests that the sun

    a. is a hot object surrounded by a hot atmosphere

    b. is a hot object surrounded by a cool atmosphere

    c. is a cool object surrounded by a hot atmosphere

    d. is a cool object surrounded by a cool atmosphere

23. The quantum number n of the lowest energy state of a hydrogen atom

    a. is 0

    b. is 1

    c. depends on the orbit size

    d. depends on the electron speed

24. Most excited states of an atom have lifetimes of about

    a. 10^{-8} s   **b.** 10^{-3} s

    c. 1 s   **d.** 10 s

25. A large value of the probability density ψ^2 of an atomic electron at a certain place and time signifies that the electron

    a. is likely to be found there

    b. is certain to be found there

    c. has a great deal of energy there

    d. has a great deal of charge there

26. The number of quantum numbers needed to determine the size and shape of the probability cloud of an atomic electron is

    a. 1   **b.** 2

    c. 3   **d.** 4

27. According to the exclusion principle, no two electrons in an atom can have the same

    a. spin   **b.** speed

    c. orbit   **d.** set of quantum numbers

28. Each probability cloud in an atom can be occupied by

    a. one electron

    b. two electrons with spins in the same direction

    c. two electrons with spins in opposite directions

    d. any number of electrons

29. An atom can be raised to an excited state by which one or more of the following processes?

    a. A collision with another atom or other particle

    b. The absorption of a photon

    c. The spontaneous emission of a photon

    d. The induced emission of a photon

30. The operation of the laser is based on

    a. the exclusion principle

    b. the interference of de Broglie waves

    c. induced emission of radiation

    d. spontaneous emission of radiation

31. Which one or more of the following properties are characteristic of the light waves from a laser?

    a. The waves all have the same frequency.

    b. The waves are all in step with one another.

    c. The waves form a narrow beam.

    d. The waves have higher photon energies than light waves of the same frequency from an ordinary source.

32. Light of wavelength 500 nm consists of photons whose energy is

    a. 1.1×10^{-48} J   **b.** 1.3×10^{-27} J

    c. 4.0×10^{-19} J   **d.** 1.7×10^{-15} J

33. An X-ray photon has an energy of 6.6×10^{-15} J. The frequency that corresponds to this energy is

    a. 4.4×10^{-48} Hz   **b.** 1.0×10^{-19} Hz

    c. 1.0×10^{15} Hz   **d.** 1.0×10^{19} Hz

34. A lamp emits light of frequency 5.0×10^{15} Hz at a power of 25 W. The number of photons given off per second is

    a. 1.3×10^{-19} **b.** 8.3×10^{-17}

    c. 7.6×10^{18} **d.** 1.9×10^{50}

35. The de Broglie wavelength of an electron whose speed is 1.0×10^8 m/s is

    a. 5.9×10^{-56} m   **b.** 1.5×10^{-19} m

    c. 7.3×10^{-12} m   **d.** 1.4×10^{11} m

36. The speed of an electron whose de Broglie wavelength is 1.0×10^{-10} m is

    a. 6.6×10^{-24} m/s   **b.** 3.8×10^{3} m/s

    c. 7.3×10^{6} m/s   **d.** 1.0×10^{10} m/s

37. Burning a ton of coal liberates 3×10^{10} J of energy. The mass lost by a ton of coal when it burns is

    a. 0.33 μg   **b.** 0.67 μg

    c. 1 μg   **d.** 100 kg

38. Which of the following transitions in a hydrogen atom emits the photon of highest frequency?

    a. $n = 1$ to $n = 2$ **b.** $n = 2$ to $n = 1$

    c. $n = 2$ to $n = 6$ **d.** $n = 6$ to $n = 2$

39. Which of the following transitions in a hydrogen atom absorbs the photon of hihgest frequency?

    a. $n = 1$ to $n = 2$ **b.** $n = 2$ to $n = 1$

    c. $n = 2$ to $n = 6$ **d.** $n = 6$ to $n = 2$

40. Which of the following transitions in a hydrogen atom emits the photon of lowest frequency?

    a. $n = 1$ to $n = 2$ **b.** $n = 2$ to $n = 1$

    c. $n = 2$ to $n = 6$ **d.** $n = 6$ to $n = 2$

QUESTIONS

1. Why do you think the wave aspect of light was discovered earlier than its particle aspects?

2. If Planck's constant were smaller than it is, would quantum phenomena be more or less conspicuous than they are now in everyday life?

3. What happens to the frequency and wavelength of the X-rays emitted form a metal surface when the speed of the electrons that strike it is increased? What happens to the speed, energy, and number per second of the X-ray photons?

4. How does the speed of the wave packet that corresponds to a moving particle compare with (a) the particle's speed and (b) the speed of light?

5. Must a particle have an electric charge in order for matter waves to be associated with its motion?

6. Can the rest mass of a moving particle be determined by measuring its de Broglie wavelength?

7. A proton and an electron have the same de Broglie wavelength. How do their speeds compare?

8. What is the simplest experimental procedure that can distinguish between a gamma ray whose wavelength is 10^{-11} m and an electron whose de Broglie wavelength is also 10^{-11} m?

9. The potential energy of a golf ball in a hole is negative relative to the ground. Under what circumstances (if any) is its kinetic energy negative? Its rest energy?

10. In the wave model of the atom the electron is in constant motion. How can such an electron have a negative amount of energy?

11. What kind of spectrum is observed in (a) light from the hot filament of a light bulb; (b) light from a sodium-vapor highway lamp; (c) light from an electric light bulb surrounded by cool sodium vapor?

12. When radiation with a continuous spectrum is passed through a volume of hydrogen gas whose atoms are all in the ground state, which spectral series will be present in the resulting absorption spectrum?

13. Explain why the spectrum of hydrogen has many lines, although a hydrogen atom contains only one electron.

14. Would you expect the fact that the atoms of an excited gas are in rapid random motion to have any effect on the sharpness of the spectral lines they produce?

15. Under what circumstances can two electrons share the same probability cloud in an atom?

16. According to the wave model, the radius of an atomic electron's orbit is proportional to n^2. Is there any aspect of the probability-cloud model that might correspond to this relationship?

17. Why is the length of the optical cavity of a laser so important?

EXERCISES

25.1 Quantum Theory of Light

1. Find the energy of a 700-nm photon.

2. Find the frequency and wavelength that correspond to a 5.0-MeV photon.

3. How many photons per second are emitted by a 150-W amateur radio transmitter operating on a wavelength of 20 m?

4. A detached retina is being ''welded'' back by using 20-ms pulses from an 0.50-W laser operating at a wavelength of 643 nm. How many photons are in each pulse?

5. What is the maximum wavelength of light that will lead to photoelectric emission from platinum, whose work function is 5.6 eV? In what part of the spectrum is such light?

6. The threshold frequency for photoelectric emission in calcium is 7.7×10^{14} Hz. Find the maximum energy in electron volts of the photoelectrons when light of frequency 1.20×10^{15} Hz is directed on a calcium surface.

7. A silver ball is suspended by a string in a vacuum chamber, and ultraviolet light of wavelength 200 nm is directed at it. What electric potential will the ball acquire as a result? The work function of silver is 4.7 eV.

25.2 X-Rays

8. The operating voltage of an X-ray tube is 50 kV. Find the highest frequency found in the X-rays it emits.

9. Electrons are accelerated in television tubes through potential differences of about 10 kV. Find the highest frequency of the em waves that are emitted when these electrons strike the screen of the tube. What type of waves are these?

10. What voltage must be applied to an X-ray tube for it to emit X-rays with a minimum wavelength of 3.0×10^{-11} m?

25.3 Matter Waves

11. What is the de Broglie wavelength of a 1.0-mg grain of sand blown by the wind at a speed of 20 m/s?

12. Find the de Broglie wavelength of an electron whose speed is 1.0×10^7 m/s.

13. Find the de Broglie wavelength of a 1.0-MeV proton.

14. Show that the de Broglie wavelength of an oxygen molecule in thermal equilibrium in the atmosphere at 20°C is smaller than its diameter of about 4×10^{-10} m.

15. Green light has a wavelength of about 550 nm. Through what potential difference must an electron be accelerated to have this wavelength?

25.4 Mass and Energy

16. The combustion of 1.00 kg of gasoline liberates 47.3 MJ of energy. How much mass is lost when 1.00 kg of gasoline is burned in the engine of a car? Would you expect to be able to measure the change in mass?

17. The source of the sun's energy (and therefore, directly or indirectly, of nearly all energy available on earth) is the conversion of hydrogen to helium. As described in Chapter 27, the nuclei of four hydrogen atoms, each of mass 1.673×10^{-27} kg, join together in a series of separate reactions to yield a helium nucleus of mass 6.646×10^{-27} kg. How much energy is liberated each time a helium nucleus is formed? How many helium nuclei are formed to produce the 1.00×10^7 J a moderately active person requires per day?

18. Typical chemical reactions absorb or release energy at the rate of several eV per molecular change. What change in mass is associated with the absorption or release of 1.00 eV?

19. Find the mass of an object whose rest mass is 1000 g when it is traveling at 10%, 90%, and 99% of the speed of light.

20. An electron is moving at 2.98×10^8 m/s. What is its mass?

21. A man has a mass of 100 kg on the ground. When he is in a spacecraft in flight, an observer on the earth measures his mass to be 101 kg. How fast is the spacecraft moving?

22. What speed must a particle have if its mass is to be triple its rest mass?

25.5 The Hydrogen Atom

25.6 Energy Levels

23. How much energy is needed to ionize a hydrogen atom when it is in the $n = 4$ state?

24. Find the average kinetic energy per molecule in a gas at room temperature (20°C) and show that this is much less than the energy required to raise a hydrogen atom from its ground state ($n = 1$) to its first excited state ($n = 2$).

25. To what temperature must a hydrogen gas be heated if the average molecular kinetic energy is to equal the binding energy of the hydrogen atom?

26. Find the quantum number of the orbit in a hydrogen atom that is 1 mm in radius. What is the energy (in electron volts) of an atom is this state? Why is such an orbit unlikely to be occupied?

27. (a) Find the de Broglie wavelength of the earth. (b) What is the quantum number that characterizes the earth's orbit about the sun? (The earth's mass is 6.0×10^{24} kg, its orbital radius is 1.5×10^{11} m, and its orbital speed is 3.0×10^4 m/s.)

25.7 Atomic Spectra

25.9 Atomic Excitation

28. Find the wavelength of the spectral line that corresponds to a transition in hydrogen from the $n = 6$ state to the $n = 3$ state. In what part of the spectrum is this?

29. Find the wavelength of the spectral line that corresponds to a transition in hydrogen from the $n = 10$ state to the ground state. In what part of the spectrum is this?

30. A proton and an electron, far apart and at rest initially, combine to form a hydrogen atom in the ground state. A single photon is emitted in this process. What is its wavelength?

31. What is the shortest wavelength present in the Brackett series of spectral lines?

32. What is the shortest wavelength present in the Paschen series of spectral lines?

33. A beam of electrons is used to bombard gaseous hydrogen. What is the minimum energy in electron volts the electrons must have if the second line of the Paschen series, corresponding to a transition from the $n = 5$ state to the $n = 3$ state, is to be emitted?

34. (a) An object of mass 0.10 g undergoes simple harmonic motion with a frequency of 20 Hz and an amplitude of 2.0 mm. If this oscillation were to emit a quantum of radiation at the same frequency, by what percentage would its energy be reduced? (b) An object of mass 1.0×10^{-26} kg undergoes simple harmonic motion with a frequency of 2.0×10^{15} Hz and an amplitude of 2.0×10^{-12} m. If this oscillation were to emit a quantum of radiation at the same frequency, by what percentage would its energy be reduced? (c) What conclusion can we draw from these results?

PROBLEMS

35. Light from the sun reaches the earth at the rate of about 1.4×10^3 W/m^2 of area perpendicular to the direction of the light. Assume sunlight is monochromatic with a frequency of 5.0×10^{14} Hz. (a) How many photons fall per second on each square meter of the earth's surface directly facing the sun? (b) How many photons are present in each cubic meter near the earth on the sunlit side?

36. Light of wavelength 420 nm falls on the cesium surface of a photoelectric cell at the rate of 5.0 mW. Given that one photoelectron is emitted for every 10^4 incident photons, find the current produced by the cell. The work function of cesium is 1.9 eV.

Answers to Multiple Choice

1. a	11. d	21. c	31. a, b, c
2. b	12. c	22. b	32. c
3. d	13. d	23. b	33. d
4. d	14. d	24. a	34. c
5. d	15. d	25. a	35. c
6. a	16. c	26. c	36. c
7. c	17. c	27. d	37. a
8. b	18. c	28. c	38. c
9. d	19. b	29. a, b	39. a
10. b	20. c	30. c	40. d

ATOMS IN COMBINATION

26.1 The Hydrogen Molecule

26.2 Covalent Bond

26.3 Solids and Liquids

26.4 Covalent and Ionic Solids

26.5 Van der Waals Bonds

26.6 Metallic Bond

26.7 Energy Bands

26.8 Impurity Semiconductors

26.9 Semiconductor Devices

Individual atoms are rare on the earth and in its lower atmosphere; only the inert gas atoms (such as those of helium and argon) occur by themselves. All other atoms are found linked in small groups called molecules or in larger ones as liquids or solids. Some of these groups consist of atoms of the same element; others consist of atoms of different elements. In every case the arrangement is favored over separate atoms because of interactions between the atoms that reduce the total energy of the system. The quantum theory of the atom accounts for these interactions in a natural way, with no special assumptions, which is further testimony to the power of this approach.

26.1 THE HYDROGEN MOLECULE

Electrical forces hold atoms together to form molecules.

A **molecule** is a group of atoms that stick together strongly enough to act as a single particle. A molecule always has a certain definite structure. Hydrogen molecules always have two hydrogen atoms each, for instance, and water molecules always consist of one oxygen atom and two hydrogen atoms each. A piece of iron is not a molecule, because even though its atoms stay together, any number of them do so to form an object of any size or shape.

A molecule of a given kind is complete in itself with little tendency to gain or lose atoms. If one of its atoms is somehow removed or another atom is somehow attached, the result is a molecule of a different kind with different properties. A liquid or a solid, on the other hand, can gain or lose additional atoms of the kinds already there without changing its character.

Atoms are bound together to form molecules by sharing electrons. An example is the hydrogen molecule, H_2, which consists of two hydrogen atoms. In this molecule the two protons are 7.42×10^{-11} m apart, and the two electrons, one from each atom, belong to the entire molecule rather than to their parent protons. Because the electrons spend more time on the average between the protons than they do on the outside, there is effectively a net negative charge between the protons (Fig. 26.1).

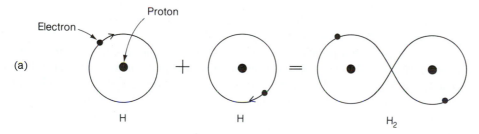

(a)

Figure 26.1

(a) Orbit model of the hydrogen molecule. (b) Probability-cloud model of the hydrogen molecule. In both models the shared electrons spend more time on the average between the protons, which leads to an attractive force.

(b)

The attractive force of this charge on the protons is more than enough to counterbalance the repulsion between them. If the protons are too close together, however, their repulsion becomes dominant.

The balance between attractive and repulsive forces occurs at a separation of 7.42×10^{-11} m, where the total energy of the H_2 molecule is -4.5 eV. Hence 4.5 eV of work must be done to break an H_2 molecule into two H atoms:

$$H_2 + 4.5\,\text{eV} \rightarrow H + H$$

By comparison, the binding energy of the hydrogen atom is 13.6 eV:

$$H + 13.6\,\text{eV} \rightarrow p^+ + e^-$$

This is an example of the general rule that it is easier to break up a molecule than to break up an atom.

Molecules and the Exclusion Principle

Why do only two hydrogen atoms join together to form a molecule? Why not three, or four, or a hundred? The basic reason for the limited size of molecules is the exclusion principle, which does not allow more than one electron in an atomic system to have the same set of quantum numbers. If a third H atom were to be brought up to an H_2 molecule, its electron would have to leave the $n = 1$ shell and go to the $n = 2$ shell in order for an H_3 molecule to be formed, since only two electrons can occupy the $n = 1$ shell (see Fig. 25.25). But an $n = 2$ electron in the hydrogen atom has over 10 eV more energy than an $n = 1$ electron, and the binding energy in H_2 is only 4.5 eV. Hence an H_3 molecule, if it could somehow be put together, would immediately break apart into $H_2 + H$ with the release of energy. Similar considerations hold for other molecules.

The exclusion principle is also responsible for the fact that the inert gases do not occur as molecules. The electron shells of the inert gas atoms are all filled to capacity, so in order for two of them to share an electron pair, one of the electrons would have to go into an empty shell of higher energy. The increase in energy involved would be more than the energy decrease produced by sharing the electrons, and therefore no such molecules as He_2 or Ar_2 occur.

26.2 COVALENT BOND

An electron pair shared by two atoms.

The mechanism by which electron sharing holds atoms together to form molecules is known as **covalent bonding.** We can think of the atoms as being held together by **covalent bonds,** with each shared pair of electrons making up a bond.

More complex atoms than hydrogen also join together to form molecules by sharing electrons. Depending on the electron structures of the atoms involved, there may be one, two, or three covalent bonds—shared electron pairs—between the atoms. For example, in the oxygen molecule O_2 there are two bonds between the O atoms, and in the nitrogen molecule N_2 there are three bonds between the N atoms. Covalent bonds are often represented by a dash for each shared pair of electrons. Thus the H_2, O_2, and N_2 molecules can be represented as follows:

$$H - H \qquad O = O \qquad N \equiv N$$

Covalent bonds are not limited to atoms of the same element or to only two atoms per molecule. Here are two examples of the bonding in more complicated molecules:

$$\text{Water, } H_2O \qquad \begin{matrix} H \\ | \\ O - H \end{matrix} \qquad \text{Ammonia, } NH_3 \quad \begin{matrix} H \\ | \\ N - H \\ | \\ H \end{matrix}$$

We see that oxygen participates in two bonds in H_2O and nitrogen in three bonds in HN_3, whereas H participates in only one bond in both cases. This behavior agrees with the fact that the hydrogen, oxygen, and nitrogen molecules have one, two, and three bonds, respectively, between their atoms.

Carbon atoms tend to form four covalent bonds at the same time, because they have four electrons in their outer shells and these shells lack four electrons to be complete. The structures of the common covalent molecules methane, carbon dioxide, and acetylene illustrate the different bonds in which carbon atoms can participate to form molecules:

$$\begin{matrix} & H & \\ & | & \\ H - &C& - H \\ & | & \\ & H & \end{matrix} \qquad O = C = O \qquad H - C \equiv C - H$$

$$\text{Methane} \qquad \text{Carbon dioxide} \qquad \text{Acetylene}$$

Carbon atoms are so versatile in forming covalent bonds with each other as well as with other atoms that millions of carbon compounds are known. The molecules in some carbon compounds contain tens of thousands of atoms. Such compounds were once thought to originate only in living things, and their study is accordingly known even today as organic chemistry.

In a molecule composed of different atoms, the shared electrons favor one or another of the kinds of atoms. The resulting molecule may have a nonuniform distribution of electric charge, with some parts of the molecule being positively charged and other parts negatively charged. A molecule of this kind is called a **polar**

molecule, whereas a molecule whose charge distribution is symmetric is called a **nonpolar molecule.** Thus the water molecule is highly polar because the two O—H bonds are 104.5° apart and the two shared electron pairs spend more time near the O atom than near the H atoms.

26.3 SOLIDS AND LIQUIDS

Long- and short-range order.

The covalent bonds that can tie a fixed number of atoms together into a molecule can also tie an unlimited number of them together into a solid or a liquid. Other bonding mechanisms are found in solids and liquids as well, one of which is responsible for the ability of metals to conduct electric current. Although only a tiny part of the universe is in the solid state and even less is in the liquid state, matter in these forms makes up much of the physical world of our experience. Modern technology is to a large extent based on the unique characteristics of various solid materials.

Most solids are crystalline in nature, with their atoms or molecules arranged in regular, repeated patterns. A **crystal** is thus characterized by the presence of **long-range order** in its structure. Every crystal of a given kind, large or small, has the same geometrical form. The word *crystal* suggests salt and sugar grains, mineral samples, and sparkling gem stones. But metals and snowflakes are crystalline, as are graphite (the soft ''lead'' in pencils), the fibers of asbestos, and the clear, flat plates of mica. Clay is composed of tiny crystals that can trap water between them to give an easily shaped material.

Other solids lack the definite arrangements of atoms and molecules so conspicuous in crystals. They can be thought of as liquids whose stiffness is due to an exaggerated viscosity. Examples of such **amorphous** (''without form'') solids are pitch, glass, and many plastics. The structures of amorphous solids exhibit **short-range order** only.

Some substances—for instance, B_2O_3—can exist in either crystalline or amorphous forms. In both cases each boron atom is surrounded by three larger oxygen atoms, which represents a short-range order. In a B_2O_3 crystal a long-range order is also present, as shown in a two-dimensional representation in Fig. 26.2. Amorphous B_2O_3, a glassy material, lacks this additional regularity.

The lack of long-range order in amorphous solids means that the bonds vary in strength. When an amorphous solid is heated, the weakest bonds break at lower temperatures than the others, and the solid softens gradually. In a crystalline solid the bonds break simultaneously, and melting has a sudden onset.

Liquids have more in common with solids than with gases, even though liquids share with gases the ability to flow from place to place. Because the density of a given liquid is never far from that of the same substance in solid form, the

○ Oxygen

• Boron   (a)   (b)

Figure 26.2

Structure of B_2O_3. (a) Amorphous B_2O_3, a glass, has short-range order only in its structure. (b) Crystalline B_2O_3 has long-range order as well.

bonding mechanism must be similar in both cases. When a solid is heated to its melting point, its atoms or molecules pick up enough energy to shift the bonds holding them together so that they form into separate clusters. Not until the liquid is heated to its vaporization point are the atoms or molecules able to break loose completely and form a gas. This interpretation of the liquid state is confirmed by X-ray studies that reveal clusters of atoms or molecules in a liquid (that is, short-range order like that in amorphous solids). Unlike the permanent arrangements in a solid, however, the clusters in a liquid are constantly shifting their arrangements.

The unusual behavior of water near the freezing point that was mentioned in Section 13.2 can be traced to the effect just described. Ice crystals have very open structures because each H_2O molecule can participate in only four bonds with other H_2O molecules (Fig. 26.3). In other solids each atom or molecule may have as many as 12 nearest neighbors. This allows the assemblies to be more compact. Because clusters of molecules are smaller and less stable in the liquid state, water molecules are on the average packed more closely together than are ice molecules, and water has the higher density. Hence ice floats. The density of water increases from 0°C to a maximum at 4°C as large clusters of H_2O molecules break up into smaller ones that occupy less space. Only above 4°C does the normal thermal expansion of a liquid show up as a decreasing density with increasing temperature (see Fig. 13.6).

Figure 26.3

The structure of an ice crystal, showing the open hexagonal arrangement of the H_2O molecules. There is less order in liquid water, which allows the molecules to be closer together on the average than they are in ice. Thus density of ice is less than that of water, and ice floats.

26.4 COVALENT AND IONIC SOLIDS

Electron sharing and electron transfer in crystals.

Molecular bonds are all essentially covalent, but in solids four bonding mechanisms are found: covalent, ionic, van der Waals, and metallic.

Diamond is an example of a covalent crystal. Figure 26.4 shows the array of carbon atoms in diamond, with each carbon atom sharing electron pairs with the four carbon atoms nearest it. All the electrons in the outer shells of the carbon atoms participate in the bonding, and as a result diamond is extremely hard and must be heated to over 3500°C before its crystal structure is disrupted and it melts. Purely covalent crystals are relatively few in number. In addition to diamond, some examples are silicon, germanium, and silicon carbide ("Carborundum"). Like diamond, all are hard and have high melting points.

In covalent bonding, two atoms share one or more pairs of electrons. In **ionic bonding,** one or more electrons from one atom transfer to another atom, producing a positive ion and a negative ion that attract each other. Let us see how ionic bonding functions in the case of NaCl, whose crystals constitute ordinary table salt. We will consider a hypothetical NaCl "molecule" for convenience. Salt crystals are actually aggregates of Na and Cl atoms that, although they do not pair off into individual molecules, do interact through ionic bonds whose nature can be most easily examined in terms of a single Na-Cl unit.

Figure 26.5 shows schematically an electron shifted from a sodium atom, which becomes an Na^+ ion, to a chlorine atom, which becomes a Cl^- ion. The ions then attract each other electrically. The combination is stable because more energy is needed to pull the Na^+ and Cl^- ions apart than will be supplied by the return of the shifted electron from the Cl^- ion to the Na^+ ion. Not all pairs of atoms can interact by electron transfer to form stable combinations. Let us see why Na and Cl can do so.

Figure 26.4

Covalent structure of diamond. Each carbon atom shares electron pairs with four other carbon atoms.

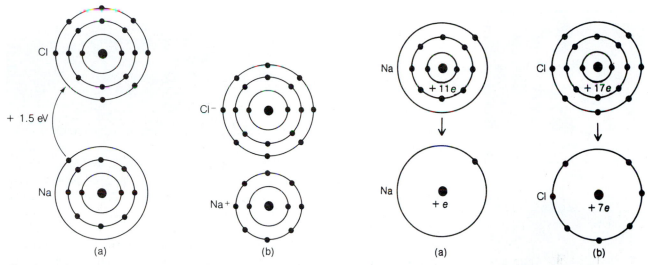

Figure 26.5

(a) Electron transfer in NaCl. (b) The resulting ions attract each other electrically.

Figure 26.6

(a) A sodium atom. The presence of ten electrons in the $n = 1$ and $n = 2$ shells effectively shields the outer $n = 3$ electron from all but $+e$ of the nuclear charge. In consequence an Na atom tends to lose its outer electron to become an Na^+ atom. (b) A chlorine atom. The inner electrons leave unshielded $+7e$ of the nuclear charge. In consequence a Cl atom tends to pick up another electron to become a Cl^- ion.

   The single outer electron of the Na atom is relatively easy to detach because the ten inner electrons shield it from all but $+e$ of the actual nuclear charge of $+11e$ (Fig. 26.6(a)). In the Cl atom, on the other hand, the ten inner electrons leave $+7e$ of the nuclear charge unshielded, so the attractive force on the outer electrons is much greater (Fig. 26.6(b)). Accordingly, a Cl atom tends to pick up an additional electron to become a Cl^- ion. Thus Na and Cl are an ideal match: One readily loses an electron; the other readily gains an electron. The same is true for the partners in other ionic solids, which owe these tendencies to their electron structures.

   The structure of a NaCl crystal is shown in Fig. 26.7. Each ion behaves much like a point charge and thus tends to attract to itself as many ions of opposite

Figure 26.7

An NaCl crystal is composed of Na^+ and Cl^- ions in an array such that each ion is surrounded by six ions of the other kind. Na^+ ions are small because they have lost their outer electrons.

The covalent bonds between its carbon atoms give diamond its characteristic hardness and high melting point. The mass of a diamond gemstone is customarily given in carats, where 1 carat is about 0.2 g. This stone has a mass of 253.7 carats.

The mineral calcite is the chief constituent of limestone and marble. Calcite crystals consist of calcium (Ca^{2+}) and carbonate (CO_3^{2-}) ions held together by ionic bonds.

sign as can fit around it. In a NaCl crystal each Na^+ ion is surrounded by six Cl^- ions, and vice versa. In crystals having different structures, the number of "nearest neighbors" around each ion may be 3, 4, 6, 8, or 12. Ionic bonds are usually fairly strong, and consequently ionic crystals are strong and hard and have high melting points.

Many crystalline bonds are partly ionic and partly covalent in origin. An example of such mixed bonding is quartz, SiO_2.

26.5 VAN DER WAALS BONDS

Weak but everywhere.

A number of molecules and nonmetallic atoms have electron structures that do not lend themselves to either covalent or ionic bonding. In this category are the inert gas atoms, which have filled outer shells, and molecules such as methane (Fig. 26.8), whose outer electrons are fully involved in covalent bonds. Even these virtually noninteracting substances, however, condense into liquids and solids at low enough temperatures through the action of **van der Waals forces,** which were proposed a century ago by the Dutch physicist Johannes van der Waals. Such familiar effects as friction, surface tension, adhesion, and cohesion also arise from these forces.

A type of van der Waals force arises between polar molecules, whose charge distributions are asymmetric. When one such molecule comes near another, the ends of opposite polarity attract each other to hold the molecules together (Fig. 26.9).

A somewhat similar effect occurs when a polar molecule is near a nonpolar molecule. The electric field of the polar molecule distorts the initially symmetric charge distribution of the nonpolar molecule, as in Fig. 26.10, and the two then

Figure 26.8

Model of the methane molecule, CH_4.

Nonpolar
molecule

Figure 26.9

Polar molecules attract each other
electrically.

Figure 26.10

A polar molecule attracts a nonpolar one by first distorting the
latter's originally symmetric
charge distribution.

attract each other in the same way as any other pair of polar molecules. This
phenomenon is like that involved in the attraction of bits of paper by a charged rubber
comb shown in Fig. 16.7.

Nonpolar molecules attract one another in much the same way that polar
molecules attract nonpolar molecules. In a nonpolar molecule the electrons are distributed symmetrically *on the average*. However, *at any moment* one part of the
molecule contains more electrons than usual and the rest of the molecule contains
fewer. Thus every molecule (and atom) behaves as though it were polar, though the
direction and magnitude of the polarization vary constantly. The fluctuations in the
charge distributions of nearby nonpolar molecules keep in step through the action of
electrical forces, and these forces also hold the molecules together (Fig. 26.11).

In general, van der Waals bonds are considerably weaker than ionic, covalent,
and metallic bonds. Usually less than 1% as much energy is needed to remove an
atom or a molecule from a van der Waals solid as is required in the case of ionic or
covalent crystals. As a result the inert gases and compounds with symmetric molecules
liquefy and vaporize at rather low temperatures. For example, the boiling point of
argon is $-186°C$, and the boiling point of methane is $-161°C$. Molecular solids
such as ordinary ice and dry ice (solid CO_2), which consist of whole molecules rather
than of atoms or ions, generally have little mechanical strength.

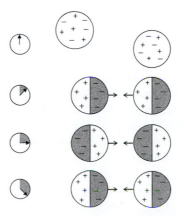

Figure 26.11

Nonpolar molecules have, on the
average, symmetric charge distributions, but at any instant the distributions are not necessarily symmetric. The fluctuations in the
charge distributions of adjacent
nonpolar molecules keep in step,
which leads to an attractive force
between them.

The water molecules in a snow-flake are held together by van der Waals bonds. The characteristic six-fold symmetry of ice crystals is the result of the way electric charges in water molecules are arranged.

26.6 METALLIC BOND

The electron "gas" that bonds metals also makes them good conductors.

Metal atoms have only a few electrons in their outer shells (see Fig. 25.25), and these electrons are not very securely attached, as we saw in the discussion of ionic bonding. When metal atoms come together to form a solid, their outer electrons are given up to a common "gas" of electrons that move relatively freely through the resulting assembly of metal ions. The negatively charged electron gas acts to hold together the positively charged metal ions.

The electron-gas picture of the metallic bond accounts nicely for the properties of metals. A metal conducts heat and electricity well because of the ease with which the free electrons can move about. (In other kinds of solids, all the electrons are bound to particular atoms or pairs of atoms.) Because the free electrons respond readily to electromagnetic waves, metals are opaque to light and have reflective surfaces. Also, because the atoms in a metal are not linked by specific bonds, the properties of alloys—mixtures of different metals—do not depend critically on the relative proportions of each kind of atom, provided their sizes are similar. The copper and tin in bronze, for instance, need not be present in an exact ratio. In sodium chloride, on the other hand, the ratio between sodium and chlorine is always the same.

Table 26.1 summarizes the characteristics of the four basic kinds of crystalline solids.

Table 26.1 Types of crystalline solids. The cohesive energy E_c is the work needed to remove an atom (or molecule) from the crystal and so indicates the strength of the bonds holding it in place.

Type	Covalent	Ionic	Molecular	Metallic
Lattice	Shared electrons	Negative ion / Positive ion	Instantaneous charge separation in molecule	Metal ion / Electron gas
Bond	Shared electrons	Electric attraction	Van der Waals forces	Electron gas
Properties	Very hard; high melting point; soluble in very few liquids	Hard; high melting point; may be soluble in polar liquid such as water	Soft; low melting and boiling points; soluble in covalent liquids	Ductile; metallic luster; ability to conduct heat and electric current readily
Example	Diamond, C $E_c = 7.4$ eV/atom	Sodium chloride, NaCl $E_c = 3.3$ eV/atom	Methane, CH_4 $E_c = 0.1$ eV/molecule	Sodium, Na $E_c = 1.1$ eV/atom

26.7 ENERGY BANDS

Its energy band structure determines whether a solid will be a conductor, an insulator, or a semiconductor.

The notion of energy bands provides a useful framework for understanding the electrical behavior of solids. The nature and properties of semiconductors in particular are clarified with the help of an energy-band analysis.

When atoms are brought as close together as those in a crystal, they interact with one another to such an extent that their outer electron shells constitute a single system of electrons common to the entire array of atoms. The exclusion principle forbids more than two electrons (one with each spin) in any energy level of a system. This principle is obeyed in a crystal because the energy levels of the outer electron shells of the various atoms are all slightly altered by their mutual interaction. (The inner shells do not interact and therefore do not change.) As a result of the shifts in the energy levels, an **energy band** exists in a crystal in place of each sharply defined

Figure 26.12

Energy bands replace energy levels of outer electrons in an assembly of atoms that are close together.

Figure 26.13

(a) Overlapping energy bands. (b) The gap between energy bands that do not overlap is called a forbidden band.

Figure 26.14

Energy levels and bands in solid sodium, which is a metal. (Not to scale.)

energy level of its component atoms (Fig. 26.12). Although these bands are actually a multitude of individual energy levels, as many as there are atoms in the crystal, the levels are so near one another as to form a nearly continuous distribution.

The energy bands in a crystal correspond to energy levels in an atom, and an electron in a crystal can only have an energy that falls within one of these bands. The various energy bands in a crystal may or may not overlap, depending on the composition of the crystal (Fig. 26.13). If they do not overlap, the gaps between them represent energies that electrons in the crystal cannot have. The gaps are accordingly known as **forbidden bands.**

The energy bands contain all the energies that an outer electron in a crystalline solid can have. The electrical properties of such a solid depend both on its energy-band structure and on the way in which the bands are normally occupied by electrons.

Figure 26.14 shows the energy bands of solid sodium. A sodium atom has only one electron in its outer shell. This means that the upper energy band in a sodium crystal is only half filled with electrons, because each level within the band, like each level in the atom, is capable of containing *two* electrons. When an electric field is set up in a sodium crystal, electrons readily pick up the small additional energy they need to move up in their energy band. The additional energy is in the form of kinetic energy, and the moving electrons constitute an electric current. Sodium is therefore a good conductor, as are other crystalline solids with energy bands that are only partly filled. Such solids are metals.

Figure 26.15 shows the energy-band structure of diamond. The two lower energy bands are completely filled, and there is a gap of 6 eV between the top of the

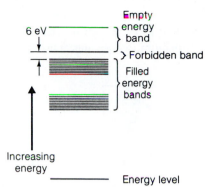

Figure 26.15

Energy levels and bands in diamond, which is an electrical insulator. (Not to scale.)

higher of these bands and the empty band above it. Hence a minimum of 6 eV of additional energy must be given to an electron in a diamond crystal if it is to be capable of free motion, because it cannot have an energy lying in the forbidden band. Such an energy boost cannot be readily given to an electron in a crystal by an electric field. Diamond, like other solids with similar energy-band structures, is therefore an electrical insulator.

Silicon has a crystal structure similar to that of diamond, and, as in diamond, a gap separates the top of a filled energy band from an empty higher band. However, whereas the gap is 6 eV wide in diamond, it is only 1.1 eV wide in silicon. At very low temperatures silicon is hardly better than diamond as a conductor, but at room temperature a few of its electrons have enough kinetic energy of thermal origin to exist in the higher band. These few electrons are enough to permit a small amount of current to flow when an electric field is applied. Thus silicon has a resistivity intermediate between those of conductors (such as sodium) and those of insulators (such as diamond), and is classified as a **semiconductor.**

Optical Properties of Solids

The optical properties of solids and their energy-level structures are closely related. Photons of visible light have energies from about 1 to 3 eV. A free electron in a metal can readily absorb such a photon because its allowed energy band is only partly filled. Metals are accordingly opaque, as mentioned before. The characteristic luster of a metal is due to the reradiation of light absorbed by its free electrons. If the metal surface is smooth, the reradiated light appears as a reflection of the original incident light.

For an electron in an insulator to absorb a photon, on the other hand, the photon energy must be more than 3 eV if the electron is to jump across the forbidden band to the next allowed band. Insulators therefore cannot absorb photons of visible light and are transparent. To be sure, most samples of insulating materials do not appear transparent, but this is usually due to the scattering of light by irregularities in their structures. Insulators are opaque to ultraviolet light, whose higher frequencies mean high enough photon energies to allow electrons to cross the forbidden band.

Because the forbidden bands in semiconductors are comparable in width to the photon energies of visible light, they are usually opaque to visible light. However, they are transparent to infrared light whose lower frequencies mean photon energies too low to be absorbed. Thus infrared lenses can be made from the semiconductor germanium, whose appearance is that of a solid metal.

26.8 IMPURITY SEMICONDUCTORS

Electrons carry current in n-type semiconductors; holes carry current in p-type semiconductors.

Small amounts of impurity can have a marked effect on the conductivity of a semiconductor. Suppose we mix a little arsenic with silicon. Silicon atoms have four electrons in their outer shells; arsenic atoms have five. When an arsenic atom replaces a silicon atom in a silicon crystal, four of its outer electrons join in covalent bonds with its nearest neighbors. The extra electron needs very little energy to be detached and move about freely in the crystal. Such a solid is called an ***n*-type** semiconductor because electric current in it is carried by the motion of negative charges (Fig. 26.16). In an energy-band diagram, as in Fig. 26.17, the effect of

Figure 26.16

Current in an *n*-type semiconductor is carried by surplus electrons that do not fit into the electron-bond structure of the crystal.

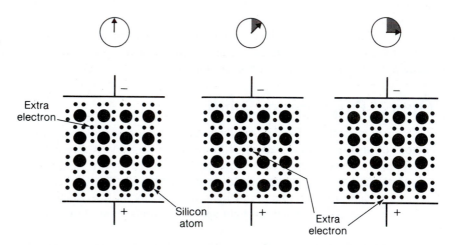

Figure 26.17

Donor levels due to arsenic atoms in a silicon crystal.

arsenic as an impurity is to supply occupied energy levels just below an empty energy band. These levels are called **donor levels.**

If we add gallium atoms to a silicon crystal, a different effect occurs. Gallium atoms have only three electrons in their outer shells, and their presence leaves vacancies called **holes** in the electron structure of the crystal. An electron requires little energy to move into a hole, but as it does so, it leaves a new hole behind. When an electric field is applied to a silicon crystal that has gallium as an impurity, electrons move toward the positive electrode by successively filling holes. The flow of current here is best described in terms of the motion of the holes, which behave as though they were positive charges since they move toward the negative electrode (Fig. 26.18). A material of this kind is therefore called a **p-type** semiconductor. In the energy-band diagram of Fig. 26.19 we see that the effect of gallium as an impurity is to provide energy levels, called **acceptor levels,** just above the highest filled band. Electrons that enter these levels leave behind unoccupied levels in the formerly filled band that allow the conduction of current.

Adding an impurity to a semiconductor is called **doping.** Phosphorus, antimony, and bismuth as well as arsenic have atoms with five outer electrons and so can be used as donor impurities in doping silicon and germanium to yield an *n*-type semiconductor. Similarly indium and tellurium as well as gallium have atoms with three outer electrons and so can be used as acceptor impurities. A tiny amount of impurity can lead to a dramatic change in the conductivity of a semiconductor. For

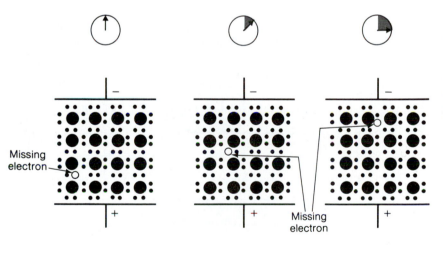

Figure 26.18

Current in a *p*-type semiconductor is carried by the motion of "holes," which are sites of missing electrons. Holes move toward the negative electrode as a succession of electrons move into them.

Figure 26.19

Acceptor levels due to gallium atoms in a silicon crystal.

instance, 1 part of a donor impurity per 10^8 parts of germanium increases its conductivity by a factor of 12. Silicon and germanium are not the only semiconducting materials with practical applications. Another important class of semiconductors consists of such compounds as GaAs, GaP, InSb, InAs, and InP.

26.9 SEMICONDUCTOR DEVICES

The properties of the p-n *junction are responsible for the microelectronics industry.*

The operation of most of the semiconductor devices that have revolutionized modern electronics is based on the nature of junctions between *p*- and *n*-type materials. Such junctions can be made in several ways. A method widely used to manufacture integrated circuits involves diffusing impurities in vapor form into a semiconductor wafer in regions defined by masks. A series of diffusion steps using donor and acceptor impurities is part of the procedure that results in circuits with as many as 2 million resistors, capacitors, diodes, and transistors on a chip 5 mm square.

Diodes

A characteristic property of a *p-n* semiconductor junction is that electric current can pass through it much more readily in one direction than in the other. In the crystal shown in Fig. 26.20, the left-hand end is a *p*-type region (current carried

Figure 26.20

(a) Current is carried in a *p*-type semiconductor by the motion of holes and in an *n*-type semiconductor by the motion of electrons. (b) Reverse bias, little current flows. (c) Forward bias, much current flows.

by holes) and the right-hand end is an *n*-type region (current carried by electrons). When a voltage is applied across the crystal so that the *p*-end is negative and the *n*-end positive, the holes in the *p*-region migrate to the left and the electrons in the *n*-region migrate to the right. Only a limited number of holes and electrons are in the respective regions, and new ones appear spontaneously at only a very slow rate. The current through the entire crystal is therefore negligible. This case is called **reverse bias.**

Figure 26.20(c) shows the same crystal with the connections changed so that the *p*-end is positive and the *n*-end negative, which is **forward bias.** Now new holes are created all the time by the removal of electrons at the *p*-end while new electrons are fed into the *n*-end of the crystal. The holes migrate to the right and the electrons to the left to produce a net positive current flowing from + to −. The electrons and holes meet at the junction between the *p*- and *n*-regions and recombine there: A hole is a missing electron, and when electrons and holes come together, they disappear into the regular structure of the crystal and can no longer act as current carriers. Thus current can flow readily through a semiconductor junction from the *p*- to the *n*-region, but hardly at all in the opposite direction. Figure 26.21(a) shows how *I* varies with *V* at a *p-n* junction. Devices using such junctions, called **diodes,** are widely employed.

Although the reverse current in many semiconductor diodes remains very small, in certain diodes the reverse current rises sharply when a particular voltage is reached, as in Fig. 26.21(b). Such diodes are called **Zener diodes** and are particularly useful in voltage regulation circuits.

One effect that contributes to the sudden increase in reverse current in a Zener diode is the pulling away of electrons on the *p* side of the junction to the *n* side by the reverse voltage. Another is the acceleration of electrons near the junction by the electric field there, which gives them enough energy to knock out other electrons from atoms in their paths and thus create new electron-hole pairs. The voltage at which Zener breakdown occurs depends on the construction of the diode and can be as little as 6 V. Figure 26.22 shows a simple voltage regulation circuit. No current passes through the diode until the breakdown voltage is reached. After this, the diode continues to maintain the breakdown voltage across its terminals, so the output voltage has this value even if the input voltage varies.

Energy is needed to create an electron-hole pair, and this energy is released when an electron and a hole recombine. In silicon and germanium the recombination energy is absorbed by the crystal as heat, but in certain other semiconductors, notably gallium arsenide, a photon is emitted when recombination occurs. This is the basis of the **light-emitting diode.** Solid-state lasers have also been made that utilize this phenomenon.

(a)

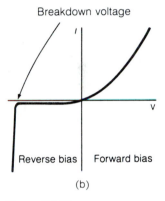

(b)

Figure 26.21

(a) The relationship between voltage and current in an ordinary semiconductor diode. The reverse current is extremely small. (b) Voltage-current relationship in a Zener diode. The reverse current increases sharply at a particular voltage.

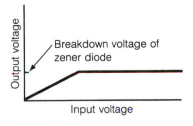

Figure 26.22

A simple voltage-regulator circuit using a Zener diode. The resistor is needed to prevent too great a current in the diode in the breakdown region.

Transistors

Transistors are semiconductor devices that can amplify weak signals into strong ones. Figure 26.23 shows an *n-p-n* junction transistor, which consists of a *p*-type material sandwiched between *n*-type materials. The *p*-type region is called the **base,** and the two *n*-type regions are the **emitter** and the **collector.**

In the figure the transistor is connected as a simple amplifier. There is a forward bias across the emitter-base junction in this circuit, so electrons pass readily from the emitter to the base. Depending on the rate at which holes are produced in the base, which in turn depends on its potential relative to the emitter and hence on the signal input, a certain proportion of the electrons from the emitter will recombine there. The rest of the electrons migrate across the base to the collector to complete the emitter-collector circuit. If the base is at a high positive potential relative to the emitter, many holes are produced in the base to recombine with electrons from the emitter. Little current can then flow through the transistor. If the relative positive potential of the base is low, the electrons arriving from the emitter will outnumber the holes being formed in the base, and the surplus electrons will continue across the base to the collector. Changes in the input-circuit current are thus mirrored by changes in the output-circuit current, which is only a few percent smaller.

The ability of the transistor in Fig. 26.23 to produce amplification comes about because the reverse bias across the base-collector junction permits a much higher voltage in the output circuit than that in the input circuit. Because electric power = (current)(voltage), the power of the output signal can greatly exceed the power of the input signal.

One problem with the junction transistor is that it is difficult to incorporate large numbers of them in an integrated circuit. The **field-effect transistor** (FET) lacks this disadvantage and is widely used today. As in Fig. 26.24, an *n*-channel field-effect transistor consists of a strip of *n*-type material with contacts at each end together with a strip of *p*-type material, called the **gate,** on one side. When connected as shown, electrons move from the **source** terminal to the **drain** terminal through the *n*-type channel. The *p-n* junction is given a reverse bias, and as a result both the *n* and *p* materials near the junction are depleted of charge carriers (see Fig. 26.20(b)).

Figure 26.23

A simple transistor amplifier.

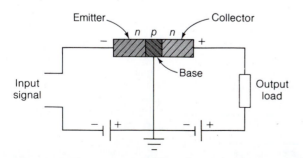

Figure 26.24

A field-effect transistor.

This silicon chip, about 6 mm square, is a microprocessor designed for telephone applications that contains tens of thousands of resistors, capacitors, diodes, and transistors.

The higher the reverse potential on the gate, the larger the depleted region in the channel, and the fewer the electrons available to carry the current. Thus the gate voltage controls the channel current.

In a metal oxide semiconductor field-effect transistor (MOSFET), the semiconductor gate is replaced by a metal film separated from the channel by an insulating layer of silicon dioxide. The metal film is thus capacitively coupled to the channel, and its potential controls the drain current through the number of induced charges in the channel. A MOSFET is easier to make than a FET and occupies only a few percent of the area needed for a junction transistor.

Important Terms

A **molecule** is a group of atoms that stick together strongly enough to act as a single particle. A molecule of a given compound always has a certain definite structure and is complete in itself, with little tendency to gain or lose atoms.

In a **covalent bond** between atoms in a molecule or a solid, the atoms share one or more electron pairs.

Solids whose constituent atoms or molecules are arranged in regular, repeated patterns are called **crystalline.** When only short-range order is present, the solid is **amorphous.**

In an **ionic bond,** electrons are transferred from one atom to another, and the two then attract each other electrically.

Van der Waals forces arise from the electric attraction between asymmetric charge distributions in atoms and molecules.

The **metallic bond** that holds metal atoms together in the solid state arises from a ''gas'' of freely moving electrons that pervades the entire metal.

Because the atoms in a crystal are so close together, the energy levels of their outer electron shells are altered slightly to produce **energy bands** characteristic of the entire crystal, in place of the individual sharply defined energy levels of the separate atoms. Gaps between energy bands represent energies that electrons in the crystal cannot have and are called **forbidden bands.**

Semiconductors are intermediate between conductors and insulators in their ability to carry electric current. An **n-type semiconductor** is one in which electric current is carried by the motion of electrons. A **p-type semiconductor** is one in which electric current is carried by the motion of **holes,** which are vacancies in the electron structure that behave like positive charges.

A semiconductor **diode** permits current to flow through it in only one direction. Signals can be amplified by means of a **transistor,** another semiconductor device.

MULTIPLE CHOICE

1. Relative to the energy needed to separate the electrons of an atom from its nucleus, the energy needed to separate the atoms of a molecule is
- **a.** smaller
- **b.** about the same
- **c.** larger
- **d.** much larger

2. The hydrogen atoms in a hydrogen molecule stick together because
- **a.** their electrons spend all their time between the nuclei
- **b.** their electrons spend more time between the nuclei than outside them
- **c.** their electrons spend more time outside the nuclei than between them
- **d.** their nuclei attract each other

3. A hydrogen molecule contains only two hydrogen atoms because
- **a.** of the exclusion principle
- **b.** of the uncertainty principle
- **c.** only two such atoms can fit together
- **d.** with more atoms, the electrons would collide with one another

4. In a molecule other than H_2 that consists of two atoms,
- **a.** only a single pair of electrons can be shared
- **b.** more than one pair of electrons can be shared
- **c.** one of the atoms must be a hydrogen atom
- **d.** one of the atoms must be a carbon atom

5. In a covalent bond,
- **a.** electrons are shifted from one atom to another
- **b.** only atoms of the same element are present
- **c.** there must be at least one carbon atom
- **d.** adjacent atoms share electron pairs

6. Covalent bonds never involve the sharing of
- **a.** 2 electrons
- **b.** 3 electrons
- **c.** 4 electrons
- **d.** 6 electrons

7. The number of covalent bonds a hydrogen atom forms when it combines chemically is
- **a.** 1
- **b.** 2
- **c.** 3
- **d.** 4

8. The number of covalent bonds a carbon atom usually forms when it combines chemically is
- **a.** 1
- **b.** 2
- **c.** 3
- **d.** 4

9. When two or more atoms join to form a stable molecule,
- **a.** energy is absorbed
- **b.** energy is given off
- **c.** there is no energy change
- **d.** any of the above, depending on the circumstances

10. Short-range order is never found in
- **a.** crystalline solids
- **b.** amorphous solids
- **c.** liquids
- **d.** gases

11. A crystalline solid always
- **a.** is transparent
- **b.** is held together by covalent bonds
- **c.** is held together by ionic bonds
- **d.** has long-range order in its structure

12. An amorphous solid is closest in structure to
- **a.** a covalent crystal
- **b.** an ionic crystal
- **c.** a semiconductor
- **d.** a liquid

13. An amorphous solid
- **a.** has its particles arranged in a regular pattern
- **b.** is held together by ionic bonds
- **c.** does not melt at a definite temperature but softens gradually
- **d.** consists of nonpolar molecules

14. The lowest melting points are usually found in solids held together by
- **a.** covalent bonds
- **b.** ionic bonds
- **c.** metallic bonds
- **d.** van der Waals bonds

15. Van der Waals forces arise from
- **a.** electron transfer
- **b.** electron sharing
- **c.** symmetric charge distributions
- **d.** asymmetric charge distributions

16. A polar molecule can attract
   a.   only ions
   b.   only other polar molecules
   c.   only nonpolar molecules
   d.   all of the above

17. The particles that make up the lattice of the van der Waals crystal of a compound·are
   a.   electrons   b.   atoms
   c.   ions   d.   molecules

18. The particles that make up the lattice of an ionic crystal are
   a.   electrons   b.   atoms
   c.   ions   d.   molecules

19. The particles that make up the lattice of a covalent crystal are
   a.   electrons   b.   atoms
   c.   ions   d.   molecules

20. A property of metals that is not due to the electron "gas" that pervades them is their unusual ability to
   a.   conduct electricity   b.   conduct heat
   c.   reflect light   d.   form oxides

21. Ice is an example of
   a.   a covalent solid   b.   an ionic solid
   c.   a metallic solid   d.   a molecular solid

22. Diamond is an example of
   a.   a covalent solid   b.   an ionic solid
   c.   a metallic solid   d.   a molecular solid

23. A crystal whose upper energy band is partly occupied by electrons is
   a.   a conductor
   b.   an insulator
   c.   an n-type semiconductor
   d.   a p-type semiconductor

24. A crystal with a wide forbidden band between a filled lower band and an empty upper band is
   a.   a conductor
   b.   an insulator
   c.   an n-type semiconductor
   d.   a p-type semiconductor

25. A solid with occupied donor levels below an empty energy band is
   a.   an n-type semiconductor
   b.   a p-type semiconductor
   c.   a transistor
   d.   a diode

26. Diamond is transparent because
   a.   its atoms are far apart
   b.   it is held together by covalent bonds
   c.   the forbidden band above its highest filled energy band is narrow relative to the photon energies in visible light
   d.   the forbidden band above its highest filled energy band is wide relative to the photon energies in visible light

27. A hole in a p-type semiconductor is
   a.   a surplus electron
   b.   a missing electron
   c.   a missing atom
   d.   a donor level

28. Current in a p-type semiconductor is carried by
   a.   electrons   b.   holes
   c.   positive ions   d.   negative ions

29. Current in an n-type semiconductor is carried by
   a.   electrons   b.   holes
   c.   positive ions   d.   negative ions

30. A junction between n- and p type semiconductors
   a.   acts as an insulator for currents in both directions
   b.   conducts current in both directions equally well
   c.   conducts current readily when the p-end is positive and the n-end is negative
   d.   conducts current readily when the p-end is negative and the n-end is positive

31. A light-emitting diode gives off a photon of light when
   a.   two holes collide
   b.   two electrons collide
   c.   a hole is created by the freeing of an electron
   d.   a hole is filled by recombination with an electron

QUESTIONS

1. What is wrong with the model of a hydrogen molecule in which the two electrons are supposed to follow figure-eight orbits that encircle the two protons?

2. What must be true of the spins of the two electrons in the H_2 molecule?

3. The energy needed to detach the electron from a hydrogen atom is 13.6 eV, but the energy needed to detach an electron from a hydrogen molecule is 15.7 eV. Why do you think the latter energy is greater?

4. Why do the inert gas atoms almost never participate in covalent bonds?

5. What property of carbon atoms allows them to form so many varied and complex molecules?

6. The atoms in a molecule are said to share electrons, yet some molecules are polar. Explain.

7. Why are electrons much more readily liberated from lithium when it is irradiated with ultraviolet light than from fluorine?

8. Why are Cl atoms more active chemically than Cl^- ions?

9. Why are Na atoms more active chemically than Na^+ ions?

10. The separation between Na^+ and Cl^- ions in a NaCl crystal is 2.8×10^{-10} m, whereas it is 2.4×10^{-10} m in a NaCl molecule such as might exist in the gaseous state. Why is it reasonable for these separations to be different?

11. Van der Waals forces can hold inert gas atoms together to form solids, but they cannot hold such atoms together to form molecules in the gaseous state. Why not?

12. The temperature of a gas falls when it passes slowly from a full container to an empty one through a porous plug. Since the expansion is into a rigid container, no mechanical work is done. What is the origin of the fall in temperature?

13. Lithium atoms, like hydrogen atoms, have only a single electron in their outer shells, yet lithium atoms do not join together to form Li_2 molecules the way hydrogen atoms form H_2 molecules. Instead, lithium is a metal with each atom part of a crystal structure. Why?

14. What is the connection between the ability of a metal to conduct electricity and its ability to conduct heat?

15. Does the ''gas'' of freely moving electrons in a metal include all the electrons present? If not, which electrons are members of the ''gas''?

16. What is the basic reason that energy bands rather than specific energy levels exist in a solid?

17. What kind of solid is one whose upper energy band is partly filled with electrons?

18. Why are some solids transparent to visible light and others opaque?

19. Semiconductors are usually opaque to visible light but transparent to infrared light. Why is this so?

20. The energy gap in silicon is 1.1 eV, and in diamond it is 6 eV. How transparent would you expect these substances to be to visible light?

21. How does the energy-band structure of a solid determine whether it is a conductor, a semiconductor, or an insulator of electricity?

22. The forbidden energy band in germanium that lies between the highest filled band and the empty band above it has a width of 0.7 eV. Compare the conductivity of germanium with that of silicon, in which the gap is 1.1 eV, at (a) very low temperatures and (b) room temperature.

23. What is the distinction between n- and p-type semiconductors?

24. Does the addition of a small amount of indium to germanium result in the formation of an n- or a p-type semiconductor? An indium atom has three electrons in its outermost shell; a germanium atom has four.

25. At a p-n junction which direction of current represents forward bias and which reverse bias? In which direction does current flow more readily?

26. Is energy absorbed or given off when an electron and a hole recombine? If energy is absorbed, where does it come from? If energy is given off, where does it go?

Answers to Multiple Choice

1. a	9. b	17. d	25. a
2. b	10. d	18. c	26. d
3. a	11. d	19. b	27. b
4. b	12. d	20. d	28. b
5. d	13. c	21. d	29. a
6. b	14. d	22. a	30. c
7. a	15. d	23. a	31. d
8. d	16. d	24. b	

27

THE NUCLEUS

27.1   Nuclear Composition

27.2   Binding Energy

27.3   Radioactive Decay

27.4   Detecting Radiation

27.5   Half-Life

27.6   Radiometric Dating

27.7   Radiation Hazards

27.8   Fission

27.9   Nuclear Reactors

27.10   Nuclear Fusion

The behavior of atomic electrons is responsible for the chief properties (except mass) of atoms, molecules, solids, and liquids. For this reason, until now we have not had to know more about the nucleus of an atom other than that it is a tiny, positively charged bit of matter that provides the atom with most of its mass and holds its electrons in place. The nucleus nevertheless turns out to be of supreme importance. To begin with, the various elements exist only because nuclei can contain more than one proton despite their mutual repulsion. And the continuing evolution of the universe is powered in part by energy that comes from nuclear reactions and transformations in the sun and other stars.

27.1 NUCLEAR COMPOSITION

Atomic nuclei of an element have the same numbers of protons but may have different numbers of neutrons.

The electron structure of the atom was understood before even the composition of its nucleus was known. The reason is that the nucleus is held together by forces vastly stronger than the electrical forces that hold the electrons to the nucleus, and it is correspondingly harder to break apart a nucleus to find out what is inside. Changes in the electron structure of an atom, such as those that occur in the emission of photons or in the formation of chemical bonds, involve energies of only several eV. Changes in nuclear structure, however, involve energies of several MeV, a million times more.

As we know, the simplest nucleus, that of the hydrogen atom, usually consists of a single proton, whose charge is $+e$ and whose mass is

$$m_{\text{proton}} = 1.673 \times 10^{-27} \text{ kg}$$

The proton mass is 1836 times that of the electron, so nearly all of the hydrogen atom's mass resides in its nucleus. This is true of all other atoms as well.

Elements more complex than hydrogen have nuclei that contain **neutrons** as well as protons. The neutron, as its name suggests, is uncharged. The neutron mass is slightly more than that of the proton:

$$m_{\text{neutron}} = 1.675 \times 10^{-27} \text{ kg}$$

Neutrons and protons are jointly called **nucleons.**

Every neutral atom contains the same number of protons and electrons. This number is the **atomic number** of the element. Thus the atomic number of hydrogen is 1, of helium 2, of lithium 3, and of uranium 92. Atomic numbers of the elements are listed in Appendix C. Except in the case of ordinary hydrogen atoms, the number of neutrons in a nucleus equals or, more often, exceeds the number of protons. The compositions of atoms of the four lightest elements are illustrated in Fig. 27.1.

Figure 27.1

The electronic and nuclear compositions of hydrogen, helium, lithium, and beryllium atoms.

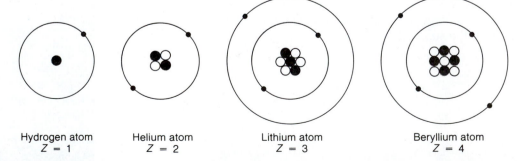

● Proton
○ Neutron
• Electron

Hydrogen atom
$Z = 1$

Helium atom
$Z = 2$

Lithium atom
$Z = 3$

Beryllium atom
$Z = 4$

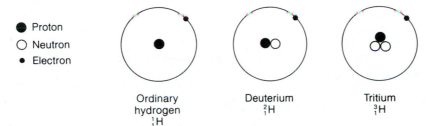

Ordinary hydrogen ^1_1H

Deuterium ^2_1H

Tritium ^3_1H

Figure 27.2

The three isotopes of hydrogen.

Although all atoms of a given element have nuclei with the same numbers of protons, the numbers of neutrons may not be the same. For instance, even though over 99.9% of hydrogen nuclei are just single protons, a few also contain a neutron as well, and a very few two neutrons (Fig. 27.2). The varieties of an element that differ in the numbers of neutrons their nuclei contain are called its **isotopes.** Table 27.1 describes the isotopes of hydrogen and chlorine. The **atomic mass unit** u used in the table has the value

$$1\ u = 1.660 \times 10^{-27}\ kg$$

The hydrogen isotope deuterium is stable, but tritium is radioactive (Section 27.3), and a sample of it gradually changes to an isotope of helium. The flux of cosmic rays from space continually replenishes the earth's tritium by nuclear reactions in the atmosphere. Only about 2 kg of tritium of natural origin is present on the earth's surface, nearly all of it in the oceans.

All elements have isotopes. The atomic number of an element determines how many electrons its atoms have and how they are arranged, which in turn governs the chemical behavior of the element. The isotopes of the element have the same atomic number, so it is not surprising that the two isotopes of chlorine, for instance, have the same yellow color, the same suffocating odor, and the same efficiency as poisons and bleaching agents. Because boiling and freezing points depend somewhat on atomic mass, they are slightly different for the two isotopes, as is their density.

Table 27.1 The isotopes of hydrogen and chlorine found in nature

	Properties of element		Properties of isotope			
Element	Atomic number	Average atomic mass, u	Protons in nucleus	Neutrons in nucleus	Atomic mass, u	Relative abundance
Hydrogen	1	1.008	1	0	1.008	99.985%
			1	1	2.014	0.015%
			1	2	3.016	Very small
Chlorine	17	35.46	17	18	34.97	75.53%
			17	20	36.97	24.47%

Other physical properties of isotopes may vary more drastically with their compositions. An example is the radioactivity of tritium, in contrast to the stability of ordinary hydrogen and deuterium.

The conventional symbols for nuclear species, or **nuclides,** all follow the pattern $^A_Z X$, where

X = chemical symbol of the element

Z = atomic number of the element
= number of protons in the nucleus

A = mass number of the nuclide
= number of protons and neutrons in the nucleus

Hence ordinary hydrogen is designated $^1_1 H$, since its atomic number and mass number are both 1, while tritium is designated $^3_1 H$. The two isotopes of chlorine in Table 27.1 are designated $^{35}_{17} Cl$ and $^{37}_{17} Cl$, respectively.

27.2 BINDING ENERGY

The energy needed to break up a nucleus into its neutrons and protons.

The nucleus of a deuterium atom consists of a proton and a neutron. Thus we would expect that the mass of a deuterium atom, $^2_1 H$, should be equal to the mass of an ordinary hydrogen atom, $^1_1 H$, plus the mass of a neutron. However, it turns out that the mass of $^2_1 H$ is 0.0024 u *less* than the combined masses of a $^1_1 H$ atom and a neutron (Fig. 27.3).

Figure 27.3

The mass of every atom is less than the total of the masses of its constituent neutrons, protons, and electrons. This phenomenon is illustrated here for the deuterium atom.

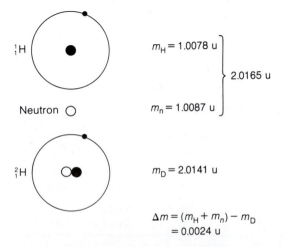

The case of deuterium is an example of a general observation:

Atomic nuclei always have less mass than the combined masses of their constituent particles.

The energy equivalent of the "missing" mass of a nucleus is called its **binding energy.** In order to break up the nucleus, this amount of energy must be supplied either by colliding with another particle or by absorbing a photon. Binding energies arise from the action of the nuclear forces that hold nuclei together, just as ionization energies of atoms, which must be supplied to remove electrons from them, are due to the action of the electrical forces that hold them together.

The usual energy unit in nuclear physics is the MeV, where

$$1 \text{ MeV} = 10^6 \text{ eV} = 1.60 \times 10^{-13} \text{ J}$$

In terms of MeV, the energy equivalent of 1 u of mass is

$$E_\text{u} = 931.5 \text{ MeV} \qquad\qquad \textit{Energy per atomic mass unit}$$

Table 27.2 gives the rest masses and their energy equivalents for several particles. When calculating binding energies, the masses of neutral atoms and of electrons, protons, and neutrons are used. The ionization energies of the electrons, which are very small compared with nuclear binding energies, usually can be neglected.

Since the "missing" mass in a deuteron (as the deuterium nucleus is called) is 0.0024 u, its binding energy is

$$(0.0024 \text{ u})(931.5 \text{ MeV/u}) = 2.2 \text{ MeV}$$

This figure is confirmed by experiments that show that the minimum energy a photon must have in order to disrupt a deuteron is 2.2 MeV (Fig. 27.4).

Nuclear binding energies are strikingly high. The range for stable nuclei is from 2.2 MeV for $^2_1$H (deuterium) to 1640 MeV for $^{209}_{83}$Bi (an isotope of the metal bismuth). Larger nuclei are all unstable and decay radioactively, as described in

Table 27.2   Rest masses and energy equivalents for selected particles

Particle	Rest mass, u	Equivalent energy, MeV
Electron	0.000549	0.511
Proton	1.007277	938.3
Neutron	1.008665	939.6
Hydrogen atom	1.007825	938.8
Deuterium atom	2.014102	1876.1
Helium ($^4_2$He) atom	4.002603	3728.4

Figure 27.4

The binding energy of the deuteron is 2.2 MeV, which is the energy difference between the deuteron's mass and the combined mass of its constituent neutron and proton. Absorbing 2.2 MeV—for instance, by being struck by a 2.2-MeV gamma-ray photon—gives a deuteron enough additional mass to split into a neutron and a proton.

Section 27.3. To appreciate how high binding energies are, we can compare them with more familiar energies in terms of kilojoules of energy per kilogram of mass. In these units a typical binding energy is 8×10^{11} kJ/kg—800 billion kJ/kg. By contrast, to boil water involves a heat of vaporization of a mere 2260 kJ/kg, and even the heat given off by burning gasoline is only 4.7×10^4 kJ/kg, 17 million times smaller.

Binding Energy per Nucleon

The **binding energy per nucleon** of a nucleus is its binding energy divided by the number A of neutrons and protons it contains. Figure 27.5 is a plot of binding energy per nucleon versus mass number A. Except for the high peak for $^4_2$He, the curve is a quite regular one. Nuclei of intermediate size have the highest binding energies per nucleon, which means that their nucleons are held together more securely than the nucleons in both heavier and lighter nuclei. The maximum in the curve is 8.8 MeV/nucleon at $A = 56$, which corresponds to the iron nucleus $^{56}_{26}$Fe.

A remarkable feature of nuclear structure is illustrated by this curve. Suppose that we split the nucleus $^{235}_{92}$U, whose binding energy is 7.6 MeV/nucleon, into two parts. Each fragment will be the nucleus of a much ligher element and therefore will

Figure 27.5

Binding energy per nucleon versus mass number. The higher the binding energy per nucleon, the more stable the nucleus. When a heavy nucleus is split into two lighter ones, a process called *fission*, the greater binding energy of the product nuclei causes the liberation of energy. When two very light nuclei join to form a heavier one, a process called *fusion*, the greater binding energy of the product nucleus again causes the liberation of energy.

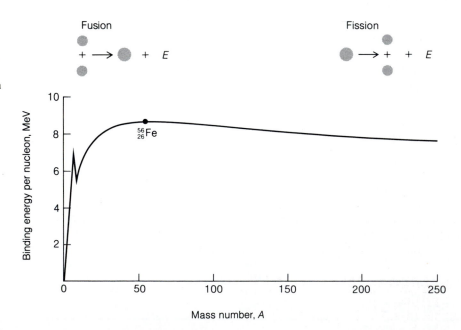

have a higher binding energy per nucleon than the uranium nucleus. The difference is about 0.8 MeV/nucleon, and so, if such **nuclear fission** were to take place, an energy of

$$\left(0.8\,\frac{\text{MeV}}{\text{nucleon}}\right)(235\text{ nucleons}) = 188\text{ MeV}$$

would be given off per splitting. This is a truly immense amount of energy to be produced in a single atomic event. As a comparison, chemical processes involve energies of the order of magnitude of 1 electron volt per reacting atom, 10^{-8} the energy involved in fission.

Figure 27.5 also shows that if two light nuclei join to form a heavier one, the higher binding energy of the resulting nucleus will also result in the evolution of energy. For instance, if two deuterons combined to make a ^4_2He nucleus, more than 23 MeV would be released. This process is known as **fusion,** and, together with fission, it promises to be the source of more and more of the world's energy as reserves of fossil fuels are depleted. Nuclear fusion is the means by which the sun and stars obtain their energy.

The graph of Fig. 27.5 has a good claim to being the most significant in all of science. The fact that binding energy exists at all means that nuclei more complex than the single proton of hydrogen can be stable. Such stability in turn accounts for the existence of the various elements and so for the existence of the many and diverse forms of matter around us. Because the curve peaks in the middle, we have the explanation for the energy that powers, directly or indirectly, the evolution of much of the universe—it comes from the fusion of protons and neutrons to form complex nuclei. And the harnessing of nuclear fission in reactors and weapons has irreversibly changed modern civilization.

27.3 RADIOACTIVE DECAY

How unstable nuclei change into stable ones.

Not all atomic nuclei are stable. At the beginning of the twentieth century research by Henri Becquerel, Marie and Pierre Curie, and others, showed that some nuclei spontaneously transform themselves into other nuclear species with the emission of radiation. Such nuclei are said to be **radioactive.**

Unstable nuclei emit three kinds of radiation:

1. **Alpha particles,** which are the nuclei of ^4_2He atoms;
2. **Beta particles,** which are electrons or positrons (positively charged electrons);
3. **Gamma rays,** which are photons of high energy.

Magnetic field directed into paper

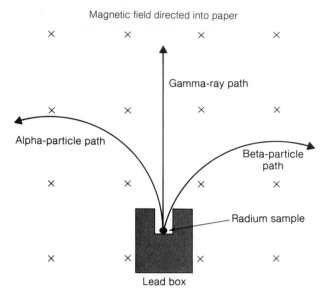

Gamma-ray path

Alpha-particle path

Beta-particle path

Radium sample

Lead box

Figure 27.6

The radiations from a radium sample may be analyzed with the help of a magnetic field. In the figure the direction of the field is into the paper; hence the positively charged alpha particles (which are helium nuclei) are deflected to the left and the negatively charged beta particles (which are electrons), to the right. Gamma rays (which are energetic photons) carry no charge and are not affected by the magnetic field.

Figure 27.7

Alpha particles are stopped by a piece of cardboard. Beta particles penetrate the cardboard but are stopped by a sheet of aluminum. Even a thick slab of lead may not stop all the gamma rays.

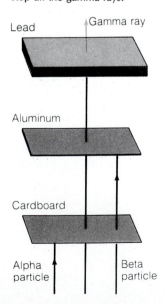

Lead

Gamma ray

Aluminum

Cardboard

Alpha particle

Beta particle

The early experimenters identified these radiations with the help of a magnetic field, as in Fig. 27.6. Gamma rays carry no charge and are not affected by the magnetic field. Figure 27.7 indicates the penetrating powers of the different radiations.

To understand why nuclei decay, let us look at Fig. 27.8, which shows the number of neutrons versus the number of protons in stable nuclei. Clearly an element of a given proton number Z has only a very narrow range of possible neutron numbers if it is to be stable.

Suppose now that a nucleus has too many neutrons for stability relative to the number of protons present. If one of the excess neutrons changes into a proton, this will both reduce the number of neutrons and increase the number of protons (Fig. 27.9). To conserve electric charge, such a change means that a negative electron must be emitted. Therefore

$$n^0 \rightarrow p^+ + e^- \qquad \qquad \textit{Electron emission} \quad (27.1)$$

The electron leaves the nucleus and is detectable as a "beta particle." The final nucleus may be left with extra energy due to its shifted binding energy, and this energy is given off in the form of gamma rays. Sometimes more than one such **beta decay** is needed for a particular unstable nucleus to become stable.

Should the nucleus have too few neutrons, the inverse process

$$p^+ \rightarrow n^0 + e^+ \qquad \qquad \textit{Positron emission} \quad (27.2)$$

may take place. Here a proton becomes a neutron with the emission of a **positron** (positive electron). This is also called beta decay, because it resembles the emission of negative electrons.

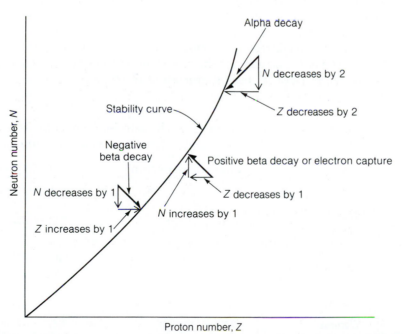

Figure 27.8

The number of neutrons versus the number of protons in stable nuclei. The larger the nucleus, the greater the proportion of neutrons.

Figure 27.9

How alpha and beta decays tend to bring an unstable nucleus to a stable configuration.

Radionuclides can be traced in living tissue by the radiation they emit. In this image of the hands of a person suffering from rheumatoid arthritis different levels of gamma ray intensity are shown by different shades of gray. The higher the intensity, the lighter the shade. The radionuclides injected into the bloodstream of the person were absorbed most readily by the inflamed tissue around the arthritic joints. The left hand and wrist are the most severely affected.

A process that competes with positron emission is the capture by a nucleus with too small a neutron/proton ratio of one of the electrons in its innermost atomic shell. The electron is absorbed by a nuclear proton, which then becomes a neutron. This process can be expressed as

$$p^+ + e^- \rightarrow n^0 \qquad\qquad \textit{Electron capture} \quad (27.3)$$

Electron capture does not lead to the emission of a particle. However, it can be detected by the X-ray photon produced when one of the atom's outer electrons falls into the vacancy left by the absorbed electron.

Another way certain nuclei can achieve stability is by **alpha decay,** in which an alpha particle consisting of two neutrons and two protons is emitted. Thus negative beta decay increases the number of protons by one and decreases the number of neutrons by one; positive beta decay and electron capture decrease the number of protons by one and increase the number of neutrons by one; and alpha decay decreases both the number of protons and the number of neutrons by two. These processes are summarized in Table 27.3 and shown schematically in Fig. 27.10.

Table 27.3   Radioactive decay. The asterisk (*) denotes an excited nuclear state and γ denotes a gamma-ray photon.

Decay	Nuclear transformation	Example
Alpha	$^A_Z X \rightarrow ^{A-4}_{Z-2} Y + ^4_2 He$	$^{238}_{92}U \rightarrow ^{234}_{90}Th + ^4_2 He$
Electron emission	$^A_Z X \rightarrow _{Z+1}^{A} Y + e^-$	$^{14}_6 C \rightarrow ^{14}_7 N + e^-$
Positron emission	$^A_Z X \rightarrow _{Z-1}^{A} Y + e^+$	$^{64}_{29}Cu \rightarrow ^{64}_{28}Ni + e^+$
Electron capture	$^A_Z X + e^- \rightarrow _{Z-1}^{A} Y$	$^{64}_{29}Cu + e^- \rightarrow ^{64}_{28}Ni$
Gamma	$^A_Z X^* \rightarrow ^A_Z X + \gamma$	$^{87}_{38}Sr^* \rightarrow ^{87}_{38}Sr + \gamma$

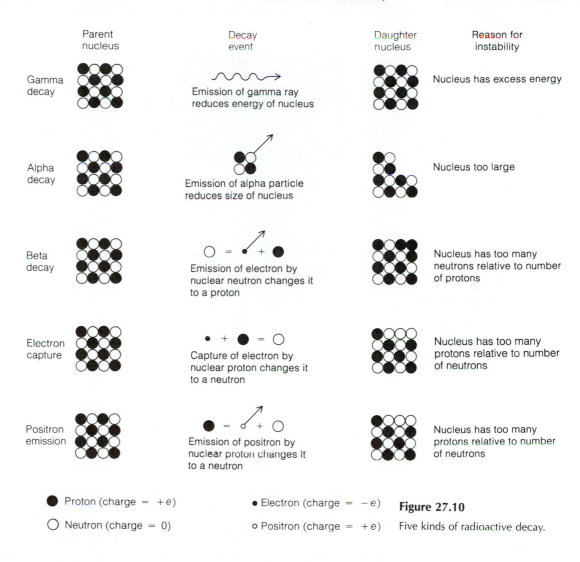

	Parent nucleus	Decay event	Daughter nucleus	Reason for instability
Gamma decay		Emission of gamma ray reduces energy of nucleus		Nucleus has excess energy
Alpha decay		Emission of alpha particle reduces size of nucleus		Nucleus too large
Beta decay		Emission of electron by nuclear neutron changes it to a proton		Nucleus has too many neutrons relative to number of protons
Electron capture		Capture of electron by nuclear proton changes it to a neutron		Nucleus has too many protons relative to number of neutrons
Positron emission		Emission of positron by nuclear proton changes it to a neutron		Nucleus has too many protons relative to number of neutrons

● Proton (charge $= +e$)   • Electron (charge $= -e$)

○ Neutron (charge $= 0$)   ○ Positron (charge $= +e$)

Figure 27.10

Five kinds of radioactive decay.

Why do heavy nuclei give off alpha particles instead of, for example, individual protons or $_2^3$He nuclei? The reason is the high binding energy of the alpha particle. As a result, an alpha particle has significantly less mass than four separate nucleons. Because of this small mass, an alpha particle can be ejected by a heavy nucleus with energy to spare. Thus the alpha particle released in the decay of $_{92}^{232}$U has a kinetic energy of 5.4 MeV, whereas 6.1 MeV would have to be supplied from the outside for this nucleus to release a proton and 9.6 MeV to release a $_2^3$He nucleus.

Often a succession of alpha and beta decays, plus gamma decays to carry off excess energy, is needed before a nucleus reaches stability. An example of such a **decay series** is shown in Fig. 27.11. Here the unstable uranium isotope $_{92}^{238}$U becomes the stable lead isotope $_{82}^{206}$Pb in 14 steps. The series branches at $_{83}^{214}$Bi, which can decay either by alpha emission followed by beta emission or in the reverse order.

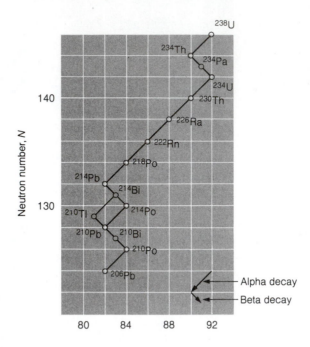

Figure 27.11

The uranium isotope $^{238}_{92}$U undergoes 14 alpha and beta decays to reach the stable lead isotope $^{206}_{82}$Pb. The decay of $^{214}_{83}$Bi can proceed either by alpha emission and then beta emission or in the reverse order.

◆ **E X A M P L E   27.1** _____

The nucleus $^6_2$He is unstable. What kind of decay would you expect it to undergo?

SOLUTION   The most stable helium nucleus is $^4_2$He, as shown by its position on the binding-energy curve of Fig. 27.5 where it is responsible for the peak at the left of the curve. Because $^6_2$He has four neutrons whereas $^4_2$He has only two, the instability of $^6_2$He must arise from an excess of neutrons. Thus it would seem likely that $^6_2$He undergoes negative beta decay to become $^6_3$Li, whose neutron/proton ratio is more consistent with stability:

$$^6_2\text{He} \rightarrow {}^6_3\text{Li} + e^-$$

This is in fact the manner in which $^6_2$He decays.   ◆

◆ **E X A M P L E   27.2** _____

The polonium isotope $^{210}_{84}$Po is unstable and emits a 5.30-MeV alpha particle. The atomic mass of $^{210}_{84}$Po is 209.9829 u and that of $^4_2$He is 4.0026 u. Identify the daughter nuclide and find its atomic mass.

SOLUTION   (a) The daughter nuclide will have an atomic number of $84 - 2 = 82$ and a mass number of $210 - 4 = 206$. From Table C.1 in the Appendix we see that $Z = 82$ corresponds to lead, so the symbol of the daughter nuclide is $^{206}_{82}\text{Pb}$.

(b) The mass equivalent of 5.30 MeV is

$$m_E = \frac{5.30\ \text{MeV}}{931.5\ \text{MeV/u}} = 0.0057\ \text{u}$$

The mass lost by $^{210}_{84}\text{Po}$ in its decay equals the atomic mass of the alpha particle ^4_2He plus m_E, which is $(4.0026 + 0.0057)\ \text{u} = 4.0083\ \text{u}$. Hence

$$\text{Mass of }^{210}_{84}\text{Po} - (\text{mass of }^4_2\text{He} + m_E) = \text{mass of }^{206}_{82}\text{Pb}$$
$$209.9829\ \text{u} - 4.0083\ \text{u} = 205.9746\ \text{u} \qquad \blacklozenge$$

27.4 DETECTING RADIATION

Ions are the clue.

Photons and subatomic particles such as those emitted in radioactive decay can be detected in a variety of ways. Nearly all of them are based on the ionization that results, directly or indirectly, from the passage of a photon or charged particle through matter.

A heavy charged particle, such as a proton or an alpha particle, loses energy chiefly by electric interactions with atomic electrons. The electrons either are raised to excited states or, more often, are pulled away entirely from their parent atoms. The ejected electrons often ionize other atoms themselves. The incoming particle slows down gradually until it comes to a stop or reacts with a nucleus in its path.

Electrons, too, lose energy by ionization, but for them another mechanism may also be important. As we know, electromagnetic radiation is given off whenever an electric charge is accelerated. Energy loss by radiation is more significant for electrons than for heavier particles because they are more violently accelerated when passing near nuclei in their paths. The faster the electron and the greater the nuclear charge of the atoms it encounters, the more it radiates.

The three principal ways in which the energetic photons of X- and gamma rays lose energy when they pass through matter are the **photoelectric effect, Compton scattering,** and **pair production** (Fig. 27.12). In each case photon energy is conveyed to electrons, which in turn lose energy mainly by exciting or ionizing atoms in the absorbing material.

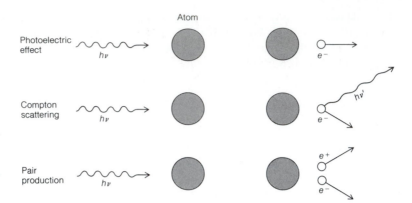

Figure 27.12

X- and gamma rays interact with matter chiefly through the photoelectric effect, Compton scattering, and pair production. Pair production requires a photon energy of at least 1.02 MeV.

At low photon energies the photoelectric effect, which was discussed in Section 25.1, is the principal mechanism of energy loss. At higher energies, a photon may transfer only part of its energy $h\nu$ to an electron it collides with. The scattered photon has the lower energy $h\nu'$, and the electron flies off with the kinetic energy $KE = h\nu = h\nu'$. This process is called Compton scattering.

At still higher energies, it is possible for a photon to materialize into a positron and an electron when it passes near a nucleus. Since the energy equivalent of the mass of either particle is 0.51 MeV, the minimum photon energy needed for such pair production is 1.02 MeV. In pair production, the electromagnetic energy of a photon is turned into matter, with any photon energy in excess of 1.02 MeV appearing as kinetic energy of the newly created positron and electron. Electric charge as well as energy is conserved in pair production since the positron ($Q = +e$) and electron ($Q = -e$) have opposite charges. Gamma rays from radioactive decay and X-rays interact with matter largely through Compton scattering.

Bubble-chamber photograph of electron-position pair formation in liquid hydrogen. A high-energy gamma ray photon created the pair and also gave energy to an atomic electron. A magnetic field in the chamber caused the paths of the electrons to curve in opposite directions.

Output

Figure 27.13

A gas-filled detector of this kind can be used either as an ionization chamber or as a Geiger counter. When it is to be an ionization chamber, the applied voltage is low, and the output pulse that occurs when a photon or charged particle passes through the tube is a measure of the amount of ionization produced in the gas. When it is to be a Geiger counter, the applied voltage is high enough to give the initial ions enough energy of their own to cause further ionization, and the result is a large output pulse that is the same for all incident photons or particles.

A number of devices can be used to detect the ionization produced by an energetic photon or charged particle. Some, such as the **ionization chamber** and the **Geiger counter,** are gas-filled tubes with electrodes that attract the electrons and ions to produce voltage pulses in external circuits (Fig. 27.13). Another device, the **scintillation counter,** is based on the flash of light emitted by certain substances (such as those used to make the screen of a television picture tube) when struck by ionizing radiation. The light is then picked up by a sensitive photomultiplier tube (Fig. 27.14). **Semiconductor detectors** make use of the fact that a thin layer on both sides of a *p-n* junction is normally depleted in charge carriers when the crystal has an external voltage across it. Electrons and holes produced at the junction by an ionizing particle lead to a measurable pulse.

Neutrons do not themselves produce ionizaton and so cannot be detected directly. However, a counter whose sensitive volume is rich in hydrogen will respond to fast neutrons through the ionization produced by protons recoiling after collisions with the neutrons. In the case of slow neutrons, one method is to employ the nuclear reaction

$$\,^{1}_{0}n + \,^{10}_{5}B \rightarrow \,^{4}_{2}He + \,^{7}_{3}Li$$

in which a neutron reacts with a nucleus of the boron isotope $^{10}_{5}B$ to produce an alpha particle and a nucleus of the lithium isotope $^{7}_{3}Li$. If $^{10}_{5}B$ is in the sensitive volume of a suitable counter (the gas BF_3 is often used), ionization that corresponds to an alpha particle signifies the capture of a slow neutron.

Figure 27.14

Essential features of a scintillation counter. The scintillator may be a crystalline or plastic solid or a liquid. The number of dynodes in the photomultiplier tube may be ten or more, with each dynode several hundred volts higher in potential than the one before it. Each dynode produces 2 to 5 secondary electrons per incident electron for a total amplification of as much as 10^6 or 10^7. A scintillation counter is faster in its operation than an ionization chamber or a Geiger counter.

27.5 HALF-LIFE

The shorter the half-life, the more likely a given nucleus will decay in a given time.

The rate R at which a sample of a radioactive nuclide decays is called its **activity.** If ΔN nuclei decay in the time Δt, then

$$R = \frac{\Delta N}{\Delta t} \qquad\qquad\qquad \textit{Activity} \quad (27.4)$$

The SI unit of activity is named after the French physicist Henri Becquerel, who discovered radioactivity in 1896:

1 becquerel = 1 Bq = 1 event/s

The traditional unit of activity, still in common use, is the **curie.** Originally the curie was defined as the activity of 1 g of radium ($^{226}_{88}$Ra), and its precise value accordingly changed as measuring techniques improved. For this reason the curie is now defined arbitrarily as

1 curie = 3.70 × 10^{10} events/s = 37 GBq

The activity of 1 g of radium is a few percent smaller. A luminous watch dial contains several microcuries of $^{226}_{88}$Ra; ordinary potassium has an activity of about 1 millicurie/kg owing to the presence of the radioactive isotope $^{40}_{19}$K; ''cobalt-60'' sources of 1 or more curies are widely used in medicine for radiation therapy and industrially for the inspection of metal castings and welded joints.

One of the characteristics of all types of radioactivity is that the rate at which the nuclei in a given sample decay always follows a curve whose shape is like that shown in Fig. 27.15. If we start with a sample whose activity is, say, 100 Bq, it will not continue to decay at that rate but instead fewer and fewer disintegrations will occur in each successive second.

Some radioactive nuclides decay faster than others, but in each case a certain definite time is required for half of an original sample to decay. This time is called the **half-life** of the nuclide. For instance, the radon isotope $^{222}_{86}$Rn undergoes alpha decay to the polonium isotope $^{218}_{84}$Po with a half-life of 3.8 days. Should we start with 1 mg of radon in a closed container (since it is a gas), $\frac{1}{2}$ mg will remain undecayed after 3.8 days; $\frac{1}{4}$ mg will remain undecayed after 7.6 days; $\frac{1}{8}$ mg will remain undecayed after 11.4 days; and so on (Fig. 27.16).

Figure 27.15

The rate at which a sample of a radioactive substance decays is not constant but varies with time in the manner shown in the curve.

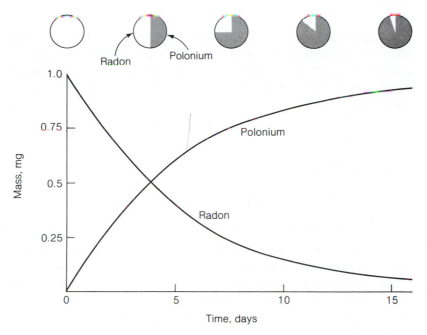

Figure 27.16

The alpha decay of radon ($^{222}_{86}$Rn) to polonium ($^{218}_{84}$Po) has a half-life of 3.8 days. The sample of radon whose decay is graphed here had an initial mass of 1.0 mg.

◆ **E X A M P L E   27.3** ─────────────────

The hydrogen isotope tritium, $^3_1$H, is radioactive and emits an electron with a half-life of 12.5 years. (a) What does $^3_1$H become after beta decay? (b) What percentage of an original sample of tritium will remain undecayed 25 years after its preparation?

SOLUTION   (a) When a nucleus emits an electron, its atomic number increases by 1 (corresponding to an increase in nuclear charge of $+e$) and its mass number is unchanged. Helium has the atomic number 2, and so

$$^3_1\text{H} \rightarrow \, ^3_2\text{He} + e^-$$

(b) Twenty-five years represents two half-lives of tritium. Hence $(\frac{1}{2})(\frac{1}{2}) = \frac{1}{4} = 25\%$ of the original sample remains undecayed.   ◆

Half-lives range from billionths of a second to billions of years. Samples of radioactive isotopes decay in the manner shown in Fig. 27.15 because a great many nuclei are involved, each with a certain probability of decaying. The fact that radon has a half-life of 3.8 days signifies that every radon nucleus has a 50% chance of decaying in any 3.8-day period. Because a nucleus does not have a memory, this does *not* mean that a radon nucleus has a 100% chance of decaying in 7.6 days: The likelihood of decay of a given nucleus stays the same until it actually does decay. Thus a half-life of 3.8 days means a 75% probability of decay in 7.6 days, an 87.5% probability of decay in 11.4 days, and so on, because in each interval of 3.8 days the probability is 50%.

This discussion suggests that radioactive decay involves events that take place within individual nuclei, rather than collective processes that involve more than one nucleus in interaction. This idea is confirmed by experiments that show that the half-life of a particular isotope is invariant under changes of pressure, temperature, electric and magnetic fields, and so on, which might, if strong enough, affect internuclear phenomena.

27.6 RADIOMETRIC DATING

A clock based on radioactivity.

Radioactivity makes it possible to find the ages of many geological and biological specimens. The ratio between the amounts of a radionuclide and of its stable daughter in a specimen depends on the half-life of the nuclide and the age of the specimen. The greater the proportion of the daughter nuclide, the older the specimen. Let us see how **radiocarbon,** the beta-active carbon isotope $^{14}_6$C, can be used to date objects of biological origin.

Cosmic rays are high-energy atomic nuclei, mainly protons, that circulate through the Milky Way galaxy of which the sun is a member. About 10^{18} of them arrive at the earth's atmosphere each second, where they disrupt the nuclei of atoms they encounter to produce showers of secondary particles. Among these secondaries are neutrons that can react with nitrogen nuclei in the atmosphere to form radiocarbon with the emission of a proton:

$$^{14}_7\text{N} + ^1_0 n \rightarrow ^{14}_6\text{C} + ^1_1\text{H}$$

The proton picks up an electron and becomes a hydrogen atom. The $^{14}_6$C isotope has too many neutrons for stability and beta-decays to become $^{14}_7$N with a half-life of 5760 years:

$$^{14}_6\text{C} \rightarrow ^{14}_7\text{N} + e^-$$

About 90 tons of radiocarbon is distributed around the world, an amount kept constant by the cosmic-ray bombardment.

Shortly after their formation, radiocarbon atoms combine with oxygen molecules to form carbon dioxide (CO_2) molecules. Green plants convert water and carbon dioxide into carbohydrates by photosynthesis, which means that every plant contains some radiocarbon. Animals eat plants, and so they too contain radiocarbon. Thus every living thing on earth has a very small proportion of radiocarbon in its tissues. Because the mixing of radiocarbon is relatively efficient, living plants and animals all have the same ratio of radiocarbon to ordinary carbon, which is $^{12}_6$C.

When a plant or animal dies, it no longer takes in radiocarbon, and the radiocarbon already present continues to decay. After 5760 years a dead organism

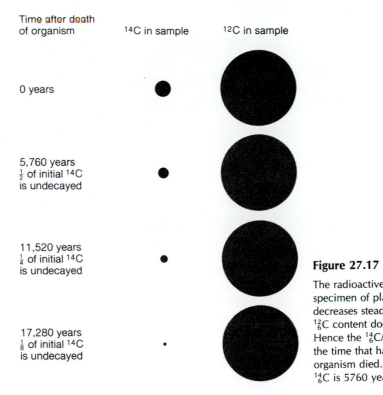

Time after death of organism	^{14}C in sample	^{12}C in sample
0 years		
5,760 years $\frac{1}{2}$ of initial ^{14}C is undecayed		
11,520 years $\frac{1}{4}$ of initial ^{14}C is undecayed		
17,280 years $\frac{1}{8}$ of initial ^{14}C is undecayed		

Figure 27.17

The radioactive $^{14}_{6}$C content of a specimen of plant or animal tissue decreases steadily, whereas its $^{12}_{6}$C content does not change. Hence the $^{14}_{6}$C/$^{12}_{6}$C ratio indicates the time that has passed since the organism died. The half-life of $^{14}_{6}$C is 5760 years.

has left only half as much radiocarbon as it had while alive, after 11,520 years only a quarter as much, and so on. The ratio of radiocarbon to ordinary carbon therefore provides the ages of ancient objects and remains of organic origin (Fig. 27.17). This technique permits the dating of mummies, wooden implements, cloth, leather, charcoal from campfires, and similar artifacts from ancient civilizations as much as 70,000 years old, about 12 half-lives of $^{14}_{6}$C.

Because the earth's history goes back 4.5 or so billion years, geologists use radioactive nuclides of much longer half-lives than that of radiocarbon to date rocks. The oldest rocks whose ages have been found in this way are in Greenland and are believed to have solidified 3.8 billion years ago. Meteorites and lunar rocks as well as terrestrial rocks have been dated by radiometric methods.

27.7 RADIATION HAZARDS

Invisible but dangerous.

Like X-rays, the various radiations from radioactive nuclei ionize matter through which they pass. All ionizing radiation is harmful to living tissue, although if the damage is slight, the tissue can often repair itself with no permanent effect. It

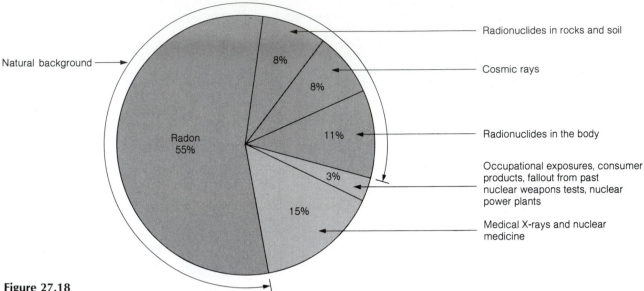

Natural background

8%

Radon
55%

Radionuclides in rocks and soil

Cosmic rays

8%

11%

Radionuclides in the body

3%

Occupational exposures, consumer products, fallout from past nuclear weapons tests, nuclear power plants

15%

Medical X-rays and nuclear medicine

Figure 27.18

Sources of radiation dosage for an average person in the United States. The total is about equivalent to 25 chest X-rays per year. Actual dosages vary widely. For instance, radon concentrations are not the same everywhere; some people receive more medical X-rays than others; cosmic rays are more intense at high altitudes; and so on. Nuclear power stations are responsible for only 0.08% of the total, although accidents can raise the amount in affected areas to dangerous levels.

is easy to underestimate radiation hazards because there is usually a delay, sometimes of many years, between an exposure and some of its possible consequences, which include cancer, leukemia, and genetic changes that may lead to children handicapped in various ways.

Radiation dosage is measured in **sieverts** (Sv), where 1 Sv is the amount of any radiation that has the same biological effects as that produced when 1 kg of body tissue absorbs 1 joule of X- or gamma rays. (An older unit, the **rem,** is equal to 0.01 Sv.) Although radiobiologists disagree about the exact relationship between radiation exposure and the likelihood of developing cancer, there is no question that such a link exists. Natural sources of radiation lead to a dosage rate per person of about 3 mSv/y averaged over the U.S. population (1 mSv = 0.001 Sv). Other sources of radiation add 0.6 mSv/y, with medical X-rays prominent. The total is about 3.6 mSv/y, about the dose received from 25 chest X-rays.

Figure 27.18 shows the relative contributions to the radiation dosage received by an average person in the United States. The most important single source is the radioactive gas radon, a decay product of radium whose own origin traces back to the decay of uranium. Uranium is found in many common rocks, notably granite. Hence radon, colorless and odorless, is present nearly everywhere, though usually in amounts too small to endanger health. Problems arise when houses are built in uranium-rich regions, because it is impossible to prevent radon from entering such houses from the ground under them. Surveys show that millions of American homes have radon concentrations high enough to pose a small but definite cancer risk. As a cause of lung cancer, radon is second only to cigarette smoking. Of the various approaches to reducing radon levels in an existing house in a hazardous region, the most effective seems to be to extract air from underneath the ground floor and disperse it into the atmosphere before it can enter the house.

Other natural sources of radiation dosage include cosmic rays and radionuclides in rocks and soil. The human body itself contains small amounts of radioisotopes of such elements as potassium and carbon.

Many useful processes involve ionizing radiation. Some employ such radiation directly, as in the X-rays and gamma rays used in medicine and industry. Smoke detectors and false teeth are among the consumer products that contain traces of radionuclides. Sometimes the radiation is unwanted but inescapable, notably in the operation of nuclear reactors and in the disposal of their wastes. It is not always easy to find an appropriate balance between risk and benefit where radiation is concerned. This seems particularly true for medical X-rays, which constitute the second largest source of radiation dosage for an average person in the United States.

It seems to be an unfortunate fact that some X-ray exposures are made for no strong reason and do more harm than good. In this category are "routine" chest X-rays upon hospital admission, "routine" X-rays as part of regular physical examinations, and "routine" dental X-rays. The once-routine mass screening by X-ray of symptomless young women for breast cancer is now generally thought to have increased, not decreased, the overall death rate due to cancer. Especially dangerous is the X-raying of pregnant women, until recently another "routine" procedure, which dramatically increases the chances of cancer in their children.

Since the carcinogenic properties of X-rays have been known since 1902, all this is hard to excuse. X-rays unquestionably are valuable in medicine. The point is that every exposure should have a definite justification that outweighs the risk.

27.8 FISSION

Divide and conquer.

As we saw earlier, a great deal of binding energy will be released if we can break a large nucleus into smaller ones. But nuclei are usually not at all easy to split apart. What we need is a way to disrupt a heavy nucleus without using more energy than we get back from the process.

The solution came in 1938 with the discovery that absorbing a neutron causes a nucleus of the uranium isotope $^{235}_{92}U$ to undergo fission. It is not the impact of the neutron that has this effect. What happens is that the $^{235}_{92}U$ nucleus becomes $^{236}_{92}U$, which is so unstable that it explodes almost at once into two **fission fragments** (Fig. 27.19). Later several other heavy nuclides were found to be fissionable in similar processes.

Because stable light nuclei have proportionally fewer neutrons than do heavy nuclei, the fragments are unbalanced when they are formed and release one or two neutrons each. Usually the fragments are still unstable and undergo radioactive decay to achieve appropriate neutron-proton ratios. The products of fission, such as the wastes from a nuclear reactor and the fallout from a nuclear bomb burst, are accord-

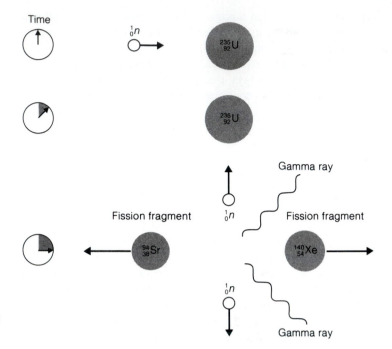

Figure 27.19

In nuclear fission an absorbed neutron causes a heavy nucleus to split into two parts, with the emission of several neutrons and gamma rays.

ingly highly radioactive. Since some of these products have long half-lives, their radioactivity will persist for many generations and their disposal remains a problem without a really satisfactory solution as yet.

A variety of nuclides may appear as fission fragments. A typical fission reaction is that shown in Fig. 27.19:

$$^{235}_{92}U + {}^{1}_{0}n \rightarrow {}^{236}_{92}U \rightarrow {}^{140}_{54}Xe + {}^{94}_{38}Sr + {}^{1}_{0}n + {}^{1}_{0}n + 200 \text{ MeV}$$

About 84% of the total energy liberated during fission appears as kinetic energy of the fission fragments, about 2.5% as kinetic energy of the neutrons, and about 2.5% in the form of instantaneous emitted gamma rays, with the remaining 11% being given off in the decay of the fission fragments.

Because each fission event liberates neutrons, a self-sustaining series of fissions is possible (Fig. 27.20). For such a **chain reaction** to occur, at least one neutron produced in each fission must, on the average, cause another fission. If too few neutrons initiate fissions, the reaction will slow down and stop. If exactly one neutron per fission causes another fission, energy will be given off at a constant rate, which is the case in a **nuclear reactor.** If the rate of fissioning increases, the energy release will be so rapid that an explosion will occur, which is the case in an **atomic bomb.** If two neutrons from each fission induce further fissions and take 10^{-8} s to do so, a chain reaction starting with a single fission will give off 2×10^{13} J of energy in less than 10^{-6} s.

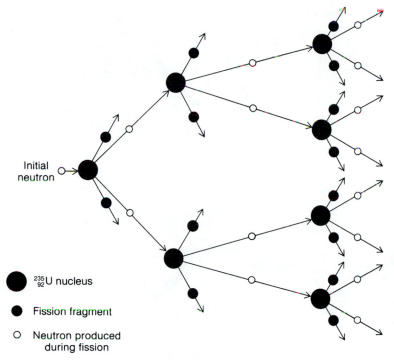

Figure 27.20

Sketch of à chain reaction. If at least one neutron from each fission on the average induces another fission, the reaction is self-sustaining. If more than one neutron per fission on the average induces another fission, the reaction is explosive.

● $^{235}_{92}$U nucleus

● Fission fragment

○ Neutron produced during fission

The destructiveness of nuclear weapons does not stop with their detonation but continues long afterward through the radioactive debris that is produced and dispersed. Tens of thousands of such weapons now exist in a number of countries. The explosion of even a fraction of them might, by causing dust clouds that block sunlight, lead to months of darkness and bitter cold that could doom many of those living things that survived the blasts themselves and the radioactive fallout from them.

27.9 NUCLEAR REACTORS

From uranium to electricity.

A nuclear reactor is a very efficient source of energy. An output of about 1 MW (1000 kW) is produced by the fission of 1 g of $^{235}_{92}$U per day, as compared with the burning of 2.6 tons of coal per day for this output in a conventional power plant.

The energy liberated in a nuclear reactor appears as heat, and it is extracted by circulating a coolant liquid or gas. The hot coolant can then be used as the heat source of a conventional steam turbine, which in turn may power an electric generator,

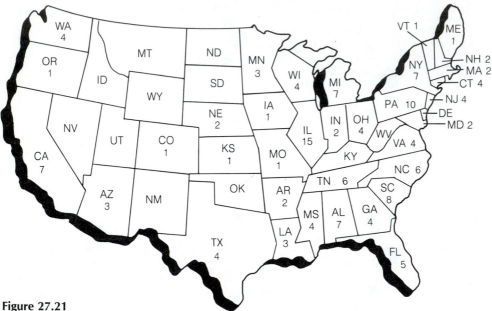

Figure 27.21

Nuclear power plants in the
United States. The total number is
142. No new plants have been
ordered since 1979, when failures
in its cooling system disabled one
of the reactors at Three Mile Island in Pennsylvania, and a certain amount of radioactive material escaped. Public unease with
the safety record of the nuclear
industry, together with a leveling
off of demand for electricity, led
to a halt in plans to expand nuclear energy further in the United
States. A much worse accident
destroyed a reactor at Chernobyl
in the Soviet Union in 1986, producing many fatalities and widespread contamination in many
parts of Europe.

a ship, or a submarine. In 1990 a total of 441 reactors in 26 countries were generating
330 GW of electric power (1 GW = 1 billion watts)—enough to supply about 160
million homes. Eighty-one more reactors were then under construction, which will
bring the total nuclear generating capacity to 400 GW. A number of countries, such
as France, obtained over half their electricity from reactors in that year. In the United
States in 1990, nuclear energy was responsible for about 20% of the electricity
generated, not far from the world average, in second place behind coal (Fig. 27.21).

Each fission in $^{235}_{92}$U frees an average of 2.5 neutrons, so no more than 1.5
neutrons per fission can be lost in a reactor. Natural uranium, however, contains only
0.7% of the fissionable isotope $^{235}_{92}$U. The rest is $^{238}_{92}$U, which readily captures the fast
neutrons emitted during fission but does not undergo fission afterward. The neutrons
absorbed by $^{238}_{92}$U are therefore wasted, and a chain reaction cannot take place in a
solid lump of natural uranium.

There is a way around this problem. As it happens, $^{238}_{92}$U has little ability to
capture *slow* neutrons, whereas slow neutrons are more likely to induce fission in
$^{235}_{92}$U than fast ones. Slowing down the fast neutrons from fission will thus prevent
them from being lost to $^{238}_{92}$U and at the same time promote further fissions in $^{235}_{92}$U.
To slow down fission neutrons, the uranium fuel in a reactor is mixed with a
moderator, a substance whose nuclei absorb energy from fast neutrons in collisions.
From Section 5.7 we know that the more nearly equal in mass colliding particles are,
the more energy is transferred. Since hydrogen nuclei are protons with nearly the
same mass as neutrons, hydrogen is widely used as a moderator in the form of water,
H_2O, whose molecules each contain two hydrogen atoms. The water also serves as
the coolant in such reactors.

Unfortunately a neutron that strikes a proton has a certain chance of sticking to it to form a deuteron, $^2_1$H. Neutron losses in a water-moderated reactor make it necessary to use **enriched uranium** whose $^{235}_{92}$U content has been increased to about 3%. Most enriched uranium is produced by the gaseous diffusion process in which uranium hexafluoride (UF$_6$) gas is exposed to a succession of semipermeable barriers. Because of their smaller mass, molecules of $^{235}_{92}$UF$_6$ are slightly more likely to diffuse through each barrier than molecules of $^{238}_{92}$UF$_6$, and any desired degree of enrichment can be achieved in this way. Another method is to use gas centrifuges for the separation. Either procedure needs a large plant and much energy, so it is expensive.

The fuel for a water-cooled reactor consists of uranium oxide (UO$_2$) pellets sealed in long, thin zirconium alloy tubes assembled into a core that is enclosed in a steel pressure vessel. A typical pressure vessel is 10 m high, has an inside diameter of 3 m, and has walls 20 cm thick. In a **pressurized-water reactor** (PWR), the most common type, the water that circulates through the core is kept at a sufficiently high pressure, about 155 atm, to prevent boiling. The water enters the pressure vessel at perhaps 280°C and leaves at 320°C, passing through a heat exchanger to produce steam that drives a turbine (Fig. 27.22). Such a PWR used to produce electricity might contain 70 tons of UO$_2$ and operate at 2700 MW to yield perhaps 1000 MW of electric power. To control the rate of the chain reaction, the reactor has movable rods of cadmium or boron, which are good absorbers of slow neutrons. As these rods are inserted farther and farther into the reactor, the reaction rate is progressively reduced.

Another type of reactor uses "heavy" water, whose molecules contain deuterium ($^2_1$H) atoms, instead of "light" water because deuterons are less likely than protons to capture neutrons. Natural uranium is the fuel in a reactor of this kind. When graphite, a form of pure carbon, is the moderator, the reactor can be cooled by a gas such as helium or carbon dioxide and operate at much higher temperatures for greater thermodynamic efficiency.

Figure 27.22

The basic design of a pressurized-water nuclear reactor. Water serves both as moderator and as coolant for the core.

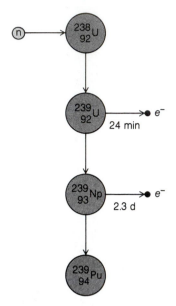

Figure 27.23

The nonfissionable uranium isotope $^{238}_{92}U$ becomes the fissionable plutonium isotope $^{239}_{94}Pu$ after absorbing a neutron and undergoing two beta decays. Transformations like this are the basis of the breeder reactor, which produces more fissionable material than it uses up.

Breeder Reactors

Some nonfissionable nuclides can become fissionable ones by absorbing neutrons. A notable example is $^{238}_{92}U$, which becomes $^{239}_{92}U$ when it captures a fast neutron. This uranium isotope then beta-decays into $^{239}_{93}Np$, an isotope of the element **neptunium,** which is also beta-active. The decay of $^{239}_{93}Np$ yields $^{239}_{94}Pu$, a fissionable isotope of **plutonium** that can support a chain reaction. The entire sequence is shown in Fig. 27.23.

Both neptunium and plutonium are **transuranic elements,** none of which are found on the earth because their half-lives are too short for them to have survived even if they had been present when the earth came into being 4.5 billion years ago. Transuranic elements up to $Z = 105$ have been produced by the neutron bombardment of lighter nuclides.

A **breeder reactor** is one designed to produce as much or more fissionable material than the $^{235}_{92}U$ it consumes, either from $^{238}_{92}U$ or from some similarly "fertile" nuclide. On the one hand, the widespread use of breeder reactors would mean that known reserves of nuclear fuel would last for centuries to come. On the other hand, since plutonium (unlike the slightly enriched uranium that fuels ordinary reactors) can also be used in nuclear weapons, such reactors would also complicate arms control.

27.10   NUCLEAR FUSION

How the sun and stars get their energy.

The basic energy-producing process in stars—and hence the source, direct or indirect, of nearly all the energy in the universe—is the fusion of hydrogen nuclei into helium nuclei. This can take place under stellar conditions in two different series of nuclear reactions. In stars like the sun, the **proton-proton** cycle predominates:

$$^1_1H + {}^1_1H \rightarrow {}^2_1H + e^+ + 0.4\,\text{MeV}$$
$$^1_1H + {}^2_1H \rightarrow {}^3_2He + 5.5\,\text{MeV} \qquad \textit{Proton-proton cycle}$$
$$^3_2He + {}^3_2He \rightarrow {}^4_2He + 2\,{}^1_1H + 12.9\,\text{MeV}$$

The first two of these reactions must each occur twice for every synthesis of ^4_2He, so the total energy liberated is 24.7 MeV.

The other reaction sequence, the **carbon cycle,** predominates in stars hotter than the sun. In this cycle, a $^{12}_6C$ nucleus absorbs three protons, with two positive beta decays along the way, until it becomes the nitrogen nucleus $^{15}_7N$. When a $^{15}_7N$ nucleus reacts with a fourth proton, the result is the formation of a ^4_2He nucleus and the reappearance of a $^{12}_6C$ nucleus. Thus the original $^{12}_6C$ nucleus acts as a kind of catalyst for the process, since it returns at the end ready to start another cycle.

The energy liberated by nuclear fusion is often called **thermonuclear energy.** High temperatures and densities are needed for fusion reactions to occur often enough for a substantial amount of thermonuclear energy to be produced. The high temperature means that the initial light nuclei have enough thermal energy to overcome their mutual electric repulsion and come close enough together to react. The high density means that such collisions are frequent. A further condition for these cycles is a large reacting mass, such as that of a star, because several steps are involved in each cycle and much time may go by between the initial fusion of one proton with another and the ultimate production of an alpha particle.

Here on the earth, where any reacting mass must be very limited in size, an efficient fusion process cannot involve more than a single step. The fusion reaction that is the basis of current research is the combination of a deuterium nucleus and a tritium nucleus to form a helium nucleus and a neutron:

$$^2_1\text{H} + ^3_1\text{H} \rightarrow ^4_2\text{He} + ^1_0 n + 17.6\,\text{MeV}$$

Most of the energy given off is carried away by the neutron. To recover this energy, the reaction chamber could be surrounded with liquid lithium, which would absorb energy from the neutrons and carry it to a steam generator. The steam would then power a turbine connected to an electric generator, as in fossil-fuel and fission power plants.

About 1 part in 5000 of the hydrogen in the waters of the world is deuterium, which adds up to over 10^{15} tons. (A gallon of seawater has the potential for fusion energy equal to the chemical energy in 300 gallons of gasoline.) Seawater contains too little tritium for economic recovery, but neutrons react with both isotopes of natural lithium to give tritium and helium:

$$^6_3\text{Li} + ^1_0 n \rightarrow ^3_1\text{H} + ^4_2\text{He}$$
$$^7_3\text{Li} + ^1_0 n \rightarrow ^3_1\text{H} + ^4_2\text{He} + ^1_0 n$$

Once a fusion reactor is given an initial load of tritium, then, it can make enough for its further operation.

The big problem in exploiting fusion energy is to achieve the required combination of temperature, density, and containment time for a deuterium-tritium mixture to react sufficiently to produce a net energy yield. Such a combination occurs in the explosion of a fission ("atomic") bomb, and including the ingredients for fusion in such a bomb leads to a much more destructive weapon, the "hydrogen" bomb.

Two approaches to the controlled release of fusion energy are being explored. In one, strong magnetic fields are used to contain the reacting nuclei, which are in the form of a hot, dense plasma (fully ionized gas). Four decades of research have led to larger and larger experimental magnetic fusion reactors that have brought to light no reasons why eventual success should not be possible. But the engineering problems to be overcome are substantial.

The other approach involves using energetic beams to both heat and compress tiny deuterium-tritium pellets to produce what are, in effect, miniature hydrogen-bomb explosions. A number of beams would strike each pellet from all sides to keep it in place as it is squeezed together. If ten pellets the size of a grain of sand are

The Tokamak reactor at Princeton University uses strong magnetic fields to confine a very hot ionized gas of deuterium and tritium while nuclear fusion reactions take place. A larger-scale reactor of this kind may produce more energy than its operation consumes.

ignited per second, the energy output could provide the electric power needed by a city of 175,000 people. Laser beams are being tried for this purpose as well as beams of particles such as electrons and protons.

Although practical fusion reactors are still in the future (how far, nobody knows), the advantages of thermonuclear energy—cheap, essentially limitless fuel; much less hazard than for fission energy; relatively minor production of radioactive wastes; no possibility of military use—are such that eventually it will almost certainly supply a large part of the world's energy needs.

Important Terms

The **atomic number** of an element is the number of electrons in each of its atoms or, equivalently, the number of protons in each of its atomic nuclei.

The **neutron** is an electrically neutral particle, slightly heavier than the proton, that is present in nuclei together with protons. Neutrons and protons are jointly called **nucleons.** The **mass number** of a nucleus is the number of nucleons it contains.

Isotopes of an element have the same atomic number but different mass numbers. Symbols of **nuclides** (nuclear species) follow the pattern

$$_Z^A X$$

where X is the chemical symbol of the element, Z its atomic number, and A the mass number of the particular nuclide.

The **binding energy** of a nucleus is the energy equivalent of the difference between its mass and the sum of the masses of its individual constituent nucleons. This amount of energy must be supplied to the nucleus if it is to be completely disintegrated.

Radioactive nuclei spontaneously transform themselves into other nuclear species with the emission of radiation. The radiation may consist of **alpha particles,** which are the nuclei of helium atoms, or **beta particles,** which are positive or negative electrons. Positive electrons are known as **positrons.** Electron capture by a nucleus is an alternative to positron emission. The emission of **gamma rays,** which are energetic photons, permits an excited nucleus to lose its excess energy.

The time required for half of a given sample of a radioactive nuclide to decay is called its **half-life.**

In **nuclear fission** the absorption of neutrons by certain heavy nuclei causes them to split into smaller **fission fragments.** Because each fission also liberates several neutrons, a rapidly multiplying sequence of fissions called a **chain reaction** can occur if a sufficient amount of the proper material is assembled. A **nuclear reactor** is a device in which a chain reaction can be initiated and controlled.

In **nuclear fusion** two light nuclei combine to form a heavier one with the emission of energy. The energy liberated in the process when it takes place on a large scale is called **thermonuclear energy.**

MULTIPLE CHOICE

1. The chemical behavior of an atom is determined by its
   a. atomic number
   b. mass number
   c. binding energy
   d. number of isotopes

2. All isotopes of a given element have the same
   a. number of protons
   b. number of neutrons
   c. number of nucleons
   d. binding energy

3. The atomic number of an atom equals the number of its
   a. protons
   b. neutrons
   c. protons and neutrons
   d. protons and electrons

4. The mass number of an atom equals the number of its
   a. protons
   b. neutrons
   c. protons and neutrons
   d. protons and electrons

5. The number of neutrons in the nucleus of the aluminum isotope $_{13}^{27}$Al is
   a. 13
   b. 14
   c. 27
   d. 40

6. Which of the following is not an isotope of hydrogen?
   a. $_0^1$H
   b. $_1^1$H
   c. $_1^2$H
   d. $_1^3$H

7. The number of electrons in a neutral atom of the zinc isotope $_{30}^{66}$Zn is
   a. 6
   b. 30
   c. 36
   d. 66

8. Relative to the sum of the masses of its constituent nucleons, the mass of a nucleus is
   a. greater
   b. the same
   c. smaller
   d. sometimes greater and sometimes smaller

9. The ionization energy of an atom relative to the binding energy of its nucleus is
   a. greater
   b. the same
   c. smaller
   d. sometimes greater and sometimes smaller

10. The binding energy per nucleon is
    a. the same for all nuclei
    b. greatest for very small nuclei
    c. greatest for nuclei of intermediate size
    d. greatest for very large nuclei

11. The element whose nuclei contain the most tightly bound nucleons is
    a. helium
    b. carbon
    c. iron
    d. uranium

12. Nuclear fission and fusion reactions give off energy because
    a. the binding energy per nucleon is least for nuclei of intermediate size
    b. the binding energy per nucleon is greatest for nuclei of intermediate size
    c. they result in the production of neutrons
    d. they result in the production of protons

13. When a nucleus undergoes radioactive decay, its new mass number is
    a. always less than its original mass number
    b. always more than its original mass number
    c. never less than its original mass number
    d. never more than its original mass number

14. A nucleus with an excess of neutrons may decay radioactively with the emission of
    a. a neutron
    b. a proton
    c. an electron
    d. a positron

15. Electron capture has the same effect on a nucleus as
    a. alpha decay
    b. negative beta decay
    c. positive beta decay
    d. gamma decay

16. Which of these types of radiation has the greatest ability to penetrate matter?
    a. alpha particles
    b. beta particles
    c. gamma rays
    d. X-rays

17. Which of these types of radiation has the least ability to penetrate matter?
    a. alpha particles
    b. beta particles
    c. gamma rays
    d. X-rays

18. Gamma rays do not lose energy in passing through matter by
    a. materializing in electron-positron pairs
    b. giving up all their energy to atomic electrons in the photoelectric effect
    c. giving up part of their energy to atomic electrons in the Compton effect
    d. undergoing radioactive decay

19. The half-life of a radioactive substance is
    a. half the time needed for a sample to decay entirely

   b. half the time a sample can be kept before it begins to decay

   c. the time needed for half a sample to decay

   d. the time needed for the remainder of a sample to decay after half of it has already decayed

20. As a sample of a radioactive nuclide decays, its half-life

   a. decreases

   b. remains the same

   c. increases

   d. any of the above, depending on the nuclide

21. A suitable object for radiocarbon dating would be one whose age is

   a. 20,000 years   **b.** 200,000 years

   c. 2 million years   **d.** 20 million years

22. The largest amount of radiation received by an average person in the United States comes from

   a. medical X-rays

   b. nuclear reactors

   c. fallout from past weapons tests

   d. natural sources

23. By ''chain reaction'' is meant

   a. the joining together of protons and neutrons to form atomic nuclei

   b. the joining together of light nuclei to form heavy ones

   c. the successive fissions of heavy nuclei induced by neutrons emitted in the fissions of other heavy nuclei

   d. the burning of uranium in a special type of furnace called a nuclear reactor

24. In a nuclear power plant the nuclear reactor itself is used to supply

   a. neutrons   **b.** electricity

   c. steam   **d.** heat

25. Most of the energy liberated during fission appears

   a. as KE of the neutrons that are emitted

   b. as KE of the fission fragments

   c. in the decay of the fission fragments

   d. as gamma rays

26. Enriched uranium is a better fuel for nuclear reactors than natural uranium because it has a greater proportion of

   a. slow neutrons   **b.** deuterium

   c. $^{235}_{92}U$ **d.** $^{238}_{92}U$

27. The purpose of the moderator in a nuclear reactor is to

   a. slow down the neutrons emitted during fission

   b. absorb the neutrons emitted during fission

   c. absorb the gamma rays emitted during fission

   d. control the rate at which energy is produced

28. The sun's energy comes from

   a. nuclear fission

   b. radioactivity

   c. the conversion of hydrogen to helium

   d. the conversion of helium to hydrogen

29. Fusion reactions on the earth are likely to use as fuel

   a. ordinary hydrogen   **b.** deuterium

   c. plutonium   **d.** uranium

30. The mass of a $^{7}_{3}Li$ nucleus is 0.042 u less than the sum of the masses of 3 protons and 4 neutrons. The binding energy per nucleon in $^{7}_{3}Li$ is

   a. 5.6 MeV   **b.** 10 MeV

   c. 13 MeV   **d.** 39 MeV

31. The product of the alpha decay of the bismuth isotope $^{214}_{83}Bi$ is

   a. $^{210}_{79}Au$ **b.** $^{210}_{81}Tl$

   c. $^{210}_{83}Bi$ **d.** $^{218}_{85}At$

32. The helium isotope $^{6}_{2}He$ undergoes beta decay with the emission of an electron. The product of the decay is

   a. $^{6}_{1}H$ **b.** $^{5}_{2}He$

   c. $^{7}_{2}He$ **d.** $^{6}_{3}Li$

33. The product of the positive beta decay of the nitrogen isotope $^{13}_{7}N$ is

   a. $^{12}_{6}C$ **b.** $^{13}_{6}C$

   c. $^{14}_{7}N$ **d.** $^{13}_{8}O$

34. When a nucleus of the copper isotope $^{64}_{29}Cu$ captures an electron, it becomes

   a. $^{63}_{28}Ni$ **b.** $^{64}_{28}Ni$

   c. $^{63}_{29}Cu$ **d.** $^{64}_{30}Zn$

35. The product of the gamma decay of the aluminum isotope $^{27}_{13}Al$ is

   a. $^{27}_{12}Mg$ **b.** $^{26}_{13}Al$

   c. $^{27}_{13}Al$ **d.** $^{27}_{14}Si$

36. When the bismuth isotope $^{213}_{83}Bi$ decays into the polonium isotope $^{213}_{84}Po$, it emits

   a. an alpha particle   **b.** an electron

   c. a positron   **d.** a gamma ray

37. The half-life of a certain radionuclide is 6 h. If we start out with 10.0 g of the isotope, after 1 day there will be

   a. none left   **b.** 0.6 g left

   c. 1.6 g left   **d.** 2.5 g left

38. After 10 years 75 g of an original sample of 100 g of a certain radionuclide has decayed. The half-life of the isotope is

   a. 5 years   **b.** 7.5 years

   c. 20 years   **d.** 40 years

QUESTIONS

1. In what ways are the isotopes of an element similar to one another? In what ways are they different?

2. What happens to the atomic number and mass number of a nucleus that emits a gamma ray? What happens to the actual mass of the nucleus?

3. Radium undergoes spontaneous decay into helium and radon. Why is radium regarded as an element rather than as a chemical compound of helium and radon?

4. (a) Under what circumstances does a nucleus emit an electron? A positron? (b) The nuclei $^{14}_{8}O$ and $^{19}_{8}O$ both undergo beta decay in order to become stable nuclei. Which would you expect to emit a positron and which an electron?

5. What is the chief way in which heavy charged particles, such as protons and alpha particles, lose energy in passing through matter?

6. In what ways do electrons lose energy in passing through matter?

7. In what ways do gamma rays lose energy in passing through matter?

8. (a) What is the basis of the radiocarbon dating procedure? (b) What steps would you take to find the age of an ancient piece of wood by radiocarbon dating?

9. What are the two basic conditions that must be met by a radionuclide in order that it be useful in dating a particular kind of rock?

10. What is the limitation on the fuel that can be used in a reactor whose moderator is ordinary water? Why is the situation different if the moderator is heavy water?

11. Why might a gas-cooled reactor with a graphite moderator be safer than a reactor that uses water as coolant and moderator?

12. What are the differences and similarities between nuclear fission and nuclear fusion? Where does the evolved energy come from in each case?

13. Why are the conditions in the interior of a star favorable for nuclear fusion reactions?

EXERCISES

27.1 Nuclear Composition

1. State the number of neutrons and protons in each of the following nuclei: $^{10}_{5}Be$; $^{22}_{10}Ne$; $^{36}_{16}S$; $^{88}_{38}Sr$; $^{180}_{72}Hf$.

2. State the number of neutrons and protons in each of the following nuclei: $^{6}_{3}Li$; $^{13}_{6}C$; $^{31}_{15}P$; $^{94}_{40}Zr$; $^{137}_{56}Ba$.

3. Complete the following nuclear reactions:

$$^{6}_{3}Li + ? \rightarrow ^{7}_{4}Be + ^{1}_{0}n$$
$$^{10}_{5}B + ? \rightarrow ^{7}_{3}Li + ^{4}_{2}He$$
$$^{35}_{17}Cl + ? \rightarrow ^{32}_{16}S + ^{4}_{2}He$$

4. A nucleus of $^{15}_{7}N$ is struck by a proton. A nuclear reaction can take place with the emission of either a neutron or an alpha particle. Give the atomic number, mass number, and chemical name of the remaining nucleus in each of these cases.

5. A reaction often used to detect neutrons occurs when a neutron strikes a $^{10}_{5}B$ nucleus, with the subsequent emission of an alpha particle. What are the atomic number, mass number, and chemical name of the remaining nucleus?

6. Ordinary boron is a mixture of the $^{10}_{5}B$ and $^{11}_{5}B$ isotopes and has a composite atomic mass of 10.82 u. Find the percentage of each isotope present in ordinary boron. The atomic masses of $^{10}_{5}B$ and $^{11}_{5}B$ are, respectively, 10.01 u and 11.01 u.

7. Using the figures in Table 27.1, verify that the average atomic mass of chlorine is 35.46 u.

27.2 Binding Energy

8. The binding energy of $^{20}_{10}Ne$ is 160.64 MeV. Find its atomic mass.

9. The atomic mass of $^{56}_{26}Fe$ is 55.934939 u. Find its binding energy and binding energy per nucleon.

10. The neutron decays in free space into a proton and an electron. What must be the minimum binding energy contributed by a neutron to a nucleus in order that the neutron not decay inside the nucleus? How does this figure compare with the observed binding energies per nucleon in stable nuclei?

11. The nuclear reaction

$$^{6}_{3}Li + ^{2}_{1}H \rightarrow 2^{4}_{2}He$$

evolves 22.4 MeV. Calculate the atomic mass of $^{6}_{3}Li$ in u.

27.3 Radioactive Decay

12. What is the daughter nuclide of the alpha decay of the polonium isotope $^{210}_{84}Po$?

13. The nucleus $^{233}_{90}$Th undergoes two negative beta decays in becoming an isotope of uranium. What is the symbol of the isotope?

14. The copper isotope $^{64}_{29}$Cu can decay by either electron capture or positron emission. What does it become in each case?

15. A $^{80}_{35}$Br nucleus can decay by electron emission, positron emission, or electron capture. What is the daughter nucleus in each case?

16. The lead isotope $^{214}_{82}$Pb decays into the bismuth isotope $^{214}_{83}$Bi. What kind of decay is involved?

17. The thorium isotope $^{232}_{90}$Th decays through the successive emission of 6 alpha particles and 4 electrons. What stable nuclide does it finally become?

18. The nuclide $^{232}_{92}$U (mass 232.0372 u) alpha decays into $^{228}_{90}$Th (mass 228.0287 u). (a) Find the amount of energy released in the decay. (b) Is it possible for $^{232}_{92}$U to decay into $^{231}_{92}$U (mass 231.0364 u) by emitting a neutron? Why? (c) Is it possible for $^{232}_{92}$U to decay into $^{231}_{91}$Pa (mass 231.0359 u) by emitting a proton? Why? (The above masses refer to neutral atoms.)

19. The atomic mass of $^{226}_{88}$Ra is 226.0254 u, and the energy liberated in its alpha decay is 4.87 MeV. (a) Identify the daughter nucleus and find its atomic mass. (b) The alpha particle emitted in the decay is observed to have a KE of 4.78 MeV. Where do you think the other 0.09 MeV goes?

27.5   Half-Life

20. How many disintegrations per second occur in a 25-millicurie sample of thorium?

21. The half-life of radium is 1600 years. How long will it take for $\frac{15}{16}$ of a given sample of radium to decay?

22. The activity of a sample of $^{227}_{90}$Th is observed to be 12.5% of its original amount 54 days after the sample was prepared. What is the half-life of $^{227}_{88}$Th?

23. The half-life of the alpha-emitter $^{210}_{84}$Po is 138 days. This nuclide is used to power a thermoelectric cell whose initial output is 1.00 W. What will the output be after 3.00 years?

27.8   Fission

27.9   Nuclear Reactors

24. (a) How much mass is lost per day by a nuclear reactor operated at a 1.0-GW (10^9-W) power level? (b) If each fission in $^{235}_{92}$U releases 200 MeV, how many fissions must occur per second to yield a power level of 1.0 GW?

25. $^{235}_{92}$U loses about 0.1% of its mass when it undergoes fission. (a) How much energy is released when 1 kg of $^{235}_{92}$U undergoes fission? (b) One ton of TNT releases about 4×10^9 J when it is detonated. How many tons of TNT are equivalent in destructive power to a bomb that contains 1 kg of $^{235}_{92}$U?

27.10   Nuclear Fusion

26. Fusion reactions involving deuterons can take place when the deuterons have energies of about 10 keV or more. Find the temperature of a deuterium plasma in which the average KE of the deuterons is 10 keV.

27. Old stars obtain part of their energy by the fusion of three alpha particles to form a $^{12}_{6}$C nucleus, whose atomic mass is 12.0000 u. How much energy is evolved in each such reaction?

Answers to Multiple Choice

1.	a	11.	c	21.	a	31.	b
2.	a	12.	b	22.	d	32.	d
3.	a	13.	d	23.	c	33.	b
4.	c	14.	c	24.	d	34.	b
5.	b	15.	c	25.	b	35.	c
6.	a	16.	c	26.	c	36.	b
7.	b	17.	a	27.	a	37.	b
8.	c	18.	d	28.	c	38.	a
9.	c	19.	c	29.	b		
10.	c	20.	b	30.	a		

APPENDIX A
USEFUL MATHEMATICS

A–1 ALGEBRA

Algebra is a generalized arithmetic in which symbols are used in place of numbers. Algebra thus provides a language in which general relationships can be expressed among quantities whose numerical values need not be known in advance.

The arithmetical operations of addition, subtraction, multiplication, and division have the same meanings in algebra. The symbols of algebra are normally letters of the alphabet. If we have two quantities a and b and add them to give the sum c, we write

$$a + b = c$$

If we subtract b from a to give the difference d, we write

$$a - b = d$$

Multiplying a and b together to give e may be written in any of these ways:

$$a \times b = e \qquad ab = e \qquad (a)(b) = e$$

Whenever two algebraic quantities are written together with nothing between them, it is understood that they are to be multiplied.

Dividing a by b to give the quotient f is usually written

$$\frac{a}{b} = f$$

but it may sometimes be more convenient to write

$$a/b = f$$

which has the same meaning.

Parentheses and brackets are used to show the order in which various operations are to be performed. Thus

$$\frac{(a + b)c}{d} - e = f$$

means that, in order to find f, we are first to add a and b together, then multiply their sum by c and divide by d, and finally subtract e.

A–2 EQUATIONS

An *equation* is simply a statement that a certain quantity is equal to another one. Thus

$$7 + 2 = 9$$

which contains only numbers, is an arithmetical equation, and

$$3x + 12 = 27$$

which contains a symbol as well, is an algebraic equation. The symbols in an algebraic equation usually cannot have any arbitrary values if the equality is to hold. Finding the possible values of these symbols is called **solving** the equation. The **solution** of the second equation above is

$$x = 5$$

since only when x is 5 is it true that $3x + 12 = 27$.

In order to solve an equation, a basic principle must be kept in mind:

Any operation performed on one side of an equation must be performed on the other.

An equation remains valid when the same quantity is added to or subtracted from both sides, or when the same quantity is used to multiply or divide both sides. Other operations, for instance squaring or taking the square root, also do not alter the equality if the same thing is done to both sides. As an example, to solve $3x + 12 = 27$, we first subtract 12 from both sides:

$$3x + 12 - 12 = 27 - 12$$
$$3x = 15$$

To complete the solution we divide both sides by 3:

$$\frac{3x}{3} = \frac{15}{3}$$
$$x = 5$$

To check a solution, we substitute it back in the original equation and see whether the equality is still true. Thus we can check that $x = 5$ is correct by reducing the original algebraic equation to an arithmetical one:

$$3x + 12 = 27$$
$$(3)(5) + 12 = 27$$
$$15 + 12 = 27$$
$$27 = 27$$

Two helpful rules follow directly from the principle stated above. The first is,

Any term on one side of an equation may be transposed to the other side by changing its sign.

To verify this rule, we subtract b from each side of the equation

$$a + b = c$$

to obtain

$$a + b - b = c - b$$
$$a = c - b$$

Thus b has disappeared form the left-hand side and $-b$ is now on the right-hand side.

The second rule is,

A quantity which multiplies one side of an equation may be transposed in order to divide the other side, and vice versa.

To verify this rule, we divide both sides of the equation

$$ab = c$$

by b. The result is

$$\frac{ab}{b} = \frac{c}{b}$$
$$a = \frac{c}{b}$$

We see that b, a multiplier on the left-hand side, is now a divisor on the right-hand side.

EXAMPLE Solve the following equation for x:

$$4(x - 3) = 7$$

SOLUTION The above rules are easy to apply here:

$$4(x - 3) = 7$$
$$x - 3 = \frac{7}{4}$$
$$x = \frac{7}{4} + 3 = 1.75 + 3 = 4.75$$

When each side of an equation consists of a fraction, all we need to do to remove the fractions is to **cross multiply:**

$$\frac{a}{b} = \frac{c}{d}$$

$$\frac{a}{b} \diagdown\!\!\!\!\!\diagup \frac{c}{d}$$

$$ad = bc$$

What was originally the denominator (lower part) of each fraction now multiplies the numerator (upper part) of the other side of the equation.

EXAMPLE Solve the following equation for y:

$$\frac{5}{y + 2} = \frac{3}{y - 2}$$

SOLUTION First we cross multiply to get rid of the fractions, and then solve in the usual way:

$$5(y - 2) = 3(y + 2)$$
$$5y - 10 = 3y + 6$$
$$5y - 3y = 6 + 10$$
$$2y = 16$$
$$y = 8$$

EXERCISES

1. Solve each of the following equations for x. The answers are given at the end of Appendix A.

(a) $\dfrac{x + 7}{6} = x + 2$

(b) $\dfrac{3x - 42}{9} = 2(7 - x)$

(c) $\dfrac{1}{x + 1} = \dfrac{1}{2x - 1}$

(d) $\dfrac{8}{x} = \dfrac{1}{4 - x}$

(e) $\dfrac{x}{2x - 1} = \dfrac{5}{7}$

A–3 SIGNED NUMBERS

The rules for multiplying and dividing positive and negative quantities are straightforward. First, perform the indicated operation on the absolute value of the quantity (the absolute value of -7 is 7, for instance). If the quantities are both positive or both negative, the result is positive:

$$(+a)(+b) = (-a)(-b) = +ab$$

$$\frac{+a}{+b} = \frac{-a}{-b} = +\frac{a}{b}$$

If one quantity is positive and the other negative the result is negative:

$$(-a)(+b) = (+a)(-b) = -ab$$

$$\frac{-a}{+b} = \frac{+a}{-b} = -\frac{a}{b}$$

A few examples might be helpful:

$$(-6)(-3) = 18 \qquad \frac{-20}{-5} = 4$$

$$(4)(-5) = -20 \qquad \frac{6}{-2} = -3$$

$$(-10)(7) = -70 \qquad \frac{-30}{15} = -2$$

EXAMPLE Find the value of

$$z = \frac{xy}{x - y}$$

when $x = -12$ and $y = 4$.

SOLUTION We begin by evaluating xy and $x - y$, which are

$$xy = (-12)(4) = -48$$
$$x - y = (-12) - 4 = -16$$

Hence

$$z = \frac{xy}{x - y} = \frac{-48}{-16} = 3$$

EXERCISES

2. Evaluate the following:

(a) $\dfrac{3(x + y)}{2}$ when $x = 5, y = -2$

(b) $\dfrac{1}{x - y} - \dfrac{1}{x + y}$ when $x = 3, y = -5$

(c) $\dfrac{3(x + 7)}{y + 2}$ when $x = 3, y = -6$

(d) $\dfrac{x + y}{2z} + \dfrac{z}{x - y}$ when $x = -2, y = 2, z = 4$

(e) $\dfrac{x + z}{y} - \dfrac{xy}{2}$ when $x = 2, y = -8, z = 10$

A–4 SIMULTANEOUS EQUATIONS

Sometimes a problem has two unknown quantities, not just a single one. To solve such a problem we need two equations. For instance, the equation

$$2x + 1 = y$$

gives one relationship between x and y, but it holds for an infinite number of pairs of x and y values. However, if x and y must also obey the second equation

$$x + 2y = 12$$

then only when $x = 2$ and $y = 5$ will both equations be satisfied. The two equations

$$2x + 1 = y$$
$$x + 2y = 12$$

are called **simultaneous equations** because together they specify x and y.

Two methods can be used to solve a pair of simultaneous equations. The most direct is simply to solve one of the equations for one of the unknowns and then to substitute the expression for this unknown into the other equation, which can then be solved for the second unknown. An example will make this clear.

EXAMPLE Solve the simultaneous equations

$$2x + 1 = y$$
$$x + 2y = 12$$

SOLUTION From the first equation

$$y = 2x + 1$$

Now we substitute this value of y into the second equation and solve for x:

$$x + 2y = 12$$
$$x + 2(2x + 1) = 12$$
$$x + 4x + 2 = 12$$
$$5x = 10$$
$$x = 2$$

With this value of x we can solve either of the original equations for y:

$$y = 2x + 1 \qquad\qquad x + 2y = 12$$
$$= 2(2) + 1 \qquad\qquad 2 + 2y = 12$$
$$= 5 \qquad\qquad\qquad 2y = 10$$
$$y = 5$$

In the second method, which is sometimes easier, we first solve both equations for one of the unknowns. Then we set the two expressions equal to each other. This gives one equation in one unknown which can be solved in the usual way.

EXAMPLE   Solve the simultaneous equations

$$2x + 1 = y$$
$$x + 2y = 12$$

SOLUTION   We begin by solving both equations for y:

$$y = 2x + 1$$
$$y = \frac{12 - x}{2}$$

Now we proceed as follows to find x:

$$y = y$$
$$2x + 1 = \frac{12 - x}{2}$$
$$4x + 2 = 12 - x$$
$$5x = 10$$
$$x = 2$$

With this value of x we find as before that $y = 5$.

EXERCISES

3. Solve each pair of the following equations for x and y.

(a) $x + 2y = 3$, $2x - 2y = 3$

(b) $2x + 3y = 2$, $6x - 3y = 2$

(c) $\dfrac{x}{2} + y = 0$, $3x - 4y = 5$

(d) $2x + 3y = -8$, $3x - 5y = 7$

(e) $x + y = 20$, $\dfrac{x}{2} + 2y = 25$

A–5   EXPONENTS

It is often necessary to multiply a quantity by itself a number of times. This process is indicated by a superscript number called the exponent, according to the following scheme:

$$A = A^1$$
$$A \times A = A^2$$
$$A \times A \times A = A^3$$
$$A \times A \times A \times A = A^4$$
$$A \times A \times A \times A \times A = A^5$$

We read A^2 as "A squared" because it is the area of a square of length A on a side; similarly A^3 is called "A cubed" because it is the volume of a cube each of whose sides is A long. More generally we speak of A^n as "A to the nth power." Thus A^5 is read as "A to the fifth power."

When we multiply a quantity raised to some particular power (say A^n) by the same quantity raised to another power (say A^m), the result is that quantity raised to a power equal to the sum of the original exponents. That is,

$$A^n A^m = A^{(n + m)}$$

For example,

$$A^2 A^5 = A^7$$

which we can check by writing out the terms:

$$(A \times A)(A \times A \times A \times A \times A)$$
$$= A \times A \times A \times A \times A \times A \times A = A^7$$

From the above result we see that when a quantity raised to a particular power (say A^n) is to be multiplied by itself a total of m times, we have

$$(A^n)^m = A^{nm}$$

For example,

$$(A^2)^3 = A^6$$

since

$$(A^2)^3 = A^2 \times A^2 \times A^2 = A^{(2 + 2 + 2)} = A^6$$

Reciprocal quantities are expressed in a similar way with the addition of a minus sign in the exponent, as follows:

$$\frac{1}{A} = A^{-1} \quad \frac{1}{A^2} = A^{-2} \quad \frac{1}{A^3} = A^{-3} \quad \frac{1}{A^4} = A^{-4}$$

Exactly the same rules as before are used in combining quantities raised to negative powers with one another and with some quantity raised to a positive power. Thus

$$A^5 A^{-2} = A^{(5-2)} = A^3$$
$$(A^{-1})^{-2} = A^{-1(-2)} = A^2$$
$$(A^3)^{-4} = A^{-4 \times 3} = A^{-12}$$
$$A A^{-7} = A^{(1-7)} = A^{-6}$$

It is important to remember that any quantity raised to the zeroth power, say A^0, is equal to 1. Hence

$$A^2 A^{-2} = A^{(2-2)} = A^0 = 1$$

This is more easily seen if we write A^{-2} as $1/A^2$:

$$A^2 A^{-2} = A^2 \times \frac{1}{A^2} = \frac{A^2}{A^2} = 1$$

A–6 ROOTS

The **square root** of A, $\sqrt{A}$, is that quantity which, when multiplied by itself, is equal to A:

$$\sqrt{A} \times \sqrt{A} = A$$

The square root has that name because the length of each side of a square of area A is given by $\sqrt{A}$. Here are some examples of square roots:

$$\sqrt{1} = 1 \qquad \text{because } 1 \times 1 = 1$$
$$\sqrt{16} = 4 \qquad \text{because } 4 \times 4 = 16$$
$$\sqrt{100} = 10 \qquad \text{because } 10 \times 10 = 100$$
$$\sqrt{42.25} = 6.5 \qquad \text{because } 6.5 \times 6.5 = 42.25$$
$$\sqrt{9A^2} = 3A \qquad \text{because } 3A \times 3A = 9A^2$$

The square root of a number less than 1 is larger than the number itself:

$$\sqrt{0.01} = 0.1 \qquad \text{because } 0.1 \times 0.1 = 0.01$$
$$\sqrt{0.49} = 0.7 \qquad \text{because } 0.7 \times 0.7 = 0.49$$

The square root of a quantity may be either positive or negative because $(+A)(+A) = (-A)(-A) = A^2$. Whether the positive or negative root (or even both) is correct in a given problem depends upon the details of the problem and is usually obvious.

Using exponents we see that, because

$$(A^{1/2})^2 = A^{2 \times (1/2)} = A^1 = A$$

we can express square roots by the exponent $\frac{1}{2}$:

$$\sqrt{A} = A^{1/2}$$

Other roots may be expressed similarly. The **cube root** of a quantity A, written $\sqrt[3]{A}$, when multiplied by itself twice equals A. That is,

$$\sqrt[3]{A} \times \sqrt[3]{A} \times \sqrt[3]{A} = A$$

which may be more conveniently written

$$(A^{1/3})^3 = A$$

where $\sqrt[3]{A} = A^{1/3}$.

In general the nth root of a quantity, $\sqrt[n]{A}$, may be written $A^{1/n}$, which is a more convenient form for most purposes. Some examples may be helpful:

$$\sqrt{A^4} = (A^4)^{1/2} = A^{1/2 \times 4} = A^2$$
$$(\sqrt{A})^4 = (A^{1/2})^4 = A^{4 \times 1/2} = A^2$$
$$\sqrt{\sqrt{A}} = (A^{1/2})^{1/2} = A^{1/2 \times 1/2} = A^{1/4}$$
$$(A^{1/4})^{-7} = A^{-7 \times (1/4)} = A^{-7/4}$$
$$(A^3)^{-1/3} = A^{-(1/3) \times 3} = A^{-1}$$
$$(A^{1/4})^{1/4} = A^{(1/4) \times (1/4)} = A^{1/16}$$

Although procedures exist for finding roots arithmetically with pencil and paper, in practice it is much easier to use an electronic calculator. Calculators for science and engineering usually have a square root [$\sqrt{}$] key. For other roots, such as $A^{1/3}$, and for nonintegral exponents in general, such as $A^{3.27}$, logarithms must be used, as described in Appendix B-5. A calculator with [LOG] and [10^x] or [INV LOG] keys makes such quantities easy to evaluate.

A–7 EQUATIONS WITH POWERS AND ROOTS

An equation that involves powers or roots or both is subject to the same basic principle that simpler equations obey: Whatever is done to one side must be done to the other. Hence the following rules:

An equation remains valid when both sides are raised to the same power, that is, when each side is multiplied by itself the same number of times as the other side.

An equation remains valid when the same root is taken of both sides.

EXAMPLE Newton's law of gravitation states that the force F between two bodies the distance r apart whose masses are m_A and m_B is given by the formula

$$F = G\frac{m_A m_B}{r^2}$$

where G is a universal constant. Solve this formula for r.

SOLUTION We begin by transposing the r^2 to give

$$Fr^2 = Gm_A m_B$$

Next we transpose the F to give

$$r^2 = \frac{Gm_A m_B}{F}$$

and finally take the square root of both sides:

$$r = \sqrt{r^2} = \sqrt{\frac{Gm_A m_B}{F}}$$

The positive value of the root is correct here since a negative separation r between the bodies is meaningless.

Quadratic equations, which have the general form

$$ax^2 + bx + c = 0$$

are often encountered in algebraic calculations. In such an equation, a, b, and c are constants. A quadratic equation is satisfied by the values of x given by the formulas

$$x_+ = \frac{-b + \sqrt{b^2 - 4ac}}{2a}$$

$$x_- = \frac{-b - \sqrt{b^2 - 4ac}}{2a}$$

By looking at these formulas, we can see that the nature of the solution depends upon the value of the quantity $b^2 - 4ac$. When $b^2 = 4ac$, $\sqrt{b^2 - 4ac} = 0$ and the two solutions are equal to just $x = -b/2a$. When $b^2 > 4ac$, $\sqrt{b^2 - 4ac}$ is a real number and the solutions x_+ and x_- are different. When $b^2 < 4ac$, $\sqrt{b^2 - 4ac}$ is the square root of a negative number, and the solutions are different. The square root of a negative number is called an **imaginary number** because squaring a real number, whether positive or negative, always gives a positive number: $(+2)^2 = (-2)^2 = +4$, hence $\sqrt{-4}$ cannot be either $+2$ or -2. Imaginary numbers do not occur in the sort of physical problems treated in this book.

EXAMPLE Solve the quadratic equation

$$2x^2 - 3x - 9 = 0$$

SOLUTION Here $a = 2$, $b = -3$, and $c = -9$, so that

$$b^2 - 4ac = (-3)^2 - 4(2)(-9) = 9 + 72 = 81$$

The two solutions are

$$x_+ = \frac{-b + \sqrt{b^2 - 4ac}}{2a}$$

$$= \frac{-(-3) + \sqrt{81}}{2(2)} = \frac{3 + 9}{4} = \frac{12}{4} = 3$$

$$x_- = \frac{-b - \sqrt{b^2 - 4ac}}{2a}$$

$$= \frac{-(-3) - \sqrt{81}}{2(2)} = \frac{3 - 9}{4} = \frac{-6}{4} = -\frac{3}{2}$$

EXAMPLE Solve the quadratic equation

$$x^2 - 4x + 4 = 0$$

SOLUTION Here $a = 1$, $b = -4$, and $c = 4$, so that

$$b^2 - 4ac = (-4)^2 - 4(1)(4) = 16 - 16 = 0$$

Hence the only solution is

$$x = \frac{-b}{2a} = \frac{-(-4)}{2} = 2$$

EXERCISES

4. Solve each of the following equations for x.

    (a) $9x^2 + 12x + 4 = 0$

    (b) $x^2 + x - 2 = 0$

    (c) $2x^2 - 5x + 2 = 0$

    (d) $x^2 + x - 20 = 0$

    (e) $4x^2 - 4x - 11 = 0$

A–8 TRIGONOMETRIC FUNCTIONS

A **right triangle** is a triangle two sides of which are perpendicular. Such triangles frequently occur in physics, and we must know how their sides and angles are related. The **hypotenuse** of a right triangle is the side opposite the right angle, as in Fig. A–1; it is always the longest side. The

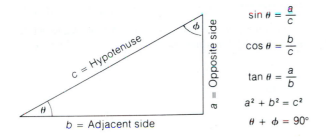

$$\sin\theta = \frac{a}{c}$$

$$\cos\theta = \frac{b}{c}$$

$$\tan\theta = \frac{a}{b}$$

$$a^2 + b^2 = c^2$$

$$\theta + \phi = 90°$$

FIG. A–1 A right triangle

three basic trigonometric functions, the **sine, cosine,** and **tangent** of an angle, are defined as follows:

$$\sin\theta = \frac{a}{c} = \frac{\text{opposite side}}{\text{hypotenuse}}$$

$$\cos\theta = \frac{b}{c} = \frac{\text{adjacent side}}{\text{hypotenuse}}$$

$$\tan\theta = \frac{a}{b} = \frac{\text{opposite side}}{\text{adjacent side}}$$

From these definitions we see that

$$\frac{\sin\theta}{\cos\theta} = \frac{a/c}{b/c} = \frac{a}{b} = \tan\theta$$

Numerical values of $\sin\theta$, $\cos\theta$, and $\tan\theta$ for angles form 0° to 90° are given in Appendix C. These figures may be used for angles from 90° to 360° with the help of this table:

θ	$90° + \theta$	$180° + \theta$	$270° + \theta$
sin	$\sin\theta$	$\cos\theta$	$-\sin\theta$	$-\cos\theta$
cos	$\cos\theta$	$-\sin\theta$	$-\cos\theta$	$\sin\theta$
tan	$\tan\theta$	$-\dfrac{1}{\tan\theta}$	$\tan\theta$	$-\dfrac{1}{\tan\theta}$

For example, if we require the value of $\sin 120°$, we first note that $120° = 90° + 30°$. Then, since

$$\sin(90° + \theta) = \cos\theta$$

we have

$$\sin 120° = \sin(90° + 30°) = \cos 30° = 0.866$$

Similarly, to find $\tan 342°$, we begin by noting that $342° = 270° + 72°$. Then, since

$$\tan(270° + \theta) = -\frac{1}{\tan\theta}$$

we have

$$\tan 342° = \tan(270° + 72°) = -\frac{1}{\tan 72°} = -\frac{1}{3.078}$$

$$= -0.325$$

The **inverse** of a trigonometric function is the angle whose function is given. For instance, the inverse of $\sin\theta$ is the angle θ. If $\sin\theta = x$, then the angle θ may be designated either as $\theta = \arcsin x$ or as $\theta = \sin^{-1} x$. It is important to keep in mind that an expression such as $\sin^{-1} x$ does *not* mean $1/\sin x$. The inverse trigonometric functions are as follows:

$$\sin\theta = x$$
$$\theta = \arcsin x = \sin^{-1} x = \text{angle whose sine is } x$$
$$\cos\theta = y$$
$$\theta = \arccos y = \cos^{-1} y = \text{angle whose cosine is } y$$
$$\tan\theta = z$$
$$\theta = \arctan z = \tan^{-1} z = \text{angle whose tangent is } z$$

To find an inverse function, say the angle whose cosine is 0.907, the procedure with most calculators is to enter 0.907 and then press in succession the [ARC] or [INV] and [COS] keys. The result is $\theta = 25°$ to the nearest degree. With a table of cosines, we look for the value nearest to 0.907 in the body of the table, which is 0.906, and then read across to find the corresponding angle.

EXERCISES

5. Express the following functions in terms of angles between 0 and 90°.

(a) $\sin 100°$

(b) $\sin 300°$

(c) $\cos 150°$

(d) $\cos 350°$

(e) $\tan 180°$

(f) $\tan 220°$

A–9 SOLVING A RIGHT TRIANGLE

To solve a given triangle means to find the values of any unknown sides or angles in terms of the values of the known sides and angles. A triangle has three sides and three angles, and we need the values of at least three of these six quantities, including one of the sides, to solve the triangle for the others. In a right triangle, one of the angles is always 90°, and so all we need here are the lengths of any two of its sides or the length of one side and the value of one of the other angles to find the remaining sides and angles.

Suppose we know the length of the side b and the angle θ in the right triangle of Fig. A–1. From the definitions of sine and tangent we see that

$$\tan \theta = \frac{a}{b} \qquad a = b \tan \theta$$

$$\sin \theta = \frac{a}{c} \qquad c = \frac{a}{\sin \theta}$$

This gives us the two unknown sides a and c. To find the unknown angle ϕ, we can use any of these formulas:

$$\phi = \sin^{-1}\frac{b}{c} \qquad \phi = \cos^{-1}\frac{a}{c} \qquad \phi = \tan^{-1}\frac{b}{a}$$

Alternatively, we can use the fact that the sum of the angles in any triangle is 180°. Because one of the angles in a right triangle is 90°, the sum of the other two must be 90°:

$$\theta + \phi = 90°$$

Hence $\phi = 90° - \theta$ here.

Another useful relationship in a right triangle is the Pythagorean theorem: The sum of the squares of the sides of such a triangle adjacent to the right angle is equal to the square of its hypotenuse. For the triangle of Fig. A–1,

$$a^2 + b^2 = c^2$$

Thus we can always express the length of any of the sides of a right triangle in terms of the other sides:

$$a = \sqrt{c^2 - b^2}$$
$$b = \sqrt{c^2 - a^2}$$
$$c = \sqrt{a^2 + b^2}$$

EXAMPLE In the triangle of Fig. A–1, $a = 7$ cm and $b = 10$ cm. Find c, θ, and ϕ.

SOLUTION From the Pythagorean theorem,

$$c = \sqrt{a^2 + b^2} = \sqrt{7^2 + 10^2}\,\text{cm}$$
$$= \sqrt{149}\,\text{cm} = 12.2\,\text{cm}$$

Since $\tan \theta = a/b$,

$$\theta = \tan^{-1}\frac{a}{b} = \tan^{-1}\frac{7\,\text{cm}}{10\,\text{cm}}$$
$$= \tan^{-1}0.7 = 35°$$

The value of the other angle ϕ is given by

$$\phi = 90° - \theta = 90° - 35° = 55°$$

EXERCISES

6. Find the values of the unknown sides and angles in the right triangles for which the following data are known:

   (a) $\theta = 45°$, $a = 10$

   (b) $\theta = 15°$, $b = 4$

   (c) $\theta = 25°$, $c = 5$

   (d) $a = 3$, $b = 4$

   (e) $a = 5$, $c = 13$

ANSWERS

1. (a) -1
     (b) 8
     (c) 2
     (d) 32/9
     (e) 5/3

2. (a) 4.5
     (b) 0.625
     (c) -3
     (d) -1
     (e) 6.5

3. (a) $x = 2, y = \frac{1}{2}$
     (b) $x = \frac{1}{2}, y = \frac{1}{3}$
     (c) $x = 1, y = -\frac{1}{2}$
     (d) $x = -1, y = -2$
     (e) $x = 10, y = 10$

4. (a) $-2/3$
     (b) $1, -2$
     (c) $2, 0.5$
     (d) $4, -5$
     (e) $0.5 \pm \sqrt{3} = 2.232, -1.232$

5. (a) $\cos 10°$
     (b) $-\cos 30°$
     (c) $-\sin 60°$
     (d) $\sin 80°$
     (e) $-1/\tan 90°$ or $\tan 0$. Since $\tan 90° = \infty$ and $\tan 0 = 0$, both expressions give the same value.
     (f) $\tan 40°$

6. (a) $b = 10, c = 14.1, \phi = 45°$
     (b) $a = 1.07, c = 4.14, \phi = 75°$
     (c) $a = 2.11, b = 4.53, \phi = 65°$
     (d) $c = 5, \theta = 37°, \phi = 53°$
     (e) $b = 12, \theta = 23°, \phi = 67°$

APPENDIX B
POWERS OF TEN
AND LOGARITHMS

B–1 POWERS OF TEN

Very small and very large numbers are common in physics. For example, the mass of an electron is 0.000, 000,000,000,000,000,000,000,000,910,9 kilogram, and the mass of the earth is 5,983,000,000,000, 000,000,000,000 kilograms. Such numbers in ordinary decimal form are clumsy to write and to make calculations with, and it is hard to appreciate their precise magnitudes because of the sea of zeros.

A better way to express these numbers makes use of powers-of-ten notation. This method is based on the fact that all numbers may be represented by a number between 1 and 10 multiplied by a power of 10. In powers-of-ten notation the mass of an electron is written simply as 9.109×10^{-31} kilogram and the mass of the earth is written as 5.983×10^{24} kilogram.

Table B-1 contains powers of 10 from 10^{-10} to 10^{10}. Evidently positive powers of 10 (which cover numbers greater than 1) follow this pattern:

TABLE B-1

$10^{-10} = 0.000,000,000,1$	$10^{0} = 1$
$10^{-9} = 0.000,000,001$	$10^{1} = 10$
$10^{-8} = 0.000,000,01$	$10^{2} = 100$
$10^{-7} = 0.000,000,1$	$10^{3} = 1000$
$10^{-6} = 0.000,001$	$10^{4} = 10,000$
$10^{-5} = 0.000,01$	$10^{5} = 100,000$
$10^{-4} = 0.000,1$	$10^{6} = 1,000,000$
$10^{-3} = 0.001$	$10^{7} = 10,000,000$
$10^{-2} = 0.01$	$10^{8} = 100,000,000$
$10^{-1} = 0.1$	$10^{9} = 1,000,000,000$
$10^{0} = 1$	$10^{10} = 10,000,000,000$

$$10^0 = 1. \qquad = 1 \qquad = 1 \text{ with decimal point moved 0 places,}$$
$$10^1 = 1.0 \qquad = 10 \qquad = 1 \text{ with decimal point moved 1 place to the right,}$$
$$10^2 = 1.00 \qquad = 100 \qquad = 1 \text{ with decimal point moved 2 places to the right,}$$
$$10^3 = 1.000 \qquad = 1,000 \qquad = 1 \text{ with decimal point moved 3 places to the right,}$$
$$10^4 = 1.0000 \qquad = 10,000 \qquad = 1 \text{ with decimal point moved 4 places to the right,}$$
$$10^5 = 1.00000 \qquad = 100,000 \qquad = 1 \text{ with decimal point moved 5 places to the right,}$$
$$10^6 = 1.000000 = 1,000,000 = 1 \text{ with decimal point moved 6 places to the right, and so on.}$$

The exponent of the 10 indicates how many places the decimal point is moved *to the right* from $1.000 \cdots$

A similar pattern is followed by negative powers of 10, whose values always lie between 0 and 1:

$$10^0 \;\; = 1. \qquad\qquad = 1 \qquad\quad = 1 \text{ with decimal point moved 0 places,}$$

$$10^{-1} = 01. \qquad\quad = 0.1 \qquad\; = 1 \text{ with decimal point moved 1 place to the left,}$$

$$10^{-2} = 001. \qquad\quad = 0.01 \qquad = 1 \text{ with decimal point moved 2 places to the left,}$$

$$10^{-3} = 0001. \qquad\quad = 0.001 \qquad = 1 \text{ with decimal point moved 3 places to the left,}$$

$$10^{-4} = 00001. \qquad = 0.000,1 \quad = 1 \text{ with decimal point moved 4 places to the left,}$$

$$10^{-5} = 000001. \qquad = 0.000,01 \; = 1 \text{ with decimal point moved 5 places to the left,}$$

$$10^{-6} = 0000001. = 0.000,001 = 1 \text{ with decimal point moved 6 places to the left, and so on.}$$

The exponent of the 10 now indicates how many places the decimal point is moved *to the left* from 1.

Here are a few examples of powers-of-ten notation:

$$600 = 6 \times 100 = 6 \times 10^2$$
$$7940 = 7.94 \times 1000 = 7.94 \times 10^3$$
$$93{,}000{,}000 = 9.3 \times 10{,}000{,}000 = 9.3 \times 10^7$$
$$0.023 = 2.3 \times 0.01 = 2.3 \times 10^{-2}$$
$$0.000{,}035 = 3.5 \times 0.000{,}01 = 3.5 \times 10^{-5}$$

EXERCISES

1. Express the following numbers in decimal notation.

(a) 2×10^5

(b) 8×10^{-2}

(c) 7.819×10^2

(d) 4.51×10^8

(e) 1.003×10^{-6}

(f) 10^{-10}

(g) 9.56×10^{-5}

2. Express the following numbers in powers-of-ten notation.

(a) 70

(b) 0.14

(c) 3.81

(d) 8400

(e) 1,000,000

(f) 0.007,890

(g) 351,600

B–2 USING POWERS OF TEN

Let us see how to make calculations using numbers written in powers-of-ten notation. To add or subtract numbers written in powers-of-ten notation, they must be expressed in terms of the *same* power of ten.

$$7 \times 10^4 + 2 \times 10^5 = 0.7 \times 10^5 + 2 \times 10^5$$
$$= 2.7 \times 10^5$$
$$5 \times 10^{-2} + 3 \times 10^{-4} = 5 \times 10^{-2} + 0.03 \times 10^{-2}$$
$$= 5.03 \times 10^{-2}$$
$$8 \times 10^{-3} - 7 \times 10^{-4} = 8 \times 10^{-3} - 0.7 \times 10^{-3}$$
$$= 7.3 \times 10^{-3}$$
$$4 \times 10^5 - 1 \times 10^6 = 4 \times 10^5 - 10 \times 10^5$$
$$= -6 \times 10^5$$

To multiply powers of ten together, add their exponents:

$$(10^n)(10^m) = 10^{n + m}$$

Be sure to take the sign of each exponent into account.

$$(10^2)(10^3) = 10^{2 + 3} = 10^5$$
$$(10^7)(10^{-3}) = 10^{7 - 3} = 10^4$$
$$(10^{-2})(10^{-4}) = 10^{-2-4} = 10^{-6}$$

To multiply numbers written in powers-of-ten notation, multiply the decimal parts of the numbers together and add the exponents to find the power of ten of the product:

$$(A \times 10^n)(B \times 10^m) = AB \times 10^{n + m}$$

If necessary, rewrite the result so the decimal part is a number between 1 and 10.

$$(3 \times 10^2)(2 \times 10^5) = (3 \times 2) \times 10^{2 + 5}$$
$$= 6 \times 10^7$$
$$(8 \times 10^{-5})(3 \times 10^7) = (8 \times 3) \times 10^{-5+7}$$
$$= 24 \times 10^2 = 2.4 \times 10^3$$
$$(1.3 \times 10^{-3})(4 \times 10^{-5}) = (1.3 \times 4) \times 10^{-3-5}$$
$$= 5.2 \times 10^{-8}$$
$$(-9 \times 10^{17})(6 \times 10^{-18}) = (-9 \times 6) \times 10^{17 - 18}$$
$$= -54 \times 10^{-1} = -5.4$$

To divide one power of ten by another, subtract the exponent of the denominator from the exponent of the numerator:

$$\frac{10^n}{10^m} = 10^{n-m}$$

Be sure to take the sign of each exponent into account.

$$\frac{10^5}{10^3} = 10^{5-3} = 10^2$$

$$\frac{10^{-2}}{10^4} = 10^{-2-4} = 10^{-6}$$

$$\frac{10^{-3}}{10^{-7}} = 10^{-3-(-7)} = 10^{-3+7} = 10^4$$

To divide a number written in powers-of-ten notation by another number written that way, divide the decimal parts of the numbers in the usual way and use the above rule to find the exponent of the power of ten of the quotient:

$$\frac{A \times 10^n}{B \times 10^m} = \frac{A}{B} \times 10^{n-m}$$

If necessary, rewrite the result so the decimal part is a number between 1 and 10:

$$\frac{6 \times 10^5}{3 \times 10^2} = \frac{6}{3} \times 10^{5-2} = 2 \times 10^3$$

$$\frac{2 \times 10^{-7}}{8 \times 10^4} = \frac{2}{8} \times 10^{-7-4} = \frac{1}{4} \times 10^{-11}$$

$$= 0.25 \times 10^{-11} = 2.5 \times 10^{-12}$$

$$\frac{-7 \times 10^5}{10^{-2}} = -7 \times 10^{5-(-2)} = -7 \times 10^{5+2}$$

$$= -7 \times 10^7$$

$$\frac{5 \times 10^{-2}}{-2 \times 10^{-9}} = -\frac{5}{2} \times 10^{-2-(-9)} = -2.5 \times 10^{-2+9}$$

$$= -2.5 \times 10^7$$

To find the reciprocal of a power of ten, change the sign of the exponent:

$$\frac{1}{10^n} = 10^{-n}$$

$$\frac{1}{10^{-m}} = 10^m$$

$$\frac{1}{10^5} = 10^{-5}$$

$$\frac{1}{10^{-3}} = 10^3$$

Hence the prescription for finding the reciprocal of a number written in powers-of-ten notation is

$$\frac{1}{A \times 10^n} = \frac{1}{A} \times 10^{-n}$$

For example,

$$\frac{1}{2 \times 10^3} = \frac{1}{2} \times 10^{-3} = 0.5 \times 10^{-3} = 5 \times 10^{-4}$$

$$\frac{1}{4 \times 10^{-8}} = \frac{1}{4} \times 10^8 = 0.25 \times 10^8 = 2.5 \times 10^7$$

The powers-of-ten method of writing large and small numbers makes arithmetic involving such numbers relatively easy to carry out. Here is a calculation that would be very tedious if each number were kept in decimal form.

$$\frac{(3800)(0.0054)(0.000,001)}{(430,000,000)(73)}$$

$$= \frac{(3.8 \times 10^3)(5.4 \times 10^{-3})(10^{-6})}{(4.3 \times 10^8)(7.3 \times 10^1)}$$

$$= \frac{(3.8)(5.4)}{(4.3)(7.3)} \times \frac{(10^3)(10^{-3})(10^{-6})}{(10^8)(10^1)}$$

$$= 0.65 \times 10^{(3-3-6-8-1)} = 0.65 \times 10^{-15}$$

$$= 6.5 \times 10^{-16}$$

EXERCISES

3. Perform the following additions and subtractions.

(a) $3 \times 10^2 + 4 \times 10^3$
(b) $7 \times 10^{-2} + 2 \times 10^{-3}$
(c) $4 \times 10^{-5} + 5 \times 10^{-3}$
(d) $6.32 \times 10^2 + 5$
(e) $4 \times 10^3 - 3 \times 10^2$
(f) $3.2 \times 10^{-4} - 5 \times 10^{-5}$
(g) $7 \times 10^4 - 2 \times 10^{-5}$
(h) $4.76 \times 10^{-3} - 4.81 \times 10^{-3}$

4. Evaluate the following reciprocals.

(a) $\dfrac{1}{10^2}$ (c) $\dfrac{1}{10^{-2}}$

(b) $\dfrac{1}{2 \times 10^2}$ (d) $\dfrac{1}{4 \times 10^{-4}}$

5. Perform the following calculations.

(a) $\dfrac{(500,000)(18,000)}{9,000,000}$

(b) $\dfrac{(30)(80,000,000,000)}{0.0004}$

(c) $\dfrac{(30{,}000)(0.000{,}000{,}6)}{(1000)(0.02)}$

(d) $\dfrac{(0.002)(0.000{,}000{,}05)}{0.000{,}004}$

(e) $\dfrac{(0.06)(0.0001)}{(0.000{,}03)(40{,}000)}$

(f) $\dfrac{(3 \times 10^4)(5 \times 10^{-12})}{10^3}$

(g) $\dfrac{9 \times 10^{12}}{9 \times 10^{-12}}$

(h) $\dfrac{(8 \times 10^{10})(3)}{6 \times 10^{-4}}$

(i) $\dfrac{10^{-3}}{(5 \times 10^4)(2 \times 10^2)}$

(j) $\dfrac{(5 \times 10^5)(2 \times 10^{-18})}{4 \times 10^4}$

B–3 POWERS AND ROOTS

To square a power of ten, multiply the exponent by 2; to cube a power of ten, multiply the exponent by 3:

$$(10^n)^2 = 10^{2n}$$
$$(10^n)^3 = 10^{3n}$$

In general, to raise a power of ten to the mth power, multiply the exponent by m:

$$(10^n)^m = 10^{m \times n}$$

Be sure to take the sign of each exponent into account, as in these examples:

$$(10^3)^2 = 10^{2 \times 3} = 10^6$$
$$(10^{-2})^5 = 10^{5 \times -2} = 10^{-10}$$
$$(10^{-4})^{-2} = 10^{-2 \times -4} = 10^8$$

To raise a number written in powers-of-ten notation to the mth power, multiply the decimal part of the number by itself m times and multiply the exponent of the power of ten by m:

$$(A \times 10^n)^m = A^m \times 10^{m \times n}$$

If necessary, rewrite the result so the decimal part is a number between 1 and 10:

$$(2 \times 10^5)^2 = 2^2 \times 10^{2 \times 5} = 4 \times 10^{10}$$

$$(3 \times 10^{-3})^3 = 3^3 \times 10^{3 \times -3} = 27 \times 10^{-9}$$
$$= 2.7 \times 10^{-8}$$
$$(5 \times 10^{-2})^{-4} = 5^{-4} \times 10^{-4 \times -2} = \frac{1}{5^4} \times 10^8$$
$$= \frac{1}{625} \times 10^8 = 0.0016 \times 10^8$$
$$= 1.6 \times 10^5$$

To take the mth root of a power of ten, divide the exponent by m:

$$\sqrt[m]{10^n} = (10^n)^{1/m} = 10^{n/m}$$

Thus

$$\sqrt{10^4} = (10^4)^{1/2} = 10^{4/2} = 10^2$$
$$\sqrt[3]{10^9} = (10^9)^{1/3} = 10^{9/3} = 10^3$$
$$\sqrt[3]{10^{-9}} = (10^{-9})^{1/3} = 10^{-9/3} = 10^{-3}$$

In powers-of-ten notation, the exponent of the 10 must be an integer. Hence in taking the mth root of a power of ten, the exponent should be an integral multiple of m. Instead of, for example,

$$\sqrt{10^5} = (10^5)^{1/2} = 10^{2.5}$$

which, while correct, is hardly useful, we would write

$$\sqrt{10^5} = \sqrt{10^1 \times 10^4}$$
$$= \sqrt{10} \times \sqrt{10^4}$$
$$= 3.16 \times 10^2$$

Here are two other examples:

$$\sqrt{10^{-3}} = \sqrt{10^1 \times 10^{-4}} = \sqrt{10} \times \sqrt{10^{-4}}$$
$$= 3.16 \times 10^{-2}$$
$$\sqrt[3]{10^8} = \sqrt[3]{10^2 \times 10^6} = \sqrt[3]{100} \times \sqrt[3]{10^6}$$
$$= 4.64 \times 10^2$$

To take the mth root of a number expressed in powers-of-ten notation, first write the number so the exponent of the 10 is an integral multiple of m. Then take the mth root of the decimal part of the number and divide the exponent by m to find the power of ten of the result:

$$(A \times 10^n)^{1/m} = \sqrt[m]{A} \times 10^{n/m}$$

Thus

$$\sqrt{9 \times 10^4} = \sqrt{9} \times 10^{4/2} = 3 \times 10^2$$
$$\sqrt{9 \times 10^{-6}} = \sqrt{9} \times 10^{-6/2} = 3 \times 10^{-3}$$

$$\sqrt{4 \times 10^7} = \sqrt{40 \times 10^6} = \sqrt{40} \times 10^{6/2}$$
$$= 6.32 \times 10^3$$
$$\sqrt[3]{3 \times 10^{-5}} = \sqrt[3]{30 \times 10^{-6}} \quad \sqrt[3]{30} \times 10^{-6/3}$$
$$= 3.11 \times 10^{-2}$$

EXERCISES

6. Evaluate the following powers.

(a) $(2 \times 10^7)^2$ (c) $(3 \times 10^{-8})^2$

(b) $(2 \times 10^7)^{-2}$ (d) $(5 \times 10^{-4})^{-3}$

7. Evaluate the following roots. Assume $\sqrt{4} = 2$, $\sqrt{40} = 6.3$, $\sqrt[3]{4} = 1.6$, $\sqrt[3]{40} = 3.4$, and $\sqrt[3]{400} = 7.4$.

(a) $(4 \times 10^6)^{1/2}$ (f) $(4 \times 10^{13})^{1/3}$

(b) $(4 \times 10^7)^{1/2}$ (g) $(4 \times 10^{14})^{1/3}$

(c) $(4 \times 10^{-4})^{1/2}$ (h) $(4 \times 10^{-6})^{1/3}$

(d) $(4 \times 10^{-5})^{1/2}$ (i) $(4 \times 10^{-7})^{1/3}$

(e) $(4 \times 10^{-12})^{1/3}$ (j) $(4 \times 10^{-8})^{1/3}$

B–4 SIGNIFICANT FIGURES

An advantage of powers-of-ten notation is that it gives no false impression of the degree of accuracy with which a number is stated. For instance, the equatorial radius of the earth is 6378 km, but it is often taken as 6400 km for convenience in making rough calculations. To show the approximate nature of the latter figure, all we have to do is write

$$r = 6.4 \times 10^3 \text{ km}$$

whose meaning is

$$r = (6.4 \pm 0.05) \times 10^3 \text{ km}$$

With this method, how large the number is and how accurate it is are both clear. The accurately known digits, plus one uncertain digit, are called **significant figures.** In the above case, r has two significant figures, 6 and 4. If we require grater accuracy, we would write

$$r = 6.38 \times 10^3 \text{ km}$$

which has three significant figures, or

$$r = 6.378 \times 10^3 \text{ km}$$

which has four significant figures.

Sometimes one or more zeros in a number are significant figures, and it is proper to retain them when expressing the number in powers-of-ten notation. There is quite a difference between 3×10^5 and 3.00×10^5:

$$3 \times 10^5 = (3 \pm 0.5) \times 10^5$$
$$3.00 \times 10^5 = (3 \pm 0.005) \times 10^5$$

When quantities are combined arithmetically, the result is no more accurate than the quantity with the largest uncertainty. Suppose a 75-kg person picks up a 0.23-kg apple. The total mass of person plus apple is still 75 kg because all we know of the person's mass is that it is somewhere between 74.5 and 75.5 kg, which means an uncertainty greater than the apple's mass. If the person's mass is instead quoted as 75.0 kg, the mass of person plus apple is 75.2 kg; if it is quoted as 75.00 kg, the mass of person plus apple is 75.23 kg. Thus

$$75 \text{ kg} + 0.23 \text{ kg} = 75 \text{ kg}$$
$$75.0 \text{ kg} + 0.23 \text{ kg} = 75.2 \text{ kg}$$
$$75.00 \text{ kg} + 0.23 \text{ kg} = 75.23 \text{ kg}$$

Significant figures must be taken into account in multiplication and division also. For example, if we divide 1.4×10^5 by 6.70×10^3, we are not justified in writing

$$\frac{1.4 \times 10^5}{6.70 \times 10^3} = 20.89552 \cdots$$

We may properly retain only two significant figures, corresponding to the two significant figures in the numerator, and so the correct answer is just 21.

In a calculation with several steps, however, it is a good idea to keep an extra digit in the intermediate steps and to round off the result only at the end. As an example,

$$\frac{5.7 \times 10^4}{3.3 \times 10^{-2}} + \sqrt{1.8 \times 10^{12}} = 1.73 \times 10^6 + 1.34 \times 10^6$$
$$= 3.07 \times 10^6 = 3.1 \times 10^6$$

If the intermediate results had been rounded off to two digits, however, the result would have been the incorrect

$$1.7 \times 10^6 + 1.3 \times 10^6 = 3.0 \times 10^6$$

B–5 LOGARITHMS

The logarithm of a number N is the exponent n to which a given base number a must be raised in order that $a^n = N$. That is, if

$$N = a^n, \quad \text{then} \quad n = \log_a N$$

Here are some examples using the base $a = 10$:

$1000 = 10^3$,   therefore $\log_{10} 1000 = 3$;

$5 = 10^{0.699}$,   therefore $\log_{10} 5 = 0.699$;

$0.001 = 10^{-3}$, therefore $\log_{10} 0.001 = -3$.

Since the decimal system of numbers has the base 10, this is a convenient number to use as the base of a system of logarithms. Logarithms to the base 10 are called **common logarithms** and are denoted simply as "$\log N$." Another widely used system of logarithms uses $e = 2.718 \cdots$ as the base. Such logarithms are called **natural logarithms** (because they arise in a natural way in calculus) and are denoted "$\ln N$." To go from one system to the other these formulas are needed:

$\log N = 0.43429 \ln N$

$\ln N = 2.3026 \log N$

Logarithms are defined only for positive numbers because the quantity a^n is positive whether n is positive, negative, or zero. Since n is the logarithm of a^n, n can describe only a positive number.

Let us consider a number N that is the product of two numbers x and y, so that $N = xy$. If $x = 10^n$ and $y = 10^m$, then

$N = xy = 10^n \times 10^m = 10^{n+m}$

$\log N = n + m$

Since $n = \log x$ and $m = \log y$, $n + m = \log x + \log y$, and so

$\log N = \log x + \log y$

Thus we have the general rule for the logarithm of a product:

$\log xy = \log x + \log y$

Similar reasoning gives the additional rules

$\log \dfrac{x}{y} = \log x - \log y$

$\log x^n = n \log x$

Before the days of electronic calculators, logarithms were widely used to simplify arithmetical work because they permit replacing multiplication and division by addition and subtraction, which are easier to do and less prone to error. Today, of course, calculators are used for such routine arithmetic. However, logarithms are still needed for finding powers and roots, with calculators replacing tables of logarithms. For example, to find $(2.13)^4$ we proceed as follows:

$\log (2.13)^4 = 4 \log 2.13 = (4)(0.3284) = 1.314$

$(2.13)^4 = \log^{-1} 1.314 = 20.6$

To obtain $\log 2.13$ with a calculator, enter 2.13 and press the [LOG] key. To obtain $\log^{-1} 1.314$ (the **antilogarithm** of 1.314, which is the number whose logarithm is 1.314), the 1.314 is entered and the [10^x] key pressed, since 10^x is the number whose logarithm is x. This key is sometimes designated [INV LOG].

The same procedure is followed for negative exponents, for instance $1/\sqrt[4]{7}$:

$$\frac{1}{\sqrt[4]{7}} = \frac{1}{7^{1/4}} = 7^{-1/4}$$

$\log 7^{-1/4} = -\tfrac{1}{4} \log 7 = -\tfrac{1}{4} \times 0.8451 = -0.2113$

$7^{-1/4} = \log^{-1}(-0.2113) = 0.615$

The logarithms of numbers written in powers-of-ten notation can be found with a calculator without first converting them to decimal notation. Let us consider 6.04×10^9. Since

$\log xy = \log x + \log y$

we have

$\log (6.04 \times 10^9) = \log 6.04 + \log 10^9$

But from the definition of logarithms to the base 10, $\log 10^9 = 9$. We therefore have

$\log (6.04 \times 10^9) = 0.7810 + 9 = 9.7810$

The same procedure holds for a number smaller than 1, for instance 2.4×10^{-5}:

$\log (2.4 \times 10^{-5}) = \log 2.4 + \log 10^{-5}$
$= 0.3802 - 5 = -4.6198$

EXERCISES

7. Evaluate the following with the help of logarithms.

(a) $0.0181^{1.5}$ (d) $\sqrt[5]{156}$

(b) $62.2^{7.13}$ (e) $(6.24 \times 10^{-4})^{1/3}$

(c) $(8.15 \times 10^{14})^6$ (f) $(2.71 \times 10^5)^{1/8}$

ANSWERS

1. (a) 200,000   (e) 0.000,001,003

(b) 0.08   (f) 0.000,000,000,1

(c) 781.9   (g) 0.000,0956

(d) 451,000,000

2. (a) 7×10^1
   (b) 1.4×10^{-1}
   (c) 3.81
   (d) 8.4×10^3
   (e) 1×10^6
   (f) 7.890×10^{-3}
   (g) 3.516×10^5

3. (a) 4.3×10^3
   (b) 7.2×10^{-2}
   (c) 5.04×10^{-3}
   (d) 6.37×10^2
   (e) 3.7×10^3
   (f) 2.7×10^{-4}
   (g) -1.3×10^5
   (h) -5×10^{-5}

4. (a) $10^{-2} = 0.01$
   (b) $5 \times 10^{-3} = 0.005$
   (c) $10^2 = 100$
   (d) $2.5 \times 10^3 = 2500$

5. (a) 1×10^3
   (b) 6×10^{15}
   (c) 9×10^{-4}
   (d) 2.5×10^{-5}
   (e) 5×10^{-6}
   (f) 1.5×10^{-10}
   (g) 1×10^{24}
   (h) 4×10^{14}
   (i) 1×10^{-10}
   (j) 2.5×10^{-17}

6. (a) 2×10^3
   (b) 6.3×10^3
   (c) 2×10^{-2}
   (d) 6.3×10^{-3}
   (e) 1.6×10^{-4}
   (f) 3.4×10^4
   (g) 7.4×10^4
   (h) 1.6×10^{-2}
   (i) 7.4×10^{-3}
   (j) 3.4×10^{-3}

7. (a) 2.44×10^{-3}
   (b) 6.16×10^{12}
   (c) 2.93×10^{89}
   (d) 3.53
   (e) 8.55×10^{-2}
   (f) 4.78

TABLE C–1

The Elements

Atomic number	Element	Symbol	Average atomic mass*	Atomic number	Element	Symbol	Average atomic mass*
1	Hydrogen	H	1.008	36	Krypton	Kr	83.80
2	Helium	He	4.003	37	Rubidium	Rb	85.47
3	Lithium	Li	6.941	38	Strontium	Sr	87.62
4	Beryllium	Be	9.012	39	Yttrium	Y	88.91
5	Boron	B	10.81	40	Zirconium	Zr	91.22
6	Carbon	C	12.01	41	Niobium	Nb	92.91
7	Nitrogen	N	14.01	42	Molybdenum	Mo	95.94
8	Oxygen	O	16.00	43	Technetium	Tc	(97)
9	Fluorine	F	19.00	44	Ruthenium	Ru	101.1
10	Neon	Ne	20.18	45	Rhodium	Rh	102.9
11	Sodium	Na	22.99	46	Palladium	Pd	106.4
12	Magnesium	Mg	24.31	47	Silver	Ag	107.9
13	Aluminum	Al	26.98	48	Cadmium	Cd	112.4
14	Silicon	Si	28.09	49	Indium	In	114.8
15	Phosphorus	P	30.97	50	Tin	Sn	118.7
16	Sulfur	S	32.06	51	Antimony	Sb	121.8
17	Chlorine	Cl	35.46	52	Tellurium	Te	127.6
18	Argon	Ar	39.95	53	Iodine	I	126.9
19	Potassium	K	39.10	54	Xenon	Xe	131.3
20	Calcium	Ca	40.08	55	Cesium	Cs	132.9
21	Scandium	Sc	44.96	56	Barium	Ba	137.3
22	Titanium	Ti	47.90	57	Lanthanum	La	138.9
23	Vanadium	V	50.94	58	Cerium	Ce	140.1
24	Chromium	Cr	52.00	59	Praseodymium	Pr	140 9
25	Manganese	Mn	54.94	60	Neodymium	Nd	144.2
26	Iron	Fe	55.85	61	Promethium	Pm	(145)
27	Cobalt	Co	58.93	62	Samarium	Sm	150.4
28	Nickel	Ni	58.70	63	Europium	Eu	152.0
29	Copper	Cu	63.55	64	Gadolinium	Gd	157.3
30	Zinc	Zn	65.38	65	Terbium	Tb	158.9
31	Gallium	Ga	69.72	66	Dysprosium	Dy	162.5
32	Germanium	Ge	72.59	67	Holmium	Ho	164.9
33	Arsenic	As	74.92	68	Erbium	Er	167.3
34	Selenium	Se	78.96	69	Thulium	Tm	168.9
35	Bromine	Br	79.90	70	Ytterbium	Yb	173.0

*The unit of mass is the u. Elements whose atomic masses are given in parentheses have not been found in nature but have been produced by nuclear reactions in the laboratory. The atomic mass in such a case is the mass number of the longest-lived radioactive isotope of the element.

TABLE C–1
The Elements (continued)

Atomic number	Element	Symbol	Average atomic mass*	Atomic number	Element	Symbol	Average atomic mass*
71	Lutetium	Lu	175.0	89	Actinium	Ac	(227)
72	Hafnium	Hf	178.5	90	Thorium	Th	232.0
73	Tantalum	Ta	180.9	91	Protactinium	Pa	231.0
74	Tungsten	W	183.9	92	Uranium	U	238.0
75	Rhenium	Re	186.2	93	Neptunium	Np	(237)
76	Osmium	Os	190.2	94	Plutonium	Pu	(244)
77	Iridium	Ir	192.2	95	Americium	Am	(243)
78	Platinum	Pt	195.1	96	Curium	Cm	(247)
79	Gold	Au	197.0	97	Berkelium	Bk	(247)
80	Mercury	Hg	200.6	98	Californium	Cf	(251)
81	Thallium	Tl	204.4	99	Einsteinium	Es	(254)
82	Lead	Pb	207.2	100	Fermium	Fm	(257)
83	Bismuth	Bi	209.0	101	Mendelevium	Md	(258)
84	Polonium	Po	(209)	102	Nobelium	No	(255)
85	Astatine	At	(210)	103	Lawrencium	Lr	(260)
86	Radon	Rn	222	104	Rutherfordium	Rf	(257)
87	Francium	Fr	(223)	105	Hahnium	Ha	(260)
88	Radium	Ra	226.0				

TABLE C–2
Natural Trigonometric Functions

Angle Degree	Angle Radian	Sine	Cosine	Tangent	Angle Degree	Angle Radian	Sine	Cosine	Tangent
0°	.000	0.000	1.000	0.000					
1°	.017	.018	1.000	.018	16°	.279	.276	.961	.287
2°	.035	.035	0.999	.035	17°	.297	.292	.956	.306
3°	.052	.052	.999	.052	18°	.314	.309	.951	.325
4°	.070	.070	.998	.070	19°	.332	.326	.946	.344
5°	.087	.087	.996	.088	20°	.349	.342	.940	.364
6°	.105	.105	.995	.105	21°	.367	.358	.934	.384
7°	.122	.122	.993	.123	22°	.384	.375	.927	.404
8°	.140	.139	.990	.141	23°	.401	.391	.921	.425
9°	.157	.156	.988	.158	24°	.419	.407	.914	.445
10°	.175	.174	.985	.176	25°	.436	.423	.906	.466
11°	.192	.191	.982	.194	26°	.454	.438	.899	.488
12°	.209	.208	.978	.213	27°	.471	.454	.891	.510
13°	.227	.225	.974	.231	28°	.489	.470	.883	.532
14°	.244	.242	.970	.249	29°	.506	.485	.875	.554
15°	.262	.259	.966	.268	30°	.524	.500	.866	.577

TABLE C–2
Natural Trigonometric Functions (continued)

Angle		Sine	Cosine	Tangent	Angle		Sine	Cosine	Tangent
Degree	Radian				Degree	Radian			
31°	.541	.515	.857	.601	61°	1.065	.875	.485	1.804
32°	.559	.530	.848	.625	62°	1.082	.883	.470	1.881
33°	.576	.545	.839	.649	63°	1.100	.891	.454	1.963
34°	.593	.559	.829	.675	64°	1.117	.899	.438	2.050
35°	.611	.574	.819	.700	65°	1.134	.906	.423	2.145
36°	.628	.588	.809	.727	66°	1.152	.914	.407	2.246
37°	.646	.602	.799	.754	67°	1.169	.921	.391	2.356
38°	.663	.616	.788	.781	68°	1.187	.927	.375	2.475
39°	.681	.629	.777	.810	69°	1.204	.934	.358	2.605
40°	.698	.643	.766	.839	70°	1.222	.940	.342	2.747
41°	.716	.658	.755	.869	71°	1.239	.946	.326	2.904
42°	.733	.669	.743	.900	72°	1.257	.951	.309	3.078
43°	.751	.682	.731	.933	73°	1.274	.956	.292	3.271
44°	.768	.695	.719	.966	74°	1.292	.961	.276	3.487
45°	.785	.707	.707	1.000	75°	1.309	.966	.259	3.732
46°	.803	.719	.695	1.036	76°	1.326	.970	.242	4.011
47°	.820	.731	.682	1.072	77°	1.344	.974	.225	4.331
48°	.838	.743	.669	1.111	78°	1.361	.978	.208	4.705
49°	.855	.755	.656	1.150	79°	1.379	.982	.191	5.145
50°	.873	.766	.643	1.192	80°	1.396	.985	.174	5.671
51°	.890	.777	.629	1.235	81°	1.414	.988	.156	6.314
52°	.908	.788	.616	1.280	82°	1.431	.990	.139	7.115
53°	.925	.799	.602	1.327	83°	1.449	.993	.122	8.144
54°	.942	.809	.588	1.376	84°	1.466	.995	.105	9.514
55°	.960	.819	.574	1.428	85°	1.484	.996	.087	11.43
56°	.977	.829	.559	1.483	86°	1.501	.998	.070	14.30
57°	.995	.839	.545	1.540	87°	1.518	.999	.052	19.08
58°	1.012	.848	.530	1.600	88°	1.536	.999	.035	28.64
59°	1.030	.857	.515	1.664	89°	1.553	1.000	.018	57.29
60°	1.047	.866	.500	1.732	90°	1.571	1.000	.000	∞

TABLE C–3
The Greek Alphabet

A	α	Alpha
B	β	Beta
Γ	γ	Gamma
Δ	δ	Delta
E	ϵ	Epsilon
Z	ζ	Zeta
H	η	Eta
Θ	θ	Theta
I	ι	Iota
K	κ	Kappa
Λ	λ	Lambda
M	μ	Mu
N	ν	Nu
Ξ	ξ	Xi
O	o	Omicron
Π	π	Pi
P	ρ	Rho
Σ	σ	Sigma
T	τ	Tau
Υ	υ	Upsilon
Φ	ϕ	Phi
X	χ	Chi
Ψ	ψ	Psi
Ω	ω	Omega

ANSWERS TO ODD-NUMBERED QUESTIONS, EXERCISES, AND PROBLEMS

CHAPTER 1

Questions

1. Centimeter; yard; kilometer. 3. Yes; an object dropped from a great height compared with one dropped from a smaller height. 5. C; D. 7. No.

Exercises and Problems

1. $(368 \text{ ft})(0.305 \text{ m/ft}) = 112 \text{ m} = 0.112 \text{ km}$. 3. $v = (50 \text{ km/h})(0.621 \text{ mi/km}) = 31 \text{ mi/h}$. 5. (a) $(4840 \text{ yd}^2)(3 \text{ ft/yd})^2(0.3048 \text{ m/ft})^2 = 4047 \text{ m}^2$. (b) $(1 \text{ km}^2)(10^3 \text{ m/km})^2/(4047 \text{ m}^2/\text{acre}) = 247 \text{ acres/km}^2$. 7. (a) $1 \text{ bd ft} = (12 \text{ in})^2(1 \text{ in}) = 144 \text{ in}^3$. (b) $1 \text{ ft}^3 = (12 \text{ in})^3 = 1728 \text{ in}^3$, $1 \text{ bd ft} = 144 \text{ in}^3/(1728 \text{ in}^3/\text{ft}^3) = 0.0833 \text{ ft}^3$. (c) $1 \text{ in}^3 = (2.54 \text{ cm})^3 = 16.4 \text{ cm}^3$, $1 \text{ bd ft} = (144 \text{ in}^3)(16.4 \text{ cm}^3/\text{in}^3) = 2.36 \times 10^3 \text{ cm}^3$. 9. $v = 132 \text{ ft/s}$, $t = x/v = 0.45 \text{ s}$. 11. $t = x/v = 0.029 \text{ s}$. 13. $x = vt = 3 \times 10^{-15} \text{ m}$. 15. $x_1 = v_1 t_1 = 1200 \text{ km}$, $x_2 = v_2 t_2 = 1000 \text{ km}$; $\bar{v} = (x_1 + x_2)/(t_1 + t_2) = 338 \text{ km/h}$. 17. $t = 1/60 \text{ h}$, $\Delta x = (v_1 - v_2)t = 0.50 \text{ km}$. 19. Let x_H be the distance the Honda has gone in the time t when it catches up with the Ford. Then $x_H = v_H t = x_0 + v_F t$, $t = x_0/(v_H - v_F) = 2.0$ h and $x_H = v_H t = 200 \text{ km}$. 21. (a) $a = (v_f - v_0)/t = 1.2 \text{ m/s}^2$. (b) -2.0 m/s^2. 23. $v = v_0 + at$, $t = (v - v_0)/a = 5.0 \text{ s}$. 25. 0.60 m/s^2; -0.80 m/s^2. 27. The airplane's speed is 550 km/h from 4:00 P.M. to 5:41 P.M., 620 km/h from 5:41 P.M. to 7:54 P.M., and 520 km/h from 7:54 P.M. to 10:00 P.M. 29. $a = v_f t$; $x = \frac{1}{2}at^2 = v_f t/2 = 90.4 \text{ m}$. 31. (a) $a = (80 \text{ m/s})/35 \text{ s}$; $t = (v_f - v_0)/a = 8.8 \text{ s}$; $x = \frac{1}{2}at^2 = 88 \text{ m}$. (b) $t = (v_f - v_0)/a = 8.8 \text{ s}$; $x = v_0 t + \frac{1}{2}at^2 = 0.61 \text{ km}$. (c) $x = \frac{1}{2}at^2 = 1.4 \text{ km}$. 33. $v_f^2 = 2ax$, $a = v_f^2/2x = 314 \text{ m/s}^2 = 32g$. 35. $v_f^2 = 2ax$, $x = v_f^2/2a = 1.0 \times 10^3 \text{ km}$. 37. (a) At a constant speed of 20 m/s, after 60 s the distance would be $x = vt = 1200 \text{ m} = 1.2 \text{ km}$; hence the train's speed must have increased. (b) $x = v_0 t + \frac{1}{2}at^2$, $a = 2(x - v_0 t)/t^2 = 0.444 \text{ m/s}^2$ and $v_f\sqrt{v_0^2 + 2ax} = 47 \text{ m/s}$. 39. Yes. 41. $v = \sqrt{2gh} = 27 \text{ m/s}$. 43. $h = \frac{1}{2}gt^2 = 24 \text{ ft}$. 45. The time of rise and the time of fall are both $t = 1.5 \text{ s}$. Hence $v = gt = 48 \text{ ft/s}$ is the value of both the initial and final speeds. 47. (a) $v_f = v_0 + gt = 19.8 \text{ m/s}$ downward. (b) $v_f = 29.6 \text{ m/s}$ downward. 49. $h = v^2/2g = 13 \text{ m}$. 51. $h = v_0 t + \frac{1}{2}gt^2$, $v_0 = h/t - gt/2 = 14 \text{ m/s}$. 53. (a) $v = \sqrt{2g_M h} = 11 \text{ m/s}$. (b) $T = 2t = 2\sqrt{2h/g_M} = 5.7 \text{ s}$. 55. (a) $h = \frac{1}{2}gt^2$, $t = \sqrt{2h/g} = 1.2 \text{ s}$. (b) $v = gt = 12 \text{ m/s}$. 57. Man: $t_1 = h/v$; monocle: $t_2 = \sqrt{2h/g}$; $\Delta t = t_1 - t_2 = 19 \text{ s}$. 59. Since the coconuts met halfway down the cliff, the one thrown upward must have had the same initial speed as the speed of the dropped coconut at the place of meeting, which is $v = \sqrt{2gh} = 20 \text{ m/s}$. 61. (a) Calling up $+$ and down $-$, the pump falls $h = v_0 t - \frac{1}{2}gt^2 = -46 \text{ m}$, so the helicopter's altitude was 46m when it dropped the pump. (b) $\Delta h = v_0 t = 32 \text{ m}$, so the helicopter was $h + \Delta h = 78 \text{ m}$ high 4 s later. 63. The snake covers 45.0 m in $t = x/v = 30 \text{ s}$. Therefore your speed must be greater than $v = 40.0 \text{ m}/30 \text{ s} = 1.3 \text{ m/s}$. 65. If $D = $ total distance, $x_1 = D/3$, $x_2 = 2D/3$, $t_1 = x_1/v_1 = D/3v_1$, $t_2 = x_2v_2 = 2D/3v_2$; $\bar{v} = D/(t_1 + t_2) = 1/[(1/3v_1) + (2/3v_2)]$, $2/3v_2 = 1/\bar{v} - 1/3v_1$, $1/v_2 = 3/2\bar{v} - 1/2v_1 = 0.022857 \text{ h/mi}$, $v_2 = 44 \text{ mi/h}$. 67. The car had an initial speed of 21.6 mi/h and accelerated at 21.6 (mi/h)/min-(1296 mi/h²) for 3 min to 86.4 mi/h. It stayed at 86.4 mi/h for 4 min and then slowed down to a stop at 43.2 (mi/h)/min-(2592 mi/h²) in 2 min. 69. If t is the interval between drops, $x_1 = \frac{1}{2}at^2$, $x_4 = \frac{1}{2}a(4t)^2$, $x_4 = 16x_1 = 12.8$ m. 71. $t_1 = v/a_1 = 6.0 \text{ s}$, $x_1 = v^2/2a_1 = 27 \text{ m}$; $t_3 = v/a_3 = 4.5 \text{ s}$, $x_3 = v^2/2a_3 = 20.25 \text{ m}$; $x_2 = 4000 \text{ m} - x_1 - x_3 = 352.75 \text{ m}$, $t_2 = x_2/v = 39.2 \text{ s}$; $t = t_1 + t_2 + t_3 = 50 \text{ s}$. 73. (a) $h_1 = v_0^2/2g = 5.10 \text{ m}$, $t_1 = \sqrt{2h_1/g} = 1.02 \text{ s}$; $h_2 = 20 \text{ m} + 5.1 \text{ m} = 25.1 \text{ m}$; $t_2 = \sqrt{2h_2/g} = 2.26 \text{ s}$; $t = t_1 + t_2 = 3.3 \text{ s}$. (b) $v = gt_2 = 22 \text{ m/s}$. 75. $t_1 = v/a = 4 \text{ s}$; $x_1 = \frac{1}{2}at^2 = 12 \text{ m}$; $t_3 = t_1 = 4 \text{ s}$; $x_3 = x_1 = 12 \text{ m}$. Hence constant-speed distance is $x_2 = 50 \text{ m} - 24 \text{ m} = 26 \text{ m}$ and $t_2 = x_2/v = 4.3 \text{ s}$; $t = t_1 + t_2 + t_3 = 12 \text{ s}$. 77. See Fig. A.1.

(a)

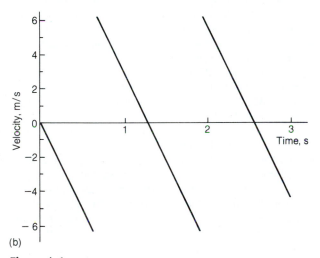

(b)

Figure A.1

Exercise 77, Chapter 1.

CHAPTER 2

Questions

1. No. The distinction between vector and scalar quantities is simply that vector quantities have directions associated with them. Both kinds of quantity are found in the physical world.   **3.** Yes.   **5.** 30 N; 0.   **7.** (a) No. B and C reach the ground together; A reaches it later. (b) No. B has

the lowest speed; A and C have the same higher speed.
9. The squirrel should stay where it is. If it lets go, it will fall with the same acceleration as the bullet and so will be struck.

Exercises and Problems

1. $F = \sqrt{(100\ \text{N})^2 + (40\ \text{N})^2} = 108\ \text{N}$.   **3.** $s = \sqrt{x^2 + y^2} = 81\ \text{m}$; $\theta = \tan^{-1} y/x = 30°$ above the horizontal.   **5.** (a) $s = \sqrt{(v_N t)^2 + (v_E t)^2} = 43\ \text{mi}$. (b) 172 mi.   **7.** The two velocities are perpendicular, so $v = \sqrt{(7.0\ \text{kn})^2 + (3.0\ \text{kn})^2} = 7.6\ \text{kn}$. If θ is the angle between $\mathbf{v}$ and NW, then $\tan \theta = 3.0\ \text{kn}/7.0\ \text{kn} = 0.43$, $\theta = 23°$. Since NW is itself $45°$ W of N, the direction of $\mathbf{v}$ is $45° + 23° = 68°$ W of N.   **9.** (a) $t = 210\ \text{s}$. $v = $ distance/time $= (100\ \text{m} + 160\ \text{m} + 100\ \text{m})/210\ \text{s} = 1.7\ \text{m/s}$. (b) $v = $ displacement/time $= 160\ \text{m}/210\ \text{s} = 0.76\ \text{m/s}$ to the east.   **11.** $F_y = 50\ \text{N} + (25\ \text{N})(\sin 45°) = 68\ \text{N}$.   **13.** $F_x = F \cos \theta = 478\ \text{lb}$.   **15.** $v_{\text{rain}} = v_{\text{car}}/\tan \theta = 8.4\ \text{m/s}$.   **17.** (a) $v_y = v \sin 25° = 85\ \text{km/h}$. (b) $v_x = v \cos 25° = 181\ \text{km/h}$. (c) $v_{\text{south}} = v_x \sin 45° = 128\ \text{km/h}$.   **19.** The angle between the path of ball 1 and the $+x$ direction is specified by $\tan \theta_1 = v_y/v_x = 2.0$, $\theta_1 = 63°$. Similarly $\tan \theta_2 = v_y/v_x = 1.5$, $\theta_2 = 56°$. The difference in angle is $7°$.   **21.** Calling E the $+x$ direction and N the $+y$ direction, $s_x = 200\ \text{km} - (100\ \text{km})(\sin 45°) = 129\ \text{km}$ and $s_y = -200\ \text{km} + (100\ \text{km})(\cos 45°) = -129\ \text{km}$. Hence $s = \sqrt{s_x^2 + s_y^2} = 182\ \text{km}$. The direction is NW since s_x and s_y are equal in length.   **23.** $t = s/v$, $h = \frac{1}{2}gt^2 = \frac{1}{2}gs^2/v^2 = 0.10\ \text{m} = 10\ \text{cm}$.   **25.** $t = \sqrt{2h/g} = 14.3\ \text{s}$, $v_x = 400\ \text{km/h} = 111\ \text{m/s}$, $v_y = gt = 140\ \text{m/s}$, $v = \sqrt{v_x^2 + v_y^2} = 179\ \text{m/s}$.   **27.** $h = \frac{1}{2}gt^2$, $t = \sqrt{2h/g}$, $v = s/t = s\sqrt{g/2h} = 12\ \text{m/s}$.   **29.** The time of fall is $t = \sqrt{2h/g} = 10.1$ s. In this time the parcel will have moved horizontally toward the ship $s_1 = v_1 t$ and the ship will have moved toward the parcel $s_2 = v_2 t$. Hence the required distance is $s_1 + s_2 = (v_1 + v_2)t = 707\ \text{m} = 0.71\ \text{km}$.   **31.** (a) $v_y = \sqrt{2gh} = 62\ \text{ft/s}$; $v_x = 80\ \text{ft/s}$; $v = \sqrt{v_x^2 + v_y^2} = 101\ \text{ft/s}$. (b) $\theta = \tan^{-1} v_y/v_x = 38°$ below the horizontal.   **33.** At maximum range, $\theta = 45°$ and $v_0 = \sqrt{Rg} = 10\ \text{m/s}$.   **35.** $R = (v_0^2 \sin 2\theta)/g$, $v_0 = \sqrt{Rg/\sin 2\theta}$. Here $R = 120\ \text{ft}$ and $\sin 2\theta = \sin 100° = \cos 10°$ since from Appendix A8 $\sin (90° + x) = \cos x$. Hence $v_0 = 62\ \text{ft/s}$.   **37.** (a) $v_0 = \sqrt{2gh}$, $R_{\text{max}} = v_0^2/g = 2h = 200\ \text{m}$. (b) $v = \sqrt{2gh} = 44\ \text{m/s}$; 44 m/s.   **39.** Let north be the $+y$ direction and east the $+x$ direction. Then $F_{1x} = 0$, $F_{1y} = 10\ \text{N}$; $F_{2x} = 20\ \text{N}$, $F_{2y} = 0$, $F_{3x} = (10\ \text{N})(\cos 40°) = 7.66\ \text{N}$, $F_{3y} = (-10\ \text{N})(\sin 40°) = -6.43\ \text{N}$; $F_{4x} = (-20\ \text{N})(\sin 50°) = -15.32\ \text{N}$, $F_{4y} = (-20\ \text{N})(\cos 50°) = -12.86\ \text{N}$. Hence $R_x = F_{1x} + F_{2x} + F_{3x} + F_{4x} = 12.34\ \text{N}$, $R_y = F_{1y} + F_{2y}$

$+ F_{3y} + F_{4y} = -9.29$ N, $R = \sqrt{R_x^2 + R_y^2} = 15$ N. R is south of east by the angle $\theta = \tan^{-1} R_y/R_x = 37°$.
41. $A_x = 0$, $A_y = 6.0$ lb; $B_x = -10.0$ lb, $B_y = 0$; $C_x = (8.0 \text{ lb}) \cos 45° = 5.66$ lb, $C_y = -(8.0 \text{ lb}) \sin 45° = -5.66$ lb. $R_x = A_x + B_x - C_x = -15.66$ lb, $R_y = A_y + B_y - C_y = 11.66$ lb. $R = \sqrt{R_x^2 + R_y^2} = 19.6$ lb, $\theta = \tan^{-1} R_y/R_x = 37°$ clockwise from $-x$ direction.
43. Here $A_x = A \sin 37° = 6$ cm; $A_y = A \cos 37° = 8$ cm; $B_x = B \cos 37° = 8$ cm; $B_y = -B \sin 37° = -6$ cm. (a) $R_x = A_x + B_x = 14$ cm, $R_y = A_y + B_y = 2$ cm; $R = \sqrt{R_x^2 + R_y^2} = 14$ cm. **R** points θ clockwise from the $+y$ direction where $\theta = \tan^{-1} R_x/R_y = 82°$. (b) $R_x = A_x - B_x = -2$ cm, $R_y = A_y - B_y = 14$ cm; $R = \sqrt{R_x^2 + R_y^2} = 14$ cm. **R** points θ counterclockwise from the $+y$ direction where $\theta = \tan^{-1} R_x/R_y = 8°$. (c) $R_x = B_x - A_x = 2$ cm, $R_y = B_y - A_y = -14$ cm; $R = \sqrt{R_x^2 + R_y^2} = 14$ cm. **R** points θ clockwise from the $+x$ direction where $\theta = \tan^{-1} R_y/R_x = 8°$. **45.** Let $A = 4.0$ km, $B = 2.0$ km, and $C = 3.0$ km with A in the $+y$ direction and θ the angle between A and B. Then $A_x = 0$, $A_y = A$; $C_x = B_x = B \sin \theta$; $C_y = A + B_y = A - B \cos \theta$; $C^2 = C_x^2 + C_y^2 = A^2 + B^2 (\sin^2 \theta + \cos^2 \theta) - 2AB \cos \theta$ (which is the law of cosines from trigonometry) and $\theta = \cos^{-1} (C^2 - A^2 - B^2)/(-2AB) = 133°$. **47.** (a) $v_x = \sqrt{v_N^2 + v_W^2} = 12.8$ m/s, $v_y = gt = 19.6$ m/s, $v = \sqrt{v_x^2 + v_y^2} = 23$ m/s. (b) $\theta_1 = \tan^{-1} v_x/v_y = 33°$. (c) $\theta_2 = \tan^{-1} v_W/v_N = 51°$. **49.** $T = (2v_0/g) \sin \theta$, $R = (2v_0^2/g) \sin \theta \cos \theta$, hence $v_0 = R/(T \cos \theta)$ and $T = (2R/gT) \tan \theta$, $\theta = \tan^{-1}(gT^2/2R) = 51°$. (a) $v_0 = R/(T \cos \theta) = 32$ m/s. (b) $h = \frac{1}{2}g(T/2)^2 = 31$ m. (c) $v_x = v_0 \cos \theta$, $v_y = v_0 \sin \theta - gt$, $v = \sqrt{v_x^2 + v_y^2} = 21$ m/s. (d) $h = v_0 y t - \frac{1}{2}gt^2 = v_0 t \sin \theta - \frac{1}{2}gt^2 = 29$ m. **51.** (a) $R = (v_0^2 g) \sin 2\theta_1 = 7953$ m $= 8.0$ km. $T_1 = (2v_0/g) \sin \theta_1 = 31$ s; $h_1 = v_y t - \frac{1}{2}gt^2 = (v_0 \sin \theta_1)(T_1/2) - \frac{1}{2}g(T_1/2)^2 = 1148$ m $= 1.1$ km. (b) $\theta_2 = 90° - \theta_1 = 60°$; $T_2 = (2v_0/g) \sin \theta_2 = 53$ s; $h_2 = (v_0 \sin \theta_2) (T_2/2) - \frac{1}{2}g (T_2/2)^2 = 3444$ m $= 3.4$ km.

CHAPTER 3

Questions

1. No. Other forces may come into being that cancel out the applied force. **3.** The deceleration of a person falling on loose earth is more gradual than if he falls on concrete, hence the force acting on him is less. **5.** $a = g$. **7.** Yes, by applying a downward force in addition to the downward force of gravity. **9.** No. Action and reaction forces act on different objects, so a single force can certainly act on an object. **11.** A propeller works by pushing backward on the air, whose reaction force in turn pushes the

propeller itself forward. No air, no reaction force, so the idea is no good. **13.** (a) For it to fly, the bird's wings must exert a downward force on the air in the box equal to the bird's weight, and this force is transmitted to the box. Hence the scale reading is the same whether the bird stands or flies. (b) Now the force exerted on the air by the bird's wings are not completely transmitted to the cage, and the scale reading will be smaller when the bird flies. Because some force is still transmitted to the bottom of the cage, the reading when the bird flies will be somewhat greater than that corresponding to the empty cage. **15.** Yes; no.

Exercises and Problems

1. $F = ma = 1.5 \times 10^{-5}$ N. **3.** $F_{max} = m_0 a_0 = 2000$ N. Hence $a = F_{max}/m = 0.67$ m/s². **5.** $a = F/m$, $v = at = Ft/m = 29$ m/s. **7.** (a) $v_f = 55.6$ m/s; $a = (v_f - v_0)/t = 18.5$ m/s² $= 1.9g$. (b) $F = ma = 2.2 \times 10^5$ N. **9.** $a = -v^2/2s = -4.8$ m/s²; $F = ma = -9600$ N $= -9.6$ kN. **11.** $\Delta v = 4.17$ m/s, $F = ma = m\Delta v/t = 500$ N $= 0.50$ kN. **13.** $a = -v_0^2/2s = -2500$ m/s²; $F = ma = -10^4$ N $= -10$ kN (over a ton); the goat will be killed. **15.** Lower string: $T = m_1 g = 98$ N. Upper string: $T = (m_1 + m_2)g = 98$ N. **17.** (a) $m = w/g = 0.82$ kg. (b) $a = F/m = 25$ m/s². **19.** $a = (m_1 - m_2)g/(m_1 + m_2) = 2.0$ m/s², the first upward and the second downward. **21.** (a) $F = w - T = ma$, $T = w - ma = mg - ma = 8.8$ kN. (b) $T = mg + ma = 10.8$ kN. **23.** $v_f^2 = v_0^2 + 2as$, $a = (v_f^2 - v_0^2)/2s = 30.6$ m/s²; $F = ma + mg = 4.0$ kN; $F/mg = 4.1$. **25.** (a) Down. (b) No. (c) Yes; $F = 100$ N, $m = w/g = 71.4$ kg, $a = F/m = 1.4$ m/s². **27.** $F = 3w$, $F_x = F \cos 50° = 1.93w$; $m = w/g$; $a = F_x/m = 1.93g = 19$ m/s². **29.** (a) $w = mg = 48$ lb. (b) $a = F/m = 67$ ft/s². **31.** $F = w + (w/g)a = 6.8$ lb. **33.** $F = w_1 - w_2$, $m = (w_1 + w_2)/g$, $a = F/m = (w_1 - w_2)g/(w_1 + w_2) = 6.4$ ft/s. **35.** $N = mg$, $F = \mu N = \mu mg = 294$ N. **37.** $F = \mu N = 2$ N. **39.** $a = v_0^2/2s$; $\mu = F/mg = ma/mg = a/g = v_0^2/2sg = 0.041$. **41.** $a = F/m = \mu mg/m = \mu g$, $v_0 = \sqrt{2as} = \sqrt{2\mu gs} = 60$ ft/s $= 41$ mi/h. **43.** (a) $F_{max} = \mu(0.4 \, m)g = ma_{max}$, $a_{max} = 0.4 \, \mu g$. (b) $F_{max} = (0.6 \, m)g = ma_{max}$, $a_{max} = 0.6 \, \mu g$. (c) $F_{max} = \mu mg = ma_{max}$, $a_{max} = \mu g$. **45.** (a) $a = v_0^2/2s = 6.75$ ft/s²; $\mu = F/mg = ma/mg = a/g = 0.21$. (b) $F = \mu w = (0.21) (5 \text{ lb}) = 1.1$ lb. **47.** (a) $F = \mu mg = ma$, $a = \mu g = 6.9$ m/s². (b) $v = 27.8$ m/s, $v^2 = 2as$, $s = v^2/2a = 56$ m. **49.** $\mu = \tan \theta = 0.18$. **51.** The component of the crate's weight parallel to the plane is $F = w \sin \theta = 171$ lb. The frictional force is $\mu N = \mu w \cos \theta = 141$ lb. Hence a force of 30 lb must be applied upward along the plane. **53.** Here $y =$

0.6 m, $h_1 = 2.0$ m, $h_2 = 1.8$ m, $m_1 = 40$ kg, $m_2 = ?$ Kangaroo by herself: $v_1^2 = 2gh_1$, $a_1 = v_1^2/2y = gh_1/y$, $F = m_1g + m_1a_1 = m_1g(1 + h_1/y) = 1699$ N. With baby: $v_2^2 = 2gh_2$, $a_2 = v_2^2/2y = gh_2/y$, $F = (m_1 + m_2)(g + a_2)$, $m_1 + m_2 = F/(g + a_2) = F/g(1 + h_2y) = 43.3$ kg, $m_2 = 43.3$ kg $- 40$ kg $= 3.3$ kg. **55.** The total mass of the system is $m = m_2 + m_3 + m_4 = 9.0$ kg, and the net force on it is $F_{net} = 120$ N $- mg = 31.8$ N. Hence the acceleration of the system is $a = F_{net}/m = 3.5$ m/s². (a) The tension T_1 in the upper string is the force that gives the acceleration a to the two lower blocks, so that $T_1 - (m_3 + m_4)g = (m_3 + m_4)a$, $T_1 = 93$ N. (b) For the lower string, $T_2 - m_4g = m_4a$, $T_2 = 53$ N. **57.** The net downward force on the object is $F = mg - R = mg - kv^2$. When the terminal speed is reached, the object falls at constant speed, hence $F = 0 = mg - kv_t^2$, $v_t = \sqrt{mg/k}$. **59.** (a) $F = m_2g = (m_1 + m_2)a$, $a = m_2g/(m_1 + m_2) = 4.9$ m/s². (b) $F = m_2g - \mu m_1g = (m_1 + m_2)a$, $a = (m_2 g - \mu m_1g)/(m_1 + m_2) = 3.9$ m/s². **61.** (a)$m_1 = 2$kg, $m_2 = 3$ kg; $F = \mu(m_1 + m_2)g + \mu m_1g = 21$ N. (b) $F = \mu(m_1 + m_2)g + \mu m_1g + \mu m_1g = 26$ N. **63.** The frictional force to be overcome is $F_f = \mu N = \mu mg \cos\theta = 2.144 \times 10^7$ N. The component of the ship's weight parallel to the ways is $F_w = mg \sin\theta = 2.049 \times 10^7$ N. Hence an additional force of $F_f - F_w = 9.5 \times 10^5$ N is needed. **65.** $N = mg \cos\theta$; $F = mg \sin\theta - \mu N = mg(\sin\theta - \mu \cos\theta) = ma$; $a = g(\sin\theta - \mu \cos\theta) = 4.11$ m/s². Since $s = \frac{1}{2}at^2$, $t = \sqrt{2s/a} = 2.1$ s. **67.** (a) From solution to Exercise 65, $a = g(\sin\theta - \mu \cos\theta) = 5.55$ m/s²; $v = \sqrt{2as} = 24$ m/s. (b) $F = \mu mg = ma$, $a = \mu g = 0.98$ m/s², $s = v^2/2a = 283$ m $= 0.28$ km.

CHAPTER 4

Questions

1. No work is done by a net force acting on a moving object when the force is perpendicular to the direction of the object's motion. **3.** Yes; no. **5.** The work done is the same for both twins. **7.** The center of mass of the stored water drops to half its original height, hence half the original PE is lost.

Exercises and Problems

1. No work is done because the force exerted on the earth is perpendicular to its orbital motion. **3.** (a) $W = Fx = 5.0$ kJ. (b) $F_f = \mu w = 200$ N, $W_f = F_f x = 4.0$ kJ; $W_a = W - W_f = 1.0$ kJ. **5.** (a) $F = w = 24$ lb. (b) $W = Fx$

$= 1.2 \times 10^3$ ft · lb. **7.** The center of mass of the post must be raised by 15 ft, so $W = wh = 7.5 \times 10^3$ ft · lb. **9.** (a) $W = Fy = 1.0$ kJ. (b) $W = Fx = 1.0$ kJ. **11.** $W = Fx \cos\theta = 24$ J. **13.** $P = mgh/t = 4.0$ kW. **15.** $P = Fv = mgv = 3.9$ kW. **17.** $F = P/v = 23$ kN. **19.** $v = P_{out}/F = (Eff)P_{in}/F = 8.0$ m/s. **21.** Eff $= P_{out}/P_{in} = (mgh/t)/P = 0.78 = 78\%$. **23.** $P = W/t = wh/t = 1.75 \times 10^4$ ft · lb/s $= 31.8$ hp. Since twice this power is required, $2P = 64$ hp. **25.** $P = (0.2)(60$ kg$)(6$ W/kg$) = 72$ W. Since $P = mgh/t$ where $m = 72$ kg here, $h/t = P/mg = 0.102$ m/s $= 367$ m/h $= 0.37$ km/h. **27.** KE $= \frac{1}{2}mv^2$, $v = \sqrt{2KE/m} = 3.2$ m/s. **29.** KE $= \frac{1}{2}(w/g)v^2 = 1.7 \times 10^5$ ft · lb. **31.** $v = s/t = 8.89$ m/s, KE $= \frac{1}{2}mv^2 = 2.8$ kJ. **33.** $Fs = \frac{1}{2}mv^2$, $v = \sqrt{2Fs/m} = 7.1$ m/s. **35.** $P = \frac{1}{2}m(v_2^2 - v_1^2)/t$, $t = m(v_2^2 - v_1^2)/2P = 8.8$ s. **37.** $P = W/t = (KE_2 - KE_1)/t = m(v_2^2 - v_1^2)/2t = 36$ kW. **39.** $mgh_2 = 2mgh_1/3$, $h_2 = 2h_1/3 = 40$ m. **41.** $Fs = mgh$, $F = mgh/s = 1.3 \times 10^5$ kN. **43.** $mgh_2 = 0.9$ mgh_1, $h_2 = 0.9$ $h_1 = 0.9$ m. **45.** (a) $P = W/t = mgh/t = 2.94$ MW. (b) 2.94×10^4 bulbs. **47.** KE/PE $= v^2/2gh = 0.1875$, hence 81\% of the initial PE is dissipated. **49.** $\frac{1}{2}mv^2 = mg\Delta h$, $\Delta h = v^2/2g = 3.3$ m; $h = 3.3$ m $+ 1.1$ m $+ 0.6$ m $= 5.0$ m. **51.** (a) KE $= W = Fs = 24$ J. (b) KE $= W - $ PE $= Fs - mgh = 9.3$ J. **53.** $\Delta E = \Delta KE + \Delta PE = \frac{1}{2}m(v_2^2 - v_1^2) + (0 - mgh) = -168$ kJ. The minus sign means that the car lost this amount of energy. **55.** $W = W_f + \Delta PE = F_f x + mgh = \mu xmg \cos\theta + mgx \sin\theta = 0.14$ kJ. **57.** The center of mass is $h = 0.75$ m high initially and $h' = \sqrt{(0.30 m)^2 + (0.75 m)^2}$ high when tilted. Hence $W = mg(h' - h) = 40$ J. **59.** $R_{max} = v_0^2/g$, $v_0^2 = R_{max}g$; KE $= \frac{1}{2}(w/g)v_0^2 = \frac{1}{2}wR_{max} = 405$ ft · lb. **61.** KE $= W - W_f - $ PE $= Fx - \mu xmg \cos\theta - mgx \sin\theta = 0.216$ J $= \frac{1}{2}mv^2$, $v = \sqrt{2W/m} = 0.46$ m/s. **63.** $h = L(1 - \cos\theta) = 1.17$ ft; $\frac{1}{2}mv^2 = mgh$, $v = \sqrt{2gh} = 8.7$ ft/s. **65.** (a) $W = mgh = 5.5$ kJ; $t = L/v = 14$ s, $P = W/t = 0.39$ kW. (b) $W = 5.5$ kJ; $v' = 1.8$ m/s, $t' = L/v' = 7.78$ s, $P' = W/t' = 0.71$ kW. (c) $W = 5.5$ kJ; $v'' = 0.2$ m/s, $t'' = L/v'' = 70$ s, $P'' = W/t'' = 78$ W. **67.** (a) $F_1 = P_1/v = 2.7$ kN. (b) $F_2 = F_1 + mg \sin\theta = 4335$ N, $P_2 = F_2v = 48$ kW. **69.** $P = P_0 + wv \sin\theta$, $\theta = \sin^{-1} (P - P_0)/wv = 2.2°$.

CHAPTER 5

Questions

1. Yes; yes. **3.** $p_2 = p_1/\sqrt{2}$ **5.** The definition of mass and the first law of motion are both included in the principle of conservation of linear momentum. This principle goes further because it can be applied to systems consisting of any number of objects that interact with one another in any manner at all. **7.** (a) Yes. (b) In the opposite direction to that in which the man walks. (c) The car also comes to a stop.   **9.** In such a collision the total momentum of the objects is initially zero and the objects stick together afterward. The total momentum remains zero.

Exercises and Problems

1. $p = \sqrt{KE/2m} = 5.0$ kg $\cdot$ m/s.   **3.** $p = mv = wv/g$ $= 4400$ slug $\cdot$ ft/s.   **5.** $KE_1 = \frac{1}{2}m_1v_1^2 = 1.4 \times 10^5$ J, $KE_2 = \frac{1}{2}m_2v_2^2 = 6.9 \times 10^4$ J $= KE_1/2$; $p_1 = m_1v_1 = p_2 = m_2v_2 = 1.7 \times 10^4$ kg $\cdot$ m/s.   **7.** (a) $v = 242$ m/s, $mv = 3.87 \times 10^7$ kg $\cdot$ m/s. (b) $t = mv/F = 114$ s.   **9.** $F = \Delta p/\Delta t = m(v_1 + v_2)/\Delta t = 7.8 \times 10^5$ N.   **11.** (a) $F = v(\Delta m/\Delta t) = \sqrt{2gh}(\Delta m/\Delta t) = 0.54$ N, $m_1 = m_0 + (\Delta m/\Delta t)t + F/g = 8.1$ kg. (b) $m_2 = m_0 + (\Delta m/\Delta t) = 8.0$ kg.   **13.** Number of bullets $= (mv_{\text{leopard}}/(mv)_{\text{bullet}} = 7.4$, so 8 bullets are needed.   **15.** Girl: $m_gv_g = m_pv_p$, $v_g = (m_p/m_g)v_p = 1.0$ m/s. Boy: $(m_b + m_p)v_b = m_pv_p$, $v_b = [m_p/(m_b + m_p)]v_p = 0.9$ m/s.   **17.** $m_1v_1 = (\Delta m/\Delta t)tv_2$, $t = m_1v_1/(\Delta m/\Delta t)v_2 = 13$ s.   **19.** The second piece continues in the original direction. $mv = mv_2/2 - mv_1/2$, $v_2 = 2v + v_1 = 24$ km/s.   **21.** (a) $v_d = m_mv_m/(m_d + m_w) = 1.9$ m/s; hence the man cannot swim back to the dinghy. (b) $\Delta v_d = m_wv_w/m_d = 2.5$ m/s; hence the dinghy now moves toward the man at 0.6 m/s and the woman can catch up with it since her own speed is 1.0 m/s.   **23.** $\Delta m/\Delta t = F/v = 720$ kg/s.   **25.** The rocket's thrust is $F = v(\Delta m/\Delta t) = 75$ kN. Since the initial weight of the rocket is m_0g, the initial force available for its acceleration is $F - m_0g$ and so $a_0 = (F - m_0g)/m_0 = 5.2$ m/s². The final acceleration is $a_f = (F - m_fg)/m_f = 40$ m/s².   **27.** $m_1v_1 = (m_1 + m_2)V$, $V = m_1v_1/(m_1 + m_2) = 3.3 \times 10^4$ km/s.   **29.** $m_1(\Delta v) = m_2v_2$, $v_2 = m_1(\Delta v)/m_2 = 1.4$ m/s.   **31.** (a) $m_1v_1 + m_2v_2 = (m_1 + m_2)v_3$, $v_3 = (m_1v_1 + m_2v_2)/(m_1 + m_2) = 1.3$ m/s. (b) $\Delta KE = KE_1 + KE_2 - KE_3 = \frac{1}{2}m_1v_1^2 + \frac{1}{2}m_2v_2^2 - \frac{1}{2}(m_1 + m_2)v_3^2 = 2.0$ J.   **33.** (a) $(m_1 + m_2)V = m_1v_1 + m_2v_2$, $V = (m_1v_1 + m_2v_2)/(m_1 + m_2) = 56$ km/h. (b) $KE_{\text{initial}} - KE_{\text{final}} = (\frac{1}{2}m_1v_1^2 + \frac{1}{2}m_2v_2^2) - \frac{1}{2}(m_1 + m_2)V^2 = 37$ kJ.   **35.** $m_1\Delta v_1 = m_2\Delta v_2$, $\Delta v_2 = \Delta v_1(m_1/m_2) = 17$ m/s, $v_f = v_i - \Delta v = 28$ m/s.   **37.** (a) Let 1 be the iron ball and 2 be the wooden ball. Then $v_1' = (m_1 - m_2)v_1/(m_1 + m_2) = 1.6$ m/s; $v_2' = 2m_1v_1/(m_1 + m_2) = 3.6$ m/s. (b) $KE_2'/KE_1 = 4(m_2/m_1)/(1 + m_2/m_1)^2 = 0.33 = 33\%$.   **39.** $h' = e^2h = 1.9$ m.   **41.** (a) $m_1v_1 = m_1v_1' + m_2v_2'$, $v_2' = (m_1v_1 - m_1v_1')/m_2 = 6.67$ m/s in the same direction as 5.0-kg ball. (b) $e = (v_2' - v_1')/(v_1 - v_2) = 0.78$.   **43.** (a) $w_1v_1 = w_1v_1' + w_2v_2'$, $v_2' = (w_1v_1 - w_1v_1')/w_2 = 0$. (b) $e = (v_2' - v_1')/(v_1 - v_2) = 0.4$.   **45.** (a) Newton's second

law of motion: $a = v_f^2/2s$, $F = ma = mv_f^2/2s = 6.8 \times 10^5$ N. (b) Work-energy: $Fs = \frac{1}{2}mv_f^2$, $F = mv_f^2/2s = 6.8 \times 10^5$ N. (c) Impulse-momentum: $t = 2s/v_f$, $Ft = mv_f$, $F = mv_f/t = mv_f^2/2s = 6.8 \times 10^5$ N.   **47.** (a) $F - w_0 = m_0a_0$, $a_0 = (g/w_0)(F - w_0) = 13$ ft/s². (b) $a_f = (g/w_f)(F - w_f) = 118$ ft/s².   **49.** The third fragment moves off in the same plane as the others in the direction that bisects the angle between the backward extensions of their paths. Letting $v = v_1 = v_2$, $v_3 = 2v \cos(\theta/2) = 17$ m/s.   **51.** If **V** is final velocity, $V_N = m_1v_1/(m_1 + m_2) = 0.444$ m/s; $V_w = m_2v_2/(m_1 + m_2) = 0.889$ m/s; $V = \sqrt{V_N^2 + V_w^2} = 0.99$ m/s. If θ is angle between **V** and north, $\tan \theta = V_w/V_N = 2$, $\theta = 63°$ west of north.   **53.** $v_2' - v_1' = e(v_1 - v_2) = 6.3$ m/s, $v_2' = v_1' + 6.3$ m/s; $m_1v_1 + m_2v_2 = m_1v_1' + m_2v_2'$, $v_2' = [(-3.0$ kg $\cdot$ m/s$) - v_1$ kg$]/(2.0$ kg$)$; setting equal the two formulas for v_2' and solving for v_1' gives $v_1' = -5.2$ m/s and so $v_2' = 1.1$ m/s. Both balls reverse their directions as a result of the collision.   **55.** (a) If V is initial velocity of wooden block, $mv = (M + m)V$ and $V = mv/(M + m)$. From PE $=$ KE, $(M + m)gh = \frac{1}{2}(M + m)V^2 = m^2v^2/2(M + m)$ and $v = (1 + M/m)\sqrt{2gh}$. (b) 561 m/s.   **59.** From conservation of momentum, $V_{1i} = v_{1f} \cos \theta_1 + v_{2f} \cos \theta_2$ and $0 = v_{1f} \sin \theta_1 - v_{2f} \sin \theta_2$, $v_{2f} = v_{1f}(\sin \theta_1/\sin \theta_2)$. Hence $v_{1i} = v_{1f} \cos \theta_1 + v_{1f} (\sin \theta_1/\sin \theta_2)\cos \theta_2$, $v_{1f} = v_{1i}/[\cos \theta_1 + (\sin \theta_1/\sin \theta_2)\cos \theta_2] = 3.8$ m/s and $v_{2f} = v_{2f}(\sin \theta_1/\sin \theta_2) = 3.2$ m/s. Alternatively, conservation of KE, here $v_{1f} = v_{1f}^2 + v_{2f}^2$, could have been used with either of the momentum equations.

CHAPTER 6

Questions

1. Under no circumstances.   **3.** At the bottom of the circle, where the string must support all the ball's weight as well as provide centripetal force.   **5.** The mass of a sprinter and the force his or her legs can exert are both unchanged on the moon. Hence the acceleraton is unchanged and the time needed would not be different.   **7.** In both cases the centripetal force is provided by the object's weight mg.   **9.** Shorter.   **11.** Such a vehicle does not have to depend only on its initial KE for the work needed to leave the earth, hence it need not reach the escape speed.

Exercises and Problems

1. (a) The record's circumference is $2\pi r = 3.14$ ft, so $v = (33.3 \times 3.14$ ft$)/60$ s $= 1.7$ ft/s. (b) $a_c = v^2/r = 6.1$ ft/s².   **3.** $v^2/r = 4g$, $r = v^2/4g = 2.3$ km.   **5.** $v^2/r = 4g$, $v = 2\sqrt{rg}$; $f = 1/T = v/2\pi r = \sqrt{g/r}/\pi = 0.38$ rev/s.   **7.** $F_c = mv^2/r = 0.38$ kN.   **9.** $F_c = mv^2/r = (w/g)v^2/r = 33$ lb.   **11.** $v = 2\pi nr/T = 28.3$ m/s; $F_c = mv^2/r = 8.9$

kN.   **13.** $F_c = mv^2/r$, $v = \sqrt{F_c r/m} = 20$ m/s.
15. $mv_{max}^2/r = \mu mg$, $v_{max} = \sqrt{\mu gr}$. Since $f = v/2\pi r$, $f_{max} = \sqrt{\mu gr}/2\pi r = 0.863$ rev/s $= 52$ rev/min. Hence the dime will stay where it is at $33\frac{1}{2}$ rev/min but will fly off at 78 rev/min.   **17.** $v = 139$ m/s; $\tan\theta = v^2/gr$, $r = v^2/(g\tan 45°) = 1971$ m $= 1.97$ km.   **19.** $\theta = \tan^{-1} v^2/gr = 6.75°$; $h = (8\text{ m})(\sin 6.75°) = 0.94$ m.   **21.** (a) The tension in the string at the top of the circle is equal to the centripetal force of mv^2/r minus the weight mg of the stone, so that $T = mv^2/r - mg$. If the string is to be just taut, $T = 0$ and $mv^2r = mg$, $v = \sqrt{rg} = 2.8$ m/s. (b) Since the mass of the stone does not matter here, the speed is again 2.8 m/s.   **23.** $mv^2/r = mg$, $v = \sqrt{rg} = 11$ m/s.
25. $v^2/r = g$, $v = \sqrt{rg}$; $T = 2\pi r/v = 2\pi\sqrt{r/g} = 2.2$ s.
27. At top, $mv^2/r = mg$, KE(top) $= \frac{1}{2}mv^2 = \frac{1}{2}mrg$ and PE(top) $= 2mrg$. At bottom, KE(bottom) $=$ KE(top) $+$ PE(top) $= \frac{5}{2}mrg = \frac{1}{2}mv'^2$, $T = mv'^2/r = 5mg$.   **29.** (a) $F = Gm_1m_2/r^2 = 6.7 \times 10^{-10}$ N. (b) $F = 6.7 \times 10^{-10}$ N. (c) The acceleration of the 2-kg mass is $a = F/m = 3.3 \times 10^{-10}$ m/s^2. The acceleration of the 5-kg mass is $a = F/m = 1.3 \times 10^{-10}$ m/s^2.   **31.** If $M =$ sun's mass and $m =$ earth's mass, $GMm/r^2 = mv^2/r$, $M = v^2r/G = 2.0 \times 10^{30}$ kg.   **33.** The moon is $r_m = 3.84 \times 10^8$ m from the earth, so from Eq. (6.10) the acceleration of gravity there is $g_m = (r_e/r_m)^2 g_0 = 2.72 \times 10^{-3}$ m/s^2. When $t = 1.00$ s, $h = \frac{1}{2}gt^2 = 1.4$ mm. The moon never reaches the earth because its "falling" causes it to move in an orbit around the earth.   **35.** (a) $w = (r_e/r)^2 w_e = 20$ lb. (b) $m = w/g = w_e/g = 18$ slugs.   **37.** $g = (r_e/r)^2 g_0$, $r = r_e\sqrt{g_0/g} = \sqrt{10 g_0}$.   **39.** $g = (r_e/r)^2 g_0 = 6.3$ m/s^2.   **41.** (a) $v = \sqrt{r_e^2 g_0/r} = 7.7$ km/s. (b) $T = 2\pi r/v = 5.5 \times 10^3$ s $= 92$ min.   **43.** From Example 6.9, $r = (r_{ee}^2 g_e T^2/4\pi^2)^{1/3} = (190 \times 10^{20})^{1/3} = (19 \times 10^{21})^{1/3} = (19)^{1/3} (10^{21})^{1/3} = 2.67 \times 10^7$ m.   **45.** Here $\mathbf{a} = \mathbf{a}_x + \mathbf{a}_y$, where $a = 2g$ and $a_y = g$. Hence $a_x = \sqrt{a^2 - a_y^2} = \sqrt{3}g$.   **47.** The horizontal component T_x of the tension is given by $T_x = F_c = mv^2/r = 4\pi^2 mr/T^2$. If the length of the wire is L and θ is the angle between the wire and the horizontal, the radius of the circle is $r = L\cos\theta$ and $T = T_x/\cos\theta = 4\pi^2 mL/T^2 = 7.9$ N.
49. $F = \sqrt{F_x^2 + F_y^2} = \sqrt{(mv^2/r)^2 + (mg)^2} = 0.94$ kN.
51. Let $R =$ radius of moon's orbit, $r =$ distance of object from moon, and m the object's mass. Then $Gmm_m/r^2 = Gmm_e/(R - r)^2$, $(R - r)/r = \sqrt{m_e/m_m}$, $r = R/(1 + \sqrt{m_e/m_m}) = 3.8 \times 10^7$ m.

CHAPTER 7

Questions

1. Such a mass distribution maximizes I and so increases the flywheel's KE at a given angular speed.   **3.** The rec-

tangle.   **5.** The solid cylinder reaches the bottom first because its moment of inertia is smaller and hence less of its initial PE becomes KE of rotation.   **7.** They will reach the bottom at the same time.   **9.** The location of the pivot point.   **11.** Reducing I means that a given torque leads to a faster turn, making it possible to complete more somersaults before reaching the water.   **13.** The length of the day will increase. The earth's moment of inertia will be greater when the water from the icecaps becomes uniformly distributed and hence, from conservation of angular momentum, the angular speed must decrease.

Exercises and Problems

1. 2π rad/9 $= 0.698$ rad.   **3.** (a) $1' = 1°/60 = 0.01745$ rad/60 $= 2.91 \times 10^{-4}$ rad. (b) With such a small angle, the chord and the arc are very nearly equal. Hence $s = \theta r = 7.27 \times 10^{-3}$ cm $= 0.073$ mm.   **5.** (a) $\omega = 2\pi$ rad/12 hr $= 1.45 \times 10^{-4}$ rad/s. (b) $\omega = 2\pi$ rad/60 min $= 1.75 \times 10^{-3}$ rad/s. (c) $\omega = 2\pi$ rad/60 s $= 0.105$ rad/s.
7. $\omega = \theta/t = (2\pi \text{ rad/turn})(49,299 \text{ turns})/20,220$ s $= 15.3$ rad/s.   **9.** (a) $v = \omega r = 94.5$ m/s. (b) $\omega = (315 \text{ rad/s})/[0.1047 \text{ (rad/s)/rpm}] = 3.0 \times 10^3$ rpm.   **11.** $v = \omega r = 226$ km/h.   **13.** $a_c = \omega^2 r = 6g$, $\omega = \sqrt{6g/r} = 11$ rad/s.   **15.** $I = \frac{2}{5}MR^2 = 9.8 \times 10^{37}$ kg $\cdot$ m^2. The angular velocity of the earth is 7.3×10^{-5} rad/s, corresponding to one rotation per day. Hence KE $= \frac{1}{2}I\omega^2 = 2.6 \times 10^{29}$ J.
17. $I = 2mR^2/5 = 0.063$ kg $\cdot$ m^2, $\omega = v/r = 33.3$ rad/s; KE $= \frac{1}{2}I\omega^2 + \frac{1}{2}mv^2 = 0.12$ kJ.   **19.** KE (initial) $=$ PE(final), $\frac{1}{2}I\omega^2 = mgh$, $h = I\omega^2/2mg$.   **21.** $mgh = \frac{1}{2}mv^2 + \frac{1}{2}I\omega^2 = \frac{1}{2}mv^2 + \frac{1}{2}(I/R^2)v^2 = \frac{1}{2}(m + I/R^2)v^2$, $v = \sqrt{2gh/(1 + I/mR^2)}$. For the solid sphere, $I/mR^2 = 2/5$, and for the hollow sphere, $I/mR^2 = 2/3$. Hence $v(\text{solid})/v(\text{hollow}) = \sqrt{(1 + 2/3)/(1 + 2/5)} = 1.09$.   **23.** $\alpha = a_T/r = 0.5$ rad/s^2.   **25.** It is easiest to use the revolution as the angular unit here. Since $\omega_f^2 = \omega_0^2 + 2\alpha\theta$, $\theta = (\omega_f^2 - \omega_0^2)/2\alpha = 60$ revolutions.   **27.** $t = (20°/33\frac{1}{3}\text{ rpm})(1 \text{ rev}/360°)(60 \text{ s/min}) = 0.10$ s, $\alpha = \omega_f/t = 35$ rad/s^2.
29. $\tau = 9.0$ N $\cdot$ m.   **31.** $I = \frac{1}{2}(w/g)r^2 = 0.9375$ slug ft^2; $\omega_f = 3.14$ rad/s. (a) $\alpha = \tau/I$, $t = \omega_f/\alpha = \omega_f I/\tau = 0.59$ s. (b) KE $= \frac{1}{2}I\omega_f^2 = 4.6$ ft $\cdot$ lb.   **33.** $\alpha = (\omega_f - \omega_0)/t = 5$ rad/s^2, $I = \tau/\alpha = 100$ kg $\cdot$ m^2.   **35.** (a) $\alpha = (\omega_f - \omega_0)/t = 2.0$ rad/s^2, $\tau = I\alpha = 16$ lb $\cdot$ ft. (b) $\theta = \omega_0 t - \frac{1}{2}\alpha t^2 = 600$ rad. (c) KE $= \frac{1}{2}I(\omega_0^2 - \omega_f^2) = 9.6 \times 10^3$ ft $\cdot$ lb.   **37.** $P = 550$ ft $\cdot$ lb/s, $\omega = 126$ rad/s; $P = \tau\omega$, $\tau = P/\omega = 4.4$ lb $\cdot$ ft.   **39.** (a) $P = Fv = mgv = 235.2$ kW. (b) $\omega = v/r = 6.7$ rad/s.   **41.** $\omega = 210$ rad/s; $\tau = P/\omega = 23.8$ N $\cdot$ m; $F = \tau/r = 0.24$ kN. Each tooth exerts half this force, or 0.12 kN.   **43.** The net force on the belt is $F = 180$ N $- 70$ N $= 110$ N and $\omega = (30 \text{ rev/s})(2\pi \text{ rad/rev}) = 188$ rad/s. The torque exerted by the motor is τ

$= Fr$ so $P = \tau\omega = Fr\omega = 2.1$ kW.   **45.** (a) $I_1/I_2 = \omega_2/\omega_1 = 0.40$. (b) $KE_1/KE_2 = \frac{1}{2}I_1\omega_1^2/\frac{1}{2}I_2\omega_2^2 = 2.5$. **47.** (a) $L = I\omega = (mr^2)(v/r) = mvr$. (b) Since $r_e = 6380$ km, $r_{min} = 6780$ km and $r_{max} = 8380$ km. The maximum speed occurs at the minimum altitude and the minimum speed at the maximum altitude. Since angular momentum is conserved in orbital motion, $mv_{min}r_{max} = mv_{max}r_{min}$, $v_{max}/v_{min} = r_{max}/r_{min} = 1.24$.   **49.** The time needed for the neutron to move at the speed v through the distance L between the disks is the same as the time needed for the second disk to turn at the angular speed ω through the angle θ. Hence $t = L/v = \theta/\omega$, $v = L\omega/\theta = 7.2 \times 10^6$ m/s.   **51.** When the rod is released, its center of gravity falls by $h = L/2$. Hence its initial potential energy is $PE = mgh = mgL/2$. The rod's KE when it has fallen is $\frac{1}{2}I\omega^2$, where $I = mL^2/3$. Since $\frac{1}{2}I\omega^2 = mgh$, $ML^2\omega^2/6 = mgL/2$, $\omega = \sqrt{3g/L}$. **53.** The easiest approach here is to use conservation of energy: PE(before) = KE(after), $mgh = \frac{1}{2}mv^2 + \frac{1}{2}I\omega^2$. Since $I = \frac{1}{2}MR^2$ and $\omega = v/R$, $mgh = \frac{1}{2}(m + M/2)v^2$, $v = \sqrt{2gh/(1 + M/2m)}$.   **55.** $\theta = \frac{1}{2}\alpha t^2$, $\alpha = 2\theta/t^2$. If $L =$ width of door, $\tau = FL$ and $I = \tau/\alpha = FLt^2/2\theta$. Since $I = ML^2/3$, $M = 3I/L^2 = 3Ft^2/2L\theta$. Here $\theta = 90° = \pi/2$ so $M = 537$ kg.   **57.** (a) $I_1 = mR_1^2 = 0.2$ kg $\cdot$ m^2, $I_2 = mR_2^2 = 0.288$ kg $\cdot$ m^2, $\omega_1 = 31.4$ rad/s, $\omega_2 = I_1\omega_1/I_2 = 21.8$ rad/s $= 3.5$ rev/s. (b) $KE_1 = \frac{1}{2} \times \frac{1}{2}I_1\omega_1^2 = 49$ J, $KE_2 = \frac{1}{2} \times \frac{1}{2}I_2\omega_2^2 = 34$ J; the decrease in KE is due to the work that had to be done to stretch the string.   **59.** (a) $I_2 = 2I_1$, $\omega_2 = I_1\omega_1/I_2 = \omega_1/2 = 50$ rad/s. (b) $KE_1 = \frac{1}{2}I_1\omega_1^2 = 5.0 \times 10^3$ ft $\cdot$ lb, $KE_2 = \frac{1}{2}I_2\omega_2^2 = 2.5 \times 10^3$ ft $\cdot$ lb. The difference represents energy dissipated as heat when the moving disk was pressed against the stationary one before the two stuck together. (c) 50 rad/s. (d) No KE was lost.

CHAPTER 8

Questions

1. (a) The forces that act on the object. (b) The forces the object exerts on anything else. (c) Yes. (d) Yes.   **3.** (a) Perpendicular to the wall. (b) The force on the ground is always greater in magnitude than the weight of the ladder because the force has a horizontal component equal to the reaction force of the wall on the ladder as well as a vertical component equal to the ladder's weight.   **5.** Negative; positive.   **7.** About whatever point makes the calculations easiest; the results will be the same regardless of where the point is.   **9.** Greater than 1; less than 1.   **11.** 4.

Exercises and Problems

1. $T_y = T \sin \theta = w$, $T = w/\sin \theta = 0.16$ kN.   **3.** The mass each rope must support is $m = 25$ kg. Since the verti-

cal component of the tension in each rope is $F = T \cos \theta$, $T = F/\cos \theta = mg/\cos \theta = 0.28$ kN.   **5.** The tension in each string is the weight mg of each 5.0-kg mass. The vertical component of each tension force on A is $T_y = mg \cos 30°$ and so $w_A = m_Ag = 2T_y = 2mg \cos 30°$, $m_A = 2m \cos 30° = 8.7$ kg.   **7.** $F = \sqrt{F_x^2 + w^2} = 94$ lb; $\theta = \tan^{-1} w/F_x = 32°$ above the horizontal.   **9.** $\theta_1 = 20°$, $\theta_2 = 30°$; $\Sigma F_x = -T_1 \sin \theta_1 + T_2 \sin \theta_2 = 0$, $T_2 = T_1 \sin \theta_1/\sin \theta_2$; $\Sigma F_y = T_1 \cos \theta_1 + T_2 \cos \theta_2 - w = 0$, $T_1 = (w - T_2\cos \theta_2)/\cos \theta_1 = w/(\cos \theta_1 + \sin \theta_1/\tan \theta_2) = 392$ lb; $T_2 = T_1 \sin \theta_1/\sin \theta_2 = 268$ lb.   **11.** $mg = F \cos \theta$, $m = (F \cos \theta)/g = 197$ kg.   **13.** Let T_1 and T_2 be the tensions in the two parts of the wire, where $T_1 = T_2 = T$. If θ is the angle either part makes with the horizontal, $\theta = \tan^{-1} (0.5$ m)/(10 m) $= 2.86°$. For the bird to be in equilibrium, $\Sigma F_y = T_{1y} + T_{2y} - mg = 2T \sin \theta - mg = 0$, $m = 0.92$ kg.   **15.** (a) $F_y = F \sin \theta = mg$, $F = mg/\sin \theta = 69$ kN. (b) $T = F_x = F \cos \theta = 59$ kN.   **17.** Let $m_1 = 15$ kg, $m_2 = 20$ kg, and $m_3 = 25$ kg. Calculating torques about the pivot with $x_1 = 2.0$ m and $x_3 = 2.0$ m gives $m_1gx_1 + m_2gx_2 = m_3gx_3$, $x_2 = 1.0$ m. **19.** Calculating torques about the end of the board on the floor, $\Sigma\tau = F_L - w(L/2) \cos \theta = 0$, $F = [w(L/2) \cos \theta]/L = (w/2) \cos \theta = 23$ lb.   **21.** Calculating torques about the base of the rod, $(F \sin \theta)(1$ m) $= (mg \cos \theta)(3$ m), $F = 3mg/\tan \theta = 0.25$ kN.   **23.** Let x be the distance from the painter to the end of the plank. Calculating torques about this end of the plank, $(700$ N)$x + (300$ N)(2.5 m) $- (1000$ N)(1.0 m) $= 0$, $x = 36$ cm.   **25.** At the point of tipping over, the CG is directly above the line of the lower wheels. Hence $h = 0.9$ m/tan 30° $= 1.6$ m.   **27.** (a) $F = \mu_s mg = 0.31$ kN. (b) $x = 0.5$ m, $mgx - Fh = 0$, $h = mgx/F = 1.3$ m.   **29.** Let x be the distance of the two men from the far end of the ladder and w be the ladder's weight. Calculating torques about the CG of the ladder yields $(w/3)(2.00$ m) $= (2w/3)(2.00$ m $- x)$, $x = 1.00$ m.   **31.** If w is the weight supported by the front axle, then $(24$ kN $- w)$ is the weight supported by the rear axle. Calculating torques about the CG of the truck, $(w)(2.5$ m) $= (24$ kN $- w)(1.5$ m), $w = 9.0$ kN.   **33.** Let the force shared by both hind paws be F_1 and that shared by both forepaws be F_2. Calculating torques about the hind paws gives $F_2L_2 = wL_1$, $F_2 = 167$ lb, so the force on each forepaw is $F_2/2 = 83$ lb. Since $F_1 = w - F_2 = 133$ lb, the force on each hind paw is $F_1/2 = 67$ lb.   **35.** The angle between the boom and the rope is $\theta = \tan^{-1} (1$ m)/(3 m) $= 18.4°$ and the CG of the boom is at its midpoint. Calculating torques about the hinge pin of the boom, $(40$ kg)$(g)(1.5$ m) $= (T)(3$ m)$(\sin 18.4°)$, $T = 0.62$ kN.   **37.** If x is the distance of the CG from the forward axle, calculating torques about the CG yields $m_1gx = m_2g(2.2$ m $- x)$, $x = 0.88$ m.   **39.** There is no horizontal force acting on the dumbbell; hence its CG must drop vertically as the dumbbell turns. The 1-kg ball therefore moves to the left. The CG is located 20 cm from the 1-kg ball, and so the final position of this ball is 20 cm to the left of its original position. **41.** Let $x = 0$ be at the right-hand end of the object. We imagine the object to consist of the cross member of mass $M_1 = 10$ whose CG is at $x_1 = -20$ cm and the horizontal member of mass $m_2 = 20$ whose CG is at $x_2 = -10$ cm. The CG of the object is at $X = (m_1x_1 + m_2x_2)/(m_1 + m_2) = -13.3$ cm, which is 6.7 cm to the right of the cross member. **43.** MA $= 5$, $F_{out} = $ MA $\times F_{in} = 4.5$ kN. **45.** $F_{in} = F_{out}L_{out}/L_{in} = mgL_{out}/L_{in} = 2.6$ kN $= 582$ lb. **47.** (a) The output RPM does not depend upon the efficiency, so $RPM_{out}/RPM_{in} = N_{in}/N_{out}$, $RPM_{out} = (N_{in}/N_{out})RPM_{in} = 2$ rev/min. (b) $\tau_{out} = (Eff)(N_{out}/N_{in})(\tau_{in}) = 96$ lb · ft. **49.** $\tau_{in} = P/\omega$, $\tau_{out} = (Eff)(\tau_{in})(N_{out}/N_{in}) = 0.30$ kN · m. **51.** AMA $=$ IMA $= (IMA_1)(IMA_2) = (2)(L/h) = mg/F_{in}$, $F_{in} = mgh/2L = 0.29$ kN. **53.** $F_{out} = (Eff)(2\pi L/p)(F_{in}) = 31$ N. **55.** Let L be the length of each rod, F_y the vertical force of the table on it, and F_x the horizontal frictional force of the table on it. Calculating torques about the pivot gives $\tau = F_yL \cos \theta - w(L/2) \cos \theta - F_xL \sin \theta$. If w is the weight of each rod, $F_y = w$ and $F_x = [F_yL \cos \theta - w(L/2)\cos \theta]/L \sin \theta = F_y/(2 \tan \theta)$. Since $F_x \le \mu_sF_y$, $F_y/(2 \tan \theta) \le \mu_sF_y$, hence $\mu_s \ge 1/(2 \tan \theta)$. **57.** For each leg, $F_y = mg/2 = 490$ N and $\theta = \cos^{-1}(1.2$ m/3 m$) = 66.4°$. Since $F_y = F \sin \theta$, $F = F_y/\sin \theta = 0.53$ kN. **59.** The angle between the supporting cable and the boom is 70° and the angle between the mast and the boom is 50°. If L is the length of the boom, calculating torques about the base of the mast gives $TL \sin 70° = mgL \sin 50°$, $T = 12$ kN. **61.** We calculate torques about the inner end of the beam. The torque exerted by the load is $\tau_1 = (1000$ lb$)(12.0$ ft$)$, the torque exerted by the beam's weight is $\tau_2 = (200$ lb$)(6.0$ ft$)$, and the torque exerted by the tension T in the rope is $\tau_3 = -(T \sin 40°)(12.0$ ft$)$. From $\tau_1 + \tau_2 + \tau_3 = 0$, $T = 1.7 \times 10^3$ lb. **63.** Here $F_y = F \sin 15°$, $w_1 = m_1g = 29.4$ N, $w_2 = m_2g = 9.8$ N, $L_1 = 0.13$ m, $L_2 = 0.28$ m, $L_3 = 0.60$ m. (a) Calculating torques about the shoulder joint yields $F_yL_1 = w_1L_2$, $F = w_1L_2/(L_1 \sin 15°) = 0.24$ kN. (b) $F_y'L_1 = w_1L_2 + w_2L_2$, $F' = (w_1L_2 + w_2L_3)/(L_1 \sin 15°) = 0.42$ kN. **65.** Since the wall is frictionless, $F_y = w = mg = 147$ N and F_x is equal in magnitude to the force the ladder exerts on the wall. Calculating torques about the foot of the ladder, $(F_x)(2.4$ m$) = (mg)\frac{1}{2} \sqrt{(3$ m$)^2 - (2.4$ m$)^2} = \frac{1}{2}(mg)(1.8$ m$)$, $F_x = 55$ N. **67.** (a) Let **F** be the force the upper hinge exerts on the door and $w = 200$ N be the door's weight acting from its CG. Since the door's weight is supported by the upper hinge, $F_y = w = 200$ N. To find F_x, we calculate torques about the lower hinge: $(F_x)(2.4$ m$)$

$= (w)(0.4$ m$)$, $F_x = 33.3$ N. Hence $F = \sqrt{F_x^2 + F_y^2} = 203$ N. If θ is the angle between **F** and the vertical, $\theta = \tan^{-1} F_x/F_y = 9.5°$. The force the door exerts on the upper hinge is equal and opposite to **F**, hence 203 N at 9.5° away from vertically downward. (b) Since the lower hinge exerts a horizontal force on the door, it must be equal in magnitude to $F_x = 33$ N. **69.** $\omega_{saw} = 3\omega_{motor} = 550$ rad/s; $v_{teeth} = \omega_{saw}r_{saw} = 115$ ft/s; $F = P/v = 2.4$ lb.

CHAPTER 9

Questions

1. The second procedure is more likely to break the string because here the tension in the string is twice as great, the reaction force exerted by the tree being equal and opposite to the pull of the two boys. **3.** All are under the same stress. The strain varies as $1/Y$, hence the aluminum leg experiences the greatest strain and the steel leg the least strain. **5.** B stretches twice as much as A. **7.** Inside the building, on the lower part of the floor; on the balcony, on the upper part of the floor.

Exercises and Problems

1. $V = $ (length)(width)(height) $= 60$ m^3, $m = \rho V = 78$ kg. **3.** (a)$\rho_{water} = 1$ kg/liter, so $V = m/\rho = 55$ L. (b) $V = m/\rho = 140$ m^3. **5.** (a) $\rho = m/V = m/(4\pi r^3/3) = 5.52 \times 10^3$ kg/m^3. (b) The interior must consist of denser materials than those at the surface; no. **7.** The weights of the liquid columns above the bottom of the oil column must be equal. Since the tube has the same cross-sectional area on both sides, $L_{water}\rho_{water} = L_{oil}\rho_{oil}$, $\rho_{oil} = L_{water}\rho_{water}/L_{oil} = 0.77$ g/cm^3. **9.** $k = F/s = mg/s = 2.94$ kN/m. **11.** $x = mg/k = 0.80$ ft. **13.** The breaking strength F is proportional to the cross-sectional area $A = \pi d^2/4$ of the rope. Hence (a) $F = (25$ kN$)(6$ mm/10 mm$)^2 = 9$ kN and (b) $F = (25$ kN$)(14$ mm/10 mm$)^2 = 49$ kN. **15.** The stress due to the acrobat's weight is $F/A = mg/A = 1.47 \times 10^7$ Pa; hence the safety factor is $(1.7 \times 10^8)/(1.47 \times 10^7) = 11.6$. **17.** $F/A = mg/\pi r^2 = 1.9 \times 10^8$ Pa, which is just a hair less than the ultimate strength of aluminum. **19.** $Y = L_0F/A\Delta L = 1.2 \times 10^{11}$ Pa. **21.** $A = 10^{-6}$ m^2. $\Delta L = L_0F/YA = L_0mg/YA = 0.29$ mm. **23.** $\Delta L = L_0F/YA = 5.8 \times 10^{-4}$ ft $= 0.0070$ in. **25.** $A = FL_0/Y\Delta L = 66.2$ in$^2 = \pi d^2/4$, $d = 2\sqrt{A/\pi} = 9.2$ in. **27.** $\Delta L = L_0F/YA = L_0F/Y\pi(r_{outside}^2 - r_{inside}^2) = 1.44 \times 10^{-4}$ m $= 0.14$ mm. **29.** The volume of lead above the lowest brick is $V = 0.9$ m $\times (0.1$ m$)^2 = 0.009$ m^3 and the weight of this lead is $w = V\rho g = 970$ N. Since $F = w$, $L_0 = 0.1$ m, and $A = 0.01$ m^2, the final height of the lowest brick is shorter by $\Delta L = L_0F/YA = 6.1 \times 10^{-7}$ m.

31. $S = Fd/sA = 3.3$ k Pa.   **33.** $L = 0.01$ m, $A =$ $4LH$, $(F/A)_{max} = F/4Lh$, $h = F/4L(F/A)_{max} = 7.1 \times 10^{-3}$ m $= 7.1$ mm.   **35.** $\Delta V = pV_0/B = 4 \times 10^{-8}$ m^3 = 40 mm^3.   **37.** $\Delta V/V_0 = p/B = 1.4 \times 10^{-6} = 1.4 \times 10^{-4}\%$.   **39.** $P = mgh/(t)$ (Eff) $= dVgh/(t)$ (Eff), $V/t =$ (P) (Eff)$/dgh = 0.0303$ m^3/s $= 30.5$ L/s $= 1.8 \times 10^3$ L/ min.   **41.** The upward force $F = mg$ on the helicopter's rotor blades equals the rate of change of the momentum of the air forced downward by them. If the area swept by the blades is $A = \pi r^2$, then $F = v(\Delta m/\Delta t) = v\rho(\Delta V/\Delta t) =$ $v\rho A(\Delta y/\Delta t)$. But $\Delta y/\Delta t$ is the speed v at which the air under the rotor blades is pushed downward, so $F = mg = v^2 \rho A$ $= v^2\rho\pi r^2$ and $v = \sqrt{mg/\rho\pi}\, r = 9.3$ m/s.   **43.** F(max) $= (0.2)(F/A)_{max} (A) = 1.5 \times 10^4$ N. The net force available for an upward acceleration is $F = F$(max) $- mg = 5200$ N, where m is the elevator's mass, and $a = F/m = 5.2$ m/ s^2.   **45.** (a) The force each wire supports is proportional to its Young's modulus, since each wire stretches by the same amount. Hence $F_s/(F_s + F_c) = Y_s/(Y_s + Y_c) = 0.65$ $= 65\%$ of the load is supported by the steel wire and 35% by the copper wire. (b) In order that each wire support an equal part of the load, the product YA must have the same value for both, and $Y_sA_s = Y_cA_c$. Since $A = \pi d^2/4$, $Y_sd_s^2 = Y_cd_c^2$ and $d_c/d_s = \sqrt{Y_s/Y_c} = 1.35$.   **47.** If θ is the angle between each half of the wire and the horizontal, $\theta = \tan^{-1}$ (20 mm/500 mm) $= 2.3°$. If L_0 is the horizontal distance between the wire's ends, $\Delta L = L_0/\cos \theta - L_0 = L_0(1/\cos \theta - 1)$, $L_0/\Delta L = \cos \theta/(1 - \cos \theta) = 1250$. From Example 8.1 we see that the tension in the wire is $F = mg/(2 \sin \theta)$, so $Y = (L_0/\Delta L)(F/A) = (L_0/\Delta L) [mg/(2A \sin \theta)] = 1.5 \times 10^{11}$ Pa.

CHAPTER 10

Questions

1. As the steam inside the can condenses, the internal pressure falls below atmospheric pressure.   **3.** Atmospheric pressure.   **5.** Airliners fly at high altitudes where there is a large pressure difference between their pressurized interiors and the outside atmosphere. Light aircraft do not have pressurized interiors, so the pressures on both sides of their windows are the same; the windows have no net force to withstand in normal flight even if they are large.   **7.** The pressures are the same.   **9.** There is no buoyant force because there is no water under the block to provide an upward force.   **11.** The load on the bridge does not change because the boat displaces a volume of water whose weight equals its own weight.   **13.** The water level falls. The floating canoe displaced a volume of water V whose weight

equalled its own weight. The sunken canoe, however, displaces a volume of water equal to the volume of its aluminum shell, which is smaller than V because aluminum is denser than water.   **15.** A ship is a hollow shell, so its average density is less than that of water. If the ship fills with water, its average density increases until it is greater than that of water and then it sinks.   **17.** Ice on the wings interferes with the intended flow of air past them and so decreases the lift they produce. Ice on the fuselage increases the weight of the airplane and increases air resistance, but these effects are usually less significant than the reduction in lift.   **19.** No. For example, water at 20°C has a greater density than ethyl alcohol but a smaller viscosity.

Exercises and Problems

1. $F = mg = 19.6$ N, $A_1 = (0.075 \times 0.15)$ m^2, $A_2 =$ (0.075×0.3) m^2, $A_3 = (0.15 \times 0.3)$ m^2. Hence $p_1 = F/$ $A_1 = 1.74$ kPa, $p_2 = 0.87$ kPa, $p_3 = 0.44$ kPa.   **3.** F $= pA = p\pi d^2/4 = 1.4$ N.   **5.** $\Delta p = F/A = 90.9$ lb/ft^2. **7.** $p = F/A = F/\pi r^2 = 6.37$ MPa $= 63$ atm.   **9.** $p =$ 15 torr $= 2 \times 10^3$ Pa $= F/A$, $F = pA = p(\pi r^2) = 0.16$ N. The needle's diameter is irrelevant.   **11.** (a) $F_{out} =$ $F_{in}(A_{out}/A_{in}) = 12.5$ N. (b) $L_{out} = L_{in}(A_{in}/A_{out}) = 8.0$ cm. **13.** $IMA_1 = (d_{out}/d_{in})^2 = 28.4$, $IMA_2 = 8.0$, IMA $=$ $(IMA_1)(IMA_2) = 228$; $F_{in} = F_{out}/($Eff$)($IMA$) = 18$ lb. **15.** $p = p_{atm} + \rho gh = 1.10 \times 10^8$ Pa.   **17.** $\Delta p = 0.1$ atm $= 1.013 \times 10^4$ Pa, $h = \Delta p/\rho g = 1.03$ m. **19.** Here $h = 0.7$ m so $\Delta p = \rho gh = 7.27$ kPa $=$ 55 torr. **21.** $F = pA = DhA = 2.6 \times 10^4$ lb.   **23.** If V_{water} is the volume of water displaced by the submerged part of the board, $V_{water}\rho_{water}g = V_{board}\rho_{wood}g$, $\rho_{wood} = \rho_{water} (V_{water}/$ $V_{board}) = \rho_{water} (h_{submerged}/h_{total}) = 0.70$ g/cm^3.   **25.** $V =$ $m/\rho_{water} = 0.07$ m^3, $F = V\rho_{air}g = 0.89$ N.   **27.** The volume of water whose mass is 60 kg is 60 kg/(1.0 $\times 10^3$ kg/m^3) $= 0.060$ m^3. By Archimedes' principle the raft rises by a height h such that its submerged volume decreases by 0.06 m^3 when the woman dives off. Hence $h = (0.060$ m$^3)/$ $(3.0$ m$)(2.0$ m$) = 0.010$ m $= 10$ mm.   **29.** $F = Vg(\rho_{air}$ $- \rho_{hydrogen}) - mg = 0.89$ kN.   **31.** $V\rho_{air}g = (m +$ $V\rho_{hydrogen})g$, $m = V(\rho_{air} - \rho_{hydrogen}) = 2.4 \times 10^5$ kg. **33.** $D_{water}V_{tank} = 8.0$ lb $+ D_{liquid}A_{tank}h_{liquid}$, $h_{liquid} = 6.48$ in. **35.** The volume of the bracelet is $V = \Delta m/\rho_{water} = 4.0$ cm^3 and its density is $\rho = m/V = 12.5$ g/cm^3. The density of pure gold is 19 g/cm^3 so the bracelet is not pure gold. **37.** Apparent mass when immersed $= m' = m - \rho V$. Here m' (in water) $= 70.0$ g $= m - \rho_{water}V$ and m' (in oil) $=$ 76.0 g $= m - \rho_{oil}V$. Subtracting the first equation from the second gives 4.0 g $= (\rho_{water} - \rho_{oil})V$, $V = 20$ cm^3. Using this value of V in either equation for m' gives $m = 90$ g.

39. $v_1A_1 = v_2A_2$, $A_2 = v_1A_1/v_2$, $D_2 = \sqrt{v_1D_1^2/v_2} = 4.5$ mm.   **41.** $v = R/A = 45$ ft/s.   **43.** The power output of the left ventricle is $P_L = pR = 6.2$ W. Since $P_R = 0.2$ $P_L = 1.2$ W, the total power output is $P_L + P_R = 7.4$ W.   **45.** $R = \Delta V/\Delta t = \sqrt{2gh}\,A$, $\Delta m/\Delta t = \rho(\Delta V/\Delta t) = \rho A\sqrt{2gh} = 1.9$ kg/s.   **47.** $p = 120$ torr $= 1.60 \times 10^4$ Pa. This is gauge pressure, so $h = p/dg = 1.5$ m.   **49.** $h = R^2/2gA^2 = 4.2$ ft.   **51.** $R = 16$ L/min $= 2.67 \times 10^{-4}$ m²/s, $p/L = 8\eta R/\pi r^4 = 0.52$ kPa/m.   **53.** (a) $\Delta p = 2.66 \times 10^3$ Pa, $R = vA = v(\pi r^2)$, $v = r^2\Delta p/8\eta L = 0.67$ mm/s. (b) In each capillary, $R = vA = 8.36 \times 10^{-15}$ m³/s $= 8.36 \times 10^{-9}$ cm³/s, so $N = (80\text{ cm}^3/\text{s})/R = 9.6 \times 10^9$ capillaries.   **55.** If V is the volume of the sinker and m is its mass, $F = mg - V\rho_{\text{oil}}g = V\rho_{\text{lead}}g - V\rho_{\text{oil}}g = Vg(\rho_{\text{lead}} - \rho_{\text{oil}})$, $V = F/g(\rho_{\text{lead}} - \rho_{\text{oil}})$; $m = \rho_{\text{lead}}V = \rho_{\text{lead}}F/g(\rho_{\text{lead}} - \rho_{\text{oil}}) = 0.022$ kg $= 22$ g.   **57.** The volume of lead is $V = m/\rho_{\text{lead}}$. The downward force exerted by the lead is $F = mg - F_{\text{buoyant}} = mg - V\rho_{\text{water}}g = mg(1 - \rho_{\text{water}}/\rho_{\text{lead}}) = 71.06$ kN. The lever arm of this force equals the horizontal displacement x of the keel's CG from being vertically below the boat, so $x = L\sin\theta = 0.513$ m and $\tau = Fx = 36$ kN · m.   **59.** From Eq. (5.7), $F = v(\Delta m/\Delta t) = v(\Delta V/\Delta t) = v\rho A(\Delta x/\Delta t)$. Since $\Delta x/\Delta t = v$ and $v^2 = 2gh$, $F = v^2\rho A = 2gh\rho A = 0.20$ kN.   **61.** $h = \frac{1}{2}gt^2$, $t = \sqrt{2h/g} = 0.45$ s, $v = x/t = 4.43$ m/s. $p = \frac{1}{2}\rho v^2 = 9.8$ kPa.   **63.** (a) $v_2 = v_1A_1/A_2 = v_1r_1^2/r_2^2 = 1.3$ m/s. Here $h_1 = h_2$ so $p_2 = p_1 + \frac{1}{2}\rho(v_1^2 - v_2^2) = 2.3 \times 10^5$ N/m² $= 2.3$ bars. (b) $v_2 = v_1A_1/A_2 = v_1r_1^2/r_2^2 = 1.3$ m/s. Here $h_1 - h_2 = -4$ m, so $p_2 = p_1 + \rho g(h_1 - h_2) + \frac{1}{2}\rho(v_1^2 - v_2^2) = 1.9 \times 10^5$ N/m² $= 1.9$ bars.   **65.** (a) $R = v_1A_1 = v_1\pi r_1^2$, $v_1 = R/\pi r_1^2 = 0.88$ m/s. (b) $v_2 = v_1A_1/A_2 = v_1(r_1/r_2)^2 = 3.5$ m/s. (c) From Bernoulli's equation, $p_2 = p_1 + \frac{1}{2}\rho(v_1^2 - v_2^2) + \rho g(h_1 - h_2) = 55$ kPa.

CHAPTER 11

Questions

1. At either extreme of its motion the energy is entirely PE; at the midpoint the energy is entirely KE.   **3.** When the object has vertical sides, so the restoring force is proportional to the displacement of the object above or below its equilibrium level.   **5.** $PE_{\max}$ and $KE_{\max}$ for a harmonic oscillator are always equal.   **7.** When the string is vertical.   **9.** 0; $1/T$; $1/T$.   **11.** In the ax's head so that there will be no reaction force on the hands of the user.

Exercises and Problems

1. $k = 2W/x^2 = 1.0$ kN/m.   **3.** $k = F/x = 20$ N/m; $\frac{1}{2}mv^2 = \frac{1}{2}kx^2$, $v = \sqrt{k/m}\,x = 1.4$ m/s.   **5.** (a) $k = F/x$,

$PE = \frac{1}{2}kx^2 = Fx/2 = 0.73$ ft · lb. (b) $KE = \frac{1}{2}mv^2 = PE$, $v = \sqrt{2PE/m} = \sqrt{2gPE/w} = 9.7$ ft/s.   **7.** (a) $A = 10$ mm. (b) $f = 1/T = \sqrt{k/m}/2\pi = 3.6$ Hz.   **9.** $T = 2\pi\sqrt{w/gk}$, $w = gkT^2/4\pi^2 = 4.4$ lb.   **11.** $T_1 = 2\pi\sqrt{m_1/k}$, $k = 4\pi^2m_1/T^2 = 0.877$ N/m. $T_2 = 2\pi\sqrt{m_2/k} = 2.3$ s.   **13.** Let $s = $ maximum sag. Then PE(gravitational) = PE(elastic), $mg(h + s) = \frac{1}{2}ks^2$, $k = 2mg(h + s)/s^2 = 8575$ N/m; $T = 2\pi\sqrt{m/k} = 0.57$ s.   **15.** $f = 1/T$, $v_{\max} = 2\pi fA = 2\pi A/T = 314$ cm/s $= 3.1$ m/s.   **17.** $a_{\max} = 4\pi^2f^2A$, $f = 1/2\pi\sqrt{a_{\max}/A} = 11$ Hz.   **19.** (a) $v_{\max} = 2\pi fA = 4.1$ m/s. (b) $a_{\max} = 4\pi^2f^2A = 1.7 \times 10^4$ m/s² $= 1.8 \times 10^3$ g.   **21.** (a) $k = F/x = 8.00$ lb/ft; $f = 1/T = \sqrt{gk/w}/2\pi = 1.80$ Hz. (b) $PE_{\max} = kA^2/2 = 0.0625$ ft · lb.   **23.** $F_{\max} = ma_{\max} = 4\pi^2wf^2A/g$, $f = \sqrt{F_{\max}g/wA}/2\pi = 48$ Hz.   **25.** (a) $T = 2\pi\sqrt{m/k}$, $k = 4\pi^2m/T^2 = 39$ N/m. (b) $v_{\max} = 2\pi fA = 0.31$ m/s. (c) $a_{\max} = 4\pi^2f^2A = 2.0$ m/s².   **27.** $T = 2\pi\sqrt{L/g} = 4.9$ s.   **29.** $g_M = 3.72$ m/s²; $T = 2\pi\sqrt{L/g}$, $L = T^2g/4\pi^2 = 0.094$ m $= 94$ mm.   **31.** (a) $T = 2\pi\sqrt{L/g} = 1.11$ s. (b) The motion of the pendulum bob is unaffected by motion of its support at constant velocity, hence $T = 1.11$ s. (c) The downward force on the bob here is $m(g - a)$, hence $T = 2\pi\sqrt{L/(g - a)} = 1.25$ s. (d) As in part (b), $T = 1.11$ s. (e) The downward force on the bob here is $m(g + a)$, hence $T = 2\pi\sqrt{L/(g + a)} = 1.01$ s.   **33.** PE (at angle θ) = KE (at $\theta = 0$), $mgL(1 - \cos\theta) = \frac{1}{2}mv_{\max}^2$, $v_{\max} = \sqrt{2gL(1 - \cos\theta)}$.   **35.** (a) $I = mL^2/3$; $h = L/2$; $T = 2\pi\sqrt{I/mgh} = 2\pi\sqrt{2L/3g} = 1.5$ s. (b) $L' = I/mh = 2L/3 = 53$ cm.   **37.** (a) $I = \frac{1}{3}mL^2$, $h = L/2$, $T = 2\pi\sqrt{I/mgh} = 2\pi\sqrt{2L/3g} = 1.60$ s. (b) Each step represents only half of a complete cycle, so such a leg swinging freely takes 0.80 s per stride. The rate of walking is therefore 1.25 strides/s and $v = (1.25)(0.80)$ m/s $= 1.0$ m/s $= 3.6$ km/h. To walk faster or slower than this would involve more effort.   **39.** (a) If x_1 and x_2 are the lengths by which the springs are stretched, then $x_1 + x_2 = 50$ cm $- 10$ cm $- 20$ cm, $x_2 = 20$ cm $- x_1$. Because $F_1 = F_2$, $k_1x_1 = k_2x_2 = k_2(20$ cm $- x_1)$, $x_1 = 8$ cm, $x_2 = 12$ cm. (b) Displacing the mass by x gives rise to a restoring force of $k_1x + k_2x = (k_1 + k_2)x$ since each spring is moved through x (one spring is compressed, the other stretched). Hence the force constant of the combination is $k' = k_1 + k_2 = 250$ N/m and $T = 2\pi\sqrt{m/k'} = 0.25$ s.   **41.** (a) $T_1 = 2\pi\sqrt{L_1/g}$, $T_2 = 2\pi\sqrt{L_2/g}$, $L_2/L_1 = (T_2/T_1)^2$, $L_2 = L_1(T_2/T_1)^2 = L_1 + \Delta L$, $\Delta L = L_1[(T_1/T_2)^2 - 1]$, $L_1 = \Delta L/[(T_1/T_2)^2 - 1] = 1586$ mm $= 159$ cm (to 3 significant figures). (b) With $L_1 = 1.586$ m, $g = 4\pi^2L_1/T_1^2 = 9.55$ m/s².   **43.** (a) On the far side of the wheel's axis from the piston (the 3 o'clock position). (b) $A = 20$ cm (the wheel's radius); $\omega = 2\pi f = 25$ rad/s. (c) $v_{\max} = 2\pi fA = 5.0$ m/s; $a_{\max} = 4\pi^2f^2A = 1.3 \times 10^2$ m/s².

CHAPTER 12

Questions

1. The energy dissipates as heat, and the string becomes warmer as a result. **3.** It increases fourfold because the rate of flow of energy is proportional to A^2. **5.** In solids, because their constituent particles are more tightly coupled together than those of gases or liquids. **9.** Frequency; amplitude; harmonic structure. **11.** The other fork could also have a frequency of 246 Hz to give 10 beats per second. **13.** (a) The short-wavelength sound will be 9 times more intense than the other one because the intensity of a mechanical wave is proportional to f^2. (b) The short-wavelength sound will seem louder, but less than 9 times louder, because the response of the ear is roughly logarithmic.

Exercises and Problems

1. The linear density of the rope is m/L. The tension at height h equals the weight of the rope below h, so $T = (m/L)hg$ and $v = \sqrt{T/(m/L)} = \sqrt{hg}$. Hence v increases as $\sqrt{h}$. **3.** $f = v/\lambda = 3.0$ kHz. **5.** (a) $f = v/\lambda = 0.86$ Hz. (b) $T = 1/f = 1.2$ s. **7.** (a) $v(\text{waves}) = f\lambda = \lambda/T = 8.8$ m/s. (b) $v(\text{molecules}) = 2\pi r/T = 1.3$ m/s. **9.** $f_1 = v/2L$, $v = 2Lf_1 = 255$ m/s. **11.** (a) Wavelength of fundamental frequency $= \lambda = 2L = 2m$, hence $v = \lambda f = 600$ m/s. (b) $f_2 = 2f_1 = 600$ Hz; $f_3 = 3f_1 = 900$ Hz; $f_4 = 4f_1 = 1200$ Hz. **13.** From Eq. (12.8), $(m/L)_B = (m/L)_A f_A^2/f_B^2 = 0.60$ g/m. **15.** (a) From Eq. (12.8), $f_A/f_B = L_B/L_A$, so with $L_B = 270$ mm, $f_B = f_A L_A/L_B = 805$ Hz. $f_2 = 2f_1 = 1611$ Hz; $f_3 = 3f_1 = 2416$ Hz. **17.** $\lambda = v/f = 29$ cm, which is slightly larger than the distance between the ears of an adult. **19.** $\Delta t = x/v_{\text{air}} - x/v_{\text{steel}} = 1.24$ s. **21.** $t = h/v_s$, $s = vt = vh/v_s = 1.6$ km. **23.** (a) $f = 440$ Hz, $\lambda = v_{\text{water}}/f = 3.4$ m. (b) $f = 440$ Hz, $\lambda = v_{\text{air}}/f = 0.78$ m. **25.** Since $v_s/v = \sin\theta$, the Mach number is $v/v_s = 1/\sin\theta = 1.15$. **27.** $\lambda/4 = v/4f = 26$ cm. **29.** $L = v/2f = 3.4$ m. **31.** The difference tone is $(396 - 264)$ Hz $= 132$ Hz, which is one octave below 264 Hz (that is, half of 264 Hz) and so is consonant. **33.** With $T = $ tension here, $f_{\text{beat}} = f_2 - f_1 = f_1(f_2/f_1 - 1) = f_2(\sqrt{T_2/T_1} - 1) = 9.8$ Hz. **35.** $f_{\text{beat}} = f_2 - f_1 = v/\lambda_2 - v\lambda_1 = v (1/\lambda_2) - (1/\lambda_1) = 408$ Hz. **37.** (a) $f_L = f_S v/(v - v_S) = 518$ Hz. (b) $f_L = f_S v/(v + v_S) = 483$ Hz. **39.** Here $v_S = 0$, so $f_L = f_S(v + v_L)/v = f_S + f_S v_L/v$ and $v_L = v(f_L - f_S)/f_S = 1.3$ m/s. **41.** Since $\lambda = c/f$ and the source is receding, $\lambda = \lambda_S \sqrt{(1 + v/c)(1 - v/c)} = 5.78 \times 10^{-7}$ m. **43.** $\beta = 10 \log I/I_0 = 10 \log[(5 \times 10^6)/10^{-12}] = 10 \log (5 \times 10^6) = 67$ dB. **45.** $I = PA = 10^{-5}$ W/m^2, $\beta = 10 \log (I/I_0) = 10 \log 10^7 = 70$ dB. **47.** $f_1 = v/2L = $

$\sqrt{T/(m/L)}/2L = \frac{1}{2}\sqrt{T/mL}$. Since $m = \rho LA$, $f_1 = \frac{1}{2}\sqrt{T/\rho L^2 A}$. But $T/A = Y\Delta L/L_0$ where $\Delta L/L_0 = 0.0100$, so $f_1 = \frac{1}{2}\sqrt{Y(\Delta L/L_0)/\rho L^2} = 253$ Hz. **49.** (a) The frequency of the waves that reach a moving cell is $f = f_s(v - v_b)/v$. The frequency of the waves that reach the stationary receiver is $f_r = fv/(v + v_b) = f_s(v - v_b)/(v + v_b)$. (b) Multiplying out and rearranging terms, $v_b(f_r + f_s) = v(f_s - f_r)$, $v_b = v(f_s - f_r)/(f_s + f_r)$ (c) Here $f_s - f_r = 40$ Hz and $f_s + f_r \approx 2 \times 10^6$ Hz, so $v_b = 0.031$ m/s $= 31$ mm/s. **51.** $\beta_{20} = 70$ dB $= 10 \log (20I_1/I_0) = 10 \log 20 + 10 \log (I_1/I_0) = 13$ dB $+ \beta_1$, $\beta_1 = (70 - 13)$ dB $= 57$ dB. **53.** $I = P/4\pi r^2 = 3.18 \times 10^{-4}$ W/m^2. $\beta = 10 \log(I/I_0) = 10 \log (3.18 \times 10^{-4}/10^{-12}) = 10 \log (3.18 \times 10^8) = 10(\log 3.18 + \log 10^8) = 10(0.5 + 8) = 85$ dB.

CHAPTER 13

Questions

1. The glass bulb heats up and expands before the mercury inside does. **3.** Glass expands less than copper with a rise in temperature, so the bulb would crack as it heats up. Lead-in wires must have the same coefficient of thermal expansion as glass. **5.** Because of the volume occupied by its molecules. **7.** No. The only temperature scale on which such comparisons might make any sense is the absolute scale. **9.** Gas molecules undergo frequent collisions with one another, which considerably increases the time needed for a particular molecule to get from one place to another. **11.** A relative humidity of 60% at 22°C means that the air contains 60% as much water vapor as the maximum possible at that temperature. Increasing T will lower the relative humidity and decreasing T will raise it. **13.** The moisture content of the room is the same after it has been heated, therefore the relative humidity has decreased.

Exercises and Problems

1. $T_F = \frac{9}{5}T_C + 32°$, hence 330°C $= 626°$F and 1170°C $= 2138°$F. **3.** If $T = T_F = T_C$, $\frac{9}{5}T + 32° = \frac{5}{9}(T - 32°)$, $T = -40°$. **5.** $T_C = \frac{5}{9}(T_F - 32°) = -80°$C. **7.** $\Delta L_c - \Delta L_s = (a_c - a_s)L_0\Delta T = 9 \times 10^{-5}$ m. **9.** $a = \Delta L/L_0\Delta T = 8.1 \times 10^{-6}/°$C. **11.** $\Delta L = aL_0\Delta T = 0.0563$ ft. The true width is 119.944 ft; to the nearest inch, 119 ft 11 in. **13.** Glycerin: $\Delta V = bV_0\Delta T = 1.275$ cm^3; beaker: $\Delta V = bV_0\Delta T = 0.023$ cm^3; thus $(1.275 - 0.023)$ cm$^3 = 1.25$ cm^3 overflows. **15.** $b_a = 3a$; $\Delta V = \Delta V_e - \Delta V_a = (b_e - b_a)V\Delta T$, $\Delta T = \Delta V/(b_e - b_a)V = 41°$F, $T = T_0 + \Delta T = 81°$F. **17.** $b = 3a = 9 \times 10^{-5}/°$C and ΔT

= 180°C. Since $m = \rho_1 V_1 = \rho_2 V_2$ and $V_2 = V_1(1 + b\Delta T)$, $\rho_2 = \rho_1/(1 + b\Delta T) = 10.8$ g/cm³. **19.** The force is $F = YA\Delta L/L_0$. Here $\Delta L/L_0 = a\Delta t$, so $F = YAa\Delta T$. **21.** (a) $\Delta L = aL_0\Delta T = 0.0312$ ft. (b) $A = 7.2$ in², $F = YA\Delta L/L_0 = 3.74 \times 10^4$ lb. **23.** $V_2 = p_1 V_1/p_2 = 2.5$ m³. **25.** At atmospheric pressure, using absolute values for p_1 and p_2, the volume of oxygen is $V_2 = p_1 V_1/p_2 = 218$ ft³. Hence its weight is $w = DV = 20$ lb. **27.** The volume of air at 1 bar absolute pressure needed to fill the tire to an absolute pressure of 3 bars is $v_1 = p_2 V_2/p_1 = 15$ L. Since 5 L of air are already in the tire, 10 L more are needed, hence 20 strokes. **29.** $p_1 = (24.0 + 14.7)$lb/in² $= 38.7$ lb/in², $T_1 = 500°R$, $T_2 = 540°R$; $p_2 = p_1 T_2/T_1 = 41.8$ lb/in². The corresponding gauge pressure is $(41.8 - 14.7)$lb/in² $= 27.1$ lb/in². **31.** (a) $p_2 = p_1 V_1/V_2 = 4.0 \times 10^5$ Pa. (b) $V_2 = p_1 V_1/p_2 = 2.7$ m³. (c) $V_2 = V_1 T_2/T_1 = 2.7$ m³. **33.** $T_1 = 293$ K, $T_2 = (p_2/p_1)(V_2/V_1)T_1 = 806$ K $= 533°C$. **35.** $6(12.0$ u$) + 12(1.008$ u$) + 6(16.00$ u$) = 180.16$ u; m (C₆H₁₂O₆) $= (180.16$ u$)(1.66 \times 10^{-27}$ kg/u$) = 2.99 \times 10^{-25}$ kg. **37.** $m_{Fe} = 55.85$ u $= 9.27 \times 10^{-26}$ kg; $n = 0.100$ kg/$m_{Fe} = 1.08 \times 10^{24}$ atoms. **39.** $v_{av} = (v_1 + v_2)/2 = 2.0$ m/s; $v_{rms} = \sqrt{(v_1^2 + v_2^2)/2} = 2.2$ m/s. **41.** $27°C = 300$ K; KE_{av} is proportional to the absolute temperature, so to triple KE_{av} the absolute temperature should be raised to $3(300$ K$) = 900$ K $= 627°C$. **43.** At the same temperature, $\frac{1}{2}m_H v_H^2 = \frac{1}{2}m_O v_O^2$, $v_O = v_H \sqrt{m_H/m_O} = 0.40$ km/s. **45.** The average time interval between collisions is $t = s/v = 2.5 \times 10^{-10}$ s. Hence each molecule makes an average of $1/t = 4 \times 10^9$ collisions per second. **47.** A clearance of 0.10 mm all around the piston means a difference in diameter of $\Delta L = 0.20$ mm $= 2.0 \times 10^{-4}$ m. Here $\Delta L = \Delta L_s - \Delta L_a = (a_s - a_a)L_0\Delta T$, $\Delta T = \Delta L/(a_s - a_a)L_0 = 209°C$ and $T = T_0 + \Delta T = 229°C$. **49.** $L = L_0 + \Delta L = L(1 + a\Delta T) = 1.00216$ m. If the apex of the triangle is h from the line of the cold bar, $L = 2\sqrt{(L_0/2)^2 + h^2} = \sqrt{L_0^2 + 4h^2}$, $L^2 = L_0^2 + 4h^2$, $h = \frac{1}{2}\sqrt{L^2 - L_0^2} = 0.033$ m $= 33$ mm. **51.** (a) For concrete $b_c = 3a_c$ is smaller than b_{water} and hence the water level rises as the temperature increases. (b) The water begins to overflow when its volume equals the pool's volume. If V_w is the water volume and V_p is the pool's total volume at 10°C, then $V_w + b_w V_w\Delta T = V_p + b_c V_p\Delta T$, $\Delta T = (V_p - V_w)/(b_w V_w - b_c V_p) = 13°C$. The overflow temperature is 10°C + 13°C = 23°C. **53.** Here $V_2 = 2V_1$ and $p_2 = p_a$ = atmospheric pressure. Hence $p_1 V_1 = p_2 V_2$, $p_1 = 2p_a$. The pressure due to the water of depth h is $p_1 - p_2 = p_a$, so $p_a = dgh$, $h = p_a/dg = 10$ m. **55.** $p_1 = p_{atm} + \rho gh = 1.993 \times 10^5$ Pa, $p_2 = p_{atm} = 1.013 \times 10^5$ Pa, $T_1 = 278$ K, $T_2 = 293$ K; $V_2/V_1 = p_1 T_2/p_2 T_1 = 2.07 = D_2^3/D_1^3$, hence $D_2 = \sqrt[3]{2.07}\, D_1 = 13$ mm. **57.** Let V_1 be the

initial volume of the gas (the volume of the cylinder) at p_1, T_1 and let V_2 be the volume it would have at p_2, T_2. Then the fraction of the original mass that escapes is $\Delta m/m_1 = \Delta V/V_1 = (V_2 - V_1)/V_1 = V_2/V_1 - 1$. Since $p_1 V_1/T_1 = p_2 V_2/T_2$, $\Delta m/m_1 = p_1 T_2/p_2 T_1 - 1 = 0.037$ and $\Delta m = (0.037)(10$ kg$) = 0.37$ kg.

CHAPTER 14

Questions

1. The specific heat capacity of alcohol is less than that of water, hence its ability to carry heat from the engine to the radiator is less. **3.** The ice is more effective because of its heat of fusion. **5.** See Fig. A.2. **7.** A liquid. **9.** The wood particles in sawdust are mixed with air and so are not in as close contact with one another as they are in solid wood. **11.** In this location convection is facilitated, which increases the rate of heat transfer to the water. **13.** (a) Conduction, convection, and radiation. (b) The same mechanisms, but radiation is less significant in this case. **15.** The rate of heat loss will decrease because polished copper has a lower emissivity. **17.** The vacuum prevents heat transfer by conductor or convection and the silvered surfaces are poor radiators.

Exercises and Problems

1. $Q = mc\Delta T = 1.2 \times 10^3$ Btu. **3.** $c = Q/m\Delta T = 1.7$ kJ/kg · °C. **5.** $Q = mc\Delta T = 5.1$ kJ, $(5.1$ kJ$)/(1.5$ MJ$) = 0.0034 = 0.34\%$. **7.** $Q/t = (m/t)c\Delta T = 113$ kJ/s $= 4.07 \times 10^5$ kJ/h; $Q'/t = (Q/t)/$Eff $= 6.10 \times 10^5$ kJ/h; $(6.10 \times 10^5$ kJ/h$)/(45$ MJ/kg$) = 14$ kg/h. **9.** $m = Q/c\Delta T = Pt/c\Delta T = 8.9 \times 10^3$ kg. (b) The mass of 1 L of

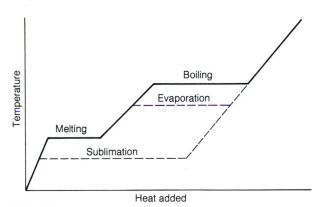

Figure A.2

Question 5, Chapter 14.

water is 1 kg, hence $(8.9 \times 10^3 \text{ L})/(200 \text{ L/drum}) = 45$ drums are needed.   **11.** $\Delta T = 80°C$; $t = Q/P = (m_1c_1 + m_2c_2 + m_3c_3)\Delta T/P = 151 \text{ s} = 2.5 \text{ min}$.   **13.** Heat gained = heat lost, $(70 \text{ kg})(c)(T - 26°C) = (10 \text{ kg})(c)(90°C - T)$, $T = 34°C$.   **15.** Heat gained = heat lost, $m_{cop}c_{cop} (T - 20°C) + m_{water}c_{water} (T - 20°C) = m_{iron}c_{iron} (120°C - T)$, $T = 21°C$.   **17.** $m = Pt/L_v = 178 \text{ g/h}$.   **19.** $Q = mc_w\Delta T + mc_i\Delta T + mL_f = 1.6 \text{ MJ}$. **21.** To heat the water to 100°C requires $Q_1 = mc\Delta T = 209.5 \text{ kJ}$. Thus $Q_2 = 1000 \text{ kJ} - Q_1 = 790.5 \text{ kJ}$ is available to convert water at 100°C to steam at 100°C and $m_{steam} = Q_2/L_v = 0.35 \text{ kg}$.   **23.** (a) From Table 9.1 the mass of 1 L of seawater is 1.03 kg; hence $Q = mc\Delta T + mL_v = 2.7 \text{ MJ}$. (b) $P = (2.7 \text{ MJ/L})(10^5 \text{ L})/(8.64 \times 10^4 \text{ s}) = 3.1$ MW.   **25.** $mL_f = \frac{1}{2}mv^2$, $v = \sqrt{2JL_f} = 0.82 \text{ km/s}$. **27.** $mc\Delta T + mL_f = \frac{1}{2}mv^2$, $v = \sqrt{2(c\Delta T + L_f)} = 0.33$ km/s, where ΔT is the difference between the melting point of lead and the initial temperature of 100°C.   **29.** For the same insulating ability the ratio k/d must be the same, so here $d_2 = d_1k_2/k_1$. Hence (a) 2.0 m, (b) 1.5 m, (c) 33 cm. **31.** $Q/t = kA\Delta T/d = 1.9 \text{ kW}$.   **33.** $A = \pi r^2 = \pi \times 10^{-2} \text{ m}^2$; $Q/t = kA\Delta T/d = 6.8 \times 10^5 \text{ W}$. The rate decreases as the milk gets warmer since this reduces ΔT. **35.** $Q/t = 300 \text{ Btu/h}$; $\Delta T = d(Q/t)/kA = 341°F$ so $T_2 = T_1 + \Delta T = 541°F$.   **37.** $t = 1 \text{ h} = 3600 \text{ s}$; $Q = kAt\Delta T/d = 5.42 \text{ MJ}$; $m = Q/(Q/m) = 0.136 \text{ kg} = 136 \text{ g}$.   **39.** $k = d/R = 6.25 \text{ Btu}/(\text{ft}^2 \cdot \text{h} \cdot °\text{F/in})$.   **41.** $R = 2d_1/k_1 + d_2/k_2 = 0.24 \text{ m}^2 \cdot °\text{C/W}$, $Q/tA = \Delta T/R = 13 \text{ W/m}^2$. **43.** The average temperature difference in the first case is 42.4°C and in the second case it is 22.5°C, so from Eq. (14.5) $t_2 = t_1\Delta T_1/\Delta T_2 = 9.4 \text{ min}$.   **45.** Here $T_1 = 773$ K. Since $R = e\sigma T^4$, $R_2/R_1 = 2 = (T_2/T_1)^4$, and $T_2 = 2^{\frac{1}{4}}T_1 = \sqrt{\sqrt{2}}\, T_1 = 919 \text{ K} = 646°C$.   **47.** $R_1/R_2 = (T_1/T_2)^4 = 0.55$.   **49.** (a) $Q/t = hA\Delta T = 84 \text{ W}$. (b) The net power radiated is $P = Ae\sigma(T_2^4 - T_1^4) = 75 \text{ W}$ since $T_1 = 298 \text{ K}$ and $T_2 = 306 \text{ K}$.   **51.** $h = s \sin \theta$; $Mgh = Mgs \sin \theta = mc\Delta T$, $\Delta T = 22°C$.   **53.** If M is the mass of the block of ice and m is the mass that melts and if $M \gg m$ (as it is here), then $mL_f = Fs = \mu Mgs$, $\mu = mL_f/Mgs = 0.85$.   **55.** Heat gained by ice = heat lost by alcohol-water mixture, $m_{ice}c_{ice} [0°C - (-10°C)] + m_{ice}L_f + m_{ice}c_{water} (5°C - 0°C) = m_{water}c_{water} (20°C - 5°C) + m_{alcohol}c_{alcohol} (20°C - 5°C)$, $m_{ice} = 0.0263 \text{ kg} = 26 \text{ g}$. **57.** The boiling point of nitrogen is $-196°C$, hence $\Delta T = 196°C$; heat lost by water = heat gained by nitrogen, $m_wL_{fw} + m_wc_{ice}\Delta T - m_nL_{fn}$, $m_n = 0.74 \text{ kg}$.   **59.** To establish the physical state of the mixture, we note that if all the steam is cooled to 100°C, the heat lost is $Q_1 = m_sc_s\Delta T_1 = 80.4 \text{ kJ}$; if all the water warms to 100°C, the heat gained is $Q_2 = m_wc_w\Delta T_2 = 293.3 \text{ kJ}$; and if all the steam condenses

to water, $Q_3 = m_sL_v = 22,600 \text{ kJ}$. We conclude that the final temperature is 100°C and that some of the steam condenses. If m is the mass of steam that condenses, $Q_1 + mL_v = Q_2$, $m = (Q_2 - Q_1)/L_v = 0.094 \text{ kg} = 94 \text{ g}$.   **61.** Let m be the mass of the substance and P the rate at which heat is given to it. Then $Q_1 = mc_{solid}\Delta T_1 = Pt_1$, $Q_2 = mL_f = Pt_2$, and $Q_3 = mc_{liquid}\Delta T_3 = Pt_3$. Routine algebra yields $L_f = c_{solid}\Delta T_1(t_2/t_1) = 30 \text{ kJ/kg}$ and $c_{liquid} = c_{solid} (\Delta T_1t_3/\Delta T_3t_1) = 3.0 \text{ kJ/kg} \cdot °\text{C}$.   **63.** We consider an area of $A = 1.00$ m^2 of the lake and ignore the increase in ice thickness in calculating the heat flow through the ice during the night, which on this basis is $Q = kAt\Delta T/d = 3456 \text{ kJ}$. The mass of newly formed ice is $m = Q/L_f = 10.3 \text{ kg}$. If h is the thickness of the newly formed ice and ρ is its density, $m = \rho Ah$, $h = m/\rho A = 0.011 \text{ m} = 1.1 \text{ cm}$ and the total thickness is 6.1 cm.   **65.** Solar energy absorption rate = $(1.2 \times 0.5 \times 0.7) \text{ kW} = 0.42 \text{ kW}$, so the heat loss rate is $(0.42 + 0.08) \text{ kW} = 0.5 \text{ kW} = 1800 \text{ kJ/h}$. Hence $(1800 \text{ kJ/h})(2430 \text{ kJ/kg}) = 0.74 \text{ kg/h}$ of sweat must be evaporated.

CHAPTER 15

Questions

1. There is no suitable low-temperature reservoir in space. **3.** The work done at constant pressure is $p(V_2 - V_1)$ and is a maximum; the final temperature would exceed the initial temperature. At constant temperature, since $pV/T =$ constant, the pressure would drop during the expansion and less work would be done. If the expansion were adiabatic, the temperature would drop, and accordingly the final pressure would be lower than in either of the other cases. Hence the least work is done during an adiabatic expansion.   **5.** Oil, coal, and natural gas. Advantages: All are easy to use and relatively cheap; in the case of coal, there are considerable resources. Disadvantages: Mining and burning coal damages the environment; burning all three adds CO_2 to the atmosphere and enhances the greenhouse effect; oil and gas resources are limited.   **7.** Wood as a fuel: very polluting and not enough for demand. The others: larger capital investment per watt needed than for fossil or nuclear power plants; for solar cells, wind, and biomass, large area of land needed; for falling water, wind, waves, and tides, only a limited number of sites are available; for solar cells, wind, waves, and tides, some means of energy storage is needed because their supply of energy is not steady.

Exercises and Problems

1. $p = 2 \times 10^5 \text{ Pa}$, $\Delta V = 10^{-3} \text{ m}^3$; $\Delta U = Q - W = Q - p\Delta V = 300 \text{ J}$. Since ΔU is positive, the internal energy increases. **3.** $p - 2160$ lb/ft^2, $\Delta V - AL - (\pi d^2/4)L$; $W = p\Delta V = \pi p d^2 L/4 = 4.4 \times 10^4$ ft · lb. **5.** (a) $a \to d$, $W = 0$; $d \to c$, $W = p\Delta V = 0.60$ MJ (by the engine); $c \to b$, $W = 0$; $b \to a$, $W = p\Delta V = -1.08$ MJ (on the engine). (b) -0.48 MJ (on the engine). **7.** Eff $= 1 - T_2/T_1 = 44\%$. **9.** Maximum efficiency is $1 - T_2/T_1 = 65\%$, hence actual efficiency is 62% of maximum. **11.** $T_2/T_1 = Q_2/Q_1$, $T_2 = T_1 Q_2/Q_1 = 375$ K. **13.** $T_1 = 600$ K, $T_2 = 400$ K, $W = $ (Eff)$(Q_1) = (1 - T_2/T_1)(Q_1) = 0.33$ MJ. **15.** $T_1 = 960°$R, Eff $= W/Q_1 = 1 - T_2/T_1$, $T_2 = T_1 (1 - W/Q_1) = 713°$R $= 253°$F. **17.** $p = P/4LAN = 651$ kPa. **19.** Eff $= P_{out}/P_{in} = 746$ W/(0.25 kg)(4.6 $\times$ 10^7 J/kg)(1/3600 s) $= 0.23 = 23\%$. **21.** (a) $Q/t = $ (18 kg/h)(44.8 MJ/kg)(10^3 kJ/MJ)/(3600 s/h) $= 224$ kW, Eff$_1 = P/(Q/t) = 0.27 = 27\%$. (b) Similarly Eff$_2 = 0.30 = 30\%$. **23.** Since 0.25 of the input heat Q is wasted, $(1 - 0.25)Q = mc\Delta T = 18$ MJ, $Q = 24$ MJ. Hence $Q/(Q/V) = 24$ MJ / 86.2 MJ/m$^3 = 0.28$ m^3 of propane is required. **25.** (a) 1 refrigeration ton $= $ (2000 lb/day)(0.454 kg/lb)(335 kJ/kg)/(8.64 $\times$ 10^4 s/day) $= 3.52$ kW. (b) The available capacity is (0.5)(0.75)(3.52 kW) $= 1.32$ kW. Each person + lamp dissipates 0.195 kW, hence 1.32 kW/(0.195 kW/person) $= 6.8$ person $\approx$ 7 persons. **27.** $T_1 = 268$ K, $T_2 = 298$ K, $Q = 1.0$ J; $W = Q[(T_2/T_1) - 1] = 0.11$ J. **29.** $W = Q_1/CP = 1.0$ J/2.4 $= 0.42$ J. **31.** $T_1 = 465°$R, $T_2 = 545°$R, CP (ideal) $= T_1/(T_2 - T_1) = 5.8$. **33.** $mgh = mc\Delta T$, $\Delta T = gh/c = 1.1°$C; $T_1 = 11.1°$C $= 284.1$ K, $T_2 = 10°$C $= 283$ K, Eff $= 1 - T_2/T_1 = 0.0039 = 0.39\%$. **35.** (a) $T_1 = 250$ K, $T_2 = 300$ K, $Q_2 = 250$ kJ. CP $= \frac{1}{2}$CP(Carnot) $= \frac{1}{2}T_1/(T_2 - T_1) = 2.5$. Since CP $= Q_1/(Q_2 - Q_1) = 1/(Q_2/Q_1 - 1)$, $Q_1 = Q_2/(1/CP + 1) = 179$ kJ. (b) $W = Q_1/CP = 72$ kJ; $P = W/t = 72$ kJ/60 s $= 1.2$ kW. **37.** $m = 1.0 \times 10^5$ lb, $Q_1 = mc_{water}(60°$F $- 32°$F$) = 2.8 \times 10^6$ Btu, $Q_2 = mL_f = 1.44 \times 10^7$ Btu, $Q_3 = mc_{ice}(32°$F $- 15°$F$) = 8.5 \times 10^5$ Btu. $Q = Q_1 + Q_2 + Q_3 = 1.805 \times 10^7$ Btu. Because 1 day $= 24$ h, $Q/t = 7.5 \times 10^5$ Btu/h.

CHAPTER 16

Questions

1. No; yes. **3.** An equal amount of charge of the opposite sign appears on the fur. **5.** It will not move; it will rotate until aligned with the field. **7.** (a) The ability of certain of the electrons to move freely through the metal, rather like molecules in a gas. (b) None of the electrons is free to move through the insulator. **9.** All metals are good electrical conductors, and most nonmetallic solids are good insulators. Liquids may be either. Gases are good insu-

lators if not ionized. Among solids, good electrical conductors are good heat conductors and vice-versa. **11.** Compare the forces that each field exerts on a charged and on an uncharged object. **13.** The direction of the force is given by the direction of the field lines at that point. The magnitude of the force is proportional to the density of field lines there.

Exercises and Problems

1. $n = m/m_p$, $Q = ne = me/m_p = 96$ C. **3.** $F_1/F_2 = (r_2/r_1)^2$, $r_2 = r_1\sqrt{F_1/F_2} = 20$ mm. **5.** $F = ke^2/r^2 = 4.2 \times 10^{-8}$ N. **7.** $r = \sqrt{kQ_AQ_B/F} = 10$ mm. **9.** $F = mg = ke^2/r^2$, $r = e\sqrt{k/mg} = 5.08$ m. **11.** Each outer charge exerts on the middle charge a force of magnitude $F = kQ_AQ_B/r^2 = 0.135$ N. Because both forces are directed toward the negative charge, the total force is 0.27 N. **13.** $E = kQ/r^2 = 3.9 \times 10^6$ N/C. **15.** $E = 0$ at the center of the square, since a charge placed there will experience no force. **17.** $eE = mg$, $E = mg/e = 1.02 \times 10^{-7}$ N/C. **19.** $E = 0$ at a point between the charges and the distance s from the charge $Q_1 = 1.5 \times 10^{-6}$ C. At this point $kQ_1/s^2 = kQ_2/(0.20$ m $- s)^2$, $(0.20$ m $- s)/s = \sqrt{Q_2/Q_1}$, $s = 0.20$ m/$(\sqrt{Q_2/Q_1} + 1) = 0.083$ m $= 83$ mm. **21.** (a) If r_1 is the distance from the test charge Q to $Q_1 = -1 \times 10^{-6}$ C, then $r_2 = (0.40$ m $- r_1)$ is the distance to $Q_2 = -3 \times 10^{-6}$ C. The forces on Q will be equal when $kQQ_1/r_1^2 = kQQ_2/r_2^2$, $r_2/r_1 = \sqrt{Q_2/Q_1} = (0.40$ m $- r_1)/r_1$, $r_1 = 0.40$ m/$\sqrt{Q_2/Q_1} + 1) = 0.15$ m $= 15$ cm. (b) Since r_1 is independent of the magnitude and sign of Q, $r_1 = 15$ cm here also. **23.** $mv^2/r = ke^2/r^2$, $v = \sqrt{ke^2/mr} = 2.2 \times 10^6$ m/s.

CHAPTER 17

Questions

1. The charges have different signs. **3.** The electron speed increases by a factor of $\sqrt{2}$. **5.** No.

Exercises and Problems

1. KE $= 7.7$ MeV $= 1.23 \times 10^{-12}$ J $= k(2e)(79e)/r$, $r = 158ke^2/$KE $= 3.0 \times 10^{-14}$ m. **3.** $V = W/Q = 6.0$ V. **5.** $d = V/E = 0.0050$ m $= 5.0$ mm. **7.** $W = QV = 3.2 \times 10^8$ J. **9.** $E = F/e = V/d$, $V = Fd/e = 2.5 \times 10^3$ V $= 2.5$ kV. **11.** (a) $W = QV = 2.8$ MJ. (b) $h = W/mg = 14$ km. (c) $v = \sqrt{2W/m} = 0.53$ km/s. **13.** KE $= 8 \times 10^{-18}$ J, $\frac{1}{2}mv^2 = $ KE, $v = \sqrt{2$KE$/m} = 4.2 \times 10^6$ m/s. **15.** KE $= \frac{1}{2}mv^2 = 3.25 \times 10^{-14}$ J/1.6 $\times 10^{-19}$ J/eV $= 2.0 \times 10^5$ eV. **17.** $E = V/d$, $w = E^2/$

$8\pi k = V^2/8\pi kd^2 = 4.0 \times 10^{-3}$ J/m³.   **19.** (a) $Q = CV$ = 0.025 C. (b) $W = \frac{1}{2}CV^2 = 12.5$ J.   **21.** $W = \frac{1}{2}CV^2$, $V = \sqrt{2W/C} = 0.45$ kV.   **23.** (a) Here $K = 1$, $A = 5 \times 10^{-3}$ m², $d = 10^{-3}$ m, so $C = A/4\pi kd = 4.4 \times 10^{-11}$ F = 44 pF. (b) $Q = CV = 2.0 \times 10^{-9}$ C. (c) $W = \frac{1}{2}CV^2$ = 4.5×10^{-8} J.   **25.** (a) $V = \sqrt{2W/C} = 707$ V. (b) $Q = CV = 0.14$ C.   **27.** $C = KC_0$, $K = C/C_0 = 6.3$. **29.** (a) From the solution to Exercise 23, $C = 44$ pF; hence $C' = KC = 172$ pF. (b) $Q = C'V = 7.8 \times 10^{-9}$ C. (c) 1.7×10^{-7} J.   **31.** $C = C_1 + C_2 = 70\ \mu$F.   **33.** $1/C = 1/C_1 + 1/C_2 + 1/C_3$, $C = 3.125\ \mu$F, $Q = Q_1 = Q_2$ = $Q_3 = CV = 3.75 \times 10^{-5}$ C; $V_1 = Q_1/C_1 = 7.5$ V, V_2 = $Q_2/C_2 = 3.75$ V, $V_3 = Q_3/C_3 = 0.75$ V.   **35.** $V_1 = V_2 = V_3 = 12$ V; $Q_1 = C_1 = C_1V_1 = 6 \times 10^{-5}$ C, $Q_2 = C_2V_2 = 1.2 \times 10^{-4}$ C, $Q_3 = C_3V_3 = 6 \times 10^{-4}$ C. **37.** (a) $1/C' = 1/2\ \mu$F + 1/4 μF, $C' = 1.33\ \mu$F; $C = C'$ + 6 μF = 7.33 μF. (b) $C = 4\ \mu$F + 6 μF = 10 μF. (c) $C = 0$.   **39.** $E = V/d = 500$ V/m. The acceleration of the electron is $a = F/m = eE/m = 8.78 \times 10^{13}$ m/s². Since $d = 0.01$ m $= \frac{1}{2}at^2$, the electron strikes the plate $t = \sqrt{2d/a} = 1.51 \times 10^{-8}$ s later. In this time it travels $x = vt$ = 0.15 m = 15 cm.   **41.** Because for a parallel-plate capacitor $C = KA/4\pi kd$, $C' = C/2$; $W_1 = \frac{1}{2}Q^2/C$, $W_2 = \frac{1}{2}Q^2/C' = Q^2/C$, hence the work needed is $\Delta W = W_2 - W_1$ = $\frac{1}{2}Q^2/C$.   **43.** At first, $Q_{01} = C_1V_0 = 1.2 \times 10^{-3}$ C, $Q_{02} = C_2V_0 = 2.4 \times 10^{-3}$ C. Because terminals of opposite sign are joined, the total charge that remains on the combination is $Q = Q_{02} - Q_{01} = 1.2 \times 10^{-3}$ C. The charge distributes itself on the capacitors so there is the same new potential difference across each. Hence $V_1 = V_2$, Q_1/C_1 = Q_2/C_2, and $Q_2 = Q - Q_1$; $Q_1 = Q/(1 + C_2/C_1) = Q/3$ = 4×10^{-4} C, $Q_2 = Q - Q_1 = 8 \times 10^{-4}$ C.

CHAPTER 18

Questions

1. With a single wire, charge would be permanently transferred from one end to the other, and soon so large a charge separation would occur that the electric field that produced the current would be canceled out. With two wires, charge can be sent on a round trip, so to speak, and a net flow of energy from one place to another can occur without a net transfer of charge.   **3.** The increase in resistivity cannot be due to the expansion of the metal because its cross-sectional area expands faster than its length and so the resistivity should decrease on this basis. Also, the coefficient of linear expansion is too much smaller than the temperature coefficient of resistivity for the expansion of the metal to have any major effect on resistivity.   **5.** If the heater is to be connected to a power source of fixed voltage, which is normally the case, then $P = V^2/R$ is the appropriate formula and a small R will give the greater rate of heat output. **7.** Some of the power produced by the cell of high emf will be dissipated in the cell of low emf.

Exercises and Problems

1. $I = Q/t = 60e/s = 9.6 \times 10^{-18}$ A.   **3.** $R = V/I = 15\ \Omega$.   **5.** $V = IR = 75$ V.   **7.** Since R is proportional to $1/A$, the resistance per meter of the 5-mm² wire is $\frac{2}{5}$ = 40% that of the 2-mm² wire.   **9.** $R = \rho L/A = 2.0\ \Omega$. **11.** $R = \rho L/\pi r^2$, $r = \sqrt{\rho L/\pi R}$, $d = 2r = 0.29$ mm. **13.** (a) The volume of a cylinder is $V = \pi d^2L/4$; hence $d_1^2 L_1 = d_2^2 L_2$, $L_2 = d_1^2L_1/d_2^2 = 100$ m. (b) Since the wire is 100 times longer than the rod and its cross-sectional area is 0.01 times as great, the wire's resistance is 10^4 times the rod's resistance because $R = \rho L/A$.   **15.** $d = 40.3$ mils, $A_{cmil} = (d_{mil})^2 = 1624$ cmil.   **17.** $A_{cmil} = (d_{mil})^2 = 400$ cmil, $L = RA/\rho = 192$ ft.   **19.** $L = 140$ m, $R = V/I = 0.15\ \Omega$; $A = \rho L/R = 16$ mm².   **21.** R and L are the same for both wires, so $\rho_A/A_A = \rho_C/A_C$. If V is the volume of one of the wires, its cross-sectional area is $A = V/L = m/Ld$, where d is the density and m the mass. Hence $\rho_A Ld_A/m_A = \rho_C Ld_C/m_C$, $m_A/m_C = \rho_A d_A/\rho_C d_C = 0.46$.   **23.** (a) $\Delta T = -20°C$, so $\Delta R = \alpha R\Delta T = -8\ \Omega$ and $R = 100\ \Omega$ $- 8\ \Omega = 92\ \Omega$. (b) $\Delta T = 60°C$, so $\Delta R = 24\ \Omega$ and $R = 124\ \Omega$.   **25.** $\alpha = \Delta R/R\Delta T = 0.0051/°C$.   **27.** $I = P/V = 1.67$ A.   **29.** Eff = (0.5 hp)(746 W/hp)/IV = 0.78 = 78%.   **31.** $P_{out} = $ (Eff)(P_{in}) = (Eff)(IV) = 1.9 kW. **33.** $W = Pt = IVt = 2.7$ kWh = 9.7 MJ.   **35.** (a) I = $P/V = 2$ A, $Q = IT = 7.2 \times 10^3$ C. (b) The battery can supply (2 Ah)(1.5 V) = 3 Wh = 0.003 kWh of electric energy. The cost is therefore $0.50/0.003 kWh = $167/ kWh, which is several thousand times more expensive than the electric energy supplied to homes and industry.   **37.** I = $V_1/R_1 = 0.02$ A, $V_2 = IR_2 = 5.0$ V.   **39.** (a) Because $R = 2R_0$ and $V = 2V_0$, $P = 2P_0$ for the pair of bulbs and each bulb operates at its original power rating of P_0. Here the idea is good. (b) Because $P = V^2/R$, $R_{10} = 1440$ Ω and $R_{100} = 144\ \Omega$. $R = R_{10} + R_{100} = 1584\ \Omega$ so at 240 V, $I = 0.1515$ A and $P_{10} = 33$ W and $P_{100} = 3.3$ W. The small bulb will be brighter than before but will soon burn out since it must dissipate 3.3 times more power; the large bulb will be very much dimmer than before. Here the idea is not so good.   **41.** (a) $R = R_1 + R_2 = 15$ A, $I = V/R$ = 0.80 A in each bulb. (b) Across the 5-Ω bulb, $V = IR_1$ = 4.0 V. Across the 10-Ω bulb, $V_2 = IR_2 = 8.0$ V. (c) P_1 = $IV_1 = 3.2$ W; $P_2 = IV_2 = 6.4$ W; $P = P_1 + P_2 = 9.6$

W,   **43.** The current in the 5.0-Ω resistor is $I = \sqrt{P/R}$ = 5.0 A. Since this is the current in both resistors, $V = IR_1 + IR_2 = 75$ V.   **45.** (a) All in parallel: $1/R = 1/R_1 + 1/R_2 + 1/R_3$, $R = 33\ \Omega$. (b) Two in series, the combination in parallel with the third: $R' = R_1 + R_2 = 200\ \Omega$, $R = \dfrac{R_3 R'}{R_3 + R'} = 67\ \Omega$. (c) Two in parallel, the combination in series with the third: $R' = \dfrac{R_1 R_2}{R_1 + R_2} = 50\ \Omega$, $R = R' + R_3 = 150\ \Omega$. (d) All in series: $R = R_1 + R_2 + R_3 = 300\ \Omega$. **47.** (a) $1/R = 1/R_1 + 1/R_2 + 1/R_3$, $R = 10.5\ \Omega$. (b) $I_1 = V/R_1 = 3.0$ A; $I_2 = V/R_2 = 1.5$ A; $I_3 = V/R_3 = 1.2$ A. **49.** (a) 5-Ω bulb: $I_1 = V/R_1 = 2.4$ A; 10-Ω bulb: $I_2 = V/R_2 = 1.2$ A. (b) 12 V, 12 V. (c) $P_1 = V^2/R_1 = 28.8$ W; $P_2 = V^2/R_2 = 14.4$ W; $P = P_1 + P_2 = 43.2$ W.   **51.** (a) $R' = (5\ \Omega \times 8\ \Omega)/(5\ \Omega + 8\Omega) = 3.08\ \Omega$; $R'' = 3.08\ \Omega + 5\ \Omega = 8.08\ \Omega$; $R = (8.08\ \Omega \times 3\ \Omega)/(8.08\ \Omega + 3\ \Omega) = 2.19\ \Omega$. (b) The current through the upper branch is $I'' = V''/R'' = 1.49$ A, hence the potential difference across the 8-Ω resistor is $V = I''r' = 4.59$ V and the current through it is $I = V/R = 0.57$ A.   **53.** $R_1 = (10\ \Omega \times 15\Omega)/(10\ \Omega + 15\ \Omega) = 6\ \Omega$; $R_2 = 4\ \Omega + 6\ \Omega + 4\ \Omega = 14\ \Omega$; $R_3 = (14\ \Omega \times 14\ \Omega)/(14\ \Omega + 14\ \Omega) = 7\ \Omega$; $R_4 = 5\ \Omega + 7\ \Omega + 3\ \Omega = 15\ \Omega$. Total current is $I_4 = 60$ V/15 $\Omega = 4$ A. $V_3 = IR_3 = 28$ V, hence $I_2 = V_3/R_2 = 2$ A. The voltage across R_1 is $V_1 = I_2 R_1 = 12$ V and so the current in the 10-Ω resistor is 12 V/10 $\Omega = 1.2$ A.   **55.** (a) $R' = r + R = 9.5\ \Omega$, $P = V^2/R' = 15$ W. (b) $I = V/R = 1.26$ A, hence the power dissipated within the battery is $P' = I^2 r = 2.4$ W; $P'/P = 0.16 = 16\%$.   **57.** (a) From $I = \mathcal{E}/(R + r)$, $r = \mathcal{E}/I - R = 0.9\ \Omega$. (b) $V = \mathcal{E} - Ir = 22$ V. **59.** $I_1 = \mathcal{E}/(R + r) = 5.0$ A, $I_2 = 3.3$ A. Since $P = I^2 R$, $P_2/P_1 = I_2^2/I_1^2 = 0.44$.   **61.** The equivalent resistance of the internal resistances of the batteries is $r/4$. Hence $I = \mathcal{E}/(R + r/4) = 2.9$ A. **63.** $W = I^2 Rt = 4608$ J; $Q = W = mc\Delta T$, $\Delta T = W/mc = 0.28°C$.   **65.** (a) $C = 2W/V^2 = 988\ \mu F$. (b) $t = RC = 15$ ms.   **67.** (a) $t = RC = 0.025$ s. (b) $I = V/R = 0.2$ A. (c) $I = Q/t$, $t = Q/I = CV/I = 0.025$ s. **69.** The heat needed is $Q = mc\Delta T + mL_f = m(c\Delta T + L_f)$ where $m = 2dAL$ is the mass of the wires, whose density is d, cross-sectional area is A, and total length is $2L$. Hence $Q = 2dAL(c\Delta T + L_f)$. The rate at which heat is supplied to the wires is $P = V^2/R = V^2/(2\rho L/A) = V^2 A/2\rho L$. The time needed to melt the wires is therefore $t = Q/P = 4\rho dL^2(c\Delta T + L_f)/V^2 = 92$ ms, independent of A. **71.** Because the currents in the upper and lower branches are the same and both flow clockwise, the current in the middle branch is 0. Hence there is no potential difference across the upper and lower branches, and $V_1 = I_1 R_1 = 10$

V, $V_2 = I_2 R_2 = 20$ V.   **73.** (a) $V = \mathcal{E}_1 + Ir = 37.5$ V. (b) $I = (V - \mathcal{E}_2)/r = 32$ A. (c) The charge passed through the battery while it is being charged is $Q = (0.25)(250$ Ah)(3600 s/h) $= 2.25 \times 10^5$ C. The energy supplied is therefore $W = Q\mathcal{E}_{av} = 7.63 \times 10^6$ J. The average current is $I_{av} = (V - \mathcal{E}_{av})/r = 36$ A, so the time required is $t = Q/I = 6.25 \times 10^3$ s and the energy dissipated as heat is $W' = I^2 rt = 8.10 \times 10^5$ J, which is 11% of the total; hence 89% of the energy supplied is stored as chemical energy. **75.** Let I_1 be the current to the right through the upper battery, I_2 be the current to the left through the lower battery, and I_3 be the current to the right through the resistor R. At the left-hand junction $I_2 = I_1 + I_3$. Proceeding clockwise in the upper loop, $\mathcal{E}_2 - \mathcal{E}_1 = I_1 r_1 + I_2 r_2$. Proceeding counterclockwise in the outermost loop, $\mathcal{E}_1 = I_3 R - I_1 r_1$. Since $I_2 = I_1 + I_3$, the first loop equation becomes $\mathcal{E}_2 - \mathcal{E}_1 = I_1 r_1 + I_1 r_2 + I_3 r_2$. Solving this equation and the second loop equation for I_1 and setting the two expressions for I_1 equal gives $(I_3 R - \mathcal{E}_1)/r_1 = (\mathcal{E}_2 - \mathcal{E}_1 - I_3 r_2)/(r_1 + r_2)$, from which $I_3 = (\mathcal{E}_2 r_1 + \mathcal{E}_1 r_2)/(Rr_1 + Rr_2 + r_1 r_2) = 0.53$ A. **77.** I_2 is downward and given by $I_2 = I_1 - I_3 = 3$A. Hence $R_2 = V_3/I_2 = 6.7\ \Omega$. To find R_1, apply Kirchhoff's second rule to the left-hand loop, proceeding clockwise. This gives $V_1 = I_1 R_1 + I_2 R_2$, hence $R_1 = (V - I_2 R_2)/I_1 = 6.0\ \Omega$.   **79.** Assume currents downward in each resistor. At the top junction, $I_1 + I_2 + I_3 = 0$. Proceeding counterclockwise in the left-hand loop, 10 V + 6 V $= I_1 \times 10\ \Omega - I_2 \times 5\ \Omega$. Proceeding clockwise in the right-hand loop, 10 V + 10 V $= I_3 \times 20\ \Omega - I_2 \times 5\ \Omega$. Solving yields $I_1 = 0.86$ A, $I_2 = -1.49$ A, $I_3 = 0.63$ A.   **81.** (a) To find the potential difference between a and b we can disregard the 8-V battery and the 6-Ω resistor. Applying Kirchhoff's second rule to the outside loop, going clockwise and assuming a clockwise current of I, yields 10 V $-$ 5 V $= I(1\ \Omega + 1\ \Omega + 3\ \Omega + 9\ \Omega)$, $I = 0.357$ A. In the upper part of the loop a potential drop occurs in both the battery and the 3-Ω resistor, hence $V = 5$ V $+ I$ (1 Ω + 3 Ω) $= 6.43$ V. The lower part of the loop gives the same result. Here there is a potential increase of 10 V in the battery and a potential drop in the battery's internal resistance and in the 9-Ω resistor, so $V = 10$ V $- I$ (1 Ω + 9 Ω) $= 6.4$ V. (b) Point b is at a potential of -6.4 V relative to a. No current passes through the 6-Ω resistor or the 8-V battery, so there is no IR drop in either. The 8-V battery's positive terminal is nearest to c, hence the potential of c relative to a is 8 V $- 6.4$ V $= 1.6$ V.   **83.** (a) $1/C = 1/C_1 + 1/C_2$, $C = 1.5\ \mu F$, $t = RC = 1.5 \times 10^{-4}$ s. (b) When fully charged, $Q = CV = 3.6 \times 10^{-5}$ C, $V_1 = Q/C_1 = 18.0$ V and $V_2 = Q/C_2 = 6.0$ V. Here $V_1' = 0.63V_1 = 11.3$ V and $V_2' = 3.8$ V.

CHAPTER 19

Questions

1. All three fields are detected in both cases.   3. C.
5. No. Because of hysteresis in the iron core, B is not proportional to I, and so the variations of B with time do not exactly follow the variations of I with time.   7. South.
9. $0°$; $90°$.   11. The electric force between the streams is attractive and the magnetic force is repulsive. Which force predominates depends on the speeds of the particles because the magnetic force increases with speed, whereas the electric force does not change.   13. The electron beam is deflected away from the wire.   15. A circle, because the forces that act are repulsive and a circle is the configuration in which each part of the loop is as far as possible from the other parts.   17. There is never a net force on the loop in such a field. There is no torque on the loop when its plane is perpendicular to the field.   19. See Fig. A.3.   21. A magnetic compass is more accurate near the equator, where the horizontal component of the earth's magnetic field is strongest. In the polar regions the field has only a very small horizontal component and so can exert relatively little torque on a compass needle.

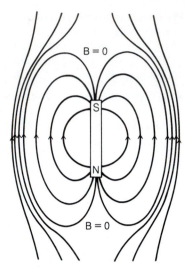

Figure A.3

Question 19, Chapter 19.

Exercises and Problems

1. $s = \mu_0 I/2\pi B = 0.080$ m $= 8.0$ cm.   3. $B = 0$ midway between the wires because the fields of the currents are opposite in direction and of the same magnitude there.
5. $B_1 = \mu_0 I/2\pi s = 2.0 \times 10^{-5}$ T, $B_2 = 0.67 \times 10^{-5}$ T. Because the fields are in opposite directions, $B = B_1 - B_2 = 1.3 \times 10^{-5}$ T.   7. $I = 2Br/\mu_0 = 8.0$ A.   9. $N/L = 3000$ turns/m, $B = \mu_0 I(N/L) = 0.014$ T.   11. The number of turns stays the same as the spring's length doubles, so B drops to $B/2$.   13. The fields of the two layers are in opposite directions, hence the result is the same as if the solenoid had 200 turns. Thus $B = \mu_0 NI/L = 1.3 \times 10^{-4}$ T.   15. $B_{\text{loop}} = B_{\text{solenoid}}$, $\mu_0 I_1/2r = \mu_0 I_2 (N/L)$, $I_1 = 2rI_2 (N/L) = 20$ A since $(N/L) = 2 \times 10^3$ turns/m.
17. $R = mv/eB = 2.8$ mm.   19. $\theta = 90°$, $\sin \theta = 1$; $B = \mu_0 I/2\pi s$; $F = QvB \sin \theta = \mu_0 IQv \sin \theta/2\pi s = 2.0 \times 10^{-9}$ N; the direction of $\mathbf{F}$ is toward the wire.   21. (a) $B = \mu_0 I/2\pi s = 1.6 \times 10^{-4}$ T, $F = evB = 1.5 \times 10^{-15}$ N. The force is directed away from the wire. (b) The force has the same magnitude but is directed toward the wire.
23. (a) The force has the same magnitude as in Exercise 21 (a), but is in the direction of the current. (b) The force has the same magnitude but is opposite to the direction of the current.   25. $E = V/s = vB$, $V = vBs = 2.5$ kV.
27. (a) $\theta = 90°$, $\sin \theta = 1$, $F = I\Delta LB = 1.0$ N. (b) $\theta = 0$, $\sin \theta = 0$, $F = 0$.   29. (a) $\theta = 45°$, $F = I\Delta L/B \sin \theta = 0.28$ N; the force is downward according to the right-hand rule. (b) $F = 0.28$ N; the force is upward.   31. $P = IV$, $I = P/V = 1.67$ A $= I_1 = I_2$; $F/l = \mu_0 I_1 I_2/2\pi s = 2.8 \times 10^{-4}$ N/m; the force is repulsive, since the currents are in opposite directions.   33. Consider a triangle in a plane perpendicular to the wires whose vertexes are the wires and the point P. The base of the triangle is s and its other legs have the same length L. If θ is the angle between the base and either leg, $\cos \theta = s/2L$ and $L^2 = x^2 + (s/2)^2$. The components of the field at P parallel to s cancel out to leave a net field perpendicular to s whose magnitude is $B = B_1 \cos \theta + B_2 \cos \theta = 2B(s/2L) = 2(\mu_0 I/2\pi L)(s/2L) = \mu_0 Is/2\pi(x^2 + (s/2)^2) = 2\mu_0 Is/\pi(4x^2 + s^2)$.   35. (a) The time per electron orbit is $t = 2\pi r/v$ so the electron's motion is equivalent to a current of $I = Q/t = ev/2\pi r$. Hence $B = \mu I/2r = \mu ev/4\pi r^2 = 13$ T. (b) $QvB = evB = mv^2/R$, $B = mv/eR = 2.4 \times 10^5$ T.   37. There the length of each turn is $2\pi r$ and $\sin \theta = 1$, so $F = 2\pi NrIB = 0.040$ N.   39. (a) Looking along the axis in Fig. 19.43, the force $\mathbf{F}_1$ on the left-hand arm is upward and the force $\mathbf{F}_2$ on the right-hand arm is downward. Each has the magnitude $F = I\Delta LB = IyB$ and has a moment arm of $x/2$, so the net torque on the loop is $\tau = 2F(x/2) = Fx = IxyB = IAB$. (b) Now the moment arm is $(x/2) \cos \theta$ so $\tau = IAB \cos \theta$.

CHAPTER 20

Questions

1. The left-hand wheels have positive charges and the right-hand wheels have negative charges.   **3.** Counter-clockwise; no current; clockwise.   **5.** As the diaphragm vibrates in response to sound waves, the coil moves back and forth in the field of the magnet. This induces a current in the coil whose frequency and amplitude correspond to those of the sound wave.   **7.** When the generator is connected to an outside circuit, work must be done to turn its shaft as the current produced delivers energy to the circuit.   **9.** The induced emf is zero when the plane of the coil is perpendicular to the magnetic field, which happens twice per rotation. Hence $T = 0.005$ s.   **11.** The emfs are the same.   **12.** (a) The emf decreases because the greater I is, the greater the IR drop in the armature and the less the potential difference across the generator. Hence the current in the magnet windings drops and, because the emf is proportional to B, it too decreases. (b) A similar argument explains the increase in the emf when the current drops.   **15.** The changing magnetic field produced by an alternating current in the primary winding.   **17.** The air gap in the recording head permits the magnetic field of its coil to reach the tape. The air gap in the playback head permits the magnetic field of the tape to enter the iron core of its coil.   **19.** (a) From Lenz's law, the self-induced emf will be opposite to the direction of the current. (b) For the same reason, the emf here will be in the same direction as that of the current.

Exercises and Problems

1. $\mathscr{E}_i = BLv = 0.60$ mV.   **3.** $\mathscr{E}_i = BLv = 1.8$ mV.   **5.** (a) $A = 5 \times 10^{-3}$ m^2, $\Phi = BA = 5 \times 10^{-6}$ Wb. (b) $\mathscr{E} = \Delta\Phi/\Delta t = 1.7$ μV.   **7.** $A = \pi r^2$, $\Delta B/\Delta t = \mathscr{E}/A = 5.1 \times 10^2$ T/s.   **9.** $A = \pi r^2 = 7.07 \times 10^{-4}$ m^2, $\mathscr{E} = -NA\Delta B/\Delta t = -NA\Delta B/\Delta t = 17$ mV.   **11.** $\mathscr{E} = IR = A(\Delta B/\Delta t)$, $\Delta B/\Delta t = IR/A = 1.0 \times 10^3$ T/s.   **13.** $V = \mathscr{E} - IR$, $I = (\mathscr{E} - V)/R = 25$ A, $P = IV = 2.9$ kW.   **15.** $V_2 = V_1 N_2/N_1 = 600$ V; $I_2 = I_1 N_1/N_2 = 0.60$ A.   **17.** $N_1/N_2 = V_1/V_2 = 21$; $I_2 = P/V_2 = 42$ A.   **19.** $L = \mathscr{E}\Delta t/\Delta I = 0.050$ H $= 50$ mH.   **21.** $L = \mu N^2 A/l = 3.9$ mH.   **23.** The length of the equivalent solenoid is $l = \pi D$ and $\mu = 400$ μ_0, so $L = (400$ $\mu_0)N^2 A/\pi D = 0.32$ H.   **25.** $t = L/R$, $L = Rt = 1.4$ H.   **27.** (a) $W = wV = B^2 V/2\mu = 4.0$ MJ. (b) $E = \sqrt{8\pi k w} = \sqrt{8\pi k B^2/2\mu} = 3.0 \times 10^{10}$ V/m.   **29.** (a) $L = \mu N^2 A/l = 4.09$ mH. (b) $I = V/(R + r) = 0.286$ A. (c) $W = \frac{1}{2}LI^2 = 1.67 \times 10^{-4}$ J.

31. (a) $W = \frac{1}{2}LI^2 = 0.25$ J. (b) $\Delta t = L\Delta I/\mathscr{E} = 0.010$ s.   **33.** $f = 1/2\pi\sqrt{LC} = 0.80$ MHz.   **35.** $C = 1/4\pi^2 Lf^2 = 0.32$ μF.   **37.** (a) $f = 1/2\pi\sqrt{LC} = 1.6$ kHz. (b) $Q = CV = 1.0 \times 10^{-5}$ C. (c) $\frac{1}{2}CV^2 = \frac{1}{2}LI^2$, $I = V\sqrt{C/L} = 0.10$ A.   **39.** (a) For an air-core solenoid, $B = \mu_0 N_1 I/l$ and $A = \pi r^2$, so $\Delta\Phi/\Delta t = A\Delta B/\Delta t = (\pi\mu_0 r^2 N_1/l)\Delta I/\Delta t = 3.95 \times 10^{-6}$ Wb/s. Hence $\mathscr{E}_0 = -N_2\Delta\Phi/\Delta t = -0.12$ mV. (b) Here $\mu = 400$ μ_0 so $\mathscr{E} = 400$ $\mathscr{E}_0 = 0.47$ V.   **41.** (a) Total current is $I = P/V = 3.5$ A, field current is $I_f = V/R_f = 0.6$ A, armature current is $I_a = 3.5$ A $- 0.6$ A $= 2.9$ A. Back emf is $\mathscr{E}_b = V - I_a R_a = 117$ V, power output is $\mathscr{E}_b I_a = 340$ W, efficiency is 340 W/420 W $= 81\%$. (b) $R = R_a R_f/(R_a + R_f) = 1.0$ Ω, I (start) $= V/R = 120$ A. (c) Total resistance needed is 120 V/25 A $= 4.8$ Ω, so series resistor should be 4.8 Ω $- 1.0$ Ω $= 3.8$ Ω.   **43.** The entire current flows through both windings, whose total resistance is $R = 0.50$ Ω. The back emf is $\mathscr{E}_b = V - IR = 14$ V and the power output is $P = I\mathscr{E}_b = 280$ W. The power input is $P' = IV = 480$ W, and the efficiency is $P/P' = 58\%$.   **45.** $\mathscr{E} = \mathscr{E}_1 + \mathscr{E}_2 + \mathscr{E}_3 = -L_1\Delta I/\Delta t - L_2\Delta I/\Delta t - L_3\Delta I/\Delta t = -L\Delta I/\Delta t$; hence $L = L_1 + L_2 + L_3$.   **47.** (a) $I = 0$, $\Delta I/\Delta t = \mathscr{E}/L = 4.2 \times 10^3$ A/s. (b) $\Delta I/\Delta t = (\mathscr{E} - IR)/L$, $I = (\mathscr{E} - L\Delta I/\Delta t)R = 3.3$ A. (c) $I = \mathscr{E}/R = 6.3$ A, $\Delta I/\Delta t = 0$.   **49.** (a) $t = T/8 = 1/8f = \pi\sqrt{LC}/4 = 2.5 \times 10^{-4}$ s. (b) $W = \frac{1}{2}CV^2 = 5.76 \times 10^{-4}$ J; $Q^2/2C = W/2$, $Q = \sqrt{WC} = 34$ μC; $LI^2/2 = W/2$, $I = \sqrt{W/L} = 0.11$ A.

CHAPTER 21

Questions

1. It increases.   **3.** $A/2$; $2A$.   **5.** The resistance of an ac circuit determines how much of the electrical energy that passes through it is dissipated as heat. The capacitive reactance and inductive reactance, respectively, determine the extent to which the capacitors and inductors in the circuit oppose the flow of current without dissipating energy. X_L and X_C vary with frequency whereas R does not, and their combined effects depend on $X_L - X_C$, so their analogy with resistance is very limited. The impedance determines the current in the circuit when a potential difference V of a certain frequency is applied according to $I = V/Z$; thus Z has the same role as R does in a dc circuit with respect to I, though the rate of energy dissipation is equal to $I^2 R$ in both cases.   **7.** $R = R_0$, $X_L = X_{L0}/2$, $X_C = 2X_C$. In order to determine what happens to Z, the values of X_{L0} and X_{C0} must be known.   **9.** The power factor of a circuit is the ratio between its resistance and impedance and governs the rate at which power is absorbed by the circuit for a given I and V; it varies with frequency. The power factor can be 0 only when $R = 0$; it is 100% at resonance, when $Z = R$. **11.** High-pass filter.

Exercises and Problems

1. $T = 0.02$ s, $f = 1/T = 50$ Hz. **3.** (a) $I_{max} = \sqrt{2} I_{eff} = 14$ A. (b) $V_{max} = \sqrt{2} V_{eff} = 85$ V. (c) No; they are coincident only when the frequency is the resonant frequency. **5.** (a) $2\pi ft = 2\pi/8 = 45°$, $V = V_{max} \sin 2\pi ft = V_{max} \sin 45° = 71$ V. (b) $2\pi ft = 2\pi/4 = 90°$, $V = V_{max} \sin 90° = 100$ V. (c) $2\pi ft = 6\pi/8 = 135°$, $V = V_{max} \sin 135° = V_{max} \sin 45° = 71$ V. (d) $2\pi ft = 2\pi/2 = 180°$, $V = V_{max} \sin 180° = 0$. **7.** (a) $X_L = 2\pi fL = 0.31$ Ω. (b) 0.31 kΩ. **9.** $X_L = 2\pi fL$, $I = V/X_L = V/2\pi fL = 0.16$ A. **11.** At $f = 200$ Hz, $L = X_L/2\pi f = 0.0955$ H. At $f = 60$ Hz, $X_L = 2\pi fL = 36$ Ω, $I = V/X_L = 6.7$ A. **13.** $C = 1/2\pi fX_C = 1.6 \times 10^{-5}$ F $= 16$ μF. **15.** At $f = 200$ Hz, $C = \frac{1}{2\pi fX_C} = 6.63$ μF. At $f = 60$ Hz, $X_C = \frac{1}{2\pi fC} = 400$ Ω, $I = V/X_C = 0.60$ A. This result is different from that of Exercise 11 because X_L and X_C vary differently with frequency. **17.** (a) $X_C = 1/2\pi fC = 637$ Ω; $Z = \sqrt{R^2 + X_C^2} = 704$ Ω; $I = V/Z = 0.17$ A. (b) $P = I^2R = 8.7$ W. **19.** $R = V/I = 120$ Ω. To halve the current, we require $Z = 240$ Ω. Since $Z = \sqrt{R^2 + X_L^2}$, $X_L = 2\pi fL = \sqrt{Z^2 - R^2}$ and $L = \sqrt{Z^2 - R^2}/2\pi f = 0.66$ H. **21.** (a) $Z = V/I = 240$ Ω; $X_C = \sqrt{Z^2 - R^2} = 226$ Ω; $C = 1/2\pi fX_C = 12$ μF. (b) No energy is dissipated in a capacitor. (c) $P = I^2R = 20$ W. **23.** (a) $R = V_1/I_1 = 5$ Ω, $Z = V_2/I_2 = 8$ Ω; $Z = \sqrt{R^2 + X_L^2}$, $L = \sqrt{Z^2 - R^2}/2\pi f = 17$ mH. (b) $P_1 = I_1^2 R = 320$ W; $P_2 = I_2^2 R = 125$ W. **25.** Since $V_C = V_L$, the circuit is in resonance, and $V_R = 60$ V. **27.** (a) Since $X_L = X_C$, $X_L - X_C = 0$ and $Z = R = 20$ Ω. (b) $\tan \phi = 0$, $\phi = 0$. (c) $V = IZ = 400$ V. **29.** (a) $X_L = 2\pi fL = 18.8$ Ω, $X_C = 1/2\pi fC = 531$ Ω; $Z = \sqrt{R^2 + (X_L - X_C)^2} = 516$ Ω, $I = V/Z = 0.23$ A. (b) $P = I^2R = 3.3$ W. (c) $V_C = IX_C = 124$ V; $V_L = IX_L = 4.4$ V; $V_R = IR = 14$V. **31.** $L = X_L/2\pi f$, $C = 1/2\pi fX_C$, $f_0 = 1/2\pi \sqrt{LC} = 1/\sqrt{X_L/f^2 X_C} = 0.63$ kHz. **33.** $V = \sqrt{V_R^2 + (V_L - V_C)^2} = 120$ V; $\phi = \tan^{-1}(V_L - V_C)/V_R = +37°$. At resonance, $V_L = V_C$. Raising the frequency f above the resonance frequency increases X_L and hence V_L, and also decreases X_C and hence V_C. Since what we have here is indeed $V_L > V_C$, the resonance frequency must be less than f. **35.** (a) $Z = V/I = 24$ Ω, $R = P/I^2 = 18$ Ω, so $X_L = \sqrt{Z^2 - R^2} = 15.9$ Ω. For a power factor of 100%, $X_C = X_L$, $C = 1/2\pi fX_C = 1/2\pi fX_L = 1.7 \times 10^{-4}$ F. (b) $P = V^2/R = 0.80$ kW. **37.** (a) $X_L = 2\pi fL = 113$ Ω, $X_C = 1/2\pi fC = 44$ Ω, $Z = \sqrt{R^2 + (X_L - X_C)^2} = 85$

Ω. (b) $I = V/Z = 1.4$ A. (c) $\tan \phi = (X_L - X_C)/R = 1.38$, $\phi = 54°$, power factor $= \cos \phi = 0.59 = 59\%$. (d) $P = I^2R = 99$ W. (e) $IV = P/\cos \phi = 0.17$ kVA. (f) $I = V/R = 2.4$ A. **39.** $I_R = V/R = 15$ A, $I_L = V/X_L = 3.0$ A, $I_C = V/X_C = 2.0$ A; $I = \sqrt{I_R^2 + (I_L - I_C)^2} = 1.8$ A. **41.** (a) $X_L = 2\pi fL = 12.6$ Ω; $I_L = V/X_L = 0.80$ A; $I_R = V/R = 1.0$ A. (b) $I = \sqrt{I_R^2 + I_L^2} = 1.3$ A. (c) $Z = V/I = 7.7$ Ω. (d) $\phi = \tan^{-1}(-I_L/I_R) = -39°$. The current lags behind the voltage. (e) $P = I_R^2 R = 10$ W (alternatively, $P = IV \cos \phi = 10$ W). **43.** (a) $X_C = 1/2\pi fC = 19.9$ Ω, $I_C = V/X_C = 1.21$ A; $X_L = 2\pi fL = 50.3$ Ω, $I_L = V/X_L = 0.48$ A; $I_R = V/R = 0.96$ A. (b) $I = \sqrt{I_R^2 + (I_C - I_L)^2} = 1.2$ A. (c) $Z = V/I = 20$ Ω. (d) $\phi = \tan^{-1}(I_C - I_L)/I_R = 30°$; the current leads the voltage. (e) $P = I_R^2 R = 23$ W. **45.** (a) Let Z_1 be the impedance at $f_1 = 100$ Hz and Z_2 be the impedance at $f_2 = 500$ Hz. Then $Z_1^2 = R^2 + X_{L1}^2 = R^2 + 4\pi^2 f_1^2 L^2$, $Z_2^2 = R^2 + 4\pi^2 f_2^2 L^2$, and $L = \sqrt{(Z_2^2 - Z_1^2)/4\pi^2(f_2^2 - f_1^2)} = 28$ mH. (b) $R = \sqrt{Z_1^2 - 4\pi^2 f_1^2 L^2} = 47$ Ω. **47.** (a) $\cos \phi = P/IV = 0.625 = 63\%$. (b) Since $\cos \phi = 0.625$, $\phi = 51°$, $\tan \phi = 1.235 = X_L/R$; $R = P/I^2 = 75$ Ω, $X_L = 1.235R = 92.6$ Ω; if $X_C = X_L$, $C = 1/2\pi fX_L = 29$ μF. (c) $I = V/R = 1.6$ A. (d) $P = I^2R = 0.19$ kW. (e) 0.19 kVA since V and I are in phase at the resonance frequency. **49.** (a) If $X_L = R$ at $f = 20$ kHz, then for the same applied voltage, $I_L > I_R$ for $f < 20$ kHz and $I_L < I_R$ for $f > 20$ kHz. Hence $X_L = R = 2\pi fL$, $L = R/2\pi f = 0.4$ mH. (b) Assume an applied voltage of 100 V for convenience. (Any other voltage could equally well be used, or we could just call the applied voltage V.) Then $X_L = 2\pi fL = 5$ Ω, $I_L = V/X_L = 20$ A, $I_R = V/R = 2$ A, $I = \sqrt{I_R^2 + I_L^2} = 20.1$ A and the proportion of I that passes through the load is $I_R/I = 0.0995 = 9.95\%$. Less than 10% of the total current passes through the load. (c) When $f = 200$ kHz we proceed in exactly the same way, with the result that $I_R/I = 0.995 = 99.5\%$. Now nearly all the current passes through the load.

CHAPTER 22

Questions

1. The electric fields produced by electromagnetic induction are readily detected through the currents they produce in metal wires, but no such simple means is available for detecting the weak magnetic fields produced by changing electric fields. **3.** Light waves consist of coupled electric and magnetic field fluctuations and need no material medium in which to occur. Sound waves consist of pressure fluctuations in a material medium and hence cannot occur in a vacuum. **5.** Perpendicular to each other. **7.** A vertical antenna will give better reception because the direction of the electric field of the radiation is vertical; therefore the induced emf in a vertical antenna will be a maximum. **9.** No elaborate mathematics is needed to use Huygens' principle. **11.** Red; blue. **13.** All red; as a black cross on a blue background. **15.** A real image is formed by light rays that pass through it; a screen placed at the location of the image would show the image. A virtual image is formed by the backward extension of light rays that were diverted by refraction or reflection. The light rays that seem to come from it do not actually pass through a virtual image, and so it cannot appear on a screen although it can be seen by the eye. **17.** The flashes of color are the result of dispersion. In red light the flashes would be red only. **19.** The water is deeper.

Exercises and Problems

1. $\lambda = c/f = 68.3$ m. **3.** (a) $f = 1/T = 1.0 \times 10^9$ Hz. (b) $\lambda = c/f = 30$ cm. (c) Microwaves. **5.** The image is always twice as far from the man as the mirror; hence it must be moving twice as fast as the mirror, namely at 60 km/h. **7.** $\sin r = (n_1/n_2)\sin i = 0.789$, $r = 52°$. **9.** $n_2 = n_1 \sin i/\sin r = 1.51$. **11.** $n_1 = 1.00$. Red light: $\sin r = (n_1/n_2)\sin i = 0.4257$, $r = 25.19°$. Violet light: $\sin r = (n_1/n_2)\sin i = 0.4201$, $r = 24.84°$; $\Delta r = 0.35°$. **13.** (a) By the same procedure as in Example 22.4 the final angle of refraction is 70°. (b) 70°. **15.** By the same procedure as in Example 22.5, the final angle of refraction is 68°. **17.** $h = h'n_1/n_2 = 54$ mm. **19.** $h' = hn_2/n_1 = 59$ cm. **21.** (a) $\sin i_c = n_2/n_1 = 0.413$, $i_c = 24°$. (b) $\sin i_c = 0.550$, $i_c = 33°$. **23.** (a) Because $v_{water} > v_{air}$ for sound, sound waves are refracted away from the normal when they reach an air-water interface from the air side and so can be totally reflected. On the other hand, sound waves are refracted toward the normal when incident from the water side, and they cannot be totally reflected in this case. (b) $i_c = \sin^{-1}(v_{air}/v_{water}) = 13°$. **25.** $\Omega = A/r^2 = 3.6$ sr. **27.** $F = EA = I\Omega = IA/r^2$, $I = r^2E = 3.6 \times 10^5$ cd. **29.** $\Omega = A/r^2 = 0.05$ sr, $F = I/\Omega = 125$ lm, Eff $= 125$ lm/20 W $= 6.3$ lm/W. **31.** $F = 4\pi I = 2513$ lm, $A = \pi d^4/4 = 0.785$ m^2, $E = F/A = 3.2 \times 10^3$ lx. **33.** $A = 1800$ m^2 so $F_{total} = EA = 5.4 \times 10^5$ lm. The useful flux emitted by each lamp is $F = \frac{2}{3}(4\pi I) = 2.09 \times 10^4$ lm/lamp so $N = F_{total}/F = 26$ lamps. **35.** Here (angle of reflection) $+ 90° +$ (angle of refraction) $= 180°$. Because the angle of reflection equals the angle of incidence, the angle of refraction must be $r = 35°$ and $n_2 = n_1 (\sin i/\sin r) = 1.43$. **37.** The angle of incidence must be 45°, so the critical angle of the glass must be 45° at most.

Hence the minimum value of n is $1/\sin i_c = 1/\sin 45° = 1.41$. **39.** Below the bulb, $E = I/r^2$, $I = Er^2 = 768$ cd. At 1.00 m to one side, $\theta = \tan^{-1}(1.00 \text{ m}/0.80 \text{ m}) = 51.3°$ and $r = \sqrt{(0.80 \text{ m})^2 + (1.00 \text{ m})^2} = 1.28$ m, $E = I \cos \theta/r^2 = 293$ lx.

CHAPTER 23

Questions

1. No; no. **3.** Crown glass has a smaller index of refraction than benzene. The immersed lens will act as a diverging lens because wavefronts passing through it are retarded less going through its center where the path length in the medium of higher light speed is greater. Thus the wavefronts are affected as in the right-hand diagram of Fig. 23.3 even though the lens is convex. **5.** Under no circumstances; when the object is between the lens and the focal point. **7.** No; yes, if the object is placed between the lens and the focal point. **9.** The lens should be moved toward the slide. **11.** (a) Inverted and larger than the object. (b) Inverted and smaller than the object. (c) Erect and smaller than the object. (d) Erect and larger than the object. **13.** (a) No. (b) A concave mirror forms a virtual image of any object that is between the mirror and its focal point.

Exercises and Problems

1. $R_1 = R_2 = -40$ cm; $1/f = (n - 1)(1/R_1 + 1/R_2)$, $f = -364$ mm. **3.** $R_1 = -30$ cm, $R_2 = +20$ cm, $1/f = (n - 1)(1/R_1 + 1/R_2)$, $f = 120$ cm. **5.** $1/f = (n - 1)(1/R)$, $f = R/(n - 1) = 382.4$ mm for blue light and 386.8 mm for red light. Hence $\Delta f = 4.4$ mm. **7.** $n - 1 = 1/f(1/n_1 + 1/n_2) = 0.56$, $n = 1.56$. **9.** (a) Converging. (b) The index of refraction of water is 1.33, so the relative index of refraction is $n' = 1.60/1.33 = 1.203$. From the lensmaker's equation, $f = f' (n' - 1)/(n - 1) = 443$ mm. **11.** (a) Converging. (b) $p = \infty$, $q = f = 80$ mm. **13.** $q = pf/(p - f) = -52$ cm. The image is virtual and 52 cm from the lens on the same side as the duck. **15.** $f = pq/(p + q) = 12$ cm; the lens is converging since its focal length is positive. **17.** No image is formed when the object is at the focal point. **19.** $q = pf/(p - f) = 30$ cm, $m = -q/p = -1$. The image is real, inverted, and the same size as the object. **21.** $m = 1/10 = -q/p$, $q = -p/10 = -0.5$ m; $f = pq/(p + q) = -556$ mm. **23.** $p = 7000$ m, $f = 1$ m, $q = pf/(p - f) \approx 1$ m; $m = -q/p = -1.43 \times 10^{-4}$; $h' = mh = -1.29 \times 10^{-3}$ m $= -1.3$ mm. **25.** $q = pf/(p - f) = -40$ cm, $m = -q/p = +5$, $h = h'/m = 2.0$ mm. **27.** $d = f/8 =$

37.5 mm. **29.** Since the shutter is open for only $\frac{1}{4} = (\frac{1}{2})$ $(\frac{1}{2})$ as long, the aperture should be opened two f-numbers to $f/4$. **31.** For $p_1 = \infty$, $f_1 = q = 20$ mm. For $p_2 = 20$ cm $= 200$ mm, $f_2 = p_2 q/(p_2 + q) = 18$ mm. **33.** (a) Converging lens. (b) $p = 25$ cm, $q = -100$ cm, $f = pq/(p + q) = +33$ cm. **35.** $q = -0.60$ m and $f = (1\text{ m})/(-1.5) = -\frac{2}{3}$ m; hence $p = qf/(q - f) = 6.0$ m. **37.** (a) $f_{eye} = f_{obj}/m_{ang} = 30$ mm. (b) $\beta = L/25$ cm, $\alpha = \beta/m_{ang} = L_0/p$, $L_0 = Lp/(25\text{ cm})(m_{ang}) = 2.0$ m. **39.** $q = pf/(p - f) = -40$ cm, $m = -q/p = 2$; the image is therefore erect, virtual, and larger than the object. **41.** $q = pf/(p - f) = 200$ cm, $m = -q/p = -4$; the image is therefore inverted, real, and larger than the object. **43.** $q = pf/(p - f) = 67$ cm, $m = -q/p = -0.67$; the image is therefore inverted, real, and smaller than the object. **45.** $q = pf/(p - f) = -6.43$ cm, $m = -q/p = 3.6$. **47.** $m = +2$, $q = (-m)p = -30.0$ cm; $f = pq/(p + q) = +30.0$ cm. Since f is positive, the mirror is concave with $R = 2f = 60.0$ cm. **49.** $f = -R/2 = -15$ cm, $q = -10$ cm, $p = qf/(q - f) = 30$ cm. **51.** $m = \frac{1}{5}$, $p = 150$ cm, $q = -mp = -30$ cm, $f = pq/(p + q) = -37.5$ cm; $R = -2f = 75$ cm. **53.** The original image distance is $q = pf/(p - f) = 0.1523$ m and so the new image distance is $q' = q + 5$ mm $= 0.1573$ m. The new object distance is $p' = q'f/(q' - f) = 3.2$ m. **55.** (a) Since $p = q$ for the convex lens, its focal length is $f = q/2 = p/2 = +15$ cm. (b) The object distance for the concave lens is $p' = 15$ cm $+ 10$ cm $= 25$ cm and its image is at the original position of the candle, so $q' = -15$ cm. Hence $f' = p'q'/(p' - q') = -37.5$ cm. **57.** (a) $p + q = 60$ cm, $p = 60$ cm $- q$; $m = -4 = -q/p$, $p = q/4 = 60$ cm $- q$, $q = 48$ cm, $p = 12$ cm; $f = pq/(p + q) = +9.6$ cm. (b) $m = -2 = -q/p$, $p = q/2$; $1/f = 1/p + 1/q = 3/q$, $q = 3f = +28.8$ cm, $p = q/2 = 14.4$ cm, $p + q = 43.2$ cm. **59.** (a) The object distance that corresponds to $q = -25$ cm is $p = -q/m = 25$ cm/m. Hence $1/f = 1/p + 1/q = m/25$ cm $- 1/25$ cm $= (m - 1)/25$ cm and $m = (25\text{ cm}/f) + 1$. (b) $m = (25\text{ cm}/5\text{ cm}) + 1 = 6.0$. **61.** Here $f = 100$ cm/$2.5 = 40$ cm. The lens forms an image q away from the eye when $p = 25$ cm. Hence $q = pf/(p - f) = -67$ cm, where the minus sign means the image is in front of the eye. The near point without the lenses is therefore 67 cm. **63.** $m_{ang} = f_{obj}/f_{eye} = 30$. (b) Objective: $p = \infty$, $f = 750$ mm, $1/q = 1/f - 1/p = 1/f - 0$, $q = f = 750$ mm. Eyepiece: $q' = -250$ mm, $f' = 25$ mm, $p' = q'f/(q' - f) = 23$ mm; $d = q + p' = 773$ mm. **65.** $f = R/2 = 50$ cm; $m = \pm h'/h = \pm 4 = -q/p$, $q = \pm 4p = pf/(p - f)$. Using the $+$ sign, $p = 1.25f = 62.5$ cm; this image is real and inverted. Using the $-$ sign, $p = 0.75f = 37.5$ cm; this image is virtual and erect.

CHAPTER 24

Questions

1. Light from incoherent sources can interfere, but the resulting interference pattern shifts continually because there is no definite phase relationship between the beams from the sources. When the sources are coherent, there is such a relationship, and the pattern is stable and therefore discernible. **3.** The bright lines formed by a diffraction grating are narrower and brighter than those formed by a double slit. **5.** The first-order spectrum contains bright lines for which $\sin \theta = \lambda/d$, where d is the spacing of the slits, and the second-order spectrum contains bright lines for which $\sin \theta = 2\lambda/d$. The second-order spectrum is deviated by more than the first-order spectrum, and is wider as well (twice as wide, in fact). **7.** The wavelengths in visible light are too long relative to the atom spacing in a crystal to permit a diffraction pattern to give any information about the crystal structure. X-ray wavelengths are comparable with such atom spacings. **9.** The wavelengths in visible light are very small relative to the dimensions of a building, whereas those in radio waves are more nearly comparable. **11.** Higher resolving power; greater light-gathering ability, hence ability to form images of faint objects. **13.** The individual waves in an ordinary light beam are polarized, but their planes of polarization are random so the beam itself is unpolarized. **15.** All; all except polarization.

Exercises and Problems

1. (a) $d = 2.5 \times 10^{-6}$ m, $\sin \theta = \lambda/d = 0.30$, $\theta = 17°$. (b) $\sin \theta = 3\lambda/d = 0.90$, $\theta = 64°$. **3.** Here $d = 2.5 \times 10^{-6}$ m. Since $\sin \theta = n\lambda/d$ and the largest value $\sin \theta$ can have is 1, the maximum number of images possible is $n = d/\lambda = 4$. **5.** Here $1/d = 3.00 \times 10^5$ m^{-1}, so $\theta_1 = \sin^{-1} \lambda_1/d = 6.89°$, $\theta_2 = \sin^{-1} \lambda_2/d = 12.12°$; $h_1 = L \tan \theta_1 = 0.242$ m, $h_2 = L \tan \theta_2 = 0.430$ m; $h_2 - h_1 = 0.19$ m $= 19$ cm. **7.** $\theta = \sin^{-1} \lambda/2a = 2.9°$. **9.** $d_0 = 1.22 \lambda L/D = 46$ m. In practice, irregularities in the earth's atmosphere prevent this resolution from being attained. **11.** (a) $\theta_0 = 1.22 \lambda/D = 3.37 \times 10^{-3}$ rad $= 0.193°$; the angular diameter is $2\theta_0 = 0.39°$. (b) The apparent angular diameter of the moon is 3476 km/3.8×10^5 km $= 0.0091$ rad $= 0.52°$. **13.** (a) $L = d_0 D/1.22\lambda = 16$ km. (b) 8 km. **15.** $\lambda = c/f = 20$ m and $d = 15$ m, so $\lambda > d$ and the only intensity maxima are in the north and south directions. **17.** For destructive interference, the path length $2d$ in a film d thick must be an odd number of half wavelengths, the smallest of which is $2d = \lambda/2$; the thinnest coating gives maximum cancellation over the greatest angle of incidence, so it is normally used. The wavelength λ in the film is λ_0/n, where λ_0 is the wavelength in air, so $d = \lambda/4$ $= \lambda_0/4n = 99.6$ nm. **19.** For a distant object, the object distance L equals the focal length f of the lens. The ratio f/D for the required resolution, with $d_0 = 1.0 \times 10^{-5}$ m, is therefore $d_0/1.22 \lambda = 15$. This is an aperture of $f/15$.

CHAPTER 25

Questions

1. Such wave phenomena as diffraction and interference are easier to demonstrate than quantum phenomena. **3.** The frequency increases and the wavelength decreases; the speed is unchanged, the energy increases, and the number per second is unchanged. **5.** No. **7.** The electron has the higher speed. **9.** Kinetic and rest energies are always positive quantities. **11.** (a) Continuous emission spectrum. (b) Emission line spectrum. (c) Absorption line spectrum. **13.** A hydrogen sample contains a great many atoms, each of which can undergo a variety of possible transitions. **15.** When their spins are opposite. **17.** The length must be an integral number of half-wavelengths in order that a standing wave be set up in the cavity that causes further emissions to be exactly in phase with the previous ones, thus producing a coherent beam.

Exercises and Problems

1. $E = hf = hc/\lambda = 1.77$ eV. **3.** $E = hf = hc/\lambda = 9.9 \times 10^{-27}$ J/photon, $P/E = 1.5 \times 10^{28}$ photons/s. **5.** $\lambda = c/f = ch/E = 220$ nm; ultraviolet. **7.** $E = hf = hc/\lambda = 6.2$ eV, so electrons will leave the ball until its potential equals $(6.2 - 4.7) = 1.5$ V. **9.** $f = eV/h = 2.4 \times 10^{18}$ Hz; X rays. **11.** $\lambda = h/mv = 3.3 \times 10^{-29}$ m. **13.** $\lambda = h/mv = h/\sqrt{2mKE} = 2.9 \times 10^{-14}$ m. **15.** $\lambda = h/mv = h/\sqrt{2mKE} = h/\sqrt{2mVe}$, $V = h^2/2me\lambda^2 = 5.0$ μV. **17.** (a) $(4m_H - m_{He})c^2 = 4.14 \times 10^{-12}$ J. (b) 2.42×10^{18} nuclei. **19.** (a) $m_0 = 1000$ g, $v/c = 0.01$, $m = m_0/\sqrt{1 - v^2/c^2} = 1.005m_0 = 1005$ g. (b) $v/c = 0.9$, $m = 2.294m_0 = 2294$ g. (c) $v/c = 0.99$, $m = 7.089m_0 = 7089$ g. **21.** $m = 1.01m_0$, $\sqrt{1 - v^2/c^2} = 1/1.01$, $v^2/c^2 = 0.0197$, $v = 4.2 \times 10^7$ m/s. **23.** $E = -E_1/n^2 = 0.85$ eV. **25.** $\frac{3}{2}kT = E_1$, $T = 2E_1/3k = 1.05 \times 10^5$ K. **27.** (a) $\lambda = h/mv = 3.7 \times 10^{-63}$ m. (b) $n\lambda = 2\pi r$, $n = 2\pi r/\lambda = 2.6 \times 10^{74}$. **29.** $1/\lambda = R(1 - 1/10^2)$, $\lambda = 9.21 \times 10^{-8}$ m = 92.1 mm; ultraviolet. **31.** The shorter the wavelength, the higher the frequency, which in turn means the greater the energy change. From Fig. 25.22 the maximum energy change in the Bracket series corresponds to a transition from $n = \infty$ to $n = 4$, so that $hf = E_i - E_f = -E_1/4^2 = 1.36 \times 10^{-19}$ J, $f = 2.06 \times 10^{14}$

Hz, and $\lambda = c/f = 1.46 \times 10^{-6}$ m = 1.46 μm. **33.** The H atoms must be raised to the $n = 5$ state from the ground state, which requires $\Delta E = E_1(1/n_i^2 - 1/n_f^2) = E_1(1 - 1/5^2) = 13.06$ eV. **35.** (a) $P/hf = 4.2 \times 10^{21}$ photons/m$^2 \cdot$ s. (b) The photon speed is c, hence there are $(4.2 \times 10^{21}$ photons/m$^2 \cdot$ s$)/c = 1.4 \times 10^{13}$ photons/m^3.

CHAPTER 26

Questions

1. The existence of such definite orbits is not consistent with the uncertainty principle. **3.** The additional attractive force of the two protons exceeds the mutual repulsion of the electrons to increase the binding energy per electron. **5.** Their ability to bond with each other as well as with other atoms. **7.** Li atoms have a single electron outside a closed shell, and this electron is held relatively weakly since the inner electrons shield most of the full nuclear charge. In fluorine, the outer shells lack an electron of completion, and the electrons in these shells are held very tightly to the nucleus since little of the nuclear charge is shielded from them. **9.** Na$^+$ ions have closed shells, whereas each Na atom has a single outer electron that can be detached relatively easily. **11.** Van der Waals forces are too weak to hold inert gas atoms together against the forces exerted during collisions in the gaseous state. **13.** A maximum of two electrons can exist in the first shell of an atom, and so only two H atoms can join at a time. A maximum of eight electrons can exist in the second shell of an atom, so there is no limit to the number of Li atoms that can join to form an array of atoms. As a result lithium is a metallic solid under ordinary conditions, whereas hydrogen is a diatomic gas. **15.** Only the outermost electrons in the atoms of a metal are members of its "gas" of free electrons. **17.** A metal. **19.** The forbidden bands in semiconductors are comparable in width to the photon energies in visible light, hence semiconductors absorb visible light and are opaque to it. The lower frequencies in infrared light mean that photon energies are too low to be absorbed, hence semiconductors are transparent to this light. **21.** A solid with a partly filled upper energy band is a conductor because electrons in this band can readily absorb small amounts of energy from an electric field and move through the solid. A solid with a filled upper band and a wide forbidden band above it is an insulator because electrons in the filled band require very large amounts of energy in order to jump the forbidden band to the empty band above it. A solid with a filled upper band and a narrow forbidden band is a semiconductor because a few electrons have enough thermal energy to enter the empty band above the forbidden band, and these electrons can move through the solid under the influence of an electric field.   **23.** In an *n*-type semiconductor, current is carried by electrons. In a *p*-type semiconductor, current is carried by holes, which are sites of missing electrons.   **25.** (a) *p-n*, forward bias; *n-p*, reverse bias. (b) Forward bias.

CHAPTER 27

Questions

1. The isotopes have the same atomic number and hence the same electron structures, therefore they have the same chemical behavior. They have different numbers of neutrons, hence different atomic masses.   **3.** Helium and radon cannot be combined chemically to form radium, nor can radium be broken down into helium and radon by chemical means.   **5.** Such particles lose energy mainly by electric interactions with the electrons of atoms in their paths. Usually the electrons are pulled away from their parent atoms to produce ion-electron pairs.   **7.** Photoelectric effect, Compton scattering, and pair production.   **9.** The radionuclide must be found in the rock, and the radionuclide's half-life must be within a few orders of magnitude of the age of the rock.   **11.** If there is a leak in a water-cooled reactor, the coolant escapes and the fuel rods may then melt with potentially disastrous results. In a gas-cooled reactor, the heat capacity of the graphite moderator will reduce the temperature rise if the coolant gas stops circulating.   **13.** The

density in the interior of a star is extremely high, hence there are many collisions between nuclei there; the temperature is also extremely high, hence many of the colliding nuclei have high enough energies to interact.

Exercises and Problems

1. $5n$, $5p$; $12n$, $10p$; $20n$, $16p$; $50n$, $38p$; $108n$, $72p$.
3. ^2_1H; ^1_0n; ^1_1H.   **5.** $^{10}_5\text{B} + ^1_0n \rightarrow ^4_2\text{He} + ^7_3\text{Li}$; hence $Z = 3$, $A = 7$, lithium.   **7.** $(0.7553)(34.97 \text{ u}) + (0.2447)(36.97 \text{ u}) = 35.46 \text{ u}$.   **9.** $\Delta m = 26m_H + 30m_n - 55.93493 \text{ u} = 0.528461 \text{ u}$, which corresponds to an energy of $E = (\Delta m)(931.5 \text{ MeV/u}) = 492.3 \text{ MeV}$. The binding energy per nucleon is $E/56 = 8.79 \text{ MeV/nucleon}$.
11. $E = 22.4 \text{ MeV} = 0.02406 \text{ u}$; $m_{\text{Li}} = 2m_{\text{He}} + E - m_H = 6.0152 \text{ u}$.   **13.** $^{233}_{92}\text{U}$.   **15.** Beta decay, $^{80}_{36}\text{Kr}$; positron emission, $^{80}_{34}\text{Se}$; electron capture, $^{80}_{34}\text{Se}$.   **17.** $^{208}_{82}\text{Pb}$.
19. (a) $^{222}_{86}\text{Rn}$; $m_E = (4.87 \text{ MeV})/(931 \text{ MeV/u}) = 0.0052 \text{ u}$; hence $m(\text{Rn}) = m(\text{Ra}) - [m_E + m(\text{He})] = 222.0176 \text{ u}$. (b) The "missing" energy goes into KE of the recoiling nucleus.   **21.** One-sixteenth is left, and since $\frac{1}{2} \times \frac{1}{2} \times \frac{1}{2} \times \frac{1}{2} = \frac{1}{16}$, this means four half lives or 6400 years.
23. $3y = 1095 \text{ d} \approx 8$ half-lives, so 0.039 of the activity remains and the power output is 0.0039 W.   **25.** (a) $m = 10^{-3} \text{ kg}$, $E = mc^2 = 9 \times 10^{13} \text{ J}$. (b) 2×10^4 tons of TNT.   **27.** $\Delta m = 3m_a - m_c = 0.007809 \text{ u} = 7.274 \text{ MeV}$.

PHOTO CREDITS

Chapter 1

p. 21, AP/Wide World Photos

Chapter 2

p. 52, AP/Wide World Photos

Chapter 3

p. 61, AP/Wide World Photos; p. 66, AP/Wide World Photos

Chapter 4

p. 95, AP/Wide World Photos; p. 109, AP/Wide World Photos

Chapter 5

p. 123, Tim Davis, Photo Researchers, Inc.; p. 130, NASA

Chapter 6

p. 155, Abram G. Schoenfeld, Photo Researchers, Inc.; p. 160, Six Flags Over Georgia; p. 162, NASA; p. 167, NASA

Chapter 7

p. 177, Black & Decker U.S. Inc., Hunt Valley, MD 21030; p. 192, AP/Wide World Photos

Chapter 8

p. 229, Pamela Clark Photography

Chapter 9

p. 248, Courtesy of the Geological Survey of Canada (GSC-120457); p. 261, Manchester (NH) Historical Associations

Chapter 10

p. 274, John Deere 590D hydraulic excavator. Photo courtesy Deere & Company; p. 277, © Ron Church, Photo Researchers, Inc.; p. 282, Sears/Zemansky/Young, *College Physics,* 7th Ed., © 1991, Addison-Wesley Publishing Co., Inc., Reading, MA. Fig 13-27(c). Reprinted with permission of publisher; p. 291, The Bettmann Archive

Chapter 12

p. 342, U.S. Coast Guard Photograph; p. 347 (left), AP/Wide World Photos; p. 347 (right), AP/Wide World Photos; p. 351, 1970 © Estate of Harold E. Edgerton. Courtesy of Palm Press; p. 353, © Sherry Suris, Photo Researchers, Inc.

Chapter 13

p. 374, Courtesy of the Triborough Bridge and Tunnel Authority; p. 381, © Bruce Roberts, Photo Researchers, Inc.

Chapter 14

p. 422, © Dr. R. Clark & M. Goff, Photo Researchers, Inc.

Chapter 15

p. 447, Courtesy of Electric Power Research Institute; p. 451, Courtesy of United Technologies, Pratt of Whitney

Chapter 16

p. 474, Fundamental Photographs, New York

Chapter 17

p. 498, AP/Wide World Photos

Chapter 19

p. 576, Bill Bachman, Photo Researchers, Inc.; p. 592, Transrapid International/ © 1988 Discover Publications

Chapter 20

p. 619, AIP/Niels Bohr Library

Chapter 22

p. 676 (left), Raytheon Company; p. 678 (right), Raytheon Company

Chapter 23

p. 731, Eugene Hecht, © 1987, Addison-Wesley Publishing Co., Inc., Reading, Massachusetts. Fig. 5.110. Reprinted with permission of publisher

Chapter 24

p. 747, by permission of DC Heath & Co.; p. 754, Fundamental Photographs, New York; p. 761, Fundamental Photographs, New York

Chapter 25

p. 775, © Guy Gillette, Photo Researchers, Inc.; p. 778 (left), Rapho, Photo Researchers, Inc.; p. 778 (right), Biophoto Associates, Photo Researchers, Inc.; p. 789, Jack Finch, Photo Researchers, Inc.; p. 799, Courtesy of the Lawrence Livermore National Laboratory

Chapter 26

p. 814 (left), Smithsonian Institution Photo No. 312-A; p. 814 (right), Jerome Wyckoff; p. 816, © Carl Zeiss, Photo Researchers, Inc.; p. 825, Courtesy of AT&T Archives

Chapter 27

p. 838, © CNRI/Science Photo Library, Photo Researchers, Inc.; p. 856, Princeton University; p. 842, Lawrence Berkeley Laboratory, University of California

INDEX

Aberrations
  of lens, 729
  of mirror, 735
Absolute pressure, 269
Absolute temperature, 380
Absolute zero, 380
Absorption, induced, 796
Absorption spectrum, 789, 794
ac circuit
  current in, 649
  impedance of, 649
  parallel, 656
  phase relations in, 640, 649
  power in, 653
  resonance in, 651
  voltage in, 647
Acceleration, 10
  angular, 183
  centripetal, 149
  constant, 11
  displacement and, 14, 16
  force and, 64
  of gravity, 22, 163
  of harmonic oscillator, 318
  tangential, 184
  velocity and, 12, 17
Acceptor level, 821
Accommodation of eye, 722
Achromatic lens, 729
Action force, 76
Activity of radionuclide, 844
Adhesion, 296
Adiabatic process, 443
Aerodynamic drag, 290
Aerodynamic lift, 290
Air pollution, 453
Air resistance, 52
Alpha decay, 838
Alpha particle, 835

Alternating current, 592, 610
Alternator, 610
Amorphous solid, 810
Ampere, 523, 589
Amplitude
  of oscillator, 311
  of wave, 339
Amplitude modulation, 672
Aneroid, 269
Angle
  critical, 690
  of incidence, 679, 682
  phase, 647
  of reflection, 679
  of refraction, 682
  of shear, 257
  solid, 694
Angular acceleration, 183
  and torque, 187
Angular displacement, 185, 194
Angular frequency, 320
Angular magnification, 727
Angular measure, 174
Angular momentum, 191
  conservation of, 192
Angular motion
  comparison with linear, 185, 194
  kinetic energy of, 178
  work and power in, 189
Angular speed, 176
Antinode, 352
Apollo 11, 129
Apparent depth, 688
Apparent power, 654
Apparent weight, 167
Arch, 261
Archimedes' principle, 279
Aspherical lens, 706

Astigmatism, 724
Atherosclerosis, 294
Atmospheric pressure, 268, 271
Atom, 384, 472
  electron clouds in, 792
  energy levels of, 786, 791
  excitation of, 794
  hydrogen, 782
  quantum numbers of, 792
  quantum theory of, 782, 791
  standing waves in, 784
  structure of, 793
Atomic bomb, 850
Atomic mass unit, 385, 831
Atomic nucleus; see Nucleus
Atomic number, 830
Atomic spectra, 788
Aurora, 789
Axis of rotation, 177

Back emf, 613
Band, energy, 817
  forbidden, 818
Banked turn, 155
Bar, 268
Batteries, series and parallel connections of, 547
Beats, 355
Becquerel, 844
Belt drive system, 226
Bernoulli's equation, 285
Beta decay, 836
Beta particle, 835
Binding energy, 833
Binoculars, 729
Biomass energy, 462
Blackbody, 422
Block and tackle, 221

Blood, circulation of, 284
Blood pressure, 277, 294
Blowtorch, 292
Boiling, 390
Boltzmann's constant, 388
Bond
  covalent, 808, 812
  ionic, 812
  metallic, 816
  van der Waals, 814
Bourdon tube, 269
Boyle's law, 377
  origin of, 387
Brake, hydraulic, 274
Branch, of circuit, 550
Breeder reactor, 834
British thermal unit, 401
Brush, in electric motor, 591
Bulk modulus, 259
Buoyant force, 278

Calorie, 401
Camera, 717
Candela, 693
Cantilever, 261
Capacitance, 506
Capacitive reactance, 644
Capacitive time constant, 553
Capacitor(s), 504
  construction of, 512
  energy of, 508
  in parallel, 515
  phase relations in, 642
  in series, 513
Capillarity, 296
Carbon cycle, 854
Carbon dioxide, in atmosphere, 424
Carnot efficiency, 445
Carnot engine, 442, 445

CAT scanner, 775
Cathode-ray tube, 501
Cell, electrochemical, 546
Cells in combination, 547
Celsius scale, 372
Center of gravity, 211
  how to calculate, 216
Center of mass, 111
Center of oscillation, 324
Center of percussion, 324
Centigrade scale, 372
Centripetal acceleration, 149
Centripetal force, 148, 152
Chain reaction, 850
Change of state, 405
Charge, electric; *see* Electric
  charge
Charles's law, 379
  origin of, 387
Chlorine, isotopes of, 831
Chlorofluorocarbons (CFCs),
  456
Chromatic aberration, 729
Circuit breaker, 527
Circle, motion in vertical, 156
Circular mil, 530
Circular motion, period of,
  151
Coated lens, 754
Coefficient of friction, 80
Coefficient of linear expan-
  sion, 373
Coefficient of performance,
  457
Coefficient of restitution, 138
Coherent light, 744
Cohesion, 296
Coil, inductance of, 622
  magnetic field of, 572
Collisions, 131
  elastic, 133
  energy transfer in, 136
  inelastic, 131
Combustion, heat of, 452
Commutator, 591
Compass, 594
Component of vector, 40
Compound, chemical, 384
Compression, uniform, 259
Compressive stress, 246, 253
Compton scattering, 841
Concave mirror, 730
Concurrent forces, 208
Condensation of vapor, 391
Conduction

of electricity, 476
of heat, 413, 414, 417
Conductivity
  electrical, 476
  thermal, 414
Conservation
  of angular momentum, 192
  of charge, 473
  of energy, 109, 112
  of linear momentum, 125
Constructive interference, 344
Continuity, equation of, 283
Convection, 413, 419
Converging lens, 706
Convex mirror, 730
Cosmic rays, 846, 849
Coulomb, 475
Coulomb's law, 473
Counter emf, 613
Covalent bonding, 808
Covalent crystal, 812
Critical angle, 690
Critical point, 411
Critical speed, 160
Crystalline solid, 250, 810
  defects in, 250, 820
Curie, 844
Current loop
  magnetic field of, 572
  magnetic torque on, 589
Current, electric, 476, 523
  in ac circuit, 649
  alternating, 592, 610
  eddy, 619
  effective, 637
  magnetic field of, 570
  magnetic force on, 583
Cycle, 309

Damped oscillator, 325
De Broglie wavelength, 777
Decay, radioactive, 835
Decay series, 839
Decibel, 360
Degrees and radians, 174
Density, 243
  measurement of, 281
  probability, 779
  relative, 246
  of water vapor, 392
Depth, apparent, 688
Depth of field, 719
Destructive interference, 344
Detergent, 296
Deuterium, 831

Dew point, 392
Diamagnetism, 575
Diamond, 812
  dispersion in, 686
  energy bands in, 819
Diastolic pressure, 277
Dielectric constant, 510
Dielectric heating, 512
Dielectric strength, 512
Diesel engine, 449
Diffraction
  of light, 755
  of X-rays, 751
Diffraction grating, 748
Diffuse reflection, 678
Diode
  light-emitting, 823
  semiconductor, 822
  Zener, 823
Diopter, 723
Direct current, 592
Dislocation in crystal, 250
Dispersion, 686
Displacement, 9
  in accelerated motion, 14,
  16
  angular, 185, 194
Dissipative force, 114
Diverging lens, 706
Domain, magnetic, 576
Donor level, 821
Doping a semiconductor, 821
Doppler effect
  in light, 359
  in sound, 357
Double-slit interference pat-
  tern, 746
Drag, aerodynamic, 290
Ductility, 248

Ear, response to sound of, 362
Earth
  magnetic field of, 594
  shape of, 162
Earth satellite, 164
Eddy current, 619
Effective current, 637
Effective voltage, 638, 647
Efficiency
  Carnot, 445
  luminous, 695
  of machine, 222
  of refrigerator, 457
Elastic collision, 133
Elastic limit, 248

Elastic potential energy, 109,
  307
Elasticity, modulus of, 252
Electric charge, 470
  conduction of, 476
  conservation of, 473
  electric force on, 473, 482
  of electron, 475
  field of, 483
  magnetic force on, 578
  radiation from, 671
  unit of, 475
Electric current, 476, 523
Electric field, 481
  of charged object, 487, 488
  energy of, 506
  inside capacitor, 505
Electric field lines, 485
Electric force, 473, 482
Electric hazards, 527
Electric motor, 591, 613
Electric oscillations, 627
Electric potential energy, 494
  of two charges, 496
Electric power, 532
Electricity and matter, 476
Electromagnet, 576
Electromagnetic induction,
  603
Electromagnetic interaction,
  566
Electromagnetic waves, 421,
  668
  radiation of, 422
  types of, 422
Electromotive force (emf),
  545
  back, 613
  induced, 603, 607
  motional, 605
  self-induced, 619
Electron, 472
  charge of, 475
  mass of, 473
  positive, 836
  spin of, 575, 792
Electron capture by nucleus,
  838
Electron gas in metal, 816
Electron microscope, 778
Electron shells in atom, 793
Electron volt, 503
Electron waves in atom, 784
Elements, 384
  transuranic, 854

Emission spectrum, 788, 794
Emission, induced, 797
  spontaneous, 797
Emissivity, 422
Energy, 93, 101
  binding, 833
  biomass, 462
  of capacitor, 508
  and civilization, 459
  conservation of, 109
  elastic potential, 109, 307
  electric potential, 494, 496,
    506
  fusion, 460, 854
  geothermal, 462
  gravitational potential, 106,
    108
  of harmonic oscillator, 316
  hydroelectric, 462
  of inductor, 625
  internal, 399
  ionization, 786
  kinetic, 101, 103
  magnetic, 108, 625
  and mass, 101, 780
  molecular, 388
  nuclear, 460
  ocean thermal, 463
  potential, 101, 105
  quantization of, 788
  renewable sources of, 461
  rest, 101, 780
  rotational kinetic, 178
  solar, 461
  thermonuclear, 854
  tidal, 462
  of wave, 462
  of wind, 462
  and work, 112
Energy band in solid, 817
Energy density
  of electric field, 506
  of magnetic field, 626
Energy efficiency ratio, 457
Energy levels of atom, 786,
    791
Energy problem, 459
Energy quantization, 788
Energy transfer in collision,
    136
Engine
  Carnot, 442, 445
  diesel, 449
  efficiency of, 445
  gasoline, 448, 453

heat, 433
  internal combustion, 448
  jet, 451
Equilibrium, 63
  neutral, 210
  rotational, 208
  stable and unstable, 210
  translational, 202
Equivalent resistance, 535
Escape speed, 166
Evaporation, 390
Excited state of atom, 786
Exclusion principle, 793
  and crystals, 817
  and molecules, 808
Expansion
  thermal, 373
  volume, 375
Explosions and momentum
    conservation, 126
Eye, 720
  defects of, 722

F-number of lens, 718
Fahrenheit scale, 372
Falling bodies, 21
Farad, 507
Faraday's law of induction,
    607
Farsightedness, 722
Fatigue of material, 249
Ferrites, 574
Ferromagnetism, 574, 576
Fiber optics, 692
Field
  depth of, 719
  electric, 481, 487, 488
  force, 479
  magnetic, 567, 570, 572
Field lines
  electric, 485
  magnetic, 595
Field-effect transistor, 824
Filters
  band-pass and band-reject,
    658
  high- and low-pass, 659
Fission, nuclear, 835, 849
Fission fragments, 849
Fluid flow, 281
  conservation of energy in,
    285
  in pipe, 294
  power in, 284
  rate of, 282

Fluorescent lamp, 695
Flux
  luminous, 693
  magnetic, 607
Focal length
  of lens, 708
  of spherical mirror, 730
Focal point, 706, 709
Foot-pound, 96
Footcandle, 697
Forbidden band in solid, 818
Force, 62
  and acceleration, 64
  action, 76
  between two currents, 587
  buoyant, 278
  centripetal, 148, 152
  conservative, 114
  on current, 583
  dissipative, 114
  drag, 290
  electric, 473, 482
  electromotive, 545
  frictional, 78
  fundamental, 566
  gravitational, 161
  magnetic, 566, 578, 583
  nonconservative, 114
  normal, 80
  reaction, 76
  restoring, 305
  and torque, 186
  units of, 65, 73
  and work, 94
Force field, 479
Forces, concurrent, 208
Fourier's theorem, 340
Free fall, 24
Free-body diagram, 69, 202
Freeze drying, 412
Freon 12, 454, 456
Frequency
  angular, 320
  of electric oscillator, 628
  fundamental, 346
  of harmonic oscillator, 312
  resonance, 651
  of wave, 337
Frequency modulation, 672
Friction, 78
  coefficient of, 80
  kinetic, 79
  rolling, 80
  sliding, 79
  starting, 79

static, 79
  and walking, 83
Fulcrum, 224
Fundamental frequency, 346
Fuse, 527
Fusion
  curve of, 411
  heat of, 406
  nuclear, 835, 854

Gain, power, 363
Galvanometer, 590
Gamma ray, 835
  interactions of, 841
Gas
  ideal, 382
  kinetic theory of, 386
Gas turbine, 450
Gasoline engine, 448, 453
Gauge pressure, 269
Gauss, 569
Gear drive, 229
Geiger counter, 843
Generator, 610
Geostationary orbit, 166
Geothermal energy, 462
Golf, 138
Grating, diffraction, 748
Gravitation, 161
Gravitational potential energy,
    106, 108
Gravity
  acceleration of, 22, 163
  center of, 211, 216
  specific, 246
Greenhouse effect, 424
Ground, electrical, 478
Ground state of atom, 786
Gyroscope, 194

Half life, 844
Harmonic motion, 309, 316
  amplitude of, 311
  damped, 325
  energy of, 316
  model of, 313
  period of, 309
Heat, 400
  of combustion, 452
  conduction of, 413, 414,
    417
  convection of, 413, 419
  of fusion, 406

Heat *(continued)*
  quantity of, 403
  specific, 402
  of vaporization, 406, 438
Heat capacity, 402
Heat engine, 433
Heat pump, 455
Heat transfer, 413
Heavy water, 853
Helium-neon laser, 798
Henry, 620
Hertz, 337
Hole in crystal structure, 821
Hooke's law, 247, 306
Horsepower, 100
Humidity, relative, 392
Huygens' principle, 677
Hydraulic brake, 274
Hydraulic jack, 275
Hydraulic press, 272
Hydroelectric energy, 462
Hydrogen
  as fuel, 453
  isotopes of, 831
  spectrum of, 791
Hydrogen atom, 782
  energy levels of, 786
  standing waves in, 784
Hydrogen bomb, 855
Hydrogen molecule, 807
Hydrometer, 281
Hyperopia, 722
Hysteresis, 577

Ice, structure of, 811
Iceberg, 280
Ideal gas law, 382
Ideal refrigerator, 458
Illumination, 696
Image
  real, 680
  virtual, 680
Image formation
  by lens, 709
  by mirror, 731
Impedance, 649
Impedance matching, 138,
    548, 655
Impulse, 123
Incidence, angle of, 679
Inclined plane, 230
Incoherent light, 745
Index of refraction, 684
Indicator diagram, 439

Induced absorption and emis-
    sion, 796
Induced emf, 603, 607
Inductance, 620
Induction, electromagnetic,
    603
Inductive reactance, 642
Inductive time constant, 622
Inductor, 620
  energy of, 625
  phase relations in, 641
Inelastic collision, 131
Inertia, 60
  moment of, 179
Infrasound, 362
Instantaneous speed, 6
Instantaneous velocity, 10
Insulator, electrical, 476
Intensity
  luminous, 693
  sound, 360
Interference
  of light, 744, 752
  of sound, 352
  of waves, 344
Internal combustion engine,
    448
Internal energy, 399
Internal reflection, 690
Internal resistance, 546
Ion, 477
Ionic bonding, 812
Ionization chamber, 843
Ionization energy, 786
Ionosphere, 478, 675
Isobaric process, 437
Isotherm, 481
Isothermal process, 442
Isotope, 831
Isotropic source, 693

Jack, hydraulic, 275
Jet engine, 451
Joule, 95
Junction, in circuit, 550

Kelvin, 380
Kilocalorie, 400
Kilogram, 62
Kilojoule, 403
Kilowatt-hour, 99
Kinetic energy, 101, 103
  of rotation, 178
Kinetic theory

of gases, 386
of liquids, 390
Kirchhoff's rules, 549
Knot, 45

Laminar flow, 281
Lamps, types of, 695
Laser, 795
Lens, 705
  aberrations of, 729
  achromatic, 729
  aspherical, 706
  coated, 754
  converging, 706
  diverging, 706
  focal length of, 708
  image formation by, 709
  magnification of, 715
  meniscus, 706
  power of, 723
  resolving power of, 756
  sign conventions for, 716
  telephoto, 719
  thin, 708
  wide-angle, 719
Lens equation, 712
Lensmaker's equation, 708
Lenz's law, 607
Lever, 224
Lift of airplane wing, 290
Light
  coherent, 744
  diffraction of, 755
  Doppler effect in, 359
  frequencies of, 673
  incoherent, 745
  interference of, 744, 752
  nature of, 668, 773
  polarized, 758
  quantum theory of, 771
  reflection of, 678
  refraction of, 681
  scattering of, 763
  speed of, 668
  wavelengths of, 673
Light-emitting diode, 823
Light pipe, 692
Light rays, 678
Lightning, 498
Linear expansion, 373
Linear momentum, 122
  conservation of, 125
Lines, field, 485
Liquids

kinetic theory of, 390
structure of, 810
Longitudinal waves, 341
Loop, in circuit, 550
Loudspeaker, 601
Lubricants, 80
Lumen, 693
Luminous efficiency, 695
Luminous flux, 693
Luminous intensity, 693
Lux, 697

Mach number, 351
Machine, 220, 224, 272
  efficiency of, 222
Maglev train, 587
Magnet, permanent, 576, 593
Magnetic domain, 576
Magnetic field, 567
  of current, 570
  of current loop, 572
  of earth, 594
  energy of, 625
  of solenoid, 572
Magnetic field lines, 595
Magnetic flux, 607
Magnetic force, 566
  between two currents, 587
  on current, 583
  on moving charge, 578
Magnetic mirror, 582
Magnetic permeability, 571
Magnetic poles, 593
Magnetic field energy, 626
Magnetic properties of matter,
    574
Magnetic torque on current
    loop, 589
Magnetism, nature of, 566
Magneto, 611
Magnetosphere, 583
Magnification
  of lens, 715
  of microscope, 726
  of mirror, 733
  of telescope, 727
Magnifying glass, 716
Malleability, 249
Manometer, 269
Mass, 60
  atomic, 832
  center of, 111
  and energy, 780
  relativity of, 780

units of, 65, 73
and weight, 69
Mass number, 832
Mass spectrometer, 580
Matter waves, 776
Maxwell's hypothesis, 668
Mechanical advantage, 220
actual, 220
of hydraulic press, 273
ideal, 220
Meniscus lens, 706
Metallic bond, 816
Metastable state, 796
Meter, 2
Microphone, 507
Microscope, 725
electron, 778
Millibar, 268
Mirror
concave, 730
convex, 730
image formation by, 731
magnetic, 582
magnification of, 733
parabolic, 735
plane, 680
spherical, 730
Mirror equation, 733
Moderator, 852
Modulation
amplitude, 672
frequency, 672
Modulus
bulk, 259
of elasticity, 252
of rigidity, 257
shear, 257
Young's, 253
Molecular bonding, 808
Molecular crystal, 814
Molecular energy, 388
Molecular speeds, 388
Molecule, 384
hydrogen, 807
polar, 509, 809
water, 509
Moment arm of force, 186
Moment of inertia, 178
Momentum
angular; *see* Angular momentum
linear, 122
Motion
accelerated, 10
angular, 173

circular, 147
at constant speed, 7
first law of, 60, 63
in free fall, 24
harmonic, 309
molecular, 386
projectile, 48
second law of, 64
third law of, 75
in vertical circle, 156
in vertical plane, 45
Motional emf, 605
Motor, electric, 591, 613
Musical sounds, 354
Myopia, 722

n-type semiconductor, 820
Near point of eye, 724
Nearsightedness, 722
Neptunium, 854
Neutral equilibrium, 210
Neutron, 472, 830
detection of, 843
slow, 852
Newton, 65
Newton's first law of motion,
60, 63
Newton's law of gravitation,
161
Newton's second law of mo-
tion, 64
Newton's third law of motion,
75
Node in standing wave, 344
Normal force, 80
Normal to a surface, 680
Nuclear fission, 835, 849
Nuclear fusion, 835, 854
Nuclear reactor, 850, 851
Nucleon, 830
Nucleus,
binding energy of, 833
composition of, 830
decay of, 835
Nuclide, 832

Ocean thermal energy, 463
Octave, 354
Ohm, 525
Ohm's law, 525
Optical properties of solids,
819
Optical pumping, 797
Optics

geometrical, 743
physical, 743
Orbit
of atomic electron, 784
of satellite, 164
Organic chemistry, 809
Oscillation, center of, 324
Oscillator
damped, 325
electric, 627
energy of, 316
harmonic, 309, 316
period, 309
Oscilloscope, 502
Otto cycle, 448
Overtone, 346
Ozone, in atmosphere, 456

p-type semiconductor, 821
Pair production, 841
Parabolic mirror, 735
Parallel ac circuits, 656
Paramagnetism, 575
Particle detection, 841
Pascal, 252
Pascal's principle, 271
Pendulum, 321
physical, 322
torsion, 330
Percussion, center of, 324
Performance, coefficient of,
457
Period
of circular motion, 151
of harmonic motion, 309,
311
of pendulum, 322, 323
of wave, 339
Periodic waves, 336
Permanent magnet, 576
Permeability, 571
Permittivity, 475
Phase angle in ac circuit, 647
Phasor, 639
Photoelectric effect, 770, 841
Photon, 771
mechanisms of energy loss,
841
Physical pendulum, 322
Piano, 346
Pitch of screw, 231
Planck's constant, 772
Plane
inclined, 230
of polarization, 758

Plasma, 476
Plastic deformation, 249
Plutonium, 854
Poise, 293
Poiseuille, 293
Poiseuille's law, 294
Polar molecule, 509, 809
Polarized light, 758
Polaroid, 760
Pole, magnetic, 593
Pollution, air, 453
Population inversion, 797
Positron, 836
Potential difference, 498
in ac circuit, 647
in dc circuit, 545
effective, 638, 647
and electric field, 499
Potential energy, 101, 105
elastic, 109, 307
electric, 494, 506
gravitational, 106
of two charges, 496
Pound, 73
Power,
in ac circuit, 653
apparent, 654
electric, 532
in fluid flow, 284
of lens, 723
in rotational motion,
189
Power factor, 653
Power gain, 363
Power transfer, 548
Presbyopia, 724
Press, hydraulic, 272
Pressure, 259, 268
absolute, 269
atmospheric, 268, 271
blood, 277
and depth, 276
gauge, 269
Pressure gauges, 269
Probability cloud, 792
Probability density, 779
Problems, how to solve, 18
Projectile flight, 48
spin stabilization in, 194
Projector, 717
Proton, 472
Proton-proton cycle, 854
Pulses in a string, 332
Pump, 272
heat, 455

Quanta, 771
Quantization in atomic world, 787
Quantum mechanics, 776
Quantum number, 785, 792
Quantum theory
    of atom, 782, 791
    of light, 771

Radar, 338, 676
Radian, 174
Radiation, 413, 421
    detection of, 841
Radiation hazards, 847
Radio communication, 672
Radioactivity, 835
Radiocarbon dating, 846
Radiometric dating, 846
Radiotelescope, 757
Radon, 845, 848
Rainbow, 687
Raindrops, 22
Rankine scale, 382
Reactance
    capacitive, 644
    inductive, 642
Reaction force, 76
Reactor
    breeder, 854
    nuclear, 850, 851
Real image, 680
Recombination, 478
Rectifier, 612
    semiconductor, 822
Reflecting telescope, 735
Reflection
    angle of, 679
    diffuse, 678
    of light, 678
    of pulse, 333, 334
    specular, 678
    total internal, 690
    of wave, 333
Refraction, 681
    angle of, 682
    index of, 684
    law of, 682, 685
Refrigerant, 454
Refrigerator, 435, 454
    ideal, 458
    thermodynamics of, 457
Relative density, 246
Relative humidity, 392
Relativity, theory of, 779
Relativity of mass, 780

Relay, 577
Rem, 848
Resistance
    electrical, 525
    equivalent, 535
    internal, 546
    thermal, 417
Resistance thermometer, 531
Resistivity, 529
    temperature variation of, 531
Resistors
    in parallel, 539
    in series, 535
Resolution of a vector, 40
Resolving power, 756
Resonance
    in ac circuit, 651
    mechanical, 347
Resonance frequency, 651
Rest energy, 101, 780
Restitution, coefficient of, 138
Restoring force, 305
Resultant vector, 37
Right-hand rule
    for induced current, 603
    for magnetic field, 567
    for magnetic force, 570, 585
Rigidity, modulus of, 257
Rms current and voltage, 639
Rocket propulsion, 128
Rolling friction, 80
Rotation
    axis of, 177
    kinetic energy of, 178
Rotational equilibrium, 208
Rotational kinetic energy, 178
Rotational power, 189
Rotational work, 189
Rpm and rps, 176
Ruby laser, 797

Satellite orbits, 164
Saturated vapor, 391
Scalar field, 481
Scalar quantity, 35
Scattering of light, 763
Scintillation counter, 843
Screw as machine, 231
Scuba diving, 276
Self-induced emf, 619
Semiconductor, 477, 819
    n-type, 820
    p-type, 821

Semiconductor detectors, 843
Semiconductor diodes, 822
Shear, 246, 256
    angle of, 257
    modulus of, 257
Shell, electron, 793
Shock wave, 350
SI units, 2
Sievert, 848
Sign conventions
    for lens, 716
    for mirror, 733
Silicon, energy bands in, 819
Simple harmonic motion, 309, 311
    amplitude of, 311
    energy of, 316
    model of, 313
    period of, 309
Sinusoidal variations, 311
Sinusoidal waves, 336
Slit, double, 746
Sky, blue color of, 763
Slug, 73
Snell's law, 682, 685
Sodium, energy bands in, 818
Sodium chloride crystal, 813
Solar energy, 461
Solenoid
    inductance of, 622
    magnetic field of, 572
Solid angle, 694
Solids
    amorphous, 810
    covalent, 812
    crystalline, 810
    dislocations in, 250
    elastic properties of, 246
    ionic, 810
    metallic, 816
    optical properties of, 819
    structure of, 250
    van der Waals, 814
Sound, 348
    Doppler effect in, 357
    intensity of, 360
    interference in, 352
    musical, 354
    speed of, 349
Sound barrier, 350
Sound recording, 604
Spark, 488
Specific gravity, 246
Specific heat capacity, 402
Spectral series, 790

Spectrometer, mass, 580
Spectrum
    absorption, 789, 794
    continuous, 789
    of diffraction grating, 748
    electromagnetic, 672
    emission, 788, 794
    of hydrogen, 791
    line, 788, 790
    of prism, 686
Specular reflection, 678
Speed, 4
    angular, 176
    average, 4
    constant, 7
    critical, 160
    escape, 166
    of harmonic oscillator, 316
    instantaneous, 6, 10
    of light, 668
    maximum possible, 782
    molecular, 389
    of sound, 349
    subsonic, 350
    supersonic, 350
    and velocity, 9
    of waves, 332, 337
    terminal, 22
Spherical aberration
    of lens, 729
    of mirror, 735
Spherical mirror, 730
Spin, electron, 575, 792
Spin stabilization, 194
Spontaneous emission, 797
Spring, oscillating, 309
Standing waves, 343, 345
    in atom, 784
    in laser, 798
State, change of, 405
Steam turbine, 446
Stefan-Boltzmann law, 422
Strain, 252
Streamline flow, 281
Strength, ultimate, 249
Stress, 246, 251
Stress concentration, 249
Sublimation, 410
    curve of, 412
Subsonic speed, 350
Sun
    energy production in, 854
    radiation from, 424
Superconductivity, 478

Superposition principle, 335, 344
Supersonic speed, 350
Surface tension, 295
Surfactant, 296
Synovial fluid, 80
Système International d'Unités, 2
Systolic pressure, 277

Tangential acceleration, 184
Tape recorder, 604
Telephoto lens, 719
Telescope
    reflecting, 735
    refracting, 726
Television, 503
Temperature,
    absolute, 380
    Celsius, 372
    centigrade, 372
    Fahrenheit, 372
    and molecular energy, 388
    Rankine, 382
Tensile stress, 246, 253
Tension in cable, 69
Tension, surface, 295
Terminal speed, 22
Terminal voltage, 546
Tesla, 568
Thermal conductivity, 414
Thermal expansion, 373
Thermal resistance, 417
Thermodynamics
    first law of, 434
    of refrigeration, 457
    second law of, 440, 446
Thermograph, 422
Thermometer, 371
    resistance, 531
Thermonuclear energy, 854
Thermostat, 371, 577
Thin film, interference in, 752
Thrust of rocket, 129
Thunderstorm, 498
Tidal energy, 462
Time constant

of capacitive circuit, 553
of inductive circuit, 622
Tonne, 107
Torque, 186
    and angular acceleration, 187
    on current loop, 589
    and equilibrium, 208
Torque transmission, 226
Torr, 277
Torricelli's theorem, 288
Torsion pendulum, 330
Total internal reflection, 690
Transformer, 615
    and impedance matching, 655
Transistor, 824
Translational equilibrium, 202
Transuranic elements, 854
Transverse waves, 341
Trigonometric notation in harmonic motion, 320
Trigonometry, 38
Triple point, 412
Tritium, 831
    decay of, 845
Truss, 261
Turbine
    gas, 450
    steam, 447
Turbojet engine, 451
Turbulent flow, 282

Ultimate strength, 249
Ultrasound, 349, 362
Uniform circular motion, 148, 151
Units
    conversion of, 3
    systems of, 2
Uranium
    enriched, 853
    fission of, 849

Van der Waals bond, 814
Vapor, 390
Vapor density of water, 392

Vaporization
    curve of, 411
    heat of, 406, 408
Vector, 36
    components of, 40
Vector addition, 37, 43
Vector field, 481
Vector quantity, 35
Vector resolution, 40
Vector subtraction, 39
Velocity, 9
    average, 9
    instantaneous, 10
Venturi meter, 292
Vertical circle, motion in, 156
Vertical plane, motion in, 45
Virtual image, 680
Viscosity, 292
Vision, defects of, 722
Volt, 498
    electron, 503
Voltage, effective, 638, 647
    terminal, 546
Volt-ampere, 654
Volume expansion, 375
Vortex, 290

Walking, physics of, 83
Water
    expansion of, 376, 811
    heavy, 853
    thermal properties of, 400, 405
Water molecule, 810
Water vapor density, 391
Water waves, 341
    refraction of, 683
Watt, 99
Wattmeter, 654
Wave(s)
    amplitude of, 339
    de Broglie, 776
    electromagnetic, 421, 668
    energy of, 462
    frequency of, 337
    interference of, 344
    light, 668

longitudinal, 341
    matter, 776
    period of, 339
    periodic, 336
    radio, 672
    reflection of, 333, 678
    refraction of, 681
    shock, 350
    sinusoidal, 336
    sound, 348
    speed of, 337
    standing, 343, 345
    in a string, 332
    superposition of, 335
    transverse, 341
    ultrasonic, 349
    water, 341
Wave function, 778
Waveform, 336, 354
Wavefront, 676
Wavelength, 337
    de Broglie, 777
Weber, 607
Wedge, 231
Weight, 69
    apparent, 167
Weight density, 243
Weightlessness, 167
Wetting agent, 296
Wide-angle lens, 719
Wind energy, 462
Wing, lift on, 290
Work, 94
    electrical, 494
    and gravity, 97
    in rotational motion, 189
Work-energy theorem, 112
Work function, 772
Work hardening, 250

X-rays, 774
    diffraction of, 751
    hazards of, 848
    interactions of, 841

Young's modulus, 253

Zener diode, 823